T0180085

Lecture Notes in Computer Science 12904

More information about this subseries at http://www.springer.com/series/7412

Marleen de Bruijne · Philippe C. Cattin ·
Stéphane Cotin · Nicolas Padoy ·
Stefanie Speidel · Yefeng Zheng ·
Caroline Essert (Eds.)

Medical Image Computing and Computer Assisted Intervention – MICCAI 2021

24th International Conference
Strasbourg, France, September 27 – October 1, 2021
Proceedings, Part IV

 Springer

Editors
Marleen de Bruijne 🆔
Erasmus MC - University Medical Center
Rotterdam
Rotterdam, The Netherlands

University of Copenhagen
Copenhagen, Denmark

Stéphane Cotin 🆔
Inria Nancy Grand Est
Villers-lès-Nancy, France

Stefanie Speidel 🆔
National Center for Tumor Diseases
(NCT/UCC)
Dresden, Germany

Caroline Essert 🆔
ICube, Université de Strasbourg, CNRS
Strasbourg, France

Philippe C. Cattin 🆔
University of Basel
Allschwil, Switzerland

Nicolas Padoy 🆔
ICube, Université de Strasbourg, CNRS
Strasbourg, France

Yefeng Zheng 🆔
Tencent Jarvis Lab
Shenzhen, China

ISSN 0302-9743   ISSN 1611-3349   (electronic)
Lecture Notes in Computer Science
ISBN 978-3-030-87201-4   ISBN 978-3-030-87202-1   (eBook)
https://doi.org/10.1007/978-3-030-87202-1

LNCS Sublibrary: SL6 – Image Processing, Computer Vision, Pattern Recognition, and Graphics

This Springer imprint is published by the registered company Springer Nature Switzerland AG
The registered company address is: Gewerbestrasse 11, 6330 Cham, Switzerland

Preface

The 24th edition of the International Conference on Medical Image Computing and Computer Assisted Intervention (MICCAI 2021) has for the second time been placed under the shadow of COVID-19. Complicated situations due to the pandemic and multiple lockdowns have affected our lives during the past year, sometimes perturbing the researchers work, but also motivating an extraordinary dedication from many of our colleagues, and significant scientific advances in the fight against the virus. After another difficult year, most of us were hoping to be able to travel and finally meet in person at MICCAI 2021, which was supposed to be held in Strasbourg, France. Unfortunately, due to the uncertainty of the global situation, MICCAI 2021 had to be moved again to a virtual event that was held over five days from September 27 to October 1, 2021. Taking advantage of the experience gained last year and of the fast-evolving platforms, the organizers of MICCAI 2021 redesigned the schedule and the format. To offer the attendees both a strong scientific content and an engaging experience, two virtual platforms were used: Pathable for the oral and plenary sessions and SpatialChat for lively poster sessions, industrial booths, and networking events in the form of interactive group video chats.

These proceedings of MICCAI 2021 showcase all 531 papers that were presented at the main conference, organized into eight volumes in the Lecture Notes in Computer Science (LNCS) series as follows:

- Part I, LNCS Volume 12901: Image Segmentation
- Part II, LNCS Volume 12902: Machine Learning 1
- Part III, LNCS Volume 12903: Machine Learning 2
- Part IV, LNCS Volume 12904: Image Registration and Computer Assisted Intervention
- Part V, LNCS Volume 12905: Computer Aided Diagnosis
- Part VI, LNCS Volume 12906: Image Reconstruction and Cardiovascular Imaging
- Part VII, LNCS Volume 12907: Clinical Applications
- Part VIII, LNCS Volume 12908: Microscopic, Ophthalmic, and Ultrasound Imaging

These papers were selected after a thorough double-blind peer review process. We followed the example set by past MICCAI meetings, using Microsoft's Conference Managing Toolkit (CMT) for paper submission and peer reviews, with support from the Toronto Paper Matching System (TPMS), to partially automate paper assignment to area chairs and reviewers, and from iThenticate to detect possible cases of plagiarism.

Following a broad call to the community we received 270 applications to become an area chair for MICCAI 2021. From this group, the program chairs selected a total of 96 area chairs, aiming for diversity — MIC versus CAI, gender, geographical region, and a mix of experienced and new area chairs. Reviewers were recruited also via an open call for volunteers from the community (288 applications, of which 149 were selected by the program chairs) as well as by re-inviting past reviewers, leading to a total of 1340 registered reviewers.

We received 1630 full paper submissions after an original 2667 intentions to submit. Four papers were rejected without review because of concerns of (self-)plagiarism and dual submission and one additional paper was rejected for not adhering to the MICCAI page restrictions; two further cases of dual submission were discovered and rejected during the review process. Five papers were withdrawn by the authors during review and after acceptance.

The review process kicked off with a reviewer tutorial and an area chair meeting to discuss the review process, criteria for MICCAI acceptance, how to write a good (meta-)review, and expectations for reviewers and area chairs. Each area chair was assigned 16–18 manuscripts for which they suggested potential reviewers using TPMS scores, self-declared research area(s), and the area chair's knowledge of the reviewers' expertise in relation to the paper, while conflicts of interest were automatically avoided by CMT. Reviewers were invited to bid for the papers for which they had been suggested by an area chair or which were close to their expertise according to TPMS. Final reviewer allocations via CMT took account of reviewer bidding, prioritization of area chairs, and TPMS scores, leading to on average four reviews performed per person by a total of 1217 reviewers.

Following the initial double-blind review phase, area chairs provided a meta-review summarizing key points of reviews and a recommendation for each paper. The program chairs then evaluated the reviews and their scores, along with the recommendation from the area chairs, to directly accept 208 papers (13%) and reject 793 papers (49%); the remainder of the papers were sent for rebuttal by the authors. During the rebuttal phase, two additional area chairs were assigned to each paper. The three area chairs then independently ranked their papers, wrote meta-reviews, and voted to accept or reject the paper, based on the reviews, rebuttal, and manuscript. The program chairs checked all meta-reviews, and in some cases where the difference between rankings was high or comments were conflicting, they also assessed the original reviews, rebuttal, and submission. In all other cases a majority voting scheme was used to make the final decision. This process resulted in the acceptance of a further 325 papers for an overall acceptance rate of 33%.

Acceptance rates were the same between medical image computing (MIC) and computer assisted interventions (CAI) papers, and slightly lower where authors classified their paper as both MIC and CAI. Distribution of the geographical region of the first author as indicated in the optional demographic survey was similar among submitted and accepted papers.

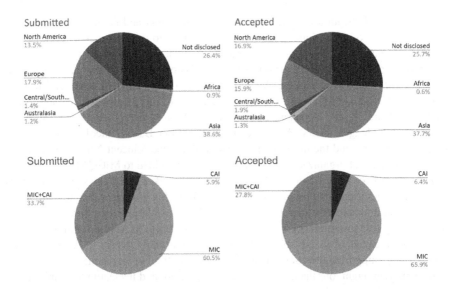

New this year, was the requirement to fill out a reproducibility checklist when submitting an intention to submit to MICCAI, in order to stimulate authors to think about what aspects of their method and experiments they should include to allow others to reproduce their results. Papers that included an anonymous code repository and/or indicated that the code would be made available were more likely to be accepted. From all accepted papers, 273 (51%) included a link to a code repository with the camera-ready submission.

Another novelty this year is that we decided to make the reviews, meta-reviews, and author responses for accepted papers available on the website. We hope the community will find this a useful resource.

The outstanding program of MICCAI 2021 was enriched by four exceptional keynote talks given by Alyson McGregor, Richard Satava, Fei-Fei Li, and Pierre Jannin, on hot topics such as gender bias in medical research, clinical translation to industry, intelligent medicine, and sustainable research. This year, as in previous years, high-quality satellite events completed the program of the main conference: 28 workshops, 23 challenges, and 14 tutorials; without forgetting the increasingly successful plenary events, such as the Women in MICCAI (WiM) meeting, the MICCAI Student Board (MSB) events, the 2nd Startup Village, the MICCAI-RSNA panel, and the first "Reinforcing Inclusiveness & diverSity and Empowering MICCAI" (or RISE-MICCAI) event.

MICCAI 2021 has also seen the first edition of CLINICCAI, the clinical day of MICCAI. Organized by Nicolas Padoy and Lee Swanstrom, this new event will hopefully help bring the scientific and clinical communities closer together, and foster collaborations and interaction. A common keynote connected the two events. We hope this effort will be pursued in the next editions.

We would like to thank everyone who has contributed to making MICCAI 2021 a success. First of all, we sincerely thank the authors, area chairs, reviewers, and session chairs for their dedication and for offering the participants and readers of these pro-ceedings content of exceptional quality. Special thanks go to our fantastic submission platform manager Kitty Wong, who has been a tremendous help in the entire process from reviewer and area chair selection, paper submission, and the review process to the preparation of these proceedings. We also thank our very efficient team of satellite events chairs and coordinators, led by Cristian Linte and Matthieu Chabanas: the workshop chairs, Amber Simpson, Denis Fortun, Marta Kersten-Oertel, and Sandrine Voros; the challenges chairs, Annika Reinke, Spyridon Bakas, Nicolas Passat, and Ingerid Reinersten; and the tutorial chairs, Sonia Pujol and Vincent Noblet, as well as all the satellite event organizers for the valuable content added to MICCAI. Our special thanks also go to John Baxter and his team who worked hard on setting up and populating the virtual platforms, to Alejandro Granados for his valuable help and efficient communication on social media, and to Shelley Wallace and Anna Van Vliet for marketing and communication. We are also very grateful to Anirban Mukhopadhay for his management of the sponsorship, and of course many thanks to the numerous sponsors who supported the conference, often with continuous engagement over many years. This year again, our thanks go to Marius Linguraru and his team who supervised a range of actions to help, and promote, career development, among which were the mentorship program and the Startup Village. And last but not least, our wholehearted thanks go to Mehmet and the wonderful team at Dekon Congress and Tourism for their great professionalism and reactivity in the management of all logistical aspects of the event.

Finally, we thank the MICCAI society and the Board of Directors for their support throughout the years, starting with the first discussions about bringing MICCAI to Strasbourg in 2017.

We look forward to seeing you at MICCAI 2022.

September 2021

Marleen de Bruijne
Philippe Cattin
Stéphane Cotin
Nicolas Padoy
Stefanie Speidel
Yefeng Zheng
Caroline Essert

Organization

General Chair

Caroline Essert   Université de Strasbourg, CNRS, ICube, France

Program Chairs

Marleen de Bruijne	Erasmus MC Rotterdam, The Netherlands, and University of Copenhagen, Denmark
Philippe C. Cattin	University of Basel, Switzerland
Stéphane Cotin	Inria, France
Nicolas Padoy	Université de Strasbourg, CNRS, ICube, IHU, France
Stefanie Speidel	National Center for Tumor Diseases, Dresden, Germany
Yefeng Zheng	Tencent Jarvis Lab, China

Satellite Events Coordinators

Cristian Linte	Rochester Institute of Technology, USA
Matthieu Chabanas	Université Grenoble Alpes, France

Workshop Team

Amber Simpson	Queen's University, Canada
Denis Fortun	Université de Strasbourg, CNRS, ICube, France
Marta Kersten-Oertel	Concordia University, Canada
Sandrine Voros	TIMC-IMAG, INSERM, France

Challenges Team

Annika Reinke	German Cancer Research Center, Germany
Spyridon Bakas	University of Pennsylvania, USA
Nicolas Passat	Université de Reims Champagne-Ardenne, France
Ingerid Reinersten	SINTEF, NTNU, Norway

Tutorial Team

Vincent Noblet	Université de Strasbourg, CNRS, ICube, France
Sonia Pujol	Harvard Medical School, Brigham and Women's Hospital, USA

Clinical Day Chairs

Nicolas Padoy   Université de Strasbourg, CNRS, ICube, IHU, France
Lee Swanström   IHU Strasbourg, France

Sponsorship Chairs

Anirban Mukhopadhyay   Technische Universität Darmstadt, Germany
Yanwu Xu   Baidu Inc., China

Young Investigators and Early Career Development Program Chairs

Marius Linguraru   Children's National Institute, USA
Antonio Porras   Children's National Institute, USA
Daniel Racoceanu   Sorbonne Université/Brain Institute, France
Nicola Rieke   NVIDIA, Germany
Renee Yao   NVIDIA, USA

Social Media Chairs

Alejandro Granados   King's College London, UK
  Martinez
Shuwei Xing   Robarts Research Institute, Canada
Maxence Boels   King's College London, UK

Green Team

Pierre Jannin   INSERM, Université de Rennes 1, France
Étienne Baudrier   Université de Strasbourg, CNRS, ICube, France

Student Board Liaison

Éléonore Dufresne   Université de Strasbourg, CNRS, ICube, France
Étienne Le Quentrec   Université de Strasbourg, CNRS, ICube, France
Vinkle Srivastav   Université de Strasbourg, CNRS, ICube, France

Submission Platform Manager

Kitty Wong   The MICCAI Society, Canada

Virtual Platform Manager

John Baxter   INSERM, Université de Rennes 1, France

Program Committee

Ehsan Adeli	Stanford University, USA
Iman Aganj	Massachusetts General Hospital, Harvard Medical School, USA
Pablo Arbelaez	Universidad de los Andes, Colombia
John Ashburner	University College London, UK
Meritxell Bach Cuadra	University of Lausanne, Switzerland
Sophia Bano	University College London, UK
Adrien Bartoli	Université Clermont Auvergne, France
Christian Baumgartner	ETH Zürich, Switzerland
Hrvoje Bogunovic	Medical University of Vienna, Austria
Weidong Cai	University of Sydney, Australia
Gustavo Carneiro	University of Adelaide, Australia
Chao Chen	Stony Brook University, USA
Elvis Chen	Robarts Research Institute, Canada
Hao Chen	Hong Kong University of Science and Technology, Hong Kong SAR
Albert Chung	Hong Kong University of Science and Technology, Hong Kong SAR
Adrian Dalca	Massachusetts Institute of Technology, USA
Adrien Depeursinge	HES-SO Valais-Wallis, Switzerland
Jose Dolz	ÉTS Montréal, Canada
Ruogu Fang	University of Florida, USA
Dagan Feng	University of Sydney, Australia
Huazhu Fu	Inception Institute of Artificial Intelligence, United Arab Emirates
Mingchen Gao	University at Buffalo, The State University of New York, USA
Guido Gerig	New York University, USA
Orcun Goksel	Uppsala University, Sweden
Alberto Gomez	King's College London, UK
Ilker Hacihaliloglu	Rutgers University, USA
Adam Harrison	PAII Inc., USA
Mattias Heinrich	University of Lübeck, Germany
Yi Hong	Shanghai Jiao Tong University, China
Yipeng Hu	University College London, UK
Junzhou Huang	University of Texas at Arlington, USA
Xiaolei Huang	The Pennsylvania State University, USA
Jana Hutter	King's College London, UK
Madhura Ingalhalikar	Symbiosis Center for Medical Image Analysis, India
Shantanu Joshi	University of California, Los Angeles, USA
Samuel Kadoury	Polytechnique Montréal, Canada
Fahmi Khalifa	Mansoura University, Egypt
Hosung Kim	University of Southern California, USA
Minjeong Kim	University of North Carolina at Greensboro, USA

Ender Konukoglu	ETH Zürich, Switzerland
Bennett Landman	Vanderbilt University, USA
Ignacio Larrabide	CONICET, Argentina
Baiying Lei	Shenzhen University, China
Gang Li	University of North Carolina at Chapel Hill, USA
Mingxia Liu	University of North Carolina at Chapel Hill, USA
Herve Lombaert	ÉTS Montréal, Canada, and Inria, France
Marco Lorenzi	Inria, France
Le Lu	PAII Inc., USA
Xiongbiao Luo	Xiamen University, China
Dwarikanath Mahapatra	Inception Institute of Artificial Intelligence, United Arab Emirates
Andreas Maier	FAU Erlangen-Nuremberg, Germany
Erik Meijering	University of New South Wales, Australia
Hien Nguyen	University of Houston, USA
Marc Niethammer	University of North Carolina at Chapel Hill, USA
Tingying Peng	Technische Universität München, Germany
Caroline Petitjean	Université de Rouen, France
Dzung Pham	Henry M. Jackson Foundation, USA
Hedyeh Rafii-Tari	Auris Health Inc, USA
Islem Rekik	Istanbul Technical University, Turkey
Nicola Rieke	NVIDIA, Germany
Su Ruan	Laboratoire LITIS, France
Thomas Schultz	University of Bonn, Germany
Sharmishtaa Seshamani	Allen Institute, USA
Yonggang Shi	University of Southern California, USA
Darko Stern	Technical University of Graz, Austria
Carole Sudre	King's College London, UK
Heung-Il Suk	Korea University, South Korea
Jian Sun	Xi'an Jiaotong University, China
Raphael Sznitman	University of Bern, Switzerland
Amir Tahmasebi	Enlitic, USA
Qian Tao	Delft University of Technology, The Netherlands
Tolga Tasdizen	University of Utah, USA
Martin Urschler	University of Auckland, New Zealand
Archana Venkataraman	Johns Hopkins University, USA
Guotai Wang	University of Electronic Science and Technology of China, China
Hongzhi Wang	IBM Almaden Research Center, USA
Hua Wang	Colorado School of Mines, USA
Qian Wang	Shanghai Jiao Tong University, China
Yalin Wang	Arizona State University, USA
Fuyong Xing	University of Colorado Denver, USA
Daguang Xu	NVIDIA, USA
Yanwu Xu	Baidu, China
Ziyue Xu	NVIDIA, USA

Zhong Xue	Shanghai United Imaging Intelligence, China
Xin Yang	Huazhong University of Science and Technology, China
Jianhua Yao	National Institutes of Health, USA
Zhaozheng Yin	Stony Brook University, USA
Yixuan Yuan	City University of Hong Kong, Hong Kong SAR
Liang Zhan	University of Pittsburgh, USA
Tuo Zhang	Northwestern Polytechnical University, China
Yitian Zhao	Chinese Academy of Sciences, China
Luping Zhou	University of Sydney, Australia
S. Kevin Zhou	Chinese Academy of Sciences, China
Dajiang Zhu	University of Texas at Arlington, USA
Xiahai Zhuang	Fudan University, China
Maria A. Zuluaga	EURECOM, France

Reviewers

Alaa Eldin Abdelaal
Khalid Abdul Jabbar
Purang Abolmaesumi
Mazdak Abulnaga
Maryam Afzali
Priya Aggarwal
Ola Ahmad
Sahar Ahmad
Euijoon Ahn
Alireza Akhondi-Asl
Saad Ullah Akram
Dawood Al Chanti
Daniel Alexander
Sharib Ali
Lejla Alic
Omar Al-Kadi
Maximilian Allan
Pierre Ambrosini
Sameer Antani
Michela Antonelli
Jacob Antunes
Syed Anwar
Ignacio Arganda-Carreras
Mohammad Ali Armin
Md Ashikuzzaman
Mehdi Astaraki
Angélica Atehortúa
Gowtham Atluri

Chloé Audigier
Kamran Avanaki
Angelica Aviles-Rivero
Suyash Awate
Dogu Baran Aydogan
Qinle Ba
Morteza Babaie
Hyeon-Min Bae
Woong Bae
Junjie Bai
Wenjia Bai
Ujjwal Baid
Spyridon Bakas
Yaël Balbastre
Marcin Balicki
Fabian Balsiger
Abhirup Banerjee
Sreya Banerjee
Shunxing Bao
Adrian Barbu
Sumana Basu
Mathilde Bateson
Deepti Bathula
John Baxter
Bahareh Behboodi
Delaram Behnami
Mikhail Belyaev
Aicha BenTaieb

Camilo Bermudez
Gabriel Bernardino
Hadrien Bertrand
Alaa Bessadok
Michael Beyeler
Indrani Bhattacharya
Chetan Bhole
Lei Bi
Gui-Bin Bian
Ryoma Bise
Stefano B. Blumberg
Ester Bonmati
Bhushan Borotikar
Jiri Borovec
Ilaria Boscolo Galazzo
Alexandre Bousse
Nicolas Boutry
Behzad Bozorgtabar
Nathaniel Braman
Nadia Brancati
Katharina Breininger
Christopher Bridge
Esther Bron
Rupert Brooks
Qirong Bu
Duc Toan Bui
Ninon Burgos
Nikolay Burlutskiy
Hendrik Burwinkel
Russell Butler
Michał Byra
Ryan Cabeen
Mariano Cabezas
Hongmin Cai
Jinzheng Cai
Yunliang Cai
Sema Candemir
Bing Cao
Qing Cao
Shilei Cao
Tian Cao
Weiguo Cao
Aaron Carass
M. Jorge Cardoso
Adrià Casamitjana
Matthieu Chabanas

Ahmad Chaddad
Jayasree Chakraborty
Sylvie Chambon
Yi Hao Chan
Ming-Ching Chang
Peng Chang
Violeta Chang
Sudhanya Chatterjee
Christos Chatzichristos
Antong Chen
Chang Chen
Cheng Chen
Dongdong Chen
Geng Chen
Hanbo Chen
Jianan Chen
Jianxu Chen
Jie Chen
Junxiang Chen
Lei Chen
Li Chen
Liangjun Chen
Min Chen
Pingjun Chen
Qiang Chen
Shuai Chen
Tianhua Chen
Tingting Chen
Xi Chen
Xiaoran Chen
Xin Chen
Xuejin Chen
Yuhua Chen
Yukun Chen
Zhaolin Chen
Zhineng Chen
Zhixiang Chen
Erkang Cheng
Jun Cheng
Li Cheng
Yuan Cheng
Farida Cheriet
Minqi Chong
Jaegul Choo
Aritra Chowdhury
Gary Christensen

Daan Christiaens
Stergios Christodoulidis
Ai Wern Chung
Pietro Antonio Cicalese
Özgün Çiçek
Celia Cintas
Matthew Clarkson
Jaume Coll-Font
Toby Collins
Olivier Commowick
Pierre-Henri Conze
Timothy Cootes
Luca Corinzia
Teresa Correia
Hadrien Courtecuisse
Jeffrey Craley
Hui Cui
Jianan Cui
Zhiming Cui
Kathleen Curran
Claire Cury
Tobias Czempiel
Vedrana Dahl
Haixing Dai
Rafat Damseh
Bilel Daoud
Neda Davoudi
Laura Daza
Sandro De Zanet
Charles Delahunt
Yang Deng
Cem Deniz
Felix Denzinger
Hrishikesh Deshpande
Christian Desrosiers
Blake Dewey
Neel Dey
Raunak Dey
Jwala Dhamala
Yashin Dicente Cid
Li Ding
Xinghao Ding
Zhipeng Ding
Konstantin Dmitriev
Ines Domingues
Liang Dong

Mengjin Dong
Nanqing Dong
Reuben Dorent
Sven Dorkenwald
Qi Dou
Simon Drouin
Niharika D'Souza
Lei Du
Hongyi Duanmu
Nicolas Duchateau
James Duncan
Luc Duong
Nicha Dvornek
Dmitry V. Dylov
Oleh Dzyubachyk
Roy Eagleson
Mehran Ebrahimi
Jan Egger
Alma Eguizabal
Gudmundur Einarsson
Ahmed Elazab
Mohammed S. M. Elbaz
Shireen Elhabian
Mohammed Elmogy
Amr Elsawy
Ahmed Eltanboly
Sandy Engelhardt
Ertunc Erdil
Marius Erdt
Floris Ernst
Boris Escalante-Ramírez
Maria Escobar
Mohammad Eslami
Nazila Esmaeili
Marco Esposito
Oscar Esteban
Théo Estienne
Ivan Ezhov
Deng-Ping Fan
Jingfan Fan
Xin Fan
Yonghui Fan
Xi Fang
Zhenghan Fang
Aly Farag
Mohsen Farzi

Lina Felsner
Jun Feng
Ruibin Feng
Xinyang Feng
Yuan Feng
Aaron Fenster
Aasa Feragen
Henrique Fernandes
Enzo Ferrante
Jean Feydy
Lukas Fischer
Peter Fischer
Antonio Foncubierta-Rodríguez
Germain Forestier
Nils Daniel Forkert
Jean-Rassaire Fouefack
Moti Freiman
Wolfgang Freysinger
Xueyang Fu
Yunguan Fu
Wolfgang Fuhl
Isabel Funke
Philipp Fürnstahl
Pedro Furtado
Ryo Furukawa
Jin Kyu Gahm
Laurent Gajny
Adrian Galdran
Yu Gan
Melanie Ganz
Cong Gao
Dongxu Gao
Linlin Gao
Siyuan Gao
Yixin Gao
Yue Gao
Zhifan Gao
Alfonso Gastelum-Strozzi
Srishti Gautam
Bao Ge
Rongjun Ge
Zongyuan Ge
Sairam Geethanath
Shiv Gehlot
Nils Gessert
Olivier Gevaert

Sandesh Ghimire
Ali Gholipour
Sayan Ghosal
Andrea Giovannini
Gabriel Girard
Ben Glocker
Arnold Gomez
Mingming Gong
Cristina González
German Gonzalez
Sharath Gopal
Karthik Gopinath
Pietro Gori
Michael Götz
Shuiping Gou
Maged Goubran
Sobhan Goudarzi
Dushyant Goyal
Mark Graham
Bertrand Granado
Alejandro Granados
Vicente Grau
Lin Gu
Shi Gu
Xianfeng Gu
Yun Gu
Zaiwang Gu
Hao Guan
Ricardo Guerrero
Houssem-Eddine Gueziri
Dazhou Guo
Hengtao Guo
Jixiang Guo
Pengfei Guo
Xiaoqing Guo
Yi Guo
Yulan Guo
Yuyu Guo
Krati Gupta
Vikash Gupta
Praveen Gurunath Bharathi
Boris Gutman
Prashnna Gyawali
Stathis Hadjidemetriou
Mohammad Hamghalam
Hu Han

Liang Han
Xiaoguang Han
Xu Han
Zhi Han
Zhongyi Han
Jonny Hancox
Xiaoke Hao
Nandinee Haq
Ali Hatamizadeh
Charles Hatt
Andreas Hauptmann
Mohammad Havaei
Kelei He
Nanjun He
Tiancheng He
Xuming He
Yuting He
Nicholas Heller
Alessa Hering
Monica Hernandez
Carlos Hernandez-Matas
Kilian Hett
Jacob Hinkle
David Ho
Nico Hoffmann
Matthew Holden
Sungmin Hong
Yoonmi Hong
Antal Horváth
Md Belayat Hossain
Benjamin Hou
William Hsu
Tai-Chiu Hsung
Kai Hu
Shi Hu
Shunbo Hu
Wenxing Hu
Xiaoling Hu
Xiaowei Hu
Yan Hu
Zhenhong Hu
Heng Huang
Qiaoying Huang
Yi-Jie Huang
Yixing Huang
Yongxiang Huang

Yue Huang
Yufang Huang
Arnaud Huaulmé
Henkjan Huisman
Yuankai Huo
Andreas Husch
Mohammad Hussain
Raabid Hussain
Sarfaraz Hussein
Khoi Huynh
Seong Jae Hwang
Emmanuel Iarussi
Kay Igwe
Abdullah-Al-Zubaer Imran
Ismail Irmakci
Mobarakol Islam
Mohammad Shafkat Islam
Vamsi Ithapu
Koichi Ito
Hayato Itoh
Oleksandra Ivashchenko
Yuji Iwahori
Shruti Jadon
Mohammad Jafari
Mostafa Jahanifar
Amir Jamaludin
Mirek Janatka
Won-Dong Jang
Uditha Jarayathne
Ronnachai Jaroensri
Golara Javadi
Rohit Jena
Rachid Jennane
Todd Jensen
Won-Ki Jeong
Yuanfeng Ji
Zhanghexuan Ji
Haozhe Jia
Jue Jiang
Tingting Jiang
Xiang Jiang
Jianbo Jiao
Zhicheng Jiao
Amelia Jiménez-Sánchez
Dakai Jin
Yueming Jin

Bin Jing
Anand Joshi
Yohan Jun
Kyu-Hwan Jung
Alain Jungo
Manjunath K N
Ali Kafaei Zad Tehrani
Bernhard Kainz
John Kalafut
Michael C. Kampffmeyer
Qingbo Kang
Po-Yu Kao
Neerav Karani
Turkay Kart
Satyananda Kashyap
Amin Katouzian
Alexander Katzmann
Prabhjot Kaur
Erwan Kerrien
Hoel Kervadec
Ashkan Khakzar
Nadieh Khalili
Siavash Khallaghi
Farzad Khalvati
Bishesh Khanal
Pulkit Khandelwal
Maksim Kholiavchenko
Naji Khosravan
Seyed Mostafa Kia
Daeseung Kim
Hak Gu Kim
Hyo-Eun Kim
Jae-Hun Kim
Jaeil Kim
Jinman Kim
Mansu Kim
Namkug Kim
Seong Tae Kim
Won Hwa Kim
Andrew King
Atilla Kiraly
Yoshiro Kitamura
Tobias Klinder
Bin Kong
Jun Kong
Tomasz Konopczynski

Bongjin Koo
Ivica Kopriva
Kivanc Kose
Mateusz Kozinski
Anna Kreshuk
Anithapriya Krishnan
Pavitra Krishnaswamy
Egor Krivov
Frithjof Kruggel
Alexander Krull
Elizabeth Krupinski
Serife Kucur
David Kügler
Hugo Kuijf
Abhay Kumar
Ashnil Kumar
Kuldeep Kumar
Nitin Kumar
Holger Kunze
Tahsin Kurc
Anvar Kurmukov
Yoshihiro Kuroda
Jin Tae Kwak
Yongchan Kwon
Francesco La Rosa
Aymen Laadhari
Dmitrii Lachinov
Alain Lalande
Tryphon Lambrou
Carole Lartizien
Bianca Lassen-Schmidt
Ngan Le
Leo Lebrat
Christian Ledig
Eung-Joo Lee
Hyekyoung Lee
Jong-Hwan Lee
Matthew Lee
Sangmin Lee
Soochahn Lee
Étienne Léger
Stefan Leger
Andreas Leibetseder
Rogers Jeffrey Leo John
Juan Leon
Bo Li

Chongyi Li
Fuhai Li
Hongming Li
Hongwei Li
Jian Li
Jianning Li
Jiayun Li
Junhua Li
Kang Li
Mengzhang Li
Ming Li
Qing Li
Shaohua Li
Shuyu Li
Weijian Li
Weikai Li
Wenqi Li
Wenyuan Li
Xiang Li
Xiaomeng Li
Xiaoxiao Li
Xin Li
Xiuli Li
Yang Li
Yi Li
Yuexiang Li
Zeju Li
Zhang Li
Zhiyuan Li
Zhjin Li
Gongbo Liang
Jianming Liang
Libin Liang
Yuan Liang
Haofu Liao
Ruizhi Liao
Wei Liao
Xiangyun Liao
Roxane Licandro
Gilbert Lim
Baihan Lin
Hongxiang Lin
Jianyu Lin
Yi Lin
Claudia Lindner
Geert Litjens

Bin Liu
Chi Liu
Daochang Liu
Dong Liu
Dongnan Liu
Feng Liu
Hangfan Liu
Hong Liu
Huafeng Liu
Jianfei Liu
Jingya Liu
Kai Liu
Kefei Liu
Lihao Liu
Mengting Liu
Peng Liu
Qin Liu
Quande Liu
Shengfeng Liu
Shenghua Liu
Shuangjun Liu
Sidong Liu
Siqi Liu
Tianrui Liu
Xiao Liu
Xinyang Liu
Xinyu Liu
Yan Liu
Yikang Liu
Yong Liu
Yuan Liu
Yue Liu
Yuhang Liu
Andrea Loddo
Nicolas Loménie
Daniel Lopes
Bin Lou
Jian Lou
Nicolas Loy Rodas
Donghuan Lu
Huanxiang Lu
Weijia Lu
Xiankai Lu
Yongyi Lu
Yueh-Hsun Lu
Yuhang Lu

Imanol Luengo
Jie Luo
Jiebo Luo
Luyang Luo
Ma Luo
Bin Lv
Jinglei Lv
Junyan Lyu
Qing Lyu
Yuanyuan Lyu
Andy J. Ma
Chunwei Ma
Da Ma
Hua Ma
Kai Ma
Lei Ma
Anderson Maciel
Amirreza Mahbod
S. Sara Mahdavi
Mohammed Mahmoud
Saïd Mahmoudi
Klaus H. Maier-Hein
Bilal Malik
Ilja Manakov
Matteo Mancini
Tommaso Mansi
Yunxiang Mao
Brett Marinelli
Pablo Márquez Neila
Carsten Marr
Yassine Marrakchi
Fabio Martinez
Andre Mastmeyer
Tejas Sudharshan Mathai
Dimitrios Mavroeidis
Jamie McClelland
Pau Medrano-Gracia
Raghav Mehta
Sachin Mehta
Raphael Meier
Qier Meng
Qingjie Meng
Yanda Meng
Martin Menten
Odyssée Merveille
Islem Mhiri

Liang Mi
Stijn Michielse
Abhishek Midya
Fausto Milletari
Hyun-Seok Min
Zhe Min
Tadashi Miyamoto
Sara Moccia
Hassan Mohy-ud-Din
Tony C. W. Mok
Rafael Molina
Mehdi Moradi
Rodrigo Moreno
Kensaku Mori
Lia Morra
Linda Moy
Mohammad Hamed Mozaffari
Sovanlal Mukherjee
Anirban Mukhopadhyay
Henning Müller
Balamurali Murugesan
Cosmas Mwikirize
Andriy Myronenko
Saad Nadeem
Vishwesh Nath
Rodrigo Nava
Fernando Navarro
Amin Nejatbakhsh
Dong Ni
Hannes Nickisch
Dong Nie
Jingxin Nie
Aditya Nigam
Lipeng Ning
Xia Ning
Tianye Niu
Jack Noble
Vincent Noblet
Alexey Novikov
Jorge Novo
Mohammad Obeid
Masahiro Oda
Benjamin Odry
Steffen Oeltze-Jafra
Hugo Oliveira
Sara Oliveira

Arnau Oliver
Emanuele Olivetti
Jimena Olveres
John Onofrey
Felipe Orihuela-Espina
José Orlando
Marcos Ortega
Yoshito Otake
Sebastian Otálora
Cheng Ouyang
Jiahong Ouyang
Xi Ouyang
Michal Ozery-Flato
Danielle Pace
Krittin Pachtrachai
J. Blas Pagador
Akshay Pai
Viswanath Pamulakanty Sudarshan
Jin Pan
Yongsheng Pan
Pankaj Pandey
Prashant Pandey
Egor Panfilov
Shumao Pang
Joao Papa
Constantin Pape
Bartlomiej Papiez
Hyunjin Park
Jongchan Park
Sanghyun Park
Seung-Jong Park
Seyoun Park
Magdalini Paschali
Diego Patiño Cortés
Angshuman Paul
Christian Payer
Yuru Pei
Chengtao Peng
Yige Peng
Antonio Pepe
Oscar Perdomo
Sérgio Pereira
Jose-Antonio Pérez-Carrasco
Fernando Pérez-García
Jorge Perez-Gonzalez
Skand Peri

Matthias Perkonigg
Mehran Pesteie
Jorg Peters
Jens Petersen
Kersten Petersen
Renzo Phellan Aro
Ashish Phophalia
Tomasz Pieciak
Antonio Pinheiro
Pramod Pisharady
Kilian Pohl
Sebastian Pölsterl
Iulia A. Popescu
Alison Pouch
Prateek Prasanna
Raphael Prevost
Juan Prieto
Sergi Pujades
Elodie Puybareau
Esther Puyol-Antón
Haikun Qi
Huan Qi
Buyue Qian
Yan Qiang
Yuchuan Qiao
Chen Qin
Wenjian Qin
Yulei Qin
Wu Qiu
Hui Qu
Liangqiong Qu
Kha Gia Quach
Prashanth R.
Pradeep Reddy Raamana
Mehdi Rahim
Jagath Rajapakse
Kashif Rajpoot
Jhonata Ramos
Lingyan Ran
Hatem Rashwan
Daniele Ravì
Keerthi Sravan Ravi
Nishant Ravikumar
Harish RaviPrakash
Samuel Remedios
Yinhao Ren

Yudan Ren
Mauricio Reyes
Constantino Reyes-Aldasoro
Jonas Richiardi
David Richmond
Anne-Marie Rickmann
Leticia Rittner
Dominik Rivoir
Emma Robinson
Jessica Rodgers
Rafael Rodrigues
Robert Rohling
Michal Rosen-Zvi
Lukasz Roszkowiak
Karsten Roth
José Rouco
Daniel Rueckert
Jaime S. Cardoso
Mohammad Sabokrou
Ario Sadafi
Monjoy Saha
Pramit Saha
Dushyant Sahoo
Pranjal Sahu
Maria Sainz de Cea
Olivier Salvado
Robin Sandkuehler
Gianmarco Santini
Duygu Sarikaya
Imari Sato
Olivier Saut
Dustin Scheinost
Nico Scherf
Markus Schirmer
Alexander Schlaefer
Jerome Schmid
Julia Schnabel
Klaus Schoeffmann
Andreas Schuh
Ernst Schwartz
Christina Schwarz-Gsaxner
Michaël Sdika
Suman Sedai
Anjany Sekuboyina
Raghavendra Selvan
Sourya Sengupta

Youngho Seo
Lama Seoud
Ana Sequeira
Maxime Sermesant
Carmen Serrano
Muhammad Shaban
Ahmed Shaffie
Sobhan Shafiei
Mohammad Abuzar Shaikh
Reuben Shamir
Shayan Shams
Hongming Shan
Harshita Sharma
Gregory Sharp
Mohamed Shehata
Haocheng Shen
Li Shen
Liyue Shen
Mali Shen
Yiqing Shen
Yiqiu Shen
Zhengyang Shen
Kuangyu Shi
Luyao Shi
Xiaoshuang Shi
Xueying Shi
Yemin Shi
Yiyu Shi
Yonghong Shi
Jitae Shin
Boris Shirokikh
Suprosanna Shit
Suzanne Shontz
Yucheng Shu
Alberto Signoroni
Wilson Silva
Margarida Silveira
Matthew Sinclair
Rohit Singla
Sumedha Singla
Ayushi Sinha
Kevin Smith
Rajath Soans
Ahmed Soliman
Stefan Sommer
Yang Song

Youyi Song
Aristeidis Sotiras
Arcot Sowmya
Rachel Sparks
William Speier
Ziga Spiclin
Dominik Spinczyk
Jon Sporring
Chetan Srinidhi
Anuroop Sriram
Vinkle Srivastav
Lawrence Staib
Marius Staring
Johannes Stegmaier
Joshua Stough
Robin Strand
Martin Styner
Hai Su
Yun-Hsuan Su
Vaishnavi Subramanian
Gérard Subsol
Yao Sui
Avan Suinesiaputra
Jeremias Sulam
Shipra Suman
Li Sun
Wenqing Sun
Chiranjib Sur
Yannick Suter
Tanveer Syeda-Mahmood
Fatemeh Taheri Dezaki
Roger Tam
José Tamez-Peña
Chaowei Tan
Hao Tang
Thomas Tang
Yucheng Tang
Zihao Tang
Mickael Tardy
Giacomo Tarroni
Jonas Teuwen
Paul Thienphrapa
Stephen Thompson
Jiang Tian
Yu Tian
Yun Tian

Aleksei Tiulpin
Hamid Tizhoosh
Matthew Toews
Oguzhan Topsakal
Antonio Torteya
Sylvie Treuillet
Jocelyne Troccaz
Roger Trullo
Chialing Tsai
Sudhakar Tummala
Verena Uslar
Hristina Uzunova
Régis Vaillant
Maria Vakalopoulou
Jeya Maria Jose Valanarasu
Tom van Sonsbeek
Gijs van Tulder
Marta Varela
Thomas Varsavsky
Francisco Vasconcelos
Liset Vazquez Romaguera
S. Swaroop Vedula
Sanketh Vedula
Harini Veeraraghavan
Miguel Vega
Gonzalo Vegas Sanchez-Ferrero
Anant Vemuri
Gopalkrishna Veni
Mitko Veta
Thomas Vetter
Pedro Vieira
Juan Pedro Vigueras Guillén
Barbara Villarini
Satish Viswanath
Athanasios Vlontzos
Wolf-Dieter Vogl
Bo Wang
Cheng Wang
Chengjia Wang
Chunliang Wang
Clinton Wang
Congcong Wang
Dadong Wang
Dongang Wang
Haifeng Wang
Hongyu Wang

Hu Wang
Huan Wang
Kun Wang
Li Wang
Liansheng Wang
Linwei Wang
Manning Wang
Renzhen Wang
Ruixuan Wang
Sheng Wang
Shujun Wang
Shuo Wang
Tianchen Wang
Tongxin Wang
Wenzhe Wang
Xi Wang
Xiaosong Wang
Yan Wang
Yaping Wang
Yi Wang
Yirui Wang
Zeyi Wang
Zhangyang Wang
Zihao Wang
Zuhui Wang
Simon Warfield
Jonathan Weber
Jürgen Weese
Dong Wei
Donglai Wei
Dongming Wei
Martin Weigert
Wolfgang Wein
Michael Wels
Cédric Wemmert
Junhao Wen
Travis Williams
Matthias Wilms
Stefan Winzeck
James Wiskin
Adam Wittek
Marek Wodzinski
Jelmer Wolterink
Ken C. L. Wong
Chongruo Wu
Guoqing Wu

Ji Wu
Jian Wu
Jie Ying Wu
Pengxiang Wu
Xiyin Wu
Ye Wu
Yicheng Wu
Yifan Wu
Tobias Wuerfl
Pengcheng Xi
James Xia
Siyu Xia
Wenfeng Xia
Yingda Xia
Yong Xia
Lei Xiang
Deqiang Xiao
Li Xiao
Yiming Xiao
Hongtao Xie
Lingxi Xie
Long Xie
Weidi Xie
Yiting Xie
Yutong Xie
Xiaohan Xing
Chang Xu
Chenchu Xu
Hongming Xu
Kele Xu
Min Xu
Rui Xu
Xiaowei Xu
Xuanang Xu
Yongchao Xu
Zhenghua Xu
Zhoubing Xu
Kai Xuan
Cheng Xue
Jie Xue
Wufeng Xue
Yuan Xue
Faridah Yahya
Ke Yan
Yuguang Yan
Zhennan Yan

Changchun Yang
Chao-Han Huck Yang
Dong Yang
Erkun Yang
Fan Yang
Ge Yang
Guang Yang
Guanyu Yang
Heran Yang
Hongxu Yang
Huijuan Yang
Jiancheng Yang
Jie Yang
Junlin Yang
Lin Yang
Peng Yang
Xin Yang
Yan Yang
Yujiu Yang
Dongren Yao
Jiawen Yao
Li Yao
Qingsong Yao
Chuyang Ye
Dong Hye Ye
Menglong Ye
Xujiong Ye
Jingru Yi
Jirong Yi
Xin Yi
Youngjin Yoo
Chenyu You
Haichao Yu
Hanchao Yu
Lequan Yu
Qi Yu
Yang Yu
Pengyu Yuan
Fatemeh Zabihollahy
Ghada Zamzmi
Marco Zenati
Guodong Zeng
Rui Zeng
Oliver Zettinig
Zhiwei Zhai
Chaoyi Zhang

Daoqiang Zhang
Fan Zhang
Guangming Zhang
Hang Zhang
Huahong Zhang
Jianpeng Zhang
Jiong Zhang
Jun Zhang
Lei Zhang
Lichi Zhang
Lin Zhang
Ling Zhang
Lu Zhang
Miaomiao Zhang
Ning Zhang
Qiang Zhang
Rongzhao Zhang
Ru-Yuan Zhang
Shihao Zhang
Shu Zhang
Tong Zhang
Wei Zhang
Weiwei Zhang
Wen Zhang
Wenlu Zhang
Xin Zhang
Ya Zhang
Yanbo Zhang
Yanfu Zhang
Yi Zhang
Yishuo Zhang
Yong Zhang
Yongqin Zhang
You Zhang
Youshan Zhang
Yu Zhang
Yue Zhang
Yueyi Zhang
Yulun Zhang
Yunyan Zhang
Yuyao Zhang
Can Zhao
Changchen Zhao
Chongyue Zhao
Fenqiang Zhao
Gangming Zhao

He Zhao
Jun Zhao
Li Zhao
Qingyu Zhao
Rongchang Zhao
Shen Zhao
Shijie Zhao
Tengda Zhao
Tianyi Zhao
Wei Zhao
Xuandong Zhao
Yiyuan Zhao
Yuan-Xing Zhao
Yue Zhao
Zixu Zhao
Ziyuan Zhao
Xingjian Zhen
Guoyan Zheng
Hao Zheng
Jiannan Zheng
Kang Zheng
Shenhai Zheng
Yalin Zheng
Yinqiang Zheng
Yushan Zheng
Jia-Xing Zhong
Zichun Zhong

Bo Zhou
Haoyin Zhou
Hong-Yu Zhou
Kang Zhou
Sanping Zhou
Sihang Zhou
Tao Zhou
Xiao-Yun Zhou
Yanning Zhou
Yuyin Zhou
Zongwei Zhou
Dongxiao Zhu
Hancan Zhu
Lei Zhu
Qikui Zhu
Xinliang Zhu
Yuemin Zhu
Zhe Zhu
Zhuotun Zhu
Aneeq Zia
Veronika Zimmer
David Zimmerer
Lilla Zöllei
Yukai Zou
Lianrui Zuo
Gerald Zwettler
Reyer Zwiggelaar

Outstanding Reviewers

Neel Dey — New York University, USA
Monica Hernandez — University of Zaragoza, Spain
Ivica Kopriva — Rudjer Boskovich Institute, Croatia
Sebastian Otálora — University of Applied Sciences and Arts Western Switzerland, Switzerland
Danielle Pace — Massachusetts General Hospital, USA
Sérgio Pereira — Lunit Inc., South Korea
David Richmond — IBM Watson Health, USA
Rohit Singla — University of British Columbia, Canada
Yan Wang — Sichuan University, China

Honorable Mentions (Reviewers)

Mazdak Abulnaga	Massachusetts Institute of Technology, USA
Pierre Ambrosini	Erasmus University Medical Center, The Netherlands
Hyeon-Min Bae	Korea Advanced Institute of Science and Technology, South Korea
Mikhail Belyaev	Skolkovo Institute of Science and Technology, Russia
Bhushan Borotikar	Symbiosis International University, India
Katharina Breininger	Friedrich-Alexander-Universität Erlangen-Nürnberg, Germany
Ninon Burgos	CNRS, Paris Brain Institute, France
Mariano Cabezas	The University of Sydney, Australia
Aaron Carass	Johns Hopkins University, USA
Pierre-Henri Conze	IMT Atlantique, France
Christian Desrosiers	École de technologie supérieure, Canada
Reuben Dorent	King's College London, UK
Nicha Dvornek	Yale University, USA
Dmitry V. Dylov	Skolkovo Institute of Science and Technology, Russia
Marius Erdt	Fraunhofer Singapore, Singapore
Ruibin Feng	Stanford University, USA
Enzo Ferrante	CONICET/Universidad Nacional del Litoral, Argentina
Antonio Foncubierta-Rodríguez	IBM Research, Switzerland
Isabel Funke	National Center for Tumor Diseases Dresden, Germany
Adrian Galdran	University of Bournemouth, UK
Ben Glocker	Imperial College London, UK
Cristina González	Universidad de los Andes, Colombia
Maged Goubran	Sunnybrook Research Institute, Canada
Sobhan Goudarzi	Concordia University, Canada
Vicente Grau	University of Oxford, UK
Andreas Hauptmann	University of Oulu, Finland
Nico Hoffmann	Technische Universität Dresden, Germany
Sungmin Hong	Massachusetts General Hospital, Harvard Medical School, USA
Won-Dong Jang	Harvard University, USA
Zhanghexuan Ji	University at Buffalo, SUNY, USA
Neerav Karani	ETH Zurich, Switzerland
Alexander Katzmann	Siemens Healthineers, Germany
Erwan Kerrien	Inria, France
Anitha Priya Krishnan	Genentech, USA
Tahsin Kurc	Stony Brook University, USA
Francesco La Rosa	École polytechnique fédérale de Lausanne, Switzerland
Dmitrii Lachinov	Medical University of Vienna, Austria
Mengzhang Li	Peking University, China
Gilbert Lim	National University of Singapore, Singapore
Dongnan Liu	University of Sydney, Australia

Bin Lou	Siemens Healthineers, USA
Kai Ma	Tencent, China
Klaus H. Maier-Hein	German Cancer Research Center (DKFZ), Germany
Raphael Meier	University Hospital Bern, Switzerland
Tony C. W. Mok	Hong Kong University of Science and Technology, Hong Kong SAR
Lia Morra	Politecnico di Torino, Italy
Cosmas Mwikirize	Rutgers University, USA
Felipe Orihuela-Espina	Instituto Nacional de Astrofísica, Óptica y Electrónica, Mexico
Egor Panfilov	University of Oulu, Finland
Christian Payer	Graz University of Technology, Austria
Sebastian Pölsterl	Ludwig-Maximilians Universität, Germany
José Rouco	University of A Coruña, Spain
Daniel Rueckert	Imperial College London, UK
Julia Schnabel	King's College London, UK
Christina Schwarz-Gsaxner	Graz University of Technology, Austria
Boris Shirokikh	Skolkovo Institute of Science and Technology, Russia
Yang Song	University of New South Wales, Australia
Gérard Subsol	Université de Montpellier, France
Tanveer Syeda-Mahmood	IBM Research, USA
Mickael Tardy	Hera-MI, France
Paul Thienphrapa	Atlas5D, USA
Gijs van Tulder	Radboud University, The Netherlands
Tongxin Wang	Indiana University, USA
Yirui Wang	PAII Inc., USA
Jelmer Wolterink	University of Twente, The Netherlands
Lei Xiang	Subtle Medical Inc., USA
Fatemeh Zabihollahy	Johns Hopkins University, USA
Wei Zhang	University of Georgia, USA
Ya Zhang	Shanghai Jiao Tong University, China
Qingyu Zhao	Stanford University, China
Yushan Zheng	Beihang University, China

Mentorship Program (Mentors)

Shadi Albarqouni	Helmholtz AI, Helmholtz Center Munich, Germany
Hao Chen	Hong Kong University of Science and Technology, Hong Kong SAR
Nadim Daher	NVIDIA, France
Marleen de Bruijne	Erasmus MC/University of Copenhagen, The Netherlands
Qi Dou	The Chinese University of Hong Kong, Hong Kong SAR
Gabor Fichtinger	Queen's University, Canada
Jonny Hancox	NVIDIA, UK

Nobuhiko Hata	Harvard Medical School, USA
Sharon Xiaolei Huang	Pennsylvania State University, USA
Jana Hutter	King's College London, UK
Dakai Jin	PAII Inc., China
Samuel Kadoury	Polytechnique Montréal, Canada
Minjeong Kim	University of North Carolina at Greensboro, USA
Hans Lamecker	1000shapes GmbH, Germany
Andrea Lara	Galileo University, Guatemala
Ngan Le	University of Arkansas, USA
Baiying Lei	Shenzhen University, China
Karim Lekadir	Universitat de Barcelona, Spain
Marius George Linguraru	Children's National Health System/George Washington University, USA
Herve Lombaert	ETS Montreal, Canada
Marco Lorenzi	Inria, France
Le Lu	PAII Inc., China
Xiongbiao Luo	Xiamen University, China
Dzung Pham	Henry M. Jackson Foundation/Uniformed Services University/National Institutes of Health/Johns Hopkins University, USA
Josien Pluim	Eindhoven University of Technology/University Medical Center Utrecht, The Netherlands
Antonio Porras	University of Colorado Anschutz Medical Campus/Children's Hospital Colorado, USA
Islem Rekik	Istanbul Technical University, Turkey
Nicola Rieke	NVIDIA, Germany
Julia Schnabel	TU Munich/Helmholtz Center Munich, Germany, and King's College London, UK
Debdoot Sheet	Indian Institute of Technology Kharagpur, India
Pallavi Tiwari	Case Western Reserve University, USA
Jocelyne Troccaz	CNRS, TIMC, Grenoble Alpes University, France
Sandrine Voros	TIMC-IMAG, INSERM, France
Linwei Wang	Rochester Institute of Technology, USA
Yalin Wang	Arizona State University, USA
Zhong Xue	United Imaging Intelligence Co. Ltd, USA
Renee Yao	NVIDIA, USA
Mohammad Yaqub	Mohamed Bin Zayed University of Artificial Intelligence, United Arab Emirates, and University of Oxford, UK
S. Kevin Zhou	University of Science and Technology of China, China
Lilla Zollei	Massachusetts General Hospital, Harvard Medical School, USA
Maria A. Zuluaga	EURECOM, France

Contents – Part IV

Surgical Data Science

Surgical Planning and Simulation

Surgical Skill and Work Flow Analysis

Surgical Visualization and Mixed, Augmented and Virtual Reality

Image Registration

Medical Image Registration Based on Uncoupled Learning and Accumulative Enhancement

Yucheng Shu[1,2], Hao Wang[1,2], Bin Xiao[1,2(✉)], Xiuli Bi[1,2], and Weisheng Li[1,2]

[1] Chongqing University of Posts and Telecommunications, Chongqing 400065, China
{shuyc,xiaobin,bixl,liws}@cqupt.edu.cn, s190201027@stu.cqupt.edu.cn
[2] Chongqing Key Laboratory of Image Cognition, Chongqing 400065, China

Abstract. As a basic building block in medical image analysis, image registration has been greatly developed since the emergence of modern deep neural networks. Compared to non-learning-based methods, the latest approaches can learn task-specific features spontaneously, thus generate the registration results with one round of inference. However, when large inter-image distortion occurs, the stability of existing methods can be strongly affected. To alleviate this problem, the iterative framework based on coarse-to-fine strategies has been introduced in recent works. However, their networks at each iteration step are relatively independent, which is not an optimal solution for the reinforcement of image features. What is more, the moving and the fixed images are often concatenated or fed to identical network layers. Consequently, the iterative learning and warping on the moving image can be entangled with the fixed image. In order to address these issues, we present a novel medical image registration framework, namely ULAE-net, to continuously enhance the spatial transformation and establish more profound contextual dependencies under a compact network layout. Extensive experiments on 3D brain MRI data sets demonstrate that our method has greatly improved the registration performance, thereby outperforms state-of-the-art methods under large-scale deformations (https://github.com/wanghaostu/ULAE-net).

Keywords: Medical image registration · Uncoupled learning · Accumulative enhancement

1 Introduction

Deformable image registration plays an essential role in the field of medical image analysis and diagnosis. By aligning the anatomical structures in the moving and the fixed image, it can lay a solid foundation for many subsequent tasks such as

This research was funded in part by the National Natural Science Foundation of China 61906024, 61976031, and 61801068, National Major Scientific Research Instrument Development Project of China 62027827, National Key R&D Program of China 2019YFE0110800, 2016YFC1000307-3.

M. de Bruijne et al. (Eds.): MICCAI 2021, LNCS 12904, pp. 3–13, 2021.
https://doi.org/10.1007/978-3-030-87202-1_1

medical image fusion [5,9], medical image reconstruction [7,13], medical image segmentation [16,20], etc.

In the traditional non-learning-based approaches [22–24], deformable registration is often performed in an iterative manner. By gradually minimizing a predefined spatial-structural energy function, they have achieved great accuracy in most of the medical image registration tasks. However, when the large-scale spatial displacement exists between the input images (e.g. 3D brain MRI), the optimization algorithms may become time-consuming, and more likely, be trapped into local minimum.

Recently, with the theoretical development of machine learning techniques, deep-learning-based methods have shown promising quality and speed in a variety of medical image registration tasks [12,21]. For instance, by employing a U-shape network to calculate the pixel-level deformation field, Voxelmorph [2,4] is able to generate 3D registration result through one round of network inference, which has tremendously sped up the medical image registration process. Nevertheless, even with the help of automatic feature learning, it is also a great challenge when two scans have large differences in appearance.

Although large data transformations can be dealt with by building long-range dependencies via pooling operation, the spatial resolution of the image features will be affected at the network bottleneck. According to our experiments, it may reduce the network's ability to generate local subtle deformations, thus resulting in the decrease of registration accuracy. To circumvent this issue, some existing methods attempt to apply the coarse-to-fine strategy with multiple registration steps. For instance, Dual-PRNet [10] was proposed to calculate multi-scale deformation fields based on two separated network branches. Hering et al. [8] proposed a hierarchical layout to align the inhale-to-exhale lung scans. Zhao et al. [25] designed the Recursive Cascaded Networks to register images iteratively.

However, the methods mentioned above still have certain limitations. Firstly, the net branches at each iteration are often relatively independent, which may directly increase the network complexity. More importantly, such discrete optimizing strategy cannot make full use of the cumulative ability of the data-driven learning, thus the spatial transformations may not be continuously enhanced. What is more, the moving and the fixed images are often concatenated as the input of a single net routine, or independently fed to similar feature encoders, which is relatively rigid and may not be suitable for the iterative learning and warping of the moving images. In order to address these issues, in this paper, we propose a novel medical image registration framework, namely ULAE-net, based on uncoupled feature learning and spatial accumulative enhancement.

The main contributions of this work can be summarized as follows:

–We present a new accumulative medical image registration framework to effectively enhance the spatial transformations from continuous optimizations under a more compact network layout, thereby construct a balanced combination based on the coarse-to-fine strategy and informative feature learning.

–By introducing a novel uncoupled feature learning algorithm, we are able to capture richer long-range dependencies and build better visual correlations between the moving and fixed images for the registration task under large deformations.

–Finally, a simple yet effective multi-window loss is designed to cooperate with our proposed framework, to further expand the network's learning potentials while avoiding the local minimum. Extensive experiments on two large scale MR brain data sets, Mindboggle101 and IXI, show that our proposed algorithm greatly improved the registration accuracy and robustness, thereby outperforms the SOTAs among traditional and deep-learning-based methods.

2   Method

2.1   Overview

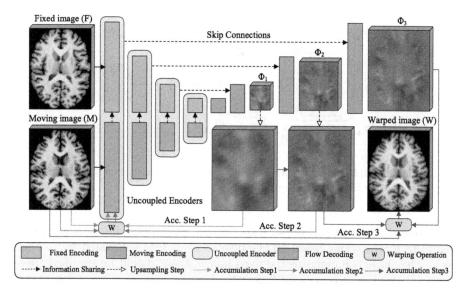

Fig. 1. Overall architecture of our proposed registration framework.

As shown in Fig. 1, given a fixed 3D image F and a moving 3D image M, the output of our registration model is a 3-channel deformation field ϕ which containing the voxel-to-voxel correspondence:

$$\phi = U_\theta(M, F) \tag{1}$$

where U corresponds to our proposed network with parameter θ. Specifically, we employ the U-shape [19] architecture with encoding routine to learn contextual features, and decoding routine to generate spatial displacements. Skip connection is also applied for gradient transduction and information sharing. Typically, the $2 \times 2 \times 2$ stride convolution is used in the encoder, and $\times 2$ trilinear up-sampling and $3 \times 3 \times 3$ convolution is applied in the decoder. As introduced in the previous section, our basic rationale is to devise a compact network layout under an unsupervised framework, thus address the large-deformation problem by learning and accumulating long-range information. Therefore, a novel uncoupled learning scheme is proposed to feed the moving and fixed image into different learning paths, and then form an integrated encoder with feature fusion. At the end of the decoder, an accumulative enhancement mechanism is proposed to cooperate the moving encoding path to continually enhance the network's potential transformation ability. Detailed introductions will be given in the following sections.

2.2   Uncoupled Spatial Encoder

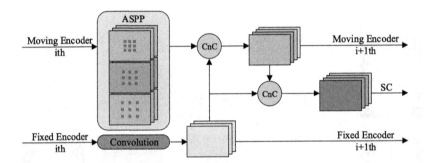

Fig. 2. The Uncoupled Spatial Encoder. CnC denotes Concatenation and Convolution, SC corresponds to Skip Connection.

During the voxel-wised registration process, the moving image will undergo spatial transformations based on the displacement field calculated by the network. Therefore, compared to the fixed image, there should be specified treatments imposed on the moving image to discover and learn more long-range dependencies. However, in the standard U-shape registration framework, the moving and fixed images are usually concatenated as the input of a same encoding routine, or independently feed to similar feature encoders, which is not an optimal solution, especially under large image deformations.

Therefore, in order to capture both long-range and short-range corresponding information between the moving and fixed image, a novel encoder structure, namely Uncoupled Spatial Encoder, is proposed for the deformable image registration. As shown in Fig. 2, the learning path of moving and fixed images are separated into two different data routines. Specifically, we apply the typical stride convolution on fixed image encoding path, and utilized the atrous spatial pyramid pooling [3] (ASPP) as the basic building block in the moving image encoding module. ASPP layer use dilated convolutions with several different sampling rates to expand the receptive field of moving routine, which is far more flexible, and can capture the context information at multi-scales. Moreover, as shown in Fig. 2, in order to learn a better contextual relationship between two images, we also apply an alternating fusion module based on concatenation and convolution, to append the fixed encoding information to the moving encoding routine. Under this feature learning mechanism, the network is able to capture much more longer-range information at each voxel and acquire a more flexible transformation field for the moving image.

2.3 Accumulative Warping Enhancement

As analyzed in the previous section, although the coarse-to-fine strategy has shown its effectiveness in large-scale deformable registration [8,10,17,25], the iterations are often performed with independent encoders and decoders, which has limitations in terms of feature learning and computational complexity. In order to continuously enhance the network's transformation ability, we propose a novel spatial accumulative enhancement mechanism. Specifically, three accumulative steps (Acc_1, Acc_2, Acc_3) are employed within the proposed network. Note that the accumulative iterations are performed compactly without the need for new network branches or extra weight sharing.

In the beginning, F and M are passed through 4 uncoupled encoder blocks to capture semantic information, and the first two decoder blocks will transmit the encoded feature to the coarsest deformation field ϕ_1. Then, ϕ_1 is up-sampled to the original shape, and the moving image M is warped accordingly to obtain roughly aligned image M_1. M_1 and fixed image F is re-input to the network as the next accumulative step. The calculation of Acc_2 and Acc_3 is similar to Acc_1, but they will utilize one and two more decoder layers, respectively. At last, the finest flow ϕ_3 will be generated at the final step, and the transformation field ϕ from M to F can be obtained by accumulating the ϕ_1, ϕ_2, and ϕ_3 as:

$$\phi = S^{2,3}(S^{1,2}(\phi_1) \circ \phi_2 + \phi_2) \circ \phi_3 + \phi_3 \qquad (2)$$

Where $\circ$ is the warping operation based on trilinear interpolation [11], S is the up-sampling function, and $+$ corresponds to the voxel-wised addition operation on the transformation tensor. The final warped image W can be acquired by $M \circ \phi$ as the final registration result.

Note that the final deformation ϕ is not simply computed by adding ϕ_1, ϕ_2, and ϕ_3, as the input at each iteration step will change accordingly. On the contrary, we calculate the final flow recursively. For example, at Acc_2, the flow ϕ_1 from Acc_1 has to be warped firstly by ϕ_2 as coordinate alignment and then added back to ϕ_2. This step will guarantee that we can finally acquire the correct flow. By closely cooperate with the uncoupled encoder, the corresponding information at different scales can be effectively recovered, and the potential abilities of the network can be continuously reinforced.

2.4 Multi-window Loss

In many deformable image registration methods [2,15,17,25], negative local normalized cross-correlation (NCC) is successfully utilized to be a similarity metric for gradient-descent optimization. However, in our framework, we propose to learn and accumulate richer contextual information with various types of spatial correlations. Thus different parts of the network should receive more specified guidance, and the direct use of NCC would be not suitable enough. Therefore, in order to prevent our model from being trapped into local optimal, we propose a new multi-window loss, which is simple yet effective, by calculate the weighted sum of NCC with different window size as the similarity metric:

$$L_{sim}(F, M \circ \phi) = -\sum_{i=1}^{K} \gamma^{i-1} NCC_{w_i}(F, M \circ \phi) \tag{3}$$

where K is the number of different windows, γ is a hyperparameter, and w_i is the window size of ith NCC part. In our experiment, we set $\gamma = 0.5$, and $K = 3$ according to the number of accumulation steps in our method, and $(w_1, w_2, w_3) = (11, 9, 7)$. We also constrain the smoothness of the deformation field ϕ with a $L2$ regularizer:

$$L_{smooth}(\phi) = \sum_{p \in \Omega} \|\nabla \phi(p)\|^2 \tag{4}$$

In order to provide more spatial transformation flexibility and transit the gradient through hierarchical aggregation, we only constrain the smoothness at the final computed deformation field ϕ in Eq. 2. Finally, with the regularization parameter λ, the loss of our method is defined as:

$$L_{total}(F, M \circ \phi) = L_{sim}(F, M \circ \phi) + \lambda L_{smooth}(\phi) \tag{5}$$

3   Experiments and Results

Data Set. In order to demonstrate the effectiveness of our proposed method, extensive experiments were conducted on two 3D brain MRI data sets: the Mindboggle101 [14] data set and IXI data set[1]. Mindboggle101 contains 101 T1 weighted MR scans, which were annotated with 25 cortical segmentation parts. It can be used to evaluate registration accuracy for large-scale deformation. We follow [15] to combine all 25 categories into 5 areas for a better review. IXI data set contains nearly 600 MR images from normal and healthy subjects. We randomly selected 100 scans for the experiment and use FreeSurfer [6] for preprocessing, including skull stripping, affine transformation, and segmentation for evaluation.

Experimental Settings. For Mindboggle101 dataset, we followed the protocols in [15] to prepare the training and testing set. Our model was trained on NKI-RS-22 and NKI-TRT-20 subset, (42×41, 1722 pairs in total), and tested on OASIS-TRT-20 subset (20×19, 380 pairs in total). For IXI data set, 80 scans (80×79, 6320 pairs in total) were used for training and 20 scans (20×19, 380 pairs

---

[1] https://brain-development.org/ixi-dataset/.

in total) were used for testing. Each scan is cropped to 160×192×160, affinely warped to MNI152 space, and normalized by the maximum voxel intensity of each brain volume. Our model and other comparative deep learning models are all implemented with Pytorch [18] and trained on 1 T V100 GPU. We set learning rate $= 10^{-4}$, $\lambda = 1$ and batch-size $= 1$. Instance Normalization and Leaky-ReLU activation was used after each convolution except the last output.

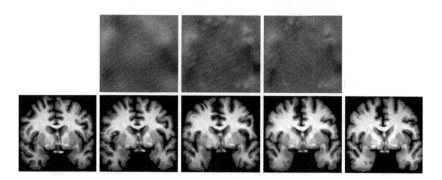

Fig. 3. Top: The generated displacement fields. Bottom (left to right): the moving image, the images warped accumulatively by three deformation flow, and the fixed image.

Visualization, Comparison, and Discussion. Firstly, the generated deformation fields and the warped images at each accumulation step are visualized in Fig. 3. It can be found that the first generated flow has the coarsest resolution and mainly focus on the alignment of the large image areas, while the last estimated flow has the finest resolution and is able to provide rich textural guidance for the registration.

We further compared our method with the state-of-the-art approaches, including SyN [1], VoxelMorph (VM) [2], RCN [25], and LapIRN [17]. VM is one of the most popular registration methods in the past three years, RCN and LapIRN are the latest methods to perform iterative large-scale deformable registration. We implemented SyN by using ANTs [1], and used the original codes of VM and LapIRN. Moreover, as the amount of parameters of default VM is smaller than other methods, we also doubled the channels of VM (VMx2), to make a relatively fair comparison. As shown in Table 1, our proposed method has demonstrated its superior registration accuracy by achieving best Dice score over all 9 categories on both Mindboggle101 and IXI data set.

As mentioned above, one merit of our method is to effectively perform registration under the accumulative layout with end-to-end training. Unlike other iterative methods, we do not have to divide the training into multiple stages, and the loss is only computed once. Therefore, the gradients at the frontend of the net can be computed by multiple times. Specifically, in our network, if a tensor is used more than one time, the gradient associated to it is the sum of the gradients corresponding to each iteration. We also evaluated the network complexity in terms of the Flops and the number of parameters (the last two rows in Table 1).

Table 1. In Mindboggle101, FL: Frontal region, PL: Parietal region, OL: Occipital region, TL: Temporal region, CL: Cingulate region; In IXI: LWm: Left white matter, RWm: Right white matter, LGm: Left gray matter, RGm: Right gray matter.

Class	Initial	SyN	VM	VMx2	RCN	LapIRN	Ours
FL	0.327	0.521	0.576	0.605	0.623	0.626	**0.664**
PL	0.312	0.452	0.526	0.555	0.564	0.566	**0.607**
OL	0.269	0.402	0.447	0.479	0.493	0.507	**0.549**
TL	0.352	0.543	0.583	0.609	0.624	0.616	**0.657**
CL	0.456	0.630	0.674	0.698	0.695	0.696	**0.722**
Avg.	0.343	0.510	0.561	0.589	0.600	0.602	**0.640**
LWm	0.664	0.788	0.839	0.841	0.843	0.847	**0.860**
RWm	0.663	0.788	0.839	0.842	0.844	0.847	**0.862**
LGm	0.495	0.661	0.792	0.806	0.808	0.813	**0.817**
RGm	0.493	0.660	0.792	0.806	0.807	0.812	**0.820**
Avg.	0.579	0.724	0.815	0.824	0.826	0.830	**0.840**
Flops	–	–	164.74G	643.67G	494.76G	588.57G	543.18G
Params.	–	–	396.45K	1.58M	1.19M	923.75K	789.56K

Moreover, to better illustrate the characteristic of our framework, we compared our accumulative enhancement process to the similar iteration steps in LapIRN, and both of them have 3 levels for large-scale deformation registration. As shown in Fig. 4, we visualize the intensity difference of the moving images between different warping stages as: $|M_1 - M|$, $|M_2 - M_1|$, and |final warping$-M_2$|. It shows that both LapIRN and our method can acquire finer flow of smaller structures at later iteration stage, which demonstrates that the iterative-based method is able to model long-range dependencies with lager receptive field, while obtain more detailed local transformations. Further more, compared to LapIRN's results (Fig. 4(b)), it can be observed that even at the finest warping process, our method (Fig. 4(c)) is still able to discover more detailed correspondences and perform the continuous transformation, thanks to uncoupled feature learning and spatially accumulated information.

Ablation Study. We conducted an additive ablation study on Mindboggle101 to illustrate the effectiveness of our proposed modules. The result is shown in Table 2. We choose the basic U-shaped transformation network with a single window NCC loss as the baseline. As shown by the statistics, after progressively

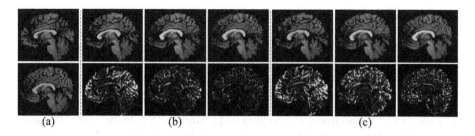

Fig. 4. (a) The moving (up) and fixed (down) image, (b) three levels of visualization of LapIRN, (c) three levels of visualization of our method. Brighter pixel corresponds to higher intensity difference.

adding Uncoupled Spatial Encoder, Accumulative Warping Enhancement, and Multi-window Loss module to the baseline, the performance of our proposed framework also improved accordingly, which implies that by actively capture richer spatial relationships within the images, all the network modules in our method are very effective for the large-scale deformable registration.

Table 2. The ablation study results on: 1) Accumulative Warping Enhancement (AWE), 2) Uncoupled Spatial Encoder (USE), and 3) Multi-window Loss (ML).

Model name	1)	2)	3)	FL	PL	OL	TL	CL	Avg.
Baseline	×	×	×	0.576	0.526	0.447	0.583	0.674	0.561
AWE	√	×	×	0.631	0.569	0.506	0.624	0.695	0.605
AWE+US	√	√	×	0.654	0.597	0.541	0.650	0.718	0.632
AWE+US+ML	√	√	√	**0.664**	**0.607**	**0.549**	**0.657**	**0.722**	**0.640**

4 Conclusion

In this paper, we presented a novel unsupervised medical image registration method, ULAE-net, for large-scale deformations. By introducing an uncoupled spatial encoder, we are able to effectively build long-range visual correlations between the moving and fixed image. During the spatial transformation stage, the proposed accumulative warping enhancement mechanism is applied to perform additive iterations integrally within the network. To provide stronger training guidance, we also proposed a simple yet effective multi-window setting for the local similarity measurement. Experiments on two 3D brain MRI data sets show that our model is able to obtain superior results compared with other state-of-the-art methods, especially in the case of large-scale deformations.

References

1. Avants, B.B., Tustison, N., Song, G.: Advanced normalization tools (ants). Or Insight **1–35** (2008)
2. Balakrishnan, G., Zhao, A., Sabuncu, M.R., Guttag, J., Dalca, A.V.: Voxelmorph: a learning framework for deformable medical image registration. IEEE Trans. Med. Imaging **38**(8), 1788–1800 (2019)
3. Chen, L.C., Papandreou, G., Kokkinos, I., Murphy, K., Yuille, A.L.: Deeplab: Semantic image segmentation with deep convolutional nets, atrous convolution, and fully connected CRFS. IEEE Trans. Pattern Anal. Mach. Intell. **40**(4), 834–848 (2018)
4. Dalca, A.V., Balakrishnan, G., Guttag, J., Sabuncu, M.R.: Unsupervised learning for fast probabilistic diffeomorphic registration. In: International Conference on Medical Image Computing and Computer-Assisted Intervention, pp. 729–738. Springer (2018). https://doi.org/10.1007/978-3-030-00928-1_82
5. Du, J., Li, W., Lu, K., Xiao, B.: An overview of multi-modal medical image fusion. Neurocomputing **215**, 3–20 (2016)
6. Fischl, B.: Freesurfer. Neuroimage **62**(2), 774–781 (2012)
7. García, H.F., Torres, C.A., Cardona, H.D.V., Álvarez, M.A., Orozco, Á.Á., Padilla, J.B., Arango, R.: 3d brain atlas reconstruction using deformable medical image registration: Application to deep brain stimulation surgery. In: 2014 XIX Symposium on Image, Signal Processing and Artificial Vision, pp. 1–5. IEEE (2014)
8. Hering, A., van Ginneken, B., Heldmann, S.: mlvirnet: Multilevel variational image registration network. In: International Conference on Medical Image Computing and Computer-Assisted Intervention. pp. 257–265. Springer (2019). https://doi.org/10.1007/978-3-030-32226-7_29
9. Hou, R., Zhou, D., Nie, R., Liu, D., Ruan, X.: Brain CT and MRI medical image fusion using convolutional neural networks and a dual-channel spiking cortical model. Medical and Biological Engineering and Computing (2019)
10. Hu, X., Kang, M., Huang, W., Scott, M.R., Wiest, R., Reyes, M.: Dual-stream pyramid registration network. In: International Conference on Medical Image Computing and Computer-Assisted Intervention. pp. 382–390. Springer (2019). https://doi.org/10.1007/978-3-030-32245-8_43
11. Jaderberg, M., Simonyan, K., Zisserman, A., Kavukcuoglu, K.: Spatial transformer networks. In: Proceedings of the 28th International Conference on Neural Information Processing Systems - Volume 2. p. 2017–2025. NIPS'15, MIT Press, Cambridge, MA, USA (2015)
12. Jingfan, F., Xiaohuan, C., Pew-Thian, Y., Dinggang, S.: Birnet: Brain image registration using dual-supervised fully convolutional networks. Medical Image Analysis (2019)
13. Kaur, H., Kumar, S.: A review on decomposition/reconstruction methods for fusion of medical images (2020)
14. Klein, A., Tourville, J.: 101 labeled brain images and a consistent human cortical labeling protocol. Front. Neurosci. **6**, 171 (2012)
15. Kuang, D., Schmah, T.: Faim-a convnet method for unsupervised 3d medical image registration. In: International Workshop on Machine Learning in Medical Imaging. pp. 646–654. Springer (2019). https://doi.org/10.1007/978-3-030-32692-0_74
16. Milletari, F., Navab, N., Ahmadi, S.A.: V-net: Fully convolutional neural networks for volumetric medical image segmentation. In: 2016 Fourth International Conference on 3D Vision (3DV) (2016)

17. Mok, T.C., Chung, A.C.: Large deformation diffeomorphic image registration with laplacian pyramid networks. In: International Conference on Medical Image Computing and Computer-Assisted Intervention, pp. 211–221. Springer (2020). https://doi.org/10.1007/978-3-030-59716-0_21

18. Paszke, A., et al.: Pytorch: An imperative style, high-performance deep learning library. arXiv:1912.01703 (2019)

19. Ronneberger, O., Fischer, P., Brox, T.: U-net: Convolutional networks for biomedical image segmentation. In: International Conference on Medical Image Computing and Computer-Assisted Intervention, pp. 234–241. Springer (2015). https://doi.org/10.1007/978-3-319-24574-4_28

20. Shu, Y., Wu, X., Li, W.: Lvc-net: Medical image segmentation with noisy label based on local visual cues. In: International Conference on Medical Image Computing and Computer-Assisted Intervention, pp. 558–566. Springer (2019). https://doi.org/10.1007/978-3-030-32226-7_62

21. Sokooti, H., De Vos, B., Berendsen, F., Lelieveldt, B.P., Išgum, I., Staring, M.: Nonrigid image registration using multi-scale 3d convolutional neural networks. In: International Conference on Medical Image Computing and Computer-Assisted Intervention, pp. 232–239. Springer (2017). https://doi.org/10.1007/978-3-319-66182-7_27

22. Sommer, S., Nielsen, M., Lauze, F., Pennec, X.: A multi-scale kernel bundle for lddmm: Towards sparse deformation description across space and scales. In: Székely, G., Hahn, H.K. (eds.) Information Processing in Medical Imaging. pp. 624–635. Springer, Berlin Heidelberg, Berlin, Heidelberg (2011). https://doi.org/10.1007/978-3-642-22092-0_51

23. Vercauteren, T., Pennec, X., Perchant, A., Ayache, N.: Diffeomorphic demons: efficient non-parametric image registration. NeuroImage **45**(1), S61–S72 (2009)

24. Wu, G., Kim, M., Wang, Q., Shen, D.: Hierarchical attribute-guided symmetric diffeomorphic registration for MR brain images. In: International Conference on Medical Image Computing and Computer-Assisted Intervention, pp. 90–97. Springer (2012). https://doi.org/10.1007/978-3-642-33418-4_12

25. Zhao, S., Dong, Y., Chang, E.I., Xu, Y., et al.: Recursive cascaded networks for unsupervised medical image registration. In: Proceedings of the IEEE/CVF International Conference on Computer Vision, pp. 10600–10610 (2019)

Atlas-based Segmentation of Intracochlear Anatomy in Metal Artifact Affected CT Images of the Ear with Co-trained Deep Neural Networks

Jianing Wang$^{(\boxtimes)}$, Dingjie Su, Yubo Fan, Srijata Chakravorti, Jack H. Noble, and Benoit M. Dawant

Department of Electrical and Computer Engineering, Vanderbilt University, Nashville, TN 37235, USA
jianing.wang@vanderbilt.edu

Abstract. We propose an atlas-based method to segment the intracochlear anatomy (ICA) in the post-implantation CT (Post-CT) images of cochlear implant (CI) recipients that preserves the point-to-point correspondence between the meshes in the atlas and the segmented volumes. To solve this problem, which is challenging because of the strong artifacts produced by the implant, we use a pair of co-trained deep networks that generate dense deformation fields (DDFs) in opposite directions. One network is tasked with registering an atlas image to the Post-CT images and the other network is tasked with registering the Post-CT images to the atlas image. The networks are trained using loss functions based on voxel-wise labels, image content, fiducial registration error, and cycle-consistency constraint. The segmentation of the ICA in the Post-CT images is subsequently obtained by transferring the predefined segmentation meshes of the ICA in the atlas image to the Post-CT images using the corresponding DDFs generated by the trained registration networks. Our model can learn the underlying geometric features of the ICA even though they are obscured by the metal artifacts. We show that our end-to-end network produces results that are comparable to the current state of the art (SOTA) that relies on a two-steps approach that first uses conditional generative adversarial networks to synthesize artifact-free images from the Post-CT images and then uses an active shape model-based method to segment the ICA in the synthetic images. Our method requires a fraction of the time needed by the SOTA, which is important for end-user acceptance.

Keywords: Non-rigid registration · Atlas-based segmentation · Metal artifact · Cochlear implant

1 Introduction

The cochlea (Fig. 1c) is a spiral-shaped structure that is part of the inner ear involved in hearing. It contains two main cavities: the scala tympani (ST) and the scala vestibuli (SV). The modiolus (MD) is a porous bone around which the cochlea is wrapped that hosts the

© Springer Nature Switzerland AG 2021
M. de Bruijne et al. (Eds.): MICCAI 2021, LNCS 12904, pp. 14–23, 2021.
https://doi.org/10.1007/978-3-030-87202-1_2

auditory nerves. A cochlear implant (CI) is an implanted neuroprosthetic device that is designed to produce hearing sensations in a person with severe to profound deafness by electrically stimulating the auditory nerves [1]. CIs are programmed postoperatively in a process that involves activating all or a subset of the electrodes and adjusting the stimulus level for each of these to a level that is beneficial to the recipient [2]. Programming parameters adjustment is influenced by the intracochlear position of the CI electrodes, which requires the accurate localization of the CI electrodes relative to the intracochlear anatomy (ICA) in the post-implantation CT (Post-CT) images of the CI recipients. This, in turn, requires the accurate segmentation of the ICA in the Post-CT images. Segmenting the ICA in the Post-CT images is challenging due to the strong artifacts produced by the metallic CI electrodes (Fig. 1b) that can obscure these structures, often severely. For patients who have been scanned before implantation, the segmentation of the ICA can be obtained by segmenting their pre-implantation CT (Pre-CT) image (Fig. 1a) using an active shape model-based (ASM) method [3]. The outputs of the ASM method are surface meshes of the ST, the SV, and the MD that have a predefined number of vertices. Importantly, each vertex corresponds to a specific anatomical location on the surface of the structures and the meshes are encoded with the information needed for the programming of the implant. Preserving point-to-point correspondence when registering the images is thus of critical importance in our application. The ICA in the Post-CT image of the patients can be obtained by registering their Pre-CT image to the Post-CT image and then transferring the segmentations of the ICA in the Pre-CT image to the Post-CT image using that transformation. This approach does not extend to CI recipients for whom a Pre-CT image is unavailable, which is the case for long-term recipients who were not scanned before surgery, or for recipients for whom images cannot be retrieved. To overcome this issue, Wang *et al.* have proposed a two-step method [4, 5], which we refer to as "cGANs+ASM". The method first uses conditional generative adversarial networks (cGANs) [6, 7] to synthesize artifact-free Pre-CT images from the Post-CT images and then uses the ASM method [3] to segment the ICA in the synthetic images. To the best of our knowledge, cGANs+ASM is the most accurate published automatic method for ICA segmentation in Post-CT images.

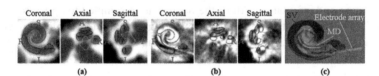

Fig. 1. A pair of registered (a) Pre-CT and (b) Post-CT images of an ear of a CI recipient. (c) An illustration of the intracochlear anatomy with an implanted CI electrode array. The meshes of the ST, the SV, and the MD are obtained by applying the ASM method to the Pre-CT image.

Here, we propose an end-to-end atlas-based method: we first generate a dense deformation field (DDF) between an artifact-free atlas image and a Post-CT image. The segmentation of the ICA in the Post-CT image can then be obtained by transferring the predefined segmentation meshes of the ICA in the atlas image to the Post-CT image using that DDF. We note that the inter-subject non-rigid registration between the atlas image and the Post-CT image is a difficult task because (1) considerable variation in cochlear anatomy across individuals has been documented [8], and (2) the artifacts in the Post-CT image change, often severely, the appearance of the anatomy, which has a significant influence on the accuracy of registration methods guided by intensity-based similarity metrics. To overcome the challenges, we propose a method to perform registrations between an atlas image and the Post-CT images that rely on deep networks. Following the idea of consistent image registration obtained by jointly estimating the forward and reverse transformations between two images that is proposed by Christensen *et al.* [9], we use a pair of co-trained networks that generate DDFs in opposite directions. One network is tasked with registering the atlas image to the Post-CT image and the other one is tasked with registering the Post-CT image to the atlas image. The networks are trained using loss functions that include voxel-wise labels, image content, fiducial registration error (FRE), and cycle-consistency constraint. We show that our model can segment the ICA and preserve point-to-point correspondence between the atlas and the Post-CT meshes, even when the ICA is difficult to localize visually.

2   Method

2.1   Data

Our dataset consists of Pre-CT and Post-CT image pairs of 624 ears. The atlas image is a Pre-CT image of an ear that is not in the 624 ears. The Pre-CT images are acquired with several conventional scanners (GE BrightSpeed, LightSpeed Ultra; Siemens Sensation 16; and Philips Mx8000 IDT, iCT 128, and Brilliance 64) and the Post-CT images are acquired with a low-dose flat-panel volumetric scanner (Xoran Technologies xCAT® ENT). The typical voxel size is $0.25 \times 0.25 \times 0.3$ mm^3 for the Pre-CT images and $0.4 \times 0.4 \times 0.4$ mm^3 for the Post-CT images. For each ear, the Pre-CT image is rigidly registered to the Post-CT image. The registration is accurate because the surgery, which consists of threading an electrode array through a small hole into the bony cavity, does not induce non-rigid deformation of the cochlea. The registered Pre-CT and Post-CT image pairs are then aligned to the atlas image so that the ears are roughly in the same spatial location and orientation. All of the images are resampled to an isotropic voxel size of 0.2 mm. Images of $64 \times 64 \times 64$ voxels that contain the cochleae are cropped from the full-sized images, and our networks are trained to process such cropped images.

2.2   Learning to Register the Artifact-Affected Images and the Atlas Image with Assistance of the Paired Artifact-Free Images

Figure 2a shows a list of images, meshes, and masks used to train our networks. For simplicity, we use O_{xSpc} to denote an object O in the x space. For example, $AtlasImg_{atlasSpc}$ is our atlas image in the atlas space. Similarly, $PostImg_{postSpc}$ is a Post-CT image in the Post-CT space. $Mesh_{atlasSpc}$ is the segmentation mesh of the ICA in $AtlasImg_{atlasSpc}$ generated by applying the ASM method to $AtlasImg_{atlasSpc}$. $PreImg_{postSpc}$ is the paired Pre-CT image of $PostImg_{postSpc}$ registered to the original Post-CT. $Mesh_{postSpc}$ is the segmentation mesh of the ICA in $PostImg_{postSpc}$. It has been generated by applying the ASM method to $PreImg_{postSpc}$ and then transferring the meshes to $PostImg_{postSpc}$. $Mask_{atlasSpc}$ and $Mask_{postSpc}$ are segmentation masks of the ST, SV, and MD. They are generated by converting $Mesh_{atlasSpc}$ and $Mesh_{postSpc}$ to masks.

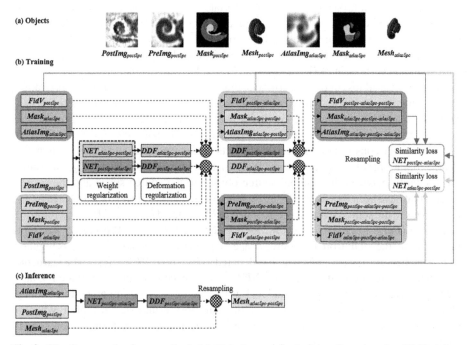

Fig. 2. The framework of our method. (a) Objects used for training the networks. (b) Training phase. (c) Inference phase.

As shown in Fig. 2b, the input of our networks is the concatenation of $AtlasImg_{atlasSpc}$ and $PostImg_{postSpc}$. The networks consist of a first network ($NET_{atlasSpc\text{-}postSpc}$) that generates a DDF from the atlas space to the Post-CT space ($DDF_{atlasSpc\text{-}postSpc}$) and a second network ($NET_{postSpc\text{-}atlasSpc}$) that generates a DDF from the Post-CT space to the atlas space ($DDF_{postSpc\text{-}atlasSpc}$). $FidV_{atlasSpc}$ and $FidV_{postSpc}$ are fiducial vertices randomly sampled from $Mesh_{atlasSpc}$ and $Mesh_{postSpc}$ on the fly for calculating FRE during training.

Assuming that $sSpc$ is the source space and $tSpc$ is the target space. The Pre-CT image, the segmentation masks, and the fiducial points in $sSpc$ are warped to $tSpc$ by using the corresponding DDFs (note that one DDF is used for the images and masks and the other for the fiducial points), and the results are denoted as $PreImg_{sSpc\text{-}tSpc}$, $Mask_{sSpc\text{-}tSpc}$, and $FidV_{sSpc\text{-}tSpc}$. Then, $PreImg_{sSpc\text{-}tSpc}$, $Mask_{sSpc\text{-}tSpc}$, and $FidV_{sSpc\text{-}tSpc}$ are transferred back to $sSpc$ using the corresponding DDF, and the results are denoted as $PreImg_{sSpc\text{-}tSpc\text{-}sSpc}$, $Mask_{sSpc\text{-}tSpc\text{-}sSpc}$, and $FidV_{sSpc\text{-}tSpc\text{-}sSpc}$, respectively. The training objective for $NET_{sSpc\text{-}tSpc}$ can be constructed by using similarity measurements between the target object in $tSpc$ (denoted as O_{tSpc}) and the source object that has been transferred to *tSpc* from *sSpc* (denoted as $O_{sSpc\text{-}tSpc}$). Specifically, we use the multiscale soft probabilistic Dice (MSPDice) [10] between $\textbf{\textit{Mask}}_{tSpc}$ and $\textbf{\textit{Mask}}_{sSpc\text{-}tSpc}$, which is denoted as MSPDice($\textbf{\textit{Mask}}_{tSpc}$, $\textbf{\textit{Mask}}_{sSpc\text{-}tSpc}$), to measure the similarity of the segmentation masks. The multiscale soft probabilistic Dice is less sensitive to the class imbalance in the segmentation tasks and is more appropriate for measuring label similarity in the context of image registration [11]. The similarity between $\textbf{\textit{FidV}}_{tSpc}$ and $\textbf{\textit{FidV}}_{sSpc\text{-}tSpc}$ is measured by the mean fiducial registration error $\overline{\text{FRE}}(\textbf{\textit{FidV}}_{tSpc}, \textbf{\textit{FidV}}_{sSpc\text{-}tSpc})$, which is calculated as the average Euclidean distance between the vertices in $\textbf{\textit{FidV}}_{tSpc}$ and the corresponding vertices in $\textbf{\textit{FidV}}_{sSpc\text{-}tSpc}$. The Post-CT images cannot be used for calculating intensity-based loss due to the artifacts, thus we use the normalized cross-correlation (NCC) between $\textbf{\textit{PreImg}}_{tSpc}$ and $\textbf{\textit{PreImg}}_{sSpc\text{-}tSpc}$, which is denoted as NCC($\textbf{\textit{PreImg}}_{tSpc}$, $\textbf{\textit{PreImg}}_{sSpc\text{-}tSpc}$), to measure the similarity between the warped source image and the target image. A cycle-consistency loss is used for regularizing the transformations. It imposes inverse consistency between the objects in the two spaces and has been shown to reduce folding problems [12]. Our cycle-consistency loss $\textbf{\textit{CycConsis}}_{sSpc\text{-}tSpc}$ measures the similarity between the original source objects in the source space and the source objects that have been transferred from the source space to the target space and then transferred back to the source space, it is calculated as MSPDice($\textbf{\textit{Mask}}_{sSpc}$, $\textbf{\textit{Mask}}_{sSpc\text{-}tSpc\text{-}sSpc}$) $+ 2 \times \overline{\text{FRE}}(\textbf{\textit{FidV}}_{sSpc}, \textbf{\textit{FidV}}_{sSpc\text{-}tSpc\text{-}sSpc}) + 0.5 \times$NCC($\textbf{\textit{PreImg}}_{sSpc}$, $\textbf{\textit{PreImg}}_{sSpc\text{-}tSpc\text{-}sSpc}$). Furthermore, the DDF from the source space to the target space $\textbf{\textit{DDF}}_{sSpc\text{-}tSpc}$ is regularized using bending energy [13], which is denoted as BendE($\textbf{\textit{DDF}}_{sSpc\text{-}tSpc}$). The learnable parameters of the registration network $\textbf{\textit{NET}}_{sSpc\text{-}tSpc}$ (except for the biases) are regularized by an L2 term, which is denoted as L2($\textbf{\textit{NET}}_{sSpc\text{-}tSpc}$). To summarize, the training objective for our networks is the weighted sum of the loss terms listed in Table 1; wherein the weights have been selected empirically by looking at training performance on a small number of epochs.

Table 1. Loss terms that are used to train our model.

Loss	Definition	Weight
MSPDice	MSPDice($\textbf{\textit{Mask}}_{postSpc}$, $\textbf{\textit{Mask}}_{atlasSpc\text{-}postSpc}$) $+$ MSPDice($\textbf{\textit{Mask}}_{atlasSpc}$, $\textbf{\textit{Mask}}_{postSpc\text{-}atlasSpc}$)	1
Mean FRE	$\overline{\text{FRE}}(\textbf{\textit{FidV}}_{postSpc}, \textbf{\textit{FidV}}_{atlasSpc\text{-}postSpc})$ $+$ $\overline{\text{FRE}}(\textbf{\textit{FidV}}_{atlasSpc}, \textbf{\textit{FidV}}_{postSpc\text{-}atlasSpc})$	2
NCC	NCC($\textbf{\textit{PreImg}}_{postSpc}$, $\textbf{\textit{AtlasImg}}_{atlasSpc\text{-}postSpc}$) $+$ NCC($\textbf{\textit{AtlasImg}}_{atlasSpc}$, $\textbf{\textit{PreImg}}_{postSpc\text{-}atlasSpc}$)	0.5
Cycle-consistency	$\textbf{\textit{CycConsis}}_{atlasSpc\text{-}postSpc} + \textbf{\textit{CycConsis}}_{postSpc\text{-}atlasSpc}$	0.5
BendE	BendE($\textbf{\textit{DDF}}_{atlasSpc\text{-}postSpc}$) $+$ BendE($\textbf{\textit{DDF}}_{postSpc\text{-}atlasSpc}$)	0.5
L2	L2($\textbf{\textit{NET}}_{atlasSpc\text{-}postSpc}$) $+$ L2($\textbf{\textit{NET}}_{postSpc\text{-}atlasSpc}$)	0.0001

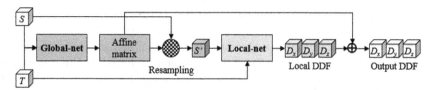

Fig. 3. Illustration of a registration network $NET_{sSpc\text{-}tSpc}$ that is tasked to generate a DDF from the source space to the target space.

2.3 Network Architecture

The registration networks in our model are adapted from the network architecture proposed by Hu *et al.* [14] and Ghavami *et al.* [15]. As shown in Fig. 3, $NET_{sSpc\text{-}tSpc}$, which is tasked with generating a DDF for warping the source image S to the target image T, is composed of a Global-net and a Local-net. After receiving the concatenation of S and T, the Global-net generates an affine transformation matrix. S is warped to T by using this affine transformation and the resulting image is denoted as S'. Then, the Local-net takes the concatenation of S' and T to generate a non-rigid local DDF. The affine transformation and the local DDF are composed to produce the output DDF. The details about the Global-net and Local-net can be found in [14].

2.4 Evaluation

As shown in Fig. 2c, at the inference phase, given a new Post-CT image $PostImg_{postSpc}$, the ICA in $PostImg_{postSpc}$ can be segmented by warping $Mesh_{atlasSpc}$ to $PostImg_{postSpc}$ using the DDF generated by the trained network. The resulting segmentation mesh of the ICA is denoted as $Mesh_{atlasSpc\text{-}postSpc}$. $Mesh_{postSpc}$, which has been described in Sect. 2.2, is used as the ground truth for comparison. As $Mesh_{atlasSpc}$ and $Mesh_{postSpc}$ are the outputs of the ASM method, both of them have a predefined number of vertices, and the vertices of $Mesh_{atlasSpc}$ and $Mesh_{postSpc}$ have a one-to-one correspondence. There are 3344, 3132, and 2852 vertices on the ST, SV, and MD mesh surfaces, respectively, for a total of 9328 vertices. Point-to-point error (P2PE), computed as the Euclidean distance in millimeters, between the corresponding vertices on $Mesh_{atlasSpc\text{-}postSpc}$ and $Mesh_{postSpc}$ are used to quantify the accuracy of the segmentation and registration. The P2PEs between the corresponding vertices on $Mesh_{postSpc}$ and the meshes generated by cGANs+ASM are calculated and serve as values that are used to compare the proposed method with the state of the art (SOTA). The method proposed in [14], which uses a unidirectional registration network trained with the MSPDice loss and the regularization loss, is used as a baseline for comparison. In addition to the MSPDice loss and the regularization loss, our training objective also includes the FRE loss, NCC loss, and the cycle-consistency loss. An ablation study is conducted to analyze how these loss terms affect the performance of our networks.

3 Experiments

The 624 ears are partitioned into 465 ears for training, 66 ears for validation, and 93 ears for testing. The partition is random, with the constraint that ears of the same object cannot be used in both training and testing. We apply augmentation to the training set by rotating each image by 6 random angles in the range of -25 and $25°$ about the x-, y-, and z-axis. The training images are blurred by applying a Gaussian filter with a kernel size selected randomly from $\{0, 0.5, 1.0, 1.5\}$ with equal probability. This results in a training set expanded to 8835 images. Each image is clipped between its 5th and 95th intensity percentiles, and the intensity values are rescaled to -1 to 1. We use a batch size of 1, at each training step, 30% of the vertices on the ICA meshes are randomly sampled and used as the fiducial points for calculating the FRE loss.

4   Results

Figure 4 shows two cases for which our method leads to (a) good and (b) poor results. For each case, the first row shows three orthogonal views of the original atlas image in the atlas space. The second row shows the Post-CT image. The third row shows the atlas image registered to the Post-CT image. The fourth row shows the paired Pre-CT image of the Post-CT image. The warped atlas image (third row) should be as similar as possible to the Pre-CT image (fourth row). The last row shows the original segmentation mesh in the atlas image (***Mesh**atlasSpc*), the segmentation mesh in the Post-CT image generated using our method (***Mesh**atlasSpc-postSpc*), and the ground truth mesh in the Post-CT image (***Mesh**postSpc*). For ***Mesh**atlasSpc* and ***Mesh**postSpc*, the ST, the SV, and the MD are shown in red, blue, and green, respectively. ***Mesh**atlas-post* is color-coded with the P2PE at each vertex on the mesh surfaces. Both these cases illustrate the severity of the artifact introduced by the implant. In the second case, the cochlea is barely visible.

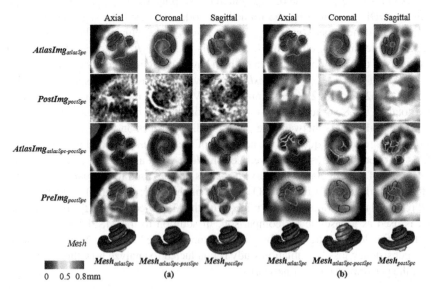

Fig. 4. Two example cases in which our method leads to (a) good and (b) poor results.

For each testing ear, we calculate the P2PEs of the vertices on the mesh surfaces of the ST, the SV, and the MD, respectively. We calculate the maximum (Max), median, and standard deviation (STD) of the P2PEs. Figure 5 shows the boxplots of these statistics for the 93 testing ears. "cGAN+ASM" denotes the results of the SOTA. "Proposed" denotes the results of our method. "Proposed-NoNCC", "Proposed-NoCycConsis", and "Proposed-NoFRE" denote the results of our proposed networks trained without using the NCC loss, the cycle-consistency loss, and the FRE loss. "Baseline" denotes the results of the baseline method. "No registration" denotes the P2PEs between the vertices on the mesh surfaces in the original atlas space and the Post-CT space. We perform two-sided and one-sided Wilcoxon signed-rank tests between the "Proposed" group and the other groups. The p-values have been corrected using the Holm-Bonferroni method [16]. The median values for each group are shown on top of the boxplots, in which red denotes that both the two-sided and the one-sided tests are significant, cyan denotes that only the two-sided test is significant, and blue denotes that the two-sided test is not significant. The results show that our networks trained using all of the proposed loss terms achieve a significantly lower segmentation error compared to the baseline method and the networks that are not trained using all of the loss terms. Our method produces results that are similar to those obtained with the SOTA in terms of the medians of the segmentation error. The Max of the segmentation error and the STD of the segmentation error for the SV and MD remain slightly superior to those obtained with the SOTA.

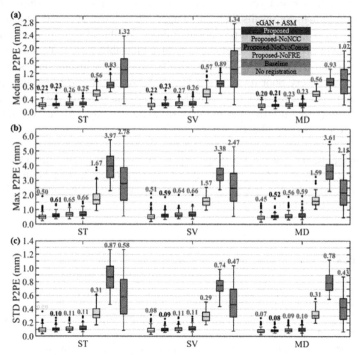

Fig. 5. Boxplots of (a) the median, (b) the Max, and (c) the STD of the P2PEs. A description of the numerical value color legend can be found in the text.

As mentioned earlier, the SOTA is a two-step process: (1) generate a synthetic Pre-CT image from a Post-CT image with cGANs trained for this purpose and (2) apply an ASM method to the synthetic image. Step 2 requires the very accurate registration of an atlas to the image to be segmented to initialize the ASM. This is achieved through an affine and then a non-rigid intensity-based registration in a volume-of-interest that includes the inner ear. Step 1 takes about 0.3s while step 2 takes on average 75s. The proposed method only requires providing a volume-of-interest that includes the inner ear to the networks and inference time is also about 0.3s. Segmentation is thus essentially instantaneous with the proposed method while it takes over a minute with the SOTA. This is of importance for clinical deployment and end-user acceptance.

5 Summary

We have developed networks capable of performing image registration between artifact-affected CT images and an artifact-free atlas image, which is a very challenging task because of the severity of the artifact introduced by the implant. Because we need to maintain point-to-point correspondence between meshes in the atlas and meshes in the segmented Post-CT images, we have introduced a point-to-point loss, which, to the best of our knowledge, has not yet been proposed. Our experiments have shown that this loss is critical to achieve results that are comparable to those obtained with the SOTA that relies on an ASM fitted to a preoperative image synthesized from a post-operative image. By design, ASM methods always produce plausible shapes. We have observed that with the point-to-point loss, our network also produces plausible shapes even when the images are of very poor quality (see Fig. 4b). We hypothesize that, thanks to the point-to-point loss, the network has been able to learn the shape of the cochlea and can fit this shape to partial information in the post-operative image. More experiments are ongoing to verify this hypothesis.

Acknowledgments. This work has been supported by NIH grants R01DC014037 and R01DC014462 and by the Advanced Computing Center for Research and Education (ACCRE) of Vanderbilt University. The content is solely the responsibility of the authors and does not necessarily represent the official views of these institutes.

References

1. What is a Cochlear Implant. https://www.fda.gov/medical-devices/cochlear-implants/what-cochlear-implant. Accessed 17 Nov 2020
2. Image-guided Cochlear Implant Programming (IGCIP). https://clinicaltrials.gov/ct2/show/NCT03306082. Accessed 17 Nov 2020
3. Noble, J.H., et al.: Automatic segmentation of intracochlear anatomy in conventional CT. IEEE Trans. Biomed. Eng. **58**(9), 2625–2632 (2011)
4. Wang, J., et al.: Metal artifact reduction for the segmentation of the intra cochlear anatomy in CT images of the ear with 3D-conditional GANs. Med. Image Anal. **58**, 101553 (2019)

5. Wang, J., et al.: Conditional generative gdversarial networks for metal artifact reduction in CT images of the ear. In: Frangi, A., et al. (eds.) Medical Image Computing and Computer Assisted Intervention – MICCAI 2018. Lecture Notes in Computer Science, vol. 11070, pp. 1–3. Springer, Cham (2018)

6. Mirza, M., Osindero, S.: Conditional generative adversarial nets. arXiv:1411.1784 (2014)

7. Isola, P., et al.: Image-to-image translation with conditional adversarial networks. In: Proceedings of the 2017 IEEE Conference on Computer Vision and Pattern Recognition (CVPR), pp. 1125–1134 (2017)

8. Pelosi, S., et al.: Analysis of intersubject variations in intracochlear and middle ear surface anatomy for cochlear implantation. Otol. Neurotol. **34**(9), 1675–1680 (2013)

9. Christensen, G.E., Johnson, H.J.: Consistent image registration. IEEE Trans. Med. Imaging **20**(7), 568–582 (2001)

10. Milletari, F., Navab, N., Ahmadi, S.: V-Net: fully convolutional neural networks for volumetric medical image segmentation. In: 2016 Fourth International Conference on 3D Vision (3DV), pp. 565–571 (2016)

11. Hu, Y., et al.: Weakly-supervised convolutional neural networks for multimodal image registration. Med. Image Anal. **49**, 1–13 (2018)

12. Kim, B., Kim, J., Lee, J.-G., Kim, D.H., Park, S.H., Ye, J.C.: Unsupervised deformable image registration using cycle-consistent CNN. In: Shen, D., et al. (eds.) MICCAI 2019. LNCS, vol. 11769, pp. 166–174. Springer, Cham (2019). https://doi.org/10.1007/978-3-030-32226-7_19

13. Rueckert, D., et al.: Nonrigid registration using free-form deformations: application to breast MR images. IEEE Trans. Med. Imaging **18**(8), 712–721 (1999)

14. Hu, Y., et al.: Label-driven weakly-supervised learning for multimodal deformable image registration. In: 2018 IEEE 15th International Symposium on Biomedical Imaging (ISBI), pp. 1070–1074 (2018)

15. Ghavami, N., et al.: Automatic slice segmentation of intraoperative transrectal ultrasound images using convolutional neural networks. In: Fei, B., Webster III, R.J. (eds.) Proceedings Medical Imaging 2018: Image-Guided Procedures, Robotic Interventions, and Modeling, vol. 10576, pp. 1057603 (2018)

16. Holm, S.: A simple sequentially rejective multiple test procedure. Scand. J. Stat. **6**(2), 65–70 (1979)

Learning Unsupervised Parameter-Specific Affine Transformation for Medical Images Registration

Xu Chen[1,3], Yanda Meng[1], Yitian Zhao[2], Rachel Williams[1],
Srinivasa R. Vallabhaneni[1,3], and Yalin Zheng[1(✉)]

[1] Institute of Life Course and Medical Sciences, University of Liverpool,
Liverpool, UK
yalin.zheng@liverpool.ac.uk
[2] Cixi Institute of Biomedical Engineering, Ningbo Institute of Industrial Technology,
Chinese Academy of Sciences, Beijing, People's Republic of China
[3] Liverpool Vascular & Endovascular Service, Royal Liverpool University Hospital
NHS Trust, Liverpool, UK

Abstract. Affine registration has recently been formulated using deep learning frameworks to establish spatial correspondences between different images. In this work, we propose a new unsupervised model that investigates two new strategies to tackle fundamental problems related to affine registration. More specifically, the new model 1) has the advantage to explicitly learn specific geometric transformation parameters (e.g. translations, rotation, scaling and shearing); and 2) can effectively understand the context between the images via cross-stitch units allowing feature exchange. The proposed model is evaluated on two two-dimensional X-ray datasets and a three-dimensional CT dataset. Our experimental results show that our model not only outperforms state-of-art approaches and also can predict specific transformation parameters. Our core source code is made available online[1]([1]https://github.com/xuuuuuuchen/PASTA).

1 Introduction

Image registration is a crucial challenge in the field of biomedical image analysis. Image registration aims to align two (or more) given images, namely, a target image $I_{tgt} : \Omega_{tgt} \subset \mathbb{R}^d \mapsto \mathbb{R}$, and a source image $I_{src} : \Omega_{src} \subset \mathbb{R}^d \mapsto \mathbb{R}$, by establishing their spatial correspondences into a common coordinate system.

Affine transformation is commonly used to correct for geometric distortions or deformations that occur with non-ideal camera angles. The planetary of surfaces, parallelism and angles between lines are all preserved in affine transformation. In general, an affine transformation is a composition of rotations, translations, scaling, and shears, which can be expressed as an energy minimization

Electronic supplementary material The online version of this chapter (https://doi.org/10.1007/978-3-030-87202-1_3) contains supplementary material, which is available to authorized users.

© Springer Nature Switzerland AG 2021
M. de Bruijne et al. (Eds.): MICCAI 2021, LNCS 12904, pp. 24–34, 2021.
https://doi.org/10.1007/978-3-030-87202-1_3

problem: $\mathbf{A}^* = argmin\{S[I_{tgt}, \mathbf{A}(I_{src})]\}$, where $\mathbf{A}$ is the affine transformation matrix[1] and S is the metrics to measure the dissimilarity between I_{tgt} and $\mathbf{A}(I_{src})$.

Image registration formulated in deep learning settings has shown promising results. Several approaches have been proposed for affine image registration and nonlinear image registration with the convolutional neural networks (CNN). The Spatial Transformer Network (STN) [8] is one of the first CNN-based methods to learn two-dimensional (2D) affine transformation for the classification of distorted MNIST digit in a supervised learning manner. The localization network of STN can regress the outputs from a CNN to produce the 2D transformation matrix $\mathbf{A}_{2D}$. $\mathbf{A}_{2D}$ has six parameters encoded from translation, scaling, rotation and shearing. *Miao et al.* proposed a supervised CNN to regress the three-dimensional (3D) transformation matrix $\mathbf{A}_{3D}$ according to the synthesized transformation parameters as the ground truth for affine registration of X-ray images [11].

On the other hand, unsupervised models for affine and nonlinear transformation learning are more desirable as transformation ground truth is no longer required, which is not always available. In a recent work by *de Vos et al.* [16], an unsupervised *Deep Learning Image Registration* (DLIR) framework for joint affine and nonlinear registration was proposed. The affine transformation framework in the *DLIR* is a multi-stage approach for the multi-temporal MRI and CT 3D image registration. The $\mathbf{A}_{3D}$ matrix is regressed by two separate CNNs. The performance of the *DLIR* outperformed conventional image registration methods when tested on a cine cardiac MRI dataset and a chest CT dataset, respectively. Similar to the DLIR, in *Hu et al.*'s work [6], segmentation labels are used as a type of ground truth and considered in a loss function to help the similarity maximisation for MRI-Ultrasound image scans registration. There are two sub-networks in this model: a CNN regressor as the *Global-Net* and a U-net-like architecture [13] as the *Local-Net* for affine and nonlinear transformation, respectively. For unsupervised 3D affine transformation learning, the AIRNet [4] was proposed to estimate the $\mathbf{A}_{3D}$ for brain MR scan alignments by training a self-supervised CNN.

Despite the recent promising progress in deep learning-based image registration, most of the existing approaches directly regress the affine transformation matrix $\mathbf{A}$, but not explicitly regress specific geometric transformation parameters in the form of translations, rotation, scaling and shearing. For **2D transformation**, there will be seven transformation parameters, namely, one rotation (θ), two translations (t_x, t_y), two scaling (sc_x, sc_y) and two shears (sh_x, sh_y). It is obvious that six parameters in the matrix form $\mathbf{A}_{2D}$ can be easily derived from these seven spatial transformation parameters (θ, t_x, t_y, sc_x, sc_y, sh_x and sh_y), but not vice versa. Similar to 2D transformation, there will be 15 transformation parameters in **3D transformation**, that is, three rotations (θ_x, θ_y, θ_z), three translation (t_x, t_y, t_z), three scaling (sc_x, sc_y, sc_z) and six shears (sh_{xy}, sh_{xz}, sh_{yx}, sh_{yz}, sh_{zx}, sh_{zy}). These 15 parameters can be used to derive the twelve parameters of the matrix $\mathbf{A}_{3D}$, but not vice versa.

---

[1] $\mathbf{A}_{2D} = \begin{bmatrix} a_1 & a_2 & a_3 \\ a_4 & a_5 & a_6 \\ 0 & 0 & 1 \end{bmatrix}$ and $\mathbf{A}_{3D} = \begin{bmatrix} a_1 & a_2 & a_3 & a_4 \\ a_5 & a_6 & a_7 & a_8 \\ a_9 & a_{10} & a_{11} & a_{12} \\ 0 & 0 & 0 & 1 \end{bmatrix}$.

Therefore, there is a dilemma: if we only determine the matrix $\mathbf{A}$ (e.g. $\mathbf{A}_{2D}$ and $\mathbf{A}_{3D}$ for 2D and 3D case, respectively) like most of the other existing deep learning-based models, we cannot infer those spatial transformation parameters as there is no unique solution and cannot explain the effect of each type of transformations. To tackle the above drawbacks and limitations, we propose a novel parameter-specific affine transformation model by explicitly learning all these spatial transformation parameters rather than learning their combinations, namely, the transformation matrix $\mathbf{A}$. Furthermore, cross-stitch units [12] have been developing for multi-task learning [3,14,15]. We introduce "cross-stitch" units [12] into our model to effectively learn an optimal combination of shared representations between image pairs.

2   Methods

In this section, we will describe our model in detail. More specifically, it has two unique features: a CNN-based parameter-specific affine transformation (PASTA) framework to formulate affine transformation, and a Cross-stitch Affine Network (CANet) to effectively learn transformation parameters in an unsupervised manner. For simplicity, the 2D affine transformation case will be presented here whist the 3D case will be shown in the supplementary material.

2.1   Parameter-Specific Affine Transformation

The affine transformation matrix $\mathbf{A}$ can be formed by composing the rotation matrix $\mathbf{M}_{ro}$, the shearing matrix $\mathbf{M}_{sh}$, the scale matrix $\mathbf{M}_{sc}$ and the translation matrix $\mathbf{M}_t$ in turn[2],

$$\mathbf{A} = \mathbf{M}_t \cdot \mathbf{M}_{sc} \cdot \mathbf{M}_{sh} \cdot \mathbf{M}_{ro} = \begin{bmatrix} a_1 & a_2 & a_3 \\ a_4 & a_5 & a_6 \\ 0 & 0 & 1 \end{bmatrix}. \tag{1}$$

Where, each of the above four 2D transformation types can be represented as: $\mathbf{M}_t = \begin{bmatrix} 1 & 0 & t_x \\ 0 & 1 & t_y \\ 0 & 0 & 1 \end{bmatrix}$ $\mathbf{M}_{ro} = \begin{bmatrix} \cos(\theta) & -\sin(\theta) & 0 \\ \sin(\theta) & \cos(\theta) & 0 \\ 0 & 0 & 1 \end{bmatrix}$ $\mathbf{M}_{sh} = \begin{bmatrix} 1 & sh_x & 0 \\ sh_y & 1 & 0 \\ 0 & 0 & 1 \end{bmatrix}$ $\mathbf{M}_{sc} = \begin{bmatrix} sc_x & 0 & 0 \\ 0 & sc_y & 0 \\ 0 & 0 & 1 \end{bmatrix}$. Our PASTA framework aims to optimise each transformation parameters, namely, one for rotation (θ), two for translation (t_x, t_y), two for scaling (sc_x, sc_y) and two for shearing (sh_x, sh_y) instead of directly optimising the transformation matrix $\mathbf{A}$. After optimising each transformation parameters, the matrix $\mathbf{A}$ as expressed in Eq. (2) can be derived according to the order in Eq. (1).

---

[2] $\mathbf{A}$ is subject to the composition order. In this work, we use the order shown in Eq. (1).

$$\mathbf{A} = \begin{bmatrix} \mathbf{a}_1 = sc_x\cos(\theta) + sh_x sc_x\sin(\theta) & \mathbf{a}_2 = -sc_x\sin(\theta) + sh_x sc_x\cos(\theta) & \mathbf{a}_3 = t_x \\ \mathbf{a}_4 = sh_y sc_y\cos(\theta) + sc_y\sin(\theta) & \mathbf{a}_5 = -sh_y sc_y\sin(\theta) + sc_y\cos(\theta) & \mathbf{a}_6 = t_y \\ 0 & 0 & 1 \end{bmatrix} \tag{2}$$

Note, the range of each of the seven transformation parameters will be empirically known for specific applications. Thus, we can linearly normalise them into the range of $[0, 1]$ where 0 and 1 corresponds to the minimum λ^i_{min} and maximum λ^i_{max} value of the ith parameter, respectively. This normalisation is beneficial: the outputs from the network will always be within $[0, 1]$, and the gradient in the optimisation will not be too small or too big. When our framework is in action, it will regress each of the seven parameters between $[0, 1]$. Each of them will then be mapped back to the actual values. Following that, these parameters will be used to generate the matrix form $\mathbf{A}$. After that, bilinear interpolation is applied when warping the I_{src} image by the $\mathbf{A}$.

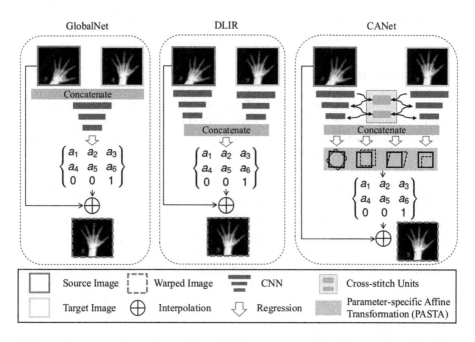

Fig. 1. Overview of our proposed model for unsupervised affine image registration

2.2   Cross-stitch Affine Network

We also propose a novel architecture for unsupervised affine transformation learning, partly motivated by cross-stitch units from multi-task learning [12]. Cross-stitch units intend to allow a model to determine how the task-specific network leverages the other task-specific network's knowledge. Cross-stitch units can learn the optimal linear combination of the output from the previous layers. We integrate these cross-stitch units into our model with our PASTA framework and provide an end-to-end learning model. We refer to it as a Cross-stitch Affine

Network (CANet), sketched in Fig. 1. For affine transformation learning, the I_{src} and I_{tgt} are fed into two separate sub-networks S and T, respectively, and the two outputs (activation maps) h_S and h_T from the S and T are concatenated to estimate the matrix $\mathbf{A}$ to warp the I_{src}. Unlike the previous works [16] that uses the hard parameter sharing, our model can learn the best-shared representations between two separate sub-networks just as in the soft parameter sharing. More specifically, the k-th layer in our CANet shares the representations via the Cross-stitch units C by learning a linear combination of the activation maps $h_{S,k}^{ij}$ and $h_{T,k}^{ij}$ at location (i,j). The outputs of the C are:

$$\begin{bmatrix} \tilde{h}_{S,k}^{ij} \\ \tilde{h}_{T,k}^{ij} \end{bmatrix} = \begin{bmatrix} C_{SS} \ C_{TS} \\ C_{ST} \ C_{TT} \end{bmatrix} \begin{bmatrix} h_{S,k}^{ij} \\ h_{T,k}^{ij} \end{bmatrix} \tag{3}$$

Where the C_{ST} and C_{TS} are the parameters weighting the activations between the sub-networks h_S and h_T, and the C_{SS} and C_{TT} are the parameters weighting the activations of the same sub-networks. Further, we demonstrate these units' effectiveness for the 2D and 3D affine transformation tasks.

3 Experiments and Results

In this section, we investigate the performance of our proposed model in both 2D and 3D applications. All models were trained on one node of a cluster with sixteen 8-core Intel CPUs, 8 TESLA V100 GPUs and 1TB memory with the spatial transformer module adapted from the open-source code in VoxelMorph [2], implemented in TensorFlow 1.14. All of the models were trained by the Adam optimizer. For a fair comparison of different models, we searched the optimal learning rate between e^{-3} and e^{-6} for each model based on the validation set.

Baseline. To evaluate the performance of the proposed model, we compared ours with the GlobalNet [6] and DLIR [16], the two most widely used networks for affine registration. To investigate the robustness and generalizability of the proposed PASTA framework, we plugged in our PASTA model into the GlobalNet [6], the DLIR [16], and our CANet for 2D and 3D tasks. All the above networks are not pre-trained on any image datasets. Two widely used metrics, normalized cross correlation score (NCC) and Dice coefficient score (DSC), were introduced for the image registration dissimilarity assessment. Mean absolute error (MAE) is introduced for evaluating the performance of individual affine transformation parameters compared with the synthetic parameters.

Datasets. In order to evaluate the performance of our models, we have applied it to three biomedical image datasets: (1) ChestMNIST of MedMNIST [18]: ChestMNIST[3] contains 10,000 frontal-view chest X-ray images based on the NIH-ChestXray14 dataset [17]. (2) HandMNIST of MedMNIST [18] contains 10,000 hand X-ray images. We used the original size 64x64 available when we downloaded them. (3) Learn2Reg[4]: 2020 MICCAI Registration Challenge (Task

---

[3] https://medmnist.github.io/#dataset.

[4] https://learn2reg.grand-challenge.org/Datasets/.

2) [5], this dataset consists of 60 3D CT thorax images taken from 30 subjects (20 for training and 10 for testing). For all the scans an automatic lung segmentation is provided to evaluate the registration methods.

Initialization. Because real-world datasets with high-quality annotation is hard to acquire and kept for evaluation only, training on synthetic data is necessary. We generated 20 synthetic transformed X-ray images for each X-ray image. In this work, the range $[\lambda_{min}, \lambda_{max}]$ of 7 transformation parameters are sh_x, $sh_y \in [-0.1, 0.1]$, $\theta \in [-30°, 30°]$, sc_x, $sc_y \in [0.9, 1.1]$ and t_x, $t_y \in [-0.2, 0.2]$. In total, 200,000 pairs of synthetic images were generated and divided 50% (n = 120,000) of images for training, 25% (n = 40,000) for validation, and the remaining 25% for testing. The corresponding seven transformations parameters were used as 'ground truth' for further evaluation and comparison. For the 3D work, each scan was used to generate 100 synthetic transformed scans to pair itself. The ranges $[\lambda_{min}, \lambda_{max}]$ of the 15 transformation parameters are θ_x, θ_y, $\theta_z \in [-5°, 5°]$, sc_x, sc_y, $sc_z \in [0.90, 1.0]$, sh_{xy}, sh_{xz}, sh_{yx}, sh_{yz}, sh_{zx}, $sh_{zy} \in [0.0, 0.1]$ and t_x, t_y, $t_z \in [-0.1, 0.1]$. We used the official data split and resized the source images into $128 \times 128 \times 128$. There are 6,000 pairs of synthetic CT scans and were divided into 2,400 pairs for training, 1,600 pairs for validation, and the remaining 2,000 pairs for testing.

Ablation Study. We investigated the number of cross-stitch units n in our proposed 2D- and 3D-CANet, respectively. We introduced a variable-controlling method to perform this ablation study to investigate the individual impact of different number of cross-stitch units.

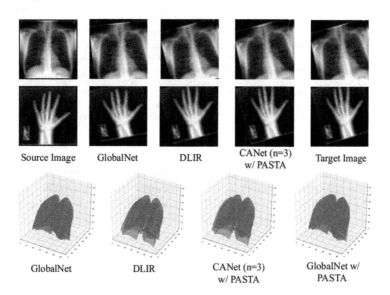

Source Image	GlobalNet	DLIR	CANet (n=3) w/ PASTA	Target Image

GlobalNet	DLIR	CANet (n=3) w/ PASTA	GlobalNet w/ PASTA

Fig. 2. 2D and 3D registration results of our proposed model for unsupervised affine image registration compared to the previous state-of-art approaches

Results on 2D Datasets. We evaluated our models on the HandMNIST dataset and ChestMNIST dataset in comparison to the GlobalNet and DLIR. The quantitative results on the both datasets are presented in Table 1. For the HandMNIST dataset, our CANet (n = 3) with PASTA framework achieved better performance in terms of NCC score of 0.966 than the GlobalNet (NCC = 0.868) and DLIR (NCC = 0.918), followed by the CANet (n = 3) (NCC = 0.964). The use of PASTA framework improved the performance in terms of NCC score of the GlobalNet from 0.868 to 0.929 and the GlobalNet from 0.918 to 0.933, respectively. On the other hand, for the HandMNIST dataset, compared to the GlobalNet (NCC = 0.859) and the DLIR (NCC = 0.957), our CANet (n = 3) with PASTA achieved the best performance in terms of NCC of 0.988, followed by the CANet (n = 3) (NCC = 0.978). For both datasets, when the number of cross-stitch units is increased from one to three, the performance in terms of NCC score are improved consistently, but the contribution from PASTA is decreasing gradually because the network is more powerful to directly regress the **A** matrix. However, the computational cost will proportionally increase.

Table 1. Quantitative 2D and 3D registration results of our proposed models compared to the others. Standard deviation is provided in the brackets.

Datasets	Models									
	PASTA	GlobalNet [8]		DLIR [16]		CANet (n = 1)		CANet (n = 2)		CANet (n = 3)	
		NCC	#Para.	NCC	#Para.	NCC	#Para.	NCC	#Para.	NCC	#Para.
HandMNIST	w/o	0.868 (0.086)	72K	0.918 (0.056)	298K	0.927 (0.051)	298K	0.920 (0.028)	302K	0.964 (0.033)	564K
	w/	0.929 (0.054)	72K	0.933 (0.049)	298K	0.913 (0.057)	298K	0.928 (0.055)	302K	**0.966** **(0.032)**	564K
ChestMNIST	w/o	0.859 (0.074)	72K	0.957 (0.029)	298K	0.962 (0.026)	298K	0.947 (0.033)	302K	0.978 (0.015)	564K
	w/	0.972 (0.025)	72K	0.945 (0.038)	298K	0.935 (0.039)	298K	0.970 (0.025)	302K	**0.988** **(0.012)**	564K
		DSC	Para.	DSC	Para.	DSC	Para.	DSC	Para.	DSC	Para.
Learn2Reg (Task 2)	w/o	0.903 (0.094)	142K	0.884 (0.030)	682K	0.889 (0.030)	683K	0.908 (0.033)	748K	0.883 (0.045)	17M
	w/	**0.938** **(0.029)**	142K	0.890 (0.037)	682K	0.886 (0.042)	683K	0.911 (0.047)	748K	0.910 (0.059)	17M

Results on 3D Dataset. We also evaluated our models on the CT lung dataset compared to the existing models, and presented the quantitative results in Table 1. The GlobalNet with our PASTA achieved the best performance among all the other models in terms of DSC score of 0.938, followed by CANet (n = 2) with the PASTA (DSC = 0.911) and CANet (n = 3) with the PASTA (DSC = 0.910). When our PASTA in action, except the GlobalNet was improved from 0.903 to 0.938, the DLIR was improved from 0.884 to 0.890, and the CANet (n = 3) was improved from 0.883 to 0.910, the performance of CANet (n = 1, 2) were similar. Figure. 2 presents the registration results of 2D and 3D registration, we can observe that the registration results of the proposed PASTA and CANet are more accurate compared with the other existing methods.

Statistical Analysis. We performed t-tests for our 2D and 3D results. Except DLIR and CANet (n = 1) for the HandMNIST and CANet (n = 1) for ChestM-NIST and 3D lung dataset, all the other networks using PASTA have shown statistically significant improvements than those without PASTA (p < 0.001). On the other hand, when PASTA is used, CANet (n = 3) performs significantly better than all the other networks (p < 0.001) but the GlobalNet for the 3D lung dataset. These results confirmed the value of PASTA and the effectiveness of CANet.

Specific Transformation Parameters Analysis. Due to the benefit of the PASTA framework, we also can investigate and evaluate the performance of individual affine transformation parameters for the 2D and 3D affine transformation compared with the synthetic parameters. The experimental results showing that the CANet (n = 3) with PASTA performs the best in terms of mean absolute error (MAE) = 0.2%, followed by CANet (n = 2) with PASTA (MAE = 0.5%), CANet (n = 1) with PASTA (MAE = 0.6%) and DLIR with PASTA (MAE = 0.6%) (Table 2).

Table 2. Quantitative results of 2D transformation in terms of (mean absolute error) of our proposed models compared with the others.

	Translation		Rotation	Shear		Scaling		Avg.
	x	y	θ	x	y	x	y	
GlobalNet w/PASTA	0.000	0.000	0.013	0.012	0.009	0.001	0.000	0.005
DLIR + PASTA	0.002	0.000	0.011	0.013	0.009	0.004	0.000	0.006
CANet (n = 1) w/PASTA	0.002	0.000	0.012	0.015	0.010	0.004	0.001	0.006
CANet (n = 2) w/PASTA	0.000	0.000	0.010	0.013	0.010	0.000	0.000	0.005
CANet (n = 3) w/PASTA	0.000	0.000	0.005	0.006	0.006	0.000	0.000	**0.002**

Results on Real Datasets. Further, we investigated and evaluated the performance between real pairs in HandMNIST dataset. 44,850 unique pairs were generated by randomly chosen from 300 different X-ray images of left hands (a ratio of 60:20:20 for training, validation and testing). The results on the testing set proved that the CANet and PASTA (NCC = 0.849) can introduce improvement in terms of NCC compared to the methods without using them or before registration (NCC = 0.655). The GlobalNet with PASTA and CANet (n = 3) with PASTA achieved better performance in terms of NCC score of 0.853 and 0.849 respectively than the GlobalNet (NCC = 0.843), the DLIR with PASTA (NCC = 0.833) and the DLIR (NCC = 0.829). Figure 3 presents the real registration results by different models.

Furthermore, we randomly chose 50 pairs of images and annotated the fingertips of thumb, middle finger and pinky finger, because 1) real pairs with true transformation parameters is not easy to acquire and 2) the error of key-points

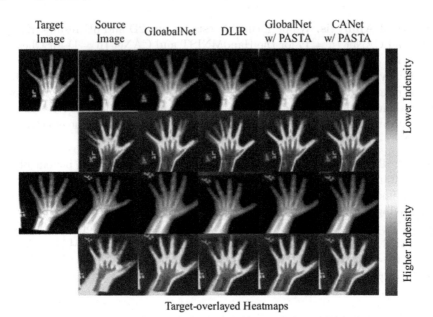

Fig. 3. 2D real registration results of our proposed model for unsupervised affine image registration compared to the previous state-of-art approaches

is more reasonable to evaluate the registration performance than the similarity between images only. Euclidean Distance is introduced to measure the registration accuracy. Before applying registration, the errors of thumb, middle finger and pinky are 7.547, 8.325 and 9.041 pixels, respectively. The quantitative results are presented in Table 3. For the thumb, middle finger alignment, our CANet (n = 3) with PASTA framework achieved the smallest distance of 4.155 and 3.877 pixels respectively, followed by the GlobalNet with PASTA (4.37 and 4.686 pixels) and the GlobalNet (4.47 and 4.458 pixels). For the pinky finger alignment, the GlobalNet with PASTA (4.607 pixel) outperformed the CANet (n = 3) with PASTA (5.420 pixel), the GlobalNet (6.604 pixel) and the DLIR (7.398 pixel).

Table 3. Quantitative results of real transformation of our proposed models compared to the others. Standard deviation is provided in the brackets.

Models	Metrics			
	Image similarity	Thumb	Middle finger	Pinky finger
	NCC	Euclidean distance of fingertips		
Before	0.655 (0.139)	7.547 (3.911)	8.325 (4.973)	9.041 (4.515)
GlobalNet [6]	0.843 (0.070)	4.470 (2.158)	4.458 (3.814)	6.604 (3.275)
DLIR [16]	0.829 (0.076)	5.655 (2.877)	6.875 (4.929)	7.398 (3.705)
GlobalNet [6] w/PASTA	0.853 (0.067)	4.372 (2.278)	4.686 (5.908)	4.607 (2.253)
CANet w/PASTA	0.849 (0.066)	4.155 (2.489)	3.877 (3.705)	5.420 (2.492)

4   Conclusion

In this work, we propose an unsupervised registration model that can explicitly learn specific geometric transformation parameters in the form of translations, rotation, scaling and shearing for both 2D and 3D transformation. we propose the CANet that can effectively learn the linear combination between the images pairs via cross-stitch units for affine transformation learning. Three public datasets: 2D ChestMNIST, HandMNIST and 3D Learn2Reg CT (task 2), are used for the evaluation of our proposed models. Our experimental results show that our models in 2D and 3D outperform the state-of-art approaches (GlobalNet and DLIR). PASTA is generic and could be compatible with other networks without increasing the computation cost. In the future, we will extend our models for joint affine and nonlinear image registration [7,16] as well as graph convolutional networks-based [9,10] or atlas-based [1,19] image segmentation problems.

Acknowledgments. Xu Chen is funded by a studentship jointly funded by the Vascular Surgery Research Fund in Liverpool and Institute of Life Course and Medical Sciences, University of Liverpool, and partially funded by The Great Britain-China Educational Trust (no.269944) administered by the Great Britain-China Centre.

References

1. Aljabar, P., Heckemann, R.A., Hammers, A., Hajnal, J.V., Rueckert, D.: Multi-atlas based segmentation of brain images: atlas selection and its effect on accuracy. Neuroimage **46**(3), 726–738 (2009)
2. Balakrishnan, G., Zhao, A., Sabuncu, M.R., Guttag, J., Dalca, A.V.: Voxelmorph: a learning framework for deformable medical image registration. IEEE Transactions on Medical Imaging (2019)
3. Beljaards, L., Elmahdy, M.S., Verbeek, F., Staring, M.: A cross-stitch architecture for joint registration and segmentation in adaptive radiotherapy. In: Medical Imaging with Deep Learning, pp. 62–74. PMLR (2020)
4. Chee, E., Wu, J.: Airnet: Self-supervised affine registration for 3d medical images using neural networks. arXiv:1810.02583 (2018)
5. Hering, A., Murphy, K., van Ginneken, B.: Lean2reg challenge: Ct lung registration - training data (2020)
6. Hu, Y., et al.: Label-driven weakly-supervised learning for multimodal deformable image registration. In: 2018 IEEE 15th International Symposium on Biomedical Imaging (ISBI 2018), pp. 1070–1074. IEEE (2018)
7. Hu, Y., et al.: Weakly-supervised convolutional neural networks for multimodal image registration. Med. Image Anal. **49**, 1–13 (2018)
8. Jaderberg, M., Simonyan, K., Zisserman, A., et al.: Spatial transformer networks. In: Advances in Neural Information Processing Systems, pp. 2017–2025 (2015)
9. Meng, Y., et al.: Regression of instance boundary by aggregated CNN and GCN. In: European Conference on Computer Vision, pp. 190–207. Springer (2020). https://doi.org/10.1007/978-3-030-58598-3_12
10. Meng, Yet al.: CNN-GCN aggregation enabled boundary regression for biomedical image segmentation. In: International Conference on Medical Image Computing and Computer-Assisted Intervention, pp. 352–362. Springer (2020). https://doi.org/10.1007/978-3-030-59719-1_35

11. Miao, S., Wang, Z.J., Liao, R.: A CNN regression approach for real-time 2d/3d registration. IEEE Trans. Med. Imaging **35**(5), 1352–1363 (2016)
12. Misra, I., Shrivastava, A., Gupta, A., Hebert, M.: Cross-stitch networks for multi-task learning. In: Proceedings of the IEEE conference on computer vision and pattern recognition, pp. 3994–4003 (2016)
13. Ronneberger, O., Fischer, P., Brox, T.: U-net: convolutional networks for biomedical image segmentation. In: International Conference on Medical Image Computing and Computer-Assisted Intervention, pp. 234–241. Springer (2015). https://doi.org/10.1007/978-3-319-24574-4_28
14. Ruder, S., Bingel, J., Augenstein, I., Søgaard, A.: Sluice networks: Learning what to share between loosely related tasks. arXiv:1705.08142 2 (2017)
15. Tissera, D., Vithanage, K., Wijesinghe, R., Kahatapitiya, K., Fernando, S., Rodrigo, R.: Feature-dependent cross-connections in multi-path neural networks. arXiv:2006.13904 (2020)
16. de Vos, B.D., Berendsen, F.F., Viergever, M.A., Sokooti, H., Staring, M., Išgum, I.: A deep learning framework for unsupervised affine and deformable image registration. Med. Image Anal. **52**, 128–143 (2019)
17. Wang, X., Peng, Y., Lu, L., Lu, Z., Bagheri, M., Summers, R.M.: Chestx-ray8: Hospital-scale chest x-ray database and benchmarks on weakly-supervised classification and localization of common thorax diseases. In: Proceedings of the IEEE Conference on Computer Vision and Pattern Recognition, pp. 2097–2106 (2017)
18. Yang, J., Shi, R., Ni, B.: Medmnist classification decathlon: a lightweight automl benchmark for medical image analysis. arXiv:2010.14925 (2020)
19. Zhao, A., Balakrishnan, G., Durand, F., Guttag, J.V., Dalca, A.V.: Data augmentation using learned transforms for one-shot medical image segmentation. arXiv preprint arXiv:1902.09383 (2019)

Conditional Deformable Image Registration with Convolutional Neural Network

Tony C. W. Mok$^{(\boxtimes)}$ and Albert C. S. Chung

Department of Computer Science and Engineering,
The Hong Kong University of Science and Technology, Hong Kong, China
{cwmokab,achung}@cse.ust.hk

Abstract. Recent deep learning-based methods have shown promising results and runtime advantages in deformable image registration. However, analyzing the effects of hyperparameters and searching for optimal regularization parameters prove to be too prohibitive in deep learning-based methods. This is because it involves training a substantial number of separate models with distinct hyperparameter values. In this paper, we propose a conditional image registration method and a new self-supervised learning paradigm for deep deformable image registration. By learning the conditional features that are correlated with the regularization hyperparameter, we demonstrate that optimal solutions with arbitrary hyperparameters can be captured by a single deep convolutional neural network. In addition, the smoothness of the resulting deformation field can be manipulated with arbitrary strength of smoothness regularization during inference. Extensive experiments on a large-scale brain MRI dataset show that our proposed method enables the precise control of the smoothness of the deformation field without sacrificing the runtime advantage or registration accuracy.

Keywords: Controllable regularization · Conditional image registration · Deformable image registration

1 Introduction

Deformable image registration and the subsequent quantitative assessment are crucial in a variety of medical imaging studies. Recent deep learning-based image registration (DLIR) methods [3,5,17,30,31] have achieved remarkable results and showed immense potential for time-sensitive medical imaging studies such as image-guided surgery and motion tracking. Unsupervised DLIR methods [3,12,23,24] circumvent costly iterative optimization in conventional image registration approaches by re-formulating the image registration problem as a

---

Electronic supplementary material The online version of this chapter (https://doi.org/10.1007/978-3-030-87202-1_4) contains supplementary material, which is available to authorized users.

M. de Bruijne et al. (Eds.): MICCAI 2021, LNCS 12904, pp. 35–45, 2021.
https://doi.org/10.1007/978-3-030-87202-1_4

learning problem with convolutional neural networks (CNN), resulting in fast image registration. While DLIR methods have a learning formulation that differs from the conventional image registration approaches [1,2,28,29], the tradeoff between registration accuracy and the smoothness of the deformation field, which is often controlled with a hyperparameter in the objective function, cannot be circumvented by DLIR methods. Typically, the optimal hyperparameter is determined using grid searching on the validation dataset [3,23]. Ironically, despite the runtime advantage of DLIR methods, searching for the optimal hyperparameter value is notoriously time-consuming and computationally intensive in DLIR methods as the hyperparameters are fixed throughout the learning and inference phase. In DLIR methods, each grid search value requires a new DLIR model trained with the distinct hyperparameter value, and each DLIR model requires up to $\sim$20 h to a few days to train from scratch [3]. As such, analyzing the effect of hyperparameters and searching for optimal regularization parameters prove to be too prohibitive in DLIR methods, leading to suboptimal registration results and limited clinical applications. Despite the computational cost of the hyperparameter searching technique, the traditional hyperparameter searching technique may not be a good solution for unsupervised DLIR methods for two thoughtful reasons. First, the optimal regularization parameter is subject to the degree of misalignment between the input images, image modality, and intensity distribution. Second, the prior knowledge of the learned model cannot be utilized in the traditional hyperparameter searching technique, resulting in a substantial computational redundancy.

In recent years, a pioneering work of Gatys et al. [9] demonstrate that CNN encodes both the content and style information of an image. Subsequent studies [4,7,14,15] further illustrate that the image information can be separated by manipulating the statistics of the feature maps with feature-wise linear modulation [6] in CNN. In this paper, motivated by these studies [7,14,15], we propose a novel conditional image registration method and a new self-supervised learning paradigm for deformable image registration to address the inefficiency of existing hyperparameter searching technique in DLIR methods. Instead of training multiple models for searching the optimal hyperparameter, we propose utilizing a single conditional model with self-supervised learning for efficient hyperparameter tuning.

Parallel to our work, Hoopes et al. [13] propose to learn the effects of registration hyperparameters on deformation field with Hypernetworks [10], which leverage a secondary network to generate the conditioned weights for the entire network layers. While the Hypernetworks-based method offers immense modulation potential, it adds an enormous number of parameters to the original image registration method. Alternatively, we propose a more parameter-efficient and scalable approach based on conditional instance normalization. Our method learns the effect of the regularization parameters and conditions on the feature statistics of high-dimensional layers such that the smoothness of the solution can be manipulated via arbitrary hyperparameter values during the inference phase. We further introduce a novel distributed mapping network to generate non-linear embedding with the condition variable. We present extensive experiments, demonstrating that our formulation enables the precise control of the smoothness of the deformation field during the inference phase and rapid grid search of an optimal hyperparameter without sacrificing the runtime advantage or the registration accuracy of the original DLIR method.

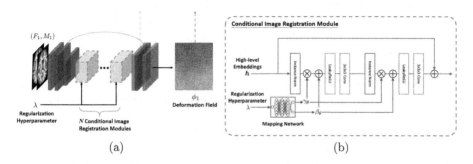

(a)   (b)

Fig. 1. Overview of the proposed (a) conditional deformable image registration method and (b) the conditional image registration module. For clarity and simplicity, we depict the first pyramid level only and illustrate the 2D formulation of our method in the figure.

2   Methods

Deformable image registration establishes a dense non-linear correspondence between a fixed image F and a moving image M, and the solution ϕ is often subject to a weighted smoothness regularization. DLIR methods often formulate the deformable image registration problem as a learning problem $\phi = f_\theta(F, M)$, in which f_θ is parameterized with CNN. Therefore, in contrast to conventional image registration approaches, the strength of the smoothness regularization is fixed throughout the training and inference phase. To address this limitation, we extend the common formulation of DLIR methods to a conditional deformable image registration setting. Instead of learning to adapt a particular weighted smoothness regularization, our proposed method learns the conditional features that correlated with arbitrary hyperparameter values. In the following sections, we describe the methodology of our proposed method.

2.1   Conditional Deformable Image Registration

Given a fixed F, a moving 3D image scan M, and a conditional variable c, we parametrize the proposed conditional image registration method as a function $f_\theta(F, M, c) = \phi$ with CNN. The proposed method works with any CNN-based DLIR methods and conditional variables. Specifically, we parametrize an example of the function f_θ with the deep Laplacian pyramid image registration network (LapIRN) and set the conditional variable to the smoothness regularization parameter λ. To condition a CNN model on a conditional variable, a concatenation-based conditioning approach [6,21,32,33] in generative models is to directly concatenate the condition variable with the input image scans. However, based on our experiments, we observed that the concatenation-based conditioning approach cannot capture a wide range of regularization parameters and bias to a limited range of hyperparameter values.

Therefore, we depart from the concatenation-based conditioning approach and extend the feature-wise linear modulation approach [4,15] instead. We condition the hidden layers on the regularization parameter directly. In particular, the network architecture of LapIRN is comprised of L CNN-based registration networks (CRN). Each CRN consists of three major components: a feature encoder, a set of N residual blocks, and a feature decoder. We replace the N residual blocks with our proposed conditional image registration modules, as shown in Fig. 1(a). The feature encoder extracts the necessary low-level features for deformable image registration, while the feature decoder upsamples and outputs the targeted displacement fields. We only condition the hidden layers in each conditional image registration module on the hyperparameter of the smoothness regularization. We set L and N to 3 and 5 in our experiments, respectively.

2.2   Conditional Image Registration Module

Based on the assumption that the characteristics of the deformation field, i.e. smoothness, can be captured and separated by CNN, we design the conditional image registration module that takes input hidden feature maps and the regularization hyperparameter as input, and outputs hidden features with shifted feature statistics based on conditional instance normalization (CIN) [7]. Specifically, the proposed conditional image registration module adopts the pre-activation structure [11] and includes two CIN layers, each followed by a leaky rectified linear unit (LeakyReLU) activation [18] with a negative slope of 0.2 and a convolutional layer with 28 filters, as depicted in Fig. 1(b). A skip connection is added to preserve the identity of the features.

Conditional Instance Normalization. While the centralized mapping network [15] generates a conditional representation with less memory consumption and computational cost, we argue that the effective representation of the hyperparameter should be diverse and adaptable to different layers in CNN. Chen et al. [4] demonstrate that modulating layers with various depths of CNN results in inconsistent performance, which implies that hidden features of different depths hold distinct feature statistics and non-linearly correspondence to the latent code.

To maintain diverse conditional representations of the hyperparameter for each hidden level, we propose to include distributed mapping networks that learn a separate intermediate non-linear latent variable for each conditional image registration module, which is shared among all the CIN layers. Formally, given a normalized regularization hyperparameter $\lambda \in \bar{\lambda}$, the distributed mapping network $g : \bar{\lambda} \to \mathcal{Z}$ first maps λ to latent code $z \in \mathcal{Z}$. Then, the CIN layers

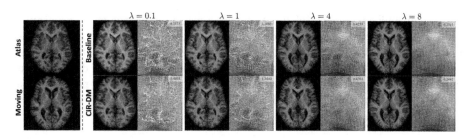

Fig. 2. Example axial MR slices of resulting warped images and deformation fields from the baseline method and our proposed method (CIR-DM) with $\lambda \in [0.1, 1, 4, 8]$. The standard deviation of the Jacobian determinant is shown at the upper-right corner of each resulting deformation fields.

learn a set of parameters that specialize z to the regularization smoothness. The distributed mapping network is parameterized with a 4-layer multilayer perceptron (MLP). For simplicity, we set the number of perceptrons in each MLP layer and the dimensionality of the latent space to 64. The middle layers in the distributed mapping network use the LeakyReLU activation to further introduce the non-linearity into the latent code. The CIN operation for each feature map h_i is defined as

$$h'_i = \gamma_{\theta,i}(z) \left(\frac{h_i - \mu(h_i)}{\sigma(h_i)} \right) + \beta_{\theta,i}(z), \tag{1}$$

where $\gamma_{\theta,i}, \beta_{\theta,i} \in \mathbb{R}$ are affine parameters learned from the latent code z, and $\mu(h_i), \sigma(h_i) \in \mathbb{R}$ are the channel-wise mean and standard deviation of feature map h_i in channel i. In other words, the control of smoothness regularization is learned by normalizing and shifting the feature statistics of the feature map with corresponding affine parameters $\gamma_{\theta,i}$ and $\beta_{\theta,i}$ for each channel in the hidden feature map h.

2.3   Self-supervised Learning

The objective of our proposed method is to compute the optimal deformation field corresponding to the hyperparameter of smoothness regularization. Formally, this task is defined as

$$\phi^* = \arg\min_{\phi} \mathcal{L}_{sim}(F, M(\phi)) + \lambda_p \mathcal{L}_{reg}(\phi), \tag{2}$$

where ϕ^* denotes the optimal displacement field ϕ, $\mathcal{L}_{sim}(\cdot, \cdot)$ denotes the dissimilarity function, $\mathcal{L}_{reg}(\cdot)$ represents the smoothness regularization function and λ_p is uniformly sampled over a predefined range. We set the predefined range of λ_p to $[0, 10]$ empirically such that the optimal deformation field with maximum λ_p is diffeomorphic in most cases. The only difference between the objective in common unsupervised DLIR methods [3,12,23,24] and our objective is that we

learn to optimize the objective function over a predefined range of hyperparameter instead of a fixed hyperparameter value. To exemplify our proposed learning
paradigm, we follow [24] and instantiate the objective function with a similar-
ily pyramid and a diffusion regularizer on the spatial gradients of displacement
fields. We also adopt a progressive training scheme to train the network in a
coarse-to-fine manner. Mathematically, the objective function for each pyramid
level $l \in L$ is defined as

$$\mathcal{L}_l(F, M(\phi), \phi, \lambda_p) = \sum_{i \in [1..l]} -\frac{1}{2^{(l-i)}} NCC_w(F_i, M_i(\phi)) + \lambda_p ||\nabla \phi||_2^2, \qquad (3)$$

where λ_p is sampled uniformly in $[0, 10]$ for each iteration and $NCC_w(\cdot, \cdot)$ denotes
the local normalized cross-correlation (NCC) with window size w, in which w is
set to $1 + 2i$. It is worth noting that our proposed learning paradigm does not
introduce extra computational cost to the original objective function and can be
easily transferred to various DLIR applications with minimum efforts.

3   Experiments

Data and Pre-processing. We evaluate our method on brain atlas registration tasks. We use 425 T1-weighted brain MR scans from the OASIS [19,20]
dataset and 40 brain MR scans from the LPBA40 [26,27] dataset. The OASIS
dataset contains subjects aged from 18 to 96, and 100 of the included subjects
were diagnosed with very mild to moderate Alzheimer's disease. We follow [24]
and perform standard pre-processing, including skull stripping, affine spatial normalization, intensity normalization, and subcortical structures segmentation, for
each MR scan using FreeSurfer [8]. For the OASIS dataset, subcortical segmentation maps of 26 anatomical structures serve as the ground truth for the evaluation
of our method. For the LPBA40 dataset, the brain MR scans in atlas space and
its subcortical segmentation map of 56 anatomical structures, which are delineated by experts, are used in our experiments. We resample all MR scans with
isotropic voxel sizes of 1^3mm and center-cropped all the pre-processed image
scans to $144 \times 192 \times 160$. We randomly split the OASIS dataset into 255, 20,
and 150 volumes and split the LPBA40 dataset into 28, 2, and 10 volumes for
training, validation, and test sets, respectively. We randomly select 3 and 2 MR
scans from the test sets as atlases in OASIS and LPBA40, respectively. Finally,
we register each subject to the chosen atlas using the baseline method and different conditional deformable image registration methods. In summary, there are
441 and 16 combinations of test scans from OASIS and LPBA40, respectively,
included in the evaluation.

Implementation. Our proposed method and the other baseline methods are
implemented with PyTorch 1.7 [25] and deployed on the same machine, equipped
with an Nvidia Titan RTX GPU and an Intel Core (i7-4790) CPU. We build our

method on top of the official implementation of LapIRN available in [22]. We adopt Adam optimizer [16] with a fixed learning rate 0.0001. We normalize $\bar{\lambda}$ to [0,1]. We train all the methods from scratch (60000 iterations in OASIS and 40000 iterations in LPBA40). The source code will be published online.

Baseline Methods. We compare our method with the original LapIRN [24] with a fixed hyperparameter (denoted as baseline). Specifically, we train seven distinct LapIRNs with different regularization hyperparameters $\lambda \in$ [0.1, 0.5, 1, 2, 4, 8, 10]. For each hyperparameter value λ, we select the top-3 models with the highest Dice score on the validation set for evaluation to alleviate the model variation. We further compare it with a concatenation-based conditioning approach (denoted as the traditional method) [6, 21, 33], which simply concatenates the regularization hyperparameter with the input scans in LapIRN to achieve conditional image registration. An ablation study of the variant of our proposed method is performed using either the 8-layer MLP centralized mapping network [15] with latent space 256 (denoted as CIR-CM) and the proposed distributed mapping network (denoted as CIR-DM). For each condition deformable image registration method, we adopt the same training scheme and select the top-3 models with the highest Dice score ($\lambda = 0.1$) on the validation set for evaluation.

Table 1. Quantitative results of the mean DSC and mean std($|J_\phi|$) over seven hyperparameter values on the OASIS and LPBA40 datasets. Initial: spatial normalization.

Method	OASIS						LPBA40					
	DSC	%DSC	std($|J_\phi|$)	%std($|J_\phi|$)	T_{train}	T_{test}	DSC	%DSC	std($|J_\phi|$)	%std($|J_\phi|$)	T_{train}	T_{test}
Initial	0.552	-	-	-	-	-	0.560	-	-	-	-	-
Baseline	0.770	-	1.157	-	200.3h	0.204s	0.729	-	0.697	-	143.2h	0.206s
Traditional	0.780	+1.41%	0.970	+4.62%	28.4h	0.212s	0.722	-1.01%	0.440	-12.35%	20.3h	0.210s
CIR-CM	0.767	-0.57%	0.900	-5.23%	28.8h	0.227s	0.721	-1.14%	0.473	-5.73%	20.5h	0.225s
CIR-DM	0.770	-0.19%	0.963	-3.78%	28.5h	0.216s	0.728	-0.17%	0.552	-3.89%	20.4h	0.218s

Measurement. We register each scan in the test set to an atlas, propagate the anatomical segmentation map of the moving image using the resulting deformation field with the nearest-neighbour interpolation, and measure the overlap of the segmentation maps using Dice similarity coefficient (DSC). We also measure the standard deviation of the Jacobian determinant on the deformation fields (std($|J_\phi|$)), representing the smoothness and local orientation consistency of the deformation field. Moreover, we compare each individual solution from all conditional methods to the solution of the corresponding test case generated from the baseline method, and measure the average difference (in percentage) of the mean Dice score (%DSC) and the standard deviation of the Jacobian determinant on the deformation fields (%std($|J_\phi|$)) over the total number of test cases. Finally, we measure the total training time in hours (T_{train}) and the average inference time per case in seconds (T_{test}) for each method. We repeat the experiment with seven distinct hyperparameter values λ. An ideal conditional image registration algorithm should achieve comparably registration accuracy and quality with the baseline method.

Results and Discussions. Table 1 presents a comprehensive summary of the results of each method in the OASIS and LPBA40 datasets. Figure 2 illustrates qualitative results compare to the baseline method and Fig. 3 shows detail results of each method over seven distinct hyperparameter values in the OASIS dataset. We demonstrate that not only does our method achieves highly consistent results with the baseline method, our method significantly reduces the total training time needed to generate solutions with diverse complexities.

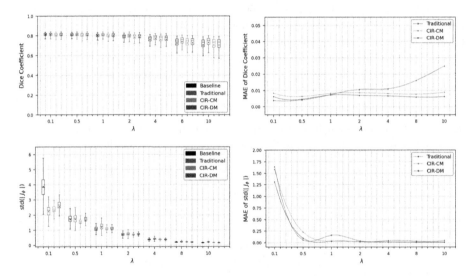

Fig. 3. Quantitative results over seven distinct hyperparameter values on the OASIS dataset. First row: the boxplot of Dice scores and the mean absolute error (MAE) of DSC compared to the baseline method. Second row: the boxplot of std($|J_\phi|$) and the MAE of std($|J_\phi|$) compared to the baseline method. The MAE of DSC (and the std($|J_\phi|$)) is computed by averaging the absolute difference of individual solutions between the targeting methods and baseline method over the total number of test cases.

Specifically, all methods under our proposed conditional framework only required one trained model to generate solutions with seven distinct hyperparameter values of the smoothness regularization λ, resulting in $\sim$7x faster total training time than the baseline method. Interestingly, we find that the complexity of the resulting deformation fields (std($|J_\phi|$)) at $\lambda = 0.1$ declines significantly (-32% to -41%) while maintains comparable Dice scores with the baseline method, indicating that our methods produce even more desirable (smoother) solutions than the baseline method. In contrast to methods based on conditional instance normalization, the traditional method achieves a consistently higher average Dice score and standard deviation of the Jacobian determinant than the baseline method on the OASIS dataset when $\lambda \geq 2$ as shown in Fig. 3, indicating the traditional method tends to bias to a limited range of λ. Compare to CIR-CM, our distributed mapping network design is in every way superior to the centralized mapping network in the context of conditional deformable image registration, as shown in Fig. 3. Importantly, our method achieves only -0.19% (-0.17% on LPBA40) difference of mean Dice score compared to the results of baseline method on OASIS, and the average inference time of CIR-DM is ~ 0.21 seconds, highlighting the fact that CIR-DM is the only method that enables precise control of the deformation field regarding diverse λ without sacrificing the registration accuracy or the runtime advantage of DLIR methods.

4   Conclusion

In summary, we have presented a novel conditional deformable image registration framework and self-supervised learning paradigm for deep learning-based deformable image registration. Our method learns the conditional features that are correlated with the regularization hyperparameter by shifting the feature statistics. It is demonstrated that our method enables precise control of the smoothness regularization in the inference phase without sacrificing the runtime advantage or the registration accuracy of the original DLIR method. Extensive experiments on brain atlas registration have been carried out, demonstrating that the results of our method consistently align with the results of the original DLIR method, and our method is superior to the common conditional approaches with diverse hyperparameter values. In principle, the proposed conditional image registration framework can be easily transferred to arbitrary CNN-based image registration approaches for controllable regularization of the deformation field and rapid hyperparameter tuning.

References

1. Ashburner, J.: A fast diffeomorphic image registration algorithm. Neuroimage **38**(1), 95–113 (2007)
2. Avants, B.B., Epstein, C.L., Grossman, M., Gee, J.C.: Symmetric diffeomorphic image registration with cross-correlation: evaluating automated labeling of elderly and neurodegenerative brain. Med. Image Anal. **12**(1), 26–41 (2008)
3. Balakrishnan, G., Zhao, A., Sabuncu, M.R., Guttag, J., Dalca, A.V.: An unsupervised learning model for deformable medical image registration. In: Proceedings of the IEEE conference on computer vision and pattern recognition, pp. 9252–9260 (2018)
4. Chen, T., Lucic, M., Houlsby, N., Gelly, S.: On self modulation for generative adversarial networks. In: International Conference on Learning Representations (2019)

5. Dalca, A.V., Balakrishnan, G., Guttag, J., Sabuncu, M.R.: Unsupervised learning for fast probabilistic diffeomorphic registration. In: International Conference on Medical Image Computing and Computer-Assisted Intervention. pp. 729–738. Springer (2018). https://doi.org/10.1007/978-3-030-00928-1_82
6. Dumoulin, V., et al.: Feature-wise transformations. Distill **3**(7), e11 (2018)
7. Dumoulin, V., Shlens, J., Kudlur, M.: A learned representation for artistic style. International Conference on Learning Representations (2017)
8. Fischl, B.: Freesurfer. Neuroimage **62**(2), 774–781 (2012)
9. Gatys, L.A., Ecker, A.S., Bethge, M.: Image style transfer using convolutional neural networks. In: Proceedings of the IEEE Conference on Computer Vision and Pattern Recognition, pp. 2414–2423 (2016)
10. Ha, D., Dai, A., Le, Q.V.: Hypernetworks. arXiv:1609.09106 (2016)
11. He, K., Zhang, X., Ren, S., Sun, J.: Identity mappings in deep residual networks. In: European Conference on Computer Vision. pp. 630–645. Springer (2016). https://doi.org/10.1007/978-3-319-46493-0_38
12. Hering, A., van Ginneken, B., Heldmann, S.: mlvirnet: multilevel variational image registration network. In: International Conference on Medical Image Computing and Computer-Assisted Intervention. pp. 257–265. Springer (2019). https://doi.org/10.1007/978-3-030-32226-7_29
13. Hoopes, A., Hoffmann, M., Fischl, B., Guttag, J., Dalca, A.V.: Hypermorph: amortized hyperparameter learning for image registration. In: International Conference on Information Processing in Medical Imaging (2021)
14. Huang, X., Belongie, S.: Arbitrary style transfer in real-time with adaptive instance normalization. In: Proceedings of the IEEE International Conference on Computer Vision, pp. 1501–1510 (2017)
15. Karras, T., Laine, S., Aila, T.: A style-based generator architecture for generative adversarial networks. In: Proceedings of the IEEE/CVF Conference on Computer Vision and Pattern Recognition, pp. 4401–4410 (2019)
16. Kingma, D.P., Ba, J.: Adam: A method for stochastic optimization. arXiv:1412.6980 (2014)
17. Krebs, J., et al.: Robust non-rigid registration through agent-based action learning. In: International Conference on Medical Image Computing and Computer-Assisted Intervention. pp. 344–352. Springer (2017). https://doi.org/10.1007/978-3-319-66182-7_40
18. Maas, A.L., Hannun, A.Y., Ng, A.Y.: Rectifier nonlinearities improve neural network acoustic models. In: International Conference on Machine Learning (ICML), vol. 30, p. 3 (2013)
19. Marcus, D.S., Wang, T.H., Parker, J., Csernansky, J.G., Morris, J.C., Buckner, R.L.: Oasis brains - open access series of imaging studies. https://www.oasis-brains.org/. Accessed on 01 March 2021
20. Marcus, D.S., Wang, T.H., Parker, J., Csernansky, J.G., Morris, J.C., Buckner, R.L.: Open access series of imaging studies (oasis): cross-sectional MRI data in young, middle aged, nondemented, and demented older adults. J. Cogn. Neurosci. **19**(9), 1498–1507 (2007)
21. Mirza, M., Osindero, S.: Conditional generative adversarial nets. arXiv:1411.1784 (2014)
22. Mok, T.C., Chung, A.: Official implementation of laplacian pyramid image registration network. https://github.com/cwmok/LapIRN. Accessed on 01 March 2021
23. Mok, T.C., Chung, A.: Fast symmetric diffeomorphic image registration with convolutional neural networks. In: Proceedings of the IEEE/CVF Conference on Computer Vision and Pattern Recognition, pp. 4644–4653 (2020)

24. Mok, T.C., Chung, A.C.: Large deformation diffeomorphic image registration with laplacian pyramid networks. In: International Conference on Medical Image Computing and Computer-Assisted Intervention, pp. 211–221. Springer (2020). https://doi.org/10.1007/978-3-030-59716-0_21

25. Paszke, A., Gross, S., Chintala, S., et al.: Automatic differentiation in pytorch. In: NIPS-W (2017)

26. Shattuck, D.W, etet al.: Lpba40 atlases download. https://resource.loni.usc.edu/resources/atlases-downloads/. Accessed 0n 01 March 2021

27. Shattuck, D.W., et al.: Construction of a 3d probabilistic atlas of human cortical structures. Neuroimage **39**(3), 1064–1080 (2008)

28. Thirion, J.P.: Image matching as a diffusion process: an analogy with maxwell's demons. Med. Image Anal. **2**(3), 243–260 (1998)

29. Vercauteren, T., Pennec, X., Perchant, A., Ayache, N.: Diffeomorphic demons: Efficient non-parametric image registration. NeuroImage **45**(1), S61–S72 (2009)

30. de Vos, B.D., Berendsen, F.F., Viergever, M.A., Staring, M., Išgum, I.: End-to-end unsupervised deformable image registration with a convolutional neural network. In: Deep Learning in Medical Image Analysis and Multimodal Learning for Clinical Decision Support, pp. 204–212. Springer (2017). https://doi.org/10.1007/978-3-319-67558-9_24

31. Yang, X., Kwitt, R., Styner, M., Niethammer, M.: Quicksilver: fast predictive image registration-a deep learning approach. NeuroImage **158**, 378–396 (2017)

32. Zhang, H., et al.: Stackgan: Text to photo-realistic image synthesis with stacked generative adversarial networks. In: Proceedings of the IEEE International Conference on Computer Vision, pp. 5907–5915 (2017)

33. Zhang, H., et al.: Stackgan++: Realistic image synthesis with stacked generative adversarial networks. IEEE Trans. Pattern Anal. Mach. Intell. **41**(8), 1947–1962 (2018)

A Deep Discontinuity-Preserving Image Registration Network

Xiang Chen[1](✉), Yan Xia[1,2], Nishant Ravikumar[1,2],
and Alejandro F. Frangi[1,2,3,4]

[1] Center for Computational Imaging and Simulation Technologies in Biomedicine,
School of Computing, University of Leeds, Leeds, UK
`scxc@leeds.ac.uk`
[2] Biomedical Imaging Department, Leeds Institute for Cardiovascular and Metabolic
Medicine, School of Medicine University of Leeds, Leeds, UK
[3] Department of Cardiovascular Sciences, KU Leuven, Leuven, Belgium
[4] Department of Electrical Engineering, KU Leuven, Leuven, Belgium

Abstract. Image registration aims to establish spatial correspondence
across pairs, or groups of images, and is a cornerstone of medical image
computing and computer-assisted-interventions. Currently, most deep
learning-based registration methods assume that the desired deformation
fields are globally smooth and continuous, which is not always valid for
real-world scenarios, especially in medical image registration (e.g. cardiac
imaging and abdominal imaging). Such a global constraint can lead to
artefacts and increased errors at discontinuous tissue interfaces. To tackle
this issue, we propose a weakly-supervised Deep Discontinuity-preserving
Image Registration network (DDIR), to obtain better registration performance and realistic deformation fields. We demonstrate that our method
achieves significant improvements in registration accuracy and predicts
more realistic deformations, in registration experiments on cardiac magnetic resonance (MR) images from UK Biobank Imaging Study (UKBB),
than state-of-the-art approaches.

Keywords: Deep learning · Image registration · Cardiac image
registration · Discontinuity-preserving image registration

1 Introduction

Image registration is a fundamental component of several applications in medical imaging. Recent years have seen a shift from traditional iterative methods to
deep learning (DL)-based registration approaches. Although training DL-based
approaches is time-consuming, inference is rapid, involving just a single forward

---

N. Ravikumar and A. F. Frangi—Joint last authorship.

Electronic supplementary material The online version of this chapter (https://
doi.org/10.1007/978-3-030-87202-1_5) contains supplementary material, which is available to authorized users.

M. de Bruijne et al. (Eds.): MICCAI 2021, LNCS 12904, pp. 46–55, 2021.
https://doi.org/10.1007/978-3-030-87202-1_5

pass through the network. Consequently, DL-based approaches offer substantial acceleration for pair-/group-wise image registration relative to traditional approaches, achieving near-real-time performance in certain applications.

Most existing DL-based registration methods constrain deformation fields to be globally smooth and continuous, through various means [3,4,7]. However, this assumption is often violated in medical image registration applications, as tissue boundaries are naturally discontinuous. This is especially pronounced in cardiac or abdominal imaging, which involve large deformations of multiple tissue-types, and organ motion/sliding at tissue boundaries. Variability in the physical properties of different tissue-types results in discontinuities at native tissue boundaries [5,6]. Hence, enforcing deformation fields to be globally smooth can generate unrealistic deformations and lead increased errors near these boundaries.

Discontinuity-preserving image registration is an active area of research in the context of traditional registration methods [6,11,13,15]. For example, Hua et al. [6] proposed a discontinuous registration approach that utilised enriched B-spline basis functions at control points near discontinuous tissue boundaries, achieving significant improvement in registration accuracy, relative to other existing discontinuity-preserving registration methods. In contrast, only one study thus far has proposed a discontinuous DL-based image registration framework. Ng et al. [10] proposed a custom discontinuity-preserving regulariser on the deformation fields (used with a typical unsupervised registration network), to preserve discontinuities, while ensuring local smoothness within specific regions. They formulated a regularisation term based on the unsigned area of the parallelogram spanned by two displacement vectors associated with moving image voxels. However, without additional boundary information for guidance, such a discontinuity regularisation term alone is insufficient to preserve strong discontinuities in deformation fields.

This paper assumes that the desired deformation fields are locally smooth, but discontinuities may exist between different regions/organs at tissue interfaces. Therefore, we generate distinct smooth deformation fields for different regions of interest and compose them to obtain the final registration field, used to warp the moving image. Such a locally-smooth and globally-discontinuous registration scheme is achieved using a novel Deep Discontinuity-preserving Image Registration network, or DDIR. The contributions of this paper are two-fold: (1) we designed a novel framework, DDIR for discontinuous DL-based image registration. This is the first study to incorporate discontinuity in DL network structure and training strategy, and not only in terms of a custom regularisation term in the loss function. (2) Our proposed DDIR achieves significant improvement in registration accuracy over state-of-the-art registration methods, and preserves key cardiac morphological indices post-registration, not afforded by the latter.

2 Method

Pair-wise image registration aims to establish spatial correspondence between the moving image $\mathbf{I}_M$ and fixed image $\mathbf{I}_F$ and is formulated as,

$$\phi(\mathbf{x}) = \mathbf{x} + u(\mathbf{x}), \tag{1}$$

where, $\mathbf{x}$ represents voxels/pixels in the moving image $\mathbf{I}_M$, $u(\mathbf{x})$ denotes the displacement field, and $\phi(\circ)$ represents the deformation function.

To generate deformation fields that are locally smooth and discontinuous at the boundaries of different organs/regions, we propose to generate deformation fields for different sub-regions, and compose them to obtain the final deformation field. Sub-regions in the images to be registered must first be segmented either manually or automatically. With short-axis (SAX) cardiac cine-magnetic resonance (CMR) images, manual and automatic segmentation results for left ventricle blood pool (LVBP), left ventricle myocardium (LVM) and right ventricle (RV) are generally available in public data sets, large-scale imaging initiatives (e.g. UK Biobank) and from previous studies on automatic CMR segmentation [2]. As the focus of this paper is on SAX-CMR image registration, we explicitly model discontinuities along cardiac boundaries by splitting the images into four sub-regions, namely, LVBP, LVM, RV, and background. These sub-regions are subsequently used to train our DDIR approach and register CMR images in manner that preserves discontinuities at their boundaries.

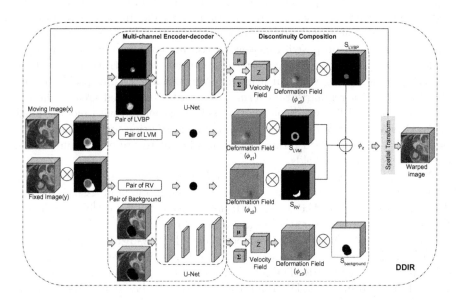

Fig. 1. Schema of DDIR. The registration network applies four different channels extracting features from pairs of LVBP, LVM, RV and background. Based on them, we obtain four sub-deformation fields for different regions. The final deformation field is obtained by composing these four deformation fields with corresponding segmentation. The cardiac MR images were reproduced by kind permission of UK Biobank ©.

Network Architecture. Most previous DL-based registration methods apply an encoder-decoder network (generally U-Net [12]) to extract feature maps from the concatenated input moving image and fixed image. However, as shown in Fig. 1, in DDIR the original moving image and fixed image (at $128 \times 128 \times 32$) are divided into four image pairs, i.e. LVBP, LVM, RV and background, using segmentation masks for the corresponding regions. In each of these pairs, voxels in corresponding regions are preserved while the rest are set at zero. Each pair is concatenated and fed as input to a distinct U-Net block, which extracts region-specific feature maps. These four U-Nets have the same architecture, including four down-sampling layers and three corresponding up-sampling layers. Using this multi-channel encoder-decoder structure, we obtain four sets of feature maps ($64 \times 64 \times 16$) corresponding to different sub-regions. We use the same U-Net architecture (with identical hyper-parameters) in all DL-based registration approaches investigated in this study.

Discontinuity Composition. Using the region-specific feature maps learned by the U-Nets, we first predict four different smooth deformation fields (corresponding to each region) and then compose them to obtain the final deformation field, to preserve local smoothness and discontinuity at the interfaces. Similar to previous papers [4,7], we assume the transformation function (denoted as ϕ_z) is parametrised by stationary velocity fields (SVF) ($z_i, i \in [0,3]$), which are sampled from a multivariate Gaussian distribution. With the predicted feature map, we compute the mean μ_i and variance Σ_i of z_i (using two different convolution layers). Based on them, four SVFs (z_0, z_1, z_2, z_3) corresponding to different regions (LVBP, LVM, RV and background) are sampled. With the corresponding integration layer and up-sampling layer, we obtain four diffeomorphic deformation fields ϕ_{z_0}, ϕ_{z_1}, ϕ_{z_2} and ϕ_{z_3}. As before, we use region-specific segmentation masks to extract each region of interest from the obtained deformation fields (setting the remaining voxels to zero) and compose them to generate the final deformation field. Denoting the segmented regions of LVBP, LVM, RV and background as $S_{LVBP}, S_{LVM}, S_{RV}$ and $S_{background}$ respectively, the composition can be formulated as,

$$\phi_z = \phi_{z_0} \times S_{LVBP} + \phi_{z_1} \times S_{LVM} + \phi_{z_2} \times S_{RV} + \phi_{z_3} \times S_{background}. \quad (2)$$

Loss Function. The loss function includes two terms, a dissimilarity and a regularisation term. The former is the distance between the warped moving image and the fixed image, while, the latter constrains the estimated deformation fields to be locally smooth (i.e. within each region), to avoid unrealistic deformations. The dissimilarity loss in DDIR captures the dissimilarity on both images and segmentations. We use normalised cross-correlation (NCC) L_{NCC} to evaluate the similarity between the warped moving image and the fixed image. As the region-wise segmentation masks are available, we also compute the region-wise dice loss, denoted L_{Dice} as in [9].

To preserve discontinuity at the interfaces of the organs/regions while ensuring local smoothness, a global smoothness constraint is not enforced on the composed deformation field. The composition of different deformation fields preserves discontinuities at interfaces, therefore, we only need to guarantee the deformation field of each sub-region smooth. This is achieved by regularising each sub-deformation field. Following Voxelmorph-diff [4], we calculate the Kullback-Leibler (KL) divergence between the approximate posterior $q_\psi(z|\mathbf{I}_F; \mathbf{I}_M)$ and the prior $p(z)$ ($p(z) = \mathcal{N}(z; 0, \Sigma_z)$) of each velocity field z, formulated as,

$$R = KL(q_\psi(z|\mathbf{I}_F; \mathbf{I}_M) || p(z|\mathbf{I}_F; \mathbf{I}_M)),$$

$$L_R = \frac{1}{4}(R_{LVBP} + R_{LVM} + R_{RV} + R_{background}), \quad (3)$$

where R denotes the regularisation for each deformation field and L_R is the combined regularisation term. The $q_\psi(z|\mathbf{I}_F; \mathbf{I}_M) = N(z; \mu_{\mathbf{z}|\mathbf{I}_F,\mathbf{I}_M}, \Sigma_{\mathbf{z}|\mathbf{I}_F,\mathbf{I}_M})$ is a multivariate normal, where, $\mu_{\mathbf{z}|\mathbf{I}_F,\mathbf{I}_M}$ and $\Sigma_{\mathbf{z}|\mathbf{I}_F,\mathbf{I}_M}$ are the mean and variance of the distribution, learned by convolution layers. The complete loss function used to train the network is, $L_{total} = \lambda_0 \times L_{NCC} + \lambda_1 \times L_{Dice} + \lambda_2 \times L_R$, where, λ_0, λ_1 and λ_2 are used to weight the importance of each loss term.

3   Experiments and Results

Data and Implementation. The registration performance of the proposed approach is evaluated on SAX-CMR images (spatial resolution at $\sim 1.8 \times 1.8 \times 10$ mm^3), available from UKBB. We chose images from 2,000 subjects at random, and used images at end-diastole (ED) and end-systole (ES) for intra-subject registration. Among these, 1,600 subjects' data was chosen at random for training DDIR, equating to 3,200 image pairs (ED-to-ES or ES-to-ED registration). Image pairs from the remaining 400 subjects were used for testing. All CMR images were resampled to $1.50 \times 1.50 \times 3.15$ mm^3 using bi-cubic interpolation, and cropped to a size of $128 \times 128 \times 32$ (with zero-padding for images with fewer than 32 slices). The region-wise segmentation masks for all CMR images were obtained automatically using the segmentation method proposed in [2]. DDIR was implemented using Python and Keras on a Tesla M60 GPU machine. The Adam optimiser was used for training, with a learning rate of $1e-4$. The batch size was set to 2, and the hyper-parameters λ_0, λ_1 and λ_2 were set to $20, 200, 0.1$ (determined empirically), respectively. The source code will be publicly available on the Github[1].

Quantitative Comparison and Analysis. To demonstrate the superiority of our approach, we compare DDIR with both traditional registration and DL-based registration methods. For the former, we choose Symmetric Normalisation (SyN) registration (3 resolution level, with 100 iterations in each sampling level) in ANTS [1], Demons (Fast Symmetric Forces Demons [14] with 800 iterations and standard deviations 1.0) in SimpleITK and B-spline registration (max iteration step is 2000, sampling 6000 random points per iteration) in SimpleElastix [8], for comparison. For the latter, DDIR is compared with Voxelmorph-diff [4]. As DDIR uses segmentation masks during training and inference, it is a weakly-supervised registration method. For fair comparison, we build three weakly-supervised versions of Voxelmorph - VM-Dice, VM(img+seg) and VM-Dice(img+seg). VM-Dice uses a Dice loss L_{Dice} term and binary cardiac segmentation masks for the fixed and moving images during training, but does not require the latter for inference. In VM(img+seg), we concatenate the fixed and

---

[1] https://github.com/cistib/DDIR.

moving images with their corresponding multi-class masks (i.e. distinct labels for each region) and use these to train the network. While, VM-Dice(img+seg) is a combination of the previous two methods. We did not compare with the DL-based discontinuity-preserving method proposed in [10], as there is no corresponding source code publicly available. This strategy to register different sub-regions and compose corresponding deformation fields is also applicable to the aforementioned networks. Hence, we also apply this strategy during inference, for trained Voxelmorph-diff and VM-Dice models (as they only require sub-images as input on the inference), for comparison with DDIR. These are denoted Voxelmorph-diff(compose) and VM-Dice(compose). These two approaches are different to DDIR as the composition of sub-deformation fields is not learned end-to-end during training (as in DDIR).

To demonstrate the advantage of incorporating discontinuity in the DL-based registration network, we also build a baseline for DDIR, DDIR(baseline), where the predicted feature maps from the four different channels are concatenated and used to compute a single diffeomorphic deformation field (instead of four sub-deformation fields, as in DDIR).

Qualitative Results. Registration results obtained using DDIR and the other methods investigated are assessed visually in Fig. 2. Here, the moving and fixed images are shown in the first column. The corresponding warped moving images, deformation fields, and Jacobian determinants (rows 1–3) obtained following registration using SyN, B-spline, Voxelmorph-diff, DDIR(baseline) and DDIR, are shown in columns 2–6. The warped moving images obtained by both traditional registration methods distinctly different to fixed image, although the B-spline result appears visually more similar than obtained by SyN. All warped moving images obtained using DL-based methods look more similar to the fixed image, than the former. The deformation fields and their corresponding Jacobian determinants estimated using each approach indicate that distinct boundaries for the left and right ventricle are retained using DDIR, not afforded by the rest.

Table 1. Quantitative comparison between DDIR and state-of-the-art methods using the DS of LVBP, LVM, RV and average Dice (denoted as Avg. DS) and HD. Statistically significant improvements in registration accuracy (DS and HD) are highlighted in bold. Besides, LVEDV and LVMM indices with no significant difference from the reference are also highlighted in bold.

Methods	LVBP DS (%)	LVM DS (%)	RV DS (%)	Avg. DS (%)	HD (mm)	LVEDV	LVMM
before Reg	57.68 ± 6.21	30.88 ± 8.68	55.13 ± 7.51	47.90 ± 6.33	12.91 ± 2.48	143.76 ± 32.13	83.67 ± 21.06
B-spline	74.44 ± 11.50	68.06 ± 7.20	61.76 ± 12.05	68.09 ± 8.76	13.72 ± 3.57	131.14 ± 40.64	81.11 ± 22.60
Demons	80.29 ± 10.00	69.96 ± 5.50	64.86 ± 9.67	71.70 ± 6.96	13.06 ± 3.12	138.00 ± 34.15	80.00 ± 21.25
SyN	70.92 ± 9.36	57.88 ± 10.59	60.30 ± 8.35	63.03 ± 8.29	12.98 ± 2.68	120.09 ± 41.83	$\mathbf{83.12 \pm 21.20}$
Voxelmorph-diff	81.73 ± 8.71	72.04 ± 4.65	65.73 ± 9.62	73.16 ± 6.26	12.96 ± 3.14	137.16 ± 32.59	78.65 ± 21.68
VM-Dice	82.28 ± 8.75	72.53 ± 4.59	66.30 ± 9.67	73.70 ± 6.28	13.00 ± 3.24	139.58 ± 32.79	78.98 ± 21.57
VM (img+seg)	82.54 ± 8.50	72.66 ± 4.80	66.69 ± 9.64	73.96 ± 6.28	12.68 ± 3.21	138.29 ± 33.00	80.83 ± 21.62
VM-Dice (img+seg)	81.97 ± 8.53	71.23 ± 4.79	70.20 ± 12.05	74.47 ± 6.79	11.28 ± 4.35	$\mathbf{144.33 \pm 32.93}$	80.17 ± 22.02
Voxelmorph-diff (compose)	78.82 ± 6.38	67.41 ± 8.80	75.10 ± 6.97	73.78 ± 6.10	11.74 ± 3.08	119.30 ± 38.71	91.39 ± 23.07
VM-Dice (compose)	79.59 ± 5.91	68.81 ± 7.81	$\mathbf{77.93 \pm 6.63}$	75.44 ± 5.36	11.14 ± 3.12	120.90 ± 38.14	94.89 ± 25.96
DDIR (baseline)	84.25 ± 8.63	75.02 ± 4.50	71.42 ± 10.32	76.90 ± 6.58	11.85 ± 3.38	$\mathbf{141.73 \pm 32.29}$	79.01 ± 21.40
DDIR	84.63 ± 8.07	75.27 ± 5.03	74.07 ± 8.73	$\mathbf{77.99 \pm 5.47}$	$\mathbf{10.65 \pm 3.51}$	$\mathbf{141.84 \pm 32.59}$	$\mathbf{81.92 \pm 21.86}$

Quantitative Results. To quantitatively evaluate the performance of our approach, we compare DDIR with previous methods using Dice score (DS) and the Hausdorff Distance (HD). DS is computed for LVBP, LVM and RV. These values and the average DS and HD across all regions are reported in Table 1. Besides, to demonstrate the clinical value of DDIR, we also compute two clinical indices, LV end-diastolic volume (LVEDV) and LV myocardial mass (LVMM). The former is computed using ED segmentations, while the latter, is computed using ED and ES segmentations, pre- and post-registration. Pre-registration, LVEDV and LVMM are computed based on the moving and fixed segmentations (used as reference values). Post-registration, we compute them based on the warped moving segmentation. Therefore, as we perform both ED-to-ES and ES-to-ED registration for each subject, the LVMM values reported in Table 1 represent the average computed at both ED and ES, across all subjects. Thus the closer LVEDV and LVMM (post-registration) are to the reference values, the better the registration performance.

DL-based approaches outperform traditional registration methods in terms of both DS and HD. The weakly-supervised variants of Voxelmorph-diff provide improvements over Voxelmorph-diff, consistent with previous research [4]. Using segmentation masks as additional input channels to the network (VM(img+seg)) yields better results than using them just to compute the loss and drive gradient updates (VM-Dice) (73.96% vs 73.70%). However, conversely the former requires

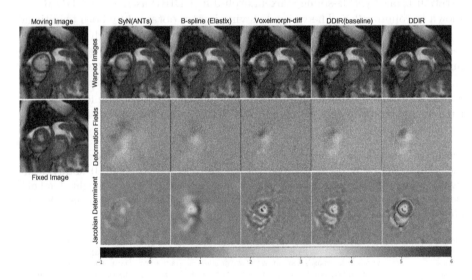

Fig. 2. Visual comparison of deformation fields estimated using DDIR and state-of-the-art methods. Left column: Moving and fixed images; Right column: corresponding warped moving image (first row), deformation fields (second row) and Jacobian Determinant (last row). Colours in the Jacobian determinant images, from blue to red represent the intensity from low to high. The cardiac MR images were reproduced by kind permission of UK Biobank ©. (Color figure online)

segmentation masks during inference, while the latter do not. The combination of these two strategies (VM-Dice(img+seg)) further improves registration performance (~0.5% in terms of average DS). Composing sub-deformation fields also improves registration accuracy of the trained networks, with Voxelmorph-diff (compose) achieving 0.6% higher average DS than Voxelmorph-diff (73.78% vs 73.16%), and VM-Dice (compose) achieving ~1.7% higher average DS than VM-Dice (75.44% vs 73.70%). We found that the DDIR(baseline) achieves ~1% higher average DS than VM-Dice(img+seg) (76.90% vs 75.93%), which highlights the advantage of using a multi-channel encoder-decoder network. Compared with DDIR, we found that incorporating discontinuity further improves the average DS (77.99% vs 76.90%). Correspondingly, DDIR also obtains the best performance in terms of the DS for LVBP, LVM and HD, while its RV DS is lower than VM-Dice(compose). We evaluated the statistical significance of these results using paired t-tests and found that DDIR significantly outperforms Voxelmorph-diff, VM-Dice, VM(img+seg) and VM-Dice(img+seg) on all DS and HD metrics (P-value < 0.05). DDIR also significantly outperforms DDIR(baseline) in terms of average DS, RV DS and HD. Each sub-deformation field generated by DDIR are smooth (without foldings). After composing, the discontinuity only exists at the interface of different sub-regions, which demonstrates that DDIR can generate locally-smooth but globally-discontinuous deformation fields.

The clinical indices, LVEDV and LVMM, show no significant differences (P-value > 0.05) post-registration using DDIR to the reference values, not afforded by other approaches. This demonstrates the superiority and clinical value of our method. To analyse the discontinuity on the deformation fields, we visualise the deformation fields generated using DDIR and DDIR (baseline) (presented in the supplementary material), where the discontinuity is observed for the former along the LV and RV boundaries. To further demonstrate the robustness and generalisability of our approach, we apply the models trained on UKBB data, to the publicly available Automatic Cardiac Diagnosis Challenge (ACDC) data set. The qualitative and quantitative results are included in the supplementary material for brevity. As cardiac motion in ACDC images is not as pronounced as in UKBB (in some cases, the images in ED are very similar to ES), only marginal differences in registration performance are observed between DDIR and the other composition-based methods in terms of DS and HD. However, as before, DDIR outperforms Voxelmorph-diff and traditional state-of-the art methods. Additionally, the clinical indices quantified (LVEDV, LVMM) post registration using DDIR show no significant differences to the reference, not afforded by any of the other methods investigated. This demonstrates the potential for applying DDIR in real clinical scenarios.

4   Conclusion

We proposed a novel weakly-supervised discontinuity-preserving registration network, DDIR, which significantly outperformed the state-of-the-art, in intra-patient CMR registration. DDIR preserves LV clinical indices post-registration, not afforded by the other approaches. This makes it compelling as a tool for use in clinical applications as it ensures that common diagnostic biomarkers for the LV are preserved post-registration.

Acknowledgements. This research was conducted using the UKBB resource under access application 11350 and was sipported by the Royal Academy of Engineering under the RAEng Chair in Emerging Technologies (CiET1919/19) scheme and EPSRC TUSCA (EP/V04799X/1).

References

1. Avants, B.B., Tustison, N.J., Song, G., Cook, P.A., Klein, A., Gee, J.C.: A reproducible evaluation of ANTs similarity metric performance in brain image registration. Neuroimage **54**(3), 2033–2044 (2011)
2. Bai, W., et al.: Automated cardiovascular magnetic resonance image analysis with fully convolutional networks. J. Cardiovasc. Magn. Reson. **20**(1), 65 (2018)
3. Balakrishnan, G., Zhao, A., Sabuncu, M.R., Guttag, J., Dalca, A.V.: Voxelmorph: a Learning Framework for Deformable Medical Image Registration. IEEE Trans. Med. Imaging **38**(8), 1788–1800 (2019)
4. Dalca, A.V., Balakrishnan, G., Guttag, J., Sabuncu, M.R.: Unsupervised learning of probabilistic diffeomorphic registration for images and surfaces. Med. Image Anal. **57**, 226–236 (2019)
5. Hua, R.: Non-rigid Medical Image Registration with Extended Free Form Deformations: Modelling General Tissue Transitions. Ph.D. thesis, University of Sheffield (2016)
6. Hua, R., Pozo, J.M., Taylor, Z.A., Frangi, A.F.: Multiresolution eXtended free-form deformations (XFFD) for non-rigid registration with discontinuous Transforms. Med. Image Anal. **36**, 113–122 (2017)
7. Krebs, J., Delingette, H., Mailhé, B., Ayache, N., Mansi, T.: learning a probabilistic model for diffeomorphic registration. IEEE Trans. Med. Imaging **38**(9), 2165–2176 (2019)
8. Marstal, K., Berendsen, F., Staring, M., Klein, S.: SimpleElastix: a user-friendly, multi-lingual library for medical image registration. In: Proceedings of the IEEE Conference on Computer Vision And Pattern Recognition Workshops, pp. 134–142 (2016)
9. Milletari, F., Navab, N., Ahmadi, S.A.: V-net: Fully convolutional neural networks for volumetric medical image segmentation. In: 2016 Fourth International Conference on 3D Vision (3DV), pp. 565–571. IEEE (2016)
10. Ng, E., Ebrahimi, M.: an unsupervised learning approach to discontinuity-preserving image registration. In: International Workshop on Biomedical Image Registration, pp. 153–162. Springer (2020). https://doi.org/10.1007/978-3-030-50120-4_15
11. Pace, D.F., Aylward, S.R., Niethammer, M.: A locally adaptive regularization based on anisotropic diffusion for deformable image registration of sliding organs. IEEE Trans. Med. Imaging **32**(11), 2114–2126 (2013)
12. Ronneberger, O., Fischer, P., Brox, T.: U-Net: convolutional networks for biomedical image segmentation. In: International Conference on Medical Image Computing and Computer-Assisted Intervention, pp. 234–241. Springer (2015). https://doi.org/10.1007/978-3-319-24574-4_28

13. Schmidt-Richberg, A., Werner, R., Handels, H., Ehrhardt, J.: Estimation of slipping organ motion by registration with direction-dependent regularization. Med. Image Anal. **16**(1), 150–159 (2012)
14. Vercauteren, T., Pennec, X., Perchant, A., Ayache, N., et al.: Diffeomorphic demons using ITK's finite difference solver hierarchy. Insight J. **1** (2007)
15. Wu, Z., Rietzel, E., Boldea, V., Sarrut, D., Sharp, G.C.: Evaluation of deformable registration of patient lung 4DCT with subanatomical region segmentations. Med. Phys. **35**(2), 775–781 (2008)

End-to-end Ultrasound Frame to Volume Registration

Hengtao Guo[1], Xuanang Xu[1], Sheng Xu[2], Bradford J. Wood[2], and Pingkun Yan[1(✉)]

[1] Department of Biomedical Engineering and Center for Biotechnology and Interdisciplinary Studies, Rensselaer Polytechnic Institute, Troy, NY 12180, USA
yanp2@rpi.edu
[2] Center for Interventional Oncology, Radiology and Imaging Sciences, National Institutes of Health, Bethesda, MD 20892, USA

Abstract. Fusing intra-operative 2D transrectal ultrasound (TRUS) image with pre-operative 3D magnetic resonance (MR) volume to guide prostate biopsy can significantly increase the yield. However, such a multimodal 2D/3D registration problem is very challenging due to several significant obstacles such as dimensional mismatch, large modal appearance difference, and heavy computational load. In this paper, we propose an end-to-end frame-to-volume registration network (FVR-Net), which can efficiently bridge the previous research gaps by aligning a 2D TRUS frame with a 3D TRUS volume without requiring hardware tracking. The proposed FVR-Net utilizes a dual-branch feature extraction module to extract the information from TRUS frame and volume to estimate transformation parameters. To achieve efficient training and inference, we introduce a differentiable 2D slice sampling module which allows gradients backpropagating from an unsupervised image similarity loss for content correspondence learning. Our experiments demonstrate the proposed method's superior efficiency for real-time interventional guidance with highly competitive registration accuracy. Source code of this work is publicly available at https://github.com/DIAL-RPI/FVR-Net.

Keywords: 2D/3D Registration · Ultrasound imaging · End-to-end · Deep learning · Prostate biopsy · Computer guided intervention

1 Introduction

Prostate cancer is a leading cause of cancer death for men in the United States [18]. Fusing transrectal ultrasound (TRUS) and magnetic resonance imaging (MRI) has been proven efficient for guiding targeted biopsies to more accurately diagnose the disease [1,16]. Real-time 2D TRUS imaging is registered to a pre-operative 3D MRI volume for joint visualization during the fusion-guided procedures. Benefited by the time-efficient imaging of TRUS and high resolution of MRI, clinicians can locate the targeted lesions, thus increase the biopsy

© Springer Nature Switzerland AG 2021
M. de Bruijne et al. (Eds.): MICCAI 2021, LNCS 12904, pp. 56–65, 2021.
https://doi.org/10.1007/978-3-030-87202-1_6

yield. The core of this technology is to register 2D TRUS images with a 3D MRI volume, which is a very challenging problem.

Existing fusion systems usually rely on external tracking devices to establish the registration [2,9,14,21]. The workflow involves reconstructing a 3D TRUS volume from a sequence of tracked 2D TRUS video frames, which is then aligned with the pre-operative MRI volume through 3D-3D image registration. During the interventional guidance stage, a tracked 2D TRUS frame is mapped to the 3D TRUS volume and then transformed into the MRI image space for fusion. These tracking-based methods require a hardware setup, which induces additional cost and human effort.

Recent advances in deep learning (DL) based image registration and volume reconstruction have enabled new opportunities to shift the MRI/TRUS fusion paradigm. Hu *et al.* [7] first proposed a weakly supervised method that uses landmark annotations as auxiliary information for training an end-to-end registration network. Haskins *et al.* [6] developed a convolutional neural network (CNN) to learn the deep similarity metric between TRUS and MRI volume for the iterative registration. Guo *et al.* [4] proposed a multi-stage registration framework that aligns a TRUS/MRI pair from coarse to fine. Deep learning has also been used for sensorless US volume reconstruction. Prevost *et al.* [17] proposed to use a CNN for directly estimating the inter-frame motion between two 2D US frames, which enables sensorless US volume reconstruction. One recent work [5] applies 3D CNN on a US video sub-sequence to better utilize the temporal context information for sensorless TRUS volume reconstruction. With the existing efforts, we are one step away from building a DL-based trackingless fusion system.

The objective of our work presented in this paper is to bridge the above research gap by developing a 2D TRUS image to 3D TRUS volume trackingless registration method. 2D/3D image registration is also often referred as slice-to-volume registration. Conventional approaches have tried to optimize the registration field according to an image matching criterion, which quantifies the alignment between the images and guides the optimization process. Classical matching criteria use pixel/voxel intensities to quantify the image similarity. For example, Wein *et al.* [19] proposed a similarity measure named linear correlation of linear combination, which reveals the correspondence between simulated US images with MRI/Computed tomography (CT). However, the iterative optimization methods are time-consuming, typically taking seconds or more to register a single pair of images, making them unsuitable for interventional guidance. Another group of 2D/3D registration methods aims to align a 2D projective image with a 3D volume, which typically applies to 2D X-ray with 3D CT volume registration. For example, Miao *et al.* [12,13] proposed to use CNN to directly predict the transformation parameters to perform 3D model to 2D X-ray registration. However, such projective 2D/3D registration is entirely different from our targeted application and thus not applicable.

In this paper, we propose a novel end-to-end frame-to-volume registration network (FVR-Net) to bridge the gap in real-time TRUS/MRI fusion for guiding prostate biopsy. The underlying framework takes one real-time 2D TRUS

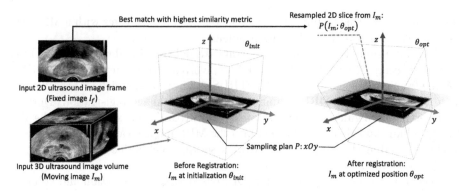

Fig. 1. Illustration of the rigid frame-to-volume registration in this work. The pink cube represents the boundaries of the input 3D ultrasound image volume.

frame and one reconstructed 3D TRUS volume as input to estimate the transformation parameters that best align these two images. We proposed a dual-branch balanced feature extraction network, which makes the model sensitive to both the frame and volume information. Besides, we introduce an auxiliary image similarity loss for end-to-end training which can significantly reduce the registration error. The experiments demonstrate that using CNN for FVR problems is highly promising, achieving competing performance to the conventional iterative registration methods while running tens of times faster.

2 Problem Definition

In this section, we give a formal definition of the mono-modal 2D TRUS frame to 3D TRUS volume registration problem. Figure 1 illustrates the overall implementation workflow. Given a 2D TRUS frame I_f (fixed image) and a 3D TRUS volume I_m (moving image) as input, we seek a mapping function θ_{opt} through minimizing the following objective function:

$$\theta_{opt} = \arg\min_{\theta} Sim(I_f, P(I_m; \theta)), \tag{1}$$

where $P(I_m; \theta)$ denotes the slice extracted from I_m and specified by the transformation θ and the sampling plane P, which is permanently set to be the xOy plane. $Sim()$ is the matching criterion, which quantifies the image similarity between the 2D frame I_f and the resampled slice $P(I_m; \theta)$. By default, for a volume transformed by an identical transformation θ_{init} as the initialization, the volume's center is placed at the coordinate system's origin. The goal of the frame-to-volume registration is to find the optimal transformation parameters θ_{opt}, such that the resampled slice $P(I_m; \theta_{opt})$ has the highest image similarity as the 2D input frame I_f.

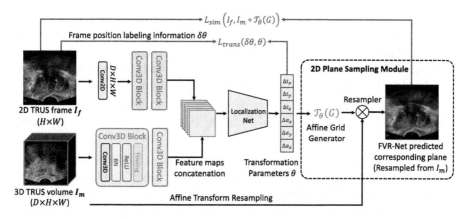

Fig. 2. Overall network structure of the proposed FVR-Net.

3   Method

This section presents the key components of the proposed method. Figure 2 depicts the proposed end-to-end frame-to-volume registration network (FVR-Net). The FVR-Net takes an 2D TRUS frame and a 3D volume as input for estimating the transformation $T(\theta)$, such that the transformed volumetric image and the sampling plane's cross-sectional area $P(I_m; \theta)$ has the highest image similarity with the 2D TRUS image frame I_f. We found the rigid registration can best suit in our application without loss of generality. Thus, the FVR-Net's output θ contains 6 degrees of freedom, i.e. $\theta = \{t_x, t_y, t_z, \alpha_x, \alpha_y, \alpha_z\}$, including the translations and rotations along the three axes.

3.1   End-to-end Slice-to-Volume Registration

In this end-to-end FVR framework, we define the real-time 2D transrectal ultrasound image frame as the fixed image I_f ($H \times W$), and the reconstructed 3D TRUS image subvolume as the moving image I_m ($D \times H \times W$).

Dual-branch Balanced Feature Extraction. The dimension gap between 2D and 3D images is a major obstacle for the registration performance. Directly concatenating these two inputs together (early-fusion) can make the network overwhelmed by the volumetric information while completely ignoring the 2D image contents. Instead, to balance the data information, we designed a dual-branch network structure to extract the image features from the frame and volume independently and then concatenate them in a late-fusion fashion.

In the frame branch, we first use a 2D convolutional layer to extract the low-level features for the input frame and extend the channel number to D, such that the feature map's size matches the input volume's size, thus achieving the data information balance. From this point, each branch is followed by two 3D convolutional blocks with the same hyper-parameters to maintain the identical feature map size. The extracted feature maps from both branches are concatenated together along the depth dimension and then serve as the input to the localization-net for estimating the transformation parameters θ. A supervised mean squared error (MSE) loss can be computed, which directly uses ground-truth labeling information to constraint the θ estimation:

$$L_{trans} = \frac{1}{N} \sum_{n=1}^{N} \|\delta\theta_n - \theta_n\|_2, \qquad (2)$$

where N denotes the total number of 2D/3D sample pairs within one training epoch, and $\delta\theta$ denotes the transformation parameters label. The localization-net uses ResNext [20] as the backbone structure. The proposed dual-branch feature extraction can ensure that the most representative image features can be learned from the images with different dimensions.

Differentiable 2D Slice Sampling. Inspired by the spatial transformer network (STN) [8], we designed a 2D slice sampling module that can introduce a new unsupervised image similarity loss for stabilizing the training process. This module has two components: the affine grid generator and the resampler. Our customized affine grid generator takes the estimated parameters θ as input and generates a transformed resampling grid $T_\theta(G)$, which has the same size as the moving image m. By applying bilinear interpolation at each point location defined by the sampling grid, the resampler can get the intensity at a particular pixel in the wrapped image. Thus, a target 2D slice $P(I_m; \theta)$, noted as $I_m \circ T_\theta(G)$ in Fig. 2, can be sampled from the 3D input volume I_m transformed by the estimated parameters θ. This FVR-Net predicted target slice denotes the results from the 2D/3D registration framework and should ideally contain the same information as the input 2D frame.

Through the partial derivative, the loss gradients can be backpropagated to the sampling grid coordinates, and furthermore, to the affine transformation parameters and the localization net. This makes the entire pipeline of the FVR-Net differentiable and can be easily trained in an end-to-end manner [8]. We further introduce an auxiliary image similarity loss, modified from Eq. 1:

$$L_{sim} = \frac{1}{N} \sum_{n=1}^{N} \|I_{f,n} - P(I_{m,n}; \theta_n)\|_2, \qquad (3)$$

Theoretically, the FVR-Net can be trained in an unsupervised fashion by using the L_{sim} alone. However, in practice, we found this yielding an unstable training process. When L_{trans} and L_{sim} are used together to update the network's parameters, the FVR-Net shows the most robust registration performance.

3.2   Implementation Details

To help the network focus on the prostate-relevant features, 2D input frames are center-cropped with a window size of 128×128 pixels. Ideally, the 3D volume should be kept intact such that the volume can always contain a full view of the 2D frame for registration. However, such an implementation encounters two difficulties. (1) The reconstructed volumes are in different sizes, making it difficult to design a network structure with fixed input and output sizes. (2) The volumes' average size is too big, making the searching space too big for the network to find the optimal transformation.

We propose a data sampling strategy to solve the above issues, which also serves as an augmentation method for network training. By referring to the clinical application, we propose to sample a smaller subvolume as the searching space based on a random initial transformation θ_{init} within a manually defined range R. For example, if we are looking for the nth frame f_n's position θ_n during the registration, one neighboring frame within the range of $[n - R, n + R]$ is randomly chosen as the initial reference frame f_{init}, which has the position of θ_{init}. A subvolume size of $128 \times 128 \times 32$ is cropped at the center of f_{init}, which serves as the 3D volume input to the network. In our experiments, we set the frame range $R = 10$ to ensure that the subvolume around the initial frame f_{init} contains the target frame f_n. Taking this subvolume as the input, the network is trained to estimate the relative transformation parameters $\delta\theta$ which is the difference between θ_n and θ_{init}, calculated through matrix manipulation.

4   Experiments and Results

4.1   Datasets and Experimental Setting

All the ultrasound volumes used in this work were collected from clinical studies using an EM-tracking enabled fusion system. The dataset contains 619 TRUS volumes reconstructed from tracked TRUS frames acquired by a Philips iU22 scanner, all from different subjects. The dataset is further divided into 488, 65, 66 cases for training, testing, and validation. Each TRUS frame has an associated positioning transformation matrix M, describing the spatial relationship between the frame and the reconstructed volume. We use this information as the ground truth label for network training and validation. Our network was trained for 150 epochs with batch size $K = 24$ using Adam optimizer [10] with an initial learning rate of 5×10^{-5}, which decays by 0.9 after every five epochs. We implemented the FVR-Net using the Pytorch library [15].

4.2   Results Evaluation

Since DL-based projective 2D/3D registration does not apply well to our problem, we compare our method with conventional iterative image registration methods as shown in Table 1. The baseline methods use mean square error (MSE) or normalized cross-correlation (NCC) as the image similarity metric, optimized with gradient descent [11] (GD) or Powell [3] optimizer. For the random guess, the transformation parameters were randomly sampled from the training set's statistics.

For evaluation, the distance error (DistErr) denotes the average distance in millimeters between the groundtruth-sampled slice and the predicted-sampled

Table 1. Performance comparison of our FVR-Net and baseline methods.

Method	DistErr (mm)	ImgSim (NCC)	Correlation							RunTime (s)
			tX	tY	tZ	aX	aY	aZ	Mean	
Random Guess	5.86	0.55	−0.01	0.05	0.03	−0.08	0.05	−0.06	0.00	–
NCC + GD	3.43	0.91	0.70	0.70	0.94	0.66	0.29	0.59	0.64	3.98
MSE + GD	3.20	0.90	0.60	0.66	0.94	0.64	0.21	0.55	0.60	3.64
NCC + Powell	3.02	0.91	0.43	0.86	0.89	0.80	0.13	0.51	0.60	7.60
MSE + Powell	2.85	0.92	0.64	0.90	0.99	0.79	0.36	0.63	**0.72**	5.85
FVR-Net (Best)	**2.73**	**0.92**	0.69	0.88	0.96	0.96	0.17	0.08	0.62	**0.07**

Table 2. Ablation studies of FVR-Net using ResNext-150 unless specifically noted.

Method	DistErr (mm)	ImgSim (NCC)	Correlation							Runtime (s)
			tX	tY	tZ	aX	aY	aZ	Mean	
ResNext-50	4.34	0.72	0.69	0.76	0.94	0.95	-0.06	0.05	0.55	**0.03**
ResNext-101	3.35	0.85	0.66	0.68	0.87	0.88	0.10	−0.01	0.53	0.04
L_{trans}	3.17	0.90	0.48	0.56	0.91	0.91	0.01	**0.14**	0.50	0.07
L_{sim}	4.86	0.61	0.01	0.05	0.03	−0.08	0.05	−0.06	0.00	0.07
EF	5.68	0.60	0.14	0.06	0.07	0.14	0.01	0.04	0.03	0.06
ULF	5.93	0.62	0.18	0.07	0.08	−0.16	0.08	−0.02	0.04	0.06
$L_{trans} + L_{sim}$	**2.73**	**0.92**	**0.69**	**0.88**	**0.96**	**0.96**	**0.17**	0.08	**0.62**	0.07

slice's corresponding corner points. The image similarity score (ImgSim) shows the quality assessment of the registration results. We also report the correlation coefficient between the groundtruth $\delta\theta$ and the estimated θ. Finally, the running time denotes the average time cost for registering a single pair of images. The average initialization error is 7.68 mm in the test set. As shown in Table 1, the iterative methods using MSE as the similarity metric can consistently outperform those using NCC. On the other hand, using Powell as the optimizer, the distance error is significantly reduced, but the running time also increased rapidly. By applying our best FVR-Net, a competitive performance (2.73 mm) has been achieved compared to the best iterative registration (2.85 mm). Although there is no significant difference in the distance error, our running time is more than 80 times faster than the traditional iterative registration. Benefitted from deep CNN's high computation efficiency, the proposed FVR-Net can perform TRUS frame-to-volume registration in approximately real-time speed. Figure 3(a) shows the results of searching for consecutive frames within one volume. Given the same input subvolume (specified by the same initialization frame), our FVR-Net can search for different frames within this subvolume. The FVR-Net results are similar to the groundtruth sampling (3rd) row.

Ablation Study. To determine whether the superior result of our network attributes to the novel (a) dual-branch balanced feature extraction and (b) auxiliary image similarity loss, we trained our FVR-Net with multiple settings, as

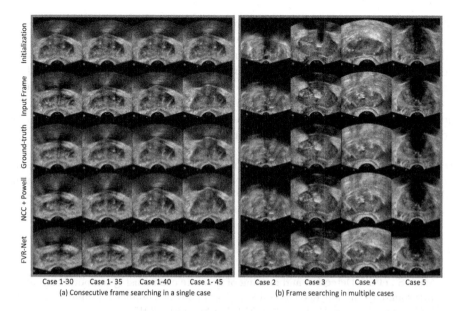

Case 1-30   Case 1- 35   Case 1-40   Case 1- 45   Case 2   Case 3   Case 4   Case 5
(a) Consecutive frame searching in a single case   (b) Frame searching in multiple cases

Fig. 3. Sample results of frame-to-volume registration. The 1st row shows the frames sampled using the initial transformation; the 2nd row includes the input frames to be registered; the 3rd row is the target slice sampled from a 3D volume using the ground-truth transformation; the 4th and 5th rows are the results produced by the baseline registration method and our proposed FVR-Net.

shown in Table 2. We tried different architectures of the localization net and found ResNext-150 produces the smallest distance error. By directly concatenating the 2D frame and 3D volume together as input, the early fusion (EF) network produces meaningless results, indicating the network learns little information for the transformation parameters estimation. A similar phenomenon happens to the unbalanced late fusion (ULF), where the frame's feature maps are still overwhelmed by the volume feature maps' large size. Applying our dual-branch balanced feature extraction, the resultant transformation parameters show a high correlation to the groundtruth. When the FVR-Net is trained in an unsupervised way using L_{sim} alone, the results show little correlation to the groundtruth. While combining two loss functions, the distance error drops significantly from 3.17 mm to 2.73 mm, demonstrating the effectiveness of the auxiliary image similarity loss's effectiveness.

5 Conclusions

This paper has introduced a novel end-to-end TRUS frame-to-volume registration network which instantly register a single 2D TRUS frame with a reconstructed 3D TRUS volume. The experimental results demonstrate our work's competitive performance and superior registration speed comparing to the conventional iterative registration. The ablation study suggests that the dual-branch balanced feature extraction and auxiliary image similarity loss can significantly reduce the registration error. Once combined with off-the-shelf 3D TRUS-MRI registration methods [4,6], the proposed FVR-Net would be able to bridge the research gap in real-time 2D-TRUS and 3D-MRI fusion for guiding the prostate biopsy. We will systematically study the entire workflow in our future work and include more ablation studies for various experimental settings.

Acknowledgements. This work was partially supported by National Institute of Biomedical Imaging and Bioengineering (NIBIB) of the National Institutes of Health (NIH) under awards R21EB028001 and R01EB027898, and through an NIH Bench-to-Bedside award made possible by the National Cancer Institute.

References

1. Azizi, S., et al.: Detection and grading of prostate cancer using temporal enhanced ultrasound: combining deep neural networks and tissue mimicking simulations. Int. J. Comput. Assisted Radiol. Surg. **12**(8), 1293–1305 (2017)
2. Bax, J., et al.: Mechanically assisted 3d ultrasound guided prostate biopsy system. Med. Phys. **35**(12), 5397–5410 (2008)
3. Fletcher, R., Powell, M.J.: A rapidly convergent descent method for minimization. Comput. J. **6**(2), 163–168 (1963)
4. Guo, H., Kruger, M., Xu, S., Wood, B.J., Yan, P.: Deep adaptive registration of multi-modal prostate images. Comput. Med. Imaging Graph. **84**, 101769 (2020)
5. Guo, H., Xu, S., Wood, B., Yan, P.: Sensorless freehand 3D ultrasound reconstruction via deep contextual learning. In: International Conference on MICCAI. pp. 463–472. Springer (2020). https://doi.org/10.1007/978-3-030-59716-0_44
6. Haskins, G., et al.: Learning deep similarity metric for 3d MR-TRUS image registration. Int. J. Comput. Assisted Radiol. Surg. **14**(3), 417–425 (2019)
7. Hu, Y., et al.: Weakly-supervised convolutional neural networks for multimodal image registration. Med. Image Anal. **49**, 1–13 (2018)
8. Jaderberg, M., Simonyan, K., Zisserman, A., Kavukcuoglu, K.: Spatial transformer networks. arXiv:1506.02025 (2015)
9. Khallaghi, S., et al.: A 2d–3d registration framework for freehand trus-guided prostate biopsy. In: International Conference on Medical Image Computing and Computer-Assisted Intervention. pp. 272–279. Springer (2015). https://doi.org/10.1007/978-3-319-24571-3_33
10. Kingma, D.P., Ba, J.: Adam: a method for stochastic optimization. arXiv:1412.6980 (2014)
11. Klein, S., Pluim, J.P., Staring, M., Viergever, M.A.: Adaptive stochastic gradient descent optimisation for image registration. Int. J. Comput. Vis. **81**(3), 227 (2009)
12. Miao, S., et al.: Dilated fcn for multi-agent 2D/3D medical image registration. In: Proceedings of the AAAI Conference on Artificial Intelligence, pp. 4694–4701 (2018)
13. Miao, S., Wang, Z.J., Liao, R.: A CNN regression approach for real-time 2d/3d registration. IEEE Trans. Med. Imaging **35**(5), 1352–1363 (2016)

14. Natarajan, S., et al.: Clinical application of a 3d ultrasound-guided prostate biopsy system. In: Urologic Oncology: Seminars and Original Investigations, vol. 29, pp. 334–342. Elsevier (2011)
15. Paszke, A., et al.: Automatic differentiation in pytorch. In: NIPS 2017 Workshop Autodiff (2017)
16. Pinto, P.A., et al.: Magnetic resonance imaging/ultrasound fusion guided prostate biopsy improves cancer detection following transrectal ultrasound biopsy and correlates with multiparametric magnetic resonance imaging. J. Urol. **186**(4), 1281–1285 (2011)
17. Prevost, R., et al.: 3D freehand ultrasound without external tracking using deep learning. Med. Image Anal. **48**, 187–202 (2018)
18. Siegel, R.L., Miller, K.D., Jemal, A.: Cancer statistics, 2019. CA: Cancer J. Clin. **69**(1), 7–34 (2019)
19. Wein, W., Khamene, A., Clevert, D.A., Kutter, O., Navab, N.: Simulation and fully automatic multimodal registration of medical ultrasound. In: International Conference on Medical Image Computing and Computer-Assisted Intervention, pp. 136–143. Springer (2007). https://doi.org/10.1007/978-3-540-75757-3_17
20. Xie, S., Girshick, R., Dollár, P., Tu, Z., He, K.: Aggregated residual transformations for deep neural networks. In: Proceedings of the IEEE Conference on Computer Vision and Pattern Recognition, pp. 1492–1500 (2017)
21. Xu, S., et al.: Real-time MRI-TRUS fusion for guidance of targeted prostate biopsies. Comput. Aided Surg. **13**(5), 255–264 (2008)

Cross-Modal Attention for MRI and Ultrasound Volume Registration

Xinrui Song[1], Hengtao Guo[1], Xuanang Xu[1], Hanqing Chao[1], Sheng Xu[2], Baris Turkbey[3], Bradford J. Wood[2], Ge Wang[1], and Pingkun Yan[1(✉)]

[1] Department of Biomedical Engineering and Center for Biotechnology and Interdisciplinary Studies, Rensselaer Polytechnic Institute, Troy, NY 12180, USA
yanp2@rpi.edu
[2] Center for Interventional Oncology, Radiology and Imaging Sciences, National Institutes of Health, Bethesda, MD 20892, USA
[3] Molecular Imaging Program, National Cancer Institute, National Institutes of Health, Bethesda, MD 20892, USA

Abstract. Prostate cancer biopsy benefits from accurate fusion of transrectal ultrasound (TRUS) and magnetic resonance (MR) images. In the past few years, convolutional neural networks (CNNs) have been proved powerful in extracting image features crucial for image registration. However, challenging applications and recent advances in computer vision suggest that CNNs are quite limited in its ability to understand spatial correspondence between features, a task in which the self-attention mechanism excels. This paper aims to develop a self-attention mechanism specifically for cross-modal image registration. Our proposed cross-modal attention block effectively maps each of the features in one volume to all features in the corresponding volume. Our experimental results demonstrate that a CNN network designed with the cross-modal attention block embedded outperforms an advanced CNN network 10 times of its size. We also incorporated visualization techniques to improve the interpretability of our network. The source code of our work is available at https://github.com/DIAL-RPI/Attention-Reg.

Keywords: Self-attention · Image feature · Image registration · Multi-modal · Prostate cancer

1 Introduction

Image-guided interventional procedures often require registering multi-modal images to visualize and analyze complementary information. For example, prostate cancer biopsy benefits from fusing transrectal ultrasound (TRUS) imaging with magnetic resonance imaging (MRI) to optimize targeted biopsy. However, image registration is a challenging task especially for multi-modal images. Traditional multi-modal image registration relies on maximizing the mutual information between images [9,16], which performs poorly when the input images have complex textural patterns, such as in the case of MRI and ultrasound

© Springer Nature Switzerland AG 2021
M. de Bruijne et al. (Eds.): MICCAI 2021, LNCS 12904, pp. 66–75, 2021.
https://doi.org/10.1007/978-3-030-87202-1_7

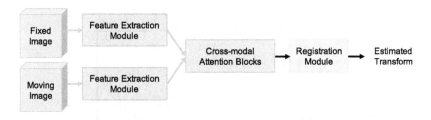

Fig. 1. Overview of the proposed registration framework with cross-modal attention.

registration. Feature based methods compute the similarity between images by representing image appearances using features [6]. However, feature engineering limits the registration performance on images in different contrasts, of complicated features, and/or with strong noise.

In the past several years, deep learning has become a powerful tool for medical image registration, starting from the early works of using neural networks for similarity metric computation to direct transformation estimation [5,17]. For example, Haskins *et al.* [4] developed a deep learning metric to measure the similarity between MRI and TRUS volumes. The correspondences between the volumes is established by optimizing the similarity iteratively, which can be computationally intensive. de Vos *et al.* [14] proposed an end-to-end unsupervised image registration method to train a spatial transform network by maximizing the normalized cross correlation. Their method can directly estimate an image transformation for registration. Balakrishnan *et al.* [1] further used mean squared voxel-wise difference and local cross-correlation to train a registration network to map image features to a spatial transformation. While the way of estimating such an image transform underwent major changes, researchers also developed novel ways to supervise the network learning process. Hu *et al.* [7] trained an image registration framework in a weakly supervised fashion by minimizing the differences between segmentation labels of the fixed image and a warped moving image. Yan *et al.* [19] developed an adversarial registration framework using a discriminator to supervise the registration estimator.

The aforementioned deep learning methods map the composite features from input images directly into a spatial transformation to align them. So far, the success comes from two primary sources. One is the ability of automatically learning image representations through training a properly designed network. The other is the capability of mapping complex patterns to an image transformation. The current methods mix these two components together for image registration. However, converting image features to a spatial relationship is extremely challenging and highly data-dependent, which is the bottleneck for further improvements of the registration performance.

In this paper, we propose a novel cross-modal attention mechanism to explicitly use the spatial correspondence to improve the performance of neural networks for image registration. By extending the non-local attention mechanism [15] to an attention operation between two images, we designed a cross-modal attention

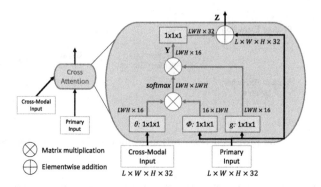

Fig. 2. The proposed cross-modal attention block.

block that is specifically oriented towards registration tasks. The attention block captures both local features and their global correspondence efficiently. Embedding this cross-modal attention block into an image registration network, as shown in Fig. 1, improves deep learning based multi-modal image registration, attaining both feature learning and correspondence establishment explicitly and synergically.

By adding the cross-modal feature correspondence, the image registration network can achieve better registration performance with a much simpler architecture. To the best of our knowledge, this is the first work to embed the non-local attention in the deep neural network for image registration.

In our experiments, we demonstrate the proposed method on the 3D MRI-TRUS fusion task, which is a very challenging cross-modality image registration problem. The proposed network was trained and tested on a dataset of 650 MRI and TRUS volume pairs. The results show that our network significantly reduced the registration error from 10.17 ± 5.75 mm to 3.71 ± 1.99 mm. The proposed method also outperformed state-of-the-art methods with only 1/10 to 1/5 of the number of parameters used by the competitors, as well as significantly reduced the run time.

2   Method

In this image registration application, the MRI volume is considered to be the fixed image, and the TRUS volume is the moving image. Our registration network consists of three main parts, as shown in Fig. 1. The feature extractor uses convolutional and max pooling layers to capture regional features, and down samples the input volume. Then we use the proposed cross-modal attention block to capture both local features and their global correspondence between modalities. Finally, this information is fed to the deep registrator that further fuses information from two modalities and infers the registration parameters.

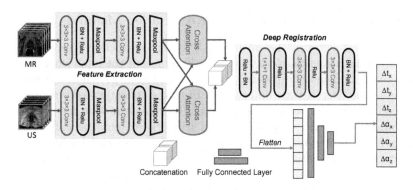

Fig. 3. Overview of the proposed network structure.

2.1 Cross-Modal Attention

The proposed cross-modal attention block takes image features extracted from MRI and TRUS volumes by the preceding convolutional layers. Unlike the non-local block [15] computing self-attention on a single image, the proposed cross-modal attention block aims to establish spatial correspondences between features from two images in different modalities. Figure 2 shows the structure of the proposed cross-modal attention block. The two input feature maps of the block are denoted as primary input $P \in \mathbb{R}^{LWH \times 32}$ and cross-modal input $C \in \mathbb{R}^{LWH \times 32}$, respectively. LWH indicates the size of each 3D feature channel after flattening. The block computes the cross-modal feature attention as

$$\mathbf{y}_i = \frac{\sum_{\forall j} f(\theta(\mathbf{c}_i)^T \phi(\mathbf{p}_j)) g(\mathbf{p}_j)}{\sum_{\forall j} f(\theta(\mathbf{c}_i)^T \phi(\mathbf{p}_j))}, \tag{1}$$

where $\mathbf{c}_i$ and $\mathbf{p}_j$ are features from $\mathbf{C}$ and $\mathbf{P}$ at location i and j, $\theta(\cdot)$, $\phi(\cdot)$ and $g(\cdot)$ are all linear embeddings, and $f(\cdot) = \exp(\cdot)$. In Eq. 1, $f(\cdot)$ computes a scalar representing correlations between the features of these two locations, $\mathbf{c}_i$ and $\mathbf{p}_j$. The result $\mathbf{y}_i$ is a normalized summary of features on all locations of $\mathbf{P}$ weighted by their correlations with the cross-modal feature on location i. Thus, the matrix $\mathbf{Y}$ composed by $\mathbf{y}_i$ integrated non-local information from $\mathbf{P}$ to every position in $\mathbf{C}$. Finally, the block's output $\mathbf{Z}$ is the sum of $\mathbf{Y}$ and $\mathbf{P}$ to allow efficient back-propagation. Therefore, the feature of a location k in $\mathbf{Z}$ summarizes non-local correlation between the entire primary feature map and location k of the cross modality feature map, as well as the information from the original primary feature map at k.

2.2 Feature Extraction and Deep Registration Modules

In the proposed network shown in Fig. 3, feature extraction modules precede the cross-modal attention block to efficiently represent the input volumes. Each feature extraction module consists of two sets of convolutional and maxpooling layers. The deep registration module fuses the concatenated outputs of the cross-modal attention blocks, and predicts the transformation for registration. Other works have used very deep neural networks to automatically learn the complex features of inputs [18]. However, since the cross-modal attention blocks help determine the spatial correspondence between the two sets of volumes, our registration module can afford to be light weighted. Thus, only three convolutional layers are used to fuse the two feature maps. The final fully connected layers convert the learnt spatial information into an estimated transformation.

2.3 Implementation Details

Due to the difficulty in representing the complex image appearances of MRI and TRUS images, surface-based and surface to volume registration methods have been investigated with considerable success [2,13,20]. That inspired us to replace the MRI volume with the prostate segmentation label volume in our work. The network remains the same and we only need to set the fixed image input as either MRI volume or segmentation label. The corresponding networks are named as Attention-Reg (image) and Attention-Reg (label), respectively. One advantage of using MRI prostate segmentation is that the binary representation is much more tolerant to image quality and device specificity than MRI volume. Moreover, using segmentation as input can readily extend the proposed method to other imaging modalities, like computed tomography. This implies that while we trained our segmentation guided model on MRI and ultrasound, it may potentially be used on any two modalities.

In this work, we focus on rigid transformation based registration. This decision is determined by the better accessibility of ground truth labels for rigid transformation, and the idea of focusing on network structure comparison only. Rigid transformations in this work are performed with 4×4 matrices generated from 6 degrees of freedom $\theta = \{\Delta t_x, \Delta t_y, \Delta t_z, \Delta a_x, \Delta a_y, \Delta a_z\}$. These 6 transformation parameters represent translations and rotations along the x, y, and z directions, respectively. We supervise the network by calculating the MSE (Mean Square Error) between the prediction and the ground truth parameters.

In our experiments, we included the recent methods of MSReg by Guo et al. [3] and DVNet by Sun et al. [12] as benchmarks. We used Adam optimizer [8] with maximum of 300 epochs to train all the networks including our proposed Attention-Reg approach. We used a step learning rate scheduler for MSReg training, with initial learning rate 5×10^{-5} which then decays to 0.9 every 5 epochs, as suggested in [3]. For DVNet, we used the same scheduler but with initial learning rate adjusted to 1×10^{-3}. The models were trained on a NVIDIA DGX-1 deep learning server with batch size of 16 for MSReg, and 8 for our proposed network. The testing phase and runtime benchmark were

Table 1. Performance comparison between Attention-Reg and similarity-based iterative registration methods.

Method	Initialization	Result SRE (mm)
Mutual information [9]	8 mm	8.96 ± 1.28
SSD MIND [6]		6.42 ± 2.86
Attention-Reg (img)		$\mathbf{3.63 \pm 1.86}$
Attention-Reg (label)		$\mathbf{3.54 \pm 1.91}$
Mutual Information [9]	16 mm	10.07 ± 1.40
SSD MIND [6]		6.62 ± 2.96
Attention-Reg (img)		$\mathbf{4.17 \pm 2.14}$
Attention-Reg (label)		$\mathbf{4.06 \pm 2.10}$

performed on a work station equipped with NVIDIA GeForce RTX 2080 Ti and AMD Ryzen 9 3900X. Both the proposed and the MSReg methods were implemented in Python using the open source PyTorch library [10]. Our implementation of the proposed Attention-Reg is available at: https://github.com/DIAL-RPI/Attention-Reg.

3 Experiments and Results

3.1 Dataset and Preprocessing

In this work, we used 528 cases of MRI-TRUS volume pair for training, 66 cases for validation, and 68 cases for testing. Each case contains a T2-weighted MRI volume and a 3D ultrasound volume. Each MRI volume has $512 \times 512 \times 26$ voxels with 0.3 mm resolution in all directions. The ultrasound is reconstructed from an electro-magnetic tracked freehand 2D sweep of the prostate. The training set was generated afresh for every training epoch to boost model robustness. On the contrary, the validation set consists of 5 pre-generated initialization matrices for each case, resulting in 330 total samples. The reason for not regenerating new validation sets every epoch is to monitor the epoch-to-epoch performance in a more stable manner. For testing, we generated 40 random initialization matrices for each case. The same test set is used for all experiments.

We measured the image registration performance using surface registration error (SRE). To accurately generate a dataset of with known SRE for training and validation, we perturbed each ground truth transformation parameter randomly within the range of 5mm of translation or 6 degrees of rotation, and then scale the perturbation to a random SRE within the desired range.

3.2 Experimental Results

We first compared our approach to classical iterative registration methods. Table 1 summarizes the comparison of our method and traditional iterative registration approaches, including mutual information [9] and MIND [6] based registration as

Table 2. Performance comparison between Attention-Reg, MSReg [3], and DVNet [12]. Both parameter count and runtime were measured per stage. SRE values are in mm.

Method	Initial.	Stage 1	Stage 2	#Parameters	Runtime
DVNet [12]	[0,20 mm]	4.77 ± 3.17	-	5,275,832	3 ms
MSReg [3]		4.75±2.63	4.04 ± 2.30	16,106,076	6 ms
Attention-Reg (image)		4.50±2.58	3.71 ± 1.99	1,248,777	3 ms
Attention-Reg (label)		**4.44±2.32**	**3.60 ± 2.01**	1,248,777	3 ms
Feature-Reg (image)	[0,20 mm]	5.14 ± 2.58	-	1,244,393	3 ms
Feature-Reg (label)		5.22 ± 2.81	-	1,244,393	3 ms

in [5]. We tested our result on two sets of initial registrations. One set is initialized at SRE = 8 mm, and the other set is initialized at SRE = 16 mm. The results of our proposed models are averaged from 6,800 test samples, with 68 cases of MRI-TRUS volume pair and 100 initialization matrices each. We used this large test set to improve the robustness of the evaluation. Attention-Reg (img) stands for the the registration result of our proposed network with MRI volume as the input of fixed image, whereas Attention-Reg (label) uses MRI prostate segmentation label as the fixed image. In both test scenarios, our methods outperformed the traditional approaches significantly (p <0.001 under t-test). It is also worth noting that when using MRI prostate segmentation as the fixed image, the performance of our network is slightly improved with statistical significance (p <0.001 under t-test).

Table 2 lists the results of our method and other end-to-end rigid registration techniques, including MSReg by Guo et $al.$ [3] and DVNet by Sun et $al.$ [12]. The ResNeXt structure that Guo et $al.$ adopted is one of the more advanced variations of CNN [18], adding more weight to this comparison. The 2D CNN network in DVNet [12] treats 3D volumes as patches of 2D images, a lighter approach in handling 3D volume registration. We tested these networks on 2,720 testing samples, which consists of 68 cases with 40 initialization positions for each case. To better compare our Attention-Reg with MSReg [3], which used two consecutive networks to boost performance, we also trained our network twice on two differently distributed training sets. The model for the 1st stage was trained and tested on a generated dataset with initial SRE uniformly distributed within the range of ([0,20 mm]), and the range for the 2nd stage was set to be $SRE \in [0, 8mm]$. The trained networks were concatenated together to form a two-stage registration network.

As shown in the top part of Table 2, our cross-modal attention network outperformed MSReg in both registration stages. Furthermore, the better result was achieved with only 1/10 the number of parameters, and half the runtime. The significantly smaller model and simpler calculation demonstrate that the proposed cross-modal attention block can efficiently capture key features of the image registration task. Again, we observed that the performance of our network with segmentation label as input was consistently better, with significantly reduced SRE when compared to MSReg (p <0.001) in both stages.

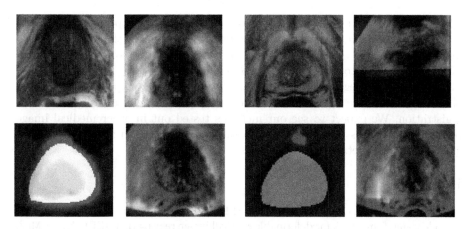

Fig. 4. Grad-CAM visualization of four pairs of feature maps resulting from the multi-modal attention blocks of (**top**) Attention-Reg (image) and (**bottom**) Attention-Reg (label). The image on the left and right in each pair are from the fixed and moving images, respectively.

To demonstrate the contribution of the proposed cross-modal attention block, we trained our Attention-Reg network without the attention block, *i.e.*, directly concatenating the outputs of feature extraction modules and feeding to the deep registration module. The results are shown in the bottom half of Table 2, which prove the importance of the proposed cross-modal attention block. Without the attention module, the registration performance under both settings was significantly reduced ($p < 0.001$ with paired t-test). Also, note that without the attention block, using segmentation label as fixed image no longer has an advantage over MRI volume. We speculate that this is also caused by the loss of attention block, which establishes a sensible spatial correlation between the MRI segmentation and the ultrasound volume, as shown in Fig. 4.

To help understand the function of cross-modal attention blocks, we employed Grad-CAM [11] to visualize the output of the two multi-modal attention blocks. Similar with Grad-CAM, we used the preceding CNN layer's weight gradient to scale the importance of each feature map channel, and thereby acquired a single volume that represents the output of the multi-modal attention block. Figure 4 shows the visualization result. It is apparent that both MRI and ultrasound features are roughly the shape and location of the corresponding ultrasound frame. This means that the network is focusing on the same region of information in both volumes.

4   Conclusion

This paper introduced a novel attention mechanism for the task of medical image registration. By comparing the proposed network with other classical methods and purely CNN-based networks up to ten times of its size, we demonstrated the effectiveness of the new cross-modal attention block. To emphasize the importance of prostate boundary, we also quantitatively evaluated the effect of replacing an MRI volume with its segmentation mask as network input. Our proposed methods have led to significant improvements in image registration accuracy over the previous registration methods. Through feature map visualization, we observed that the network indeed extracted meaningful features to guide image registration. We expect to see our methods tested out in other medical image registration settings in the future with such improvement in accuracy and efficiency, and interpretability.

References

1. Balakrishnan, G., Zhao, A., Sabuncu, M.R., Guttag, J., Dalca, A.V.: VoxelMorph: a learning framework for deformable medical image registration. IEEE Trans. Med. Imaging **38**(8), 1788–1800 (2019)
2. Bashkanov, O., et al.: Learning multi-modal volumetric prostate registration with weak inter-subject spatial correspondence (2021)
3. Guo, H., Kruger, M., Xu, S., Wood, B.J., Yan, P.: Deep adaptive registration of multi-modal prostate images. Comput. Med. Imaging Graph. **84**, 101769 (2020)
4. Haskins, G., et al.: Learning deep similarity metric for 3D MR-TRUS image registration. Int. J. Comput. Assist. Radiol. Surg. **14**(3), 417–425 (2019)
5. Haskins, G., Kruger, U., Yan, P.: Deep learning in medical image registration: a survey. Mach. Vis. Appl. **31**(1), 8 (2020)
6. Heinrich, M.P., et al.: MIND: modality independent neighbourhood descriptor for multi-modal deformable registration. Med. Image Anal. **16**(7), 1423–1435 (2012)
7. Hu, Y., et al.: Weakly-supervised convolutional neural networks for multimodal image registration. Med. Image Anal. **49**, 1–13 (2018)
8. Kingma, D.P., Ba, J.: Adam: a method for stochastic optimization. arXiv preprint arXiv:1412.6980 (2014)
9. Maes, F., Collignon, A., Vandermeulen, D., Marchal, G., Suetens, P.: Multimodality image registration by maximization of mutual information. IEEE TMI **16**(2), 187–198 (1997)
10. Paszke, A., Gross, S., Chintala, S., et al.: Automatic differentiation in pytorch. In: NIPS 2017 Workshop Autodiff, pp. 1–4 (2017)
11. Selvaraju, R.R., Cogswell, M., Das, A., Vedantam, R., Parikh, D., Batra, D.: Grad-CAM: visual explanations from deep networks via gradient-based localization. In: Proceedings of the IEEE International Conference on Computer Vision, pp. 618–626 (2017)
12. Sun, Y., Moelker, A., Niessen, W.J., van Walsum, T.: Towards robust CT-ultrasound registration using deep learning methods. In: Stoyanov, D., et al. (eds.) MLCN/DLF/IMIMIC -2018. LNCS, vol. 11038, pp. 43–51. Springer, Cham (2018). https://doi.org/10.1007/978-3-030-02628-8_5
13. Thomson, B.R., et al.: MR-to-US registration using multiclass segmentation of hepatic vasculature with a reduced 3D U-Net. In: Martel, A.L., et al. (eds.) MICCAI 2020. LNCS, vol. 12263, pp. 275–284. Springer, Cham (2020). https://doi.org/10.1007/978-3-030-59716-0_27
14. de Vos, B.D., Berendsen, F.F., Viergever, M.A., Staring, M., Išgum, I.: End-to-end unsupervised deformable image registration with a convolutional neural network. In: Cardoso, M.J., et al. (eds.) DLMIA/ML-CDS -2017. LNCS, vol. 10553, pp. 204–212. Springer, Cham (2017). https://doi.org/10.1007/978-3-319-67558-9_24

15. Wang, X., Girshick, R., Gupta, A., He, K.: Non-local neural networks. In: Proceedings of the IEEE Conference on Computer Vision and Pattern Recognition, pp. 7794–7803 (2018)
16. Wells, W.M., Viola, P., Atsumi, H., Nakajima, S., Kikinis, R.: Multi-modal volume registration by maximization of mutual information. Med. Image Anal. **1**(1), 35–51 (1996)
17. Wu, G., Kim, M., Wang, Q., Gao, Y., Liao, S., Shen, D.: Unsupervised deep feature learning for deformable registration of MR brain images. In: Mori, K., Sakuma, I., Sato, Y., Barillot, C., Navab, N. (eds.) MICCAI 2013. LNCS, vol. 8150, pp. 649–656. Springer, Heidelberg (2013). https://doi.org/10.1007/978-3-642-40763-5_80
18. Xie, S., Girshick, R., Dollár, P., Tu, Z., He, K.: Aggregated residual transformations for deep neural networks. In: Proceedings of the IEEE Conference on CVPR, pp. 1492–1500 (2017)
19. Yan, P., Xu, S., Rastinehad, A.R., Wood, B.J.: Adversarial image registration with application for MR and trus image fusion. In: Shi, Y., Suk, H.-I., Liu, M. (eds.) MLMI 2018. LNCS, vol. 11046, pp. 197–204. Springer, Cham (2018). https://doi.org/10.1007/978-3-030-00919-9_23
20. Zhang, Y., Bi, J., Zhang, W., Du, H., Xu, Y.: Recent advances in registration methods for MRI-TRUS fusion image-guided interventions of prostate. Recent Patents Eng. **11**(2), 115–124 (2017)

Bayesian Atlas Building with Hierarchical Priors for Subject-Specific Regularization

Jian Wang[1(✉)] and Miaomiao Zhang[1,2]

[1] Computer Science, University of Virginia, Charlottesville, USA
jw4hv@virginia.edu
[2] Electrical and Computer Engineering, University of Virginia, Charlottesville, USA

Abstract. This paper presents a novel hierarchical Bayesian model for unbiased atlas building with subject-specific regularizations of image registration. We develop an atlas construction process that automatically selects parameters to control the smoothness of diffeomorphic transformation according to individual image data. To achieve this, we introduce a hierarchical prior distribution on regularization parameters that allows multiple penalties on images with various degrees of geometric transformations. We then treat the regularization parameters as latent variables and integrate them out from the model by using the Monte Carlo Expectation Maximization (MCEM) algorithm. Another advantage of our algorithm is that it eliminates the need for manual parameter tuning, which can be tedious and infeasible. We demonstrate the effectiveness of our model on 3D brain MR images. Experimental results show that our model provides a sharper atlas compared to the current atlas building algorithms with single-penalty regularizations. Our code is publicly available at https://github.com/jw4hv/HierarchicalBayesianAtlasBuild.

1 Introduction

Deformable atlas building is to create a "mean" or averaged image and register all subjects to a common space. The resulting atlas and group transformations are powerful tools for statistical shape analysis of images [12,18], template-based segmentation [13,21,22], or object tracking [16,17], just to name a few. A good quality of altas heavily relies on the registration process, which is typically formulated as a regularized optimization to solve [5,7,25,30]. An issue in the current process of registration-based atlas construction is how to regularize model parameters. Having an appropriate regularization is critical to the "sharpness" of the atlas, as well as ensuring a set of desirable properties of transformations, i.e., a smooth and invertible smooth mapping between images, also known as diffeomorphisms, to preserve the topology of original images.

Current atlas building models either exhaustively search for an optimal regularization in the parameter space, or treat it as unknown variables to estimate from Bayesian models. While ad hoc parameter-tuning may yield satisfactory results, it requires expert domain knowledge to guide the tuning process [14,18,26,27].

© Springer Nature Switzerland AG 2021
M. de Bruijne et al. (Eds.): MICCAI 2021, LNCS 12904, pp. 76–86, 2021.
https://doi.org/10.1007/978-3-030-87202-1_8

Inspired by probabilistic models, several works have proposed Bayesian models of atlas building with automatically estimated regularizations [1,2,31]. These approaches define a posterior distribution that consists of an image matching term between a deformed atlas and each individual as a likelihood, and a regularization as a prior to support the smoothness of transformation fields. The regularization parameter is then jointly estimated with atlas after carefully integrating out the image deformations using Monte Carlo sampling. However, sampling in a high-dimensional transformation space (i.e., on a dense 3D image grid 128^3) is computationally expensive and often leads to a long execution time with high memory consumption. More importantly, the aforementioned methods are limited to regularizations with single-penalty for population studies. This prohibits the model's ability to adaptively search for the best regularization parameter associated with an individual subject, which is critical to images with various degrees of geometric transformations. The typical "one-fits-all" fails in cases where large geometric variations occur, i.e., brain shape changes of Alzheimer's disease group. Allowing the subject-specific (data-driven) regularization can substantially affect the sharpness and quality of the atlas [29].

In this paper, we propose a hierarchical Bayesian model of atlas building with subject-specific regularizations in the context of Large Deformation Diffeomorphic Metric Mapping (LDDMM) algorithm [7]. In contrast to previous approaches treating the regularization of individual subjects as a single-penalty function with adhoc parameters, we develop a data-adaptive algorithm to automatically adjust the model parameters accordingly. To achieve this, we introduce a novel hierarchical prior that features (i) prior distributions with multiple regularization parameters on the group transformations in a low-dimensional bandlimited space; and (ii) a hyperprior to model the regularization parameters as latent variables. We then develop a Monte Carlo Expectation Maximization (MCEM) algorithm, where the expectation step integrates over the regularization parameters using Hamiltonian Monte Carlo (HMC) sampling. The joint estimation of model parameters including atlas, registration, and hyperparameters in the maximization step successfully eliminates a massive burden of multi-parameters tuning. We demonstrate the effectiveness of our algorithm on both 2D synthetic images and 3D real brain MRIs.

To the best of our knowledge, we are the first to extend the atlas building to a data-adaptive and parameter-tuning-free framework via hierarchical Bayesian learning. Experimental results show that our model provides an efficient atlas construction of population images, particularly with large variations of geometric transformations. This paves a way for an improved quality of clinical studies where atlas building is required, for example, statistical shape analysis of brain changes for neurodegenerative disease diagnosis [12], or atlas-based segmentation for in-utero placental disease monitoring [16].

2   Background: Atlas Building with Fast LDDMM

We first briefly review an unbiased atlas building algorithm [14] based on Fourier-approximated Lie Algebra for Shooting (FLASH), a fast variant of LDDMM with geodesic shooting [30]. Given a set of images $I_1, \cdots, I_N$ with N being the number of images, the problem of atlas building is to find a template image I and transformations $\phi_1, \cdots, \phi_N$ that minimize the energy function

$$E(I, \phi_n) = \sum_{n=1}^{N} \text{Dist}(I \circ \phi_n, I_n) + \text{Reg}(\alpha, \phi_n). \tag{1}$$

The $\text{Dist}(\cdot, \cdot)$ is a distance function that measures the dissimilarity between images, i.e., sum-of-squared differences [7], normalized cross correlation [6], and mutual information [28]. The $\text{Reg}(\cdot)$ is a weighted regularization with parameter α that guarantees the diffeomorphic properties of transformation fields.

Regularization In Tangent Space of Diffeomorphisms. Given an open and bounded d-dimensional domain $\Omega \subset \mathbb{R}^d$, we use $\text{Diff}(\Omega)$ to denote a space of diffeomorphisms and its tangent space $V = T\text{Diff}(\Omega)$. The regularization of LDDMM is defined as an integral of the Sobolev norm of the time-dependent velocity field $v(t) \in V (t \in [0, 1])$ in the tangent space, i.e.,

$$\text{Reg}(\alpha, \phi_n) = \int \langle \mathcal{L}(\alpha) v_n(t), \mathcal{L}(\alpha) v_n(t) \rangle \, dt, \text{ with } \frac{d\phi_n(t)}{dt} = -D\phi_n(t) \cdot v_n(t). \tag{2}$$

Here $\mathcal{L}$ is a symmetric, positive-definite differential operator, with parameter α controling the smoothness of transformation fields. In this paper, we use the Laplacian operator $\mathcal{L} = (-\alpha\Delta + \text{Id})^3$, where Id is an identity matrix. The operator D is a Jacobian matrix and $\cdot$ denotes an element-wise matrix multiplication.

According to the geodesic shooting algorithm [25], the minimum of LDDMM is uniquely determined by solving a Euler-Poincaré differential equation (EPDiff) [3,19] with initial conditions. This inspires a recent model FLASH to reparameterize the regularization of Eq. (2) in a low-dimensional bandlimited space of initial velocity fields, which dramatically reduces the computational complexity of transformation models with little to no loss of accuracy [30].

Fourier Computation of Diffeomorphisms. Let $\widetilde{\text{Diff}}(\Omega)$ and $\tilde{V}$ denote the space of Fourier representations of diffeomorphisms and velocity fields respectively. Given time-dependent velocity field $\tilde{v}(t) \in \tilde{V}$, the diffeomorphism $\tilde{\phi}(t) \in \widetilde{\text{Diff}}(\Omega)$ in the finite-dimensional Fourier domain can be computed as

$$\tilde{\phi}(t) = \tilde{\text{Id}} + \tilde{u}(t), \quad \frac{d\tilde{u}(t)}{dt} = -\tilde{v}(t) - \tilde{\mathcal{D}}\tilde{u}(t) * \tilde{v}(t), \tag{3}$$

where $\tilde{\text{Id}}$ is the frequency of an identity element, $\tilde{\mathcal{D}}\tilde{u}(t)$ is a tensor product $\tilde{\mathcal{D}} \otimes \tilde{u}(t)$, representing the Fourier frequencies of a Jacobian matrix $\tilde{\mathcal{D}}$ with central difference approximation, and $*$ is a circular convolution [1].

The Fourier representation of the geodesic shooting equation (EPDiff) is

$$\frac{\partial \tilde{v}(t)}{\partial t} = -\tilde{\mathcal{K}} \left[(\tilde{\mathcal{D}}\tilde{v}(t))^T \star \tilde{\mathcal{L}}\tilde{v}(t) + \tilde{\nabla} \cdot (\tilde{\mathcal{L}}\tilde{v}(t) \otimes \tilde{v}(t)) \right], \tag{4}$$

---

[1] To prevent the domain from growing infinity, we truncate the output of the convolution in each dimension to a suitable finite set.

where $\star$ is the truncated matrix-vector field auto-correlation. The operator $\tilde{\nabla}\cdot$ is the discrete divergence of a vector field. Here $\tilde{\mathcal{K}}$ is an inverse operator of $\tilde{\mathcal{L}}$, which is the Fourier transform of a Laplacian operator in this paper.

The regularization in Eq. (2) can be equivalently formulated as

$$\text{Reg}(\alpha, \phi_n) = \langle \tilde{\mathcal{L}}(\alpha)\tilde{v}_n(0), \tilde{\mathcal{L}}(\alpha)\tilde{v}_n(0) \rangle, \quad \text{s.t. Eq. (3) \& Eq. (4)}.$$

We will drop off the time index in remaining sections for notational simplicity, e.g., defining $\tilde{v}_n \triangleq \tilde{v}_n(0)$.

3   Our Model: Bayesian Atlas Building with Hierarchical Priors

This section presents a hierarchical Bayesian model for atlas building that allows subject-specific regularization with no manual effort of parameter-tuning. We introduce a hierarchical prior distribution on the initial velocity fields with adaptive smoothing parameters followed by a likelihood distribution on images.

Likelihood. Assuming an independent and identically distributed (i.i.d.) Gaussian noise on image intensities, we formulate the likelihood of each observed image I_n as

$$p(I_n \,|\, I, \tilde{v}_n, \sigma^2) = \frac{1}{(\sqrt{2\pi\sigma^2})^M} \exp\left(-\frac{1}{2\sigma^2}\|I \circ \phi_n - I_n\|_2^2\right). \tag{5}$$

Here σ^2 denotes a noise variance, M is the number of image voxels, and ϕ_n is an inverse Fourier transform of $\tilde{\phi}_n$ at time point $t = 1$. It is worth mentioning that other noise models such as spatially varying noises [24] can also be applied.

Prior. To ensure the smoothness of transformation fields, we define a prior on each initial velocity field $\tilde{v}_n$ as a complex multivariate Gaussian distribution

$$p(\tilde{v}_n \,|\, \alpha_n) = \frac{1}{(2\pi)^{\frac{M}{2}} |\tilde{\mathcal{L}}_n^{-1}(\alpha_n)|} \exp\left(-\frac{1}{2}\langle \tilde{\mathcal{L}}_n(\alpha_n)\tilde{v}_n, \tilde{\mathcal{L}}(\alpha_n)\tilde{v}_n \rangle\right), \tag{6}$$

where $|\cdot|$ is matrix determinant. The Fourier coefficients of a discrete Laplacian operator is $\tilde{\mathcal{L}}_n(\xi_1, \dots, \xi_d) = \left(-2\alpha_n \sum_{j=1}^{d}(\cos(2\pi\xi_j) - 1) + 1\right)^3$, with $(\xi_1, \dots, \xi_d)$ being a d-dimensional frequency vector.

Hyperprior. We treat the subject-specific regularization parameter α_n of the prior distribution (6) as a random variable generated from Gamma distribution, which is a commonly used prior to model positive real numbers [23]. Other prior such as inverse Wishart distribution [11] can also be applied. The hyperprior of our model is formulated as

$$p(\alpha_n \,|\, k, \beta) = \frac{\alpha_n^{k-1} \exp^{(-\alpha_n/\beta)}}{\Gamma(k)\beta^k}, \tag{7}$$

with k and β being positive numbers for shape and scale parameters respectively. The Gamma function $\Gamma(k) = (k-1)!$ for all positive integers of k. We finally arrive at the log posterior of the diffeomorphic transformation and regularization parameters as

$$E(\tilde{v}_n, \alpha_n, I, \sigma, k, \beta) \triangleq \ln \prod_{n=1}^{N} p(I_n \mid I, \tilde{v}_n, \sigma^2) \cdot p(\tilde{v}_n \mid \alpha_n) \cdot p(\alpha_n \mid k, \beta)$$

$$= \sum_{n=1}^{N} \frac{1}{2} \ln |\mathcal{L}_n| - M \ln \sigma - \frac{\|I \circ \phi_n - I_n\|_2^2}{2\sigma^2} - \frac{1}{2}(\tilde{\mathcal{L}}\tilde{v}_n, \tilde{\mathcal{L}}\tilde{v}_n)$$

$$(k-1)\ln \alpha_n - \frac{\alpha_n}{\beta} - k \ln \beta - \ln \Gamma(k) + \text{const.} \qquad (8)$$

3.1 Model Inference

We develop an MCEM algorithm to infer the model parameter Θ, which includes the image atlas I, the noise variance of image intensities σ^2, the initial velocities of diffeomorphic transformations $\tilde{v}_n$, and the hyperparameters k and β. We treat the regularization parameter α_n as latent random variables and integrate them out from the log posterior in Eq. (8). Computations of two main steps (expectation and maximization) are illustrated below.

Expectation: HMC. Since the E-step does not yield a closed-form solution, we employ a powerful Hamiltonian Monte Carlo (HMC) sampling method [9] to approximate the expectation function Q with respect to the latent variables α_n. For each α_n, we draw a number of S samples from the log posterior (8) by using HMC from the current estimated parameters $\hat{\Theta}$. The Monte Carlo approximation of the expectation Q is

$$Q(\Theta|\hat{\Theta}) \approx \frac{1}{S} \sum_{n=1}^{N} \sum_{j=1}^{S} \ln p(\alpha_{nj} \mid I_n; \hat{\Theta}). \qquad (9)$$

To produce samples of α_n, we first define the potential energy of the Hamiltonian system $H(\alpha_n, \gamma) = U(\alpha_n) + W(\gamma)$ as $U(\alpha_n) = -\ln p(\alpha_n | I_n; \Theta)$. The kinetic energy $W(\gamma)$ is a typical normal distribution on an auxiliary variable γ. This gives us Hamilton's equations to integrate

$$\frac{\alpha_n}{dt} = \frac{\partial H}{\partial \gamma} = \gamma, \quad \frac{d\gamma}{dt} = -\frac{\partial H}{\partial \alpha_n} = -\nabla_{\alpha_n} U. \qquad (10)$$

Since α_n is a Euclidean variable, we use a standard "leap-frog" numerical integration scheme, which approximately conserves the Hamiltonian and results in high acceptance rates. The gradient of U with respect to α_n is

$$\nabla_{\alpha_n} U = \frac{3}{2S} \sum_{j=1}^{S} [\sum_{i=1}^{d} \frac{\tilde{A}_i}{\alpha_{nj}\tilde{A}_i + 1} - \langle 2(\alpha_{nj}\tilde{A} + 1)^5 \tilde{A}\tilde{v}_{nj}, \tilde{v}_{nj}\rangle], \qquad (11)$$

where $\tilde{\mathcal{A}} = -2\sum_{i=1}^{d}(\cos(2\pi\xi_i) - 1)$. Here $\tilde{\mathcal{A}}$ denotes a discrete Fourier Laplacian operator with a d-dimensional frequency vector.

Starting from the current point α_n and initial random auxiliary variable γ, the Hamiltonian system is integrated forward in time by Eq. (10) to produce a candidate point $(\hat{\alpha}_n, \hat{\gamma})$. The candidate point $\hat{\alpha}_n$ is accepted as a new point in the sample with probability $p(accept) = \min(1, -U(\hat{\alpha}_n) - W(\hat{\gamma}) + U(\alpha_n) + W(\gamma))$.

Maximization: Gradient Ascent. We derive the maximization step to update the parameters $\Theta = \{I, \tilde{v}_n, \sigma^2, k, \beta\}$ by maximizing the HMC approximation of the expectation Q in Eq. (9).

For updating the atlas image I, we set the derivative of the Q function with respect to I to zero. The solution for I gives a closed-form update

$$I = \frac{\sum_{j=1}^{S}\sum_{n=1}^{N}(I_n \circ \phi_{nj}^{-1}) \cdot |D\phi_{nj}^{-1}|}{\sum_{j=1}^{S}\sum_{n=1}^{N}|D\phi_{nj}^{-1}|}. \tag{12}$$

Similarly, we obtain the closed-form solution for the noise variance σ^2 after setting the gradient of Q w.r.t. σ^2 to zero

$$\sigma^2 = \frac{1}{MNS}\sum_{n=1}^{N}\sum_{j=1}^{S}\|I \circ \phi_{nj} - I_n\|_2^2. \tag{13}$$

The closed-form solutions for hyperparameters k and β are

$$k = \psi^{-1}(\frac{1}{NS}\sum_{i=1}^{N}\sum_{j=1}^{S}\ln\alpha_{nj} - \ln\beta), \quad \beta = \frac{1}{NSk}\sum_{n=1}^{N}\sum_{j=1}^{S}\alpha_{nj}. \tag{14}$$

Here ψ is a digamma function, which is the logarithmic derivative of the gamma function $\Gamma(\cdot)$. The inverse of digamma function ψ^{-1} is computed by using a fixed-point iteration algorithm [20].

As there is no closed-form update for initial velocities, we employ a gradient ascent algorithm to estimate $\tilde{v}_{nj}$. The gradient $\nabla_{\tilde{v}_{nj}}Q$ is computed by a forward-backward sweep approach. Details are introduced in the FLASH algorithm [30].

4 Experimental Evaluation

We compare the proposed model with LDDMM atlas building algorithm that employs single-penalty regularization with manually tuned parameters on 3D brain images [30]. In HMC sampling, we draw 300 samples for each subject, with initialized value of $\alpha = 10$, $k = 9.0$, $\sigma = 0.05$, and $\beta = 0.1$. An averaged image of all image intensities is used for atlas initialization.

Data. We include 100 3D brain MRI scans with segmentation maps from a public released resource Open Access Series of Imaging Studies (OASIS) for Alzheimer's disease [10]. The dataset covers both healthy and diseased subjects, aged from 55 to 90. The MRI scans are resampled to 128^3 with the voxel size of $1.25\,\mathrm{mm}^3$. All MRIs are carefully prepossessed by skull-stripping, intensity normalization, bias field correction, and co-registration with affine transformation.

Experiments. We estimate the atlas of all deformed images by using our method and compare its performance with LDDMM atlas building [30]. Final results of atlases estimated from both our model and the baseline algorithm are reported. We also compare the time and memory consumption of proposed model with the baseline that performs HMC sampling in a full spatial domain [31]. To measure the sharpness of estimated atlas I, we adopt a metric of normalized standard deviation computed from randomly selected 3000 image patches [15]. Given $N(i)$, a patch around a voxel i of an atlas I, the local measure of the sharpness at voxel i is defined as sharpness$(I(i)) = \mathrm{sd}_{N(i)}(I)/\mathrm{avg}_{N(i)}(I)$, where sd and avg denote the standard deviation and the mean of N_i.

To further evaluate the quality of estimated transformations, we perform atlas-based segmentation after obtaining transformations from our model. For a fair comparison, we fix the atlas for both methods and examine the registration accuracy by computing the dice similarity coefficient (DSC) [8] between the propagated segmentation and the manual segmentation on six anatomical brain structures, including cerebellum white matter, thalamus, brain stem, lateral ventricle, putamen, caudate. The significance tests on both dice and sharpness between our method and the baseline are performed.

Results. Figure 1 visualizes a comparison of 3D atlas on real brain MRI scans. The top panel shows that our model substantially improves the quality of atlas with sharper and better details than the baseline with different values of manually set regularization parameters, e.g., $\alpha = 0.1, 3.0, 6.0, 9.0$. Despite the observation of a smaller value of $\alpha = 0.1$ produces sharper atlas, it breaks the smoothness constraints on the transformation fields hence introducing artifacts on anatomical structures (outlined in purple boxes). The mean and standard deviation of our estimated hyperprior parameters k and β in Eq. (7) over 30 pairwise image registrations are 47.40/7.22, and 0.036/0.005. The bottom panel quantitatively reports the sharpness metric of all methods. It indicates that our algorithm outperforms the baseline by offering a higher sharpness score while preserving the topological structure of brain anatomy.

Figure 2 reports results of fixed-atlas-based segmentation by performing the baseline with various regularization parameters and our algorithm. It shows the dice comparison on six anatomical brain structures of all image pairs. Our algorithm produces better dice coefficients without the need of parameter tuning.

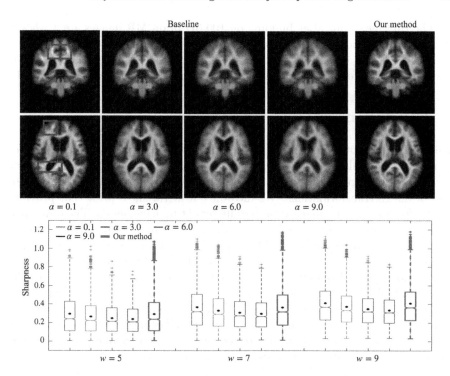

Fig. 1. Top: atlases estimated by baseline with different α and our model (artifacts introduced by small regularization are outlined in purple boxes). Bottom: sharpness measurement of atlas for all methods with different patch size w. The mean of the sharpness metric of **our method** vs. the best performance of baseline without artifacts ($\alpha = 3$) is **0.290**/0.264, **0.362**/0.323, **0.405**/0.360.

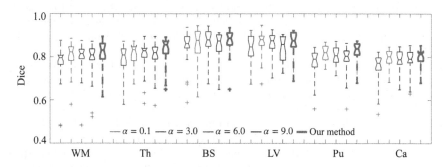

Fig. 2. A comparison of dice evaluation for fixed-atlas-based segmentation on six brain structures (cerebellum white matter (WM), thalamus (Th), brain stem (BS), lateral ventricle (LV), putamen (Pu), caudate (Ca)).

The runtime of our atlas building on 100 3D brain MR images are 4.4 hours with 0.89GB memory consumption. The p-values of significance differences test on both dice ($p = 0.002$) and sharpness ($p = 0.0034$) reject the null hypothesis that there's no differences between our model estimation and baseline algorithms.

5   Conclusion

This paper presents a novel hierarchical Bayesian model for unbiased diffeomorphic atlas building with subject-specific regularization. We design a new parameter choice rule that allows adaptive regularization to control the smoothness of image transformations. We introduce a hierarchical prior that provides prior information of regularization parameters at multiple levels. The developed MCEM inference algorithm eliminates the need of manual parameter tuning, which can be tedious and infeasible in multi-parameter settings. Experimental results show that our proposed algorithm yields a better registration model as well as an improved quality of atlas. While our algorithm is presented in the setting of LDDMM, the theoretical development is generic to other deformation models, e.g., stationary velocity fields [4]. In addition, this model can be easily extended to multi-atlas building where a much higher degree of variations exist in the population studies. Our future work will focus on conducting subsequent statistical shape analysis in the resulting atlas space.

References

1. Allassonnière, S., Amit, Y., Trouvé, A.: Towards a coherent statistical framework for dense deformable template estimation. J. Roy. Stat. Soc. Ser. B (Stat. Methodol.) **69**(1), 3–29 (2007)
2. Allassonnière, S., Kuhn, E.: Stochastic algorithm for parameter estimation for dense deformable template mixture model. arXiv preprint arXiv:0802.1521 (2008)
3. Arnol'd, V.I.: Sur la géométrie différentielle des groupes de Lie de dimension infinie et ses applications à l'hydrodynamique des fluides parfaits. Ann. Inst. Four. **16**, 319–361 (1966)
4. Arsigny, V., Commowick, O., Pennec, X., Ayache, N.: A Log-Euclidean framework for statistics on diffeomorphisms. In: Larsen, R., Nielsen, M., Sporring, J. (eds.) MICCAI 2006. LNCS, vol. 4190, pp. 924–931. Springer, Heidelberg (2006). https:// doi.org/10.1007/11866565_113
5. Ashburner, J., Friston, K.J.: Diffeomorphic registration using geodesic shooting and Gauss-newton optimisation. NeuroImage **55**(3), 954–967 (2011)
6. Avants, B.B., Epstein, C.L., Grossman, M., Gee, J.C.: Symmetric diffeomorphic image registration with cross-correlation: evaluating automated labeling of elderly and neurodegenerative brain. Med. Image Anal. **12**(1), 26–41 (2008)
7. Beg, M.F., Miller, M.I., Trouvé, A., Younes, L.: Computing large deformation metric mappings via geodesic flows of diffeomorphisms. Int. J. Comput. Vis. **61**(2), 139–157 (2005)
8. Dice, L.R.: Measures of the amount of ecologic association between species. Ecology **26**(3), 297–302 (1945)

9. Duane, S., Kennedy, A.D., Pendleton, B.J., Roweth, D.: Hybrid Monte Carlo. Phys. Lett. B **195**(2), 216–222 (1987)

10. Fotenos, A.F., Snyder, A., Girton, L., Morris, J., Buckner, R.: Normative estimates of cross-sectional and longitudinal brain volume decline in aging and AD. Neurology **64**(6), 1032–1039 (2005)

11. Gori, P., et al.: A Bayesian framework for joint morphometry of surface and curve meshes in multi-object complexes. Med. Image Anal. **35**, 458–474 (2017)

12. Hong, Y., Golland, P., Zhang, M.: Fast geodesic regression for population-based image analysis. In: Descoteaux, M., Maier-Hein, L., Franz, A., Jannin, P., Collins, D.L., Duchesne, S. (eds.) MICCAI 2017. LNCS, vol. 10433, pp. 317–325. Springer, Cham (2017). https://doi.org/10.1007/978-3-319-66182-7_37

13. Iglesias, J.E., Sabuncu, M.R., Van Leemput, K.: Incorporating parameter uncertainty in Bayesian segmentation models: application to Hippocampal subfield volumetry. In: Ayache, N., Delingette, H., Golland, P., Mori, K. (eds.) MICCAI 2012. LNCS, vol. 7512, pp. 50–57. Springer, Heidelberg (2012). https://doi.org/10.1007/978-3-642-33454-2_7

14. Joshi, S., Davis, B., Jomier, M., Gerig, G.: Unbiased diffeomorphic atlas construction for computational anatomy. NeuroImage **23**, S151–S160 (2004)

15. Legouhy, A., Commowick, O., Rousseau, F., Barillot, C.: Online atlasing using an iterative centroid. In: Shen, D., et al. (eds.) MICCAI 2019. LNCS, vol. 11766, pp. 366–374. Springer, Cham (2019). https://doi.org/10.1007/978-3-030-32248-9_41

16. Liao, R., et al.: Temporal registration in application to in-utero MRI time series. arXiv preprint arXiv:1903.02959 (2019)

17. Lorenzo-Valdés, M., Sanchez-Ortiz, G.I., Mohiaddin, R., Rueckert, D.: Atlas-based segmentation and tracking of 3D cardiac MR images using non-rigid registration. In: Dohi, T., Kikinis, R. (eds.) MICCAI 2002. LNCS, vol. 2488, pp. 642–650. Springer, Heidelberg (2002). https://doi.org/10.1007/3-540-45786-0_79

18. Ma, J., Miller, M.I., Trouvé, A., Younes, L.: Bayesian template estimation in computational anatomy. NeuroImage **42**(1), 252–261 (2008)

19. Miller, M.I., Trouvé, A., Younes, L.: Geodesic shooting for computational anatomy. J. Math. Imaging Vis. **24**(2), 209–228 (2006)

20. Minka, T.: Estimating a Dirichlet distribution (2000)

21. Pohl, K.M., Fisher, J., Grimson, W.E.L., Kikinis, R., Wells, W.M.: A Bayesian model for joint segmentation and registration. NeuroImage **31**(1), 228–239 (2006)

22. Rohlfing, T., Brandt, R., Menzel, R., Maurer, C.R., Jr.: Evaluation of atlas selection strategies for atlas-based image segmentation with application to confocal microscopy images of bee brains. NeuroImage **21**(4), 1428–1442 (2004)

23. Simpson, I.J., et al.: Probabilistic non-linear registration with spatially adaptive regularisation. Med. Image Anal. **26**(1), 203–216 (2015)

24. Simpson, I.J., Woolrich, M.W., Andersson, J.L., Groves, A.R., Schnabel, J.A.: A probabilistic non-rigid registration framework using local noise estimates. In: 2012 9th IEEE International Symposium on Biomedical Imaging (ISBI), pp. 688–691. IEEE (2012)

25. Vialard, F.X., Risser, L., Rueckert, D., Cotter, C.J.: Diffeomorphic 3D image registration via geodesic shooting using an efficient adjoint calculation. Int. J. Comput. Vis. **97**(2), 229–241 (2012)

26. Vialard, F.X., Risser, L., Holm, D.D., Rueckert, D.: Diffeomorphic atlas estimation using Karcher mean and geodesic shooting on volumetric images. In: MIUA, pp. 55–60 (2011)

27. Wang, J., Xing, W., Kirby, R.M., Zhang, M.: Data-driven model order reduction for diffeomorphic image registration. In: Chung, A.C.S., Gee, J.C., Yushkevich, P.A., Bao, S. (eds.) IPMI 2019. LNCS, vol. 11492, pp. 694–705. Springer, Cham (2019). https://doi.org/10.1007/978-3-030-20351-1_54

28. Wells, W.M., III., Viola, P., Atsumi, H., Nakajima, S., Kikinis, R.: Multi-modal volume registration by maximization of mutual information. Med. Image Anal. 1(1), 35–51 (1996)

29. Yeo, B.T., Sabuncu, M.R., Desikan, R., Fischl, B., Golland, P.: Effects of registration regularization and atlas sharpness on segmentation accuracy. Med. Image Anal. 12(5), 603–615 (2008)

30. Zhang, M., Fletcher, P.T.: Fast diffeomorphic image registration via Fourier-approximated lie algebras. Int. J. Comput. Vis. 127(1), 61–73 (2019)

31. Zhang, M., Singh, N., Fletcher, P.T.: Bayesian estimation of regularization and atlas building in diffeomorphic image registration. In: Gee, J.C., Joshi, S., Pohl, K.M., Wells, W.M., Zöllei, L. (eds.) IPMI 2013. LNCS, vol. 7917, pp. 37–48. Springer, Heidelberg (2013). https://doi.org/10.1007/978-3-642-38868-2_4

SAME: Deformable Image Registration Based on Self-supervised Anatomical Embeddings

Fengze Liu[1], Ke Yan[2(✉)], Adam P. Harrison[2], Dazhou Guo[2], Le Lu[2], Alan L. Yuille[1], Lingyun Huang[3], Guotong Xie[3], Jing Xiao[3], Xianghua Ye[4], and Dakai Jin[2]

[1] Johns Hopkins University, Baltimore, MD, USA
[2] PAII Inc., Bethesda, MD, USA
[3] Ping An Technology, Shenzhen, China
[4] The First Affiliated Hospital Zhejiang University, Hangzhou, China

Abstract. In this work, we introduce a fast and accurate method for unsupervised 3D medical image registration. This work is built on top of a recent algorithm self-supervised anatomical embedding (SAM), which is capable of computing dense anatomical/semantic correspondences between two images at the pixel level. Our method is named SAM-enhanced registration (SAME), which breaks down image registration into three steps: affine transformation, coarse deformation, and deep deformable registration. Using SAM embeddings, we enhance these steps by finding more coherent correspondences, and providing features and a loss function with better semantic guidance. We collect a multiphase chest computed tomography dataset with 35 annotated organs for each patient and conduct inter-subject registration for quantitative evaluation. Results show that SAME outperforms widely-used traditional registration techniques (Elastix FFD, ANTs SyN) and learning based VoxelMorph method by at least 4.7% and 2.7% in Dice scores for two separate tasks of within-contrast-phase and across-contrast-phase registration, respectively. SAME achieves the comparable performance to the best traditional registration method, DEEDS (from our evaluation), while being orders of magnitude faster (from 45 s to 1.2 s).

Keywords: Deformable registration · Affine registration · Unsupervised · Self-supervised anatomical embedding · Deep learning

1 Introduction

Deformable image registration is a fundamental task in medical image analysis [16]. Traditional registration methods solve an optimization problem and iteratively minimize a preset similarity measure to align a pair of images. Recently,

---

F. Liu and K. Yan—Equal contribution.

© Springer Nature Switzerland AG 2021
M. de Bruijne et al. (Eds.): MICCAI 2021, LNCS 12904, pp. 87–97, 2021.
https://doi.org/10.1007/978-3-030-87202-1_9

learning-based deformable registration, using deep networks, have been investigated [2,10,12,13,19]. Compared with their conventional counterparts, learning-based methods can incorporate more flexible losses, integrate other computing modules and are much faster in inference. VoxelMorph was a representative work [2] that learns a parameterized registration function using a convolutional neural network (CNN). Many recent methods focus on designing more sophisticated networks using pyramid [13] or cascaded structures [10,19], or connecting registration to pipelines that include synthesis and segmentation [12]. Ideally, registration should focus on aligning semantically similar/coherent voxels, e.g., the same anatomical locations. This semantic information can come in the form of extra manual annotations (e.g. organ masks) [2], but requiring prohibitive labor costs from professionals. Existing unsupervised methods instead optimize similarity measures describing local intensities as a proxy of the semantic information, such as the mean squared error (MSE) or normalized cross correlation (NCC). However, these are less reliable in settings with large deformations, complex anatomical differences, or cross-modality/cross-phase imagery.

In this paper, we exploit incorporating a novel form of semantic information in registration. Self-supervised anatomical embedding (SAM) is a recent work as a means to produce pixel-wise embeddings in radiological images by encoding anatomical semantic information [18]. It requires no annotations in training. SAM can match corresponding points between two images, which is exactly the fundamental goal of image registration. The most simple and straightforward way to register two images with SAM is to extract SAM embeddings from both fixed and moving images, match each moving pixel to the closest fixed pixel in SAM space, and calculate the corresponding coordinate offsets to generate a deformation field. However, this approach is highly inefficient, as there are millions of pixels in a typical 3D computed tomography (CT) scan. Besides, SAM would not incorporate spatial smoothness constraints [2], which is useful when the correspondences predicted by SAM contain noises.

We propose SAM-enhanced registration (SAME) to address these issues. SAME is comprised of three consecutive steps. (1) **SAM-affine**, which uses correspondence points generated from SAM on a sparse grid to compute the affine transformation matrix. Affine registration [11] has been widely used either alone or as an initialization of deformable methods [2,9]. (2) **SAM-coarse**, which uses a coarse correspondence grid to directly produce a coarse-level deformation field. These first two steps are efficient, require no additional training, and can provide a good initialization for the final step. (3) Lastly, **SAM-VoxelMorph** enhances the deep learning-based VoxelMorph registration method [2], using SAM-based correlation features [4] and a newly formulated SAM similarity loss. SAME is evaluated on a multi-phase chest CT dataset for inter-subject registration with 35 thoracic organs annotated. Quantitative experimental results show that SAM-affine significantly outperforms traditional optimization-based affine registration in both accuracy and speed. The complete SAME consistently outperforms traditional approaches [1,15] and VoxelMorph [2] in both within-contrast-phase and across-contrast-phase tasks by average Dice scores of 4.7%

and 2.7%, respectively. SAME matches DEEDS [9], as the state-of-the-art in CT registration [17], while being orders of magnitude faster (1.2 sec vs. 45 sec).

2 Method

In this section, we present the details of the proposed SAME for deformable registration and describe how SAM is integrated in each of the three steps.

2.1 Self-supervised Anatomical Embedding (SAM)

SAM is recently proposed by [18], as a novel pixel-level contrastive learning framework with a coarse-to-fine network and a hard-and-diverse negative sampling strategy. In an unsupervised manner, it predicts a global and a local embedding vector with semantic meanings per pixel in a CT volume—the same anatomical location in different images expressing similar embeddings. SAM is readily used to find correspondences between images, providing a means to solve the registration problem from a new perspective. Let $X_f, X_m \in \mathbb{R}^{D \times H \times W}$ be the fixed and moving images to be registered . For each image, we extract the global and local SAM embedding volumes and concatenate them in the channel dimension, resulting in $S_f, S_m \in \mathbb{R}^{C \times D \times H \times W}$ (C is the concatenated channel dimension). Given a point $p_f = (x, y, z)$ in X_f, we take its embedding vector $S_f(:, z, y, x)$ and convolve it with S_m to get a similarity heatmap volume. The point with the highest similarity score becomes the matched point in the moving image. Results show that matching for a single point only consumes 0.2 sec on a common chest CT scan [18].

2.2 SAM-Affine and SAM-Coarse

Matched SAM correspondences can be directly employed to estimate an affine transformation matrix [2,9,11]. First, we select a set of points on X_f for matching. Intuitively, evenly distributed points on the image may lead to a better estimation. Therefore, we use the points on a regular grid on X_f, see Fig. 1. It would be more precise to run point matching on every pixel (instead of a coarse grid) and directly generate a fine deformation field, but that would consume 0.5h for a CT with 200 slices. To balance accuracy and speed, we use a grid with stride 8. Since SAM is only designed for points inside the body, we segment the body mask of X_f using intensity thresholding and morphological post processing, and then remove grid points outside the mask. When doing point matching, we downsample S_m with spatial stride of 4 to reduce computation. After the corresponding points in X_m are located, we need to filter out low-quality matches. We examine their similarity scores and discard those lower than a threshold θ. After that, we can get k matched points in X_f, X_m, which can be represented by $3 \times k$ matrices: $\mathbf{P}_f$ and $\mathbf{P}_m$, respectively. We pad them with 1s to create

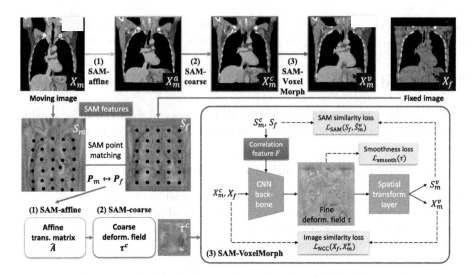

Fig. 1. SAM-enhanced registration (SAME) framework. The moving image is warped by three consecutive steps: SAM-affine, SAM-coarse, SAM-VoxelMorph, gradually approaching the fixed image. Variables X, S, and P denote the image, SAM embedding, and point coordinates, respectively. Subscripts m, f stand for moving or fixed, respectively. Superscripts a, c and v indicate the variable is generated after each of the three steps (affine, coarse deform, or VoxelMorph).

homogeneous versions of the matched points coordinates, $\tilde{\mathbf{P}}_f$, $\tilde{\mathbf{P}}_m \in \mathbb{R}^{4 \times k}$, and estimate the affine matrix $\hat{\mathbf{A}} \in \mathbb{R}^{4 \times 4}$ by a simple least squares fitting:

$$\hat{\mathbf{A}} = \arg\min_{\mathbf{A}} \| \mathbf{A}\tilde{\mathbf{P}}_m - \tilde{\mathbf{P}}_f \|_F^2. \tag{1}$$

Next, we transform X_m with $\hat{\mathbf{A}}$ to obtain X_m^a and extract new SAM embeddings S_m^a from it. Then, points in $\mathbf{P}_f$ are matched again on X_m^a to get $\mathbf{P}_m^a$. $\mathbf{P}_m^a$ and $\mathbf{P}_f$ actually represent a mapping from X_m^a to X_f on k sparse points. We can compute their difference $\Delta = \mathbf{P}_f - \mathbf{P}_m^a$, and map each point in Δ back to the original coordinates of the image to get $\tau^c \in \mathbb{R}^{3 \times D \times H \times W}$. Note, there are only k deformation in Δ that are not necessarily uniformly spaced. Thus values in τ^c are filled in using linear interpolation. This gives us the final coarsely estimated deformation map, which is applied to warp (X_m^a, S_m^a) to (X_m^c, S_m^c). Although coarsely estimated (on only k points), τ_c can effectively reduce the difference between the moving and the fixed images. Compared to a global affine alignment, this provides local warps that can serve as a better initialization for a final learning-based deformable registration step. One question is that whether we could omit SAM-affine and compute τ^c directly. We observed that before affine registration, the two images may have significant offsets, so τ^c is potentially large in magnitude, which will magnify the noises in the matched points. Thus, we first perform affine registration to reduce the magnitude of deformations.

2.3 SAM-VoxelMorph

The objective of the final step is to predict a fine deformation map $\tau \in \mathbb{R}^{3 \times D \times H \times W}$, which is a spatial transformation function that can warp the moving image to best match the fixed one. Following the framework of VoxelMorph [2], we learn a function $\Phi : (X_f, X_m^c) \to \tau$ with a CNN . The original VoxelMorph uses pure pixel intensity-based features and similarity losses. We improve them by leveraging the semantic information contained in SAM embeddings using SAM correlation features and a SAM loss (see Fig. 1).

The loss function in VoxelMorph and follow-up works includes two parts, an image similarity loss and a smoothness loss. We use the local normalized cross-correlation (NCC) loss [2] for the former, while the latter is defined as

$$\mathcal{L}_{smooth}(\tau) = \frac{1}{|\Omega|} \sum_{\mathbf{u} \in \Omega} ||\nabla \tau_{\mathbf{u}}||^2, \tag{2}$$

where Ω is the set of all pixels within the body mask. However, the NCC loss only compares local image intensities, which may not be robust under CT contrast injection, pathological changes, and large or complex deformations in the two images. On the other hand, the SAM embeddings can uncover semantic similarities between two pixels. Thus, we add a proposed SAM loss:

$$\mathcal{L}_{SAM}(S_f, S_m^v) = \frac{1}{|\Omega|} \sum_{\mathbf{u} \in \Omega} \langle S_f(\mathbf{u}), S_m^v(\mathbf{u}) \rangle, \tag{3}$$

where the superscript v indicates the feature map has been warped by τ predicted by SAM-VoxelMorph. The final loss is

$$\mathcal{L} = \mathcal{L}_{NCC}(X_f, X_m^v) + \lambda \mathcal{L}_{SAM}(S_f, S_m^v) + \gamma \mathcal{L}_{smooth}(\tau). \tag{4}$$

While the SAM loss is an effective means to more semantically align images, the *features* extracted in standard VoxelMorph still lack semantic information, which may be needed to better guide predictions. The correlation feature was originally proposed in FlowNet [4] to manage this problem for optical flow. It was also used in [7] for registration. Briefly, it computes the similarity of pixel $\mathbf{u}$ on X_f and pixel $\mathbf{u} + \mathbf{d}$ on X_m, where $\mathbf{d}$ is a small displacement. This similarity is computed for each pixel and for n possible displacement values to generate an n-channel feature map, which is then concatenated to the original feature map at some point in the network. When using SAM, the semantic similarity of two pixels can be simply computed as the inner product of two SAM vectors, $F(\mathbf{u}) = \langle S_f(\mathbf{u}), S_m^c(\mathbf{u} + \mathbf{d}) \rangle$. We empirically find that using 27 displacement values $\mathbf{d} \in \{-2, 0, 2\}^3$ yields good results. Injecting the SAM correlation features provides improved cues to the network when predicting deformations, thus brings further boosts in accuracy.

Table 1. Comparison of different registration methods. We show the average Dice score (%) of two tasks: CE-to-CE and CE-to-NC registration. VM: VoxelMorph. Best and second best performance is shown in bold and gray box, respectively.

Methods	CE-to-CE	CE-to-NC	Inference time (s)	std of $|J_\phi|$
Elastix-affine [11]	28.44	27.96	3.38	-
MIND-affine [8]	28.24	27.91	7.86	-
SAM-affine (SA)	33.80	33.77	0.48	-
SAM-coarse (SC)	44.67	43.68	0.78	-
SA + SC	46.76	45.67	1.05	0.40
SA + VM [2]	48.79	47.35	0.78	0.38
SA + SAM-VM	51.99	49.90	0.84	0.36
SA + SC + VM	54.12	50.64	1.13	0.68
SA + SC + SAM-VM (ours)	**54.42**	50.96	1.16	0.66
SyN [1]	49.75	47.95	74.34	-
FFD [15]	49.36	48.22	93.51	0.51
DEEDS [9]	52.72	**51.15**	45.35	0.40

*Paired t-tests show SAME significantly outperforms all other methods ($p < 10^{-4}$), except for DEEDS in the CE-to-NC setting. SAM-VM significantly outperforms VM ($p < 10^{-7}$).
**The average surface distance (ASD) in CE-to-CE: FFD 4.6mm, SA+VM 4.1mm, DEEDS 4.0mm, SA+SAM-VM 3.9mm, SA + SC + SAM-VM 3.8mm.

3   Experiments

Dataset and Task. To evaluate SAME, we collected a chest CT dataset containing 94 subjects, each with a contrast-enhanced (CE) and a non-contrast (NC) scan. We randomly split the patients to 74, 10, and 10 for training, validation, and testing. Each image has manually labeled masks of 35 organs (including lung, heart, airway, esophagus, aorta, bones, muscles, arteries and veins) [5]. For the validation and test sets, we construct 90 image pairs for inter-subject registration and calculate an atlas-based segmentation accuracy on the 35 organs. Performances of two tasks are evaluated: intra-phase registration (CE-to-CE) and cross-phase registration (CE-to-NC). Every image is resampled to an isotropic resolution of 2mm and cropped to $208 \times 144 \times 192$ by clipping black borders. The image intensity is normalized to $(-1, 1)$ using a window of $(-800, 400)$ HU.

Implementation Details. Our method was developed using PyTorch 1.5. It was run on a Ubuntu server with 12 CPU cores of 3.60GHz. It requires one NVIDIA Quadro RTX 6000 GPU to train and test. We trained a SAM model using the training set of the chest CT dataset. Its structure is identical with the one in [18], which outputs a 128D global embedding and a 128D local one for each pixel. This model is fixed and applied in all three steps of SAME. In SAM-affine and SAM-coarse, the similarity threshold θ is set to 0.7 to select high-confidence matches. In SAM-VoxelMorph, we use a 3D progressive holistically-nested network (P-HNN) [6] as the backbone and concatenate the correlation feature before the third convolutional block. We also tried 3D U-Net [3] but observed no significant accuracy gains. The loss weights in Eq. 4 are empirically set to $\lambda = 1, \gamma = 0.5$. We train SAM-VoxelMorph using the Adam optimizer with a learning rate of 0.001 for 10 epochs. Each training batch contains 2 image pairs with random contrast phases (CE or NC). We evaluate the registration results using average Dice score over 35 organ masks. The organ masks are not used during training.

Table 2. Ablation study for different settings on incorporating SAM to VoxelMorph (VM). The average Dice score (%) is reported. All methods are initialized by SAM-affine without SAM-coarse.

Methods	SAM loss	SAM correlation feature	CE-to-CE	CE-to-NC
VM [2]	✗	✗	48.79	47.35
	✓	✗	50.43	48.24
SAM-VM	✗	✓	51.37	48.99
	✓	✓	**51.99**	**49.90**

Quantitative Results. From Table 1 we can see that **SAM-affine** outperforms the traditional affine registration method in Elastix [11] by 5–6%, meanwhile being 6 times faster. It is also better than affine registration with the MIND [8] robust descriptor. This is because SAM can match corresponding anatomical locations between two images accurately and efficiently. Compared with other methods that iteratively optimizes the affine parameters, SAM-affine directly calculates affine matrix by least squared fitting. **SAM-coarse** surpasses SAM-affine by 10% since it allows for locally deformable warping with more degrees of freedom. Cascading these two steps further boosts the accuracy. VoxelMorph pre-aligned by SAM-affine outperforms SAM-affine + SAM-coarse moderately since the latter can only perform a coarse deformable transformation. However, note that the former is a learning-based dense registration method, while the latter does not require any extra training. It only utilizes the matching result of a pretrained SAM model on grid points. The 2% small gap demonstrates the capability of our proposed SAM-coarse.

SAM-affine + SAM-coarse can provide a good initialization to the learning-based VM in the third step, allowing it to better perform. From the 4 rows in the middle block of Table 1, we also observe consistent improvement by replacing the original VoxelMorph [2] with **SAM-VM**. The SAM embeddings contain more semantic information than the raw pixel intensities, which is incorporated to SAM-VM by the SAM-based correlation feature and SAM loss. An ablation study of SAM-VM is shown in Table 2, where the best result is achieved when both the correlation feature and SAM loss are used. On one hand, explicitly inputting the correlation feature calculated by SAM provides extra guidance for determining the deformation fields . On the other hand, the SAM loss provides a more semantically informed supervisory signal.

In the bottom block of Table 1, we evaluate several widely-used non-rigid registration methods including FFD [15], SyN [1], and DEEDS [9]. FFD was implemented using Elastix [10], where parameters matched the best performing FFD method in EMPIRE10 Challenge [14]. The only modification was an extra bending energy term with weight 0.01 to regularize the smoothness. For SyN (implemented in ANTS) and DEEDS (implemented by the original author), parameters were set according to those used in [17]. For affine transform, the

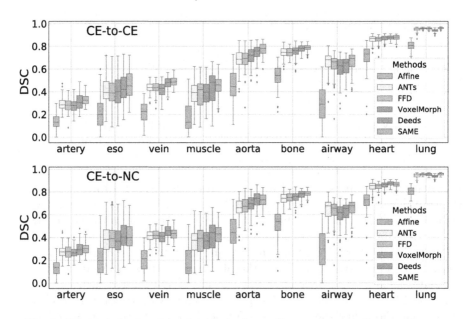

Fig. 2. Comparison of registration methods on all organ groups. Eso: esophagus.

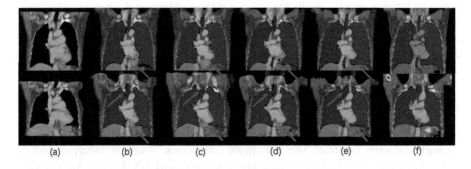

Fig. 3. Visualization of registration results from different methods. From left to right is (a) the moving image, (b) warped moving image of ANTs, (c) DEEDS, (d) SAM-affine + VoxelMorph, (e) SAME, and (f) the fixed image.

default implementation in each package was used. The proposed SAME (combination of three steps) achieves markedly better results than SyN and FFD. Compared with the best traditional method (DEEDS), it performs better in the within-phase setting and comparably in the cross-phase setting, meanwhile is 38 times faster. Cross-phase registration is more difficult because the brightness and appearance of contrast-enhanced and non-contrast CTs can be very different (see X_m and X_f in Fig. 1), and DEEDS has explicitly designed the modality independent features in its registration. SAME takes a different approach that uses the modality invariant SAM embeddings to align images.

We have computed the standard deviation of Jacobian determinants to measure the smoothness of the deformation field. In Table 1, it is observed that SAME achieves the best Dice with a certain degree of sacrifice in smoothness. This is mainly because SAME cascades two deformable methods, SAM-coarse (SC) and SAM-VM. The smoothness of SAM-VM alone is slightly better than the original VM (0.36 vs. 0.38), but SC itself brings more non-smoothness (0.40). SC generates a deformation field by directly differentiating two sets of coordinates without any constraint. This approach gives SC more flexibility to model large deformation but may also produce less smoothed results. We will study on adding constraints to improve the smoothness of SC in the future. On the other hand, if SC is not used, SA + SAM-VM can also achieve competing accuracy (52.0% Dice score) with good smoothness (0.36), where the overall performance is still comparable to DEEDS (52.7%, 0.40) while significantly better than FFD (49.4%, 0.51), and SA+VM (50.8%, 0.38).

Organ-specific results are shown in Fig. 2. For the sake of conciseness, we divide the 35 organs in our dataset into 9 groups and calculate the median and inter-quartile range of Dice score within each group. The affine in Fig. 2 is from Elastix [11], whereas the VoxelMorph refers to SAM-affine + VM [2] in Table 1. The results of SAME surpass DEEDS on 8 out of 9 groups except heart in the within-phase condition. In the cross-phase setting, SAME outperforms DEEDS on the artery, bone, airway and lung organs. In other organs, like esophagus and muscle, SAME shows results with smaller variance and comparable median performance with DEEDS. Organ groups such as artery, esophagus, vein, and muscle display lower Dice scores for all methods because they are typically small and can be confused with surrounding tissues. Qualitative examples are illustrated in Fig. 3. Manual organ masks of the fixed images are overlaid to show whether the warped moving images align well with the fixed image. Arrows pointed to regions where SAME works better than other methods.

4   Conclusion

In this paper, we propose SAME, a fast and accurate framework for unsupervised medical image registration. We expect SAM-affine and SAM-coarse to be promising alternatives of traditional optimization-based methods for registration initialization. The SAM correlation feature and SAM loss may also be combined with other learning-based algorithms [12,19] for further accuracy improvement.

References

1. Avants, B.B., Epstein, C.L., Grossman, M., Gee, J.C.: Symmetric diffeomorphic image registration with cross-correlation: evaluating automated labeling of elderly and neurodegenerative brain. Med. Image Anal. **12**(1), 26–41 (2008). https://doi.org/10.1016/j.media.2007.06.004. www.itk.org
2. Balakrishnan, G., Zhao, A., Sabuncu, M.R., Guttag, J., Dalca, A.V.: Voxel-Morph: A Learning framework for deformable medical image registration. IEEE Trans. Med. Imaging **38**(8), 1788–1800 (2019). https://doi.org/10.1109/TMI.2019.2897538. http://voxelmorph.csail.mit.edu
3. Çiçek, Ö., Abdulkadir, A., Lienkamp, S.S., Brox, T., Ronneberger, O.: 3D U-Net: learning dense volumetric segmentation from sparse annotation. In: Ourselin, S., Joskowicz, L., Sabuncu, M.R., Unal, G., Wells, W. (eds.) MICCAI 2016. LNCS, vol. 9901, pp. 424–432. Springer, Cham (2016). https://doi.org/10.1007/978-3-319-46723-8_49
4. Dosovitskiy, A., et al.: FlowNet: learning optical flow with convolutional networks. In: ICCV, vol. 2015 Inter, pp. 2758–2766 (2015). https://doi.org/10.1109/ICCV.2015.316
5. Guo, D., et al.: DeepStationing: thoracic lymph node station parsing in CT scans using anatomical context encoding and key organ auto-search. In: MICCAI (2021)
6. Harrison, A.P., Xu, Z., George, K., Lu, L., Summers, R.M., Mollura, D.J.: Progressive and multi-path holistically nested neural networks for pathological lung segmentation from CT images. In: Descoteaux, M., Maier-Hein, L., Franz, A., Jannin, P., Collins, D.L., Duchesne, S. (eds.) MICCAI 2017. LNCS, vol. 10435, pp. 621–629. Springer, Cham (2017). https://doi.org/10.1007/978-3-319-66179-7_71
7. Heinrich, M.P., Hansen, L.: Highly accurate and memory efficient unsupervised learning-based discrete CT registration using 2.5D displacement search. In: Martel, A.L., et al. (eds.) MICCAI 2020. LNCS, vol. 12263, pp. 190–200. Springer, Cham (2020). https://doi.org/10.1007/978-3-030-59716-0_19
8. Heinrich, M.P., et al.: MIND: modality independent neighbourhood descriptor for multi-modal deformable registration. Med. Image Anal. **16**(7), 1423–1435 (2012). https://doi.org/10.1016/j.media.2012.05.008. http://users.ox.ac.uk/~shil3388/
9. Heinrich, M.P., Jenkinson, M., Brady, S.M., Schnabel, J.A.: Globally optimal deformable registration on a minimum spanning tree using dense displacement sampling. In: Ayache, N., Delingette, H., Golland, P., Mori, K. (eds.) MICCAI 2012. LNCS, vol. 7512, pp. 115–122. Springer, Heidelberg (2012). https://doi.org/10.1007/978-3-642-33454-2_15
10. Hu, X., Kang, M., Huang, W., Scott, M.R., Wiest, R., Reyes, M.: Dual-stream pyramid registration network. In: Shen, D., et al. (eds.) MICCAI 2019. LNCS, vol. 11765, pp. 382–390. Springer, Cham (2019). https://doi.org/10.1007/978-3-030-32245-8_43
11. Klein, S., Staring, M., Murphy, K., Viergever, M.A., Pluim, J.P.: Elastix: a toolbox for intensity-based medical image registration. IEEE Trans. Med. Imaging **29**(1), 196–205 (2010). https://doi.org/10.1109/TMI.2009.2035616. http://elastix.isi.uu.nl/wiki.php
12. Liu, F., et al.: JSSR: A Joint Synthesis, Segmentation, and Registration System for 3D Multi-modal Image Alignment of Large-Scale Pathological CT Scans. In: Vedaldi, A., Bischof, H., Brox, T., Frahm, J.-M. (eds.) ECCV 2020. LNCS, vol. 12358, pp. 257–274. Springer, Cham (2020). https://doi.org/10.1007/978-3-030-58601-0_16

13. Mok, T.C.W., Chung, A.C.S.: Large deformation image registration with anatomy-aware Laplacian pyramid networks. In: Shusharina, N., Heinrich, M.P., Huang, R. (eds.) MICCAI 2020. LNCS, vol. 12587, pp. 61–67. Springer, Cham (2021). https://doi.org/10.1007/978-3-030-71827-5_7

14. Murphy, K., et al.: Evaluation of registration methods on thoracic CT: the empire10 challenge. IEEE Trans. Med. Imaging **30**(11), 1901–1920 (2011)

15. Rueckert, D., Sonoda, L.I., Hayes, C., Hill, D.L.G., Leach, M.O., Hawkes, D.J.: Nonrigid registration using free-form deformations: application to breast MR Images. IEEE Trans. Med. Imaging **18**(8), 712–721 (1999)

16. Rueckert, D., Schnabel, J.A.: Medical image registration, pp. 131–154. Springer, Berlin, Heidelberg (2011)

17. Xu, Z., et al.: Evaluation of six registration methods for the human abdomen on clinically acquired CT. IEEE Trans. Biomed. Eng. **63**(8), 1563–1572 (2016)

18. Yan, K., et al.: Self-supervised learning of pixel-wise anatomical embeddings in radiological images (2020). https://arxiv.org/abs/2012.02383

19. Zhao, S., Dong, Y., Chang, E., Xu, Y.: Recursive cascaded networks for unsupervised medical image registration. In: 2019 IEEE/CVF International Conference on Computer Vision (ICCV), pp. 10599–10609 (2019). https://doi.org/10.1109/ICCV.2019.01070

Weakly Supervised Registration of Prostate MRI and Histopathology Images

Wei Shao[1]([✉]), Indrani Bhattacharya[1], Simon J.C. Soerensen[2],
Christian A. Kunder[3], Jeffrey B. Wang[4], Richard E. Fan[2], Pejman Ghanouni[1],
James D. Brooks[2], Geoffrey A. Sonn[1,2], and Mirabela Rusu[1]

[1] Department of Radiology, Stanford University, Stanford, CA 94305, USA
weishao@stanford.edu
[2] Department of Urology, Stanford University, Stanford, CA 94305, USA
[3] Department of Pathology, Stanford University, Stanford, CA 94305, USA
[4] School of Medicine, Stanford University, Stanford, CA 94305, USA

Abstract. The interpretation of prostate MRI suffers from low agreement across radiologists due to the subtle differences between cancer and normal tissue. Image registration addresses this issue by accurately mapping the ground-truth cancer labels from surgical histopathology images onto MRI. Cancer labels achieved by image registration can be used to improve radiologists' interpretation of MRI by training deep learning models for early detection of prostate cancer. A major limitation of current automated registration approaches is that they require manual prostate segmentations, which is a time-consuming task, prone to errors. This paper presents a weakly supervised approach for affine and deformable registration of MRI and histopathology images without requiring prostate segmentations. We used manual prostate segmentations and mono-modal synthetic image pairs to train our registration networks to align prostate boundaries and local prostate features. Although prostate segmentations were used during the training of the network, such segmentations were not needed when registering unseen images at inference time. We trained and validated our registration network with 135 and 10 patients from an internal cohort, respectively. We tested the performance of our method using 16 patients from the internal cohort and 22 patients from an external cohort. The results show that our weakly supervised method has achieved significantly higher registration accuracy than a state-of-the-art method run without prostate segmentations. Our deep learning framework will ease the registration of MRI and histopathology images by obviating the need for prostate segmentations.

1 Introduction

Prostate cancer is the most diagnosed solid-organ cancer and the second leading cause of cancer death among American men [1]. Contemporary prostate cancer screening and diagnosis is marred by underdetection of aggressive cancers and overdiagnosis and over treatment of low-risk prostate cancers. Magnetic

© Springer Nature Switzerland AG 2021
M. de Bruijne et al. (Eds.): MICCAI 2021, LNCS 12904, pp. 98–107, 2021.
https://doi.org/10.1007/978-3-030-87202-1_10

resonance imaging (MRI) is increasingly used to improve prostate cancer diagnosis in patients with elevated PSA levels [2]. However, the interpretation of MRI suffers from low inter-reader agreement across radiologists coupled with large variations in reported sensitivity (58–96%) and specificity (23–87%) [3]. This is largely due to the lack of histological confirmation of the extent of cancer areas on MRI. One solution to this challenge is to map the ground-truth extent of prostate cancer from surgical histopathology images onto pre-operative MRI using image registration. Radiologists can use such mapping as a training tool to learn from subjects that have already undergone radical prostatectomy. Moreover, accurate cancer labels on MRI can also be used to develop and validate machine learning approaches for early detection of prostate cancer on MRI [4–7].

Previous approaches require manual prostate segmentations to facilitate MRI-histopathology registration [8–13]. While prostate segmentations on MRI are required for ultrasound-MRI fusion targeted biopsy, such segmentations are not routinely performed in men undergoing prostate after conventional prostate biopsy. Moreover, pathologists do not segment the prostate on digital histopathology. Since prostate segmentations on MRI and histopathology are not always immediately available and gland segmentation is a time-consuming task prone to errors, there is value in registration of MRI and histopathology without the need for gland segmentation.

Here, we present a weakly supervised registration approach that avoids the above shortcomings of prostate segmentation by not relying on it at inference. MRI-histopathology registration without prostate segmentations is challenging since the appearance, size, and location of the prostate on MRI and histopathology images differ substantially. To address the above challenges, we trained our registration network with manual prostate segmentations and mono-modal synthetic image pairs so that the network is able to align both prostate boundaries and local prostate features. We acknowledge that many weakly supervised registration methods have been proposed for image registration [14–16]. Our method used both label-driven and image-driven losses, unlike Hu et al. that only used label-based loss [14]. Moreover, others used multimodal losses for the training [15,16], however, these methods are not suitable for registration of MRI and histopathology due to their large differences in appearance. The paper has the following three major contributions.

- Our MRI-histopathology registration neural network does not require prostate segmentations during the testing.
- Our registration network trained with mono-modal synthetic image pairs and manual prostate segmentations can be used for multi-modal registration.
- An unsupervised intensity loss, an auxiliary segmentation loss, and a regularization loss were developed for the challenging registration problem.

2 Methods

2.1 Data Acquisition

This institutional review board-approved study included 183 patients with biopsy-confirmed prostate cancer from two cohorts. The first cohort is an internal cohort consisting of 161 patients. The second cohort is an external validation cohort consisting of 22 patients [17]. For both cohorts, each patient had a pre-operative T2-w MRI and whole-mount histopathology images from radical prostatectomy. Slice-to-slice correspondences between MRI and histopathology images were optimized using customized 3D-printed molds. An intensity standardization algorithm [18] was used to correct for misaligned MR image intensities between patients. The prostate, urethra, and anatomic landmarks on MRI and histopathology images were manually segmented. Cancerous regions on the histopathology images were also manually segmented.

2.2   Overview of Proposed Method

Figure 1 presents an overview of our weakly supervised registration pipeline. The input to our registration pipeline is a histopathology image and the corresponding axial MRI slice. First, we cropped a 100 mm × 100 mm central region from the MRI slice as the fixed image. Second, we padded the histopathology image to the same size as the cropped MRI, i.e., 100 mm × 100 mm. The histopathology image served as the moving image. The registration neural network then takes the fixed image the moving image as the input and outputs a vector θ that parameterizes a composite transformation (affine + deformable) between the two images. Finally, the predicted transformation is used to deform the ground-truth cancer labels from the histopathology image onto the MRI.

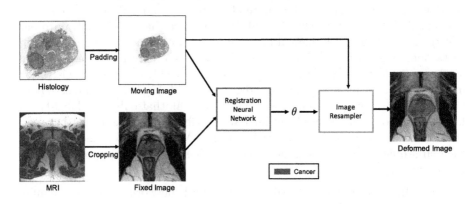

Fig. 1. An overview of our weakly supervised learning registration pipeline.

2.3   Registration Neural Network

Our registration neural network shown in Fig. 2 is considered as a function that maps a fixed image and a moving image to a vector θ that parameterizes a geometric transformation between the two images. Features in the fixed and moving images were extracted by the first few layers of the VGG16 network [19]

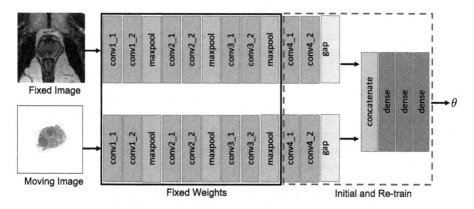

Fig. 2. Proposed registration neural network. Fixed Weights correspond to layers whose weights were frozen during the training, and Initial and Re-train correspond to layers whose weights were updated during the training.

pre-trained on the ImageNet dataset [20]. During the training, we freeze all layers of the VGG16 network except for the last two convolutional layers. Inspired by [16], the extracted image features were connected to the global average pooling (GAP) layer to transform their dimensions from $(N \times N \times C)$ to $(1 \times 1 \times C)$, where N is the size of a feature image and C is the number of feature images. We concatenated the outputs of the GAP layers into a column vector, which was connected to a stack of three dense layers to regress the parameter vector θ. One advantage of using the GAP layer is that it can convert feature images of any dimension to 1×1, and thus it allows our neural network to have fixed and moving images of different sizes. The GAP layer could also prevent the neural network from overfitting since it has reduced the complexity the model.

2.4 Transformation Model

For each pair of images, our registration network predicts an affine transformation followed by a nonrigid transformation. To parameterize a 2D affine transformation, we set the length of the output parameter vector θ from the registration network to 6. Instead of directly using θ as the affine matrix, we define the affine transformation ϕ parameterized by θ as

$$\phi(x,y) = \begin{bmatrix} 1+\alpha\theta_1 & \alpha\theta_2 \\ \alpha\theta_4 & 1+\alpha\theta_5 \end{bmatrix} \begin{bmatrix} x \\ y \end{bmatrix} + \begin{bmatrix} \alpha\theta_3 \\ \alpha\theta_6 \end{bmatrix} \tag{1}$$

where (x,y) is a pixel in the fixed image I_f, and $\alpha = 0.001$ is a constant. The purpose of scaling θ by a small constant and adding an identity matrix is to guarantee that the transformation ϕ is close to the identity at the beginning of the training, which would substantially improve the robustness of our model.

In this paper, we parameterize nonrigid transformations by a 4×4 thin-plate spline grid [21]. Therefore, the length of the parameter vector θ is set to

$2 \times 4 \times 4 = 32$. Similarly, we scale the parameter vector θ by a small constant $\alpha = 0.001$ and add an identity matrix so that the initial estimation of the nonrigid transformation ϕ is close to the identity transformation.

2.5 Loss Functions

In this paper, we propose three loss functions: an unsupervised intensity loss L_{int} computed on two synthetic mono-modal image pairs, an auxiliary segmentation loss L_{seg} computed on two prostate segmentations, and a regularization loss L_{reg} that measures the smoothness of nonrigid transformations.

Our registration framework consists of an affine registration network and a deformable registration network. We trained the two registration networks simultaneous and used the output of the affine network as an initialization of the deformable network. The loss function of the affine registration network is defined as

$$L_{affine} = L_{seg} + 0.05L_{int} \tag{2}$$

The loss function of the deformable registration network is defined as

$$L_{def} = L_{seg} + 0.05L_{int} + 0.05L_{reg} \tag{3}$$

Each training example (I_f, I_m, S_f, S_m) consists of four images, where I_f and I_m are the fixed and moving images, and S_f and S_m are the corresponding prostate segmentations. We applied two random transformations to deform I_f and I_m into I'_f and I'_m. The mono-modal image pairs (I_f, I'_f) and (I_m, I'_m) were fed into the registration network to predict two transformations ϕ^f_{mono} and ϕ^m_{mono}. We define the unsupervised intensity loss L_{int} as

$$L_{int} = MSE(I_f, I'_f(\phi^f_{mono})) + MSE(I_m, I'_m(\phi^m_{mono})) \tag{4}$$

where $MSE(\cdot, \cdot)$ is the mean squared error. The motivation of using L_{int} is to train the network to align local prostate features irrespective of image modalities.

We used manual prostate segmentations S_f and S_m to train the network to align prostate boundaries. Let ϕ_{multi} denote the transformation between I_f and I_m predicted by the registration neural network. We define the auxiliary segmentation loss L_{seg} as

$$L_{seg} = Dice(S_f, S_m(\phi_{multi})) = \frac{2|S_f \cap S_m(\phi_{multi})|}{|S_f| + |S_m(\phi_{multi})|} \tag{5}$$

where $|\cdot|$ is cardinality of a set.

We used the regularization loss L_{reg} to measure the smoothness of thin-plate spline transformations ϕ

$$L_{reg} = ||Lu||^2_{L^2} \tag{6}$$

where $u = \phi - Id$, $L = -0.75\nabla^2 - 0.25\nabla(\nabla \cdot) + 0.01I$, ∇ is the gradient operator, $\nabla \cdot$ is the divergence operator, and I is the identity matrix.

2.6   Previous Method

The ProsRegNet network [12] is the first and the only deep learning MRI-histopathology approach and was used to compare with our weakly supervised approach. We downloaded the ProsRegNet code from the public repository https://github.com/pimed//ProsRegNet. We acknowledge the following differences between the proposed method and the ProsRegNet method.

- Network architecture. ProsRegNet used ResNet101 while our method used VGG16 for feature extraction. The ProsRegNet method used a correlation layer to compute correlation between feature maps while our method used the global average pooling layer to reduce the size of each feature map.
- Loss function. ProsRegNet only used the mean squared error (MSE) loss while our method used the MSE loss and two additional losses, i.e., the Dice coefficient loss and the regularization loss.
- ProsRegNet used masked synthetic unimodal image pairs for the training. The proposed method used unmasked synthetic unimodal image pairs, unmasked multimodal image pairs, and manual prostate masks for the training.

2.7   Evaluation Metrics

We use Dice coefficient defined in Eq. 5 to measures the overlap between prostate segmentation of the fixed image (S_f) and prostate segmentation of the deformed image $(S_m(\phi))$.

We use Hausdorff distance to measure the distance between boundaries of the prostate segmentations

$$Haus = \max \left\{ \sup_{a \in S_f} \inf_{b \in S_m(\phi)} ||a - b||, \sup_{b \in S_m(\phi)} \inf_{a \in S_f} ||a - b|| \right\}. \tag{7}$$

We use the mean landmark error (MLE) to measure the accuracy of estimated point-to-point correspondences

$$MLE = \frac{1}{N} \sum_{i=1}^{N} ||p_i' - \phi(p_i)|| \tag{8}$$

where N is the number of landmarks, $p_1, \cdots, p_N$ and $p_1', \cdots, p_N'$ are landmarks in the fixed image and moving images, respectively.

We also use Eq. 8 to evaluate the distance between the center of urethra segmented on the fixed image and the deformed image, i.e., urethra deviation.

2.8   Experimental Design

We trained and validated our registration networks with 135 and 10 patients from the internal cohort, respectively. We used an initial learning rate of 0.001, a learning rate decay of 0.9, a batch size of 1, 50 epochs, and the Adam optimizer [22]. We tested the performance of our method using 16 patients from the internal cohort and 22 patients from the external cohort. We compared our weakly supervised registration approach to the previous ProsRegNet deep learning registration approach [12]. Since the goal of this paper is to develop a registration approach that does not require prostate segmentation, both our method and the competing ProsRegNet method were tested without using prostate segmentations. All experiments were performed on the NVIDIA Quadro GV100 GPU (32 GB memory, 5120 CUDA cores, 640 tensor cores).

3   Results

3.1   Qualitative Results

Figure 3 shows the registration results of a representative patient from the internal cohort. The first row of Fig. 3 shows that there is large misalignment between the MRI and histopathology images before image registration. The second row of Fig. 3 shows that the previous ProsRegNet barely improved the registration between the two images. The third row of Fig. 3 shows that our weakly supervised method has not only accurately aligned prostate boundaries, but also local features inside the prostate which includes the urethra, benign prostatic hyperplasia regions, and cancerous regions. This example demonstrates the superiority of our weakly supervised method. Accurate transformations predicted our weakly supervised registration network enables accurate mapping of the ground-truth cancer labels from the histopathology image onto MRI.

3.2   Quantitative Results

The Dice coefficients and Hausdorff distances in Table 1 show that our weakly supervised approach has achieved significantly more accurate alignment of the prostate boundaries for both cohorts ($p < 0.05$). The urethra deviations and landmark errors show that our weakly supervised approach has also achieved much more accurate alignment of local prostate features than the ProsRegNet approach for both cohorts. One major reason for the poor performance of the ProsRegNet model is that it was trained using images masked by prostate segmentations while prostate segmentations were not provided during the testing.

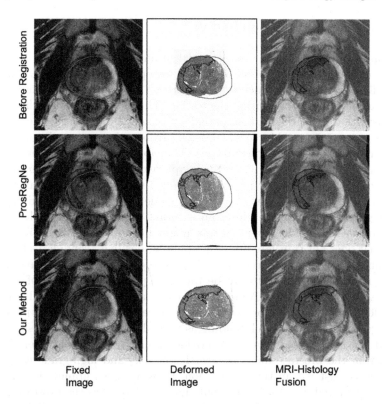

Fig. 3. Comparison of the registration results of our weakly supervised method and previous ProsRegNet method of a representative subject from the internal cohort. The green outlines correspond to manual segmentation of the prostate on the fixed MRI image. The red outlines correspond to the boundaries of the cancer labels which were obtained by deforming the ground-truth cancer labels from the histopathology images onto MRI using the resulting transformations from image registration. (Color figure online)

Table 1. Registration results. Cohort 1 is internal and Cohort 2 is external.

Dataset	Registration approach	Dice coefficient	Hausdorff distance	Urethra deviation (mm)	Landmark error (mm)
Cohort 1	Input	0.80 (± 0.06)	7.57 (± 2.21)	4.32 (± 2.29)	4.48 (± 1.77)
	ProsRegNet	0.80 (± 0.06)	7.76 (± 2.00)	4.54 (± 2.46)	4.58 (± 1.87)
	Our Method	**0.90 (± 0.02)**	**4.38 (± 0.64)**	**3.09 (± 1.06)**	**3.09 (± 0.81)**
Cohort 2	Input	0.83 (± 0.05)	7.23 (± 2.30)	3.39 (± 1.47)	3.40 (± 1.04)
	ProsRegNet	0.82 (± 0.04)	8.00 (± 2.12)	3.40 (± 1.34)	4.09 (± 1.38)
	Our Method	**0.89 (± 0.03)**	**5.01 (± 0.97)**	**2.59 (± 1.16)**	**3.13 (± 0.76)**

4   Discussion and Conclusion

This paper presents a weakly supervised learning approach for the affine and deformable image registration of MRI and histopathology images. Unlike previous fully-automated approaches, our approach does not require manual segmentations of the prostate during the testing. Therefore, our approach will ease the registration of MRI and histopathology images by obviating the need for prostate segmentations. Moreover, previous registration approaches are sensitive to errors in prostate segmentations. Therefore, fine-tuning of the prostate segmentations is a necessity for previous approaches to achieve accurate registration results. Although our model was trained with patients from an internal cohort, the results show that the trained model can be generalized to accurately register unseen images from an external cohort. We acknowledge that the use of patients undergoing prostate might introduce spectrum bias in our cohort that does not translate into other populations (e.g., candidates for active surveillance). However, we used grade as a surrogate for the aggressive disease since it has been demonstrated to be the most powerful predictor of outcome in localized prostate cancer. Moreover, grade is also used for clinical decisions (e.g., selection of patients for active surveillance). As such, radical prostatectomy cases are ideal for model building as they allow: (1) precise registration of MR and histopathology images, (2) provide ground truth cancer label at every voxel within the prostate and (3) diversity of low- and high-grade cancers as they often coexist within the same lesion. However, such granularity of labels cannot be achieved in subjects that only undergo biopsy as the prostate is only sampled at the biopsy sites. The proposed algorithm can relieve radiologists the task of manual prostate segmentation while generating accurate cancer labels on MRI. Accurate labels can facilitate clinical care and facilitate use of deep learning approaches for detection of aggressive cancers on preoperative MRI.

References

1. American Cancer Society. Facts & Figures 2021. American Cancer Society, Atlanta, GA (2021)
2. Baris Turkbey, L., Peter Choyke, L.: Multiparametric MRI and prostate cancer diagnosis and risk stratification. Curr. Opin. Urol. **22**(4), 310–315 (2012)
3. Ahmed, H.U., et al.: Diagnostic accuracy of multi-parametric MRI and TRUS biopsy in prostate cancer (PROMIS): a paired validating confirmatory study. The Lancet **389**(10071), 815–822 (2017)
4. Lovegrove, C.E., et al.: The role of pathology correlation approach in prostate cancer index lesion detection and quantitative analysis with multiparametric MRI. In: NIH (2016)
5. Bhattacharya, I., et al.: CorrSigNet: learning CORRelated prostate cancer SIGnatures from radiology and pathology images for improved computer aided diagnosis. In: Martel, A.L., et al. (eds.) MICCAI 2020. LNCS, vol. 12262, pp. 315–325. Springer, Cham (2020). https://doi.org/10.1007/978-3-030-59713-9_31
6. Seetharaman, A., et al.: Automated detection of aggressive and indolent prostate cancer on magnetic resonance imaging. Med. Phys. (2021)

7. Saha, A., Hosseinzadeh, M., Huisman, H.: End-to-end prostate cancer detection in bpMRI via 3D CNNs: effects of attention mechanisms, clinical priori and decoupled false positive reduction. Med. Image Anal. 102155 (2021)

8. Chappelow, J., et al.: Elastic registration of multimodal prostate MRI and histology via multiattribute combined mutual information. Med. Phys. **38**(4), 2005–2018 (2011)

9. Reynolds, H.M., Williams, S., Zhang, A., Chakravorty, R., Rawlinson, D., Ong, C.S., et al.: Development of a registration framework to validate MRI with histology for prostate focal therapy. Med. Phys. **42**(12), 7078–7089 (2015)

10. Wu, H.H., et al.: A system using patient-specific 3D-printed molds to spatially align in vivo MRI with ex vivo MRI and whole-mount histopathology for prostate cancer research. J. Magnet. Reson. Imaging **49**(1) (2019)

11. Rusu, M., et al.: Registration of presurgical MRI and histopathology images from radical prostatectomy via RAPSODI. Med. Phys. **47**(9), 4177–4188 (2020)

12. Shao, W., et al.: Prosregnet: a deep learning framework for registration of MRI and histopathology images of the prostate. Med. Image Anal. **68**, 101919 (2021)

13. Sood, R.R., et al.: 3D registration of pre-surgical prostate MRI and histopathology images via super-resolution volume reconstruction. Med. Image Anal. **69**, 101957 (2021)

14. Hu, Y., et al.: Label-driven weakly-supervised learning for multimodal deformable image registration. In: 2018 IEEE 15th International Symposium on Biomedical Imaging (ISBI 2018), pp. 1070–1074. IEEE (2018)

15. Balakrishnan, G., et al.: Voxelmorph: a learning framework for deformable medical image registration. IEEE Trans. Med. Imaging **38**(8), 1788–1800 (2019)

16. de Vos, B.D., Berendsen, F.F., Viergever, M.A., Sokooti, H., Staring, M., Igum, I.: A deep learning framework for unsupervised affine and deformable image registration. Med. Image Anal. **52**, 128–143 (2019)

17. Choyke, P., Turkbey, B., Pinto, P., Merino, M., Wood, B.: Data from PROSTATE-MRI (2016)

18. Nyúl, L.G., Udupa, J.K., Zhang, X.: New variants of a method of MRI scale standardization. IEEE Trans. Med. Imaging **19**(2), 143–150 (2000)

19. Simonyan, K., Zisserman, A.: Very deep convolutional networks for large-scale image recognition. arXiv.org (2015)

20. Deng, J., Dong, W., Socher, R., Li, L.-J., Li, K., Fei-Fei, L.: Imagenet: a large-scale hierarchical image database. In: 2009 IEEE Conference on Computer Vision and Pattern Recognition, pp. 248–255. IEEE (2009)

21. Donato, G., Belongie, S.: Approximate thin plate spline mappings. In: Heyden, A., Sparr, G., Nielsen, M., Johansen, P. (eds.) ECCV 2002. LNCS, vol. 2352, pp. 21–31. Springer, Heidelberg (2002). https://doi.org/10.1007/3-540-47977-5_2

22. Kingma, D., Ba, J.: Adam: a method for stochastic optimization. arXiv.org (2017)

4D-CBCT Registration with a FBCT-derived Plug-and-Play Feasibility Regularizer

Yudi Sang and Dan Ruan[(✉)]

University of California, Los Angeles, CA 90095, USA
yudisang@ucla.edu, druan@mednet.ucla.edu

Abstract. Deformable registration of phase-resolved lung images is an important procedure to appreciate respiratory motion and enhance image quality. Compared to high-resolution fan-beam CTs (FBCTs), cone-beam CTs (CBCTs) are more readily available for on-table acquisition in companion with treatment. However, CBCT registration is challenging because classic regularization energies in convention methods usually cannot overcome the strong artifacts and the lack of structural details. In this study, we propose to learn an implicit feasibility prior of respiratory motion and incorporate it in a plug-and-play (PnP) fashion into the training of an unsupervised image registration network to improve registration accuracy and robustness to noise and artifacts. In particular, we propose a novel approach to develop a feasibility descriptor from a set of deformation vector fields (DVFs) generated from FBCTs. Subsequently, this FBCT-derived feasibility descriptor was used as a spatially variant regularizer on DVF Jacobian during the unsupervised training for 4D-CBCT registration. In doing so, the higher-quality, higher-confidence information from FBCT is *transferred* into the much challenging problem of CBCT registration, without explicit FB-CB synthesis. The method was evaluated using manually identified landmarks on real CBCTs and automatically detected landmarks on simulated CBCTs. The method presented good robustness to noise and artifacts and generated physically more feasible DVFs. The target registration errors on the real and simulated data were (1.63 ± 0.98) and (2.16 ± 1.91) mm, respectively, significantly better than the classic bending energy regularization in both the conventional method in SimpleElastix and the unsupervised network. The average registration time was 0.04 s.

Keywords: Deep learning · Image registration · 4D cone-beam CT

1 Introduction

Cone-beam computed tomography (CBCT) is a modality commonly available from on-board-imaging systems integrated to image-guided radiation therapy (IGRT) machines. CBCT and can provide important information for daily or fraction-wise patient setup, verification, and on-table adaptation, with much

© Springer Nature Switzerland AG 2021
M. de Bruijne et al. (Eds.): MICCAI 2021, LNCS 12904, pp. 108–117, 2021.
https://doi.org/10.1007/978-3-030-87202-1_11

lower imaging dose than fan-beam CT (FBCT). 4D-CBCT images are computed by sorting the projections into different respiratory bins and reconstructing them separately [18]. In treatments of lung or liver tumors, where substantial respiratory motion is involved, 4D-CBCT can provide onboard 4D volumetric information of target location and motion range with good accuracy. This is especially important in stereotactic ablative radiotherapy (SABR), where a high radiation dose is used per fraction and higher accuracy is required to protect nearby organs in adaptive strategies [20].

Deformable image registration (DIR) is the process of estimating the deformation vector fields (DVFs) to align two images. DIR between phases in a 4D-CBCT is a fundamental step for motion estimation and contour propagation of target volumes. Furthermore, it can integrate information from different phases and enhance image quality through motion-compensated reconstruction [11,13]. However, the DIR of 4D-CBCT is much more challenging than fan-beam CT (FBCT) due to its low spatial resolution and low signal-to-noise ratio (SNR) caused by the artifacts[16].

Classic DIR is usually formulated in a regularized optimization framework [12,14,19] where the DVF is updated iteratively, using descent algorithms to minimize a loss function consisting of image dissimilarity that measures the disagreement between the warped moving image and the fixed image, and regularization energy that penalize non-physical deformations. Commonly used regularization encourages prescribed physical behaviors such as smoothness and diffeomorphism to enhance DVF feasibility [19].

Deep learning approaches have been developed for DIR in recent years. The registration process can be integrated into a deep network to infer DVFs directly from input image pairs with much higher efficiency. Training of the network can be supervised or unsupervised. In the supervised setting, the network learns the map between the pair of image input and the corresponding ground-truth DVF directly [22]. The unsupervised setting is very similar to the classic DIR approach in terms of the loss function, but replaces the iterative DVF solving process with fast inference from a trained network [1,21].

Currently, most studies in unsupervised learning lung image registration focus on high-quality FBCTs, and encourage DVF feasibility through regularization on the first or second order spatial gradient of the DVF [4,5,10,21]. Fechter and Baltas [4] also used a patch boundary smoothness constraint to guarantee homogeneity at patch transitions. In addition, they used the sum of all deformation vectors along voxel trajectory as a constraint for periodic motion. Fu et al. [5] used an adversarial loss on the warped image in a two-stage coarse-to-fine registration network to encourage realistic transformation.

Despite the good performances on FBCTs, the DIR of CBCT remains challenging because the lack of structural details and the presence of strong artifacts can introduce undesired local minima, which are difficult for the classic regularization energies to overcome. More powerful and robust feasibility prior is needed to deal with the lower image quality and further improve registration accuracy.

In this study, we propose to achieve efficient and accurate 4D-CBCT interphase registration by deriving an implicit feasibility descriptor for respiratory motion from high-quality 4D-FBCT and incorporate it in a plug-and-play (PnP) fashion into an unsupervised CBCT DIR network as a flexible regularizer.

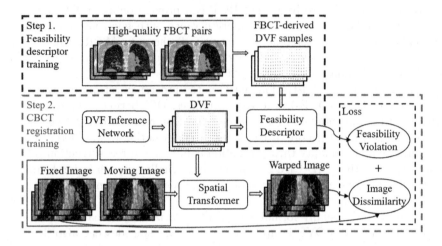

Fig. 1. Overview of the proposed method.

2 Method

2.1 Overview

Our method consists of two major modules developed sequentially. First, a feasibility descriptor in the form of a convolutional auto-encoder (CAE) is trained in a supervised setting, using FBCT-derived data, to characterize feasible respiratory motions represented by a set of physically sound DVFs. Then, an unsupervised DVF inference network is trained with CBCT, which includes the FBCT-derived CAE as a PnP regularizer to simultaneously enforce image matching and penalize DVFs that deviate from the learned implicit feasibility condition.

As shown in Fig. 1, the proposed method consists of a DVF inference network, a spatial transformer [9], and an feasibility descriptor. As in the other learning-based DIR methods, the inference network concatenates fixed and moving images as input and outputs a DVF estimation; the spatial transformer uses tri-linear interpolation for image re-sampling and warps the moving image with the DVF [1,21]. The major contribution of this work lies in the injection of the FBCT-derived feasibility descriptor, which is trained independently beforehand and then plugged into the main registration network. Loss functions defined as the combination of the dissimilarity between the fixed and warped images and the DVF feasibility violation provided by the descriptor, are used to drive the network optimization, with a back-propagation scheme.

2.2 Feasibility Descriptor of Respiratory Motion

To introduce spatially variant regularization on deformation, we use a CAE to model the DVF feasibility conditions. As shown in Fig. 2, the Jacobian matrix of an input DVF is computed and represented by a nine-channel feature map. In the encoder, four alternating layers of convolution and average-pooling are applied. Then, a fully connected layer with 256 units is used, which generates the latent representation of the DVF. The decoder path consists of a fully connected layer and three transposed convolution layers with stride of 2. All the (de)convolution layers use zero-padding, kernel size of 3, and ReLU activation, except the last layer, which uses linear activation.

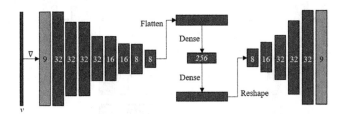

Fig. 2. Architecture of the feasibility descriptor.

The loss function used for the CAE training is the squared Frobenius norm of point-wise difference between the DVF Jacobian and the CAE output:

$$L_V = \|\nabla(v) - \text{CAE}(\nabla(v))\|_{\text{Frob}}^2, \tag{1}$$

where v is a DVF sample, and ∇ is the Jacobian operator. Once the CAE is properly trained, the deviation of a candidate DVF from the manifold, as described by the auto-encoding discrepancy of its Jacobian matrix in squared Frobenius norm, provides a measure of the physical or physiological feasibility of that DVF. The advantage of using a supervised CAE instead of conventional regularization energy is its potential to incorporate spatial variability from the training process. After the CAE training, the network parameters remain fixed during the subsequent DVF inference network training.

2.3 Unsupervised Learning for DVF Estimation

The DVF inference network takes concatenated pairs of moving and fixed images as input, and outputs a DVF. As shown in Fig. 3, the network uses a general U-net structure [15] to take advantage of the hierarchical structure and skip connections for effective learning of features at all scales. All the $3 \times 3 \times 3$ convolution layers use a stride of 1, zero-padding, and ReLU activation, except the last layer, which uses tanh activation. Average pooling and up-sampling with a scaling factor of 2 are used in the encoding and decoding paths, respectively.

The loss function for the DVF inference network training consists of an image intensity matching cost L_D and an implicit feasibility violation penalty on DVF Jacobian L_V. In this work, we used normalized cross-correlation (NCC) as the similarity metric. The loss function can be written as:

$$L = L_D + \mu L_V = -\text{NCC}(I_F, I_M \circ v) + \mu \|\nabla(v) - \text{CAE}(\nabla(v))\|^2_{\text{Frob}}, \quad (2)$$

where I_F is the fixed image, I_M is the moving image, $v(I_F, I_M)$ is the DVF estimated by the inference network, $I_W = I_M \circ v$ is the image warped with DVF v, and μ is a balancing parameter.

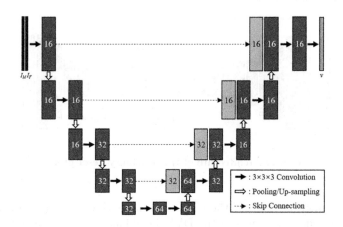

Fig. 3. Architecture of the DVF inference network.

3   Experiments and Results

The implementation was based on TensorFlow. The proposed method was compared against a classic B-spline method in SimpleElastix [12] and the DVF inference network trained with and without bending energy penalty (BP) [21]. The metrics used in SimpleElastix registration were also NCC and BP, with the weight for BP manually tuned for each specific case to optimize the performance. All the registrations were from exhalation to inhalation. The experiments were performed on a workstation equipped with an NVIDIA GTX 1080 Ti GPU and an Intel i7-6700HQ 3.5 GHz CPU.

3.1   Training of the Feasibility Descriptor

To obtain the DVF samples that represent realistic respiratory motion, we performed conventional B-spline registration on a set of ten FBCT scans from the DIR-Lab dataset [2,3]. The scans were acquired as part of the radiotherapy planning process for the treatment of thoracic malignancies. Each scan has ten breathing phases. The slice thickness was 2.5 mm and in-plane spacing was 0.97

to 1.16 mm. All images were resampled with slice thickness 2.34 mm and typical in-plane pixel spacing 1.16 mm and then cropped with a $256 \times 256 \times 64$ window that covered the lungs. Image intensities were clamped between -1000 and 500 HU and scaled between 0 and 1.

SimpleElastix was used to generate DVFs between breathing phases, with BP regularized NCC objective. In order to accommodate spatially variant regularization, we used various values of trade-off parameters λ ranging from 0.01 to 2, therefore the learned manifold could address different local trade-offs. For each scan, 15 moving and fixed image pairs were selected, all from exhalation to inhalation. Then, they were augmented by five different registrations performed with different λ trade-offs for BP regularization. As a result, 750 DVFs were generated as the training set. The CAE was trained for 200 epochs with batch size of 1. ADAM optimizer with learning rate of 10^{-4} was used.

3.2   Training of the DVF Inference Network

The 4D-CBCT data was from the 4D-Lung collection in the Cancer Imaging Archive (TCIA) [8]. They were acquired during chemoradiotherapy of 20 locally-advanced, non-small cell lung cancer patients. Each patient has multiple scans. Each scan has ten breathing phases. The reconstructed slice thickness was 3 mm and in-plane spacing was 0.98 to 1.17 mm. The images were pre-processed to the same size and pixel spacing described in Sect. 3.1. The training, validation, and testing sets contain 10, 5, and 5 patients, respectively. In the training set, 25 scans from the 10 patients were used for data augmentation purpose.

For each scan, 15 moving and fixed image pairs were selected. The network was trained for 150 epochs with batch size of 1. The balancing parameter μ in Eq. (2) was set to 10^{-6}. ADAM optimizer with learning rate of 10^{-4} was used. For comparison, U-nets with and without BP were trained and tested with the same setting. The weight for BP was set to 0.1. The hyperparameters for the networks were tuned based on the learning curves of the validation set.

3.3   Evaluation

To quantitatively assess the registration performance, we annotated ten pairs of anatomical landmarks at end-inhale (EI) and end-exhale (EE) phases in each of the testing CBCT scans.

To further obtain landmark annotation on a larger scale, we obtained simulated 4D-CBCTs from the SPARE dataset [17]. The CBCTs in the dataset were reconstructed from projections simulated from real patient 4D FBCTs using FDK algorithm. We applied an automatic landmark pair detection algorithm proposed in [6] to the original FBCTs to take advantage of its higher image quality and structural details. The locations of the landmarks were then mapped to the simulated CBCTs using a rigid registration of the FB-CB pairs. Finally, since the SPARE CBCTs were simulated with a much shorter acquisition time than the real CBCTs in the training set, we adjusted the SNR of the simulated CBCT by adding the ground-truth FBCT to it. The final adjusted CBCT was the combination of 40% FBCT and 60% simulated CBCT. As shown in Fig. 4, the adjusted image quality was similar to the real CBCT in the training set. Eventually, nine scans from the dataset were used, each with 100 landmark pairs detected in the EI and EE phases.

The Euclidean distance between the transformed and the fixed landmarks was calculated as target registration error (TRE) to indicate registration performance. Paired t-tests were used to examine statistical significance.

3.4   Results

Figure 5 shows example registration results of the proposed method in comparison to the conventional method with BP regularization in SimpleElastix. Although both methods presented a good image intensity matching, the DVFs in our method were much smoother and better characterized the respiratory motion. The motion boundaries in our method were generally in alignment with the lung contours. SimpleElastix generated locally smoothed DVFs, but failed to estimate motions in the homogeneous region where structural details were missing, and was also strongly affected by the noise and artifacts.

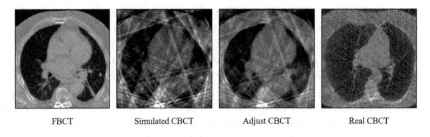

FBCT   Simulated CBCT   Adjust CBCT   Real CBCT

Fig. 4. Example CBCT data. The adjusted CBCT is generated using a weighted summation of the FBCT and the simulated CBCT and has a similar noise level to the real CBCT in the training set.

Table 1. Target registration errors based on the anatomical landmarks. Results are provided as mean ± standard deviation in millimeter (p-value from paired t-tests).

	Before	SimpleElastix	U-net	U-net BP	Our method
Real CBCT	6.12 ± 4.12	1.74 ± 1.66 (0.008)	2.45 ± 1.87 (10^{-5})	1.93 ± 1.62 (0.001)	**1.63 ± 0.98**
Simulated CBCT	7.53 ± 4.15	2.57 ± 2.51 (0.002)	3.11 ± 3.10 (10^{-8})	2.61 ± 2.48 (10^{-6})	**2.16 ± 1.91**

Table 1 shows the quantitative TRE results. Our method achieved the best TREs on both real and simulated data. Paired t-tests indicated that the TRE reductions in our method were statistically significant ($p < 0.01$) compared to all the other methods tested. The average registration time was 52 s for SimpleElastix and 0.04 s for all three networks.

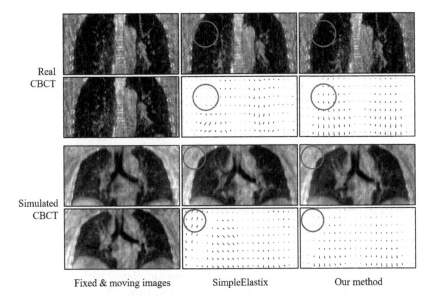

Fig. 5. Example registration results. The red circle indicates a locally homogeneous region. The blue circle indicates a region affected by a strong artifact. Note that the 3D DVFs are visualized with their 2D projection on the coronal plane.

4   Discussion and Conclusion

In this study, we have developed a deep neural network for lung 4D-CBCT DIR. The network was trained with the incorporation of a novel DVF feasibility descriptor that characterizes a learned flexible motion prior from high-quality FBCT scans. Experiments with real and simulated CBCTs showed that introducing the feasibility descriptor improved registration accuracy significantly, compared to conventional BP regularization energy in both the conventional and the deep learning methods.

The quality of CBCT in clinical settings is influenced by many factors including setup differences and detector properties. However, the FBCT quality is usually higher and more consistent across different machines or vendors. In addition, offline FBCT registration can afford to be performed with more sophisticated methods at higher time cost, even with manual interaction or guidance when needed, to ensure the quality of feasibility prior. Therefore, the proposed feasibility descriptor, once properly trained, can be used as a PnP module for the retraining of the DVF inference network as it migrates to new imaging platforms, transferring the advantages of FBCT image quality and fine-tuned methods into the fast registration of CBCT.

A similar idea of using a learning-based feasibility prior in registration has been proposed by Hu et al. [7], where adversarial deformation constraints are introduced with a discriminator network to distinguish the registration-predicted DVFs from the motion data offered by biomechanical models. Compared to this adversarial regularization method, our method is more stable and simpler in the training setup, and the MRM can be transferred more easily as a PnP module.

In our experiments, the DVF training samples were generated using a conventional registration method in SimpleElastix with various values of trade-off parameters. While this method was robust enough for estimating the general respiratory motion field from FBCT, other more sophisticated methods could also be used to further improve the motion prior. A few FBCT scans in the DIR-Lab dataset contain sorting artifacts and may affect the accuracy of estimated motion in prior generation. While the motion prior can benefit from a high-quality prior generation dataset, we have observed that the use of various trade-off values and the CAE scheme for establishing MRM is robust against moderate noise in the training DVF samples. This robustness is desirable so that users can choose to either utilize an existing MRM with applicable site or customize the prior training with their best available dataset and registration resource.

The proposed method was evaluated using anatomical landmarks in both real and simulated CBCTs. The SPARE CBCTs were simulated with a short acquisition time and had a much lower SNR. We boosted their SNR using a weighted sum of the ground-truth FBCT so that the quality matched that of a common clinical on-table or on-board CBCT. The root-mean-square error (RMSE) of the adjusted CBCT to the ground truth was 180 HU, higher than a reference value of 152 HU from a phantom-based reconstruction study in [23], indicating that the adjusted CBCT was not overly enhanced in quality. The proposed DVF inference network was trained with real CBCTs and did not require further training before tested with the simulated data. While the SNR adjustment is reasonable, we are working on simulating realistic CBCT from FBCT for a more direct performance quantification. As the rationale of PnP modality-agnostic prior generalizes, we plan to investigate its utility in other registration applications.

References

1. Balakrishnan, G., Zhao, A., Sabuncu, M.R., Guttag, J., Dalca, A.V.: Voxelmorph: a learning framework for deformable medical image registration. IEEE Trans. Med. Imag. **38**, 1788–1800 (2019)
2. Castillo, E., Castillo, R., Martinez, J., Shenoy, M., Guerrero, T.: Four-dimensional deformable image registration using trajectory modeling. Phys. Med. Biol. **55**(1), 305 (2009)
3. Castillo, R., et al.: A framework for evaluation of deformable image registration spatial accuracy using large landmark point sets. Phys. Med. Biol. **54**(7), 1849 (2009)
4. Fechter, T., Baltas, D.: One-shot learning for deformable medical image registration and periodic motion tracking. IEEE Trans. Med. Imag. **39**(7), 2506–2517 (2020)
5. Fu, Y., et al.: LungRegNet: an unsupervised deformable image registration method for 4D-CT lung. Med. Phys. **47**(4), 1763–1774 (2020)
6. Fu, Y., Wu, X., Thomas, A.M., Li, H.H., Yang, D.: Automatic large quantity landmark pairs detection in 4DCT lung images. Med. Phys. **46**(10), 4490–4501 (2019)

7. Hu, Y., et al.: Adversarial deformation regularization for training image registration neural networks. In: Frangi, A.F., Schnabel, J.A., Davatzikos, C., Alberola-López, C., Fichtinger, G. (eds.) MICCAI 2018. LNCS, vol. 11070, pp. 774–782. Springer, Cham (2018). https://doi.org/10.1007/978-3-030-00928-1_87

8. Hugo, G.D., et al.: A longitudinal four-dimensional computed tomography and cone beam computed tomography dataset for image-guided radiation therapy research in lung cancer. Med. Phys. **44**(2), 762–771 (2017)

9. Jaderberg, M., Simonyan, K., Zisserman, A., et al.: Spatial transformer networks. Adv. Neural Inf. Process. Syst. **28**, 2017–2025 (2015)

10. Jiang, Z., Yin, F.F., Ge, Y., Ren, L.: A multi-scale framework with unsupervised joint training of convolutional neural networks for pulmonary deformable image registration. Phys. Med. Biol. **65**(1), 015011 (2020)

11. Li, T., Koong, A., Xing, L.: Enhanced 4D cone-beam CT with inter-phase motion model. Med. Phys. **34**(9), 3688–3695 (2007)

12. Marstal, K., Berendsen, F., Staring, M., Klein, S.: SimpleElastix: a user-friendly, multi-lingual library for medical image registration. In: Proceedings of the IEEE Conference on Computer Vision and Pattern Recognition Workshops, pp. 134–142 (2016)

13. Nakagawa, K., et al.: 4D registration and 4D verification of lung tumor position for stereotactic volumetric modulated arc therapy using respiratory-correlated cone-beam CT. J. Radiat. Res. **54**(1), 152–156 (2013)

14. Oliveira, F.P., Tavares, J.M.R.: Medical image registration: a review. Comput. Methods. Biomech. Biomed. Eng. **17**(2), 73–93 (2014)

15. Ronneberger, O., Fischer, P., Brox, T.: U-net: convolutional networks for biomedical image segmentation. In: Proceedings of International Conference on Medical Image Computing and Computer-Assisted Interventio, pp. 234–241 (2015)

16. Schulze, R., et al.: Artefacts in CBCT: a review. Dentomaxillofac. Radiol. **40**(5), 265–273 (2011)

17. Shieh, C.C., et al.: SPARE: sparse-view reconstruction challenge for 4D cone-beam CT from a 1-min scan. Med. Phys. **46**(9), 3799–3811 (2019)

18. Sonke, J.J., Zijp, L., Remeijer, P., van Herk, M.: Respiratory correlated cone beam CT. Med. Phys. **32**(4), 1176–1186 (2005)

19. Sotiras, A., Davatzikos, C., Paragios, N.: Deformable medical image registration: a survey. IEEE Trans. Biomed. Eng. **32**(7), 1153 (2013)

20. Sweeney, R.A., et al.: Accuracy and inter-observer variability of 3D versus 4D cone-beam CT based image-guidance in SBRT for lung tumors. Radiat. Oncol. **7**(1), 1–8 (2012)

21. de Vos, B.D., Berendsen, F.F., Viergever, M.A., Sokooti, H., Staring, M., Išgum, I.: A deep learning framework for unsupervised affine and deformable image registration. Med. Image Anal. **52**, 128–143 (2019)

22. Yang, X., Kwitt, R., Styner, M., Niethammer, M.: Quicksilver: fast predictive image registration-a deep learning approach. NeuroImage **158**, 378–396 (2017)

23. Zhi, S., Kachelrieß, M., Mou, X.: High-quality initial image-guided 4D CBCT reconstruction. Med. Phys. **47**(5), 2099–2115 (2020)

Unsupervised Diffeomorphic Surface Registration and Non-linear Modelling

Balder Croquet[1,2(✉)], Daan Christiaens[1,2], Seth M. Weinberg[4],
Michael Bronstein[5,6,7], Dirk Vandermeulen[1,2], and Peter Claes[1,2,3]

[1] Medical Imaging Research Center, UZ Leuven, Leuven, Belgium
`balder.croquet@kuleuven.be`
[2] Department of Electrical Engineering, ESAT/PSI, KU Leuven, Leuven, Belgium
[3] Department of Human Genetics, KU Leuven, Leuven, Belgium
[4] Department of Oral and Craniofacial Sciences, Center for Craniofacial and Dental
Genetics, University of Pittsburgh, Pittsburgh, USA
[5] Department of Computing, Imperial College London, London, UK
[6] IDSIA, USI Lugano, Lugano, Switzerland
[7] Twitter, London, UK

Abstract. Registration is an essential tool in image analysis. Deep learning based alternatives have recently become popular, achieving competitive performance at a faster speed. However, many contemporary techniques are limited to volumetric representations, despite increased popularity of 3D surface and shape data in medical image analysis. We propose a one-step registration model for 3D surfaces that internalises a lower dimensional probabilistic deformation model (PDM) using conditional variational autoencoders (CVAE). The deformations are constrained to be diffeomorphic using an exponentiation layer. The one-step registration model is benchmarked against iterative techniques, trading in a slightly lower performance in terms of shape fit for a higher compactness. We experiment with two distance metrics, Chamfer distance (CD) and Sinkhorn divergence (SD), as specific distance functions for surface data in real-world registration scenarios. The internalised deformation model is benchmarked against linear principal component analysis (PCA) achieving competitive results and improved generalisability from lower dimensions.

Keywords: Diffeomorphic registration · Geometric deep learning · Deformation modelling

1 Introduction

3D surface scans are an increasingly popular low-cost radiation-free alternative or addition to traditional medical imaging modalities. In this work we focus on 3D

Electronic supplementary material The online version of this chapter (https://doi.org/10.1007/978-3-030-87202-1_12) contains supplementary material, which is available to authorized users.

M. de Bruijne et al. (Eds.): MICCAI 2021, LNCS 12904, pp. 118–128, 2021.
https://doi.org/10.1007/978-3-030-87202-1_12

surface scans of faces, which are for instance used in the clinic for surgical planning and auditing of craniofacial reconstructive surgery [21], or in medical research to describe and understand the effects of teratogens on facial development [22]. Traditionally, analyses were executed using sparse landmarks, more recent studies make use of non-rigid registration to transfer thousands of corresponding landmarks from a template to a target [21]. MeshMonk [28] is an open-source toolbox that was developed and validated for phenotyping of faces. It consists of a scaled rigid alignment step, based on rigid Iterative closest point (ICP), followed by an iterative non-rigid registration step that makes use of a Visco elastic model. Its major limitation is that it is iterative, with a complexity that scales with the number of vertices. Furthermore, MeshMonk does not provide any theoretical guarantees on the deformation. Contrarily, diffeomorphic non-rigid registration, enforces the deformation to be smooth, differentiable, invertible and topology preserving [3], thus, providing extra robustness in registration. Furthermore, the metric induced by the diffeomorphism can be used as an alternative over point-to-point distances. State of the art iterative techniques include DARTEL [3], Diffeomorphic Demons [26] and LDDMM [5,7] on volumetric images, with LDDMM also being generalised to surfaces [25]. Deformetrica [6] is an open-source implementation of a specific instance of LDDMM that makes use of control points, it is applicable to many representations. The main limitation of Deformetrica is that it is iterative and thus still not usable in real-time.

Recently, the success of deep learning (DL) in image tasks have sparked interest in DL based volume registration. Supervised approaches such as [29], making use of generated ground truth data can result in bias, leading to the introduction of the first unsupervised approaches [10,19]. For surface data such as point clouds or meshes, the DL literature is far less expanded. Rigid registration of point clouds has been demonstrated in robotics [31]. Le Clerc and Sun [8] proposed a convolutional neural network for a vertex classification based mesh registration. Fu et al. [17] proposed a deep neural network that performs volumetric point-cloud registration of segmented prostates in a multi-modal setting using bio-mechanical constraints. Bahri et al. [4] proposed an encoder-decoder architecture capable of non-rigid registration of faces by translation of latent geometric information. FlowNet3D [20], predicts scene flow directly from unstructured point clouds, it is trained with synthetic ground truth data. There are also a few diffeomorphic surface registration models. ResNet-LDDMM [1] uses a deep residual neural network based on LDDMM to register 3D shapes. Dalca et al. [11], introduced a probabilistic model for diffeomorphic registration of volumes that has the option to incorporate a surface deformation model. However, to the best of our knowledge, a standalone deep learning based diffeomorphic model in conjunction with a lower-dimensional probabilistic deformation model has not yet been explored for surfaces.

In this work we introduce a geometric conditional variational autoencoder (CVAE) network [18] based on [11,19]. The network is able to perform a one-step 3D surface registration by deforming the ambient space while jointly learning a lower dimensional probabilistic deformation model (PDM). Because we deform the ambient space instead of the individual vertices, the model is independent of the resolution and representation of the input. Futhermore, when trained, it could

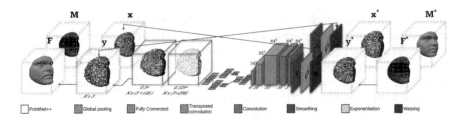

Fig. 1. Network architecture

be used for real-time patient assessment. The registration component is benchmarked against an iterative registration technique in a real-world scenario. The PDM serves as a learned active shape model, imposing a probabilistic prior on the deformations. The PDM is benchmarked against a traditional deformation model on a dataset that is in correspondence. The source code is publicly available.[1]

2   Methods

2.1   Registration Model

Assuming that the moving shape M and fixed shape F are embedded in $\Omega = \mathbb{R}^d$, shape registration involves finding a deformation function φ that can be directly applied such that the two shapes maximally overlap $F \approx \varphi \circ M$. We choose to work with a mapping $\varphi : \Omega \to \Omega$ that is a diffeomorphism of the ambient space. We used the computationally more efficient stationary velocity field (SVF) parametrisation of diffeomorphisms instead of the variable velocity fields or shooting formulations seen in LDDMM. The SVF preserves the same robustness guarantees as its alternatives [3] and is formally written as an initial value problem of the ordinary differential equation;

$$\frac{d\phi_t}{dt} = \mathbf{v}(\phi_t), \text{ with } \phi_0 = Id. \tag{1}$$

This is solved over unit time using a numerical integrator, resulting in the final deformation $\varphi = \phi_1^v$. Mathematically, the transformation φ becomes the Lie group exponential map parametrised by v, $\varphi = Exp(v)$. In this work, we used an Euler integration with scaling and squaring similar to [2].

2.2   CVAE Network

We propose a CVAE [18] that internalises a PDM, while being able to perform a one-step warp of the moving template surface to a fixed target surface. As input for the network, we used independently and randomly sampled point clouds x and y of the affine aligned moving (M) and fixed (F) mesh respectively, where

---

[1] https://gitlab.kuleuven.be/u0132345/deepdiffeomorphicfaceregistration.

$x, y \in \mathbb{R}^{N \times d}$. Point clouds for every surface were resampled at every epoch during training to improve robustness. A schematic representation of the model architecture is shown in Fig. 1. The encoder predicts a probabilistic parametrisation for the deformation of x given y. Given the fixed point cloud, the parametric encoder $g_\omega(\mu, \sigma; y)$ predicts the mean $\mu \in \mathbb{R}^Z$ and covariance $\sigma \in \mathbb{R}^Z$ of the posterior normal distribution $q_\omega(z|y) = \mathcal{N}(\mu, \sigma)$. Given the approximate posterior, an encoding $z \in \mathbb{R}^Z$ is sampled for the input. To structure the latent space, we made the approximate posterior distribution approach a prior distribution $p(z) = \mathcal{N}(0, I)$ using the Kullback-Leibler (KL) divergence. The decoder reconstructs point cloud y by warping the conditional point cloud x according to the predicted deformation parametrised by z, this results in the distribution $p_\gamma(y|x, z)$. The decoder is comprised of four blocks. First, the latent representation is decoded as a vector field. Second, the vector field is smoothed using a Gaussian smoothing layer, resulting in a velocity field $v \in \mathbb{R}^{V^d \times d}$. Third, the velocity field is integrated as explained in Sect. 2.1 using an exponentiation layer [10,19], resulting in the final deformation function $\phi_1^v = Exp(v)$. Last, the moving and fixed point clouds are warped using the warp layer $y^* = \phi_1^v \circ x$. Because we deform the ambient space instead of individual points, we can apply this same mapping to the vertices of the surface M during inference and it allows us to choose the size of point clouds x and y independent of the mesh resolution.

2.3   Objective Function

The network parameters $\theta = (\omega, \gamma)$ were trained end-to-end, minimising the variational lower bound; $-\mathbb{E}_{z \sim q_\omega(z|y)} \log p_\gamma(y|x, z) + KL(q_\omega(z|y)||p(z))$, where the first term is the reconstruction log-likelihood between the sampled points and the second term a divergence between the approximate posterior and prior. KL divergence can be computed in closed form [18] and minimising the negative log-likelihood is equivalent to minimising the mean squared error (MSE):

$$\mathcal{L}_{MSE}(x, y) = \frac{1}{N} \sum_{i=1}^{N} \|x_i - y_i\|_2^2. \tag{2}$$

However, the MSE can only be computed if correspondence between shapes is known, which is rarely the case in real-world scenarios. Therefore, we experimented with two alternative loss functions. In a first instance, the Chamfer distance (CD) takes the closest point instead of the corresponding point:

$$\mathcal{L}_{CD}(x, y) = \frac{1}{N} \sum_{i=1}^{N} \min_{c \in y} \|x_i - c\|_2^2 + \frac{1}{N} \sum_{j=i}^{N} \min_{c \in x} \|c - y_i\|_2^2. \tag{3}$$

As explained in [14], the nearest neighbour projections of a CD tend to result in low quality gradients. Therefore, in a second and alternative instance, we also looked at optimal transport theory. Intuitively, a Wasserstein distance attempts to find the transport plan π that orders the points. We approximated the Wasserstein distance using the debiased Sinkhorn divergence (SD) [9]

$$\mathcal{L}_{SD}(\alpha, \beta, \epsilon) = OT_\epsilon(\alpha, \beta) - \frac{1}{2} OT_\epsilon(\alpha, \alpha) - \frac{1}{2} OT_\epsilon(\beta, \beta), \tag{4}$$

where OT_ϵ the optimal transport distance with an entropic regularisation term.

$$OT_\epsilon(\alpha, \beta) = \min_{\pi \in \Pi} \sum_{i=1}^{N} \sum_{j=1}^{N} \pi_{i,j} \frac{1}{p} \|x_i - y_j\|_p + \epsilon KL(\pi \| \alpha \otimes \beta), \qquad (5)$$

subject to $\pi > 0, \pi 1 = \alpha, \pi^T 1 = \beta$, given π the regularised transport plan, 1 vector of ones, and, $\alpha = \sum_{i=1}^{N} \alpha_i \delta_{x_i}$, $\beta = \sum_{j=1}^{M} \beta_j \delta_{y_j}$ probability measures. The SD approaches the Wasserstein distance as $\epsilon \to 0$. In this work, we used $\epsilon = 10^{-4}$, $p = 1$ and the geometric loss python package [16]. The SD was first applied to diffeomorphic registration in [15].

Experimentally we have found that these constraints were not enough, resulting in unnecessarily complicated deformation fields and loss of correspondence. To overcome these problems, we introduced two additional regularisation terms. The first term encourages minimal deformation, the second term encourages points of the deformed shape and the original shape to stay close together, preventing drift of the landmarks:

$$\mathcal{R}_{smooth}(v) = \frac{1}{3V^3} \sum_{i=1}^{3V^3} (-\alpha \nabla^2 v + \gamma v)^2, \quad \mathcal{R}_{vert}(x, x^*) = \frac{1}{N} \sum_{i=1}^{N} \|x_i - x_i^*\|_1. \quad (6)$$

with ∇^2 the Laplacian operator approximated using central finite differences, $\alpha = 10^{-6}$ and $\gamma = 1$. The final objective function is defined as:

$$\mathcal{O}(x, y, z, v) = \frac{\lambda_1}{2} (\mathcal{L}(y, \phi_1^v \circ x) + \mathcal{L}(x, \phi_0^{-v} \circ y)) + \lambda_2 \mathcal{R}_{smooth}(v)$$
$$+ \frac{\lambda_3}{2} (\mathcal{R}_{vert}(x, \phi_1^v \circ x) + \mathcal{R}_{vert}(y, \phi_0^{-v} \circ y)) + \frac{1}{Z} KL(q_\omega(z|F) \| p(z)). \tag{7}$$

We found $\lambda_2 = 10$ and $\lambda_3 = 100$ to work best. λ_1 varies per loss function; 4×10^4, 8×10^4 and 5×10^3 for MSE, CD and SD respectively. Once the order of magnitude was found for these parameters we did not experience major differences in performance when adjusting the factor.

3   Experiments and Results

As dataset we used "3D Facial Norms" [27], which contains 3D facial surface scans of 2454 individuals (952M/1502F), the download is available through Face-Base [30], a controlled-access data respository managed by the U.S. National Institutes of Health. We excluded 47 unsuitable scans due to imaging artifacts. The other images were cleaned by manually removing the clothing and hair, pose normalised with a 3D facial template based on 5 manually indicated landmarks followed by a rigid ICP procedure as implemented in the open-source toolbox MeshMonk [28], and isotropically scaled to fit within a centered cube with side $= 2$. The dataset was split in a training, validation and test set of sizes 2007, 200 and 200 respectively.

We designed two experiments to examine different properties of the network in Fig. 1, the same split was used in every experiment. In the first experiment (Subsect. 3.2), the model was evaluated as a registration algorithm on the dataset without correspondence. As an additional preprocessing, we algoritmically cropped the meshes (removing e.g. extended neck or forehead) by finding the nearest vertex on M (template) for every vertex on F (facial surface) and removing the vertices that mapped to the border of M. We experimented with two loss functions (CD and SD) to cope with the lack of correspondence. In the second experiment (Subsect. 3.3), the model was evaluated as a PDM with the data brought into correspondence using MeshMonk [28]. This allowed us to exclude uncertainty of the loss function.

3.1 Implementation Details

We worked on $d = 3$ dimensional space, with point clouds of $N = 5000$ samples. When spatial correspondence was known, the point cloud was obtained by uniformly sampling the same vertices from moving and fixed shape; when it was not known, point cloud sampling was performed uniformly taking mesh triangle area into account. The encoder g_ω consisted of two PointNet++ [23] blocks as implemented in [13]. PointNet++ layers consist of a sampling grouping and PointNet layer and allow for hierarchical learning on point clouds [23]. During sampling, we iteratively sampled a ratio of 0.5 and 0.25 of the furthest points for the first and second layer respectively. Grouping was performed in a ball query with a radius of 0.2 and 0.4 and a maximum of 64 neighbours. For the PointNet layer, we used 3 linear blocks, consisting of a linear, ReLU and batch normalisation layer of $3 \rightarrow 64 \rightarrow 64 \rightarrow 128$ channels in the first layer and $(3 + 128) \rightarrow 128 \rightarrow 128 \rightarrow 256$ channels in the second layer. The learned hierarchical representations are shown together with the model architecture in Fig. 1. The features and positions are concatenated and fed in 3 similar linear blocks of $(256 + 3) \rightarrow 256 \rightarrow 512 \rightarrow 1024$ channels. This was followed by a global pooling layer, taking the maximum across the node dimensions, resulting in a 1024 dimensional feature vector. The bottleneck consisted of two pairs of two fully connected layers with an ELU and linear activation function resulting in σ, μ. Due to memory constraints only a single latent encoding z of predetermined size Z was sampled per epoch from q_ω. The latent encoding was decoded using two fully connected layers, two transposed convolution layers (4 kernel size, 1 padding and 2 stride), and two convolution layers (1 kernel size, 0 padding and 1 stride). Each layer was followed by an ELU activation function, the transposed convolutions were followed by a batch normalisation layer. The Gaussian smoothing layer uses a kernel of size 15 and standard deviation of 4, the predicted vector field consisted of $V = 64$ 3-dimensional vectors in every dimension. The exponentiation layer made use of $T = 7$ scaling and squaring steps. For the warp layer, we used trilinear interpolation to interpolate the deformation of every point from the deformation field. All CVAE models were trained for 300 epochs with batch size 7 using Adam optimizer with a learning rate of 10^{-5}, the model using SD was trained with a batch size 3 due to memory constraints.

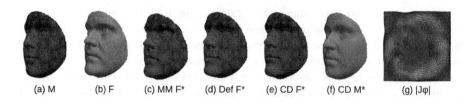

(a) M   (b) F   (c) MM F*   (d) Def F*   (e) CD F*   (f) CD M*   (g) |Jφ|

Fig. 2. Registration example on unseen data sample: a, b) moving shape, fixed shape; c) registration result of baseline MeshMonk (MM); d) registration result of Deformetrica (Def); e, f) registration result and inverse of CVAE using Chamfer distance (CD); g) slice 32 of deformation field color-coded by the Jacobian determinant (values range between 0.01 and 3.4). (Color figure online)

3.2   Validation of Registration

In the first experiment, we evaluated the registration performance of the network using a latent size of $Z = 32$. We compared the performances of the proposed model with CD (Eq. 3) and SD (Eq. 4) as distance function to MeshMonk (MM) [28] and Deformetrica (Def) [6] as baseline. We used recommended parameters for both, in Def we used a varifold distance with 10^{-6} noise-std. A registration result of MM, Def and CD are displayed together in Fig. 2. The performance was quantified as shape fit and model compactness. The shape fit is the RMSE between M^* and F. Because correspondence in this experiment was not known, we used the average squared error over the 3-nearest vertices as distance between vertices. Figure 3a shows a median RMSE by MeshMonk of 0.92, better than 1.18 for Def, 1.13 for CD and 1.14 for SD. For the Diffeomorphic techniques, the distributions of errors in shape fit tend to be closer to the median, this could be a direct consequence of the Diffeomorphic constraints. However, for the CVAE it could also be due to the internal PDM, which acts as regularisation. This same constraints could also explain the slightly lower accuracy in terms of shape fit. To validate the indication of corresponding points we looked at model compactness. In the context of surface registration, a more compact shape model typically originates from an improved (more consistent) indication of corresponding points [12]. This was quantified as the cumulative percentage of explained variance of a PCA point-distribution model on the correspondences outputted by the registration pipeline. Figure 3b shows a higher compactness with an AUC of 195.29 and 195.48 for CD and SD respectively, compared to 192.01 for MM and 194.42 for Def. This potentially indicates an improved quality of correspondences as obtained by the proposed models. However, one should always take into account the trade-off between model compactness and goodness-of-fit (and its dependence on hyperparameter choices). If the CVAE results in shapes that are closer to the template we would indeed expect a lower model complexity. Finally, we did not observe significant differences between SD and CD; this is probably due to faces already being very similar in shape. Inference took on average 0.1s on NVIDIA GeForce RTX 2080 Ti.

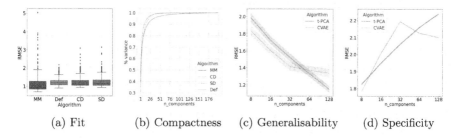

| (a) Fit | (b) Compactness | (c) Generalisability | (d) Specificity |

Fig. 3. a,b) Registration performance: shape fit and compactness of baselines Mesh-Monk (MM) and Deformetrica (Def), and, CVAE trained with Chamfer distance (CD) and Sinkhorn divergence (SD); c,d) PDM performance: generalisability and specificity for CVAE and t-PCA on data in correspondence.

3.3 Validation of Internal Probabilistic Deformation Model

In the second experiment, we additionally evaluated the performance of the proposed model as a non-linear PDM given spatially dense correspondences. Having correspondence allowed us to use exact point-to-point distances, excluding the uncertainty of shape fit, focusing solely on the degrees of freedom offered by the PDM and regularisation terms. We used the MSE (Eq. 2) as distance metric. As baseline we used iterative optimisation to find the training set SVFs

$$v = \arg\max_{v} \frac{\lambda_1}{2} (\mathcal{L}_{MSE}(y, \phi_1^v \circ x) + \mathcal{L}_{MSE}(x, \phi_0^{-v} \circ y)) + \lambda_2 \mathcal{R}_{smooth}(v), \quad (8)$$

followed by incremental PCA [24] on the velocity fields (incremental PCA has better memory scaling than traditional PCA), this is referred to as tangent-PCA (t-PCA). Similar to the CVAE, the velocity fields were also smoothed using the same Gaussian kernel. The first 5 components for both t-PCA and the CVAE are visualised in supplementary material A, Fig. 1. The performance of the deformation model was quantified as a trade-off between generalisability and specificity in terms of the shape it produced. In statistical shape analysis, generalisability is the ability of a model to fit to unseen data, measuring the RMSE between M^* and F on the test set. Figure 3c shows a better generalisability for the CVAE using a lower number of components ≤ 32 and an earlier plateauing compared to the t-PCA baseline. Specificity is the ability to generate shapes that are similar to the training set, measured by sampling the latent space according to its distribution and averaging the RMSE of the generated example to the 3-nearest neighbour shapes in the training set. Here we used 10,000 generated shapes. Figure 3d shows an upward trend as the model gets more freedom to express the entire deformation space. A drop in specificity for the proposed model is observed where the number of components > 32. This could be due to $\mathcal{R}_{vert}$ over-constraining the model, however, more experimentation is required.

4   Discussion and Conclusion

In this work, we proposed an unsupervised end-to-end one-step surface registration technique that jointly learns a probabilistic deformation model. The internalised PDM is competitive to linear PCA and registration results are competitive to iterative non-rigid registration algorithms that are widely used in the field. Furthermore because this technique deforms the ambient space and we work with a latent encoding the technique can easily be generalised to other image representations, showing potential for real time hybrid shape and deformation analysis software. A limitation of this work is that the parameters have been specifically fine tuned for face registration, applying it to a different anatomical structure would require retraining. Another limitation is the lack of ground truth data, which makes evaluation challenging. A last limitation is that this network is restricted to a single moving template shape. Future work involves replacing the vector field by control points, incorporating surface to volume registration and performing statistics on the predicted deformations. In conclusion, we propose a registration algorithm that, in contrast to traditional techniques, does not require iterative optimisation; is guaranteed to be diffeomorphic; works on any mesh resolution; and internalises deformation constraints. The performance of the proposed model is competitive to traditional techniques and can easily be integrated in other pipelines.

Acknowledgements. This project was funded by the Research Foundation in Flanders (FWO, Fonds Wetenschappelijk Onderzoek) Phd Fellowship (1SB0121N), postdoctoral fellowship (12ZV420N) and research project (G078518N), by the Research Fund KU Leuven (BOF-C1, C14/15/081 and C14/20/081) and by ERC Consolidator grant No. 724228 (LEMAN).

References

1. Amor, B.B., Arguillère, S., Shao, L.: ResNet-LDDMM: advancing the LDDMM framework using deep residual networks. arXiv preprint arXiv:2102.07951 (2021)
2. Arsigny, V., Commowick, O., Pennec, X., Ayache, N.: A log-Euclidean framework for statistics on diffeomorphisms. In: Larsen, R., Nielsen, M., Sporring, J. (eds.) MICCAI 2006. LNCS, vol. 4190, pp. 924–931. Springer, Heidelberg (2006). https://doi.org/10.1007/11866565_113
3. Ashburner, J.: A fast diffeomorphic image registration algorithm. Neuroimage **38**(1), 95–113 (2007)
4. Bahri, M., et al.: Shape my face: registering 3D face scans by surface-to-surface translation. arXiv preprint arXiv:2012.09235 (2020)
5. Beg, M.F., Miller, M.I., Trouvé, A., Younes, L.: Computing large deformation metric mappings via geodesic flows of diffeomorphisms. Int. J. Comput. Vis. **61**(2), 139–157 (2005)
6. Bône, A., Louis, M., Martin, B., Durrleman, S.: Deformetrica 4: an open-source software for statistical shape analysis. In: Reuter, M., Wachinger, C., Lombaert, H., Paniagua, B., Lüthi, M., Egger, B. (eds.) ShapeMI 2018. LNCS, vol. 11167, pp. 3–13. Springer, Cham (2018). https://doi.org/10.1007/978-3-030-04747-4_1

7. Brunn, M., Himthani, N., Biros, G., Mehl, M., Mang, A.: Fast GPU 3D diffeomorphic image registration. J. Parallel Distrib. Comput. **149**, 149–162 (2021)
8. Clerc, F., Sun, H.: Memory-friendly deep mesh registration. J. WSCG **28**, 1–10 (2020). https://doi.org/10.24132/CSRN.2020.3001.1
9. Cuturi, M.: Sinkhorn distances: lightspeed computation of optimal transport. In: Advances in Neural Information Processing Systems, vol. 26, pp. 2292–2300 (2013)
10. Dalca, A.V., Balakrishnan, G., Guttag, J., Sabuncu, M.R.: Unsupervised learning for fast probabilistic diffeomorphic registration. In: Frangi, A., Schnabel, J., Davatzikos, C., Alberola-López, C., Fichtinger, G. (eds.) MICCAI 2018. LNCS, vol. 11070, pp. 729–738. Springer, Cham (2018). https://doi.org/10.1007/978-3-030-00928-1_82
11. Dalca, A.V., Balakrishnan, G., Guttag, J., Sabuncu, M.R.: Unsupervised learning of probabilistic diffeomorphic registration for images and surfaces. Med. Image Anal. **57**, 226–236 (2019)
12. Davies, R.H., Twining, C.J., Cootes, T.F., Waterton, J.C., Taylor, C.J.: A minimum description length approach to statistical shape modeling. IEEE Trans. Med. Imaging **21**(5), 525–537 (2002)
13. Fey, M., Lenssen, J.E.: Fast graph representation learning with PyTorch Geometric. In: ICLR Workshop on Representation Learning on Graphs and Manifolds (2019)
14. Feydy, J.: Geometric data analysis, beyond convolutions. Ph.D. thesis, Université Paris-Saclay (2020)
15. Feydy, J., Charlier, B., Vialard, F.X., Peyré, G.: Optimal transport for diffeomorphic registration. In: Descoteaux, M., Maier-Hein, L., Franz, A., Jannin, P., Collins, D., Duchesne, S. (eds.) MICCAI 2017. LNCS, vol. 10433, pp. 291–299. Springer, Cham (2017). https://doi.org/10.1007/978-3-319-66182-7_34
16. Feydy, J., Séjourné, T., Vialard, F.X., Amari, S.i., Trouve, A., Peyré, G.: Interpolating between optimal transport and mmd using sinkhorn divergences. In: The 22nd International Conference on Artificial Intelligence and Statistics, pp. 2681–2690 (2019)
17. Fu, Y., et al.: Deformable MR-CBCT prostate registration using biomechanically constrained deep learning networks. Med. Phys. (2020)
18. Kingma, D.P., Mohamed, S., Jimenez Rezende, D., Welling, M.: Semi-supervised learning with deep generative models. In: Advances in Neural Information Processing Systems, vol. 27, pp. 3581–3589 (2014)
19. Krebs, J., Mansi, T., Mailhé, B., Ayache, N., Delingette, H.: Unsupervised probabilistic deformation modeling for robust diffeomorphic registration. In: Stoyanov, D., et al. (eds.) DLMIA 2018, ML-CDS 2018. LNCS, vol. 11045, pp. 101–109. Springer, Cham (2018). https://doi.org/10.1007/978-3-030-00889-5_12
20. Liu, X., Qi, C.R., Guibas, L.J.: Flownet3D: learning scene flow in 3D point clouds. In: Proceedings of the IEEE/CVF Conference on Computer Vision and Pattern Recognition, pp. 529–537 (2019)
21. Matthews, H.S., et al.: Pitfalls and promise of 3-dimensional image comparison for craniofacial surgical assessment. Plastic Reconstr. Surg. Glob. Open **8**(5) (2020)
22. Muggli, E., et al.: Association between prenatal alcohol exposure and craniofacial shape of children at 12 months of age. JAMA Pediatr. **171**(8), 771–780 (2017)
23. Qi, C.R., Yi, L., Su, H., Guibas, L.J.: Pointnet++: deep hierarchical feature learning on point sets in a metric space. In: Advances in Neural Information Processing Systems, vol. 30, pp. 5099–5108 (2017)
24. Ross, D.A., Lim, J., Lin, R.S., Yang, M.H.: Incremental learning for robust visual tracking. Int. J. Comput. Vis. **77**(1–3), 125–141 (2008)

25. Vaillant, M., Glaunes, J.: Surface matching via currents. In: Christensen, G.E., Sonka, M. (eds.) IPMI 2005. LNCS, vol. 3565, pp. 381–392. Springer, Heidelberg (2005). https://doi.org/10.1007/11505730_32

26. Vercauteren, T., Pennec, X., Perchant, A., Ayache, N.: Diffeomorphic demons: efficient non-parametric image registration. NeuroImage **45**(1), S61–S72 (2009)

27. Weinberg, S.M., et al.: The 3D facial norms database: Part 1. a web-based craniofacial anthropometric and image repository for the clinical and research community. Cleft Palate-Craniofacial J. **53**(6), 185–197 (2016)

28. White, J.D., et al.: Meshmonk: Open-source large-scale intensive 3d phenotyping. Sci. Rep. **9**(1), 1–11 (2019)

29. Yang, X., Kwitt, R., Niethammer, M.: Fast predictive image registration. In: Carneiro, G., et al. (eds.) DLMIA 2016, LABELS 2016. LNCS, vol. 10008, pp. 48–57. Springer, Cham (2016). https://doi.org/10.1007/978-3-319-46976-8_6

30. Zachary, R., Mary, L., Marazita, S.W.: 3D facial norms. FaceBase Consortium (2015). https://doi.org/10.25550/VWP

31. Zhang, Z., Dai, Y., Sun, J.: Deep learning based point cloud registration: an overview. Virtual Real. Intell. Hardw. **2**(3), 222–246 (2020)

Learning Dual Transformer Network for Diffeomorphic Registration

Yungeng Zhang⦿, Yuru Pei$^{(\boxtimes)}$⦿, and Hongbin Zha

Department of Machine Intelligence, Key Laboratory of Machine Perception (MOE),
Peking University, Beijing, China
peiyuru@cis.pku.edu.cn

Abstract. Diffeomorphic registration is widely used in medical image processing with the invertible and one-to-one mapping between images. Recent progress has been made to diffeomorphic registration by utilizing a convolutional neural network for efficient and end-to-end inference of registration fields from an image pair. However, existing deep learning-based registration models neglect to employ attention mechanisms to handle the long-range cross-image relevance in embedding learning, limiting such approaches to identify the semantically meaningful correspondence of anatomical structures. In this paper, we propose a novel dual transformer network (DTN) for diffeomorphic registration, consisting of a learnable volumetric embedding module, a dual cross-image relevance learning module for feature enhancement, and a registration field inference module. The self-attention mechanisms of DTN explicitly model both the inter- and intra-image relevances in the embedding from both the separate and concatenated volumetric images, facilitating semantical correspondence of anatomical structures in diffeomorphic registration. Extensive quantitative and qualitative evaluations demonstrate that the DTN performs favorably against state-of-the-art methods.

Keywords: Dual transformer · Diffeomorphic registration · Relevance learning

1 Introduction

Deformable registration is a fundamental and challenging problem in medical image processing, with the goal to find dense per-voxel displacement and establish alignment between a pair of images. The study of deformable registration has a variety of potential applications, especially in the analysis of multi-modal images captured from different subjects or in a longitudinal treatment with structure variations due to treatments and growths. The deformable registration produces the nonlinear voxel-wise mapping between images, facilitating the atlas-based annotation, statistical shape analysis, and shape comparison of anatomical structures. In order to accomplish the task of deformable registration effectively,

© Springer Nature Switzerland AG 2021
M. de Bruijne et al. (Eds.): MICCAI 2021, LNCS 12904, pp. 129–138, 2021.
https://doi.org/10.1007/978-3-030-87202-1_13

we need to infer the semantic correspondence of fine-grained structures. The volumetric images vary in shapes, scales, and poses, so that it is a challenging issue to identify the real matching anatomical structures.

Traditional deformable registration is known to be computationally expensive due to iterative optimization of large-scale parameters [21]. Recently, the state-of-the-art convolutional neural network (CNN)-based methods has been proposed to address the deformable registration [4,9,12,18,26,27]. The CNN performs the end-to-end inference of the displacement or velocity fields from a pair of images, using regularization, such as the smoothness and the Jacobian determinant [14,18,26], for the invertible and the diffeomorphic transformations. Moreover, the symmetric registration infers a pair of diffeomorphic maps regarding the middle of the geodesic path [18]. However, these methods conduct a straightforward inference from the CNN-based low-level local embedding with varying scales of contexts, without addressing the global relevance of the image pair. Thus, the resultant alignment may suffer implausible voxel-wise mapping, where the prior affine transformation and landmark annotation are required to circumvent the trap of local minima.

Variants of transformers have gained great success in a group of tasks in natural language processing (NLP), including cross-language translation [25] and the question-answering [22]. Recently, the transformer has been extended to computer vision community, such as object detection [5], image recognition [10], and segmentation [7,23]. The transformer facilitates the global embedding of images by the relevance modeling of image words. Attention was utilized in various image processing tasks by highlighting salient feature regions and suppressing irrelevant ones [20]. Liao et al. [15] utilized an attention-driven hierarchical strategy and a greedy supervised approach in rigid CT registration. An auto-attention mechanism was introduced to multiple regions for reliable visual cues in the registration of X-ray and CT images [17]. Nevertheless, such attention schemes addressed the long-range dependencies of a single image or the rigid transformation, which can not effectively handle the cross-image semantic correspondence and deformable registration.

In response to these difficulties, we propose a dual transformer network (DTN) for diffeomorphic registration. The proposed approach exploits the self-attention scheme to model the inter- and intra-image global contextual relevances explicitly. The dual transformer conducts the relevance modeling and the feature enhancement on two kinds of image embedding for semantically meaningful correspondences of anatomical structures. The DTN consists of a learnable image embedding module, a cross-image relevance learning module, and a registration field inference module. The combinational embedding, taking the strength of both the low-level spatial features and the high-level contextual relevance-based enhancements, is used to predict the registration fields. One difficulty in unsupervised deformable registration is to identify the semantic correspondence between anatomical structures. The proposed DTN addresses the cross-image and global relevance to improve the discriminative power of image embedding for voxel-wise correspondence. We evaluate the proposed approach on the clinically obtained brain MRI scans of the OASIS dataset [16] qualitatively and quantitatively. The atlas-based registration and segmentation demonstrate our model achieves performance improvements over the compared deep-learning-based methods. The main contributions of this work are as follows:

- We devise a novel dual transformer for volumetric diffeomorphic registration, facilitating the semantically meaningful correspondence of anatomical structures.
- We conduct the volume embedding enhancement for velocity field inference, taking advantage of both the CNN-based local features and the attention-based global and cross-image relevances.
- The proposed DTN has gained success in the diffeomorphic registration and atlas-based segmentation of multi-category anatomical structures.

2   Proposed Method

As shown in Fig. 1, the DTN can be stacked on an existing encoder-decoder network of the 3D U-net [8]. We present the dual transformer to take the strengths of both the CNN-based local features and the cross-image global contextual relevance-based enhancements for volumetric image embedding. Given the input moving and fixed volumetric images, $V_m, V_f \in \mathbb{R}^{h_0 \times w_0 \times d_0}$, the goal is to estimate the diffeomorphic registration field $\psi \in \mathbb{R}^{3 h_0 \times w_0 \times d_0}$ for the one-to-one map and the atlas-based registration. The DTN has two branches to address the relevance learning on the volumetric embedding of separate one-channel images and the two-channel image concatenation. First, the DTN extracts the CNN-based low-level image embedding of both separate and the concatenated images. Second, The image embeddings are collapsed into sequences, which are fed to the dual

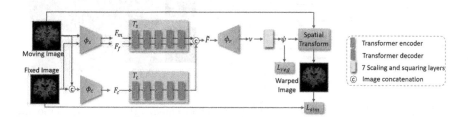

Fig. 1. Diffeomorphic registration by the proposed DTN. The framework is stacked on an existing CNN-based encoder-decoder network, such as 3D U-net, for volumetric image embedding and velocity field inference. The DTN consists of dual transformers, T_s and T_c, to handle the cross-image global relevance learning on separate single-channel images and the image concatenation. The proposed DTN takes advantage of both the CNN-based low-level local features and the attention-based global relevances for diffeomorphic registration. (The skip connections of the 3D U-net are removed for the sake of the clarity.)

transformer for global relevance-based feature enhancement. Finally, the resultant features from two branches are concatenated together to infer the velocity field $\nu \in \mathbb{R}^{3h_0 \times w_0 \times d_0}$ and the registration field ψ.

2.1   Volumetric Image Embedding

The DTN utilizes the encoder of 3D U-net for embedding of separate and concatenated volumetric images. The first branch takes single-channel images as the input, and outputs features $F_m, F_f \in \mathbb{R}^{q_s \times h \times w \times d}$, and $F_{\{m,f\}} = \phi_s(V_{\{m,f\}}, \Theta_s)$. q_s denotes the channel number and Θ_s network parameters. Aside from the embedding on single-channel volumetric image, we estimate the embedding of the image concatenation $[V_m \; V_f]$. The resultant q_c-channel feature $F_c \in \mathbb{R}^{q_c \times h \times w \times d}$, and $F_c = \phi_c([V_m \; V_f], \Theta_c)$. Θ_c denotes learnable network parameters.

2.2   Dual Transformer

We present a dual transformer to model the cross-volume dependencies to enhance the volumetric embedding, as shown in Fig. 1. We utilize an encoder-decoder transformer [6,24], consisting of concatenated three encoders and two decoders, to model the inter- and intra-volume relevance. The transformer encoder requires a sequence input, where the spatial dimensions of the features are collapsed into a vector with the resultant feature $\hat{F}_{\{m,f\}} \in \mathbb{R}^{q_s \times hwd}$ and $\hat{F}_c \in \mathbb{R}^{q_c \times hwd}$. A learnable position encoding is added to the feature sequence to retain the positional information as [24] (Fig. 2(a)).

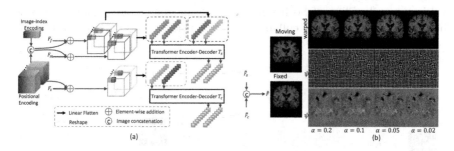

Fig. 2. (a) The dual transformer-based feature enhancements. (b) The warped moving images with varying α. The registration fields ψ are visualized in grid and color images. Red points on warped images indicate voxels with $|J(\psi)| \le 0$. (Color figure online)

The first branch addresses the inter and intra-image relevances on the separate image embedding. Transformer T_s handles the relevance of sequences from both the fixed and moving images, i.e., $\hat{F}_f$ and $\hat{F}_m$. T_s takes $(\hat{F}_m, \hat{F}_f)$ and the positional encoding as the input and outputs $\tilde{F}_s \in \mathbb{R}^{q_s \times hwd}$. Here we use the image index encoding (Fig. 2 (a)) to indicate the fixed or moving image. Since the queries and keys are defined as the embedding sequences from $\hat{F}_f$ and $\hat{F}_m$, T_s models both the intra- and inter-image dependencies. The resultant feature embedding $\tilde{F}_s$ enhances the convolutional features with increasing receptive fields and the cross-image global relevance.

In the second branch, T_c utilizes the self-attention scheme to model the global dependencies of the concatenated volumetric embedding F_c. The input of T_c is $\hat{F}_c$ and the positional encoding. The resultant $\tilde{F}_c \in \mathbb{R}^{q_c \times hwd}$. Since the entangled embedding F_c is computed from the image pair, T_c also addresses the inter-volume feature enhancements. The final enhanced feature embedding $\tilde{F}$ is computed as a concatenation of features from both T_s and T_c, and $\tilde{F} = \tilde{F}_s \copyright \tilde{F}_c$. $\copyright$ denotes the image concatenation operator. We noticed the U-Net Transformer [19] had transformer modules in the skip connections for segmentation. The multiple transformers are promising to enhance features further, though enlarging the memory and computational complexity. We implemented the self-attention learning on the voxel sequence at the bottom of the U-net.

2.3 Diffeomorphic Registration

Since a diffeomorphism has the differentiable and invertible properties theoretically, the one-to-one mapping and the topology preservation are guaranteed in the diffeomorphic registration. We conduct the diffeomorphic registration and utilize the stationary velocity field $\nu_t, t \in [0, 1]$, which satisfies

$$d\psi_t/dt = \nu_t(\psi_t) = \nu_t \circ \psi_t. \tag{1}$$

The diffeomorphic deformation field $\psi^{(0)} = I$ is an identity transformation. The deformation field represented as a member of Lie algebra can be computed as the exponential of the velocity field [1]. The exponentiated velocity field $\psi^{(1)} = \exp(\nu)$ guarantees the mapping between images to be invertible. Given the enhanced feature $\tilde{F}$, the diffeomorphic registration decoder is used to infer the invertible deformation fields ψ. The decoder network parameterizes a nonlinear mapping function $\phi_r(\tilde{F}, \Theta_r) = \psi$, as a combination of a CNN-based decoder and scaling and squaring layers (Fig. 1). Θ_r denotes the parameters of the registration inference module.

2.4 Unsupervised Learning

The proposed DTN is optimized in an unsupervised manner by the metric space alignment. Given image pair (V_m, V_f), the DTN estimates registration field ψ. The spatial transformer [13] is used to warp the moving image. The resultant warped moving image $V'_m = V_m \circ \psi$. The l_1-norm-based image similarity loss $L_{sim} = \|V'_m - V_f\|_1$. We apply the Frobenius norm-based smoothness regularizer on the velocity field, and $L_{reg} = \|\nabla \nu\|_F^2$. The loss function is defined as follows:

$$L = L_{sim} + \alpha L_{reg}, \tag{2}$$

where the hyperparameter α is used to balance the image similarity and the smoothness regularization on the velocity field. We optimize the parameters of the proposed DTN by minimizing the loss function.

3 Experiments

Dataset and Metric. We used the publicly available dataset of the OASIS with 425 T1-weighted brain MRI scans [16]. The MRI scans are re-sampled to a resolution of $256 \times 256 \times 256$ with an isotropic voxel size of 1 mm $\times$ 1 mm $\times$ 1 mm, and then cropped to $144 \times 160 \times 192$. We conducted the standard preprocessing for affine transformation and brain structure extraction using FreeSurfer [11] as [18]. The OASIS dataset provides brain segmentation with visual inspections, which are viewed as the ground truth in the evaluation process. The dataset was split into 256 and 150 scans for training and testing. The remaining 19 scans have been used for validation.

We evaluate the proposed approach using the Dice similarity coefficients (DSC), which measures the consistency between the ground truth segmentations and those estimated by the atlas-based registration. The negative Jacobian determinant $|J(\psi)| \leq 0$ is used to evaluate the registration fields. We compute the derivatives of the volumetric registration fields for the negative Jacobian determinant, which is related to structural folding without topology preservation and the violation of the diffeomorphic property. Generally, the low value of negative Jacobian determinant and the high value of the DSC suggest reliable diffeomorphic registration fields.

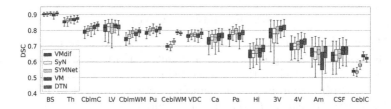

Fig. 3. Boxplots of DSCs on 16 anatomical structures, including brain stem (BS), thalamus (Th), cerebellum cortex (CblmC), lateral ventricle (LV), cerebellum white matter (CblmWM), putamen (Pu), cerebral white matter (CeblWM), ventral DC (VDC), caudate (Ca), pallidum (Pa), hippocampus (Hi), the 3rd ventricle (3V), the 4th ventricle (4V), amygdala (Am), CSF, and cerebral cortex (CeblC), by the SyN [2], the VM [4], the VM$_{dif}$ [9], the SYMNet [18], and the proposed DTN. The symmetric structures are combined into one in the plots.

Implemental Details. We built the DTN on the five-level 3D Unet, where the volumetric image embedding module and the registration field inference module are the encoder and the decoder of the U-net, respectively. The CNN-based feature embedding has $q_s = 128$ channels and $q_c = 96$ channels. The resultant embedding of the dual transformation has the resolution of $224 \times 9 \times 10 \times 12$. The hyperparameter α in the registration loss (2) is set to 0.1. The proposed framework is implemented using the PyTorch on a PC with a NVIDIA GTX TITAN xp GPU. The network parameters are optimized using the ADAM algorithm with a learning rate of $1e - 4$. The momentums are set to 0.5 and 0.999.

The mini-batch consists of 1 image. The training takes 50 h with 400 epochs and 102.4 k iterations. The online testing takes 0.653 s.

3.1 Qualitative Assessment

The proposed DTN realizes the diffeomorphic registration of volumetric images. The dual transformer-based attention enhances the cross-image volumetric embedding, facilitating the diffeomorphic registration.

Comparison with Sstate-of-the-Art. We compare with the symmetric image normalization registration method (SyN) [2] and recent deep-learning-based registration models, including the VM [4], the diffeomorphic variant VM_{dif} [9], and the SYMNet [18] (Figs. 3 and 4), using the publicly available implementations provided by authors with the suggested parameter setting. Both the VM [4] and the VMdif [9] use the MSE similarity loss, where the smooth regularization weight and the learning rate are set to 0.01 and 1e−4, respectively. The SYMNet [18] uses the NCC similarity loss, where the orientation consistency weight, the smooth regularization weight, the magnitude weight, and the learning rate are set to 1000, 10, 0.001, and 1e−4, respectively. We used the same data splitting for training, validation, and testing datasets in comparison. We use the ANTs toolbox [3] for the SyN implementations with the maximum iteration set to [100, 100, 100] in the iterative optimization. The unsupervised VM predicts the displacement vector fields directly from the input image pair without the guarantee of the diffeomorphic property. The VM_{dif}, SYMnet, and the proposed DTN infer the velocity field for the diffeomorphic registration.

Table 1 reports the DSCs and the negative Jacobian determinant by the proposed DTN and the compared methods, including the SyN, the VM, the VM_{dif}, and the SYMNet. The proposed DTN is feasible to estimate the diffeomorphic registration field with low voxel numbers of the negative Jacobian determinants, facilitating performance improvements on the average DSC of anatomical structures. The deep learning-based methods, including the VM, the SYMNet, and the proposed DTN, achieve higher DSCs than the iterative optimization-based SyN. Similar to deep learning-based diffeomorphic registration models of the VM_{dif} and the SYMNet, the proposed DTN is feasible to reduce the negative Jacobian determinant without sacrificing the atlas-based registration. Figure 3 illustrates the boxplot of DSCs on 16 anatomical structures. As we can see, the proposed DTN outperforms the compared diffeomorphic registration methods on 13 out of 16 structures. We further conducted the statistical significance tests. There existed statistically significant improvements with the p-values < 0.05 in terms of the average DSC in the t-test. The proposed DTN outperformed the compared transformer-free models with the p-values below 5e−15 (VM [4]), 5e−75 (VMdif [9]), and 5e−20 (SYMNet [18]). The dual transformer scheme effectively models the cross-image global relevances and enhances the volumetric image embedding for semantically meaningful correspondence. For instance, the boundaries of the lateral-ventricle are more consistent with the atlas than the compared methods as shown in Fig. 4. Note that the proposed approach utilizes the basic diffeo-morphic registration loss (2) on the image similarity and the regularization as the first-order gradients of the velocity field, without relying on the delicately designed regularizer.

Table 1. The DSC by the SyN [2], the VM [4], VM$_{dif}$ [9], SYMNet [18], and variants of our methods.

		SyN	VM	VM$_{dif}$	SYMNet	DTN	DTN$_s$	DTN$_c$
DSC	Avg.	0.740	0.756	0.733	0.755	0.769	0.763	0.764
	Std.	0.089	0.081	0.086	0.081	0.076	0.076	0.077
$\|J(\psi)\| \leq 0$	Avg.	0	20604	2.84	1169	0.497	0.531	0.655
	Std.	0	3015	9.64	405	2.13	2.21	2.59

Regularization. We study the effect of the regularization on velocity fields. The deep learning-based diffeomorphic registration model is not guaranteed to produce invertible one-to-one map, due to computational errors in registration fields. Figure 2 (b) shows the voxels with negative Jacobian determinants with varying α. The average voxel number with $\|J(\psi)\| \leq 0$ reduces from 1376 to 0 when α ranges from 0.02 to 0.2. By tuning the hyperparameter α, the resultant registration field balances topology preservation with plausible registration accuracies.

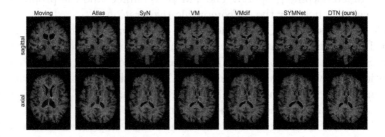

Fig. 4. The atlas-based registration of lateral-ventricle, thalamus, and hippocampus by the SyN [2], VM [4], VM$_{dif}$ [9], SYMNet [18], and ours.

Ablation Study. The proposed dual transformer network addressed the relevance modeling and the feature enhancement on two kinds of image embedding for semantically meaningful correspondences of anatomical structures. We conducted an ablation study to analyze the proposed dual transformer network (Table 1). The proposed DTN with the dual transformer architecture outperformed the DTN$_s$ and the DTN$_c$ with the only transformer of T_s on the dependencies of the separate image embedding or T_c on the relevance of the concatenated image embedding. We compared with the transformer-free deep diffeomorphic registration models, including the VMdif [9] and the SYMNet [18]. As shown in Fig. 1, the encoder-decoder architecture with ϕ_c and ϕ_r followed the VMdif [9] when removing the dual transformer. As shown in Table 1, both the DTN_s and DTN_c are helpful to improve the registration accuracy compared with transformer-free models with the DSCs of 0.763 ± 0.076 and 0.764 ± 0.077, respectively. The transformer branches are complementary, where the proposed DTN with the dual transformers outperformed the DTN_s and the DTN_c with p-values below $1e-6$ and $1e-4$, respectively. We designed the dual architecture to identify the semantically meaningful correspondence of anatomical structures. The attention-based global and cross-image relevance enhanced volumetric embedding and improved the registration accuracy compared with transformer-free models (Table 1).

4   Conclusion

This paper presents the DTN for the volumetric diffeomorphic registration, taking advantage of both the CNN-based low-level features and the attention-based global and cross-image relevances for feature enhancements. The DTN explicitly models the long-range cross-image relevance in embedding learning to identify the semantically meaningful correspondence of anatomical structures. The qualitative and quantitative evaluations of the proposed approach are conducted on the OASIS dataset with clinically obtained brain MRI scans. The improvements on the atlas-based registration suggest that the dual transformer facilitates the semantically meaningful correspondence of anatomical structures.

Acknowledgments. This work was supported in part by National Natural Science Foundation of China under Grant 61876008 and 82071172, Beijing Natural Science Foundation under Grant 7192227, and Research Center of Engineering and Technology for Digital Dentistry, Ministry of Health.

References

1. Ashburner, J.: A fast diffeomorphic image registration algorithm. NeuroImage **38**, 95–113 (2007)
2. Avants, B., Epstein, C., Grossman, M., Gee, J.: Symmetric diffeomorphic image registration with cross-correlation: evaluating automated labeling of elderly and neurodegenerative brain. Med. Image Anal. **12**(1), 26–41 (2008)
3. Avants, B., Tustison, N., Song, G., Cook, P., Klein, A., Gee, J.: A reproducible evaluation of ants similarity metric performance in brain image registration. NeuroImage **54**, 2033–2044 (2011)
4. Balakrishnan, G., Zhao, A., Sabuncu, M.R., Guttag, J., Dalca, A.V.: An unsupervised learning model for deformable medical image registration. In: IEEE Conference on Computer Vision and Pattern Recognition, pp. 9252–9260 (2018)
5. Carion, N., Massa, F., Synnaeve, G., Usunier, N., Kirillov, A., Zagoruyko, S.: End-to-end object detection with transformers. ArXiv abs/2005.12872 (2020)
6. Chen, H., et al.: Pre-trained image processing transformer. ArXiv abs/2012.00364 (2020)

7. Chen, J., et al.: TransuNet: transformers make strong encoders for medical image segmentation. ArXiv abs/2102.04306 (2021)
8. Çiçek, Ö., Abdulkadir, A., Lienkamp, S.S., Brox, T., Ronneberger, O.: 3D U-Net: learning dense volumetric segmentation from sparse annotation. In: International Conference on Medical Image Computing and Computer-Assisted Intervention, pp. 424–432 (2016)
9. Dalca, A.V., Balakrishnan, G., Guttag, J., Sabuncu, M.R.: Unsupervised learning for fast probabilistic diffeomorphic registration. In: Frangi, A., Schnabel, J., Davatzikos, C., Alberola-López, C., Fichtinger, G. (eds.) MICCAI 2018. LNCS, vol. 11070, pp. 729–738. Springer, Cham (2018). https://doi.org/10.1007/978-3-030-00928-1_82
10. Dosovitskiy, A., et al.: An image is worth 16x16 words: transformers for image recognition at scale. ArXiv abs/2010.11929 (2020)
11. Fischl, B.: Freesurfer. NeuroImage **62**, 774–781 (2012)
12. Haskins, G., Kruger, U., Yan, P.: Deep learning in medical image registration: a survey. Mach. Vis. Appl. **31**, 1–18 (2020)
13. Jaderberg, M., Simonyan, K., Zisserman, A., et al.: Spatial transformer networks. In: NeurIPS, pp. 2017–2025 (2015)
14. Leow, A., et al.: Inverse consistent mapping in 3D deformable image registration: its construction and statistical properties. Inf. Process. Med. Imaging **19**, 493–503 (2005)
15. Liao, R., et al.: An artificial agent for robust image registration. In: AAAI (2017)
16. Marcus, D., Wang, T.H., Parker, J., Csernansky, J., Morris, J., Buckner, R.: Open access series of imaging studies (OASIS): cross-sectional MRI data in young, middle aged, nondemented, and demented older adults. J. Cogn. Neurosci. **19**, 1498–1507 (2007)
17. Miao, S., et al.: Dilated FCN for multi-agent 2D/3D medical image registration. ArXiv abs/1712.01651 (2018)
18. Mok, T.C.W., Chung, A.C.S.: Fast symmetric diffeomorphic image registration with convolutional neural networks. In: IEEE/CVF Conference on Computer Vision and Pattern Recognition (CVPR), pp. 4643–4652 (2020)
19. Petit, O., Thome, N., Rambour, C., Soler, L.: U-net transformer: self and cross attention for medical image segmentation. ArXiv abs/2103.06104 (2021)
20. Schlemper, J., et al.: Attention gated networks: learning to leverage salient regions in medical images. Med. Image Anal. **53**, 197–207 (2019)
21. Sotiras, A., Davatzikos, C., Paragios, N.: Deformable medical image registration: a survey. IEEE Trans. Med. Imaging **32**(7), 1153–1190 (2013)
22. Tan, H.H., Bansal, M.: LXMERT: learning cross-modality encoder representations from transformers. In: EMNLP/IJCNLP (2019)
23. Valanarasu, J.M.J., Oza, P., Hacihaliloglu, I., Patel, V.: Medical transformer: gated axial-attention for medical image segmentation (2021)
24. Vaswani, A., et al.: Attention is all you need. ArXiv abs/1706.03762 (2017)
25. Wang, Q., et al.: Learning deep transformer models for machine translation. ArXiv abs/1906.01787 (2019)
26. Zhang, J.: Inverse-consistent deep networks for unsupervised deformable image registration. ArXiv abs/1809.03443 (2018)
27. Zhang, Y., Pei, Y., Guo, Y., Ma, G., Xu, T., Zha, H.: Fully convolutional network for consistent voxel-wise correspondence. In: AAAI (2020)

Construction of Longitudinally Consistent 4D Infant Cerebellum Atlases Based on Deep Learning

Liangjun Chen, Zhengwang Wu, Dan Hu, Yuchen Pei, Fenqiang Zhao,
Yue Sun, Ya Wang, Weili Lin, Li Wang[✉], Gang Li[✉], and the UNC/UMN
Baby Connectome Project Consortium

Department of Radiology and BRIC, University of North Carolina at Chapel Hill,
Chapel Hill, NC 27599, USA
{li_wang,gang_li}@med.unc.edu

Abstract. Longitudinal infant dedicated cerebellum atlases play a fundamental role in characterizing and understanding the dynamic cerebellum development during infancy. However, due to the limited spatial resolution, low tissue contrast, tiny folding structures, and rapid growth of the cerebellum during this stage, it is challenging to build such atlases while preserving clear folding details. Furthermore, the existing atlas construction methods typically independently build discrete atlases based on samples for each age group without considering the within-subject temporal consistency, which is critical for large-scale longitudinal studies. To fill this gap, we propose an age-conditional multi-stage learning framework to construct longitudinally consistent 4D infant cerebellum atlases. Specifically, 1) A joint affine and deformable atlas construction framework is proposed to accurately build *temporally continuous* atlases based on the entire cohort, and rapidly warp the new images to the atlas space; 2) A longitudinal constraint is employed to enforce the within-subject temporal consistency during atlas building; 3) A Correntropy based regularization loss is further exploited to enhance the robustness of our framework. Our atlases are constructed based on 405 longitudinal scans from 187 healthy infants with age ranging from 6 to 27 months, and are compared to the atlases built by state-of-the-art algorithms. Results demonstrate that our atlases preserve more structural details and fine-grained cerebellum folding patterns, which ensure higher accuracy in subsequent atlas-based registration and segmentation tasks.

Keywords: Cerebellum · 4D infant atlas · Joint affine and deformable registration

Electronic supplementary material The online version of this chapter (https://doi.org/10.1007/978-3-030-87202-1_14) contains supplementary material, which is available to authorized users.

M. de Bruijne et al. (Eds.): MICCAI 2021, LNCS 12904, pp. 139–149, 2021.
https://doi.org/10.1007/978-3-030-87202-1_14

1   Introduction

As an important bilateral neuroanatomical structure, the cerebellum is not only related to motor tasks [1–4], but also involved in cognitive functions (e.g., language, learning, and memory [5–7]) and various neurodevelopmental disorders (e.g., attention deficit/hyperactivity disorder and autism [8–10]). Furthermore, as shown in Fig. 1, during the first two postnatal years, the infant cerebellum grows dramatically (volume increasing by 280%) even more than the cerebral cortex (volume increasing by 116%) [11]. Thus, to boost the understanding of the cerebellum and related cognitive function development during infancy, it is essential to construct longitudinally consistent 4D infant cerebellum atlases with temporally continuous time points, which are able to well characterize the rapidly developing patterns of the cerebellum in size, shape, tissue contrast, and appearance. However, the existing cerebellum atlases [12–14] are mainly proposed for adult brains, and longitudinally-consistent infant-dedicated 4D cerebellum atlases are still lacking due to limited imaging resolution, low tissue contrast, tiny folding structures, and rapidly changing size and shape.

Meanwhile, the existing atlas construction methods have obvious shortcomings. Specifically, these methods usually iteratively perform pair-wise image registrations [15] to construct a population-average atlas from a group of samples, making the atlas construction prohibitively expensive in computation. Moreover, to characterize the complex development during infancy, multiple age-specific atlases [16,17] are often built by subdividing the entire collection of samples into different age groups, and then constructing the atlas for each age group independently, which typically cause longitudinal inconsistency among different age atlases. In [18], the authors proposed a 4D longitudinal atlas construction method for developing infant brains, which encourages the temporal consistency among the generated longitudinal atlases. However, due to the high computational expense, the longitudinal atlases are still built for specific time points, e.g., with the time interval of months or years, which are too sparse to well capture the rapidly developing patterns of the infant cerebellum. Recently, learning based registration methods [19–23] have achieved many successes in medical image analysis. Motivated by these works, [24] proposed a learning-based conditional deformable template construction network for adult brains with relatively subtle longitudinal changes, which can not only build atlases based on conditional information, but also align the new images to the atlas space. However, this method requires that the input images have already been pre-affinely aligned; consequently, the constructed longitudinal atlases have similar sizes across ages, thus

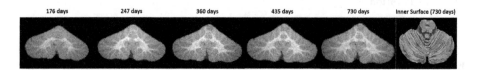

Fig. 1. T1w images and white matter surface of the infant cerebellum from the same subject at 176, 247, 360, 435, and 730 days of age.

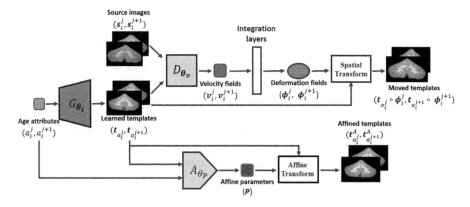

Fig. 2. A schematic illustration of our atlas construction framework, consisting of the deformable stage DACN (upper) and the affine stage AARN (lower).

cannot well reflect the dynamic size and shape development of the infant cerebellum. Besides, similar to the major cerebral cortical folding patterns, which are established at term birth and preserved during brain development [25,26], both middle cerebellar peduncles and deep white matter in the cerebellum appear myelinated from 4th postnatal month [27]. These established cerebellum structures show apparent within-subject temporal consistency (as shown in Fig. 1). In practice, we also find that it is much easier to achieve good results on within-subject registration, due to the existence of the within-subject temporal consistency. However, all aforementioned atlas construction methods ignore such temporal consistency, which is essential for large-scale longitudinal studies and helpful for enhancing the accuracy of both within-subject and inter-subject alignment and in turn generating sharper atlases.

To address these issues, we propose an age-conditional multi-stage learning framework to construct longitudinally-consistent and temporally-continuous 4D infant cerebellum atlases. Specifically, 1) To accurately build temporally continuous atlases based on the entire cohort, we develop a multi-stage atlas construction framework, which consists of a deformable atlas construction network (DACN) and an affine atlas rescaling network (AARN). The trained framework can also rapidly warp the new images to the atlas space; 2) By enforcing the within-subject temporal consistency during atlas building, we efficiently incorporate the longitudinal constraint into our framework, which further prompts the within-subject alignment and consequently helps to generate sharper atlases; 3) A Correntropy based regularization loss is exploited to robustly penalize the deviations of the estimated deformation fields, which can further enhance the training stability of our deformable atlas construction network; 4) By leveraging our framework, we construct, to the best of our knowledge, the first longitudinally-consistent and temporally-continuous 4D infant cerebellum atlases based on 405 longitudinal scans from 187 infants with age ranging from 6 months (6M) to 27 months (27M). The results indicate that the built 4D atlases preserve more details with clearer folding patterns, compared to the atlases built by state-of-the-art methods.

2   Method

As presented in Fig. 2, our framework includes two stages, i.e., a deformable atlas construction network (DACN) and an affine atlas rescaling network (AARN).

2.1   Deformable Atlas Construction Network (DACN)

We devise a DACN to construct age-conditional atlases from the entire dataset in an unsupervised way. The trained DACN can jointly produce age-conditional atlases and the corresponding deformations for registering new images.

Network: Let $S_{\text{affine}} = \left\{ s_1^1, \ldots, s_1^{n_1}, \ldots, s_i^j, \ldots, s_l^1, \ldots, s_l^{n_l} \right\}$ denotes a dataset of affinely aligned cerebellum images with total of $\mathbb{N}$ samples from l subjects and each subject $i(i = 1, ..., l)$ has n_i longitudinal scans, s_i^j represents the jth time point image of subject i and a_i^j refers to the age of subject i at j^{th} time point.

There are two sub-networks in the proposed DACN, i.e., an atlas synthesis network and a deformable registration network. Atlas synthesis network is designed to generate a conditional template $t_a = G_{\theta_t}(a)$ by taking an age a as input, and θ_t is the learnable parameters. This sub-network first decodes a using a dense layer; then the decoded a is input to four sequentially linked upsampling blocks using ReLU activation. Deformable registration network is formulated to estimate t_i^j, which is the age-conditional atlas $t_{a_i^j}$ warped to subject image s_i^j. The network D_{θ_v} is a U-Net like architecture [28] consisting of an input convolutional layer, four downsampling blocks, and three upsampling blocks, and is first used to estimate a stationary velocity field $v_i^j = D_{\theta_v}\left(s_i^j, t_{a_i^j}\right)$. Then the stationary velocity field v_i^j is further input to integration layers [29] to generate the corresponding deformation field ϕ_i^j. Finally, the warped atlas $t_i^j = t_{a_i^j} \circ \phi_i^j$ is achieved through spatial transform layers [24]. The optimal parameters $\hat{\theta} = \left\{ \hat{\theta}_t, \hat{\theta}_v \right\}$ are learned by enforcing t_i^j as similar as s_i^j.

Longitudinal Constraint: To incorporate the within-subject temporal consistency into our framework to further prompt the within-subject alignment and consequently generate sharper atlases, we first sort the scans belonging to each subject by age, and then rearrange the training data to jointly input two adjacent longitudinal scans from the same subject. For instance, the subject shown in Fig. 1 has five scans at 176, 247, 360, 435, and 730 days. Thus, we group them into 4 pairs, e.g., {176, 247}, {247, 360}, {360, 435}, and {435, 730}, respectively. Besides, for the subjects which have only one scan, we pair it with itself. At each training step, by jointly inputting two adjacent longitudinal scans from the same subject (s_i^j, s_i^{j+1}), we can achieve warped images (r_i^j, r_i^{j+1}) in atlas space from DACN. Thereby, we explicitly enforce the longitudinal correspondence between these two warped images (r_i^j, r_i^{j+1}) in atlas space as the longitudinal constraint.

Losses: We train the DACN by minimizing the following loss function:

$$\mathbb{L}_{\text{DACN}} = L_{\text{Similar}} + \lambda_1 * L_{\text{LC}} + \lambda_2 * L_{\text{Smooth}} + \lambda_3 * L_{\text{Reg}}. \tag{1}$$

a) L_{Similar}, which supervises the similarity between the warped atlases and subject images, is a localized NCC (LNCC) loss: For N sliding windows,

$$L_{\text{Similar}} = L_{\text{LNCC}}(\boldsymbol{s}, \boldsymbol{t}) = \frac{1}{N} \sum_k \frac{\sum_{m \in \zeta_k} (\boldsymbol{s}_m - \bar{\boldsymbol{s}}_k)(\boldsymbol{t}_m - \bar{\boldsymbol{t}}_k)}{\sqrt{\sum_{m \in \zeta_k} (\boldsymbol{s}_m - \bar{\boldsymbol{s}}_k)^2 \sum_{m \in \zeta_k} (\boldsymbol{t}_m - \bar{\boldsymbol{t}}_k)^2}}, \tag{2}$$

where $\boldsymbol{s}_m$ and $\boldsymbol{t}_m$ refer to the mth voxel in the subject images and warped atlases, respectively. And $\bar{\boldsymbol{s}}_k$ and $\bar{\boldsymbol{t}}_k$ are the average image intensity values over window ζ_k, which is centered at k^{th} voxel.

b) L_{LC}, which enhances the longitudinal correspondence (LC) between two warped images, is also based on LNCC: $L_{\text{LC}} = \frac{1}{\bar{N}} \sum_{i=1}^{l} \sum_{j=1}^{n_i} L_{\text{LNCC}}(\boldsymbol{r}_i^j, \boldsymbol{r}_i^{j+1})$.

c) L_{Smooth}, which provides explicit guidance of the smoothness constraint for the estimated deformation fields, is based on the Laplacian operator: $L_{\text{Smooth}} = \|\Delta\phi\|_2^2$, where $\Delta\phi = \frac{\partial^2 \phi}{\partial x^2} + \frac{\partial^2 \phi}{\partial y^2} + \frac{\partial^2 \phi}{\partial z^2}$ is the Laplacian operator in terms of ϕ.

d) L_{Reg}, which constrains the large deformations by penalizing the deviations of estimated deformation fields, is based on a Correntropy loss (Closs) [30], a more robust alternative for the mean square error (MSE): $L_{\text{Reg}} = 1 - \exp\left(-\frac{\phi^2}{2\sigma^2}\right)$, where σ is the tunable kernel bandwidth. The MSE similarity metric is too sensitive to outliers, which degrades the training stability. By leveraging the Closs based regularization, our framework is more robust in estimating the deformation fields, which significantly improves training stability.

Training: Balancing the training of the atlas synthesis network and the deformable registration network is important yet very difficult, since the deformable registration network tends to dominate the training process due to its powerful data generation capability. To train these two sub-networks more tractably, we design a training strategy comprising three steps: 1) We first individually train the deformable registration network as a scan-to-scan registration task by minimizing $L_{\text{Similar}} + L_{\text{Smooth}}$ until convergence; 2) We fix the deformable registration network, and then individually train atlas synthesis network by minimizing $L_{\text{Similar}} + L_{\text{LC}}$ until convergence; 3) We fine-tune our framework by jointly training both sub-networks to minimize the losses in Eq. (1).

2.2 Affine Atlas Rescaling Network (AARN)

Inspired by existing affine registration networks [22,31], we design an AARN to affinely rescale the size of the atlases generated in the first stage based on age.

Network, Training, and Loss: During training, AARN takes the affinely aligned images S_{affine} and their age attributes as input, and the original images

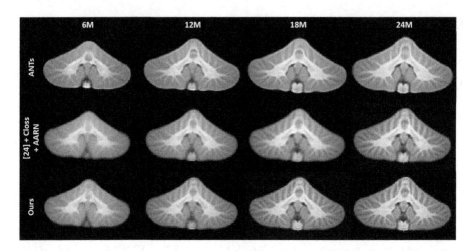

Fig. 3. Visual comparison of T1w templates, which are built by ANTs and our framework, respectively. Note that although limited time points are shown, our framework provides temporally-continuous 4D atlases at arbitrary time points.

S_{ori} are regarded as targets. We first train AARN to predict the affine transformation parameters $P_{S_{\text{affine}},a} = A_{\theta_P}(S_{\text{affine}}, a)$. Then, the affine-rescaled output images S_{output} are obtained using the affine transform [31]. Specifically, the age a is decoded by a dense layer, reshaped, and concatenated with S_{affine}. And the affine parameters P are estimated through an input convolutional layer, four downsampling blocks, and two dense layers. Herein, we still use the LNCC loss to measure the similarity between the S_{output} and S_{ori} for training.

Age-Conditional Atlas Construction: Once the training is completed, given an age attribute a, we can directly achieve an age-conditional atlas t_a from network $G_{\theta_t}(a)$. Then age-conditional atlas t_a and its age attribute a are input to the AARN to predict affine parameters $P_{t_a,a} = A_{\theta_P}(t_a, a)$. Finally, we can achieve the affine-rescaled atlas t_a^A through the affine transform, as showing in Fig. 2.

3   Experiments

Dataset and Experimental Setup: We evaluated the proposed framework on the UNC/UMN Baby Connectome Project (BCP) dataset [32]. In total, 405 T1-weight images from 187 subjects with age ranging from 6 to 27 months were included. 64% infants have 2 or more scans and 33% infants have 3 or more scans. The distribution of longitudinal scans is shown in the supplementary material. Meanwhile, 12 scans were manually delineated by experienced experts, and the achieved cerebellum tissue segmentation maps were exploited as ground-truth in testing for quantitative comparison.

Table 1. The registration accuracy comparison between different atlases using different registration methods (mean and standard deviation).

Metrics	Tissue	SyGN + SyN	[24] + Closs + SyN	Ours + SyN	Ours
CC (%)	N/A	87.8 ± 3.9	88.1 ± 3.9	88.7 ± 3.6	$\mathbf{90.3 \pm 1.3}$
DSC (%)	CSF	51.8 ± 2.5	52.6 ± 2.5	53.0 ± 2.4	$\mathbf{54.2 \pm 1.6}$
	GM	64.7 ± 1.1	64.3 ± 1.1	65.1 ± 1.0	$\mathbf{67.4 \pm 1.1}$
	WM	63.8 ± 1.5	64.2 ± 1.4	64.7 ± 1.4	$\mathbf{68.5 \pm 0.7}$
ASSD (mm)	CSF	0.37 ± 0.08	0.37 ± 0.08	0.35 ± 0.07	$\mathbf{0.31 \pm 0.02}$
	GM	0.21 ± 0.03	0.22 ± 0.03	0.20 ± 0.03	$\mathbf{0.18 \pm 0.01}$
	WM	0.34 ± 0.04	0.36 ± 0.05	0.33 ± 0.03	$\mathbf{0.28 \pm 0.02}$
HD95 (mm)	CSF	1.85 ± 0.65	1.81 ± 0.63	1.80 ± 0.62	$\mathbf{1.40 \pm 0.22}$
	GM	0.95 ± 0.26	0.97 ± 0.27	0.94 ± 0.26	$\mathbf{0.83 \pm 0.09}$
	WM	1.47 ± 0.26	1.49 ± 0.29	1.47 ± 0.25	$\mathbf{1.27 \pm 0.24}$

The resolution of the T1w images is $0.8 \times 0.8 \times 0.8\,\text{mm}^3$. The cerebellum images were affinely registered to the averaged 24-month T1w image.

Our framework was compared with the state-of-the-art symmetric group-wise normalization (SyGN) approach from ANTs [15]. Specifically, we built the atlases based on SyGN (ANTs atlases) at time points 6M, 12M, 18M, and 24M.

Our framework and ANTs atlas building approach were all performed on a Dell Alienware PC with a 6 cores Intel Core I7-8700K 3.7 GHz CPU, 16 GB NVIDIA RTX 2080Ti GPU, and 64 GB memory. We trained all networks on a platform of Keras 2.0.6 and adopted maximum epoch number to 10,00, and the initial learning rate to 0.0001, which was divided by 10 after 200 iterations. In SyGN, the greedy B-spline was chosen as the transformation model, and cross-correlation was the similarity metric for the registration, with default shrinkage factors, smoothing factors, and max iterations of $8 \times 4 \times 2 \times 1$, $3 \times 2 \times 1 \times 0$, and $100 \times 70 \times 50 \times 10$, respectively.

Parameters of our framework were experimentally set as: lr $= 0.0001$ (Adam optimizer), kernel size $= 3$, stride $= 2$, $\lambda_1 = 0.5$, $\lambda_2 = 0.1$, $\lambda_3 = 1$, and $\sigma = 0.8$.

To quantitatively evaluate our atlases, we first warped 12 test images and corresponding tissue segmentation maps (TSMs) to atlases, as performing an image normalization task. Then, the Pearson's correlation coefficient (CC) between each pair of warped test images, and Dice similarity coefficient (DSC), average symmetric surface distance (ASSD), and 95th Hausdorff Distance (HD95) between each pair of warped TSMs were calculated. Intuitively, the atlas with sharper tissue patterns would lead to better spatial normalization.

Results: We first visually compared our atlases with the ANTs-based atlases. The typical image slices of our atlases and the ANTs-based atlases at different time points are illustrated in Fig. 3. At all time points, one can easily observe that our atlases show sharper structural details comparing to ANTs-based atlases. In particular, attributing to the distinct tissue patterns in our cerebellum atlases, the critical lobules and cruses are more identifiable. Meanwhile, by taking advantage of our LC loss, our framework can estimate atlases with obviously more apparent boundaries and demarcative structures, which hints that the enhanced within-subject registration accuracy can further improve the quality of our atlases.

The values of CC, DSC, ASSD, and HD95 are summarized in Table 1 for quantitative comparison. The results of ANTs-based atlases (SyGN + SyN) and our atlases (Ours) are respectively illustrated to testify the improvement of our atlas construction framework. Moreover, to individually verify the quality of our infant cerebellum atlases, we used SyN to warp each test image to our atlases (Ours + SyN). Besides, to certify the effectiveness of the newly introduced within-subject consistency constraint, we performed experiments on [24] with Closs ([24] + Closs + SyN) and affinely aligned the generated atlases with trained AARN. From Table 1, we have at least three observations.

First, when using symmetric normalization (SyN) registration algorithm [33] as the registration method, our atlases (Ours + SyN) show obviously enhanced consistency after registration, compared to ANTs atlases. These results indicate that our atlases successfully preserve detailed structures of infant cerebellum and fine-grained cerebellum folding patterns, which is beneficial for many related typical applications, e.g., spatial normalization and tissue segmentation.

Second, compared to the atlases built without LC loss ([24] + Closs + SyN), our atlases (Ours + SyN) manifest better registration consistency, which reveals that incorporating the within-subject consistency constraint can help improve the quality of the generated atlases. This is because longitudinal constraint can obviously enhance the within-subject registration accuracy to remedy the bad influence from the low tissue contrast, and consequently provide sharper atlases.

Third, compared to the state-of-the-art SyN registration method (Ours + SyN), our framework (Ours) achieves remarkably better registration consistency in terms of all metrics, which demonstrates that our framework has improved registration accuracy. Note that the accurate registration is critical for infant cerebellum atlas construction, which suffers from the significantly degraded performance due to the limited spatial resolution, low tissue contrast, tiny folding structures, and rapid development of the infant cerebellum.

Moreover, compared to the existing methods, our framework achieved statistically significant improvement in both atlas-based registration and segmentation tasks (p-values < 0.05). Therefore, both qualitative and quantitative evaluations suggest the superior performance of our framework. Besides, the SyGN approach takes hours or days (increased linearly with the number of images) to create each atlas and additional hours to obtain the deformation for each new image. In contrast, our trained framework can simultaneously generate age-conditional atlases and warp the new image to atlas space in seconds. These results imply that our framework yields markedly improved efficiency compared to the traditional iterative atlas construction approaches.

4   Conclusion

In this paper, we propose an age-conditional multi-stage learning framework to construct longitudinally consistent 4D infant cerebellum atlases. Our framework first builds atlases in a deformable manner, and then affinely transforms the achieved atlases to the age-specific sizes. By doing this, our framework can accurately build temporally continuous atlases and efficiently warp the new image to the atlas space. To improve the longitudinal consistency of our atlases, we incorporate the within-subject temporal consistency constraint into our framework to further prompt the within-subject alignment and, in turn sharpen atlases. To enhance the training stability of our framework, instead of the commonly used mean square error (MSE), we introduce a more robust Correntropy based similarity measure as the regularization loss for the estimated deformation fields. By taking advantage of our framework, we construct the first longitudinally-consistent and temporally-continuous 4D infant cerebellum atlas based on 405 longitudinal scans from 187 infants. Experimental results demonstrate that, compared to the state-of-the-art methods, our atlas construction framework achieves enhanced quality in generating infant cerebellum atlases (preserving more structural details with clearer folding patterns) and improved registration accuracy in aligning the new images to the atlas space. The achieved 4D infant cerebellum atlases are practically useful for studying early cerebellum growth patterns and clinical applications. Moreover, our framework can also be extensively applied to other atlas construction tasks.

Acknowledgments. This work was partially supported by NIH grants (MH116225, MH109773, MH117943, and MH123202). This work also utilizes approaches developed by an NIH grant (1U01MH110274) and the efforts of the UNC/UMN Baby Connectome Project Consortium.

References

1. Robinson, E.C., et al.: MSM: a new flexible framework for multimodal surface matching. Neuroimage **100**, 414–426 (2014)
2. Kase, C., Norrving, B., Levine, S., et al.: Cerebellar infarction. Clinical and anatomic observations in 66 cases. Stroke **24**(1), 76–83 (1993)
3. Davie, C., Barker, G., Webb, S., et al.: Persistent functional deficit in multiple sclerosis and autosomal dominant cerebellar ataxia is associated with axon loss. Brain **118**(6), 1583–1592 (1995)
4. Klockgether, T.: The clinical diagnosis of autosomal dominant spinocerebellar ataxias. Cerebellum **7**(2), 101 (2008)
5. Desmond, J.E., Gabrieli, J.D., Glover, G.H.: Dissociation of frontal and cerebellar activity in a cognitive task: evidence for a distinction between selection and search. Neuroimage **7**(4), 368–376 (1998)
6. Riva, D., Giorgi, C.: The cerebellum contributes to higher functions during development: evidence from a series of children surgically treated for posterior fossa tumours. Brain **123**(5), 1051–1061 (2000)

7. Stoodley, C.J., Schmahmann, J.D.: Functional topography in the human cerebellum: a meta-analysis of neuroimaging studies. Neuroimage **44**(2), 489–501 (2009)
8. Seidman, L.J., Valera, E.M., Makris, N.: Structural brain imaging of attention-deficit/hyperactivity disorder. Biol. Psychiatry **57**(11), 1263–1272 (2005)
9. Bishop, D.V.: Cerebellar abnormalities in developmental dyslexia: cause, correlate or consequence? (2002)
10. Courchesne, E., Saitoh, O., Yeung-Courchesne, R., et al.: Abnormality of cerebellar Vermian lobules VI and VII in patients with infantile autism: identification of hypoplastic and hyperplastic subgroups with MR imaging. **162**(1), 123–130 (1994)
11. Knickmeyer, R.C., Gouttard, S., Kang, C., et al.: A structural MRI study of human brain development from birth to 2 years. J. Neurosci. **28**(47), 12176–12182 (2008)
12. Diedrichsen, J., Balsters, J.H., Flavell, J., Cussans, E., Ramnani, N.: A probabilistic MR atlas of the human cerebellum. Neuroimage **46**(1), 39–46 (2009)
13. Diedrichsen, J.: A spatially unbiased atlas template of the human cerebellum. Neuroimage **33**(1), 127–138 (2006)
14. Schmahmann, J.D., Doyon, J., McDonald, D., et al.: Three-dimensional MRI atlas of the human cerebellum in proportional stereotaxic space. Neuroimage **10**(3), 233–260 (1999)
15. Avants, B.B., et al.: The optimal template effect in hippocampus studies of diseased populations. Neuroimage **49**(3), 2457–2466 (2010)
16. Schuh, A., Makropoulos, A., Robinson, E.C., et al.: Unbiased construction of a temporally consistent morphological atlas of neonatal brain development. bioRxiv (2018) 251512
17. Shi, F., et al.: Infant brain atlases from neonates to 1- and 2-year-olds. PloS ONE **6**(4), e18746 (2011)
18. Zhang, Y., Shi, F., Wu, G., Wang, L., Yap, P.T., Shen, D.: Consistent spatial-temporal longitudinal atlas construction for developing infant brains. IEEE TMI **35**(12), 2568–2577 (2016)
19. Dalca, A.V., Balakrishnan, G., Guttag, J., Sabuncu, M.R.: Unsupervised learning for fast probabilistic diffeomorphic registration. In: Frangi, A., Schnabel, J., Davatzikos, C., Alberola-López, C., Fichtinger, G. (eds.) MICCAI 2018. LNCS, vol. 11070, pp. 729–738. Springer, Cham (2018). https://doi.org/10.1007/978-3-030-00928-1_82
20. Fan, J., Cao, X., Xue, Z., Yap, P.T., Shen, D.: Adversarial similarity network for evaluating image alignment in deep learning based registration. In: Frangi, A., Schnabel, J., Davatzikos, C., Alberola-López, C., Fichtinger, G. (eds.) MICCAI 2018. LNCS, vol. 11070, pp. 739–746. Springer, Cham (2018). https://doi.org/10.1007/978-3-030-00928-1_83
21. Fan, J., Cao, X., Wang, Q., Yap, P.T., Shen, D.: Adversarial learning for mono-or multi-modal registration. Med. Image Anal. **58**, 101545 (2019)
22. Shen, Z., Han, X., Xu, Z., Niethammer, M.: Networks for joint affine and non-parametric image registration. In: CVPR, pp. 4224–4233 (2019)
23. Wei, D., et al.: Deep morphological simplification network (MS-Net) for guided registration of brain magnetic resonance images. Pattern Recogn. **100**, 107171 (2020)
24. Dalca, A.V., Rakic, M., Guttag, J., Sabuncu, M.R.: Learning conditional deformable templates with convolutional networks. In: NeurIPS (2019)
25. Hill, J., et al.: A surface-based analysis of hemispheric asymmetries and folding of cerebral cortex in term-born human infants. J. Neurosci. **30**(6), 2268–2276 (2010)
26. Li, G., et al.: Mapping region-specific longitudinal cortical surface expansion from birth to 2 years of age. Cereb. Cortex **23**(11), 2724–2733 (2013)

27. Triulzi, F., Parazzini, C., Righini, A.: MRI of fetal and neonatal cerebellar development. In: Seminars in Fetal and Neonatal Medicine, vol. 10, pp. 411–420. Elsevier (2005)

28. Ronneberger, O., Fischer, P., Brox, T.: U-net: Convolutional networks for biomedical image segmentation. In: Navab, N., Hornegger, J., Wells, W., Frangi, A. (eds.) MICCAI 2015. LNCS, vol. 9351, pp. 234–241. Springer, Cham (2015). https://doi.org/10.1007/978-3-319-24574-4_28

29. Krebs, J., Mansi, T., Mailhé, B., Ayache, N., Delingette, H.: Unsupervised probabilistic deformation modeling for robust diffeomorphic registration. In: Stoyanov, D., et al. (eds.) DLMIA 2018, ML-CDS 2018. LNCS, vol. 11045, pp. 101–109. Springer, Cham (2018). https://doi.org/10.1007/978-3-030-00889-5_12

30. Chen, L., Qu, H., Zhao, J., Chen, B., Principe, J.C.: Efficient and robust deep learning with correntropy-induced loss function. Neural Comput. Appl. 27(4), 1019–1031 (2016)

31. de Vos, B.D., Berendsen, F.F., Viergever, M.A., Sokooti, H., Staring, M., Išgum, I.: A deep learning framework for unsupervised affine and deformable image registration. Med. Image Anal. 52, 128–143 (2019)

32. Howell, B.R., et al.: The UNC/UMN baby connectome project (BCP): an overview of the study design and protocol development. NeuroImage 185, 891–905 (2019)

33. Avants, B., Gee, J.C.: Geodesic estimation for large deformation anatomical shape averaging and interpolation. Neuroimage 23, S139–S150 (2004)

Nesterov Accelerated ADMM for Fast Diffeomorphic Image Registration

Alexander Thorley[1], Xi Jia[1], Hyung Jin Chang[1], Boyang Liu[2],
Karina Bunting[2], Victoria Stoll[2], Antonio de Marvao[3], Declan P. O'Regan[3],
Georgios Gkoutos[4], Dipak Kotecha[2], and Jinming Duan[1(✉)]

[1] School of Computer Science, University of Birmingham, Birmingham, UK
j.duan@cs.bham.ac.uk
[2] Institute of Cardiovascular Sciences, University of Birmingham, Birmingham, UK
[3] MRC London Institute of Medical Sciences, Imperial College London, London, UK
[4] Institute of Cancer and Genomic Sciences, University of Birmingham,
Birmingham, UK

Abstract. Deterministic approaches using iterative optimisation have
been historically successful in diffeomorphic image registration (DiffIR).
Although these approaches are highly accurate, they typically carry
a significant computational burden. Recent developments in stochastic
approaches based on deep learning have achieved sub-second runtimes
for DiffIR with competitive registration accuracy, offering a fast alternative to conventional iterative methods. In this paper, we attempt to
reduce this difference in speed whilst retaining the performance advantage of iterative approaches in DiffIR. We first propose a simple iterative
scheme that functionally composes intermediate non-stationary velocity
fields to handle large deformations in images whilst guaranteeing diffeomorphisms in the resultant deformation. We then propose a convex
optimisation model that uses a regularisation term of arbitrary order to
impose smoothness on these velocity fields and solve this model with a
fast algorithm that combines Nesterov gradient descent and the alternating direction method of multipliers (ADMM). Finally, we leverage
the computational power of GPU to implement this accelerated ADMM
solver on a 3D cardiac MRI dataset, further reducing runtime to less
than 2 s. In addition to producing strictly diffeomorphic deformations,
our methods outperform both state-of-the-art deep learning-based and
iterative DiffIR approaches in terms of dice and Hausdorff scores, with
speed approaching the inference time of deep learning-based methods.

Keywords: Image registration · Diffeomorphism · ADMM

1 Introduction

Over the past two decades, diffeomorphic image registration (DiffIR) has become
a powerful tool for deformable image registration. The goal of DiffIR is to find a

Electronic supplementary material The online version of this chapter (https://
doi.org/10.1007/978-3-030-87202-1_15) contains supplementary material, which is
available to authorized users.

smooth and invertible spatial transformation between two images, such that every point in one image has a corresponding point in the other. Such transformations are known as diffeomorphisms and are fundamental inputs for computational anatomy in medical imaging.

Beg et als. pioneering work developed the large-deformation diffeomorphic metric mapping (LDDMM) framework [5], formulated as a variational problem where the diffeomorphism is represented by a geodesic path (flow) parameterised by time-dependent (non-stationary) smooth velocity fields through an ordinary differential equation. Vialard et al. proposed a geodesic shooting algorithm [24] in which the geodesic path of diffeomorphisms can be derived from the Euler-Poincaré differential (EPDiff) equation given only the initial velocity field. In this case, geodesic shooting requires only an estimate of the initial velocity rather than the entire sequence of velocity fields required by LDDMM [5], thus reducing the optimisation complexity. Recently, Zhang et al. developed the Fourier-approximated Lie algebras for shooting (FLASH) [28] for DiffIR, where they improved upon the geodesic shooting algorithm [22,24] by speeding up the calculation of the EPDiff equation in the Fourier space. However, solutions of these approaches [5,24,28] reply on gradient descent, which requires many iterations to converge and is therefore very slow. To reduce the computational cost of DiffIR, Ashburner represented diffeomorphsims by stationary velocity fields (SVFs) [1], and proposed a fast DiffIR algorithm (DARTEL) [2] by composing successive diffeomorphisms using scaling and squaring. However, SVFs do not provide geodesic paths between images which may be critical to statistical shape analysis. Another approach for fast DiffIR is demons [23], although this is a heuristic method and does not have a clear energy function to minimise.

An alternative approach to improve DiffIR speed is to leverage deep learning, usually adopting convolutional neural networks (CNNs) to learn diffeomorphic transformations between pairwise images in a training dataset. Registration after training is then achieved efficiently by evaluating the network on unseen image pairs with or without further (instance-specific) optimisation [4]. Deep learning in DiffIR can be either supervised or unsupervised. Yang et al. proposed Quicksilver [26] which utilises a supervised encoder-decoder network as the patch-wise prediction model. Quicksilver is trained with the initial momentum[1] of LDDMM as the supervision signal, which does not need to be spatially smooth and hence facilitates a fast patch-wise training strategy. Wang et al. extended FLASH [28] to DeepFLASH [25] in a learning framework, which predicts the initial velocity field in the Fourier space (termed the low dimensional bandlimited space in their paper). DeepFLASH is more efficient in general as the backbone network consists of only a decoder. As Quicksilver and DeepFLASH need to be trained using the numerical solutions of LDDMM [22] as ground truth, their performance may be bounded by that of LDDMM [22]. Dalca et al. proposed diffeomorphic VoxelMorph [9] that leverages the U-Net [20] and the spatial transformer network [15]. Mok et al. developed a fast symmetric method (SymNet) [18] that guarantees topology preservation and invertibility of the transformation. Both

---

[1] Momentum is the dual form of velocity. They are connected by a symmetric, positive semi-definite Laplacian operator as defined in Eq. (6).

Voxelmorph and SymNet impose diffeomorphisms using SVFs in an unsupervised fashion. Whilst the fast inference speed of deep learning is a clear benefit, the need for large quantities of training data and the lack of robust mathematical formulation in comparison with deterministic iterative approaches means both paradigms still have their advantages.

The goal of this study is to reduce the speed gap between learning-based and iterative approaches in DiffIR, while still retaining the performance advantage of iterative approaches. We first propose an iterative scheme to decompose a diffeomorphic transformation into the compositions of a series of small non-stationary velocity fields. Next, we define a clear, general and convex optimisation model to compute such velocity fields. The data term of the model can be either $\mathcal{L}^1$ or $\mathcal{L}^2$ norm and the $\mathcal{L}^2$ regularisation term uses a derivative of arbitrary order whose formulation follows the multinomial theorem. We then propose a fast solver for the proposed general model, which combines Nesterov acceleration [19] and the alternating direction method of multipliers (ADMM) [6,13,17]. By construction, all resulting subproblem solutions are point-wise and closed-form without any iteration. As such, the solver is very accurate and efficient, and can be implemented using existing deep learning frameworks (e.g. PyTorch) for further acceleration via GPU. We show on a 3D cardiac MRI dataset that our methods outperform both state-of-the-art learning and optimisation-based approaches, with a speed approaching the inference time of learning-based methods.

2   Diffeomorphic Image Registration

Let $\phi : \mathbf{R}^3 \rightarrow \mathbf{R}^3$ denote the deformation field that maps the coordinates from one image to those in another image. Computing a diffeomorphic deformation can be treated as modelling a dynamical system [5], given by the ordinary differential equation (ODE): $\partial\phi/\partial t = \mathbf{v}_t(\phi_t)$, where $\phi_0 = $ Id is the identity transformation and $\mathbf{v}_t$ indicates the velocity field at time t ($\in [0,1]$). A straightforward approach to numerically compute the ODE is through Euler integration, in which case the final deformation field ϕ_1 is calculated from the compositions of a series of small deformations. More specifically, a series of N velocity fields are used to represent the time varying velocity field $\mathbf{v}_t$. For N uniformly spaced time steps $(0, t_1, t_2, ..., t_{N-2}, t_{N-1})$, ϕ_1 can be achieved by:

$$\phi_1 = \left(\text{Id} + \frac{\mathbf{v}_{t_{N-1}}}{N}\right) \circ \left(\text{Id} + \frac{\mathbf{v}_{t_{N-2}}}{N}\right) \circ ... \circ \left(\text{Id} + \frac{\mathbf{v}_{t_1}}{N}\right) \circ \left(\text{Id} + \frac{\mathbf{v}_0}{N}\right), \quad (1)$$

where $\circ$ denotes function composition. This greedy composite approach is seen in some DiffIR works [7,23] and if the velocity fields $\mathbf{v}_{t_i}, \forall i \in \{0, ..., N-1\}$ are sufficiently small whilst satisfying some smoothness constraint, the resulting compositions should result in a deformation that is diffeomorphic [8].

The fact that the numerical solver (1) requires small and smooth velocity fields motivates us to consider classical optical flow models, among which a simple yet effective method was proposed by Horn and Schunck (HS), who computed velocity $\mathbf{v}$ through $\min_{\mathbf{v}}\{\frac{1}{2}\|\langle\nabla I_1, \mathbf{v}\rangle + I_1 - I_0\|^2 + \frac{\lambda}{2}\|\nabla\mathbf{v}\|^2\}$ [14]. In the minimisation problem, the first data term imposes brightness constancy, where

$I_0 \in \mathbb{R}^{MNH}$ (MNH is the image size) and $I_1 \in \mathbb{R}^{MNH}$ are a pair of input images and $\langle \cdot, \cdot \rangle$ denotes inner product. The second regularisation term induces spatial smoothness on $\mathbf{v} \in (\mathbb{R}^{MNH})^3$, where λ is a hyperparameter and ∇ denotes the gradient operator. As the HS data term is based on local Taylor series approximations of the image signal, it is only capable of recovering small velocity fields, which together with the smooth regularisation makes the HS model perfect to combine with (1) for effective diffeomorphic registration.

Based on this observation, we propose a simple iterative scheme for DiffIR. For a pair of images I_0 (target image) and I_1 (source image), first we solve the HS model for $\mathbf{v}_{t_{N-1}}^*$. Of note, as both terms in the HS model are linear and quadratic, if we derive the first-order optimality condition with respect to $\mathbf{v}$ from the model, we end up with a system of linear equations (i.e. regularised normal equation), which can be solved analytically by matrix inversion. We then warp the source image I_1 to I_1^ω using $I_1^\omega = I_1 \circ (\mathrm{Id} + \mathbf{v}_{t_{N-1}}^*)$. Note that we set the denominator coefficient N to 1 in this paper. Next, we pass I_1^ω and I_0 to the HS model again and solve for $\mathbf{v}_{t_{N-2}}^*$, with which we update the warped source image I_1^ω with $I_1^\omega = I_1 \circ (\mathrm{Id} + \mathbf{v}_{t_{N-1}}^*) \circ (\mathrm{Id} + \mathbf{v}_{t_{N-2}}^*)$. We repeat this iterative process until that the final warped source image $I_1^\omega = I_1 \circ (\mathrm{Id} + \mathbf{v}_{t_{N-1}}^*) \circ (\mathrm{Id} + \mathbf{v}_{t_{N-2}}^*) \circ \ldots \circ (\mathrm{Id} + \mathbf{v}_{t_1}^*) \circ (\mathrm{Id} + \mathbf{v}_0^*)$ is close to the target image I_0 within a pre-defined, small and positive tolerance. We note that unlike [5, 22, 24, 28] there is no need for the proposed iterative scheme to pre-define the number of velocity fields as it is automatically computed based on the similarity between the warped and target images, however we can and will impose a restriction on the number of iterations to reduce computation time.

The HS model however suffers from the fact that the quadratic data term is not robust to outliers and that the first-order diffusion regularisation performs poorly against large deformations (which often requires an additional affine linear pre-registration to be successful [12]). To circumvent these shortcomings of the HS model, we propose a new, general variational model, given by

$$\min_{\mathbf{v}} \left\{ \frac{1}{s} \| \rho(\mathbf{v}) \|^s + \frac{\lambda}{2} \| \nabla^n \mathbf{v} \|^2 \right\}, \tag{2}$$

where $\rho(\mathbf{v}) = \langle \nabla I_1, \mathbf{v} \rangle + I_1 - I_0$, $s = \{1, 2\}$ and $n > 0$ is an integer number. ∇^n denotes an arbitrary order gradient. For a scalar-valued function $f(x, y, z)$, $\nabla^n f = [(\begin{smallmatrix} n \\ k_1, k_2, k_3 \end{smallmatrix}) \frac{\partial^n f}{\partial x^{k_1} \partial y^{k_2} \partial z^{k_3}}]^{\mathrm{T}}$, where $k_i = 0, ..., n$, $i \in \{1, 2, 3\}$ and $k_1 + k_2 + k_3 = n$. Moreover, $(\begin{smallmatrix} n \\ k_1, k_2, k_3 \end{smallmatrix})$ are known as multinomial coefficients which are computed by $\frac{n!}{k_1! k_2! k_3!}$. Note that if we set $s = 2$ and $n = 1$, the proposed model is recovered to the original HS model. However, if we set $s = 1$, the data term will be the sum absolute differences (SAD) [27] which is more robust to outliers in images. On the other hand, the proposed regularisation has a general form, which allows reconstruction of a smooth polynomial function of arbitrary degree for the velocity. Unfortunately, solving the proposed model becomes non-trivial due to the non-differentiality in the data term and the generality in the derivative order. We note that even if we only solve the original HS model, we still need to do matrix inversion which is impractical for 3D high-dimensional data used in this paper. To address these issues, we propose an effective algorithm in the following section to minimise the proposed convex model efficiently.

3   Nesterov Accelerated ADMM

In order to minimise the convex problem (2) efficiently, we utilise the alternating direction method of multipliers (ADMM) [6,13,17] accelerated by Nesterov's approach for gradient descent methods [19]. First, we introduce an auxiliary primal variable $\mathbf{w} \in (\mathbb{R}^{MNH})^3$, converting (2) into the following equivalent form

$$\min_{\mathbf{v}} \left\{ \frac{1}{s}\|\rho(\mathbf{v})\|^s + \frac{\lambda}{2}\|\nabla^n\mathbf{w}\|^2 \right\} \quad s.t. \; \mathbf{w} = \mathbf{v}. \tag{3}$$

The introduction of the constraint $\mathbf{w} = \mathbf{v}$ decouples $\mathbf{v}$ in the regularisation from that in the data term so that we can derive closed-form, point-wise solutions with respect to both variables regardless of the non-differentiality and generality of the problem (2). To guarantee an optimal solution, the above constrained problem can be converted to a saddle problem solved by the proposed fast ADMM. Let $\mathcal{L}_A(\mathbf{v}, \mathbf{w}; \mathbf{b})$ be the augmented Lagrange functional of (3), defined as

$$\mathcal{L}_A(\mathbf{v}, \mathbf{w}; \mathbf{b}) = \frac{1}{s}\|\rho(\mathbf{v})\|^s + \frac{\lambda}{2}\|\nabla^n\mathbf{w}\|^2 + \frac{\theta}{2}\|\mathbf{w} - \mathbf{v} - \mathbf{b}\|^2, \tag{4}$$

where $\mathbf{b} \in (\mathbb{R}^{MNH})^3$ is the augmented Lagrangian multiplier (aka. dual variable), and $\theta > 0$ is the penalty parameter which has impact on how fast the minimisation process converges. It is known that one of the saddle points for (4) will give a minimiser for the constrained minimisation problem (3). We use the Nesterov accelerated ADMM to optimise variables in (4), which is given as

$$\begin{cases} \mathbf{v}^k = \arg\min_{\mathbf{v}} \frac{1}{s}\|\rho(\mathbf{v})\|^s + \frac{\theta}{2}\|\widehat{\mathbf{w}}^k - \mathbf{v} - \widehat{\mathbf{b}}^k\|^2 \\ \mathbf{w}^k = \arg\min_{\mathbf{w}} \frac{\lambda}{2}\|\nabla^n\mathbf{w}\|^2 + \frac{\theta}{2}\|\mathbf{w} - \mathbf{v}^k - \widehat{\mathbf{b}}^k\|^2 \\ \mathbf{b}^k = \widehat{\mathbf{b}}^k + \mathbf{v}^k - \mathbf{w}^k \\ \alpha^{k+1} = \frac{1+\sqrt{1+4(\alpha^k)^2}}{2} \\ \widehat{\mathbf{w}}^{k+1} = \mathbf{w}^k + \frac{\alpha^k-1}{\alpha^{k+1}}(\mathbf{w}^k - \mathbf{w}^{k-1}) \\ \widehat{\mathbf{b}}^{k+1} = \mathbf{b}^k + \frac{\alpha^k-1}{\alpha^{k+1}}(\mathbf{b}^k - \mathbf{b}^{k-1}) \end{cases} \tag{5}$$

At the beginning of the iteration, we set $\widehat{\mathbf{w}}^{-1} = \widehat{\mathbf{w}}^0 = \mathbf{0}$, $\widehat{\mathbf{b}}^{-1} = \widehat{\mathbf{b}}^0 = \mathbf{0}$ and $\alpha^0 = 1$. k above denotes the number of iterations. As can be seen from (5), we decompose (4) into two convex subproblems with respect to $\mathbf{v}$ and $\mathbf{w}$, update the dual variable $\mathbf{b}$, and then apply Nesterov's acceleration to both primal variable $\mathbf{w}$ and dual variable $\mathbf{b}$. This accelerated iterative process is repeated until the optimal solution $\mathbf{v}^*$ is found. The convergence of this accelerated ADMM has been proved in [13] for convex problems and the authors have also demonstrated the accelerated ADMM has a convergence rate of $\mathcal{O}(1/k^2)$, which is the optimal rate for first-order algorithms. Moreover, due to the decoupling we are able to derive a closed-form, point-wise solution for both subproblems of $\mathbf{v}$ and $\mathbf{w}$

$$\begin{cases} \mathbf{v}^k = \widehat{\mathbf{w}}^k - \widehat{\mathbf{b}}^k - \frac{\hat{z}}{\max(|\hat{z}|,1)}\frac{\nabla I_1}{\theta} & \text{if } s = 1 \\ (\mathbf{J}\mathbf{J}^T + \theta\mathbb{1})\mathbf{v}^k = \theta(\widehat{\mathbf{w}}^k - \widehat{\mathbf{b}}^k) - \mathbf{J}(I_1 - I_0) & \text{if } s = 2 \;. \\ \mathbf{w}^k = \mathcal{F}^{-1}(\mathcal{F}(\mathbf{v}^k + \widehat{\mathbf{b}}^k)/(\lambda\mathcal{F}(\Delta^n) + \theta)) \end{cases} \tag{6}$$

In the case of $s = 1$, we have $\hat{z} = \theta\rho(\widehat{\mathbf{w}}^k - \widehat{\mathbf{b}}^k)/(|\nabla I_1|^2 + \epsilon)$, where $\epsilon > 0$ is a small, positive number to avoid division by zeros. In the case of $s = 2$, $\mathbf{J}\mathbf{J}^T$ (where $\mathbf{J} = \nabla I_1$) is a rank-1 outer product and $\mathbb{1}$ is an identity matrix. As the respective $\mathbf{v}$-subproblem in this case is differentiable, we can derive the Sherman–Morrison formula (i.e. see (6) middle equation) by directly differentiating this subproblem. Due to the identity matrix, the Sherman–Morrison formula [23] will lead to a closed-form, point-wise solution to $\mathbf{v}^k$. In the last equation, $\mathcal{F}$ and $\mathcal{F}^{-1}$ respectively denote the discrete Fourier transformation (DFT) and inverse DFT, and Δ^n is the nth-order Laplace operator which are discretised via the finite difference [11,17]. In this case, $\mathcal{F}(\Delta^n) = 2^n(3 - \cos(\frac{2\pi p}{M}) - \cos(\frac{2\pi q}{N}) - \cos(\frac{2\pi r}{H}))^n$, where $p \in [0, M)$, $q \in [0, N)$ and $r \in [0, H)$ are grid indices. We note that all three solutions are closed-form and point-wise, and therefore can be computed very efficiently in 3D.

4 Experimental Results

Dataset: The dataset used in our experiments consists of 220 pairs of 3D high-resolution (HR) cardiac MRI images corresponding to the end diastolic (ED) and end systolic (ES) frames of the cardiac cycle. The raw volumes have a resolution of $1.2 \times 1.2 \times 2.0\,\mathrm{mm}^3$. HR imaging requires only one single breath-hold and therefore introduces no inter-slice shift artifacts [10]. Class segmentation labels at ED and ES are also contained in the dataset for the right ventricle (RV), left ventricle (LV) and left ventricle myocardium (LVM). All images are resampled to $1.2 \times 1.2 \times 1.2\,\mathrm{mm}^3$ resolution and cropped or padded to matrix size $128 \times 128 \times 96$. To train comparative deep learning methods and tune hyperparameters in different methods, the dataset is split into $100/20/100$ corresponding to training, validation and test sets. We report final quantitative results on the test set only (Table 1).

Hyperparameter Tuning: We embed our approach into a three-level pyramid scheme to avoid convergence to local minima, resulting in a total of three loops in the final implementation. We tune two versions of hyperparameters for our model, the first with limited iterations to minimise runtime, and the second without such restrictions leading to a slower runtime but higher accuracy. For the first, we restrict the iteration number within the ADMM loop in (5) to ten in the first pyramid level and five in the following two levels. We also cap the number of warpings in (1) to ten in the first two pyramid levels and five in the final level. For the second, iterations are only terminated when a convergence threshold (range $[0.01, 0.13]$) for the ADMM loop is met and the difference between two warpings is below 2%. This range of thresholds ensures runtime is competitive with other iterative methods whilst increasing accuracy compared to our first version. For both versions, the validation set is used to tune hyperparameters λ (range $[0, 60]$) and θ (range $[0.01, 0.15]$) to maximise dice score for each combination of data term ($\mathcal{L}^2$ or $\mathcal{L}^1$) and regulariser of order 1 to 3 ($1^{st}O$, $2^{nd}O$ or $3^{rd}O$). See supplementary for the optimal values of λ and θ.

Comparative Methods: We compare our methods against a variety of state-of-the-art iterative and deep learning-based image registration methods. For iterative methods, we use the MIRTK implementation of free form deformations

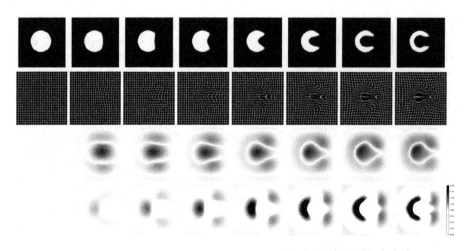

Fig. 1. A toy experiment showing evolution of diffeomophism throughout iterations using our method. Rows one to four correspond to image, grid deformation, HSV deformation and jacobian determinant, respectively. Columns one to eight correspond to increasing iterations.

(FFD) [21] with a three-level pyramid scheme. Control point spacing is selected to maximise dice score on the validation set. We also compare with the popular diffeomorpic demons algorithm [23] implemented in SimpleITK [16], using a three-level pyramid scheme. The number of iterations and smoothing $\lambda's$ are optimised on the validation set. We also compare with ANTs SyN [3] using the official ANTs implementation with mean square distance and four registration levels. The hyper-parameters such as smoothing parameters, gradient step size, and the number of iterations are selected based on the validation set. For deep learning, we train both VoxelMorph [4] and its diffeomorphic version [9] on the training set. We adopt a lighter 5-level hierarchical U-shape network [18] than the original Voxelmorph paper [4] as our 128×128×96 images require significant GPU memory. Both VoxelMorphs are optimised under an SAD loss defining image similarity, a diffusion regularisation loss to enforce smoothness of the velocity field and a local orientation consistency loss [18] that regularises the Jacobian determinant of the velocity field. For the diffeomorphic VoxelMorph, an additional 7 scaling and squaring layers are used to encourage diffeomorphic properties in the resulting deformation. The weights of smoothness regularisation and local orientation consistency are tuned to 0.1 and 1000 respectively on the validation set, and both trained for 30000 iterations with batch size 2 and learning rate 0.0001 on a GTX 1080Ti GPU. Training takes around 13 h.

Quantitative Results: To evaluate performance, we use the deformations produced by each method to warp the associated source (ES) class label to its corresponding target (ED) label. The warped source and target labels are then used to compute the dice similarity coefficient and Hausdorff distance between pairs, with Hausdorff distances averaged over each 2D slice within the 3D volumes. We

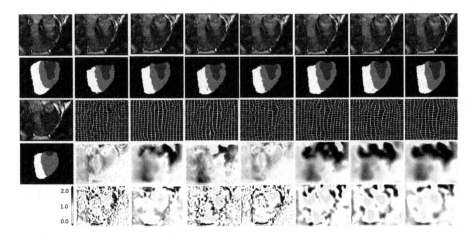

Fig. 2. Visual results of different methods. Column 1: target image, target label, source image, source label. Columns 2–8: warped source image, warped source label, grid$_\phi$, HSV$_\phi$, det(J_ϕ) for (1) Voxelmorph, (2) SyN, (3) Demons, (4) diffeomorphic voxelmorph, (5) FFD, (6) $\mathcal{L}^1+3^{rd}O$ and (7) $\mathcal{L}^2+3^{rd}O$, respectively.

compute the percentage of non-positive voxels in the Jacobian J_ϕ of each deformation ϕ to assess its diffeomorphic properties, for which a purely diffeomorphic deformation should be zero everywhere. We also compare runtime, measuring speed of iteritive methods and inference time for two Voxelmorph models. All methods using a GPU are tested on a 16GB NVIDIA Tesla V100.

Table 1. Statistics of each method averaged over test set. Dice and Hausdorff scores are denoted as an average across RV, LV and LVM. Standard deviations are shown after $\pm$. Hausdorff distance is measured in mm and runtime in seconds. Results of our methods using 2nd hyperparameter version are given in brackets.

Method	Dice	Hausdorff	% of $|J_\phi| \leq 0$	Runtime
Unreg	$.493 \pm .043$	8.404 ± 0.894	–	–
Voxelmorph	$.709 \pm .032$	7.431 ± 1.049	0.01 ± 0.01	$\mathbf{0.09 \pm 0.23}$
SyN	$.721 \pm .051$	6.979 ± 1.233	0.00 ± 0.00	77.39 ± 9.57
Demons	$.727 \pm .040$	6.740 ± 0.998	0.00 ± 0.00	13.01 ± 0.17
Diff-Voxelmorph	$.730 \pm .031$	6.498 ± 0.975	0.01 ± 0.01	$\mathbf{0.09 \pm 0.21}$
FFD	$.739 \pm .047$	7.044 ± 1.200	0.77 ± 0.30	8.98 ± 1.97
$\mathcal{L}^1+1^{st}O$	$.727 \pm .039$ ($.728 \pm .040$)	6.949 ± 1.056 (6.911 ± 1.092)	$\mathbf{0 \pm 0}$ ($\mathbf{0 \pm 0}$)	$1.63 \pm .26$ ($7.90 \pm .79$)
$\mathcal{L}^2+1^{st}O$	$.719 \pm .039$ ($.726 \pm .041$)	7.499 ± 1.227 (7.105 ± 1.199)	$\mathbf{0 \pm 0}$ ($\mathbf{0 \pm 0}$)	$2.17 \pm .50$ ($4.69 \pm .34$)
$\mathcal{L}^1+2^{nd}O$	$.749 \pm .042$ ($.761 \pm .041$)	6.891 ± 1.134 ($\mathbf{6.364 \pm 1.036}$)	$\mathbf{0 \pm 0}$ ($\mathbf{0 \pm 0}$)	$1.43 \pm .05$ ($5.02 \pm .57$)
$\mathcal{L}^2+2^{nd}O$	$.735 \pm .043$ ($.751 \pm .045$)	6.895 ± 1.212 (6.855 ± 1.231)	$\mathbf{0 \pm 0}$ ($\mathbf{0 \pm 0}$)	$1.38 \pm .04$ ($5.00 \pm .41$)
$\mathcal{L}^1+3^{rd}O$	$.753 \pm .042$ ($\mathbf{.768 \pm .042}$)	6.963 ± 1.100 (6.515 ± 1.077)	$\mathbf{0 \pm 0}$ ($\mathbf{0 \pm 0}$)	$1.57 \pm .24$ ($4.49 \pm .56$)
$\mathcal{L}^2+3^{rd}O$	$.736 \pm .042$ ($.757 \pm .046$)	7.302 ± 1.237 (6.927 ± 1.264)	$\mathbf{0 \pm 0}$ ($\mathbf{0 \pm 0}$)	$1.37 \pm .03$ ($4.72 \pm .41$)

158   A. Thorley et al.

In Fig. 2, we show the visual comparison between all methods, where grid_ϕ shows the deformation applied to a unit grid, $\det(\mathbf{J}_\phi)$ the jacobian determinant and HSV_ϕ the final composition of each iterative velocity field, with hue depicting direction and saturation magnitude. In Table 1, we show the final results of each method averaged over the test set. Results from both versions of hyperparameters are displayed with the latter in brackets. It can be seen that our accelerated ADMM outperforms all other methods by at least 1% in terms of dice score despite limiting the number of iterations and warpings, increasing to nearly 3% for our second version. There is also a clear upward tendency in dice (see Fig. 3), further demonstrating the superiority of our methods. Moreover, our methods consistently maintain diffeomorphic properties with no negative elements present in $\mathbf{J}_\phi$. Although the speed of our method is still slower than deep learning methods, the average runtime we achieved for our first hyperparameter version is barely over a second.

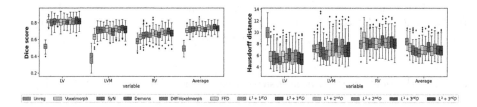

Fig. 3. Boxplots of Dice and Hausdorff for left ventricle (LV), left ventricle myocardium (LVM) and right ventricle (RV) of all methods on test set.

5 Conclusion

In this paper, we proposed an accelerated algorithm that combines Nesterov gradient descent and ADMM, which tackles a general variational model with a regularisation term of arbitrary order. Through the compositions of a series of velocity fields produced by our accelerated ADMM, we model deformations as the solution to a dynamical system to ensure diffeomorphic properties. We implement our methods in the PyTorch framework, leveraging the power of a GPU to further accelerate our methods. In addition to reducing the difference in speed between deep learning to under 2 s, our methods have achieved a new state-of-the-art performance for the DiffIR task.

Acknowledgements. The research is supported by the BHF Accelerator Award (AA/18/2/34218), the Ramsay Research Fund from the School of Computer Science at the University of Birmingham and the Wellcome Trust Institutional Strategic Support Fund: Digital Health Pilot Grant.

References

1. Arsigny, V., Commowick, O., Pennec, X., Ayache, N.: A log-Euclidean framework for statistics on diffeomorphisms. In: Larsen, R., Nielsen, M., Sporring, J. (eds.) MICCAI 2006. LNCS, vol. 4190, pp. 924–931. Springer, Heidelberg (2006). https://doi.org/10.1007/11866565_113

2. Ashburner, J.: A fast diffeomorphic image registration algorithm. NeuroImage **38**(1), 95–113 (2007)

3. Avants, B.B., Tustison, N.J., Song, G., Cook, P.A., Klein, A., Gee, J.C.: A reproducible evaluation of ANTs similarity metric performance in brain image registration. NeuroImage **54**(3), 2033–2044 (2011)

4. Balakrishnan, G., Zhao, A., Sabuncu, M.R., Guttag, J., Dalca, A.V.: Voxelmorph: a learning framework for deformable medical image registration. IEEE Trans. Med. Imaging **38**(8), 1788–1800 (2019)

5. Beg, M.F., Miller, M.I., Trouvé, A., Younes, L.: Computing large deformation metric mappings via geodesic flows of diffeomorphisms. Int. J. Comput. Vis. **61**(2), 139–157 (2005)

6. Boyd, S., Parikh, N., Chu, E.: Distributed Optimization and Statistical Learning via the Alternating Direction Method of Multipliers. Now Publishers Inc., Delft (2011)

7. Christensen, G.E., Rabbitt, R.D., Miller, M.I.: Deformable templates using large deformation kinematics. IEEE Trans. Image Process. **5**(10), 1435–1447 (1996)

8. Christensen, G.E., et al.: Topological properties of smooth anatomic maps. In: Information Processing in Medical Imaging, vol. 3, pp. 101–112. Kluwer Academic, Boston (1995)

9. Dalca, A.V., Balakrishnan, G., Guttag, J., Sabuncu, M.R.: Unsupervised learning of probabilistic diffeomorphic registration for images and surfaces. Med. Image Anal. **57**, 226–236 (2019)

10. Duan, J., et al.: Automatic 3D bi-ventricular segmentation of cardiac images by a shape-refined multi-task deep learning approach. IEEE Trans. Med. Imaging **38**(9), 2151–2164 (2019)

11. Duan, J., Qiu, Z., Lu, W., Wang, G., Pan, Z., Bai, L.: An edge-weighted second order variational model for image decomposition. Digit. Sig. Process. **49**, 162–181 (2016)

12. Fischer, B., Modersitzki, J.: Curvature based image registration. J. Math. Imaging Vis. **18**(1), 81–85 (2003)

13. Goldstein, T., O'Donoghue, B., Setzer, S., Baraniuk, R.: Fast alternating direction optimization methods. SIAM J. Imaging Sci. **7**(3), 1588–1623 (2014)

14. Horn, B.K., Schunck, B.G.: Determining optical flow. Artif. Intell. **17**(1–3), 185–203 (1981)

15. Jaderberg, M., Simonyan, K., Zisserman, A., Kavukcuoglu, K.: Spatial transformer networks. In: Proceedings of the 28th International Conference on Neural Information Processing Systems, vol. 2, pp. 2017–2025 (2015)

16. Lowekamp, B.C., Chen, D.T., Ibáñez, L., Blezek, D.: The design of SimpleITK. Front. Neuroinform. **7**, 45 (2013)

17. Lu, W., Duan, J., Qiu, Z., Pan, Z., Liu, R.W., Bai, L.: Implementation of high-order variational models made easy for image processing. Math. Methods Appl. Sci. **39**(14), 4208–4233 (2016)

18. Mok, T.C., Chung, A.: Fast symmetric diffeomorphic image registration with convolutional neural networks. In: Proceedings of the IEEE/CVF Conference on Computer Vision and Pattern Recognition, pp. 4644–4653 (2020)

19. Nesterov, Y.E.: A method of solving a convex programming problem with convergence rate $O(1/k^2)$. In: Doklady Akademii Nauk, vol. 269, pp. 543–547. Russian Academy of Sciences (1983)

20. Ronneberger, O., Fischer, P., Brox, T.: U-Net: convolutional networks for biomedical image segmentation. In: Navab, N., Hornegger, J., Wells, W., Frangi, A. (eds.) MICCAI 2015. LNCS, vol. 9351cience. Springer, Cham (2015). https://doi.org/10.1007/978-3-319-24574-4_28

21. Rueckert, D., Sonoda, L.I., Hayes, C., Hill, D.L., Leach, M.O., Hawkes, D.J.: Non-rigid registration using free-form deformations: application to breast MR images. IEEE Trans. Med. Imaging 18(8), 712–721 (1999)

22. Singh, N., Hinkle, J., Joshi, S., Fletcher, P.T.: A vector momenta formulation of diffeomorphisms for improved geodesic regression and atlas construction. In: 2013 IEEE 10th International Symposium on Biomedical Imaging, pp. 1219–1222. IEEE (2013)

23. Vercauteren, T., Pennec, X., Perchant, A., Ayache, N.: Diffeomorphic demons: efficient non-parametric image registration. NeuroImage 45(1), S61–S72 (2009)

24. Vialard, F.X., Risser, L., Rueckert, D., Cotter, C.J.: Diffeomorphic 3D image registration via geodesic shooting using an efficient adjoint calculation. Int. J. Comput. Vis. 97(2), 229–241 (2012)

25. Wang, J., Zhang, M.: DeepFlash: an efficient network for learning-based medical image registration. In: Proceedings of the IEEE/CVF Conference on Computer Vision and Pattern Recognition, pp. 4444–4452 (2020)

26. Yang, X., Kwitt, R., Styner, M., Niethammer, M.: Quicksilver: fast predictive image registration-a deep learning approach. NeuroImage 158, 378–396 (2017)

27. Zach, C., Pock, T., Bischof, H.: A duality based approach for realtime TV-L^1 optical flow. In: Hamprecht, F.A., Schnörr, C., Jähne, B. (eds.) DAGM 2007. LNCS, vol. 4713, pp. 214–223. Springer, Heidelberg (2007). https://doi.org/10.1007/978-3-540-74936-3_22

28. Zhang, M., Fletcher, P.T.: Fast diffeomorphic image registration via Fourier-approximated lie algebras. Int. J. Comput. Vis. 127(1), 61–73 (2019)

Spectral Embedding Approximation and Descriptor Learning for Craniofacial Volumetric Image Correspondence

Diya Sun[1], Yungeng Zhang[1], Yuru Pei[1(✉)], Tianmin Xu[2],
and Hongbin Zha[1]

[1] Department of Machine Intelligence, Key Laboratory of Machine Perception
(MOE), Peking University, Beijing, China
peiyuru@cis.pku.edu.cn
[2] School of Stomatology, Peking University, Beijing, China

Abstract. Deformable image correspondence is crucial in various medical image research. Existing deep learning-based registration and correspondence models mostly learn a nonlinear voxel-wise mapping function between volumetric images by metric space alignments in the spatial domain, without addressing the intrinsic structure correspondence. Thus, the registration requires prior affine transformation or landmark annotations to handle high-frequency perturbations due to pose and structural variations. This paper presents a novel and efficient correspondence framework via low-dimensional spectral mapping to handle the intrinsic correspondence of anatomical structures. We devise a novel multipath graph convolutional network (GCN)-based embedding approximation module, relieving the time complexity in the eigendecomposition-based spectral embedding of volumetric images. We present a descriptor learning module and surpass the descriptor selection or hand-crafted descriptors. Experimental results demonstrate the efficacy of the core modules, i.e., the image embedding approximation and descriptor learning, for volumetric image correspondence and the atlas-based registration of craniofacial anatomical structures. The proposed approach achieves comparable corresponding accuracies with the state-of-the-art deep registration models, being resilient to pose and shape perturbations.

Keywords: Image correspondence · Descriptor learning · Spectral embedding approximation

1 Introduction

Deformable image registration and correspondence is a fundamental issue in medical image processing and has been studied for decades [12,24]. The estimation of voxel-wise correspondence and transformations facilitates atlas-based registration and segmentation, alleviating the tedious and time-consuming manual or interactive annotations, as well as labeling variations of practitioners. Existing image registration mostly relies on metric space alignment in the spatial

© Springer Nature Switzerland AG 2021
M. de Bruijne et al. (Eds.): MICCAI 2021, LNCS 12904, pp. 161–170, 2021.
https://doi.org/10.1007/978-3-030-87202-1_16

domain, prone to be trapped in a local minimum due to pose and structural variations. The seminal functional maps have been applied to dense correspondence and shape matching [21], which has been extended to the volumetric images [18,25] to handle the intrinsic structural correspondence in the spectral domain. By virtue of the graph Laplacian eigendecomposition, the dense correspondence is formulated as a low-dimensional spectral map by aligning projected shape descriptors in the intrinsic spectral domain of anatomical structures. However, the existing functional map suffers from the descriptor selection and the computational complexity in experiments.

First, the functional map relies on descriptor alignments so that the dense correspondence is affected by noisy and inconsistent hand-crafted descriptors. Several delicately designed norms [16] and regularizer [9,23] have been introduced to handle the inconsistent and less discriminative descriptors. The learning-based methods rely on the given corresponding landmarks to find the optimal descriptors [5,8,11,13,17]. However, it is not a trivial task to identify the optimal descriptors from raw volumetric images for map estimation and image correspondence. Second, the map computation is conducted in the spectral domain spanned by the eigenvectors of the $O(n^2)$ graph Laplacian matrix, where n is the number of graph nodes associated with voxels or supervoxels. The affinity or the graph Laplacian matrix increases quadratically with growing nodes. The spectral embedding requires the worst time complexity of $O(n^3)$ for eigendecomposition, limiting the scalability performance. The learning-based methods have been applied to graph representation and embedding [3,10,14,20,22]. Recently, the spectral-free techniques utilize the Chebyshev polynomial approximation to avoid the eigendecomposition in the spectral graph convolutions [7,15]. However, there is no guarantee to get the orthonormal spectral bases by the GCN-based embedding, which can not be applied directly to the spectral embedding and mapping of volumetric images.

To circumvent the delicately-designed descriptor selection and the time complexity issues on the spectral embedding of volumetric images, We introduce a spectral mapping-based image correspondence network (SMNet) as shown in Fig. 1. we devise a novel multi-path GCN-based approximation module for image embedding, addressing multiple frequency bands to span the spectral domain for image correspondence. A novel criterion is introduced to enforce the orthonormality of approximated spectral bases. The image correspondence is formulated as a low-dimensional spectral map computed by descriptor alignment in the spectral domain as the functional map [21]. A difficulty of image correspondence is the less semantically meaningful matching due to noisy and inconsistent descriptors. Instead of relying on prior hand-crafted volumetric descriptors, the proposed approach takes advantage of the intrinsic structural embedding and presents a map-guided optimized descriptors for semantic correspondence and the atlas-based registration. The image embedding approximation and descriptor learning modules are trained by projected descriptor alignments and distortion minimization. The proposed approach is applied to craniofacial cone-beam computed tomography (CBCT) obtained in clinical orthodontics. We conduct qualitative and quantitative evaluations to demonstrate the efficacy of our method compared with existing deep learning-based registration. The main contributions of this work include:

- The proposed descriptor learning and the multi-path GCN-based embedding approximation framework surpasses the descriptor selection and relieves the time complexity in spectral embedding, facilitating the generalization of the spectral map-based correspondence of volumetric images.
- The spectral map-based image correspondence is robust to high-frequency perturbations associated with pose and structural variations, relieving the prior affine transformation and landmark annotations.
- Extensive experimental results validate the proposed SMNet on spectral map-based image correspondence and the atlas-based segmentation of craniofacial anatomical structures.

2   Method

The goal is to estimate pairwise image correspondence and multi-category segmentations of craniofacial structures when given volumetric images $X, Y \in \mathbb{R}^n$ with n voxels. Without loss of generality, we decompose the volumes by the SLIC algorithm [1] and represent the image as a supervoxel graph $G(\mathcal{S}, \mathcal{E})$. $\mathcal{S}$ denotes the set of $|\mathcal{S}| = m$ supervoxel nodes, and $\mathcal{E}$ the set of edges connecting neighboring supervoxels. The proposed SMNet has two primary modules: the spectral embedding approximation module and the volumetric image descriptor

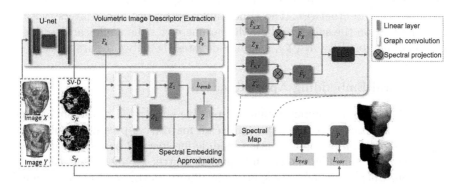

Fig. 1. Volumetric image correspondence by the proposed SMNet. Given the image pair and the supervoxel decomposition (SV-D), the framework utilizes a 3D U-net-based module for supervoxel descriptor $\hat{F}_s$. We introduce a multi-path GCN-based spectral basis approximation module, which avoids the computationally expensive spectral embedding by eigendecomposition. The SMNet is trained by descriptor alignment in the spectral space, where the supervoxel-wise correspondence is formulated as a low-dimensional spectral map C solved by the linear least squares (LLS). The spectral map is up-scaled to the supervoxel-wise matching probability P for image correspondence.

extraction module, as shown in Fig. 1. We utilize the 3D U-net-based network to extract multi-scale feature $F \in \mathbb{R}^{n \times q}$ from raw CBCTs, relieving the descriptor selection. The supervoxel descriptor $F_s \in \mathbb{R}^{m \times q}$ is defined as an average of voxel clustering according to the supervoxel decomposition. Instead of eigendecomposition of the large scale graph Laplacian matrix of the supervoxel graph, we present a multi-path GCN-based embedding approximation technique to address multiple frequency bands of the spectral bases $Z \in \mathbb{R}^{m \times k}$ spanning k-dimensional spectral space. The optimal low-dimensional spectral map $C \in \mathbb{R}^{k \times k}$ is computed by aligning projected descriptors $\tilde{F}$ in the embedding space, which is used to segment multi-category anatomical structures and to restore the matching probability matrix $P \in \mathbb{R}^{m \times m}$ of supervoxel graphs.

2.1   Volumetric Image Descriptor

We present a 3D U-net [4]-based module to extract multi-scale features from the raw one-channel volumetric image. The hierarchical encoder-decoder network with skip connections is used to extract the volumetric features. The feature maps obtained in the decoder with the resolution of 256×8^3, 128×16^3, 64×32^3, 32×64^3, and 4×128^3 are re-sampled and concatenated, resulting in $q = 484$-channel features $F \in \mathbb{R}^{n \times q}$. The supervoxel descriptor $F_s \in \mathbb{R}^{m \times q}$ are computed as an average of voxel clusters, and

$$F_{s,ij} = \frac{1}{|s_i|} \|\delta_{s,i} F_j\|_1, \tag{1}$$

where F_j denotes the j-th feature channel, $F_{s,ij}$ the j-th feature channel of supervoxel s_i. The delta function of supervoxel s_i is defined as $\delta_{s,i}(v) = 1$, if voxel $v \in s_i$ and 0 otherwise. The U-net, followed by two linear layers (with a kernel of 1×1) on the supervoxel graph, parameterizes the nonlinear descriptor extraction function $\hat{F}_s = h_\Theta(X)$, where Θ denotes the network parameters.

2.2   Spectral Embedding Approximation

In order to avoid the computationally expensive supervoxel graph embedding by eigendecomposition of the graph Laplacian matrix, we present a multi-path GCN-based embedding approximation module for spectral bases with multiple frequency bands. Consider the graph Laplacian matrix $L = D - A$ of the supervoxel graph, where A denotes the symmetric affinity matrix, and D the diagonal degree matrix. The spectra-free GCN utilizes the first-order Chebyshev polynomial approximation, where the convolution operation is defined as

$$Z^{(l+1)} = \sigma(D^{-1/2} A D^{-1/2} Z^{(l)} W^{(l)}). \tag{2}$$

$Z^{(l)} \in \mathbb{R}^{m \times k^{(l)}}$ denotes the $k^{(l)}$-channel signal at the l-th layer, which is initialized as $Z^{(0)} = F_s$ obtained in the volumetric descriptor extraction module. $W^{(l)}$ denotes the learnable weights. The multiplication with W can be viewed as the

1-hop aggregation on the supervoxel graph. σ denotes the tanh activation function. The resultant vectors $Z^{(l+1)}$ compose of the approximated spectral bases to minimize $tr(Z^T L Z)$. The difficulty is that the GCN implementation does not address the orthonormality of Z. Instead of using the QR iteration algorithm [19] to compute the orthogonal eigenvectors, we present a novel criterion on the embedding as follows:

$$\mathcal{L}_{emb} = \|Z^T Z - I_k\|_F^2 + \alpha \|off(Z^T L Z)\|_F^2, \tag{3}$$

where I_k denotes a $k \times k$ identity matrix. The first term is used to enforce the orthonormality of column vectors. The second term is used to ensure $Z^T L Z$ to be the diagonal matrix. Note that if Z is similar to eigenvectors of L, the product $Z^T L Z$ tends to be a diagonal matrix with the diagonal entry being related to eigenvalues. $off(M)$ has the off-diagonal elements equal to matrix M, and the diagonal entry is set to 0. The hyperparameter α is used to balance the orthonormal and the diagonalization constraints. As shown in Fig. 1, we utilize three GCN paths to approximate the spectral bases. The resultant embedding, i.e., Z_1, Z_2, and Z_3, are concatenated to approximate the spectral bases Z associated with the graph Laplacian matrix of the supervoxel graph.

2.3   Spectral Map-Based Correspondence

Spectral Map. When given the supervoxel descriptor $\hat{F}_{s,X}$ and $\hat{F}_{s,Y}$ (Sec. 2.1), and the approximated spectral bases Z_X and Z_Y (Sect. 2.2) of volumetric images X and Y, we compute the low-dimensional spectral map $C \in \mathbb{R}^{k \times k}$. Since $k \ll \min(m_X, m_Y)$, the supervoxel-wise correspondence computed by the spectral map is more efficient than the straightforward supervoxel-wise correspondence computation. m_X and m_Y denote the supervoxel node numbers of image X and Y, respectively. The supervoxel descriptors are projected using the approximated spectral bases. The resultant coefficient matrix $\tilde{F}_{s,X}$ and $\tilde{F}_{s,Y}$ satisfy $\hat{F}_{s,X} = Z_X \tilde{F}_{s,X}$ and $\hat{F}_{s,Y} = Z_Y \tilde{F}_{s,Y}$. The spectral map is computed by descriptor alignments in the spectral domain by minimizing $\|C^T \tilde{F}_{s,X} - \tilde{F}_{s,Y}\|_2^2$. The spectral map C can be solved by the linear least squares (LLS), where the i-th row vector $c_i = (\tilde{F}_{s,X} \tilde{F}_{s,X}^T)^{-1} \tilde{F}_{s,X} \tilde{f}_{Y,i}$. $\tilde{f}_{Y,i}$ denotes the i-th row of $\tilde{F}_{s,Y}$.

Loss. We train the volumetric image descriptor extraction module and the spectral embedding approximation module from both the real clinically obtained CBCT dataset $\mathcal{D}_r$ and the synthetic dataset $\mathcal{D}_s$ generated by random volumetric deformations. There is no prior correspondence or landmark annotation in $\mathcal{D}_r$, while the synthetic training dataset $\mathcal{D}_s$ has the ground truth map as an identity matrix. The resultant spectral map C from the SMNet is required to align the volumetric probe descriptors F_p. The probe supervoxel descriptors are defined as the intensity histograms of supervoxels, the Chi-squared distance of the intensity histograms, and the supervoxel-wise Euclidean distance vector from sampled surrounding supervoxels as [25]. Note that the probe functions are only required in the training process. We define the correspondence loss $\mathcal{L}_{cor}$ as a combination of the probe descriptor alignment on $\mathcal{D}_r$ and the supervised spectral map distance on $\mathcal{D}_s$.

$$\mathcal{L}_{cor} = \sum_{X,Y \in \mathcal{D}_r} \|PF_{p,X} - F_{p,Y}\|_F^2 + \beta \sum_{X,Y \in \mathcal{D}_s} \|P - I\|_F^2. \qquad (4)$$

P denotes the supervoxel-wise matching probability matrix, and $P = Z_X C Z_Y^\dagger$, where $\dagger$ is the pseudo-inverse operation. We utilize the orthogonal regularization on the spectral map as [21], and $\mathcal{L}_{reg} = \|C^T C - I\|^2$. The final loss

$$\mathcal{L} = \mathcal{L}_{cor} + \gamma_1 \mathcal{L}_{reg} + \gamma_2 \mathcal{L}_{emb}. \qquad (5)$$

The hyperparameters γ_1 and γ_2 trade off the correspondence and the constraints on spectral maps and bases. By minimizing $\mathcal{L}$, we optimize the volumetric descriptor extraction module and the spectral embedding approximation module.

3 Experiments

Dataset and Metric. We apply the proposed SMNet on both the synthetic and clinically obtained craniofacial CBCTs. The dataset consists of 408 clinically obtained CBCTs with a resolution of $128 \times 128 \times 128$ and the isotropic voxel size of $1.56\,\mathrm{mm} \times 1.56\,\mathrm{mm} \times 1.56\,\mathrm{mm}$, where 390 CBCTs are used for training and the remaining for testing. We retain 15000 supervoxels without considering the air background. We randomly sample 10,000 pairs of CBCTs from the training dataset. The synthetic dataset consists of 1020 volumes, split into 1000 for training and 20 for testing. The synthetic dataset is generated by perturbing a reference CBCT, where random displacements with a zero mean and a variance of 8 mm are applied to the control grid of the B-spline-based deformations.

We quantitatively evaluate the proposed SMNet on the image correspondence and multi-category segmentation using two metrics: the Dice similarity coefficient (DSC) and the supervoxel matching accuracy acc. For the synthetic dataset with known supervoxel-wise correspondence, we compute the matching accuracy acc as the ratio of the correctly matched supervoxel number N_a and the testing supervoxel number N_s, and $acc = \frac{N_a}{N_s}$. Generally, the high values of the acc and the DSC mean reasonable image correspondence.

Implemental Details. The supervoxel descriptors obtained from the U-net and the linear layers have 600 channels. The hyperparameter α in embedding constraint (3) is set to 1. β in $\mathcal{L}_{cor}$ is set to 1. γ_1 and γ_2 in $\mathcal{L}$ are set to 0.5 and 2, respectively. The dimension of the spectral space is set to $k = 48$, where Z_1, Z_2, and Z_3 each contributes 16 spectral bases. The proposed framework is implemented using the Tensorflow on a PC with two NVIDIA GTX 1080ti GPUs. The network parameters are optimized using the ADAM algorithm with a learning rate of $1e - 5$. The momentums are set to 0.5 and 0.999. The minibatch consists of two CBCTs. The training takes 16.7 h with 20 epochs. In the testing process, the descriptor extraction and spectral embedding take 0.066 s. The LLS-based spectral map and supervoxel-wise correspondence take 0.11 s.

3.1   Qualitative Assessment

The proposed SMNet realizes the supervoxel-wise correspondence by utilizing the low-dimensional spectral map and efficient spectral bases approximation. Figure 2 (a) visualizes two estimated spectral bases Z with their corresponding eigenvectors ψ of L in curves and on color supervoxels. The histogram overlaps plots are shown in Fig. 2 (b). The approximated spectral bases are consistent with those obtained by the eigendecomposition and relive the computationally expensive eigendecomposition-based spectral embedding. Figure 2 (c) shows the matrix $Z^T L Z$, which tends to be a diagonal matrix with the approximated eigenvalues of L. The multi-path GCNs with increasing numbers of graph convolutional operations handle the multiple frequency bands. We visualize the diagonal values of $Z_i^T L Z_i, i = 1, 2, 3$, which approximate the eigenvalues of matrix L as shown in Fig. 2 (d). The approximated eigenvalues related to Z_1 are in the low-frequency band, while those of Z_3 are in the high-frequency band. The above observation is consistent with our expectations in that Z_1 is approximated eigenvectors with more graph convolutional operations with the low-pass property than Z_3. The pseudo colors in Fig. 2 (e) indicate estimated corresponding supervoxels between volumetric images.

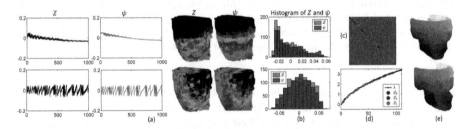

Fig. 2. (a) Visualization of two approximated spectral bases Z and corresponding eigenvectors ψ in curves (sampled 1000 nodes) and on color supervoxels. (b) The histogram overlaps of the approximated spectral bases Z and corresponding eigenvectors ψ. (c) $Z^T L Z$. (d) Comparison of $diag(Z_i^T L Z_i)|_{i=1,2,3}$ with eigenvalues λ of L. (e) Supervoxel-wise correspondence in pseudo colors.

Figure 3 (a) shows the supervoxel matching accuracy acc regarding seven anatomical structures on the synthetic testing dataset. In our experiments, u nearest supervoxels obtained by the correspondence probabilistic matrix P are viewed as the matching candidates. The acc reaches more than 80% when $u = 2$ on reported structures. The estimated spectral map facilitates multi-category segmentation of anatomical structures as shown in Fig. 3 (b), where the estimated segmentation of craniofacial structures is consistent with the ground truth.

Comparison with State-of-the-Art. Table 1 reports the DSCs of the multi-category anatomical structure segmentation by the estimated supervoxel-wise

Table 1. The DSC of the craniofacial structures segmentation on the clinical obtained CBCT dataset by the compared methods, including the VM [2] and the VM_{dif} [6], and the proposed SMNet. [‡] indicates the model without prior affine transformation.

	Max	Mad	Zyg	Fro	Sph	Occ	Tem
SMNet	**0.85 ± 0.03**	**0.90 ± 0.02**	**0.81 ± 0.12**	**0.84 ± 0.05**	**0.79 ± 0.04**	**0.77 ± 0.12**	**0.87 ± 0.03**
VM	0.75 ± 0.03	0.89 ± 0.02	0.79 ± 0.04	0.82 ± 0.03	0.64 ± 0.06	0.65 ± 0.11	0.76 ± 0.04
$VM^{‡}$	0.46 ± 0.14	0.71 ± 0.13	0.46 ± 0.23	0.66 ± 0.14	0.28 ± 0.14	0.19 ± 0.24	0.54 ± 0.12
Vm_{dif}	0.71 ± 0.02	0.85 ± 0.02	0.75 ± 0.03	0.77 ± 0.03	0.67 ± 0.03	0.70 ± 0.08	0.74 ± 0.03
$Vm_{dif}{}^{‡}$	0.39 ± 0.15	0.63 ± 0.18	0.36 ± 0.21	0.40 ± 0.21	0.26 ± 0.19	0.23 ± 0.28	0.45 ± 0.16

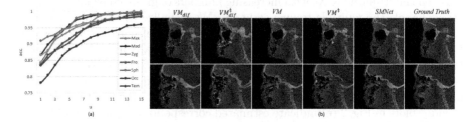

Fig. 3. (a) The supervoxel-wise matching accuracies of seven anatomical structures, including the mandible (Mad), the maxilla (Max), the zygoma (Zyg), the frontal (Fro), the sphenoid (Sph), the occipital (Occ), and the temporal (Tem). (b) Comparison with the deep learning-based registration model VM [2] and VM_{dif} [6] on the atlas-based registration. [‡] indicates the model without prior affine transformation.

correspondence. We compare the proposed SMNet with the existing deep learning-based registration model VM and its diffeomorphic variant VM_{dif} [2,6]. The proposed approach achieves the DSCs of 0.90 vs. 0.89 of the VM and 0.85 of the VM_{dif} on the mandible. Compared with the deep registration model [2,6], the correspondence computation in the spectral embedding space is robust to high-frequency perturbations due to pose and shape variations, mitigating the prior affine registration. However, the deep registration model in the spatial domain deteriorates when the image pair varies in head poses without the affine registration in the preprocessing (Fig. 3 (b)). For instance, the DSC of the mandible reduces from 0.85 to 0.63 without the affine transformation by the VM_{dif}. The proposed spectral mapping-based method is feasible to find the intrinsic structure correspondence, independent of prior transformations.

4   Conclusion

In this paper, we present the SMNet for supervoxel-wise image correspondence by virtue of low-dimensional spectral maps in the spectral domain. We address two critical issues in the spectral map-based correspondence, i.e., the optimal volumetric descriptor definition and the efficient spectral embedding of volumetric images. The proposed multi-path GCN is efficient and effective to approximate the spectral bases of volumetric images, relieving the time-consuming eigen-decomposition. We present a volumetric descriptor learning scheme to mitigate the descriptor selection. The spectral map-based image correspondence is robust to high-frequency perturbation associated with pose and structural variations, avoiding the prior affine transformation or landmark annotations. Experimental results validate the proposed SMNet on image correspondence and atlas-based craniofacial structure segmentation.

Acknowledgments. This work was supported in part by National Natural Science Foundation of China under Grant 61876008 and 82071172, Beijing Natural Science Foundation under Grant 7192227, and Research Center of Engineering and Technology for Digital Dentistry, Ministry of Health.

References

1. Achanta, R., Shaji, A., Smith, K., Lucchi, A., Fua, P., Süsstrunk, S.: Slic super-pixels compared to state-of-the-art superpixel methods. IEEE Trans. PAMI **34**, 2274–2282 (2012)
2. Balakrishnan, G., Zhao, A., Sabuncu, M.R., Guttag, J., Dalca, A.V.: An unsuper-vised learning model for deformable medical image registration. In: IEEE/CVF Conference on Computer Vision and Pattern Recognition (CVPR), pp. 9252–9260 (2018)
3. Bruna, J., Zaremba, W., Szlam, A., LeCun, Y.: Spectral networks and locally connected networks on graphs. CoRR abs/1312.6203 (2014)
4. Çiçek, Ö., Abdulkadir, A., Lienkamp, S.S., Brox, T., Ronneberger, O.: 3D U-Net: learning dense volumetric segmentation from sparse annotation. In: Ourselin, S., Joskowicz, L., Sabuncu, M., Unal, G., Wells, W. (eds.) MICCAI 2016. LNCS, vol. 9901, pp. 424–432. Springer, Cham (2016). https://doi.org/10.1007/978-3-319-46723-8_49
5. Corman, E., Ovsjanikov, M., Chambolle, A.: Supervised descriptor learning for nonrigid shape matching. In: Agapito, L., Bronstein, M., Rother, C. (eds.) ECCV 2014. LNCS, vol. 8928, pp. 283–298. Springer, Cham (2014). https://doi.org/10.1007/978-3-319-16220-1_20
6. Dalca, A.V., Balakrishnan, G., Guttag, J., Sabuncu, M.R.: Unsupervised learn-ing for fast probabilistic diffeomorphic registration. In: Frangi, A., Schnabel, J., Davatzikos, C., Alberola-López, C., Fichtinger, G. (eds.) MICCAI 2018. LNCS, vol. 11070, pp. 729–738. Springer, Cham (2018). https://doi.org/10.1007/978-3-030-00928-1_82
7. Defferrard, M., Bresson, X., Vandergheynst, P.: Convolutional neural networks on graphs with fast localized spectral filtering. In: NIPS (2016)
8. Donati, N., Sharma, A., Ovsjanikov, M.: Deep geometric functional maps: robust feature learning for shape correspondence. In: IEEE/CVF Conference on Computer Vision and Pattern Recognition (CVPR), pp. 8589–8598 (2020)
9. Ezuz, D., Ben-Chen, M.: Deblurring and denoising of maps between shapes. Com-put. Graph. Forum **36**(5), 165–174 (2017)
10. Grover, A., Leskovec, J.: node2vec: Scalable feature learning for networks. In: ACM SIGKDD International Conference on Knowledge Discovery and Data Min-ing (2016)

11. Halimi, O., Litany, O., Rodol, E., Bronstein, A.M., Kimmel, R.: Unsupervised learning of dense shape correspondence. In: IEEE/CVF Conference on Computer Vision and Pattern Recognition (CVPR), pp. 4370–4379 (2019)
12. Haskins, G., Kruger, U., Yan, P.: Deep learning in medical image registration: a survey. Mach. Vis. Appl. 1–18 (2020). https://doi.org/10.1007/s00138-020-01060-x
13. Kanavati, F., et al.: Supervoxel classification forests for estimating pairwise image correspondences. Pattern Recogn. **63**, 561–569 (2017)
14. Khosla, M., Anand, A., Setty, V.: A comprehensive comparison of unsupervised network representation learning methods. ArXiv abs/1903.07902 (2019)
15. Kipf, T., Welling, M.: Semi-supervised classification with graph convolutional networks. ArXiv abs/1609.02907 (2017)
16. Kovnatsky, A., Bronstein, M.M., Bresson, X., Vandergheynst, P.: Functional correspondence by matrix completion. In: IEEE/CVF Conference on Computer Vision and Pattern Recognition (CVPR), pp. 905–914 (2015)
17. Litany, O., Remez, T., Rodolà, E., Bronstein, A., Bronstein, M.: Deep functional maps: structured prediction for dense shape correspondence. IEEE International Conference on Computer Vision (ICCV), pp. 5660–5668 (2017)
18. Lombaert, H., Arcaro, M., Ayache, N.: Brain transfer: spectral analysis of cortical surfaces and functional maps. In: International Conference on Information Processes Medical and Imaging, pp. 474–487 (2015)
19. Lu, P.E., Chang, C.: Explainable, stable, and scalable graph convolutional networks for learning graph representation. ArXiv abs/2009.10367 (2020)
20. Ou, M., Cui, P., Pei, J., Zhang, Z., Zhu, W.: Asymmetric transitivity preserving graph embedding. In: Proceedings of the 22nd ACM SIGKDD International Conference on Knowledge Discovery and Data Mining (2016)
21. Ovsjanikov, M., Ben-Chen, M., Solomon, J., Butscher, A., Guibas, L.: Functional maps: a flexible representation of maps between shapes. ACM Trans. Graph. **31**(4), 30 (2012)
22. Perozzi, B., Al-Rfou, R., Skiena, S.: Deepwalk: online learning of social representations. In: ACM SIGKDD International Conference on Knowledge Discovery and Data Mining (2014)
23. Ren, J., Panine, M., Wonka, P., Ovsjanikov, M.: Structured regularization of functional map computations. Comput. Graph. Forum **38** (2019)
24. Sotiras, A., Davatzikos, C., Paragios, N.: Deformable medical image registration: a survey. IEEE Trans. Med. Imaging **32**(7), 1153–1190 (2013)
25. Zhang, Y., Pei, Y., Guo, Y., Ma, G., Xu, T., Zha, H.: Consistent correspondence of cone-beam CT images using volume functional maps. In: International Conference on Medical Image Computing and Computer-Assisted Intervention, pp. 801–809 (2018)

A Deep Network for Joint Registration and Parcellation of Cortical Surfaces

Fenqiang Zhao[1], Zhengwang Wu[1], Li Wang[1], Weili Lin[1], Shunren Xia[2],
Gang Li[1(✉)], and the UNC/UMN Baby Connectome Project Consortium

[1] Department of Radiology and BRIC, University of North Carolina at Chapel Hill,
Chapel Hill, NC, USA
gang_li@med.unc.edu
[2] Key Laboratory of Biomedical Engineering of Ministry of Education,
Zhejiang University, Hangzhou, China

Abstract. Cortical surface registration and parcellation are two essential steps in neuroimaging analysis. Conventionally, they are performed independently as two tasks, ignoring the inherent connections of these two closely-related tasks. Essentially, both tasks rely on meaningful cortical feature representations, so they can be jointly optimized by learning shared useful cortical features. To this end, we propose a deep learning framework for joint cortical surface registration and parcellation. Specifically, our approach leverages the spherical topology of cortical surfaces and uses a spherical network as the shared encoder to first learn shared features for both tasks. Then we train two task-specific decoders for registration and parcellation, respectively. We further exploit the more explicit connection between them by incorporating the novel parcellation map similarity loss to enforce the boundary consistency of regions, thereby providing extra supervision for the registration task. Conversely, parcellation network training also benefits from the registration, which provides a large amount of augmented data by warping one surface with manual parcellation map to another surface, especially when only few manually-labeled surfaces are available. Experiments on a dataset with more than 600 cortical surfaces show that our approach achieves large improvements on both parcellation and registration accuracy (over separately trained networks) and enables training high-quality parcellation and registration models using much fewer labeled data.

Keywords: Surface registration · Parcellation · Deep neural network

1 Introduction

Cortical surface registration and parcellation are two essential tasks in surface-based neuroimaging analysis. Cortical surface registration estimates a deformation field to align cortical features from different scans and establishes vertex-wise cortical correspondences across individuals or time points [3,17,21], thus enabling subsequent analyses, e.g., group comparison or longitudinal studies. Cortical surface parcellation learns the mapping between the cortical features

© Springer Nature Switzerland AG 2021
M. de Bruijne et al. (Eds.): MICCAI 2021, LNCS 12904, pp. 171–181, 2021.
https://doi.org/10.1007/978-3-030-87202-1_17

and the anatomically or functionally meaningful regions, thereby parcellating the cortex into different regions of interest (ROIs) [1,6,14,19]. Conventionally, these two tasks are generally performed independently or sequentially [7], ignoring their inherent close relationship. However, they both rely on learning effective cortical feature representations, one for inferring the deformation field and the other for predicting parcellation labels, which means that they explore cortical features in different aspects and thus conducting them together enables better regularization for more meaningful and robust feature representation, and therefore, they can help each other. For example, for the parcellation task, with the vertex-wise correspondence established by cortical surface registration, the manual parcellation labels from one subject or an atlas can be easily propagated to a new subject, thus helping the parcellation task to better learn the mapping from features to parcellation labels. For the registration task, the parcellated ROI boundaries can be used as an extra guidance (in addition to cortical geometric or functional features) to better learn the mapping from features to deformation field for registering two surfaces.

Therefore, in this paper, for the *first* time, we explore the idea of joint registration and parcellation of cortical surfaces through a deep spherical neural network, leveraging the spherical topology of the cerebral cortex. To extract more useful cortical features, we design a shared encoder to first learn the common features shared by both tasks. Then we train two task-specific decoders for registration and parcellation, respectively. We further exploit the more explicit connection between the two tasks which is formulated by a novel parcellation map similarity loss. This loss forces the warped predicted parcellation map of the moving surface to match the manual parcellation map of the fixed surface, e.g., an atlas, thereby providing extra supervision of the ROI boundary consistency for the registration task. Conversely, this loss can be considered as a data augmentation method that generates quasi-ground-truth parcellations for the unlabeled surfaces to help the semi-supervised learning of the parcellation network [16]. Consequently, the two tasks can mutually guide each other's training and boost each other's performance compared with separately trained networks, especially with limited labeled data, which is of significant importance for practical use cases where only few manually labeled surfaces are available.

2   Method

Our goal is to train a joint registration and parcellation network (**JRP-Net**) that generates the parcellation map of a given cortical surface based on its cortical features and simultaneously yields the individual-to-atlas deformation field, more accurately with few manually labeled surfaces. To this end, we formulate our approach as in Fig. 1. Our JRP-Net consists of three modules: a shared encoder (SE) for extracting mutual high-level features, a registration decoder (RD) for cortical surface registration and a parcellation decoder (PD) for surface parcellation. Let M, F be the moving surface and fixed surface defined on the sphere S^2 discretized using icosahedron subdivisions [3]. The SE module

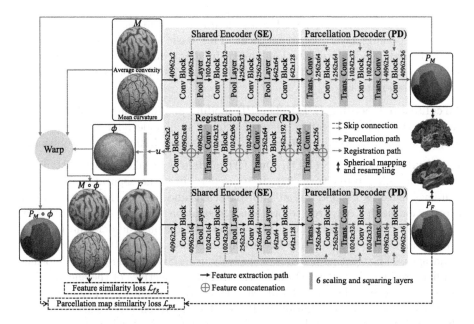

Fig. 1. Our JRP-Net for joint registration and parcellation of cortical surfaces. The input cortical surfaces are color-coded by two geometric features, i.e., average convexity (up) and mean curvature (down). The Conv Block contains repeated 1-ring-Conv+BN+ReLU. The Trans. Conv represents the spherical transposed convolution for upsampling the surface. The input and output sizes are denoted before and after each operation. See Sect. 2 for more interpretations of the math symbols.

learns the mapping f_{SE} between an input surface map and extracted features $Z : Z_F = f_{SE}(F), Z_M = f_{SE}(M)$. The RD module takes the extracted features from both F and M as input and outputs the spherical velocity field $u : u = f_{RD}(Z_F, Z_M)$ and further derives the diffeomorphic deformation field $\phi = exp(u)$ using 6 "scaling and squaring" layers as in [17]. The PD module takes the extracted features as input and outputs the parcellation map $P = f_{PD}(Z)$. We will detail the network architectures (Sect. 2.1), the specific losses (Sect. 2.2) and training strategy (Sect. 2.3) to train our JRP-Net effectively in the following.

2.1 Network Architecture

We construct our network based on the popular Spherical U-Net [22]. It leverages the spherical topology of cortical surfaces and extends convolution, pooling operations to the spherical space using 1-ring filter on regularly resampled spherical surfaces, and then constructs the network using corresponding spherical operations, achieving promising performance in various tasks, e.g., parcellation [19], registration [21] and harmonization [20].

Shared Encoder (SE). Our SE module shares a similar architecture with the encoder part of Spherical U-Net. It consists of 4 repeated 1-ring-Convolution

+Batch Normalization (BN)+ReLU layers in four resolutions with three spherical mean pooling layers between them. This basic encoder architecture has demonstrated good representation ability in many tasks [11,23] and can be trained to extract shared features at various spatial scales for both registration and parcellation tasks.

Parcellation Decoder (PD). The PD module predicts the parcellation map based on the extracted feature maps Z using a similar decoder architecture as in Spherical U-Net, with modifications to the feature channels and resolutions.

Registration Decoder (RD). Our RD module is similar to a recent unsupervised learning framework [21]. It is used to fuse the separately extracted high-level features Z_F and Z_M at multiple scales and predict the corresponding deformation field. Compared to previous registration methods [21] that directly take the concatenation of F and M as input and output the deformation field using a single network, our novel registration network (SE+RD) has two main advantages. 1) Extracting features exclusively using SE and computing deformation field exclusively using RD makes the training objectives of the encoder and the decoder more explicit and distinct and thus easier to accomplish; 2) Our deep multi-scale feature concatenation in RD enables more effective learning of the differences of the high-level boundary features of F and M, which is important for computing the deformation field aligning F and M [8].

2.2   Loss Functions

Feature Similarity Loss ($\mathcal{L}_{fs}$). $\mathcal{L}_{fs}$ is applied to the registration network to enforce the feature similarity between the warped moving surface (moved surface) and the atlas surface:

$$\mathcal{L}_{fs}(F, M, \phi) = \|F - M \circ \phi\|^2 - \lambda_{cc}\frac{cov(F, M \circ \phi)}{\sqrt{\sigma_F \cdot \sigma_{M \circ \phi}}}, \tag{1}$$

where ϕ is the learned deformation field and $M \circ \phi$ represents the moved surface maps, $cov(\cdot, \cdot)$ is the covariance, σ is the standard deviation and λ_{cc} is the weight for the correlation coefficient term.

Spherical Deformation Smoothness Loss ($\mathcal{L}_s$). We use the same operator ∇_s as in [21] on the spherical surface to approximate the spherical deformation's gradients on the sphere and accordingly penalize the gradients as:

$$\mathcal{L}_s(\phi) = \frac{1}{N}\sum_{n=1}^{N}\|\nabla_s R_{v_n}(u)\|^2, \tag{2}$$

where $R_{v_n}(u)$ denotes the local 1-ring velocity vectors of vertex v_n on a surface with N vertices. Hence, it can encourage the deformation field to be smooth.

Supervised Parcellation Loss ($\mathcal{L}_{sp}$). We use the weighted cross entropy loss to supervise the training of the parcellation network when manual parcellation maps are available:

$$\mathcal{L}_{sp}(P, P^*) = -\frac{1}{N}\sum_{n=1}^{N} W_c \log(\frac{exp(P(v_n)[c])}{\sum_{j=1}^{C} exp(P(v_n)[j])}), \tag{3}$$

where P and P^* are the predicted and manual parcellation maps respectively, W_c is the inverse proportions of the c-th ROI's area in P^*, $P(v_n)[j]$ represents the probability of vertex v_n predicted as label j and c is the manual label of v_n. Considering the real applications of cortical surface analysis, we assume F (atlas) always has manual labels and thus $\mathcal{L}_{sp}(P, P^*) = \mathcal{L}_{sp}(P_F, P_F^*)$, when only F is labeled, otherwise $\mathcal{L}_{sp}(P, P^*) = \mathcal{L}_{sp}(P_F, P_F^*) + \mathcal{L}_{sp}(P_M, P_M^*)$ when both F and M are labeled.

Parcellation Map Similarity Loss ($\mathcal{L}_{ps}$). We use the multi-class Dice loss, which addresses imbalanced parcellation labels inherently [16]:

$$\mathcal{L}_{ps}(P_F, P_M, \phi) = -\frac{1}{K}\sum_{k=1}^{K} Dice(P_F^k, P_M^k \circ \phi) = -\frac{1}{K}\sum_{k=1}^{K} \frac{2\sum_v P_F^k(v) \cdot (P_M^k \circ \phi)(v)}{\sum_v P_F^k(v) + \sum_v (P_M^k \circ \phi)(v)}, \tag{4}$$

where k indicates a ROI label (out of K ROIs) and v is the vertex location. The P_M in Eq. 4 is manually labeled parcellation map when available or predicted using f_{PD} otherwise. Hence it can provide extra ROI boundary consistency supervision for the registration task and augmented quasi-ground-truth parcellations of M for the parcellation task, when M is not labeled.

2.3 Training Strategy

To train the proposed JRP-Net, we design a progressive training strategy to learn the network parameters in an easy-to-hard manner with three steps. 1) We train the parcellation network (SE+PD) by minimizing $\mathcal{L}_{sp}$ using the strong supervision from all available manually labeled surfaces for 2,000 iterations (i.e., 200 epochs for 10 labeled surfaces when batch size is 1, which empirically pretrains the parcellation network sufficiently). 2) We jointly train registration and parcellation networks (SE+PD+RD) by optimizing $\mathcal{L}_{fs} + \lambda_s \mathcal{L}_s + \lambda_{sp}\mathcal{L}_{sp}$, where λ represents the weights, using all available surfaces for 10 epochs. 3) We incorporate $\mathcal{L}_{ps}$ and optimize the full objective $\mathcal{L}_{fs} + \lambda_s \mathcal{L}_s + \lambda_{sp}\mathcal{L}_{sp} + \lambda_{ps}\mathcal{L}_{ps}$ for 100 epochs (or early stopped when obtaining stable results). Herein, $\mathcal{L}_{ps}$ teaches PD to parcellate an unlabeled surface such that the predicted parcellation matches the inversely warped manual parcellation of the atlas via RD. Conversely, $\mathcal{L}_{ps}$ enforces RD to correctly warp the predicted parcellation map via PD to match the atlas parcellation map, thus enabling mutual learning between the two tasks.

3 Experiments and Results

3.1 Experimental Setting

We used an infant dataset with 623 cortical surfaces, which were reconstructed via iBEAT V2.0 Cloud (http://www.ibeat.cloud/) [4,7,12,13]. We mapped each surface onto the sphere with 2 features at each vertex, i.e., 'sulc' (average convexity) and 'curv' (mean curvature) [2], and labeled it into 36 gyrus-based regions based on the parcellation protocol in [1]. We used the public UNC 4D infant cortical surface atlas [5,15] as the fixed surfaces. We followed the implementation in [18] to initialize the registration process and perform the spherical interpolation. We randomly split the data into 60% for training, 10% for validation, and 30% for testing and made sure the longitudinal surfaces from the same subject are in the same set. All the results reported are then solely based on the hold-out test set.

Baselines. We used our SE+PD architecture in Fig. 1 (similar to Spherical U-Net [19]) as the baseline parcellation model (**BL-Parc**) trained via $\mathcal{L}_{sp}$, and SE+RD architecture as the baseline registration model (**BL-Reg**) trained via $\mathcal{L}_{fs} + \lambda_s \mathcal{L}_s$. Jointly training two independent BL-Parc and BL-Reg models (i.e., task-specific encoders for each task instead of the SE) is called **JRP-w/oSE-w/$\mathcal{L}_{ps}$**. Similarly, **JRP-w/SE-w/o$\mathcal{L}_{ps}$** means jointly training with SE but without $\mathcal{L}_{ps}$. We also compared with 3 available registration methods, FreeSurfer [3], Spherical Demons [17], and an unsupervised learning approach S3Reg [18,21].

Implementation Details. We implemented our method using PyTorch. We used Adam optimizer with a fixed learning rate 5e-4 for all deep learning experiments. Sulc and curv values were firstly normalized between $[-1, 1]$. The loss weights are $\lambda_{cc} = 1.2$, $\lambda_s = 10.0$, $\lambda_{sp} = 2.0$ and $\lambda_{ps} = 5.0$. Note that when computing $\mathcal{L}_{fs}$, we assign different weights for the similarities of sulc and curv maps, 0.75 for sulc and 0.25 for curv, because sulc is a more robust cortical folding measure, while curv is more variable and contains more noises for registration. These parameters are empirically determined. We used the official codes of Spherical Demons and FreeSurfer (7.1.0) for their experiments. We run all the experiments on a PC with an NVIDIA RTX2080 Ti GPU and an Intel Core i7-9700K CPU.

3.2   Results

We evaluate all the methods using the Dice overlap metric computed as in Eq. 4. For parcellation task, it is between predictions and the ground truth parcellations; for registration task, it is between moved manual parcellation maps and the fixed (atlas) parcellation map. We also provide the within-group correlation coefficient (CC) of moved cortical features as in [5,10,21] for additionally evaluating the within-group spatial normalization accuracy for registration task.

Comparison of Registration-Only Methods. As shown in Table 1, using our BL-Reg alone for cortical surface registration achieves comparable accuracy with available registration methods, but is 50+ times faster than [21], 500+ times faster than [3,17]. Note that conventional methods [3,17] do not have GPU implementation and we report the time on CPU for them in Table 1 (other methods are evaluated on GPU). We also run BL-Reg on the same CPU, it takes

Table 1. Average (standard deviation) performance of cortical surface parcellation and registration on the hold-out test set using different models trained with all manual parcellations in the training set. Parc. means parcellation, Reg. means registration.

Models	Parcellation	Registration			Run time
	Dice (%)	Dice (%)	CC_sulc	CC_curv	
Parc. only (BL-Parc)	88.48(3.68)	–	–	–	0.01 s
Reg. only (FreeSurfer [3])	–	76.69(5.52)	0.7788(0.0549)	0.2792(0.0813)	~30 min
Reg. only (Spherical Demons [17])	–	76.58(5.83)	0.7825(0.0675)	0.2720(0.1237)	~1.5 min
Reg. only (Zhao et al. [21])	–	77.03(5.66)	0.7859(0.0526)	0.2955(0.0810)	~10 s
Reg. only (BL-Reg) (Ours)	–	77.34(3.82)	0.7946(0.0374)	0.2968(0.0796)	0.16 s
JRP-w/oSE-w/$\mathcal{L}_{ps}$ (Ours)	88.48(3.68)	88.78(3.12)	0.8209(0.0339)	0.3313(0.0794)	0.01 s + 0.16 s
JRP-w/SE-w/o$\mathcal{L}_{ps}$ (Ours)	89.34(2.47)	78.18(4.39)	0.7996(0.0466)	0.3095(0.0953)	0.16 s
JRP-w/SE-w/$\mathcal{L}_{ps}$ (Ours)	**89.98(2.53)**	**90.23(2.56)**	**0.8288(0.0317)**	**0.3359(0.0676)**	0.16 s

0.4 s and thus is still much faster than [3,17]. Compared to [21] that directly concatenates F and M as input in a coarse-to-fine manner, our SE+RD architecture fuses high-level features in deep feature space, thus avoiding the time-consuming re-interpolation of deformations in original spherical space while obtaining better results. This indicates the effective learning of high-level features in SE and the deformation computation based on boundary differences in RD. To additionally validate the topology-preserving registrations, we computed the folded triangles [9] in moved surfaces and found there is no folded triangles for all methods.

Ablation Study on Shared Encoder (SE). Comparing JRP-w/SE-w/o$\mathcal{L}_{ps}$ with BL-Parc and BL-Reg in Table 1, we can see that jointly training SE improves Dice over separately training two networks by 0.86% and 0.84% for parcellation and registration tasks respectively, which indicates that SE learns and exploits the shared useful features between the two tasks successfully.

Ablation Study on Parcellation Map Similarity Loss ($\mathcal{L}_{ps}$). Table 1 shows incorporating $\mathcal{L}_{ps}$ substantially improves the performance, with >10% and 1.5% Dice improvement over separately trained networks for registration and parcellation, respectively. For JRP-w/oSE-w/$\mathcal{L}_{ps}$, the $\mathcal{L}_{ps}$ is computed between two manual parcellations according to Eq. 4, thus it does not provide any information for training the independent parcellation network but provides maximum extra supervision for the registration network and leads to huge improvement on registration accuracy. Further with SE (JRP-w/SE-w/$\mathcal{L}_{ps}$), $\mathcal{L}_{ps}$ back-propagates gradients from RD to SE and thus also enhance the performance of parcellation task by using more effective features in SE.

178   F. Zhao et al.

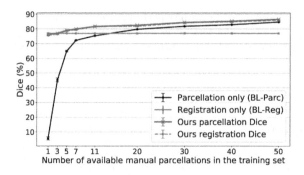

Fig. 2. Test Dice of the models trained with different numbers of manual parcellations.

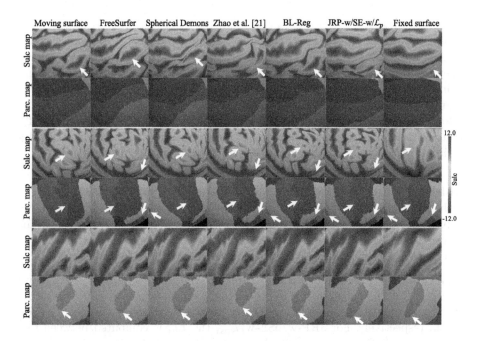

Fig. 3. For each case (left to right): the moving surfaces' sulc map (first row) and manual parcellation map (second row), corresponding moved maps by different methods, and fixed surfaces' maps. Note that all the methods fairly take the features (sulc and curv) as input and output the deformation field. The manual parcellation maps are only for validating the registration performance and were not used to drive the registration.

How Registration and Parcellation Help Each Other. We simulate the common practical use case where only few manually labeled surfaces are available. We randomly used N surfaces' manual parcellations to train BL-Parc, BL-Reg and JRP-Net (JRP-w/SE-w/$\mathcal{L}_{ps}$) on the training set. To achieve more reliable results, we repeated the experiment 10 times for each model (each time with randomly selected manual parcellations). Figure 2 shows the results on the test set averaged over 10 times experiments. We can see that in the extreme case where only atlases' parcellations are available (11 atlases in UNC 4D atlas [5]), our method still achieves good performance, with large improvements over independently trained models. This demonstrates that registration did help parcellation by providing estimated supervision from augmented parcellation maps on unlabeled surfaces. Conversely, the registration performance is also boosted by the parcellation task (when $N > 3$) even using predicted parcellations to enforce the anatomical consistency. Also note that the lines of registration Dice and parcellation Dice of our JRP-Net in Fig. 2 are highly coincident. A paired t-test with a significance level 0.05 on the 10 times experiments shows there is no significant difference between them, which could be a strong evidence that the two tasks mutually guide each other's training to successfully learn the optimal features using SE and find the optimal solution for both tasks. Figure 3 shows some practical registration cases from the test set that available registration methods all fail to align but are correctly aligned using our JRP-Net ($N = 50$). The suboptimal results of available methods are an inherent problem because they only use feature similarity as the registration objective, while our JRP-Net incorporating the parcellation map similarity can effectively solve this problem.

4   Conclusion

In this paper, we propose the JRP-Net for joint registration and parcellation of cortical surfaces, which are connected by the shared encoder and parcellation map similarity loss. By leveraging the inherent relation between the two tasks, our shared encoder is effective in extracting more meaningful features shared by both tasks. The parcellation map similarity loss further enables effective mutual training between the two tasks. Both visual and quantitative results show large improvements of our method in both registration and parcellation over separately learned networks. In future, we will release our model for enhancing cortical surface registration and parcellation for the research community.

Acknowledgements. This work was partially supported by NIH grants (MH116225, MH117943, MH109773, MH123202). This work also utilizes approaches developed by an NIH grant (1U01MH110274) and the efforts of the UNC/UMN Baby Connectome Project Consortium.

References

1. Desikan, R.S., et al.: An automated labeling system for subdividing the human cerebral cortex on MRI scans into GYRAL based regions of interest. Neuroimage **31**(3), 968–980 (2006)

2. Fischl, B.: Freesurfer. Neuroimage **62**(2), 774–781 (2012)
3. Fischl, B., Sereno, M.I., Tootell, R.B., Dale, A.M.: High-resolution intersubject averaging and a coordinate system for the cortical surface. Hum. Brain Mapp. **8**(4), 272–284 (1999)
4. Li, G., et al.: Measuring the dynamic longitudinal cortex development in infants by reconstruction of temporally consistent cortical surfaces. Neuroimage **90**, 266–279 (2014)
5. Li, G., et al.: Construction of 4D high-definition cortical surface atlases of infants: methods and applications. Med. Image Anal. **25**(1), 22–36 (2015)
6. Li, G., Wang, L., Shi, F., Lin, W., Shen, D.: Simultaneous and consistent labeling of longitudinal dynamic developing cortical surfaces in infants. Med. Image Anal. **18**(8), 1274–1289 (2014)
7. Li, G., et al.: Computational neuroanatomy of baby brains: a review. Neuroimage **185**, 906–925 (2019)
8. Liu, L., Hu, X., Zhu, L., Heng, P.-A.: Probabilistic multilayer regularization network for unsupervised 3D brain image registration. In: Shen, D., et al. (eds.) MICCAI 2019. LNCS, vol. 11765, pp. 346–354. Springer, Cham (2019). https://doi.org/10.1007/978-3-030-32245-8_39
9. Möller, T.: A fast triangle-triangle intersection test. J. Graph. Tools **2**(2), 25–30 (1997)
10. Robinson, E.C., et al.: MSM: a new flexible framework for multimodal surface matching. Neuroimage **100**, 414–426 (2014)
11. Ronneberger, O., Fischer, P., Brox, T.: U-net: convolutional networks for biomedical image segmentation. In: Navab, N., Hornegger, J., Wells, W.M., Frangi, A.F. (eds.) MICCAI 2015. LNCS, vol. 9351, pp. 234–241. Springer, Cham (2015). https://doi.org/10.1007/978-3-319-24574-4_28
12. Sun, L., et al.: Topological correction of infant white matter surfaces using anatomically constrained convolutional neural network. NeuroImage **198**, 114–124 (2019)
13. Wang, L., et al.: Volume-based analysis of 6-month-old infant brain MRI for autism biomarker identification and early diagnosis. In: Frangi, A.F., Schnabel, J.A., Davatzikos, C., Alberola-López, C., Fichtinger, G. (eds.) MICCAI 2018. LNCS, vol. 11072, pp. 411–419. Springer, Cham (2018). https://doi.org/10.1007/978-3-030-00931-1_47
14. Wu, Z., et al.: Registration-free infant cortical surface parcellation using deep convolutional neural networks. In: Frangi, A.F., Schnabel, J.A., Davatzikos, C., Alberola-López, C., Fichtinger, G. (eds.) MICCAI 2018. LNCS, vol. 11072, pp. 672–680. Springer, Cham (2018). https://doi.org/10.1007/978-3-030-00931-1_77
15. Wu, Z., Wang, L., Lin, W., Gilmore, J.H., Li, G., Shen, D.: Construction of 4D infant cortical surface atlases with sharp folding patterns via spherical patch-based group-wise sparse representation. Hum. Brain Mapp. **40**(13), 3860–3880 (2019)
16. Xu, Z., Niethammer, M.: DeepAtlas: joint semi-supervised learning of image registration and segmentation. In: Shen, D., et al. (eds.) MICCAI 2019. LNCS, vol. 11765, pp. 420–429. Springer, Cham (2019). https://doi.org/10.1007/978-3-030-32245-8_47
17. Yeo, B.T., Sabuncu, M.R., Vercauteren, T., Ayache, N., Fischl, B., Golland, P.: Spherical demons: fast diffeomorphic landmark-free surface registration. IEEE Trans. Med. Imaging **29**(3), 650–668 (2009)
18. Zhao, F., et al.: S3Reg: superfast spherical surface registration based on deep learning. IEEE Trans. Med. Imaging (2021)

19. Zhao, F., et al.: Spherical deformable U-Net: application to cortical surface parcellation and development prediction. IEEE Trans. Med. Imaging **40**(4), 1217–1228 (2021)
20. Zhao, F., et al.: Harmonization of infant cortical thickness using surface-to-surface cycle-consistent adversarial networks. In: Shen, D., et al. (eds.) MICCAI 2019. LNCS, vol. 11767, pp. 475–483. Springer, Cham (2019). https://doi.org/10.1007/978-3-030-32251-9_52
21. Zhao, F., et al.: Unsupervised learning for spherical surface registration. In: Liu, M., Yan, P., Lian, C., Cao, X. (eds.) MLMI 2020. LNCS, vol. 12436, pp. 373–383. Springer, Cham (2020). https://doi.org/10.1007/978-3-030-59861-7_38
22. Zhao, F., et al.: Spherical U-Net on cortical surfaces: methods and applications. In: Chung, A.C.S., Gee, J.C., Yushkevich, P.A., Bao, S. (eds.) IPMI 2019. LNCS, vol. 11492, pp. 855–866. Springer, Cham (2019). https://doi.org/10.1007/978-3-030-20351-1_67
23. Zhong, T., et al.: DIKA-Nets: domain-invariant knowledge-guided attention networks for brain skull stripping of early developing macaques. NeuroImage **227**, 117649 (2021)

4D-Foot: A Fully Automated Pipeline of Four-Dimensional Analysis of the Foot Bones Using Bi-plane X-Ray Video and CT

Shuntaro Mizoe[1], Yoshito Otake[1(✉)], Takuma Miyamoto[2], Mazen Soufi[1], Satoko Nakao[2], Yasuhito Tanaka[2], and Yoshinobu Sato[1]

[1] Division of Information Science, Nara Institute of Science and Technology, Ikoma, Japan
otake@is.naist.jp
[2] Department of Orthopedics, Nara Medical University, Kashihara, Japan

Abstract. We aim to elucidate the mechanism of the foot by automated measurement of its multiple bone movement using 2D-3D registration of bi-plane x-ray video and a stationary 3D CT. Conventional analyses allowed tracking of only 3 large proximal tarsal bones due to the requirement of manual segmentation and manual initialization of 2D-3D registration. The learning-based 2D-3D registration, on the other hand, has been actively studied and demonstrating a large capture range, but the accuracy is inferior to conventional optimization-based methods. We propose a fully automated pipeline using a cost function that seamlessly incorporates the reprojection error at the landmarks in CT and x-ray detected by off-the-shelf CNNs into the conventional image similarity cost, combined with the automated bone segmentation. We experimentally demonstrated that the pipeline allowed a robust and accurate 2D-3D registration to track all 12 tarsal bones, including the metatarsals at the foot arch, which is especially important in the foot biomechanics but has been unmeasurable with previous methods. We evaluated the proposed fully automated pipeline in studies using a bone phantom and real x-ray images of human subjects. The real image study showed the registration error of 0.38 ± 1.95 mm in translation and $0.38 \pm 1.20°$ in rotation for the proximal tarsal bones.

Keywords: 2D/3D registration · Automated segmentation · Automated initialization

1 Introduction

The foot consists of flexible structures of bones, joints, muscles, and soft tissues, allowing complex movements and shock absorption in human motion. We aim to accurately track the foot bones for biomechanical analysis (e.g., interaction between the small bones at multiple joints). While its importance is acknowledged, especially in injury prevention and rehabilitation of the ankle disease,

---

S. Mizoe and Y. Otake—Equal contribution.

© Springer Nature Switzerland AG 2021
M. de Bruijne et al. (Eds.): MICCAI 2021, LNCS 12904, pp. 182–192, 2021.
https://doi.org/10.1007/978-3-030-87202-1_18

most conventional methods are limited to either static anatomical analyses using CT [1,2] or skin-marker-based motion capture [3–5] which is prone to error due to the skin movement.

Some recent studies employ a 2D-3D registration between x-ray videos acquired by a biplane imaging system and a CT image for the analysis of the 3D bone movement [6,7]. The approach demonstrated a high accuracy, however, the target bones have been limited to only the proximal tarsal bones, namely talus, calcaneus, and navicular bones, and the methods required laborious manual segmentation of each bone from the CT and manual initialization of the 2D-3D registration. While Esteban et al. [8] and Grupp et al. [9] studied 2D-3D registration in the analysis of pelvis anatomy using a CNN-based landmark detection for initialization of the intensity-based registration, both works assumed manual segmentation of the target anatomy in CT, which is prohibitive especially in the clinical analysis of the foot bones. On the other hand, several attempts have been made by training CNNs for directly solving the 2D-3D alignment in an end-to-end manner (e.g. [10,11]), showing better stability due to a large capture range but inferior accuracy compared to the conventional intensity-based method. Our approach is to achieve stable and accurate registration using a cost function incorporating similarities of both intensity and landmark positions, unlike landmark only for the initialization.

We propose a fully automated pipeline of 2D-3D registration between x-ray video and CT for the motion analysis of all 12 tarsal bones (i.e., 3 proximal tarsal, 4 distal tarsals, and 5 metatarsal bones) and tibia-fibula (as one rigid object). The contribution of this paper is threefold: (1) Proposal of a 4D foot analysis system including the movement of the foot arch (metatarsal bones) which was previously unmeasurable, (2) introduction of a cost term in 2D-3D registration that incorporates reprojection error of the landmarks detected by CNNs allowing robust and accurate registration without any manual interactions, (3) quantitative evaluation of impacts of the errors in automated segmentation and landmark detection on the final registration accuracy.

2   Method

2.1   Overview of the Proposed Pipeline

Figure 1 shows the overview of the proposed pipeline. The input CT and biplane x-ray videos are first processed by CNNs, Bayesian U-net [14] for bone segmentation and landmark extraction in CT, and DeepLabCut [13] for landmark extraction in x-ray video. Then the intensity-based 2D-3D registration is performed frame-by-frame using the proposed cost terms incorporating information of the landmark and intensity similarities, resulting in a robust and accurate registration for multiple small bones in the foot.

2.2   Automated Segmentation and Landmark Detection

Segmentation of each lower leg and foot bones (2 lower leg bones, 7 tarsal bones, 5 metatarsal bones, and 14 phalanges bones) in CT is performed by

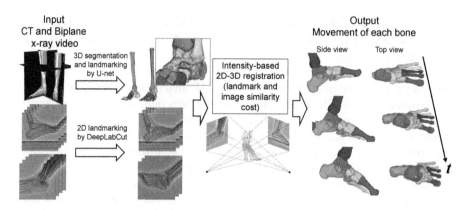

Fig. 1. Overview of the proposed automated pipeline for 4D analysis of the foot bones. The CT images and biplane x-ray video are automatically annotated (segmentation and landmarking) using CNNs and the movement of each tarsal and metatarsal bones are estimated using the proposed intensity-based 2D-3D registration.

the Bayesian U-net [14], that previously demonstrated a significantly superior accuracy than the previous multi-atlas method in segmentation of the hip and thigh muscles and bones. Our implementation including the network architecture, hyper-parameters, and pre- and post-processing follows [14][1] except for the size of convolution kernel of 7×7 for leveraging a larger receptive field. Figure 2 shows detail of the target bones.

Bone name	Color	Bone name	Color	Bone name	Color
Tibia		1st metatarsal		2nd middle phalanges	
Fibula		2nd metatarsal		3rd middle phalanges	
Talus		3rd metatarsal		4th middle phalanges	
Calcaneus		4th metatarsal		5th middle phalanges	
Navicular		5th metatarsal		1st distal phalanges	
Medial cuneiform		1st proximal phalanges		2nd distal phalanges	
Intermediate cuneiform		2nd proximal phalanges		3rd distal phalanges	
Lateral cuneiform		3rd proximal phalanges		4th distal phalanges	
Cuboid		4th proximal phalanges		5th distal phalanges	
		5th proximal phalanges			

Fig. 2. List of the foot bones annotated in this study. (note that the phalanges bones are not included in our 2D-3D registration analysis due to the limited field-of-view of the x-ray video).

---

[1] Source code was obtained from https://github.com/yuta-hi/bayesian_unet

2.3 2D-3D Registration Incorporating Landmark Reprojection Error

The intensity-based 2D-3D registration optimizes similarity between the x-ray image (fixed image) and digitally reconstructed radiograph (DRR) generated from CT (moving image). DRRs were generated using the tri-linear interpolation ray-tracing algorithm [12] implemented on the graphics processing unit (GPU). In this study, we parameterized the rigid transformation of each bone with a 6 degree-of-freedom variable (3 rotation parameters represented as Euler angle around the geometrical centroid of each bone and 3 translation parameters), resulting in a $6N$ parameter optimization problem for N bones. Following [12], we employed covarience matrix adaptation evolutionary strategy (CMA-ES) [15] for optimization and the gradient correlation similarity measure [16] for the cost function. Initialization of the translation parameters was derived by the paired point registration of the landmarks for each frame independently, assuming all bones moved rigidly. The registration of 14 bones (bones in Fig. 2 except for the 14 phalanges bones) was split into 3 stages, 1) proximal tarsal, tibia, and fibula (5 bones), distal tarsal (4 bones), and metatarsals (5 bones), to reduce the optimization parameters. The proposed cost function incorporating the landmark reprojection error derived from CNNs and the conventional image similarity is defined as follows.

$$\hat{\Theta} = \underset{\Theta}{\operatorname{argmin}}\{(1-\alpha)C_{landmark}(p_i^{2D}, p_i^{3D}, \Theta)$$

$$-\alpha GC(I^{Xp}, \sum_{k=1}^{N} I_k^{DRR}(\Theta)) + \lambda g_{rigidity}(\Theta)\} \tag{1}$$

The parameter α changes balance between the two data fitness terms, the landmark fitness and the image fitness defined by the gradient correlation (denoted by GC) between the X-ray image I^{Xp} and sum of DRRs of each bone I_k^{DRR}. The third term encourages rigidity of the target bones and λ is the weight parameter. The rigidity term was effective only for the bones with no landmark identified, such as metatarsal bones in this study. $C_{landmark}(p_i^{2D}, p_i^{3D}, \Theta) = \sum_{i=1}^{M} ||p_i^{2D} - P(T(\Theta))p_i^{3D}||$ represents the reprojection error of i_{th} landmark, where p_i^{2D} and p_i^{3D} are the landmark location identified by CNNs in 2D and 3D. $g_{rigidity}(\Theta) = \sum_{k=2}^{N} d(T_1(\Theta), T_k(\Theta))$, $T_k(\Theta)$ is the transformation of k_{th} bone, $P(T_k)$ is the projection matrix with the extrinsic parameter defined by T_k, and $d(T_1, T_k)$ indicates difference between the two transformations (in our implementation, assuming small difference, we first concatenate T_1 and T_k^{-1}, convert it to 3 translation and 3 rotation parameters, and calculate Euclidean distance between the two 6-element vectors). Our implementation of the 2D-3D registration is available at https://github.com/YoshitoOtake/4DFoot.

3 Experiment and Results

After evaluation of the accuracy of individual automated segmentation and landmark detection components by the cross-validation, accuracy of the 2D-3D registration was evaluated using; 1) the bone phantom with metallic beads attached to 14 anatomical landmarks, providing the ground truth using the radio stereometric analysis, and 2) the images from 5 volunteer subjects with fully manual annotations. Firstly, using the ground truth in the phantom image, we validate that registration using manually annotated segmentation and landmarks can be used as the quasi-ground-truth. Then, using the manually annotated quasi-ground-truth, we evaluate the accuracy of the proposed fully automated pipeline for real subjects' images.

3.1   Experimental Materials

Thirty-five CTs of the lower leg and the foot obtained from 35 patients, and 18 biplane x-ray videos of the foot during the gait obtained from 5 healthy volunteers, were used in the experiment. The phase from heel contact to toe-off was manually identified by an expert surgeon and used in the experiment. The field of view of the CTs was 323–486 mm^2, the matrix size was 512×512, and the slice interval was 0.625 mm. All individual bone regions shown in Fig. 2 and 17 anatomical landmarks (on the tibia and 3 proximal tarsal bones) in the CTs and 12 landmarks (on the same bone for each view) in all frames of the x-ray video were manually annotated by an expert orthopedic surgeon. Since we could not find a sufficient number of 3D landmarks visible in two views simultaneously, 5 landmarks were used only in one x-ray view, the other 5 were used only in the other view, and the remaining 7 were used in both views. Thus, $(5 + 7) = 12$ landmarks were used in each 2D view, which amounts to 17 in 3D. The biplane x-ray imager was equipped so that the two views are aligned to the patient's right-left direction (referred to as *lateral view*) and the oblique direction (referred to as *oblique view*). The distance between the x-ray source and detector was approximately 1200 mm for both views. The matrix size of the x-ray image was 512×512, and the pixel spacing was 0.558×0.558 mm. Geometric calibration of the two imagers was performed by obtaining 12 x-ray images of a cube-shaped calibration phantom (edge length of 110 mm) having 8 metallic spheres of 10 mm diameter at each corner. In our system, the two x-ray views were not synchronized. They record images alternately at 15 fps with half a frame (1/30 sec) phase offset. In order to obtain a *pseudo synchronized* pair of videos, a CNN-based video interpolation method, SuperSloMo [18], with a pre-trained model was used to double the frame rate of each video.

3.2   Evaluation of Automated Segmentation and Landmark Detection

Three-fold cross-validation using the 35 CTs was performed to evaluate the segmentation accuracy. In training, the right and left sides of the foot were split at the middle of the axial slice, and the right foot was flipped for the data augmentation purpose (note that the training/test split in the cross-validation was performed patient-wise since the right and left side of the same patient are similar). The dice coefficient for each bone used in the 2D-3D registration in the following experiments is summarized in Fig. 3. The dice for the lower leg, tarsal bones, and metatarsal bones were 0.990 ± 0.012, 0.971 ± 0.053, and 0.975 ± 0.022. The phalanges bones were not used in the 2D-3D registration but included in the segmentation target. The dice coefficients were 0.956 ± 0.050, 0.847 ± 0.154, and 0.794 ± 0.210 for phalanx proximalis, medialis, and distalis.

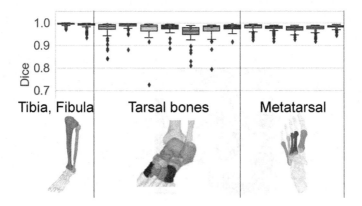

Fig. 3. Results of automated segmentation of the foot bones. Red dots indicates the five cases that were used in the 2D-3D registration experiment.

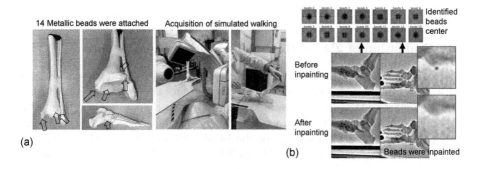

Fig. 4. Accuracy evaluation experiment using the bone phantom. (a) Experimental setting, and (b) preprocessing of the x-ray videos. The ground truth was obtained by radio stereometric analysis (RSA) using the metallic beads attached to the bones. Metallic beads in the x-ray image and CT were removed by inpainting in order to avoid the bias in the 2D-3D registration due to the strong gradient created by the beads.

The landmark detection in CT was performed by the U-net using the heatmap approach [17] with the σ (radius of the Gaussian representing the landmark) of 5 mm. As the result of three-fold cross-validation, the Euclidean distance errors of landmarks on the tibia, talus, calcaneus, and navicular were 4.27 ± 2.26, 3.65 ± 2.04, 4.07 ± 2.01, and 4.13 ± 2.41 mm, respectively.

The landmark detection in the x-ray video was performed using DeepLab-Cut [13]. We used the pre-trained Resnet-50 with fine-tuning using our own training data set. The leave-one-patient-out evaluation demonstrated the average Euclidean distance error of all the landmarks for the lateral and oblique view was 3.01 ± 2.29 and 2.73 ± 2.00 mm, respectively.

The landmark detection errors in CT and x-ray video were comparable to those reported in [17], where the authors applied their state-of-the-art method in the spine CT data set.

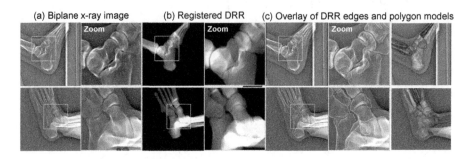

Fig. 5. A representative registration result. The intensity-based 2D-3D registration optimized the similarity between (a) the original biplane x-ray image and (b) DRR. The overlay of the DRR edges and polygon models in (c) demonstrate the accurate alignment between the two images indicating the 3D position of each bone was correctly estimated.

3.3   Evaluation of 2D-3D Registration Using Bone Phantom

The bone phantom and its x-ray videos used in the experiment were shown in Fig. 4. The phantom was moved by hands to simulate the gate. The 14 metallic beads attached to the phantom were localized in the x-ray images first manually and then refined by the Gaussian fitting search at their vicinity. To avoid the strong image gradient at the edge of the beads affecting registration accuracy, the bead regions were inpainted [19] (see Fig. 4b). The localized beads position with the geometric calibration provided the ground truth movement of each bone, while the inpainted x-ray videos were used for the 2D-3D registration. The experimental results were shown in Fig. 6a. The average absolute translation error was 0.40 ± 0.28 mm, and the rotation error was $0.66 \pm 0.59°$. Thus, we confirmed that 2D-3D registration based on manual annotation is of a level that can be used as a quasi-ground-truth in terms of the clinically required accuracy. The larger error in the navicular bone was likely attributed to its small size and rotationally symmetric shape. Relatively larger error in Y translation and smaller error in X rotation could be attributed to the sensitivity to the imaging direction (i.e., movement in the out-of-plane direction is less sensitive to the in-plane direction).

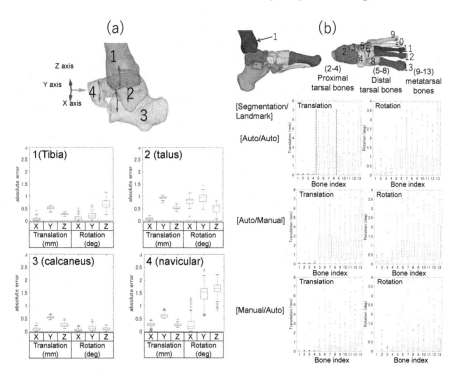

Fig. 6. Quantitative evaluation of the 3D tracking of each bone for the experiment with (a) the bone phantom and (b) real subject images.

3.4 Evaluation of 2D-3D Registration Using Images of Real Subjects

Figure 5 demonstrates a representative registration result using real subject images. DRR at the registered position correctly aligned with the x-ray image providing a visual assessment of the registration accuracy. Figure 6b and Table 1 show the quantitative results. As described above, the registration result using manual segmentation and manual landmark detection was used as the quasi-ground-truth in this experiment. In order to evaluate the effect of using automated annotation in the registration, the results in three scenarios were compared, 1) automated segmentation (Auto)/automated landmark detection (Auto), 2) Auto/Manual, 3) Manual/Auto. Overall, registration of proximal tarsal bones showed excellent accuracy (<0.5-mm translation and <0.5-degree rotation), comparable to the bone phantom experiment regardless of the annotation method. The insensitivity of the registration results to the landmark detection error suggests that the error was in an acceptable range in our 2D-3D registration application.

The distal tarsal bones and metatarsal bones showed relatively lower accuracy (~3 mm translation and ~1.5° rotation), especially when we used automated segmentation, likely due to their small size increasing sensitivity to the segmentation error. Parameters for the CMA-ES optimizer were: population size 1000, stopping criterion 0.01 (mm or deg), the two-level multi-resolution pyramid with down-sampling by a factor of 2 and 1. One registration trial required approximately 60,000 function evaluations (i.e., DRR generation and cost calculation), and the computation time was approximately 20 s on a workstation with AMD EPYC 7742 64-core processor and two nVidia GeForce RTX3090.

Table 1. Comparison of the registration accuracy using automated- and manual- segmentation and landmark (trans: translation error, rot: rotation error, bone phantom experiment used manual segmentation and manual landmark)

Segmentation/ Landmark	Proximal tarsal bones				Distal tarsal bones				Metatarsal bones			
	trans(mm)		rot(deg)		trans(mm)		rot(deg)		trans(mm)		rot(deg)	
	mean	std	mean	std	mean	std	mean	std	mean	std	mean	std
Auto/Auto	0.38	1.95	0.38	1.20	3.09	4.46	1.41	2.06	3.25	7.79	1.30	2.88
Auto/Manual	0.15	0.90	0.27	0.90	2.79	4.16	1.28	1.95	3.04	8.81	1.21	2.95
Manual/Auto	0.39	2.07	0.34	1.48	1.66	4.93	0.72	1.98	1.88	9.04	0.63	2.79
(bone phantom)	0.40	0.28	0.66	0.59	—	—	—	—	—	—	—	—

4   Discussion and Conclusion

We have presented a fully automated pipeline of segmentation and 2D-3D registration for 4D analysis of the foot bones and evaluated the accuracy with fully manually annotated data sets. Our primary contribution has been the proposal and quantitative evaluation of the registration cost incorporating reprojection error at landmarks derived by CNNs with a conventional image similarity cost. We showed that the combination of simple off-the-shelf CNN-based image recognition and the conventional intensity-based registration allowed highly accurate 4D tracking of the complex movement of small foot bones, including the foot arch, whose shock absorption function is critical in the analysis of foot biomechanics but has been unmeasurable with previous methods. Furthermore, the experiment suggested that the error in the automated segmentation had a larger impact on the registration accuracy than the landmark detection error, especially for the distal part, namely the distal tarsal and metatarsal bones, which is small in size and symmetric in shape. The lower accuracy in those bones is attributed partly to the lack of landmarks since our current landmarks are placed only on the proximal tarsal bones as shown in Fig. 1 and the distal bones are associated with landmarks placed on the bones close to them. We plan to add several landmarks on those distal parts to improve accuracy. Application in a clinical routine and the analysis of patients with ankle disease are also underway.

References

1. Stanković, K., Booth, B.G., Danckaers, F., Burg, F., Vermaelen, P., Duerinck, S., et al.: Three-dimensional quantitative analysis of healthy foot shape: a proof of concept study. J. Foot Ankle Res. **11**(1), 8 (2018)
2. Nozaki, S., Watanabe, K., Kamiya, T., Katayose, M., Ogihara, N.: Three-dimensional morphological variations of the human calcaneus investigated using geometric morphometrics. Clin. Anat. **33**(5), 751–758 (2020)
3. Eichelberger, P., Blasimann, A., Lutz, N., Krause, F., Baur, H.: A minimal markerset for three-dimensional foot function assessment: measuring navicular drop and drift under dynamic conditions. J. Foot Ankle Res. **11**(1), 15 (2018)
4. Kim, T., Park, J.C.: Short-term effects of sports taping on navicular height, navicular drop and peak plantar pressure in healthy elite athletes: a within-subject comparison. Med. (United States) **96**(46), 3–8 (2017)
5. Behling, A.V., Manz, S., von Tscharner, V., Nigg, B.M.: Pronation or foot movement - what is important. J. Sci. Med. Sports **23**(4), 366–371 (2020)
6. Cao, S., et al.: In vivo kinematics of functional ankle instability patients during the stance phase of walking. Gait Posture **73**, 262–268 (2019)
7. Lenz, A.L., et al.: Compensatory motion of the subtalar joint following tibiotalar arthrodesis. J. Bone Joint Surg. **102**(7), 600–608 (2020)
8. Esteban, J., Grimm, M., Unberath, M., Zahnd, G., Navab, N.: Towards fully automatic X-Ray to CT registration. In: Shen, D., et al. (eds.) MICCAI 2019. LNCS, vol. 11769, pp. 631–639. Springer, Cham (2019). https://doi.org/10.1007/978-3-030-32226-7_70
9. Grupp, R.B., et al.: Automatic annotation of hip anatomy in fluoroscopy for robust and efficient 2D/3D registration. Int. J. Comput. Assist. Radiol. Surg. **15**(5), 759–769 (2020)
10. Gao, C., et al.: Generalizing spatial transformers to projective geometry with applications to 2D/3D registration. In: Martel, A.L., et al. (eds.) MICCAI 2020. LNCS, vol. 12263, pp. 329–339. Springer, Cham (2020). https://doi.org/10.1007/978-3-030-59716-0_32
11. Miao, S., et al.: Dilated FCN for multi-agent 2D/3D medical image registration. In: 32nd AAAI Conference on Artificial Intelligence, AAAI 2018, pp. 4694–4701 (2018)
12. Otake, Y., et al.: Intraoperative image-based multiview 2D/3D registration for image-guided orthopaedic surgery: incorporation of fiducial-based C-arm tracking and GPU-acceleration. IEEE Trans. Med. Imaging **31**(4), 948–962 (2012)
13. Mathis, A., et al.: DeepLabCut: markerless pose estimation of user-defined body parts with deep learning. Nat. Neurosci. **21**(9), 1281–1289 (2018)
14. Hiasa, Y., Otake, Y., Takao, M., Ogawa, T., Sugano, N., Sato, Y.: Automated muscle segmentation from clinical CT using Bayesian U-Net for personalized musculoskeletal modeling. IEEE Trans. Med. Imaging **39**(4), 1030–1040 (2019)
15. Nikolaus, H.: The CMA evolution strategy: a comparing review. In: Lozano, J., et al. (ed.) Towards a New Evolutionary Computation, vol. 192, pp. 75–102. Springer, Heidelberg (2006). https://doi.org/10.1007/3-540-32494-1_4
16. Penney, G.P., et al.: A comparison of similarity measures for use in 2-D-3-D medical image registration. IEEE Trans. Med. Imaging **17**(4), 586–595 (1998)
17. Payer, C., Štern, D., Bischof, H., Urschler, M.: Integrating spatial configuration into heatmap regression based CNNs for landmark localization. Med. Image Anal. **54**, 207–219 (2019)

18. Jiang, H., Sun, D., Jampani, V., Yang, M.-H., Learned-Miller, E., Kautz, J.: Super SloMo: high quality estimation of multiple intermediate frames for video interpolation. In: Proceedings of the IEEE Conference on Computer Vision and Pattern Recognition (CVPR), pp. 9000–9008 (2018)
19. Bertalmio, M., Bertozzi, A.L., Sapiro, G.: Navier-stokes, fluid dynamics, and image and video inpainting. In: Proceedings of the 2001 IEEE Computer Society Conference on Computer Vision and Pattern Recognition, CVPR 2001, vol. 1, p. I (2001)

Equivariant Filters for Efficient Tracking in 3D Imaging

Daniel Moyer[1]([✉]), Esra Abaci Turk[2], P. Ellen Grant[2], William M. Wells[1,3], and Polina Golland[1]

[1] CSAIL, Massachusetts Institute of Technology, Cambridge, MA, USA
dmoyer@csail.mit.edu
[2] Boston Children's Hospital, Harvard Medical School, Boston, MA, USA
[3] Brigham and Women's Hospital, Harvard Medical School, Boston, MA, USA

Abstract. We demonstrate an object tracking method for 3D images with fixed computational cost and state-of-the-art performance. Previous methods predicted transformation parameters from convolutional layers. We instead propose an architecture that neither flattens convolutional features nor uses fully connected layers, but instead relies on equivariant filters to preserve transformations between inputs and outputs (e.g., rotations/translations of inputs rotate/translate outputs). The transformation is then derived in closed form from the outputs of the filters. This method is useful for applications requiring low latency, such as real-time tracking. We demonstrate our model on synthetically augmented adult brain MRI, as well as fetal brain MRI, which is the intended use-case.

Keywords: Tracking · Equivariant convolution · Fetal MRI

1 Introduction

Real-time prospective tracking and slice prescription in volumetric modalities like MRI become possible as registration speeds approach slice acquisition/safety-prescribed cooldown times [10,19]. If rigid landmarks are easily identifiable, rigid tracking in volumes reduces to automatic tagging or segmentation. However, in low signal-to-noise (SNR), low resolution applications such as fetal imaging, standard anatomic landmarks may not be identifiable from images; it is helpful in those cases to learn features that can be robustly detected. Moreover, if utility is directly proportional to speed (e.g., for navigator scans), SNR and resolution are reduced in favor of lower scan times. In this regime anatomic intuition may no longer match imaging conditions, and relevant features may be more robust if learned from images.

Recent learning based methods for rigid and affine registration have used convolutional neural networks (CNNs), which generally have faster inference-time speed than traditional image-grid iterative algorithms. CNN-based registrations almost universally regress transformation parameters directly by applying a set of stacked convolutional filters to images followed by a series of fully connected layers. Such architectures are common in other vision tasks [2].

© Springer Nature Switzerland AG 2021
M. de Bruijne et al. (Eds.): MICCAI 2021, LNCS 12904, pp. 193–202, 2021.
https://doi.org/10.1007/978-3-030-87202-1_19

Convolutions are translation equivariant: as inputs shift, output features shift accordingly. However, the fully connected network that estimates transformation parameters must learn from data a similar property between shifts and output estimates; the network structure does not naturally give rise to these symmetries. Moreover, neither structure encodes these properties naturally for rotation, and thus the equivariances must be learned. This is inefficient in data and computation, and, as we show, induces sub-optimal performance.

Instead, we constrain the feature extraction portion of networks (i.e., the convolutional filters) to be both translation and rotation equivariant. If the activation maps shift and rotate with the target object, we can avoid the fully connected regression step and instead compute least squares optimal transformations in closed form from summary statistics of the activation maps. We introduce a novel architecture for rigid tracking that leverages these two techniques. Our proposed method has the same computational costs as regular convolution at inference time, and we show that it provides more accurate tracking on human adult brain MRI and our intended use-case of fetal brain MRI.

1.1   Previous Work

Real-time object tracking (rigid registration) in volumetric imaging has a long history [5,20]. Early real-time methods in MRI and PET relied on either voxelwise intensity matching and gradient descent on parameters [5,15], or intrinsic properties of the image acquisition in k-space e.g., the oversampling of low frequencies in PROPELLER [11]. For adult human heads constrained by the physical geometry of the head coil, observed shifts and rotations are small in scale (e.g., maximally $10°$ [15]), and assumed to be global shifts of the entire non-zero acquisition [11]. Later systems for adult human heads include "navigator" scans and Kalman filters[19], but still consider a range of motion reasonable for scanning adults.

In fetal imaging, classical tracking methods such as PROPELLER fail in practice [10]. Iterative intensity matching is feasible but tends to be slower due to the extended search space [3] and multiple scales of image gridding.

Current methods in rigid and affine alignment are built almost exclusively on convolutional networks [3,13,17]. As they have a fixed number of operations which are amicable to GPU parallelization and compiler optimizations [4]. Salehi et al. 2018 [12] use a regression similar to image prediction architectures [2], where the reference and moving images are concatenated and provided as input to a commonly used architecture of alternating convolution and pooling. This leads to a flattening followed by a series of fully connected layers which estimate transformation parameters. Chee et al. 2018 [3] and de Vos et al. 2019 [17] instead use convolution layer stacks with separate inputs but shared weights.

We avoid the previous paradigm of stacked convolution to direct regression; rigid motion has known analytic symmetries, properties which are ignored by the structure of fully connected layers, and must be relearned in their parameters. These same properties we exploit in our proposed method.

We emphasize that our work is unrelated to the slice-to-volume registration problem [6], as our goal is to align volumetric 3D images.

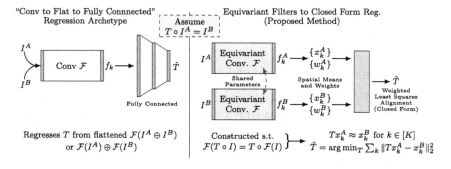

Fig. 1. Our network method, along with the convolution archetypes.

2 Method

We assume that two given images I^A and I^B are related by an unknown rigid transformation T, i.e.,

$$T \circ I^A = I^B. \tag{1}$$

A filter bank $\mathcal{F} : \mathcal{I} \to \mathcal{I}(\mathbb{R}^+)^K$ with K non-negative real-valued channels is called equivariant[1] under rigid transformations if for each channel $\mathcal{F}_k$ of $\mathcal{F}$,

$$\mathcal{F}_k(T \circ I) = T \circ \mathcal{F}_k(I). \tag{2}$$

In other words, $\mathcal{F}$ is equivariant if the outputs of $\mathcal{F}$ translate and rotate as the inputs are rotated and translated. Assuming $\mathcal{F}$ is equivariant and denoting $\mathcal{F}(I) = f$, we have

$$f_k^B = \mathcal{F}_k(I^B) = \mathcal{F}_k(T \circ I^A) = T f_k^A. \tag{3}$$

Since filter outputs $\{f_k^A\}$ and $\{f_k^B\}$ are in correspondence, we expect specific summary statistics (e.g., spatial mean activation point) to similarly match across images. Extracting the statistics from the filter outputs yields point clouds $\{x_k^A\}$ and $\{x_k^B\}$ with correspondence. $T x_k^A = x_k^B$ for $k \in \{1, \ldots, K\}$ up to discretization and resampling error. Thus, T can be estimated from $\{x_k^A\}$ and $\{x_k^B\}$ using for example the least squares formulation [1,7,8].

Our proposed method implements the above procedure. It applies equivariant filters to each volume, computes the spatial mean of each filter, and then estimates the transformation that minimizes squared error, which in turn provides the ℓ_2 optimal alignment of those spatial means between the volumes (Fig. 1). The learned parameters remain entirely in the equivariant filter construction; after filter outputs are collected all subsequent computations are analytically prescribed. The entire structure is differentiable, thus a loss function on the outputs can be backpropagated in order to train the equivariant filter parameters.

---

[1] More general forms of equivariance are defined in the literature [18], but this definition is sufficient for our purposes. We do not use properties of any other form.

Losses may be defined either on transformations or on image differences under the transformations (i.e., $d(\hat{T} \circ I^A, I^B)$ for image metric d).

2.1 Construction of Equivariant Convolutional Filters

Previously we defined $\mathcal{F}$ generally; however, $\mathcal{F}$ must be an efficient but expressive differentiable operation on image arrays. This leads us to stacked convolutional filters, which are already intrinsically equivariant to translation. In order to ensure rotational equivariance, we follow the construction in Weiler et al. [18], which retains translation equivariance.

Any rigid transformation T can be decomposed into translation t and rotation R. Given a field $F(x)$ over the spatial domain, the transformation operator is $\rho(R)F(R^{-1}(x - t))$, where the function ρ depends on the order of the field. While the images we consider and our desired outputs are both scalar fields with trivial ρ functions, for more expressive equivariant filters we require higher order fields (e.g., vector or matrix valued fields). Conventional multi-channel image operations use all scalar fields.

Weiler et al. [18] showed that convolutions between fields ("cross-correlation" in other settings) can be expressed as a block diagonal operation between specific basis components (irreducible representations), and that rotation equivariant kernels are a linear subspace of the possible kernels. The authors then construct a basis of such kernels analytically using spherical harmonics. Convolutions expressed in this basis are weighted combinations of the basis elements, which are analytically determined sub-blocks of conventional convolution kernels. These may be pre-computed for given field orders and discrete kernel widths. The weights between the basis elements form the learnable factors.

For a single channel (i.e., a scalar field), equivariant kernels must be isotropic (i.e., having spherical isoclines). However, grouping multiple channels together into higher order fields allows for non-isotropic equivariant kernels, acting on sets of channels. Higher order convolutional fields also require specialized non-linearities, but these are operationally the same as normal convolutions, and also without learned parameters. We refer the reader to [14,18] for details. We stack alternating layers of equivariant convolutions and their corresponding non-linearities to form $\mathcal{F}$, our equivariant filter bank.

2.2 Registration of Equivaritant Filters

We reduce each non-negative filter response f_k to its mean spatial position

$$x_k = \frac{1}{N_1 N_2 N_3} \sum_v v f_k(v) \tag{4}$$

for images with dimension $N_1 \times N_2 \times N_3$, where $v_{\ell,m,n}$ is the spatial coordinate of image index (ℓ, m, n), and $f_k(v_{\ell,m,n})$ is the corresponding value for the kth filter. Since $f_k^B = T f_k^A$, the spatial means of the filters will also be related similarly: $x_k^B \approx T x_k^A$. Here the error is due to discretization. This over-determined linear system is then solved for rigid transformation T, which we can express as the least squares problem:

$$\hat{T} = \arg\min_{T} \sum_{k} \|x_k^B - Tx_k^A\|_2^2. \tag{5}$$

Analytic solutions have been described multiple times [1,7,8]. We further introduce non-negative channel importance weights w_k^A and w_k^B, in order to reduce the influence of less important filters; we use the total power of each channel, e.g., $\tilde{w}_k^A = \sum_v f_k^A(v)$ as proposal weights (filter outputs are non-negative), then normalize to form $w_k^A = \tilde{w}_k^A / \sum_{k'} w_{k'}^A$, and similarly for $\{w_k^B\}$. From these we form generic channel weights $w_k = w_k^A w_k^B$, which we add to our optimization:

$$\hat{T} = \arg\min_{T} \sum_{k} w_k \|x_k^B - Tx_k^A\|_2^2. \tag{6}$$

As shown in [7], this also has an analytic solution similar to the unweighted problem. Computing weighted centroids $c^A = \sum_k x_k^A w_k$ and $c^B = \sum_k x_k^B w_k$, the optimal translation is $\hat{t} = c^B - c^A$. Re-centering each point $\bar{x}_k^A = x_k^A - c^A$ and stacking the $\bar{x}_k^A$ into matrices X_A and X_B, the optimal rotation is $\hat{R} = VU^T$, where $(\bar{X}_A)^T W \bar{X}_B = UDV^T$ is the SVD of the weighted cross correlation matrix and $W = \mathrm{diag}(w_1, \ldots, w_k)$.

2.3   Loss Functions and Implementations

Our architectures prescribe a forward operation. If the transformation parameters are known, loss functions can be defined for the output rotation and translation parameters. For translation parameters, the ℓ_2 loss is natural, but for rotation parameters more nuanced metric spaces can also be used. Salehi et al. 2018 [12] propose a geodesic loss on the rotation parameter matrix

$$\mathcal{L}_{\mathrm{geo}} = \arccos\left\{(\mathrm{tr}\,\tilde{R}R - 1)/2\right\}. \tag{7}$$

Zhou et al. [21] propose a projection and renormalization of the first two rows of each rotation matrix ("6D-loss")

$$\mathcal{L}_{6D} = \|\ (\ \tilde{R}_{:,1:2}/\|\tilde{R}_{:,1:2}\|_2\ ) - (\ R_{:,1:2}/\|R_{:,1:2}\|_2\ )\ \|_2 \tag{8}$$

which they show to be continuous for all elements in $SO(3)$.

In the absence of known transformation parameters we may also use image losses under the estimated transformation, i.e.,

$$\mathcal{L}_{\mathrm{Image}} = d(T \circ I^A, I^B). \tag{9}$$

Given a loss choice, we back-propagate errors and fit parameters by standard gradient based optimization [2].

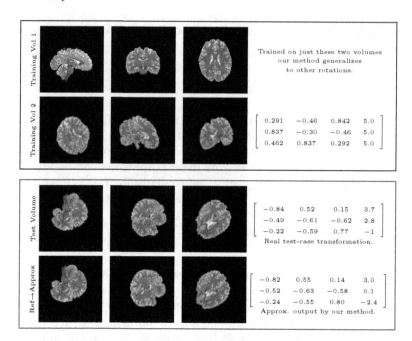

Fig. 2. The training and results for our first experiment. At top in the blue box we show central slices of the entire training dataset: a single subject in two poses. Below in the red box we first show the central slices of the same subject in a novel pose, then in the second row we show a reference volume transformed by our estimated parameters. True and estimated parameters are shown at right. (Color figure online)

Table 1. Mean and std. of error measures for the first experiment averaged over 100 novel poses. The proposed method was trained on a single pair of images. The Euler Angle error is the Mean Abs. Error in degrees averaged over the axes.

Angular error		Translational error		Dice index
Euler Angle	$\|I - R\hat{R}^T\|_F$	mm	Voxels	–
$2.0° \pm 1.6$	0.07 ± 0.03	6.1 ± 2.0	2.2 ± 0.7	0.96 ± 0.01

3   Experiments

We demonstrate our method in three separate empirical experiments, two on subsets of the Human Connectom Project (HCP) young adult cohort [16], and one on a fetal imaging dataset. The HCP dataset consists of subjects' T2-weighted brain volumes, downsampled from 0.7 mm isotropic to 2.8 mm isotropic, and then padded to a $96 \times 96 \times 96$ voxel volume. Images were masked for brain tissue and histograms normalized by percentile filter; training-testing splits as well

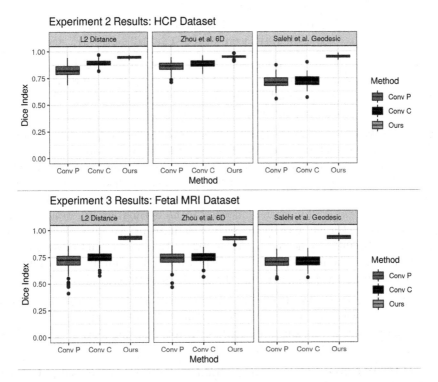

Fig. 3. Testing results for the second experiment on HCP data (**top row**) and for the third experiment with fetal MRI data (**bottom row**), measured by Dice Index overlap of masks under estimated transformations between images. Each inset indicates training under a different loss (ℓ_2, $\mathcal{L}_{6D}$ [21], and $\mathcal{L}_{geo}$ [12]), for all methods. **Higher** is better, with 1 indicating perfect overlap. **Conv P** is comparable to [3,17], while **Conv C** is comparable to [12].

as data augmentation vary by experiment. The fetal dataset consists of MRI time-series from 53 healthy mothers pregnant with healthy singletons at gestational ages ranging from 25 to 35 weeks. MRI were acquired on a 3T Skyra Scanner (Siemens Healthcare, Erlangen, Germany). Multislice single-shot gradient echo EPI sequences were acquired at 3mm isotropic resolution, with a mean matrix size (field-of-view) of $120 \times 120 \times 80$ which we crop to $64 \times 64 \times 64$, TR = 5–8, TE = 32–38, and 90° flip angles. Automatic brain masking was applied and dilated by 4 voxels to reflect uncertainty, after which images were intensity normalized.

In all three experiments training, validation, and testing datasets are completely disjoint, and in the latter two experiments subjects are not repeated in separate poses between datasets i.e., no individuals are shared between training/validation/testing sets. For both datasets and all three experiments we train both baseline methods and variations of our proposed method using the Adam optimizer [9] for 2000 epochs, taking the best performing parameters and hyperparameters with respect to a validation set. For all experiments batch size was set to 1. Three separate losses were tested: $\mathcal{L}_{geo}$ [12], $\mathcal{L}_{6D}$ [21], and ℓ_2.

Parameters: After several preliminary experiments we selected a stack of 5 equivariant convolutions, all with $5 \times 5 \times 5$ kernels, using field-appropriate ReLU non-linearities throughout, and 16, 16, and 4 fields of orders 0, 1, and 2 respectively, except for the final layer which has 64 scalar channels.

Baselines: We compare our method against two convolutional baseline architectures, one using a combined processing stream (**"Conv C"**) as found in [12], and one using separate, parallel convolutional processing streams with tied weights (**"Conv P"**), as found in both [3] and [17]. For each method, we searched over the number of layers and channels per layer for optimal configurations given compute resources. We found the following to be optimal for our tests: 3 conv layers of $3 \times 3 \times 3$ kernels with 64 channels each, interspersed with $2 \times 2 \times 2$ pooling, followed by fully connected layers of 1024, 512, and 256 units, using ReLU activations throughout except for the output layer. The initial fully-connected layer has more parameters than the entirety of our network.

Single Subject Experiment: In order to demonstrate the structural advantage of our method, we first train on *just two volumes* of one subject. As Fig. 2, and Table 1 report, from this single example of a rotation our method generalizes well to new poses of the same subject.

Group HCP Experiment: Next we train the network on 500 subjects from the HCP dataset, holding out 100 for validation and 100 for testing, in order to test the generalization of different methods across subjects. As shown in the **top row** of Fig. 3, our method performs better than both baselines, though all methods work reasonably well using either ℓ_2 or $\mathcal{L}_{6D}$ losses.

Fetal Experiment: Our third experiment trains the proposed method on the fetal MRI dataset, using 33 subjects for training (50 volumes per subject), 4 subjects held out for validation, and 16 subjects for testing (10 volumes per subject). As shown in the **bottom row** of Fig. 3, our method significantly outperforms the baseline methods for this dataset. This may be due to the smaller and more highly varying size of the fetal brains.

4   Discussion and Conclusion

We have described an equivariant filter based method for rigid tracking suitable for applications where speed is paramount. As we have shown empirically, our approach is demonstrably superior to methods based on conventional convolution. Moreover, due to the equivariant property of the filters, it generalizes to unseen poses by design. While a complete prospective motion correction system will have multiple other components (e.g., reconstruction, denoising, masking, slice proscription), tracking is a key step in this process, and accurate and efficient methods for tracking such as the one introduced here are thus of value for building such systems.

Acknowledgements. This work was supported by NIH NIBIB NAC P41EB015902, NIH NICHD R01HD100009, Wistron Corporation, and the MIT-IBM Watson AI Lab.

References

1. Arun, K.S., Huang, T.S., Blostein, S.D.: Least-squares fitting of two 3-D point sets. IEEE Trans. Pattern Anal. Mach. Intell. **5**, 698–700 (1987)
2. Bengio, Y., Goodfellow, I., Courville, A.: Deep Learning, vol. 1. MIT Press, Massachusetts (2017)
3. Chee, E., Wu, Z.: AIRNet: self-supervised affine registration for 3D medical images using neural networks. arXiv preprint arXiv:1810.02583 (2018)
4. Chen, T., et al.: TVM: an automated end-to-end optimizing compiler for deep learning. In: 13th USENIX Symposium on Operating Systems Design and Implementation (OSDI 18), pp. 578–594. USENIX Association, Carlsbad, October 2018
5. Cox, R.W., Jesmanowicz, A.: Real-time 3D image registration for functional MRI. Magnetic Reson. Med. Official J. Int. So. Magnetic Reson. Med. **42**(6), 1014–1018 (1999)
6. Ferrante, E., Paragios, N.: Slice-to-volume medical image registration: a survey. Med. Image Anal. **39**, 101–123 (2017)
7. Horn, B.K.: Closed-form solution of absolute orientation using unit quaternions. Josa a **4**(4), 629–642 (1987)
8. Kabsch, W.: A solution for the best rotation to relate two sets of vectors. Acta Crystallographica Sect. A Cryst. Phys. Diffr. Theoret. General Crystallogr. **32**(5), 922–923 (1976)
9. Kingma, D.P., Ba, J.: Adam: a method for stochastic optimization. arXiv preprint arXiv:1412.6980 (2014)
10. Malamateniou, C., et al.: Motion-compensation techniques in neonatal and fetal MR imaging. Am. J. Neuroradiol. **34**(6), 1124–1136 (2013)
11. Pipe, J.G.: Motion correction with propeller MRI: application to head motion and free-breathing cardiac imaging. Magn. Resonance Med. Official J. Int. Soc. Magn. Resonance Med. **42**(5), 963–969 (1999)
12. Salehi, S.S.M., Khan, S., Erdogmus, D., Gholipour, A.: Real-time deep pose estimation with geodesic loss for image-to-template rigid registration. IEEE Trans. Med. Imaging **38**(2), 470–481 (2018)
13. Sloan, J.M., Goatman, K.A., Siebert, J.P.: Learning rigid image registration-utilizing convolutional neural networks for medical image registration (2018)
14. Smidt, T.E., Geiger, M., Miller, B.K.: Finding symmetry breaking order parameters with Euclidean neural networks. Phys. Rev. Res. **3**(1), L012002 (2021)
15. Thesen, S., Heid, O., Mueller, E., Schad, L.R.: Prospective acquisition correction for head motion with image-based tracking for real-time fmri. Magn. Resonance Med. Official J. Int. Soc. Magn. Resonance Med. **44**(3), 457–465 (2000)
16. Van Essen, D.C., et al.: The WU-Minn human connectome project: an overview. Neuroimage **80**, 62–79 (2013)
17. de Vos, B.D., Berendsen, F.F., Viergever, M.A., Sokooti, H., Staring, M., Išgum, I.: A deep learning framework for unsupervised affine and deformable image registration. Med. Image Anal. **52**, 128–143 (2019)
18. Weiler, M., Geiger, M., Welling, M., Boomsma, W., Cohen, T.: 3D steerable CNNs: learning rotationally equivariant features in volumetric data. arXiv preprint arXiv:1807.02547 (2018)

19. White, N., et al.: Promo: real-time prospective motion correction in MRI using image-based tracking. Magn. Resonance Med. Official J. Int. Soc. Magn. Resonance Med. **63**(1), 91–105 (2010)
20. Woods, R.P., Grafton, S.T., Holmes, C.J., Cherry, S.R., Mazziotta, J.C.: Automated image registration: I. general methods and intrasubject, intramodality validation. J. Comput. Assist. Tomogr. **22**(1), 139–152 (1998)
21. Zhou, Y., Barnes, C., Lu, J., Yang, J., Li, H.: On the continuity of rotation representations in neural networks. In: Proceedings of the IEEE/CVF Conference on Computer Vision and Pattern Recognition, pp. 5745–5753 (2019)

Revisiting Iterative Highly Efficient Optimisation Schemes in Medical Image Registration

Lasse Hansen$^{(\boxtimes)}$ ⓘ and Mattias P. Heinrich ⓘ

Institute of Medical Informatics, Universität zu Lübeck, Lübeck, Germany
{hansen,heinrich}@imi.uni-luebeck.de

Abstract. 3D registration remains one of the big challenges in medical imaging, especially when dealing with highly deformed anatomical structures such as those encountered in inter- or intra-patient registration of abdominal scans. In a recent MICCAI registration challenge (Learn2Reg) deep learning based network architectures with inference times of <2 s showed great success for supervised alignment tasks. However, in unsupervised settings deep learning methods have not yet outperformed their conventional algorithmic counterparts based on continuous iterative optimisation (and probably won't as they share the same objective function (image metric)). This finding has brought us to revisit conventional optimisation schemes and investigate an iterative message passing approach that enables fast runtimes (using iterative optimisation with only few displacement candidates) and high registration accuracy. We conduct experiments on three challenging abdominal datasets ((pre-aligned) inter-patient CT, intra-patient MR-CT) and carry out an in-depth evaluation with a set of selected comparison methods. Our results clearly indicate that optimisation based methods are highly competitive both in accuracy and runtime when compared to Deep Learning methods. Moreover, we show that semantic label information (when available) can be efficiently exploited by our approach (cf. weakly supervised learning). Data and code will be made publicly available to ensure reproducibility and accelerate research in the field of 3D medical registration (https://github.com/lasseha/iter_lbp).

Keywords: Registration · Discrete optimisation · Generalisation

1 Introduction/Motivation

While the main focus of this work is on investigating the general problem of efficient medical image registration, we here specifically deal with the clinical task of inter- and intra-patient alignment of abdominal CT/MR scans. Enabling deformable multimodal fusion (of thorax and abdomen) has numerous medical applications, e.g. for aligning pre-interventional scans for image-guided (radio)therapy and multimodal diagnostic. Inter-subject CT registration can enable statistical modelling

© Springer Nature Switzerland AG 2021
M. de Bruijne et al. (Eds.): MICCAI 2021, LNCS 12904, pp. 203–212, 2021.
https://doi.org/10.1007/978-3-030-87202-1_20

of variations of abdominal organs for abnormality detection and to provide a canonical atlas space.

Medical image registration is often considered a task with substantial computation times that may prevent its application in clinical workflows. While fully-convolutional segmentation networks can produce contours almost instantly, a classical iterative registration algorithm may often take minutes to converge. Hence, deep learning based image registration has been proposed to improve run times, so far however often with a degradation in alignment quality. The recent comprehensive medical registration challenge Learn2Reg showed that especially for large internal motion and multimodal fusion tasks, conventional methods are more robust and accurate than learning based approaches at the cost of longer runtimes. Yet little effort has recently been devoted in designing new iterative optimisation strategies for registration. GPU-accelerated optimisation routines have been explored in the AirLab framework [19] and as instance refinement in the DL-based VoxelMorph and PDD [2,9]. Discrete optimisation may inherently reduce the number of iterations but has large memory requirements and either limited accuracy or a fixed capture range.

We hypothesis that a suitable combination of discrete and iterative schemes has been mostly overlooked in previous research but could provide new perspectives for medical registration without learning. We also note, that the limited availability of GPUs in clinical setups is often disregarded for run times - hence speeding up CPU computation bring real benefits. Since registration algorithms are often used as versatile tools to handle a variety of tasks an evaluation on complementary applications is desirable: here we consider monomodal inter-subject abdominal CT registration and multimodal intra-patient CT/MR fusion both in affine/rigid and nonlinear settings.

Related Work: Learning-based registration can be roughly subdivided in metric- and label-supervised approaches or combinations of them, where most algorithms employ a fully-convolutional multi-scale CNN architecture with (potentially multiple) spatial transformer layers [2,18]. Since, the alignment target can often not be determined alone by a limited number of manually segmented structures - e.g. fissures and lobes are not sufficient to guide intra patient lung registration - metric-supervised methods may be considered more general. In conventional image registration a generic similarity metric (possible based on hand-crafted features) is optimised together with a regulariser in a multi-scale and usually iterative fashion.

Two popular MRF-based discrete optimisation approaches, drop [7] and deeds [11] are related to our work. Drop employs Fast-PD as optimisation backend, limits the complexity by sampling discrete displacements only along the 3 principal axes (31 possible vectors in 3D) and iteratively refines the matching. It uses a B-spline transformation, in a multi-resolution setting and local cross-correlation as metric. Deeds relies on the faster MST-BP optimisation [5], and uses a dense discrete displacement search (up to 5000 vectors each). It employs a multi-scale approach with up to 5 warps, MIND features [12] for similarity and symmetry constraints. Both methods run in ≈1 min on multi-thread CPUs.

The closest work to our method in learning based registration is linearised multi-sampling [14], which was proposed as an extension for spatial transformer networks to tackle the noisy gradient estimation in differentiable trilinear interpolation. Different to other DL-registration approaches the gradient with respect to a sub-pixel displacement is not determined by differentiating the interpolation coefficients directly, but instead a hyperplane is fitted at a small number of random sampling points in the proximity of the current displacement. Using the slope of this plane as gradient estimation was shown to be more stable especially for difficult transformations.

Contribution: We propose an efficient strategy for dynamically solving a regularised cost function that is founded in graph-based optimisation and surpasses most conventional and learning-based registration algorithms in terms of accuracy and speed. Similar to multi-sampling we employ a randomised set of displacement candidates nearby the previous solution at each iteration. But instead of directly computing a metric-based displacement gradient, we perform a probabilistic discrete optimisation using a sparse variant of loopy belief propagation (LBP) [6] for the joint regularised cost function. In contrast to dense discrete optimisation, the space of considered displacements is reduced by orders of magnitudes and no fixed range of motion has to be pre-defined. The methods improves runtimes - with $<1\,$s GPU or $<5\,$s CPU - compared to DL registration, while matching or exceeding accuracy on three demanding tasks for inter- and intra-patient registration of CT-CT and MR-CT of the abdomen.

2   Method

Our method comprises a spatially randomly distributed sampling of control points (keypoints) and a conventional feature extraction step using hand-crafted contrast-invariant descriptors followed by the proposed iterative (dynamic) loopy belief propagation optimisation of a regularised registration cost function. The registration is performed in a small number of outer iterations in which the displacement search is dynamically changed based on the probabilistic estimate of the inner optimisation iterations for LBP. The feature descriptors can be efficiently evaluated on-line through indexing or nearest neighbour interpolation for each keypoint in the fixed image and each candidate in the moving image. Finally, either a trimmed least square or thin-plate spline fitting is employed to obtain either a linear transform or extrapolate a nonlinear displacement field.

We employ MIND with self-similarity context with 12 channels [12] as state-of-the-art descriptor and enhance the capturing of spatial context by sampling a small local patch of size $3 \times 3 \times 3$ of these descriptors at each keypoint positions $\mathbf{p}_{f_i}$, resulting in a 324-dimensional vector. We use a symmetric k-nearest neighbour (kNN) graph on the set of keypoints in the fixed scan P_f with edges $(ij) \in E$ that connect keypoints $\mathbf{p}_{f_i}$ and $\mathbf{p}_{f_j}$. To optimise a regularised cost function that minimises the dissimilarity between feature vectors in fixed and - at a displaced position - in the moving scan we use the sum of squared difference metric and define a diffusion like regulariser. In discrete optimisation the regularisation term is considered for each pair of possible candidate displacement for each edge using

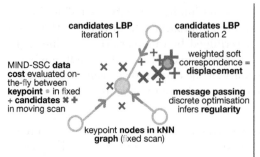

Algorithmic steps for **iterative LBP registration**
1) **Extract MIND** for fixed and moving scan
2) Compute sparse **keypoint graph** (symmetric kNN) on fixed scan (storing edge indices)
3) initialise with zero displacement **and iterate**
    1) **draw new displacement candidates**
    2) **compute MIND data term** (sampling)
    3) **iterate for loopy belief propagation**
        1) gather incoming messages
        2) subtract reverse messages
        3) **min-sum regularisation**
        4) scatter outgoing messages
    4) compute **soft**(max) **correspondences** and update displacements
4) **fit linear** (least squares) **or nonlinear** (thin-plate spline) **transformation** for dense displacements → **optionally** second transform

Fig. 1. Schematic overview and pseudo-code of our proposed iterative optimisation approach for keypoints based 3D medical image registration using loopy belief (LBP) message passing.

$r_{ij}^{pq} = \left| (\mathbf{c}_i^p - \mathbf{p}_{f_i}) - (\mathbf{c}_j^q - \mathbf{p}_{f_j}) \right|_2^2$. In contrast to most common discrete optimisation schemes, the set of candidates $\mathbf{c}_j^q$ is not pre-determined or even equally spaced for each keypoint but dynamically adapted throughout iterations. The actual optimisation is performed in a few inner iterations, where messages are passed in parallel that enable the exchange of information across the graph and a more accurate deformation estimation. The algorithm is described in detail in [6] and uses the following equation to compute outgoing messages $\mathbf{m}_{i \to j}^t$ from $\mathbf{p}_{f_i}$ to

$\mathbf{p}_{f_j}$ at iteration t: $\mathbf{m}_{i \to j}^t = \min_{1,\dots,q,\dots l} \left(\mathbf{d}_i + \alpha \mathbf{r}_{ij}^q - \mathbf{m}_{j \to i}^{t-1} + \sum_{(h,i) \in E} \mathbf{m}_{h \to i}^{t-1} \right)$.

Once each LBP optimisation has converged, new candidates are drawn from a local neighbourhood (uniformly spatially distributed) and the search region can gradually adapt to the true optimum. To extrapolate from the final correspondences to a dense displacement either thin-plate splines or affine least-squares fitting is employed. A visual overview of the concept with a brief pseudo-code is given in Fig. 1. The source-code is available at https://github.com/lasseha/iter_lbp.

Compared to conventional stochastic gradient descent methods our proposed optimisation scheme benefits from the discrete setting of the search that finds the combinatorial minimum of a wide range of displacements. In contrast to commonly used loopy belief propagation (which iterates only over steps of message passing) we introduce a second and important outer iteration over the capture ranges of displacements. This approach is to the best of our knowledge the first that couples a discrete belief propagation optimisation with a sparsely distributed and locally adaptive solution space of potential displacements for each control point.

3   Experiments

To demonstrate the effectiveness and robustness of our iterative optimisation approach, we perform a comprehensive evaluation on three challenging abdominal datasets covering inter- and intra-patient, multimodal (CT/MR) as well as pre- and non-pre-aligned registration tasks. The datasets are described in detail in Sect. 3.1. The evaluation metrics are outlined below and closely follow the evaluation design of the L2R challenge, assessing accuracy and smoothness of the displacement field as well as the runtime of the algorithm. Finally, we describe implementation and configuration details of our proposed and comparison methods. All methods and experiments (with exception of drop2) were implemented and evaluated using the latest version of PyTorch (v1.7.1).

3.1   Datasets

CT-CT (L2R 2020): For our experiments on inter-patient alignment of abdominal CT scans we use the corresponding dataset (Task 3) from the Learn2Reg (L2R) challenge [8]. It contains 30 abdominal CT scans with 13 manually labelled anatomical structures: spleen ■, right kidney ■, left kidney ■, gall bladder ■, esophagus ■, liver ■, stomach ■, aorta ■, inferior vena cava ■, portal and splenic vein ■, pancreas ■, left adrenal gland, and right adrenal gland. The image data and labels themselves stem from [20] and were pre-registered to a canonical space and resampled to same voxel resolution and spatial dimensions ($192 \times 160 \times 256$). We report evaluation results on the predefined validation set, which describes a subset of 10 CT scans and 45 registration pairs.

CT-CT: To study the effect of non-pre-aligned data and thus more complex deformations on the different registration methods we also consider the original data from [20]. The segmentation labels, resampling procedure of the CT scans and validation pairs remain the same as described above.

MR-CT: Inter-patient alignment of abdominal CT to MR is investigated on 16 selected scan pairs from different trials in The Cancer Imaging Archive (TCIA) project [1,3,4,17]. Four different sized organs (liver, spleen, left and right kidney) were manually labelled by us for each MR-CT scan pair to assess the registration accuracy. The data (with exception of a withheld test set) will be made publicly available as part of a larger dataset for abdominal registration later this year. Pre-processing comprises resampling the scan pairs to an isotropic resolution of 2 mm and cropping/padding to consistent voxel dimensions of $192 \times 160 \times 192$. We report evaluation results over all 16 MR-CT pairs.

Evaluation Metrics: The used evaluation metrics cover the three most important aspects of medical image registration: 1) registration accuracy, 2) smoothness of the displacement field and 3) runtime of the algorithm. We assess the accuracy by means of alignment of the manually labelled organ segmentations using the Dice similarity metric (F1 score). In addition, the plausibility of the deformation field is of great clinical relevance and can be estimated by the standard deviation of the (logarithmic) Jacobian determinant [15,16] (SDlogJ). Finally, we report the runtime of the registration methods, both on CPU (Apple M1) and GPU (NVIDIA RTX 2080 Ti), including all necessary computations after reading the images until predicting the final dense displacement field.

Compared Methods and Hyper-parameters: We employ a number of different related state-of-the-art learning- and optimisation-based registration methods in comparison to our proposed iterative LBP approach. Hyperparameters for ours and all comparison methods were selected on a subset of training scans for CT-CT and a small number of validation scans for MR-CT. Note, that iterative registration methods are much less sensitive to manual parameter choices than deep learning approaches. First, we substantially extended and improved upon the original VoxelMorph method [2] by replacing the simplistic U-Net with a two-stream architecture and a custom MIND-based metric-loss. Despite these advancements VoxelMorph+ does not yield satisfactory results for very large misalignments (CT-CT w/o pre-align and MR-CT). Second, the overall runner-up in the Learn2Reg challenge, PDD-net [10] again with MIND loss is used with all available extensions: multiple warps and instance optimisation for CT-CT, as well as affine least squares fitting for MR-CT. Third, drop2 a recent re-implementation of [7] is used as a related baseline for discrete optimisation. We explored a wide range of settings and found that diffusion regularisation and an initial B-spline spacing of 80mm in combination with the following settings worked best: cross-correlation (weight: 0.25) and entropy correlation (weight: 0.5) for CT-CT and MR-CT respectively with three pyramid levels each: 12 mm, 8 mm, 4 mm for CT-CT and 8 mm, 6 mm, 4 mm for MR-CT with a doubling of the iterations for the latter.

Furthermore, we evaluate two alternatives to our proposed method, all with the same metric patch-based MIND-SSC and one (nonlinear, for CT-CT (L2R 2020)) or two warps (affine + nonlinear, for CT-CT and MR-CT): dense 3D displacement sampling with LBP and continuous Adam optimisation with diffusion regularisation. For our proposed method, we chose 20 outer and 3 inner (LBP) iterations, where 12 (CT-CT) or 16 (MR-CT) displacement candidates are drawn from a uniform distribution for each of 2048 (or 8192 for nnUNet features) keypoints using capture ranges that start from ±15 voxels and linearly decrease to ±3 voxels. For Adam we employed a learning rate of 0.1 and 100 iterations. For the dense variant of LBP, we considered 11^3 displacements in parallel with a range of ±30 voxels. The regularisation is performed on a kNN graph with 10 nearest neighbours for the LBP variants and 16 neighbours for Adam. The regularisation weight (alpha) is set to 2.5 (LBP) and 0.2 (Adam), respectively. For all methods (with exception of Voxelmorph+) a coarse and convex binary mask is used to restrict registration to the region of interest (which includes approximately 50% of the original number of image voxels).

4   Results

Quantitative and qualitative results of our experiments are shown in Table 1 and Fig. 2, respectively. For the pre-aligned CT-CT dataset our proposed iterative LBP approach yields a Dice score of 40.1%, which is competitive to the best performing deep learning based comparison method, PDD-Net, with 41.5%. The sparse candidate sampling strategy of our method enables fast runtimes of

Table 1. Quantitative evaluation results of proposed and comparison methods on all three abdominal datasets considered. Runtimes are given for both GPU and CPU (GPU/CPU). Experiments in the first half (after the first double rule) are based on similarity metrics (MIND, NCC), while experiments in the second half (after the second double rule) make use of label information.

	CT-CT (L2R 2020)			CT-CT			MR-CT		
	Dice [%]	SDlogJ	Time [s]	Dice [%]	SDlogJ	Time [s]	Dice [%]	SDlogJ	Time [s]
Initial	25.1			11.6			26.5		
VoxelMorph+	35.4	.134	**0.2**/13						
PDD-Net	**41.5**	.129	1.4/73	35.1	.131	1.4/73	63.6	.210	1.0/37
drop2	37.0	.147	−/43	31.5	.132	−/43	56.2	.109	−/68
Adam	36.6	.080	1.6/27	30.9	.087	2.7/46	71.1	.074	2.7/46
Dense LBP	38.6	.119	0.8/14	36.4	.098	1.3/23	71.7	.085	1.3/23
Iterative LBP	40.1	.093	0.6/**5**	**37.7**	.092	**1.0/13**	**76.4**	.075	**1.0/13**
VoxelMorph+	43.9	.162	**0.2/13**						
Adam	54.6	.046	3.4/52						
Dense LBP	62.1	.170	2.2/33						
Iterative LBP	**65.2**	.088	1.9/22						

<1 s on GPU and approximately 5 s on CPU. Making use of label information (weakly supervised learning for Voxelmorph+, using nnUNet [13] softmax features for optimisation based approaches) consistently and significantly improves the registration results by 8.5% points (VoxelMorph+), 18% points (Adam), 22% points (dense LBP) and 25.1% (iterative LBP). The inference time of the network is included in the total runtimes of the optimisation based methods. Our methods final Dice score of 65.2% is also comparable to the reported score of 67% of the winning entry (LapIRN [18]) of the L2R challenge on the hidden test set (validation and test results are generally comparable). Results on the non-pre-aligned CT-CT data show only a moderate decrease in Dice score of 37.7% for our method (−2.4% points). Other comparison methods are less robust, e.g. drop2 or Adam, decrease by 5.5% points and 5.7% points Dice score, respectively. In this experiment, in addition to the Dice metric, we evaluate the 95th percentile of the Hausdorff Distance (HD95) to further highlight the differences between the state-of-the-art continuous optimiser (Adam) and our proposed iterative LBP approach. We find initial values of 30.14 voxels, which are reduced by Adam to 26.1 voxels. Our proposed method can substantially and statistically significantly (p < 0.0002 using a Wilcoxon signed rank test) improve upon this with an error of only 17.3 voxels. For the inter-patient MR-CT dataset our iterative LBP method clearly yields the best Dice score of 76.4% with runtimes of 1.0 s on GPU and approximately 13 s on CPU.

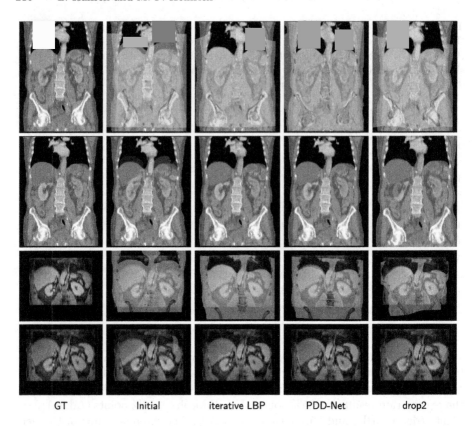

GT Initial iterative LBP PDD-Net drop2

Fig. 2. Qualitative results of selected methods (coronal view). Row 1–2: Overlayed warped moving CT scan and warped segmentation labels for inter-patient CT-CT (L2R 2020). Row 3–4: intra-patient MR-CT. PDD and our proposed iterative LBP achieve more regular transformations than drop2 and are both very close to the ground truth for inter-subject alignment. Our method yields the highest accuracy for multimodal fusion especially at organ boundaries (lungs, kidneys).

Differences to drop2 the iterative MRF registration technique proposed in [7]: while the general idea of solving registration as iterative discrete optimisation is similar our method differs in a number of important design choices that greatly improve accuracy and runtime. First, our model uses a sparse keypoint graph and computes the similarity cost of displacements separately for each node, while [7] uses a grid-based model. We employ a feature-based metric and found that spatially sampling one high-dimensional vector per keypoint sufficiently captures similarity and no spline interpolant has to be evaluated. Loopy belief propagation is used as discrete optimisation backend, which has a simpler implementation, provides probabilistic estimates and lends itself to parallelism. Finally, our displacement sampling injects more randomness by avoiding an axis parallel selection of motion vectors and thus converges faster.

5   Discussion and Conclusion

We have developed an efficient discrete optimisation framework for versatile medical image registration that contrary to recent trends does not rely on learned convolutions and improves both runtime and accuracy compared to deep learning registration. This avoids lengthy preparations for supervised training and makes our method generally applicable. In addition, separating label prediction and optimisation-based registration offers improved explainability for clinicians compared to end-to-end learning methods (black box approach). The extensive experimental validation on three different abdominal registration tasks demonstrate substantial advantages over both unsupervised learning based and previous (iterative) discrete optimisers. We believe this is due to the following reasons: 1) our method avoids regular grids and dynamically adapts the shape of the solution space, finding an optimal balance between coverage and accuracy. 2) the combination of locally discriminative image features and globally regular graph optimisation can more efficiently address the computational demands on medical image registration than convolutional network architectures. A further surprising outcome is the competitiveness of Adam optimisation using regularisation across keypoints and MIND as loss term. This shows that iterative optimisation for registration should always be considered as a strong baseline when discussing new learning-based approaches. Future work could further reduce the complexity of supervised nnUNet feature extraction through quantisation and pruning.

References

1. Akin, O., et al.: Radiology data from the cancer genome atlas kidney renal clear cell carcinoma [TCGA-KIRC] collection. Cancer Imaging Arch. (2016). https://doi.org/10.7937/K9/TCIA.2016.V6PBVTDR
2. Balakrishnan, G., Zhao, A., Sabuncu, M.R., Guttag, J., Dalca, A.V.: VoxelMorph: a learning framework for deformable medical image registration. IEEE Trans. Med. Imaging (TMI) **38**(8), 1788–1800 (2019)
3. Clark, K., et al.: The cancer imaging archive (TCIA): maintaining and operating a public information repository. J. Digit. Imaging **26**(6), 1045–1057 (2013)
4. Erickson, B., et al.: Radiology data from the cancer genome atlas liver hepatocellular carcinoma [TCGA-LIHC] collection. Cancer Imaging Arch. (2016). https://doi.org/10.7937/K9/TCIA.2016.IMMQW8UQ
5. Felzenszwalb, P.F., Huttenlocher, D.P.: Pictorial structures for object recognition. Int. J. Comput. Vis. **61**(1), 55–79 (2005)
6. Felzenszwalb, P.F., Huttenlocher, D.P.: Efficient belief propagation for early vision. Int. J. Comput. Vis. **70**(1), 41–54 (2006)
7. Glocker, B., Komodakis, N., Tziritas, G., Navab, N., Paragios, N.: Dense image registration through MRFs and efficient linear programming. Med. Image Anal. **12**(6), 731–741 (2008)
8. Hansen, L., Hering, A., Heinrich, M.P., Dalca, A., et al.: Learn2Reg: 2020 MICCAI registration challenge (2020). https://learn2reg.grand-challenge.org

9. Heinrich, M.P.: Closing the gap between deep and conventional image registration using probabilistic dense displacement networks. In: Shen, D., et al. (eds.) MICCAI 2019. LNCS, vol. 11769, pp. 50–58. Springer, Cham (2019). https://doi.org/10.1007/978-3-030-32226-7_6

10. Heinrich, M.P., Hansen, L.: Highly accurate and memory efficient unsupervised learning-based discrete CT registration using 2.5D displacement search. In: Martel, A.L., et al. (eds.) MICCAI 2020. LNCS, vol. 12263, pp. 190–200. Springer, Cham (2020). https://doi.org/10.1007/978-3-030-59716-0_19

11. Heinrich, M.P., Jenkinson, M., Brady, S.M., Schnabel, J.A.: MRF-based deformable registration and ventilation estimation of lung CT. IEEE Trans. Med. Imaging (TMI) **32**(7), 1239–48 (2013)

12. Heinrich, M.P., Jenkinson, M., Papież, B.W., Brady, S.M., Schnabel, J.A.: Towards realtime multimodal fusion for image-guided interventions using self-similarities. In: Mori, K., Sakuma, I., Sato, Y., Barillot, C., Navab, N. (eds.) MICCAI 2013. LNCS, vol. 8149, pp. 187–194. Springer, Heidelberg (2013). https://doi.org/10.1007/978-3-642-40811-3_24

13. Isensee, F., Jäger, P.F., Kohl, S.A., Petersen, J., Maier-Hein, K.H.: nnU-Net: a self-configuring method for deep learning-based biomedical image segmentation. Nat. Meth. **18**(2), 203–211 (2021)

14. Jiang, W., Sun, W., Tagliasacchi, A., Trulls, E., Yi, K.M.: Linearized multi-sampling for differentiable image transformation. In: Proceedings of the IEEE/CVF International Conference on Computer Vision, pp. 2988–2997 (2019)

15. Kabus, S., Klinder, T., Murphy, K., van Ginneken, B., Lorenz, C., Pluim, J.P.W.: Evaluation of 4D-CT lung registration. In: Yang, G.-Z., Hawkes, D., Rueckert, D., Noble, A., Taylor, C. (eds.) MICCAI 2009. LNCS, vol. 5761, pp. 747–754. Springer, Heidelberg (2009). https://doi.org/10.1007/978-3-642-04268-3_92

16. Leow, A.D., et al.: Statistical properties of Jacobian maps and the realization of unbiased large-deformation nonlinear image registration. IEEE Trans. Med. Imaging **26**(6), 822–832 (2007)

17. Linehan, M., et al.: Radiology data from the cancer genome atlas cervical kidney renal papillary cell carcinoma [KIRP] collection. Cancer Imaging Arch. (2016). https://doi.org/10.7937/K9/TCIA.2016.ACWOGBEF

18. Mok, T.C.W., Chung, A.C.S.: Large deformation diffeomorphic image registration with Laplacian pyramid networks. In: Martel, A.L., et al. (eds.) MICCAI 2020. LNCS, vol. 12263, pp. 211–221. Springer, Cham (2020). https://doi.org/10.1007/978-3-030-59716-0_21

19. Sandkühler, R., Jud, C., Andermatt, S., Cattin, P.C.: AirLab: autograd image registration laboratory. arXiv preprint arXiv:1806.09907 (2018)

20. Xu, Z., et al.: Evaluation of six registration methods for the human abdomen on clinically acquired CT. IEEE Trans. Biomed. Eng. **63**(8), 1563–1572 (2016)

Multi-scale Neural ODEs for 3D Medical Image Registration

Junshen Xu[1](✉), Eric Z. Chen[2], Xiao Chen[2], Terrence Chen[2],
and Shanhui Sun[2]

[1] Department of Electrical Engineering and Computer Science, MIT,
Cambridge, MA, USA
junshen@mit.edu
[2] United Imaging Intelligence, Cambridge, MA, USA
{zhang.chen,xiao.chen01,terrence.chen,shanhui.sun}@united-imaging.com

Abstract. Image registration plays an important role in medical image analysis. Conventional optimization based methods provide an accurate estimation due to the iterative process at the cost of expensive computation. Deep learning methods such as learn-to-map are much faster but either iterative or coarse-to-fine approach is required to improve accuracy for handling large motions. In this work, we proposed to learn a registration optimizer via a multi-scale neural ODE model. The inference consists of iterative gradient updates similar to a conventional gradient descent optimizer but in a much faster way, because the neural ODE learns from the training data to adapt the gradient efficiently at each iteration. Furthermore, we proposed to learn a modal-independent similarity metric to address image appearance variations across different image contrasts. We performed evaluations through extensive experiments in the context of multi-contrast 3D MR images from both public and private data sources and demonstrate the superior performance of our proposed methods.

Keywords: Multi-modal image registration · Neural ordinary differential equations · Disentangled representation · Self-supervised learning

1 Introduction

Image registration is an essential step in many tasks such as motion correction and atlas-based image segmentation. It is indispensable in many clinical applications such as surgical planning [24] and radiogenomics analysis [16], where 3D

J. Xu—This work was carried out during the internship of the author at United Imaging Intelligence, Cambridge, MA 02140.

Electronic supplementary material The online version of this chapter (https://doi.org/10.1007/978-3-030-87202-1_21) contains supplementary material, which is available to authorized users.

M. de Bruijne et al. (Eds.): MICCAI 2021, LNCS 12904, pp. 213–223, 2021.
https://doi.org/10.1007/978-3-030-87202-1_21

images are commonly used. Conventional image registration methods solve an optimization problem by minimizing the dissimilarity between the transformed image and the target image. Although the conventional methods often achieve high accuracy, the slow process and expensive computation due to the use of iterative non-linear optimization algorithms hinder their clinical translation.

Recently deep learning based approaches have been proposed for image registration [5, 8, 25, 30], which learn to map from image or feature space to a spatial transformation space using neural networks trained on large datasets. Since the registration during the inference is just one forward pass of the network, the deep learning methods are intrinsically faster than the conventional methods. Supervised methods [8, 25, 31] require ground truth transformations for training, which are typically difficult to obtain in clinical practice especially for deformable registration. Unsupervised methods, such as DIRNet [30] and VoxelMorph [5], directly regress deformation fields by minimizing dissimilarity between input and target images. However, for large motion, the learn-to-map based methods often do not perform well [27]. Multi-stage methods are proposed [29, 32], where several networks are cascaded to gradually refine the estimated transformation. Deep reinforcement learning is also applied to image registration, especially for rigid registration [14, 20, 28]. In terms of non-rigid registration, Krebs et al. [18] proposed an agent-based method for prostate MR registration, which is limited to low dimensional B-spline.

Multi-modal images are frequently used in clinic. The main challenge of multi-modal image registration is to find a proper dissimilarity cost function to distinguish motions from contrast changes. Some classical methods use mutual information (MI) [11, 21] and modality-independent neighborhood descriptor (MIND) [13]. Some learning-based methods convert the problem to mono-modal registration by image-to-image translation [1, 7], which are prone to synthetic artifacts. UMDIR [23] measured the difference between multi-modal images in a feature space; however, only 2D deformable image registration between two modalities was addressed.

Recently, neural ordinary differential equations (ODEs) are proposed to represent more complex dynamics over classical ODEs. Compared to the common deep learning models such as ResNet and UNet, neural ODE models are more memory and parameter efficient and provide the benefits of adaptive computation [10], which are potentially suitable for medical applications [9]. The optimization dynamics are also inherently continuous. These merits motivate us to learn the optimization in medical image registration using neural ODEs. To the best of our knowledge, our study is the first work to apply neural ODEs to image registrations.

The contributions of our proposed methods are summarized as follows: 1) We proposed a new direction of modeling image registration optimizer as a continuous optimization dynamics via neural ODEs. 2) We introduced multi-scale architecture to neural ODEs to reduce searching space by performing registration iteratively on different scales. 3) Our proposed method is a general learn-to-learn image registration framework and is not limited to specific transformations. 4) Our framework can handle multiple contrasts with a single trained network attribute to proposed contrast-independent similarity metric for $n(\geq 2)$ modalities.

2   Method

Let x_{mov} and x_{fix} denote the moving and fixed images, respectively, and let ϕ_θ be the transformation field between two images parameterized by θ. The image registration optimization problem can be written as:

$$\hat{\theta} = \arg\min_{\theta} \mathcal{L}(x_{\mathrm{mov}} \circ \phi_\theta, x_{\mathrm{fix}}) + \mathcal{R}(\theta), \tag{1}$$

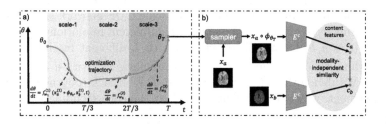

Fig. 1. Overview of our proposed method: a) Image registration optimization is modeled as a continuous optimization dynamics via neural ODEs, b) A modality-independent similarity metric is realized via a pretrained content encoder.

where $x_{\mathrm{mov}} \circ \phi_\theta$ is the transformed image, $\mathcal{L}$ is the dissimilarity cost function, and $\mathcal{R}$ is the regularization term. The form of θ is determined by the types of transformation.

We propose a novel multi-modal image registration method by learning an image registration optimizer via neural ODEs, and deriving loss function from a pretrained image content encoder. Figure 1 illustrates an overview of our method.

2.1   Learn Registration Optimizer via Neural ODEs

The optimization problem for image registration in Eq. 1 can be solved via gradient descent optimizer. Thus, the update of θ at each step is $\theta_{t+1} = \theta_t - \eta_t \partial (\mathcal{L} + \mathcal{R})/\partial \theta_t \triangleq \theta_t + f(\theta_t, t)$, where t represents t^{th} step and η_t is the step size. If the time difference is small enough, the difference equation can be re-written as an ODE function, $\frac{d\theta_t}{dt} = f(\theta_t, t), t \in [0, T]$. Given the initial parameter θ_0, the final parameter θ_T is the solution to this ODE initial value problem. Some conventional methods such as LDDMM [6] and SyN [3] describe the evolution of transformation as a differential equation. These methods solve differential equations where system dynamics are described by predefined functions, which is less flexible. Neural ODEs use trainable network to replace the predefined function, which can be considered as learning an optimizer to compute the gradient update. θ_T is the inference of the neural ODE model. We further assume that for two 3D images x_a and x_b, the evolution of θ_t follows a neural ODE:

$$\frac{d\theta_t}{dt} = f_w(x_a \circ \phi_{\theta_t}, x_b, t), t \in [0, T], \tag{2}$$

where f_w is a neural network with parameter w which takes the current warped image $x_a \circ \phi_{\theta_t}$, the target image x_b and the time variable t as inputs. Therefore, the final output at time T can be computed by integrating function f over time interval $[0,T]$. In practice, the integral is evaluated by an ODE solver, such as Runge–Kutta methods and adaptive step size solver. To train the neural ODE model, we adopt the adaptive checkpoint adjoint method [33].

Multi-scale ODE network: Empirically, we found that solving neural ODE for image registration problem requires an extensive number of function evaluations (NFE) by ODE solver, which leads to prolonged training and inference time. To address this problem, we propose a multi-scale ODE network (MS-ODENet) which performs registration by solving ODEs at different resolutions. Specifically, let $\{x_i^l\}_{l=1}^L (i = a, b)$ be an image pyramid with L different resolutions, where $x_i^{(L)} = x_i$, and $x_i^{(l-1)}$ is generated by down sampling $x_i^{(l)}$ with a factor of 2 on all 3 axes. We divide the whole time interval $[0,T]$ into L congruent segments. In each segment, we solve the ODE in Eq. 3 at the corresponding resolution as shown in Fig. 2a and b.

$$\frac{d\theta_t}{dt} = f_{w_l}^{(l)}(x_a^{(l)} \circ \phi_{\theta_t}, x_b^{(l)}, t), t \in [\frac{l-1}{L}T, \frac{l}{L}T], \tag{3}$$

where $f_{w_l}^{(l)}$ is the network at the l-th scale. The output parameter θ_T is the integral over all scales. The benefits of this design are two-fold: 1) The time cost for function evaluations is much smaller at low resolutions, allowing a larger number of steps to reach the desired accuracy. 2) The searching space for image registration is largely reduced and therefore the convergence of the ODE network is faster, which also makes the model less sensitive to local optimal.

Loss Functions: The proposed registration network is trained in an unsupervised manner by minimizing a loss function similar to Eq. 1. The utilized similarity metric $\mathcal{L}_{\text{sim}}$ (Fig. 1b) between two images (x_a and x_b) is defined in Eq. 4:

$$\mathcal{L}_{\text{sim}} = \mathbb{E}_{x_a, x_b, a, b}||E_{3D}^c(x_a \circ \phi_{\theta_T}) - E_{3D}^c(x_b)||_2^2, \tag{4}$$

where $E_{3D}^c(\cdot)$ is a 3D content feature extractor. The 3D content feature is composed of 2D content features generated from N randomly selected 2D images of different axes from the image utilizing a 2D feature extractor E^c (Sect. 2.2). In addition, we perform a random perturbation to a given image x_a such that $\tilde{x}_a = x_a \circ \phi_{\tilde{\theta}}$, where $\tilde{\theta}$ is randomly sampled from a distribution of parameter $\mathbb{P}_\theta$. The pair x_a and $\tilde{x}_a$ are fed to the registration network resulting transformation parameters $\hat{\theta}_T = \text{MS-ODENet}(\theta_0, x_a, \tilde{x}_a, L, T)$. Our network can align two images from the same modality so that we expect $\hat{\theta}_T$ approximating to the purturbation $\tilde{\theta}$ by minimizing L2 loss: $\mathcal{L}_{\text{self}} = \mathbb{E}_{x_a, a, \tilde{\theta}}||\tilde{\theta} - \hat{\theta}_T||_2^2$. The total loss function is summarized as $\mathcal{L} = \mathcal{L}_{\text{sim}} + \lambda_{\text{self}}\mathcal{L}_{\text{self}} + \lambda_{\text{reg}}\mathcal{L}_{\text{reg}}$, where $\mathcal{L}_{\text{reg}} = \mathbb{E}_{x_a, x_b, a, b}||\nabla\phi_{\theta_T}||_2^2$ is the regularization term enforcing a smooth motion field in deformable registration. λ_{self} and λ_{reg} are weighting coefficients.

2.2 Pretrained Feature Extraction Network

Different contrast images from the same patient share the same anatomical structures (content features) but have different style features. This inspires us decomposing the image into content features and style features in the latent space. The realization of this feature domain disentanglement is extended from the diverse image-to-image translation framework [19]. We trained content encoder E^c, style encoder E^s and generator G to perform image translation from one contrast to another. Note that only E^c is used in our registration task. Figure 2c illustrates an overview of our feature extraction framework utilizing image-to-image translation. Suppose two different groups $\mathcal{X}_a$ and $\mathcal{X}_b$ sampled from M modality groups ($\{\mathcal{X}_i\}_{i=1}^M$). Given x_a and x_b two image samples in these two groups, we perform cross-modality translation as follows: (a) Extract the content and style features, i.e., $s_a = E^s(x_a, a)$, $c_a = E^c(x_a)$, $s_b = E^s(x_b, b)$, $c_b = E^c(x_b)$. (b) Swap style features and generate translated images, i.e., $\hat{x}_{ab} = G(c_a, s_b, b)$, $\hat{x}_{ba} = G(c_b, s_a, a)$. (c) Encode the translated images to reconstruct content and style features of the original images, $\hat{s}_a = E^s(\hat{x}_{ba}, a)$, $\hat{c}_a = E^c(\hat{x}_{ab})$, $\hat{s}_b = E^s(\hat{x}_{ab}, b)$, $\hat{c}_b = E^c(\hat{x}_{ba})$. (d) Swap the style features again to generate the original images, $\hat{x}_{aba} = G(\hat{c}_a, \hat{s}_a, a)$, and $\hat{x}_{bab} = G(\hat{c}_b, \hat{s}_b, b)$. We trained the network using adversarial loss in [19]. Moreover, we expect that $\hat{x}_{aba}$ and $\hat{x}_{bab}$ are consistent to the original images x_a and x_b respectively. We use L1 loss $\mathcal{L}_{\text{cyc}}$ to enforce this cycle consistency. We also introduce the feature reconstruction loss for content/style features, $\mathcal{L}_{\text{rec}}^c / \mathcal{L}_{\text{rec}}^s$, which are the L1 loss between $c_a, c_b / s_a, s_b$ and $\hat{c}_a, \hat{c}_b / \hat{s}_a, \hat{s}_b$. Similarly, we define mono-modal reconstruction loss $\mathcal{L}_{\text{rec}}^x$ as the L1 loss between the reconstructed images $\hat{x}_a = G(c_a, s_a, a)$, $\hat{x}_b = G(c_b, s_b, b)$ and the original images x_a, x_b. Since training is not trivial in 3D due to high dimensionality, we use a 2D multi-channel (adjacent slices in 3D volume) network.

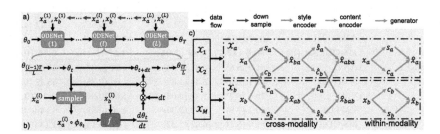

Fig. 2. Implementation of our proposed framework: a) Overview of multi-scale ODENet: several cascaded ODENets in different resolutions, b) a computational graph of a single scale ODENet with Euler's solver, c) overview of the utilized image translation towards metric learning.

3 Experiments and Results

3.1 Experiment Setup

Dataset: Currently, there is a lack of large scale medical image dataset dedicated to multi-modal image registration. We use a public dataset and a separate acquired volunteer dataset. The public dataset is the brain tumor segmentation (BraTS) 2020 dataset [22] which consists of multi-modal 3D brain MRI with four distinctive contrasts (T1, T2, T2-FLAIR, and T1Gd) from 494 subjects with glioblastomas, resulting in various multi-modal image registration tasks (12 pairs per subject). Besides, tumor masks provide a clinically meaningful evaluation metric for registration. For each subject, all modalities were normalized into 3D volumes with a size of $240 \times 240 \times 160$ and a resolution of $1 \times 1 \times 1$ mm^3. The dataset is split into 444 subjects (5,328 pairs) for training and validation and 50 subjects (600 pairs) for testing. Since different modalities have been registered, we simulate motions by applying random rigid transformation, random control point deformation, or both. Note that the simulated transformation fields are only used for evaluation, not for training. We also acquired additional multi-modal 3D MR brain data on 25 volunteers using a special MR technique that can acquire multiple contrasts simultaneously. This private dataset is used only for testing the generalizability of the proposed methods. The motion is simulated as described above. Images were preprocessed to remove skull using DeepBrain[1] and resized as the public data.

Table 1. Quantitative results for image registration on BraTS, where R, D, and B represent rigid, dense and B-spline parameterization in MS-ODENet. The mean (standard deviation) are reported.

Transformations	Methods	Dice/%	RMSE(x) $\times 10^{-2}$	RMSE (ϕ)/mm	Time/s
Rigid	ANTs	63.0 (40.5)	8.34 (7.64)	7.28 (6.66)	17.17 (4.05)
	ANTs+I2I	60.6 (40.8)	8.83 (7.81)	7.54 (6.78)	30.69 (4.70)
	MS-ODENet (R)	**90.6 (17.2)**	**3.89 (2.95)**	**3.57 (5.18)**	**0.55 (0.08)**
Deformable	ANTs	81.9 (8.8)	6.31 (1.98)	1.21 (0.37)	55.35 (7.07)
	ANTs+I2I	81.1 (8.3)	6.34 (1.82)	1.06 (2.35)	68.47 (6.18)
	VM	79.4 (8.7)	8.81 (3.03)	1.61 (0.69)	**0.24 (0.05)**
	VM+I2I	80.1 (7.8)	8.52 (2.30)	1.26 (0.31)	0.34 (0.06)
	MS-ODENet (D)	81.6 (8.1)	6.63 (2.22)	1.11 (0.18)	1.13 (0.10)
	MS-ODENet (B)	**83.0 (8.7)**	**6.17 (2.43)**	**0.99 (0.32)**	0.31 (0.06)
Rigid + deformable	ANTs	73.6 (22.2)	6.79 (2.51)	2.99 (2.95)	70.87 (7.97)
	ANTs+I2I	71.1 (22.1)	6.97 (2.26)	2.83 (2.80)	87.25 (12.2)
	RCN	64.9 (10.1)	8.74 (3.21)	5.48 (1.97)	2.49 (0.31)
	VM*	78.1 (9.8)	8.62 (3.02)	2.92 (2.31)	0.60 (0.23)
	VM*+I2I	78.4 (9.2)	7.07 (1.96)	2.04 (1.38)	1.01 (0.38)
	MS-ODENet (R+D)	79.6 (9.1)	6.70 (2.28)	1.82 (1.22)	1.39 (0.12)
	MS-ODENet (R+B)	**81.1 (9.9)**	**6.28 (2.38)**	**1.52 (1.02)**	**0.56 (0.09)**

---

[1] https://github.com/iitzco/deepbrain.

Registration Networks: We evaluated our methods on rigid, deformable and a hybrid of rigid and deformable motions. For 3D rigid motion, f includes convolution layers followed by fully connected layers. The output size of f is the same as the degree of freedom in the transformation. For deformable motion, fully convolutional network is used. One variation of the deformable registration is kernel based method (B-spline). Parameter θ is the grid of control points which can also be regarded as an image. For dense motion, the UNet as in [26] is used for voxel-wise estimation. For hybrid motion, we cascade the rigid and deformable MS-ODENets sequentially.

Feature Extraction Network: The backbone network is similar to [15]. The content encoder is a fully convolutional network while the style encoder has convolution layers followed by a global average pooling and a fully connected layer, which forces the network to extracts global style features. The generator is a CNN with deconvolution layers to generate images of the original size. Besides, to condition on modality $\mathcal{X}_i$ for style encoder and image generator, the modality code i is converted into a one-hot vector concatenated with the input tensor along the channel dimension.

Implementation Details: We follow the protocol in [2] for training GAN with number of slices set to 5. For registration, we set $\lambda_{\text{self}} = 1$, and let λ_{reg} be 10 and 2 for dense and B-spline respectively. We use $L = T = 3$ for MS-ODENet. Let $F(dt)$ be Euler's method with fixed step size dt and $A(\epsilon)$ be adaptive Heun's method with tolerance of error ϵ. We use adaptive solvers when the search space is small (Rigid: $A(10^{-3})$-$A(10^{-3})$-$F(0.1)$) to avoid extensive NFE and use fixed step size solver for large search space (B-spline: $F(0.2)$-$F(0.2)$-$F(0.25)$, Dense: $F(0.2)$-$F(0.2)$-$F(0.5)$). All networks are trained using Adam [17] optimizer with a learning rate of 10^{-4} on an NVIDIA Tesla V100 GPU.

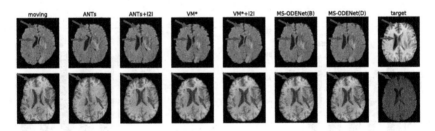

Fig. 3. Examples of compared methods on deformable registration. Top row: T2-FLAIR to T1 registration. Bottom row: T1 to T2-FLAIR. Red-arrows highlight example different areas.

Evaluation Metrics: For evaluation, Dice scores [12] are computed over tumor masks. With available synthetic transformation fields and ground truth images, root mean square errors are calculated and denoted as RMSE (ϕ) and RMSE (x), respectively. The metrics are averaged over all pairs of test data.

3.2   Results

Table 1 shows the quantitative results for rigid, deformable and hybrid registration (rigid+deformable). **Rigid:** Random rotation and translation were synthesized along all three axes and were sampled from $U(-75°, 75°)$ and $U(-20, 20)$mm, respectively. We used the rigid registration method with MI metric in Advanced Normalization Tools (ANTs) [4] as the baseline. We also used the trained GAN to perform image translation followed by mono-modal image registration with ANTs (ANTs+I2I). **Deformable:** We made synthetic image pairs through elastic transformations by perturbing B-spline control points with noise from $\mathcal{N}(0, (8\text{mm})^2)$ on three axes. For comparison, we use SyN [3] in ANTs, VoxelMorph (VM) [5] with MI metric, and also their variants with image translation (ANTs+I2I and VM+I2I). For MS-ODENet, we use two different parameterizations, namely dense deformation (D) and B-spline (B). Figure 3 shows example results. **Rigid + deformable:** Random rotations, translations and control point noise are from $U(-40°, 40°)$, $U(-10, 10)$mm, and $\mathcal{N}(0, (8\text{mm})^2)$, respectively. We compared our method to SyN, RCN [32] and VM. RCN is an iterative deep learning method consisting of an affine network and n deformable networks. We use MI metric for training RCN and set $n = 2$ due to memory limit. Since VM was proposed to solve deformable registration, we performed a rigid registration using our rigid MS-ODENet prior to applying VM. This approach is denoted as VM*.

Ablation: Figure 4(a) summarizes the results that evaluate models with no learned content loss (replaced by MI), no multi-scale ODE, no self-supervision or no multi-slice GAN respectively and compare them with the full model on rigid + B-spline deformable registration. To investigate the necessity of iterative registration for large motions, we conducted a rigid registration experiment with $L = T = 1$ and various numbers of steps in ODE solver. Figure 4(b) shows the corresponding results.

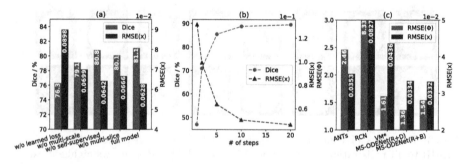

Fig. 4. a) Quantitative results for the ablation study. b) Performance on large motion for neural ODE models with different number of solver steps. c) Results for the generalizability study.

Generalization: We performed the rigid+deformable test on our private dataset. Figure 4(c) shows bar chart among compared methods.

4   Discussions and Conclusions

Table 1 shows that our proposed MS-ODENet outperforms classical methods under various transformations. In all experiments, MS-ODENet is much faster than ANTs due to the fast inference of neural networks and the smaller number of iterations needed in neural ODEs. In rigid registration (Fig. 1), MS-ODENet greatly outperforms ANTs with or without the domain translation. Classical methods only consider local gradient information and tend to get stuck at local minima. The MS-ODENet learns the optimization process via neural ODEs, which utilizes not only the current local information but also the experiences learned from the dataset, and is thus more likely to reach the global minimum. In deformable registration, our proposed methods achieve similar or better accuracy compared with classical methods. For rigid+deformable registration, ANTs suffers greatly from the additional rigid transformation, indicating that traditional optimization-based methods rely heavily on the initialization of parameters.

Our methods also outperform other deep learning methods (Table 1) on deformable registration. The iterative updates in MS-ODENet improve registration accuracy progressively (Fig. 4(c)). When the step number is one, the MS-ODENet is equivalent to a conventional deep learning model. Note that all the ablated methods (Fig. 4(a)) outperform the other deep learning methods. Furthermore, with the adjoint method, the training of our neural ODE model does not backpropagate through the operations of the solver [10,33] and thus is more memory efficient than traditional deep learning models. Unlike other coarse-to-fine methods that need to be trained in separate stages [29], our multi-scale neural ODE model can be trained end-to-end. In the generalization study, the proposed MS-ODENets show consistent improvement over the other methods. The RCN does not employ iterative networks for affine transformation and therefore has poor generalizability for large transformation.

In this work, we present a new framework for 3D multi-modal image registration. We formulate the optimization in conventional registration methods as a continuous process and learn the optimizer via a neural ODE. Furthermore, for efficient learning and inference, we propose a multi-scale architecture to narrow the searching space from coarse to fine image resolutions. In addition, we employ an image-to-image translation GAN to learn a modality-independent metric between images from different modalities. Experiment results show that our proposed framework is superior to other compared methods. For future work, we will extend our framework to other types of medical registration such as 3D-2D image registration.

References

1. Arar, M., Ginger, Y., Danon, D., Bermano, A.H., Cohen-Or, D.: Unsupervised multi-modal image registration via geometry preserving image-to-image translation. In: Proceedings of the IEEE/CVF Conference on Computer Vision and Pattern Recognition, pp. 13410–13419 (2020)

2. Arjovsky, M., Chintala, S., Bottou, L.: Wasserstein GAN. arXiv preprint arXiv:1701.07875 (2017)
3. Avants, B.B., Epstein, C.L., Grossman, M., Gee, J.C.: Symmetric diffeomorphic image registration with cross-correlation: evaluating automated labeling of elderly and neurodegenerative brain. Med. Image Anal. **12**(1), 26–41 (2008)
4. Avants, B.B., Tustison, N., Song, G.: Advanced normalization tools (ANTS). Insight J. **2**(365), 1–35 (2009)
5. Balakrishnan, G., Zhao, A., Sabuncu, M.R., Guttag, J., Dalca, A.V.: VoxelMorph: a learning framework for deformable medical image registration. IEEE Trans. Med. Imaging **38**(8), 1788–1800 (2019)
6. Beg, M.F., Miller, M.I., Trouvé, A., Younes, L.: Computing large deformation metric mappings via geodesic flows of diffeomorphisms. Int. J. Comput. Vis. **61**(2), 139–157 (2005)
7. Cao, X., Yang, J., Gao, Y., Guo, Y., Wu, G., Shen, D.: Dual-core steered non-rigid registration for multi-modal images via bi-directional image synthesis. Med. Image Anal. **41**, 18–31 (2017)
8. Cao, X., et al.: Deformable image registration based on similarity-steered CNN regression. In: Descoteaux, M., Maier-Hein, L., Franz, A., Jannin, P., Collins, D.L., Duchesne, S. (eds.) MICCAI 2017. LNCS, vol. 10433, pp. 300–308. Springer, Cham (2017). https://doi.org/10.1007/978-3-319-66182-7_35
9. Chen, E.Z., Chen, T., Sun, S.: MRI image reconstruction via learning optimization using neural ODEs. In: Martel, A.L., et al. (eds.) MICCAI 2020. LNCS, vol. 12262, pp. 83–93. Springer, Cham (2020). https://doi.org/10.1007/978-3-030-59713-9_9
10. Chen, R.T., Rubanova, Y., Bettencourt, J., Duvenaud, D.K.: Neural ordinary differential equations. In: Advances in Neural Information Processing Systems, pp. 6571–6583 (2018)
11. De Nigris, D., Mercier, L., Del Maestro, R., Louis Collins, D., Arbel, T.: Hierarchical multimodal image registration based on adaptive local mutual information. In: Jiang, T., Navab, N., Pluim, J.P.W., Viergever, M.A. (eds.) MICCAI 2010. LNCS, vol. 6362, pp. 643–651. Springer, Heidelberg (2010). https://doi.org/10.1007/978-3-642-15745-5_79
12. Dice, L.R.: Measures of the amount of ecologic association between species. Ecology **26**(3), 297–302 (1945)
13. Heinrich, M.P., et al.: MIND: modality independent neighbourhood descriptor for multi-modal deformable registration. Med. Image Anal. **16**(7), 1423–1435 (2012)
14. Hu, J., et al.: End-to-end multimodal image registration via reinforcement learning. Med. Image Anal. **68**, 101878 (2020)
15. Huang, X., Liu, M.Y., Belongie, S., Kautz, J.: Multimodal unsupervised image-to-image translation. In: Proceedings of the European Conference on Computer Vision (ECCV), pp. 172–189 (2018)
16. Incoronato, M., et al.: Radiogenomic analysis of oncological data: a technical survey. Int. J. Mol. Sci. **18**(4), 805 (2017)
17. Kingma, D.P., Ba, J.: Adam: a method for stochastic optimization. arXiv preprint arXiv:1412.6980 (2014)
18. Krebs, J., et al.: Robust non-rigid registration through agent-based action learning. In: Descoteaux, M., Maier-Hein, L., Franz, A., Jannin, P., Collins, D.L., Duchesne, S. (eds.) MICCAI 2017. LNCS, vol. 10433, pp. 344–352. Springer, Cham (2017). https://doi.org/10.1007/978-3-319-66182-7_40
19. Lee, H.Y., et al.: DRIT++: diverse image-to-image translation via disentangled representations. Int. J. Comput. Vis. **128**(10), 2402–2417 (2020). https://doi.org/10.1007/s11263-019-01284-z

20. Ma, K., et al.: Multimodal image registration with deep context reinforcement learning. In: Descoteaux, M., Maier-Hein, L., Franz, A., Jannin, P., Collins, D.L., Duchesne, S. (eds.) MICCAI 2017. LNCS, vol. 10433, pp. 240–248. Springer, Cham (2017). https://doi.org/10.1007/978-3-319-66182-7_28

21. Maes, F., Collignon, A., Vandermeulen, D., Marchal, G., Suetens, P.: Multimodality image registration by maximization of mutual information. IEEE Trans. Med. Imaging **16**(2), 187–198 (1997)

22. Menze, B.H., et al.: The multimodal brain tumor image segmentation benchmark (BRATS). IEEE Trans. Med. Imaging **34**(10), 1993–2024 (2014)

23. Qin, C., Shi, B., Liao, R., Mansi, T., Rueckert, D., Kamen, A.: Unsupervised deformable registration for multi-modal images via disentangled representations. In: Chung, A.C.S., Gee, J.C., Yushkevich, P.A., Bao, S. (eds.) IPMI 2019. LNCS, vol. 11492, pp. 249–261. Springer, Cham (2019). https://doi.org/10.1007/978-3-030-20351-1_19

24. Risholm, P., Golby, A.J., Wells, W.: Multimodal image registration for preoperative planning and image-guided neurosurgical procedures. Neurosurg. Clin. **22**(2), 197–206 (2011)

25. Rohé, M.-M., Datar, M., Heimann, T., Sermesant, M., Pennec, X.: SVF-Net: learning deformable image registration using shape matching. In: Descoteaux, M., Maier-Hein, L., Franz, A., Jannin, P., Collins, D.L., Duchesne, S. (eds.) MICCAI 2017. LNCS, vol. 10433, pp. 266–274. Springer, Cham (2017). https://doi.org/10.1007/978-3-319-66182-7_31

26. Ronneberger, O., Fischer, P., Brox, T.: U-Net: convolutional networks for biomedical image segmentation. In: Navab, N., Hornegger, J., Wells, W.M., Frangi, A.F. (eds.) MICCAI 2015. LNCS, vol. 9351, pp. 234–241. Springer, Cham (2015). https://doi.org/10.1007/978-3-319-24574-4_28

27. Shen, Z., Han, X., Xu, Z., Niethammer, M.: Networks for joint affine and non-parametric image registration. In: Proceedings of the IEEE Conference on Computer Vision and Pattern Recognition, pp. 4224–4233 (2019)

28. Sun, S., et al.: Robust multimodal image registration using deep recurrent reinforcement learning. In: Jawahar, C.V., Li, H., Mori, G., Schindler, K. (eds.) ACCV 2018. LNCS, vol. 11362, pp. 511–526. Springer, Cham (2019). https://doi.org/10.1007/978-3-030-20890-5_33

29. de Vos, B.D., Berendsen, F.F., Viergever, M.A., Sokooti, H., Staring, M., Išgum, I.: A deep learning framework for unsupervised affine and deformable image registration. Med. Image Anal. **52**, 128–143 (2019)

30. de Vos, B.D., Berendsen, F.F., Viergever, M.A., Staring, M., Išgum, I.: End-to-end unsupervised deformable image registration with a convolutional neural network. In: Cardoso, M.J., et al. (eds.) DLMIA/ML-CDS -2017. LNCS, vol. 10553, pp. 204–212. Springer, Cham (2017). https://doi.org/10.1007/978-3-319-67558-9_24

31. Yang, X., Kwitt, R., Styner, M., Niethammer, M.: Quicksilver: fast predictive image registration-a deep learning approach. NeuroImage **158**, 378–396 (2017)

32. Zhao, S., Dong, Y., Chang, E.I., Xu, Y., et al.: Recursive cascaded networks for unsupervised medical image registration. In: Proceedings of the IEEE International Conference on Computer Vision, pp. 10600–10610 (2019)

33. Zhuang, J., Dvornek, N., Li, X., Tatikonda, S., Papademetris, X., Duncan, J.: Adaptive checkpoint adjoint method for gradient estimation in neural ODE. arXiv preprint arXiv:2006.02493 (2020)

Image-Guided Interventions
and Surgery

Self-supervised Generative Adversarial Network for Depth Estimation in Laparoscopic Images

Baoru Huang[1,2(✉)], Jian-Qing Zheng[3], Anh Nguyen[1], David Tuch[4], Kunal Vyas[4], Stamatia Giannarou[1,2], and Daniel S. Elson[1,2]

[1] The Hamlyn Centre for Robotic Surgery, Imperial College London, London SW7 2AZ, UK
Baoru.Huang18@imperial.ac.uk
[2] Department of Surgery and Cancer, Imperial College London, London SW7 2AZ, UK
[3] The Kennedy Institute of Rheumatology, University of Oxford, Oxford, UK
[4] Lightpoint Medical Ltd., Chesham, UK

Abstract. Dense depth estimation and 3D reconstruction of a surgical scene are crucial steps in computer assisted surgery. Recent work has shown that depth estimation from a stereo image pair could be solved with convolutional neural networks. However, most recent depth estimation models were trained on datasets with per-pixel ground truth. Such data is especially rare for laparoscopic imaging, making it hard to apply supervised depth estimation to real surgical applications. To overcome this limitation, we propose SADepth, a new self-supervised depth estimation method based on Generative Adversarial Networks. It consists of an encoder-decoder generator and a discriminator to incorporate geometry constraints during training. Multi-scale outputs from the generator help to solve the local minima caused by the photometric reprojection loss, while the adversarial learning improves the framework generation quality. Extensive experiments on two public datasets show that SADepth outperforms recent state-of-the-art unsupervised methods by a large margin, and reduces the gap between supervised and unsupervised depth estimation in laparoscopic images.

Keywords: Depth estimation · Laparoscopic images · Generative adversarial network

1 Introduction

Robot-assisted minimally invasive surgery with stereo laparoscopic vision has become popular due to the advantages of enhanced movement range, precision, vision and proficiency [22,23]. Surgical scene depth estimation is a fundamental problem in image-guided intervention and has received substantial prior interest due to its promise for robot navigation, 3D registration between pre- and intra-operative organ models, and augmented reality [30]. Obtaining depth maps is

© Springer Nature Switzerland AG 2021
M. de Bruijne et al. (Eds.): MICCAI 2021, LNCS 12904, pp. 227–237, 2021.
https://doi.org/10.1007/978-3-030-87202-1_22

not trivial due to the inherent problems such as tissue deformation, specular reflections, and lack of photometric constancy across frames [20].

Several traditional methods used multi-view stereo algorithms such as Simultaneous Localization and Mapping (SLAM) [12] and Structure from Motion (SfM) [19], but these struggle with less textured tissues. More recently deep learning-based depth estimation has used RGB images as the training data and Convolutional Neural Networks (CNNs) for supervised learning [4,6]. To produce accurate results in less than a second of GPU time, Luo *et al.* [21] treated the problem as a multi-class classification indicating all possible disparities, and exploited a product layer to simplify the representations of a Siamese architecture. Chang *et al.* [2] proposed PSMNet, where the capacity of global context information at different scales and locations could be extracted by a spatial pyramid pooling module to form a cost volume. Duggal *et al.* [5] sped up the runtime of stereo matching and developed a differentiable PatchMatch module that could discard most disparities without the need of full cost volume evaluation.

The methods above are fully supervised and require ground truth depth during training. However, acquiring per-pixel ground truth depth data is challenging for real-world settings [18] and especially for laparoscopic vision where port space is limited, working distance is short and sterilization is required [15]. One alternative is self-supervised training of depth estimation models using image reconstruction as the supervisory signal [7]. The input is usually a set of images in the form of monocular or stereo images [32]. Godard *et al.* [9] proposed a training loss that included a left-right depth consistency term and a reconstruction term for single image depth estimation, despite the absence of ground truth depth. This was extended by [10] with full-resolution multi-scale sampling to reduce visual artifacts, and a minimum reprojection loss to robustly handle occlusions. Johnston *et al.* [17] further closed the gap with fully-supervised methods by including a self-attention mechanism and made use of contextual information. Ye *et al.* [30] proposed a deep learning framework for surgical scene depth estimation in self-supervised mode for scalable data acquisition by adopting a differentiable spatial transformer and an autoencoder.

In this paper, we present a new method for self-supervised adversarial depth estimation: SADepth. A U-Net architecture [26] was adopted as a generative structure and fed with stereo pairs as inputs to benefit from complementary information. To cope with local minima caused by classic photometric reprojection loss, we applied the disparity smoothness loss and formed the network across multiple scales. The use of a generative adversarial network (GAN) allowed us to improve the reconstructed image quality, which formed a supervisory signal for training, while keeping the overall end-to-end optimization objective.

2   Methodology

2.1   Overview

Here we describe the proposed self-supervised adversarial depth estimation framework, SADepth. Stereo depth estimation predicts depth maps $\boldsymbol{D}^l, \boldsymbol{D}^r \in \mathbb{R}_+^{h \times w}$

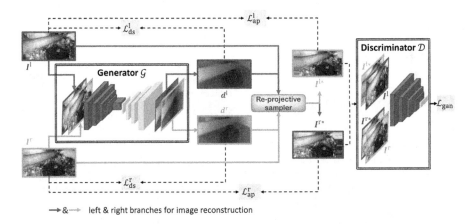

Fig. 1. Overview of the self-supervised adversarial depth estimation network, SADepth.

based on the stereo RGB images $\boldsymbol{I}^l, \boldsymbol{I}^r \in \mathbb{R}_+^{h \times w \times 3}$ of height and width h, w. A generative network $\mathcal{G}$ with stereo image pairs $\boldsymbol{I}^l$ and $\boldsymbol{I}^r$ as inputs, was used to produce two distinct left and right disparity maps $\boldsymbol{d}^l$ and $\boldsymbol{d}^r$, $i.e.$ $\boldsymbol{d}^l, \boldsymbol{d}^r = \mathcal{G}(\boldsymbol{I}^l, \boldsymbol{I}^r)$. As the two disparity maps were generated from different input images, a 'reprojection sampler' [16] could be used for photometric reprojection loss computation of mutual counter-parts, $i.e.$ reconstructed left and right images $\boldsymbol{I}^{l*}$ and $\boldsymbol{I}^{r*}$. The discriminator $\mathcal{D}$ was exploited to indicate if the reconstructed images were real or fake (original input images were regarded as real). By forcing the reconstructed image to be consistent with the original input, we could derive accurate disparity maps for depth inference, as shown in the following sections.

2.2 Network Architecture

Generator. The generator followed the general U-Net [26] architecture consisting of an encoder-decoder network, where the encoder was designed to obtain compact image representations and the decoder produced disparity maps for left and right input images, recovering them at the original scale (illustrated in Fig. 2). Encoder-decoder skip connections were applied to represent deep abstract features while preserving local information. To make the model compact - and different from less streamlined previous approaches which had two branches or two sub-networks for the encoder [2,25] - we first concatenated the left and right images into a 6-channel tensor and then fed it to a ResNet18 model [13]. The input size was $\# \ channels \times h \times w = 6 \times 192 \times 384$. Similar to [9], our decoder was formed of five cascaded blocks where each block had four parts: the first convolutional layer, an upsampling layer, a concatenation manipulation, and the second convolutional layer. In the upsampling layer, features were interpolated to twice the input size and both convolutional layers were followed by an *ELU* activation function [3]. In particular, sigmoids were applied at the output to

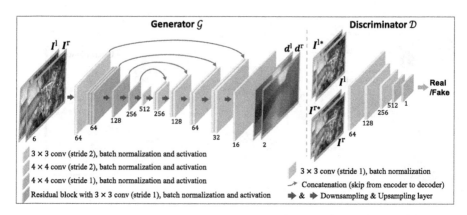

Fig. 2. The detailed architecture of the SADepth generator and discriminator. The generator was an autoencoder architecture with concatenated stereo image pairs as inputs and left and right disparity maps as outputs using a sigmoid function. These outputs were then transformed to reconstruct the counter-part camera input images using a 'reprojection sampler', and these reconstructed images were fed into the discriminator together with the original input image pair. The discriminator output a scalar indicating whether the reconstructed images generated from the 'reprojection sampler' were real or fake.

generate a 2-channel tensor representing the left and right disparity $\mathbf{d}^l$ and $\mathbf{d}^r$. Finally the sigmoid outputs were converted to depth by $\mathbf{D}^{l(r)} = 1/(a\mathbf{d}^{l(r)} + b)$, where parameters a and b were selected to constrain the depth $\mathbf{D}^{l(r)}$ between 0.1 and 100 units. The depth maps were then back-projected into point clouds by applying the intrinsic parameters and using the counter-part camera's extrinsic parameters to form reconstructed stereo images. The structural similarity between the original and reconstructed images was regarded as a supervisory signal to train the generator (see Sect. 2.3 for the generator loss).

Discriminator. Goodfellow *et al.* [11] introduced a generative adversarial learning strategy and presented impressive results for image generation tasks. GANs have been widely exploited in different tasks with different GAN models including *e.g.* DualGAN [31] and CycleGAN [33]. To improve the generation quality of the reconstructed images $\boldsymbol{I}^{l*}$ and $\boldsymbol{I}^{r*}$, and following the work in [25] for natural scenes, we applied an adversarial learning strategy for laparoscopic images to include geometry constraints during training and force the network to make a consistent depth map prediction. The original input stereo image pairs and reconstructed images $\boldsymbol{I}^{r*}$ and $\boldsymbol{I}^{l*}$ generated from the 'reprojection sampler' were fed into the discriminator $\mathcal{D}$, which consisted of convolutional, batch normalization and activation function layers and classified the input and reconstructed images as real or fake. As training progressed, the reconstructed images became more similar to the original inputs, while the discriminator also became better at distinguishing between the input and reconstructed images, resulting in an overall improvement of the associated disparity maps.

2.3   Training Losses

Generator Loss. In the depth estimation generator network $\mathcal{G}$, the loss $\mathcal{L}^{\text{r}}_{\text{rec}}$ was formed from the appearance matching loss $\mathcal{L}^{\text{r}}_{\text{ap}}$ and disparity smoothness loss $\mathcal{L}^{\text{r}}_{\text{ds}}$

$$\mathcal{L}^{\text{r}}_{\text{rec}} = \mathcal{L}^{\text{r}}_{\text{ap}} + \alpha_{\text{ds}}\mathcal{L}^{\text{r}}_{\text{ds}} \tag{1}$$

where α_{ds} balanced the loss magnitude of the two parts to stabilize the training and was set to 0.001.

Appearance-Matching Loss. Self-supervised training typically assumes that the appearance and material properties (*e.g.* brightness and Lambertian) of object surfaces are consistent between frames. A local structure-based appearance loss [9] can effectively improve the depth estimation performance compared with simple pairwise pixel differences [32]. Following [10], we exploited the appearance-matching loss as part of the generator loss which forced the reconstructed image to be similar to the corresponding training inputs. During the training, the right disparity map $\mathbf{d}^{\text{r}}$ generated by the autoencoder was then transformed to produce $\boldsymbol{I}^{\text{r}*}$ – a reconstruction of the original right input image – using RGB intensity information from the counter-part camera image $\boldsymbol{I}^{\text{l}}$ (see Fig. 1). This was achieved by first converting the disparity map $\mathbf{d}^{\text{r}}$ to a depth map $\mathbf{D}^{\text{r}}$, from which a point cloud of the surgical scene could be generated. Then the point cloud was transferred into the other camera's coordinate system and projected onto its image plane. The reconstructed input image $\boldsymbol{I}^{\text{r}*}$ was generated with bilinear interpolation for each output pixel using the weighted sum of the four neighboring intensities. In contrast to [7], this bilinear sampling was locally fully differentiable, which allowed it to be integrated into the fully convolutional architecture without requiring simplification or approximation of the cost function. To compare the reconstructed image $\boldsymbol{I}^{\text{r}*}$ and the original input image $\boldsymbol{I}^{\text{r}}$, a combination of structural similarity (SSIM) index [27] and $\mathcal{L}_1$ loss were applied as the photometric image reconstruction cost $\mathcal{L}^{\text{r}}_{ap}$:

$$\mathcal{L}^{\text{r}}_{ap} = \frac{1}{N}\sum_{i,j}\frac{\gamma}{2}(1 - \text{SSIM}(I^{\text{r}}_{ij}, I^{\text{r}*}_{ij})) + (1 - \gamma)\|I^{\text{r}}_{ij} - I^{\text{r}*}_{ij}\|_1 \tag{2}$$

where N denotes the number of pixels and γ represents the weighting for L1-norm loss term, which was set to 0.85. Similar to [9], the calculation of SSIM here was simplified to a 3×3 block filter instead of a Gaussian. The training of the depth estimation generator then involved minimizing the reconstruction loss between input and reconstructed images.

Disparity Smoothness Loss. Since disparities should be locally smooth and discontinuities usually occur at image gradients, we applied the disparity smoothness loss to penalize unexpected discontinuities in the disparity maps. Following [14], this cost was an edge-aware term weighted with the input image gradients $\partial\mathbf{I}$:

$$\mathcal{L}_{\mathrm{ds}}^{\mathrm{r}} = \frac{1}{N}\sum_{ij}|\partial_x(\mathbf{d}_{ij}^{\mathrm{r}})|e^{-|\partial_x I_{ij}^{\mathrm{r}}|} + |\partial_y(\mathbf{d}_{ij}^{\mathrm{r}})|e^{-|\partial_y I_{ij}^{\mathrm{r}}|} \qquad (3)$$

where $\mathbf{d}^{\mathrm{r}}$ represents the generated disparity map and $\boldsymbol{I}^{\mathrm{r}}$ is the original input right image.

Discriminator Loss. The adversarial objective of the generative network can be expressed as follows:

$$\mathcal{L}_{\mathrm{gan}}^{\mathrm{r}}(\boldsymbol{I}^{\mathrm{r}}, \boldsymbol{I}^{\mathrm{r*}}; \mathcal{G}, \mathcal{D}) = \mathbb{E}_{\boldsymbol{I}^{\mathrm{r}} \sim P(\boldsymbol{I}^{\mathrm{r}})}[\log(\mathcal{D}(\boldsymbol{I}^{\mathrm{r}}))] + \mathbb{E}_{\boldsymbol{I}^{\mathrm{r*}} \sim P(\boldsymbol{I}^{\mathrm{r*}})}[\log(1 - \mathcal{D}(\boldsymbol{I}^{\mathrm{r*}}))] \quad (4)$$

where a cross-entropy loss measured the expectation of the reconstructed image $\boldsymbol{I}^{\mathrm{r*}}$ against the distribution of the input image $\boldsymbol{I}^{\mathrm{r}}$. Note that both generator and discriminator losses included losses for left and right images but only the right image equations are shown.

Multi-scale Loss. One remaining issue with the above learning pipeline was that the training objective risked becoming stuck in local minima due to the application of a photometric reprojection loss [28]. The strategy introduced in [32] indicated that combining the individual losses across multiple scales in the decoder was effective, which could improve the depth estimation performance and reduce sensitivity to architectural choices. Hence, the lower resolution depth maps (from the intermediate layers) were first upsampled to the input image resolution and then reprojected and resampled, with the errors computed at the higher input resolution. This manipulation is similar to matching patches, which enables low-resolution disparity maps to warp an entire patch of pixels in a high resolution image while promoting the depth maps at every scale to reconstruct the high resolution input image as accurately as possible [10].

Joint Optimization Loss. Finally, the joint optimization loss was a combination of generator loss and adversarial loss, written as:

$$\mathcal{L}_{\mathrm{total}} = \frac{1}{m}\sum_{s=1}^{m}\frac{\mathcal{L}_s^{\mathrm{l}} + \mathcal{L}_s^{\mathrm{r}}}{2} = \frac{1}{m}\sum_{s=1}^{m}\left(\alpha(\mathcal{L}_{\mathrm{rec}}^{\mathrm{l}} + \mathcal{L}_{\mathrm{rec}}^{\mathrm{r}}) + \beta(\mathcal{L}_{\mathrm{gan}}^{\mathrm{l}} + \mathcal{L}_{\mathrm{gan}}^{\mathrm{r}})\right) \quad (5)$$

Training. The depth estimation procedure was trained based on the reconstruction supervision signal and no per-pixel depth ground truth labels were needed. The augmentation of input data was performed on the fly by flipping 50% of the input images horizontally and reorienting the stereo pairs. Parameter m was set to 4, which means that there were 4 output scales with resolutions $\frac{1}{2^0}$, $\frac{1}{2^1}$, $\frac{1}{2^2}$ and $\frac{1}{2^3}$ of the input resolution. α and β were set to 0.5.

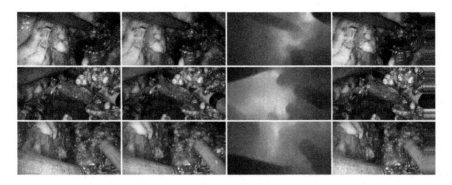

Fig. 3. Qualitative results on *dVPN* dataset. From left to right, they are left image, right image, right depth map and reconstructed right image.

3 Experiments and Results

3.1 Dataset

We evaluated SADepth on two datasets. The first was the *dVPN* dataset, collected from da Vinci partial nephrectomy, with 34320 pairs of rectified stereo images for training and 14382 pairs for testing [30]. The second was the *SCARED* dataset [1] released during the Endovis challenge at MICCAI 2019, with 17206 pairs (dataset 1, 2, 3, 6 and 7) of rectified stereo images for training and 5637 pairs for testing. To verify the generalization of our framework, we only trained on the *dVPN* dataset but tested on both *dVPN* and *SCARED* datasets.

3.2 Evaluation Metrics, Baseline, and Implementation Details

Evaluation Metrics. As the ground truth depth labels were not available for the *in vivo* surgical data in the *dVPN* dataset, we adopted the SSIM index to evaluate the similarity between the reconstructed image and the original input image (*i.e.* I^{r*} and I^r) as the evaluation metric. For the *SCARED* dataset the team at Intuitive Surgical collected the ground truth by using structured light, thus we used the absolute error to assess our SADepth model.

Table 1. SSIM score for *dVPN* test set (higher is better).

Method	Training	Mean SSIM	Std. SSIM
ELAS [8]	No training	47.3	0.079
SPS [29]	No training	54.7	0.092
V-Basic [30]	Unsupervised	55.5	0.106
V-Siamese [30]	Unsupervised	60.4	0.066
Monodepth [9]	Unsupervised	54.9	0.087
Monodepth2 [10]	Unsupervised	71.2	0.075
SADepth (ours)	Unsupervised	**79.6**	**0.049**

Baseline. We compared SADepth with several recent works. For the *dVPN* dataset, we compared our method with stereo matching-based methods: ELAS [8] and SPS [29]; Siamese-based networks: V-Basic [30] and V-Siamese [30]; and recent deep learning methods: Monodepth [9] and the stereo mode of Monodepth2 [10]. For the *SCARED* dataset, we compared our results with the methods summarized by the recent MICCAI sub-challenge paper [1].

Implementation Details. The SADepth model was implemented in PyTorch [24], with a batch size of 16 and input/output resolution of (192×384). The learning rate was set to 10^{-4} for the first 15 epochs and then dropped to 10^{-5} for the remainder. The model was trained for 20 epochs using the Adam optimizer which took about 22 h on a single NVIDIA 2080 Ti GPU.

Table 2. The mean absolute depth error for the SCARED test set 1 and 2 (unit: mm) (lower is better).

Method	Training	Test Set 1 Average	Test Set 2 Average
Lalith Sharan [1]	Supervised	43.03	48.72
Xiaohong Li [1]	Supervised	22.77	20.52
Huoling Luo [1]	Supervised	19.52	18.21
Zhu Zhanshi [1]	Supervised	9.60	21.20
Wenyao Xia [1]	Supervised	6.73	9.44
Congcong Wang [1]	Supervised	4.10	4.28
Trevor Zeffiro [1]	Supervised	3.60	3.47
J.C. Rosenthal [1]	Supervised	3.44	4.05
Dimitris Psychogyios 1 [1]	Supervised	3.00	1.67
Dimitris Psychogyios 2 [1]	Supervised	2.95	2.30
KeXue Fu [1]	Unsupervised	20.94	17.22
Monodepth [9]	Unsupervised	23.56	21.62
Monodepth2 [10]	Unsupervised	21.92	15.25
SADepth (ours)	Unsupervised	**17.42**	**11.23**

3.3   Results

The SADepth and other state-of-the-art results for the *dVPN* dataset are summarized in Table 1 using the mean and standard deviation (Std.) of the SSIM index. The SADepth model effectively outperformed other methods with an SSIM of 79.6, *i.e.* 24.7 units higher than Monodepth [9], 8.4 units higher than Monodepth2 [10], and 19.2 units higher than the Siamese architecture [30].

Table 2 presents the results of SADepth on the test set 1 and test set 2 (as defined in the *SCARED* dataset), together with the performance reported in the

MICCAI sub-challenge summary paper [1]. The results show an improvement over the unsupervised methods from the summary paper and recent baselines, while it is also competitive with some supervised approaches. This confirms that SADepth generalizes well across different datasets collected from different laparoscopes and subjects, while still producing superior performance compared with the state-of-the-art unsupervised approaches.

4   Conclusions

We have presented a new self-supervised adversarial depth estimation framework SADepth with an encoder-decoder generator and a concatenated stereo image pair as the input. The adversarial learning strategy improved the generation quality of the framework and led to the state-of-the-art performance on two public datasets. Furthermore, SADepth did not require any per-pixel depth labels and generalized well across different laparoscopes, suggesting excellent applicability to scalable data acquisition when accurate ground truth depth cannot be collected.

References

1. Allan, M., et al.: Stereo correspondence and reconstruction of endoscopic data challenge. arXiv:2101.01133 (2021)
2. Chang, J.R., Chen, Y.S.: Pyramid stereo matching network. In: Proceedings of the IEEE Conference on Computer Vision and Pattern Recognition, pp. 5410–5418 (2018)
3. Clevert, D.A., Unterthiner, T., Hochreiter, S.: Fast and accurate deep network learning by exponential linear units (ELUs). arXiv preprint arXiv:1511.07289 (2015)
4. Do, T., Nguyen, B.X., Tjiputra, E., Tran, M., Tran, Q.D., Nguyen, A.: Multiple meta-model quantifying for medical visual question answering. arXiv preprint arXiv:2105.08913 (2021)
5. Duggal, S., Wang, S., Ma, W.C., Hu, R., Urtasun, R.: DeepPruner: learning efficient stereo matching via differentiable PatchMatch. In: Proceedings of the IEEE/CVF International Conference on Computer Vision, pp. 4384–4393 (2019)
6. Eigen, D., Puhrsch, C., Fergus, R.: Depth map prediction from a single image using a multi-scale deep network. arXiv preprint arXiv:1406.2283 (2014)
7. Garg, R., B.G., V.K., Carneiro, G., Reid, I.: Unsupervised CNN for single view depth estimation: geometry to the rescue. In: Leibe, B., Matas, J., Sebe, N., Welling, M. (eds.) ECCV 2016. LNCS, vol. 9912, pp. 740–756. Springer, Cham (2016). https://doi.org/10.1007/978-3-319-46484-8_45
8. Geiger, A., Roser, M., Urtasun, R.: Efficient large-scale stereo matching. In: Kimmel, R., Klette, R., Sugimoto, A. (eds.) ACCV 2010. LNCS, vol. 6492, pp. 25–38. Springer, Heidelberg (2011). https://doi.org/10.1007/978-3-642-19315-6_3
9. Godard, C., Mac Aodha, O., Brostow, G.J.: Unsupervised monocular depth estimation with left-right consistency. In: Proceedings of the IEEE Conference on Computer Vision and Pattern Recognition, pp. 270–279 (2017)

10. Godard, C., Mac Aodha, O., Firman, M., Brostow, G.J.: Digging into self-supervised monocular depth estimation. In: Proceedings of the IEEE/CVF International Conference on Computer Vision, pp. 3828–3838 (2019)
11. Goodfellow, I.J., et al.: Generative adversarial networks. arXiv preprint arXiv:1406.2661 (2014)
12. Grasa, O.G., Bernal, E., Casado, S., Gil, I., Montiel, J.: Visual slam for handheld monocular endoscope. IEEE Trans. Med. Imaging 33(1), 135–146 (2013)
13. He, K., Zhang, X., Ren, S., Sun, J.: Deep residual learning for image recognition. In: Proceedings of the IEEE Conference on Computer Vision and Pattern Recognition, pp. 770–778 (2016)
14. Heise, P., Klose, S., Jensen, B., Knoll, A.: PM-Huber: PatchMatch with Huber regularization for stereo matching. In: Proceedings of the IEEE International Conference on Computer Vision, pp. 2360–2367 (2013)
15. Huang, B., et al.: Tracking and visualization of the sensing area for a tethered laparoscopic gamma probe. Int. J. Comput. Assist. Radiol. Surg. 15(8), 1389–1397 (2020). https://doi.org/10.1007/s11548-020-02205-z
16. Jaderberg, M., Simonyan, K., Zisserman, A., Kavukcuoglu, K.: Spatial transformer networks. arXiv preprint arXiv:1506.02025 (2015)
17. Johnston, A., Carneiro, G.: Self-supervised monocular trained depth estimation using self-attention and discrete disparity volume. In: Proceedings of the IEEE/CVF Conference on Computer Vision and Pattern Recognition, pp. 4756–4765 (2020)
18. Joung, S., Kim, S., Park, K., Sohn, K.: Unsupervised stereo matching using confidential correspondence consistency. IEEE Trans. Intell. Transp. Syst. 21(5), 2190–2203 (2019)
19. Leonard, S., et al.: Evaluation and stability analysis of video-based navigation system for functional endoscopic sinus surgery on in vivo clinical data. IEEE Trans. Med. Imaging 37(10), 2185–2195 (2018)
20. Liu, X., et al.: Dense depth estimation in monocular endoscopy with self-supervised learning methods. IEEE Trans. Med. Imaging 39(5), 1438–1447 (2019)
21. Luo, W., Schwing, A.G., Urtasun, R.: Efficient deep learning for stereo matching. In: Proceedings of the IEEE Conference on Computer Vision and Pattern Recognition, pp. 5695–5703 (2016)
22. Mack, M.J.: Minimally invasive and robotic surgery. JAMA 285(5), 568–572 (2001)
23. Nguyen, A., et al.: End-to-end real-time catheter segmentation with optical flow-guided warping during endovascular intervention. In: 2020 IEEE International Conference on Robotics and Automation (ICRA), pp. 9967–9973. IEEE (2020)
24. Paszke, A., et al.: Automatic differentiation in PyTorch (2017)
25. Pilzer, A., Xu, D., Puscas, M., Ricci, E., Sebe, N.: Unsupervised adversarial depth estimation using cycled generative networks. In: 2018 International Conference on 3D Vision (3DV), pp. 587–595. IEEE (2018)
26. Ronneberger, O., Fischer, P., Brox, T.: U-Net: convolutional networks for biomedical image segmentation. In: Navab, N., Hornegger, J., Wells, W.M., Frangi, A.F. (eds.) MICCAI 2015. LNCS, vol. 9351, pp. 234–241. Springer, Cham (2015). https://doi.org/10.1007/978-3-319-24574-4_28
27. Wang, Z., Bovik, A.C., Sheikh, H.R., Simoncelli, E.P.: Image quality assessment: from error visibility to structural similarity. IEEE Trans. Image Process. 13(4), 600–612 (2004)
28. Watson, J., Firman, M., Brostow, G.J., Turmukhambetov, D.: Self-supervised monocular depth hints. In: Proceedings of the IEEE/CVF International Conference on Computer Vision, pp. 2162–2171 (2019)

29. Yamaguchi, K., McAllester, D., Urtasun, R.: Efficient joint segmentation, occlusion labeling, stereo and flow estimation. In: Fleet, D., Pajdla, T., Schiele, B., Tuytelaars, T. (eds.) ECCV 2014. LNCS, vol. 8693, pp. 756–771. Springer, Cham (2014). https://doi.org/10.1007/978-3-319-10602-1_49

30. Ye, M., Johns, E., Handa, A., Zhang, L., Pratt, P., Yang, G.Z.: Self-supervised Siamese learning on stereo image pairs for depth estimation in robotic surgery. arXiv preprint arXiv:1705.08260 (2017)

31. Yi, Z., Zhang, H., Tan, P., Gong, M.: DualGAN: unsupervised dual learning for image-to-image translation. In: Proceedings of the IEEE International Conference on Computer Vision, pp. 2849–2857 (2017)

32. Zhou, T., Brown, M., Snavely, N., Lowe, D.G.: Unsupervised learning of depth and ego-motion from video. In: Proceedings of the IEEE Conference on Computer Vision and Pattern Recognition, pp. 1851–1858 (2017)

33. Zhu, J.Y., Park, T., Isola, P., Efros, A.A.: Unpaired image-to-image translation using cycle-consistent adversarial networks. In: Proceedings of the IEEE International Conference on Computer Vision, pp. 2223–2232 (2017)

Personalized Respiratory Motion Model Using Conditional Generative Networks for MR-Guided Radiotherapy

Liset Vázquez Romaguera[1]([✉]), Tal Mezheritsky[1], and Samuel Kadoury[1,2]

[1] MedICAL, Polytechnique Montreal, Montreal, QC, Canada
liset.vazquez@polymtl.ca
[2] CHUM Research Center, Montreal, QC, Canada

Abstract. MRI-guided radiotherapy systems enable real-time 2D cine acquisitions for target monitoring, but cannot provide volumetric information due to spatio-temporal constraints. Hence, respiratory motion models coupled with a temporal predictive mechanism are a suitable solution to enable ahead-of-time 3D tumor and anatomy tracking in combination with real-time online plan adaptation. We propose a novel subject-specific probabilistic model to enable 3D+t predictions from image-based surrogates during radiotherapy treatments. The model is trained end-to-end to simultaneously capture and learn a distribution of realistic motion fields over a population dataset. Furthermore, the distribution is conditioned on a sequence of partial observations, which can be extrapolated in time using a *seq2seq*-inspired mechanism allowing for scalable predictive horizon. Based on the generative properties of conditional variational autoencoders, it integrates anatomical features and temporal information to construct an interpretable latent space with respiratory phase discrimination. The choice of a probabilistic framework allows improving uncertainty estimation during the volume generation phase. Experimental validation on 25 subjects demonstrates the potential of the proposed model, which achieves a mean landmark error of 1.4 ± 1.1 mm, yielding statistically significant improvements over state-of-the-art methods.

Keywords: Image-guided radiotherapy · Motion modelling · 4D MRI · Generative model · Conditional variational autoencoders

1 Introduction

Shape and motion variability in abdominal and thoracic organs due to the breathing-induced deformation represents an important challenge in external beam radiation therapy. Consequently, tumor tracking and motion compensation strategies are crucial to improve control of radiation beams within the body. Technological innovations such as the MR-Linac have enabled the integration of MR imaging capabilities with linear accelerators into a single device

Supported by NSERC research grant (CRDPJ-517413-17) and by the Canada First Research Excellence Fund through the TransMedTech Institute.

M. de Bruijne et al. (Eds.): MICCAI 2021, LNCS 12904, pp. 238–248, 2021.
https://doi.org/10.1007/978-3-030-87202-1_23

enabling real-time target monitoring during treatment. However, the unobserved out-of-plane motion may degrade dosimetric benefits [10]. Furthermore, volumetric information is useful for dose recalculation and adaptive radiotherapy planning. Current solutions are based on deformable registration [19,24] and statistical motion models [5,6,9,12,20,21,27]. The former strategy applies 2D-3D deformable registration between in-room cine-MRI and a pre-treatment volume to estimate the 3D target position. However, this simple yet effective technique is limited to local motion modeling. Alternatively, several methods are based on maximizing the correlation between a surrogate and a motion model, which can be either population-based or subject-specific. The term surrogate (also known as partial observation) refers to a signal acquired during the intervention, which is directly correlated with the motion of interest [16]. In population-based models, motion data from multiple patients are combined to capture broader motion variability. For instance, Tanner et al. [29] employed a 3D breath-hold scan and interleaved MR slices to drive a statistical model. Although this type of model has shown promising results, their construction involves the challenging task of identifying correspondent landmarks. In contrast, patient-specific models do not require establishing correspondences across a population, providing an improved fit to the patient's anatomy. Typically, the motion extracted from pre-treatment 4D datasets through deformable registration is used to compute a statistical model. Subsequently, partial observations are linked to the model by maximising a similarity metric between the image surrogate and its corresponding slice in the warped reference volume [9,30]. Often, multi-layer perceptrons are employed to enable ahead-of-time prediction of the model coefficients [20]. Their main limitation is that the weights optimization relies on a pixel-wise similarity metric, which only captures the variation in a single plane [18]. Recent advancements in deep learning have opened new opportunities to relate partial observations to high-dimensional data given sufficiently large training datasets [1,17,22]. In the context of motion modelling for image-guided radiation treatments (IGRT), Giger et al. [7] leveraged a conditional generative adversarial network to create a patient-specific model able to relate ultrasound to 3D deformations. However, it lacked interpretability capabilities towards the 3D prediction. Predicting the breathing-induced deformation fields from partial observations has also been explored in 2D [23].

In this work, we propose a predictive framework for abdominal motion, leveraging the advantages of both population-based and patient-specific motion models. During training, the model learns from a population dataset, capturing the wide range in motion variability from multiple anatomies. Moreover, the model's generation capability can potentially benefit from a progressive increase in the amount of data. Once the model is created, it can be personalized to a given patient using relatively few temporal samples, tailoring the model to the subject's specific anatomy at the beginning of the IGRT. In terms of motion modelling, we introduce a novel conditional model which considers the temporal consistency of 2D surrogate images to regress multiple feature representations ahead of time. These feature vectors can be seen as conditioning variables of a

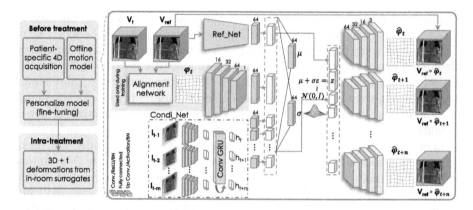

Fig. 1. Proposed motion model used for multi-time volume prediction. The 3D deformations are mapped to a probabilistic latent space, which is conditioned both on extrapolated-in-time vectors and anatomical features. The blue background indicates those components used only during training, whereas the rest are used at all stages.

low-dimensional space of breathing-induced 3D deformations. Besides, our model has additional advantages compared to related approaches, namely: a latent space capable to discriminate and visualize respiratory phases and the ability to provide uncertainty measures over the model's predictions. Both characteristics make the results more interpretable for clinical procedures.

2   Methods

2.1   Model Building

During training, our conditional probabilistic model receives as input a single pre-treatment volume gated at a reference respiratory phase and cine-MR images at times $\langle t-1, t-2, \ldots, t-m \rangle$, which act as predictive variables to recover the dense displacement vector fields (DVF) corresponding to n future respiratory phases. It also receives a set of dense 3D deformations at times $\langle t, t+2, \ldots, t+n \rangle$. Figure 1 shows a schematic representation of the training framework. It is composed of the following blocks: (1) alignment network to generate the DVF, (2) conditional variational autoencoder to learn the motion distribution, and (3) temporal predictor to generate conditioning feature vectors ahead-of-time.

Alignment Network. Since our method does not rely on any surface-based information (i.e. prior segmentations) and avoids explicit voxel generation, we work with deformations between pairs of volumes, from the same subject, over a population dataset. We use a registration function, parameterized with a neural network, which receives a specific reference volume V_{ref} and a target volume V_t at time t as inputs to generate a breathing-induced organ deformation (ϕ_t) between them. This deformation is then passed to the following block, which learns the distribution of DVF across the training dataset. In our setup, V_{ref} is taken at the exhale phase since it presents the most reproducible liver representation. We assume that both volumes were previously rigidly aligned to a common reference space. For registration, we use the U-net-like architecture proposed in [2] with pre-trained weights since this step is out of the scope of this work, however other similar configuration can be used.

Conditional Motion Field Generation. We formulate the 3D volume estimation from partial observations as a conditional manifold learning task, an extension of [26]. The predictive variables, i.e. the pre-operative volume and the cine acquisitions are integrated during optimization in the form of conditional variables which modulate the motion distribution learned by the model. Let ϕ_t, V_{ref} and $I_s = \langle I_{t-1}, I_{t-2}, \ldots I_{t-m} \rangle$ be the 3D deformation, the reference volume and the surrogate image sequence, respectively. The goal of the model is to learn the conditional distribution $P(\phi_t|I_s, V_{ref})$ to produce a displacement matrix $\hat{\phi}_t \in \mathcal{R}^{H \times W \times D \times 3}$, given the available partial information and subject anatomy, where H, W and D denote the height, width and depth of the volumes, respectively. Following the generative process of conditional variational autoencoders (CVAE), a latent variable z is generated from the prior distribution $p_\theta(z)$ which is constrained to be a Gaussian, i.e. $z \sim \mathcal{N}(0, I)$. By randomly sampling values of z, we can generate new DVF. However, computing the posterior distribution $p_\theta(z|I_s, V_{ref})$ to obtain z is analytically intractable [14]. Therefore, an encoder network is adopted to find an approximation of the posterior distribution:

$$q_\psi(z|I_s, V_{ref}) = \mathcal{N}\left(\mu(\phi_t, I_s, V_{ref}), \sigma(\phi_t, I_s, V_{ref})\right). \tag{1}$$

This network, parameterized with stacked 3D convolution layers, learns the mean $\mu \in \mathcal{R}^d$ and diagonal covariance matrix $\sigma \in \mathcal{R}^d (d \ll H \times W \times D)$ from the data, as depicted in Fig. 1. At training, the sampling of z is differentiable with respect to μ and σ by using the "reparameterization trick" [14], and defining $z = \mu + \epsilon * \sigma$, where $\epsilon \sim \mathcal{N}(0, I)$. The distance between both distributions p_θ and q_ψ can be minimized using the Kullback-Leibler (KL) divergence within a combined loss function which also seeks to minimize a reconstruction loss. The spatial warping block warps the reference volume with the transformation provided by the decoder enabling the model to calculate a similarity measure $\mathcal{L}_{sim}$ between $V_{ref} \circ \phi_t$ and the expected in-room volume V_t. We use stochastic gradient descent to find the optimal parameters $\hat{\theta}$ by minimizing the following loss function:

$$\hat{\theta} = \arg\min_\theta \left[\mathcal{L}_{sim}\left(V_{ref} \circ \phi_t, V_t\right) + \mathrm{KL}(q_\psi(z|I_s, V_{ref}) \parallel p_\theta(z)) \right] \tag{2}$$

where the KL-divergence can be computed in closed form. We adopt a negative local cross correlation as similarity loss function. In the proposed architecture, we use a multi-branch convolutional neural network composed by three sub-models that encode: (1) the 3D motion fields provided by the alignment module, (2) the pre-treatment volume ("Ref-Net" sub-network) and (3) the 2D cine image surrogates ("Condi-Net" sub-network). The first and second sub-models possess identical configurations. They are composed of successive 3D convolutions with kernel size $3 \times 3 \times 3$ and a stride of 2 followed, by ReLU activations and batch normalization (BN). On the other hand, Condi-Net acts as temporal predictor. As illustrated in Fig. 1, each branch ends in a fully connected (FC) layer. The respective outputs are further concatenated and mapped to two additional FC layers to generate μ and σ, which are combined with ϵ to construct the latent space sample z, representing the normal Gaussian distribution. The decoder, also modeled with a convolutional neural network, reconstructs the displacement vector fields given the pre-operative volume and the spatiotemporal features extracted from the 2D slices provided in real-time. The conditional dependency is explicitly modeled by the concatenation of z with the feature representation of V_{ref} and I_s. This means that our model leverages the transformation retrieved from the latent space given the conditioning feature. Finally, a differentiable layer with explicit spatial transformation capabilities [11] applies the predicted deformation on the pre-operative volume yielding the warped volume that is compared to the target volume in the first term of Eq. (2). Using this scheme, our model is able to provide volumetric information.

Temporal Predictor (Condi-Net). A last module enables multi-time surrogate extrapolation. Its design is inspired by the *seq2seq* configuration, widely applied for natural language processing and other time-series tasks [28]. It is comprised by $m = 3$ stacks of 2D convolutions with kernel size 3×3 and a stride of 2 followed by ReLU activations and BN. Each stack independently processes the channel-wise concatenation of a single temporal image with their corresponding slice in the pre-operative volume. For a single timestep horizon the result is fed to a convolutional layer. To enable multi-time predictions, the temporal representations are stacked together and fed to convolutional gated recurrent units (GRU) [3], which are arranged in an encoder-decoder configuration. The encoder processes the spatiotemporal features and summarizes the information in a context vector. This embedding is tiled and fed to the decoder, which learns how to extrapolate n feature vectors (depending on the desired number of outputs volume) corresponding to n future time steps.

2.2   Model Personalization and Application

The goal of the personalization step is to fine-tune the weights of the pre-trained model, which is learned from a population, in order to adapt for the subject-specific anatomy and motion patterns. Thus this process follows a similar methodology as during training, but using a lower initial learning rate on a single subject. During the model application (test stage), the alignment module and the motion encoder are removed. Therefore, the decoder operates as a generative network given the patient anatomy and the cine acquisition, yielding realistic ahead-of-time DVF by sampling $z \sim \mathcal{N}(0, I)$.

3   Experimental Setup and Results

A dataset of free-breathing MR images acquired from a cohort of 25 healthy volunteers, each providing their written consent, was used in this study. Sagittal slices were acquired during 20 min on a MRI clinical scanner (3T Philips Ingenia) using a 2D T2-weighted balanced turbo field echo sequence. Data frames spanning the right liver lobe and navigator slices were acquired following an interleaved scheme and subsequently sorted to create a time-resolved 4D dataset, as detailed in [25]. The in-plane and through-plane resolution was 3.4×3.4 mm^2 and 3.5 mm, respectively, and image dimension of $32 \times 64 \times 64$. For each subject, 80 different sequences of 2D navigators showing different motion amplitudes and frequencies were acquired, which portrays the considerable inter-cycle variability that must be taken into account to increase the robustness of the motion model during radiotherapy. Hence, we leverage this variability as a data augmentation strategy for model creation. The time horizon for a single time step prediction is equivalent to a temporal resolution of 450 ms. The number of volumes for each subject was 2480. We followed a leave-one-out scheme, thereby creating the models with 24 anatomies and the remaining case for personalization/test. Each subject dataset was split into fine-tuning images (5 min, 620 volumes) and test images (15 min, 1860 volumes). We assume that this fine-tuning data represents the treatment planning acquisition, as depicted in the upper left of Fig. 1. The network's parameters were optimized using the Adam optimizer [13] with an initial learning rate (lr) set at 10^{-3}. For fine-tuning, the $lr = 10^{-5}$ was progressively reduced after 3 epochs without improvements in the validation loss. Training was performed in PyTorch with a batch size of 10.

As a first experiment, between 3 and 5 expert-selected vessel annotations were used to measure the geometrical accuracy between ground-truth (GT) and predicted positions over the last minute ($\approx$12 respiratory cycles). These landmark positions were scattered out-of-plane and tracked with subpixel resolution. Two of them were tagged on the same anatomical structure (main portal trunk bifurcation and the first bifurcation of the right portal vein) across all the sub-

Table 1. Target tracking errors (in mm) measured at different respiratory phases for a predictive horizon of 450 ms. Values are mean $\pm$ std.

Model	Mid-inh	Inhale	Mid-exh	Exhale	Overall
Initial motion	7.0 ± 6.0	7.0 ± 10.34	5.3 ± 4.9	2.6 ± 2.1	5.4 ± 5.8
FM [17]	3.1 ± 2.6	3.9 ± 3.2	2.7 ± 2.2	1.9 ± 2.1	2.9 ± 2.7
ME [19]	3.0 ± 2.7	2.5 ± 2.5	2.5 ± 1.7	1.7 ± 1.4	2.4 ± 2.0
PCA [20]	1.6 ± 2.0	2.0 ± 2.6	1.6 ± 0.9	2.0 ± 1.2	1.8 ± 1.6
Proposed* (C)	2.4 ± 2.6	3.3 ± 3.0	2.0 ± 0.9	1.3 ± 1.0	2.2 ± 1.8
Proposed (S)	1.8 ± 1.3	1.9 ± 1.4	1.7 ± 1.2	1.5 ± 1.0	1.7 ± 1.2
Proposed (C)	$\mathbf{1.4 \pm 1.0}$	$\mathbf{1.8 \pm 1.6}$	$\mathbf{1.3 \pm 1.0}$	$\mathbf{1.1 \pm 0.8}$	$\mathbf{1.4 \pm 1.1}$

* Model applied as population-based, i.e. in unseen cases without fine-tuning

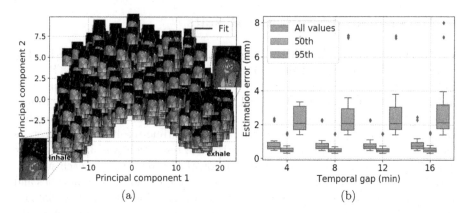

(a) (b)

Fig. 2. (a) Low-dimensional mapping of the latent representations of breathing phases. (b) Analysis of the drift effect on the estimation error when increasing the temporal gap between training and test subsets.

jects. The tracking capabilities of the proposed model, with sagittal (S) and coronal (C) orientations used for surrogates, was compared to three state-of-the-art approaches developed in the context of IGRT. Namely, Principal Component Analysis (PCA) [20], a registration-based motion extrapolation (ME) technique [19] and a deep network based on feature merging (FM) [17]. Results presented in Table 1 were tested for statistical significance using the Wilcoxon signed-rank test with significance level $\alpha = 0.01$. Effect size was measured using Cohen's d. The reference volume was excluded from the error calculation. When comparing the overall tracking errors, the accuracy with the proposed model using coronal images improved by a significant margin of 0.4 mm ($p \ll 0.01$, $d = 0.95$), 1.0 mm ($p \ll 0.01$, $d = 1.02$) and 1.5 mm ($p \ll 0.01$, $d = 0.70$) over PCA, ME and FM, respectively. Moreover, using coronal orientation showed increased performance compared to the sagittal view ($p \ll 0.01$, $d = 0.22$), which is in line with previously reported results [18,27]. We also reported the model's result when applied as population-based, i.e., in unseen cases without prior fine-tuning. Figure 2a shows the mapping of the latent vectors to a Cartesian plane via PCA. A clear phase discrimination can be observed, which is plausible for the motion modeling task. When investigating the model's tolerance to potential shifts of the surrogate location, we found little variation of the NCC between GT and predicted volumes with a shift of ≈ 20 mm away from the central slice in both directions. This suggests the training process relies primarily on encoded phase information in latent space, an important advantage over current techniques. Each acquisition was divided into 5 equally-sized subsets of 4 min in order to investigate the influence of the organ drift. The analysis was conducted by fixing the fifth subset as the testing set, while the first four subsets were used as four separate training sets. The geometrical error distributions reported in Fig. 2b were measured using 3D deformable registration between ground-truth and predicted volumes with the B-spline transformation model as implemented in the

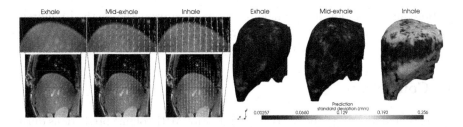

Fig. 3. Left: Most probable deformation (yellow) and reference motion fields (green). Right: Motion-based prediction uncertainty maps at several phases (Color figure online).

Elastix framework [15]. It can be seen that the error distributions show little to no degradation with the increase of the temporal gap between training and testing sets, showing superior robustness compared to existing techniques where estimation errors are ×4 higher between the extreme subsets [8].

The median [interquartile range] of the error distributions when increasing the horizon to 900 and 1350 ms are 1.8 [2.6] mm and 1.9 [2.9] mm, respectively. The quality of the obtained deformations was assessed through the Jacobian matrix determinant ($|J|$). The percentage of voxels with a non-negative $|J|$ was 99.4% across the entire dataset. Qualitative comparison of the most probable deformation field at selected phases is shown on the left of Fig. 3, where the reference and the predicted deformations are overlaid. Overall, there is a satisfactory alignment with some minor exceptions. The right part of Fig. 3 displays uncertainty maps of the predicted DVF, defined as the standard deviation of $N = 50$ different predictions generated by randomly sampling the latent space. Finally, difference maps between GT and predictions across multiple respiratory phases are shown in Fig. 4. It is noticeable that the model correctly predicts the

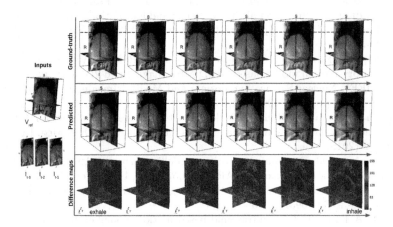

Fig. 4. Difference maps between ground-truth and predicted volumes.

spatiotemporal motion from inhale to exhale. The proposed method required a mean computation time of 7.44 ms (average from 20 measurements) for inference on a NVIDIA Titan RTX GPU with 64 Gb RAM.

4 Conclusion

We presented a novel probabilistic framework for MRI volume predictions during IGRT with a variable predictive horizon from real-time 2D surrogates. It offers several advantages over existing solutions. First, it avoids pre-processing steps such as surface segmentation or landmark annotations. Second, it provides an explainable latent space and quantitative uncertainty metrics, therefore making the results clinically interpretable by physicists. The accuracy of the tracking results are within the clinically acceptable margins (<2 mm) for motion management in modalities such as high-intensity focused ultrasound, conventional radiotherapy, or particle therapy. With a prediction horizon of (at least) 450 ms, the motion model is applicable in real-time and meets the typical temporal requirements [4]. Future studies should investigate how the model copes with inter-session variations as well as assessing the dosimetric impact.

References

1. Abdi, A.H., Pesteie, M., Prisman, E., Abolmaesumi, P., Fels, S.: Variational shape completion for virtual planning of jaw reconstructive surgery. In: Shen, D., et al. (eds.) MICCAI 2019. LNCS, vol. 11768, pp. 227–235. Springer, Cham (2019). https://doi.org/10.1007/978-3-030-32254-0_26
2. Balakrishnan, G., Zhao, A., Sabuncu, M.R., Guttag, J., Dalca, A.V.: VoxelMorph: a learning framework for deformable medical image registration. IEEE Trans. Med. Imaging 38(8), 1788–1800 (2019)
3. Ballas, N., Yao, L., Pal, C., Courville, A.C.: Delving deeper into convolutional networks for learning video representations. In: ICLR (Poster) (2016)
4. Ehrhardt, J., Lorenz, C., et al.: 4D Modeling and Estimation of Respiratory Motion for Radiation Therapy. Springer, Heidelberg (2013). https://doi.org/10.1007/978-3-642-36441-9
5. Fayad, H.J., Buerger, C., Tsoumpas, C., Cheze-Le-Rest, C., Visvikis, D.: A generic respiratory motion model based on 4D MRI imaging and 2D image navigators. In: 2012 IEEE Nuclear Science Symposium and Medical Imaging Conference Record (NSS/MIC), pp. 4058–4061. IEEE (2012)
6. Garau, N., et al.: A ROI-based global motion model established on 4DCT and 2D cine-MRI data for MRI-guidance in radiation therapy. Phys. Med. Biol. 64(4), 045002 (2019)
7. Giger, A., et al.: Respiratory motion modelling using cGANs. In: Frangi, A.F., Schnabel, J.A., Davatzikos, C., Alberola-López, C., Fichtinger, G. (eds.) MICCAI 2018. LNCS, vol. 11073, pp. 81–88. Springer, Cham (2018). https://doi.org/10.1007/978-3-030-00937-3_10
8. Giger, A.T., et al.: Liver-ultrasound based motion modelling to estimate 4D dose distributions for lung tumours in scanned proton therapy. Phys. Med. Biol. 65(23), 235050 (2020)

9. Harris, W., Yin, F.F., Cai, J., Ren, L.: Volumetric cine magnetic resonance imaging (VC-MRI) using motion modeling, free-form deformation and multi-slice undersampled 2D cine MRI reconstructed with spatio-temporal low-rank decomposition. Quant. Imaging Med. Surg. **10**(2), 432 (2020)

10. Henke, L., et al.: Phase I trial of stereotactic MR-guided online adaptive radiation therapy (SMART) for the treatment of oligometastatic or unresectable primary malignancies of the abdomen. Radiother. Oncol. **126**(3), 519–526 (2018)

11. Jaderberg, M., Simonyan, K., Zisserman, A., et al.: Spatial transformer networks. In: Advances in Neural Information Processing Systems, pp. 2017–2025 (2015)

12. Jud, C., Preiswerk, F., Cattin, P.C.: Respiratory motion compensation with topology independent surrogates. In: Workshop on Imaging and Computer Assistance in Radiation Therapy (2015)

13. Kingma, D.P., Ba, J.: Adam: a method for stochastic optimization. arXiv preprint arXiv:1412.6980 (2014)

14. Kingma, D.P., Welling, M.: Auto-encoding variational Bayes. CoRR arXiv:1312.6114 (2013)

15. Klein, S., Staring, M., Murphy, K., Viergever, M.A., Pluim, J.P.: elastix: A toolbox for intensity-based medical image registration. IEEE Trans. Med. Imaging **29**(1), 196–205 (2009)

16. McClelland, J.R., Hawkes, D.J., Schaeffter, T., King, A.P.: Respiratory motion models: a review. Med. Image Anal. **17**(1), 19–42 (2013)

17. Mezheritsky, T., Romaguera, L.V., Kadoury, S.: 3D ultrasound generation from partial 2D observations using fully convolutional and spatial transformation networks. In: 2020 IEEE 17th International Symposium on Biomedical Imaging (ISBI), pp. 1808–1811. IEEE (2020)

18. Paganelli, C., et al.: Time-resolved volumetric MRI in MRI-guided radiotherapy: an in silico comparative analysis. Phys. Med. Biol. **64**(18), 185013 (2019)

19. Paganelli, C., et al.: Feasibility study on 3D image reconstruction from 2D orthogonal cine-MRI for MRI-guided radiotherapy. J. Med. Imaging Radiat. Oncol. **62**(3), 389–400 (2018)

20. Pham, J., Harris, W., Sun, W., Yang, Z., Yin, F.F., Ren, L.: Predicting real-time 3D deformation field maps (DFM) based on volumetric cine MRI (VC-MRI) and artificial neural networks for on-board 4D target tracking: a feasibility study. Phys. Med. Biol. **64**(16), 165016 (2019)

21. Preiswerk, F., et al.: Model-guided respiratory organ motion prediction of the liver from 2D ultrasound. Med. Image Anal. **18**(5), 740–751 (2014)

22. Qin, C., et al.: Joint learning of motion estimation and segmentation for cardiac MR image sequences. In: Frangi, A.F., Schnabel, J.A., Davatzikos, C., Alberola-López, C., Fichtinger, G. (eds.) MICCAI 2018. LNCS, vol. 11071, pp. 472–480. Springer, Cham (2018). https://doi.org/10.1007/978-3-030-00934-2_53

23. Romaguera, L.V., Plantefève, R., Romero, F.P., Hébert, F., Carrier, J.F., Kadoury, S.: Prediction of in-plane organ deformation during free-breathing radiotherapy via discriminative spatial transformer networks. Med. Image Anal. **64**, 101754 (2020)

24. Seregni, M., Paganelli, C., Kipritidis, J., Baroni, G., Riboldi, M.: Out-of-plane motion correction in orthogonal cine-MRI registration. Radiother. Oncol. **123**, S147–S148 (2017)

25. von Siebenthal, M., Szekely, G., Gamper, U., Boesiger, P., Lomax, A., Cattin, P.: 4D MR imaging of respiratory organ motion and its variability. Phys. Med. Biol. **52**(6), 1547 (2007)

26. Sohn, K., Lee, H., Yan, X.: Learning structured output representation using deep conditional generative models. In: Advances in Neural Information Processing Systems, pp. 3483–3491 (2015)
27. Stemkens, B., Tijssen, R.H., De Senneville, B.D., Lagendijk, J.J., Van Den Berg, C.A.: Image-driven, model-based 3D abdominal motion estimation for MR-guided radiotherapy. Phys. Med. Biol. **61**(14), 5335 (2016)
28. Sutskever, I., Vinyals, O., Le, Q.V.: Sequence to sequence learning with neural networks. arXiv preprint arXiv:1409.3215 (2014)
29. Tanner, C., et al.: In vivo validation of spatio-temporal liver motion prediction from motion tracked on MR thermometry images. Int. J. Comput. Assist. Radiol. Surg. **11**(6), 1143–1152 (2016)
30. Zhang, Y., Yin, F.F., Pan, T., Vergalasova, I., Ren, L.: Preliminary clinical evaluation of a 4D-CBCT estimation technique using prior information and limited-angle projections. Radiother. Oncol. **115**(1), 22–29 (2015)

Multimodal Sensing Guidewire for C-Arm Navigation with Random UV Enhanced Optical Sensors Using Spatio-Temporal Networks

Andrei Svecic[1], Gilles Soulez[2], Frédéric Monet[1], Raman Kashyap[1], and Samuel Kadoury[1,2(✉)]

[1] Polytechnique Montréal, Montréal, QC, Canada
samuel.kadoury@polymtl.ca
[2] CHUM Hospital Research Center, Montréal, QC, Canada

Abstract. Percutaneous transluminal angioplasty (PTA) revascularization is a common minimally invasive treatment for occlusions in peripheral arteries, but it's success in long occlusions is limited by technical challenges associated with crossing occluded vessels and lumen re-entry. Revascularization needs to be guided closely using ionizing imaging such as fluoroscopy, while intravascular guidewires lack the capability of characterizing physiological conditions near occlusions, such as blood flow. We propose a multimodal sensing framework to infer both three-dimensional shape and vascular flow from an optical fiber device using random optical gratings enhanced with ultraviolet exposure, allowing a fully-distributed strain sensor. A two-branch spatio-temporal neural network is proposed to process a generated optical signal trajectory from scattered wavelength distributions. A shape network is first used in combination with the pre-procedural 3D angiography image to track the 3D shape related to backscattered wavelength shift, while a flow velocity network trained on 4D-MRI measurements allows to extract vascular flow. A final refinement is performed to adjust the 3D-2D projection onto C-arm images, allowing to correct for slight deviations of the sensed shape. Synthetic and porcine experiments were performed in a controlled environment setting, enabling to measure the accuracy of the 3D shape tracking and flow measurements, with errors of 2.4 ± 0.9 mm and flow differences below 2 cm/s, demonstrating the ability to provide anatomical and physiological properties during vascular procedures.

Keywords: Fiber optic sensing · Optical frequency domain reflectometry · Vascular flow · Arterial image guidance

1 Introduction

Percutaneous transluminal angioplasty (PTA) revascularization is a common minimally invasive treatment option for arterial occlusions, but several challenges

Supported by the Canada Research Chairs and Canadian Institutes of Health Research (CIHR).

M. de Bruijne et al. (Eds.): MICCAI 2021, LNCS 12904, pp. 249–258, 2021.
https://doi.org/10.1007/978-3-030-87202-1_24

related to vessel crossings and sub-lumen re-entry hinders the treatment of long occlusions with guidewires. Assessment of the severity of non-occlusive plaque remains difficult angiographically, as there is no visible residual channel between the proximal part of the occlusion to the distal part [6]. Improved visualization using pre-treatment CT or MR angiography (CTA, MRA) overlaid on X-ray angiography during revascularization is well suited for vascular lesions, as it can produce vascular maps identifying the lesion extent and characterize the surrounding vasculature [2,16] to establish a compliant pathway.

While image-guided X-ray angiography has excellent temporal and spatial resolution, 3D positioning of the microcatheter to the target lesion in the peripheral artery remains difficult, with no assessment if the recanalization is intraluminal or subintimal. Flow information is also extremely valuable for guiding PTA in femoral occlusions, as flow limiting dissection should be selectively stented [19]. Thus, the ability to measure subintimal blood flow velocity in combination with a precise localization of the device inside the arteries can complement physiological treatments and improve image-guidance to support atherosclerotic lesion therapies. Several image-based processes allow to estimate blood flow based on 2D-3D modeling with optical flow measurements [9], but lacks accuracy, and are not able to non-invasively assess flow in arteries without contrast.

Alternatively, fiber optic sensing using fiber Bragg gratings (FBG) enables the real time three-dimensional (3D) visualisation of endovascular devices [7], which was used and evaluated clinically for endovascular aortic or peripheral lesion repair [5]. Recent sensing technologies measuring light deflections along the entire length of the optical fibers enable shape and force measurements within needles, guidewires [14] and surgical robots [12], surpassing FBGs which can only perform point measurements along the device. These techniques are based on optical frequency domain reflectometry (OFDR) which measure Rayleigh backscatter in the fibers under strain to infer continuous measurements in real-time, providing navigation capabilities in complex environments [18]. Specifically, Rayleigh scattering from ultraviolet (UV) exposure causes microscopic defects in the glass structure of the fibers [10]. Still, they depend on a geometric model for 3D tracking, which requires precise calibration and are inaccurate to fabrication irregularities. Furthermore, OFDR has not been exploited for characterization of blood flow within arteries and is limited to real-time processing to infer the shape and flow measurements, thus favouring data-driven approaches as in [17].

We present an inference framework allowing to track the 3D shape combined with distributed flow measurements along a custom-built guidewire device in real-time for endovascular procedures, such as arterial interventions. We fabricate a guidewire device integrating optical fiber triplets, using random optical gratings enhanced with ultraviolet exposure on each fiber to improve sensing with OFDR to measure the local strain and Doppler wavelength shifts estimating blood flow. Wavelength optical data, in combination to pre-procedural MR angiography and intra-procedural C-arm imaging, are used in a dual-branch event-driven network, learning correspondences between 4D flow imaging and backscattered optical wavelength data to estimate shape and flow measures.

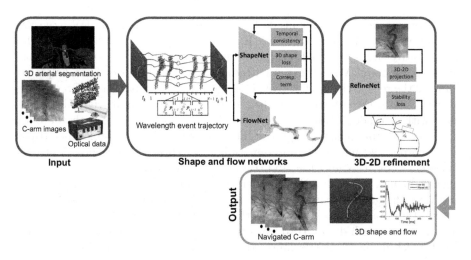

Fig. 1. The pipeline of the dual branch architecture for tracking the shape and sensing the flow from an optical random grating device. Using a single 3D angiographic volume, live 2D C-arm images and a stream of backscattered wavelength data from the interferometer, we first generate signal trajectories from asynchronous events, capturing the fluctuations of wavelength data during guidance. The first branch addresses the 3D shape recovery, with consistency, temporal and 3D correspondence constraints between targeted vessels and event trajectories. The second branch integrates a temporal convolutional network (TCN) to recover the sensed flow surrounding the continuous sensing fibers. Finally, a 3D-2D refinement stage is performed to correct local discrepancies of the shape on C-arm images while imposing stability constraints.

2 Materials and Methods

Given (1) a single input angiographic 3D image with segmented arteries S_{MRA}, (2) optical measurements, denoted as Λ, obtained in real-time from optical random optical gratings within the arteries and (3) synchronized acquired C-arm images, our goal is to generate as an output the full 3D guidewire shape $\mathbf{S}_f$, including the distal end, with blood flow measures $\mathbf{F}_f$. Figure 1 describes the steps of the dual-branch network used to recover shape and flow measurements.

2.1 Device Fabrication with UV Enhanced Random Gratings

We use noise-based Bragg gratings distributed randomly on the fiber, with the particular aspect of exhibiting a large bandwidth that are unaffected by the fiber's length, as opposed to uniform/discrete FBGs. The gratings have an enhanced backscatter, which is several orders of magnitude above typical Rayleigh backscatter of standard SMF-28 optical fibers, and are used in distributed sensing with OFDR [11]. The random gratings were written directly onto the fibers using a Talbot interferometer configuration with a phase-mask, where optimal noise frequency and displacement amplitude were used to obtain the strongest reflection magnitude for the sensors for the same UV exposure, which were set at 10 periods (5V) 20 Hz, respectively. A Q-switched laser with a 213 nm wavelength was used to expose the core of the fiber, based on a fifth harmonic. A random UV interference pattern was produced on each fiber, helping to improve the reflectivity from a given frequency range at a particular Bragg wavelength. Sets of three fibers were then extruded together in a triangular geometry, with a 120deg configuration between each fiber (Fig. 3), with the outer diameter of approximately 260 μm. The reusable and sterilizable fiber triplet was incorporated into a 0.67-mm-inner-diameter catheter (5-French Polyamide catheter, Cook, Bloomington, IN). Because spectral drift computed in the frequency domain (through a Fast Fourier Transform) is proportional to the strain, and thus the geometrical position of the fiber based on the directional strain around the rotational axis, as well as the surrounding medium's velocity, we use the raw wavelength data (Λ) to learn the relationship with these two variables. Further details on the device fabrication and setup can be found in [15].

2.2 Event Trajectory Generation of Wavelength Data

Individual events of wavelength data does not allow to fully recover the sensing information from the random optical gratings, and thus the guidewire tracking from single timepoints lacks robustness. Spatio-temporal wavelength data is extracted from the interferometer stream, which can be included within the time range $[t_k, t_{k+1}]$ (where k indicates the wavelength data). This integrates sequential wavelength distributions $\Lambda(k)$ and $\Lambda(k+1)$, which are used to asynchronously track relevant features from the guidewire, thus generating sparse event trajectories $\mathcal{T}(h)$ [4]. In this case, $h \in [1, \ldots, H]$ represents temporal sets of shifts of all the H wavelength data which is extracted for shape and flow inference. Here, N_H indicates the number of gratings (fiber sections) along the optical fiber triplet. While the sensor gratings are continuous, this discretization is needed to evaluate the loss functions in a discrete manner. Because the tracking of features can drift during device insertion, we process the feature tracking forward from the initial set $\Lambda(t_k)$ and backwards from the following set, $\Lambda(t_k+1)$. The results from this bidirectional feature tracking are combined by linking the nearest backward feature location to every forward feature local to a midpoint $(t_k + t_{k+1})/2$. To track the device from a trajectory of wavelength readings, we denote $[0, 1, \ldots, N]$ as the indices of the tracking frames from a current set of wavelength data, with N is a particular tracking frame rate.

2.3 Shape and Flow Networks

To recover the guidewire shape $S = \{\mathbf{S}_f\}, f \in [0, \ldots, N]$ for each sequential batch of wavelength data Λ, we train a shape recovery network (ShapeNet) which seeks to maximize the fit of the inferred shape with respect to the segmented anatomy S_{MRA}, combined with a minimization of drifting due to summation of tracking errors and inherent errors in wavelength shift correspondences. Based on a skeletal pose model [20], we formulate the network's overall loss term as:

$$\mathcal{L}_{\text{global}}(S) = \alpha_{\text{cor}}\mathcal{L}_{\text{cor}} + \alpha_{3D}\mathcal{L}_{3D} + \alpha_{\text{temp}}\mathcal{L}_{\text{temp}}. \tag{1}$$

A first signal correspondence term is based on the spatio-temporal deformation of the shape which is asynchronous in nature and is embedded in each stream of events. For each tracking frame i in a batch, we extract signal correspondences within the parsed event trajectories of two sequential acquisitions ($i - 1$ and $i + 1$), which are encoded in the event stream. This generates separate sets of signal correspondences $\mathcal{P}_{i,i-1}$ and $\mathcal{P}_{i,i+1}$, where $\mathcal{P}_{i,*} = \{(p_{i,h}, p_{*,h})\}$, with $h \in [1, \ldots, H]$. This favors the inferred shape's 3D coordinates to correspond with points along the centrelines extracted from the 2D detection of vessels on C-arm images [13]:

$$\mathcal{L}_{\text{cor}}(S) = \sum_{i=1}^{N-1} \sum_{j \in \{i-1, i+1\}} \sum_{l=1}^{N_H} \tau(p_{j,l}) \|\pi(P_l(\mathbf{S}_j)) - p_{j,l}\|_2^2 \tag{2}$$

where $\tau(p_{i,l})$ is a binary function $\{0, 1\}$ indicating whether there is a correspondence between the inferred 3D point $P_l(\mathbf{S}_j)$ with the detected wavelength data from tracked frame i, π is the 3D-2D projection, and $p_{j,l}$ detected vessels on the C-arm. The overall 3D term enables to align the event trajectory of the inferred shape with segmented 3D arteries (S_{MRA}). The 3D segmentation of the arteries were obtained with deformable mesh model approach [1]. The alignement term is defined as:

$$\mathcal{L}_{3D}(S) = \sum_{i=0}^{N} \sum_{l=1}^{N_H} \|P_l(\mathbf{S}_i) - (\mathbf{P}_{i,l}^{3D} + \mathbf{t}')\|_2^2 \tag{3}$$

with P_l as a 3D position of the guidewire in a section N_H and $\mathbf{P}_{i,l}^{3D} \in S_{MRA}$ represents the 3D point coordinate of the 3D centerline from the MRA, corresponding to the frame i and gratting l. Here, $\mathbf{t}' \in \mathbb{R}^3$ is a separate variable translating $\mathbf{P}_{i,l}^{3D}$ within the coordinate system of the reconstructed guidewire shape. Because only deforming sections of the fiber optic device will lead to variations in the wavelength event trajectory, random grating fluctuations were incorporated as a temporal stability constraint, allowing to capture the sections within the fiber with little motion. The term induces a penalty for severe deformations in shape between previous wavelength events and the current wavelength data:

$$\mathcal{L}_{\text{temp}}(S) = \sum_{i=0}^{N} \sum_{l=1}^{N_H} \varphi(l) \|P_l(\mathbf{S}_i) - P_l(\mathbf{S}_{i+1})\|_2^2 \tag{4}$$

with φ is a binary function $\{0, 1\}$ which provides an indication whether there is a correspondence between the inferred shape and the event trajectory.

The second branch relates to FlowNet, a data-driven model learning the relationship between Doppler wavelength data sensed in the vessel and vascular flow $F = \{\mathbf{F}_f\} = \Psi(\Lambda, \beta, S_{MRA})$, with β as the learned parameters and Ψ represent a Temporal Convolutional Network (TCN) used previously for motion-based analysis and predictions made from time-varying features [3,8]. The rationale for exploiting time series data lies from the fact the input is dynamically changing, which in the current process, is caused by blood flow variations surrounding the fiber. The input of the branch is the concatenation of the k wavelength observations along the entire sensing area in addition to the output of ShapeNet, such that $\Lambda_{cat} = [\widehat{\lambda}_{n-k}, \cdots, \widehat{\lambda}_n, \widehat{\lambda}_{n+1}, \cdots, \widehat{\lambda}_{n+k}] \in \mathbb{R}^{(2k+1)}$. The output of the network is a point series of velocity measures produced by the network.

2.4   Refinement Network

The final step of the framework comprises of an image-based refinement step of the tracked shape with respect to vessel features extracted from C-arm images, favouring the projected 3D shape to match the 2D vessel centreline detection obtained by a CNN from the C-arm images [13]. While shortest distance criterions may be sub-optimal for 2D-3D adjustments, the added stability term allows to maintain the overall 3D shape. The data term $\mathcal{L}_{proj}$ measures the point-to-line misalignment of the matching correspondences. For each point P_c along the centerline of the 3D reconstructed guidewire within the artery, we determine the closest 2D vessel border coordinate, estimated such that:

$$\mathcal{L}_{proj}(\mathbf{S}_f) = \sum_{c \in \mathcal{C}} \|\pi(P_c(\mathbf{S}_f) - u_c))\|_2^2 \qquad (5)$$

where $\mathcal{C}$ is the centerline of the 3D recovered shape and u_c the found closest vessel boundary on the 2D image. To ensure regularity of the reconstructed 3D shape with respect to the previous step, we enforce a stability term which minimizes the divergence between the shapes:

$$\mathcal{L}_{stab}(\mathbf{S}_f) = \sum_{l=1}^{N_H} \|P_l(\mathbf{S}_f) - P_l(\widehat{\mathbf{S}}_f)\|_2^2 \qquad (6)$$

where $\widehat{\mathbf{S}}_f$ is the inferred centerline determined in the previous step. The overall loss of the RefineNet is set as: $\mathcal{L}_{refine}(S) = \alpha_{proj}\mathcal{L}_{proj} + \mathcal{L}_{stab}$.

2.5   Implementation Details

For ShapeNet and RefineNet, a ResNet101 was used as the backbone. The channel size was set at $\{64, 128, 256, 512\}$, with ReLU functions linked after with 2 strides. The FlowNet network was composed of four fully connected layers of 5×5 kernels, with 2 max pooling layers and softmax layer at the end, with a stride of 1. Hyperparameters were determined using an Adam optimizer, with $5e-5$ as the learning rate, using a momentum $= 0.09$ and weight decay $= 0.00050$. The parameters were set as follows: $\{\alpha_{corr} = 50, \alpha_{3D} = 1, \alpha_{temp} = 70, \alpha_{proj} = 0.2\}$. A total of 4 NVIDIA Titan X GPU K80 dual-GPU graphics cards were used to train and test the models. The average inference time was 0.3s for a set of wavelength data.

Fig. 2. (a) Conveyer belt setup used for splicing and extruding the custom optical fiber triplet. (b) The reflectometer used to measure optical backscatter from the interrogated fiber triplet. (c) Enhancement of Rayleigh scattering with a fiber exposition process using a UV laser (XVVL-5HG; Xiton photonics). (d) Bent fiber with resulting 3D shape reconstruction.

3   Results

3.1   Experimental Setup and Training Data

An Optical Backscattering Reflectometer (OBR4600, LUNA Inc.) was used to acquire and process the backscattered data. The system's maximal framerate was used (5 Hz) for data sampling rate and an optical switch (JDSU SB series; Fiberoptic switch) with a channel transition period of 300 ms was used to scan each fiber of the triplet (Fig. 3). A dataset of 4D flow acquisitions from 18 porcine models with a 3T MRI (Achieva TX, Philips Healthcare, Best, The Netherlands), using a 16-channel surface coil for signal reception, a velocity encoding (VENC) value of 110 cm/s in each direction and peripheral pulse oximetry for cardiac synchronization, was used to train ShapeNet and FlowNet. For each porcine model, 30 different MR acquisitions were performed with the sensing triplet inserted inside the hepatic arteries of the pig, collecting the backscattered signals at several positions and synchronized with the respiration gating. Device navigation was paused when performing flow acquisitions. Acquisitions prior to contrast agent administration were made with flip angles of 4 and 20 degrees, using identical acquisition values to the ones used to generate T1 maps.

3.2   Synthetic and In-Vivo Experiments

The first set of experiments consisted in testing the framework in vascular phantoms created from patient-specific anatomies, providing ground-truth data in terms of shape and measured flows. The synthetic arteries made of polyvinyl alcohol were placed in an agar gel recipient with fiducial markers to allow for imaging and flow testing by connecting arteries to a flow pump (FloGard, Baxter) with controlled temperature and constant intravascular flow. Increasing levels of flow (5, 10, 15, 25 cm/s) were injected to the model, reproducing physiological

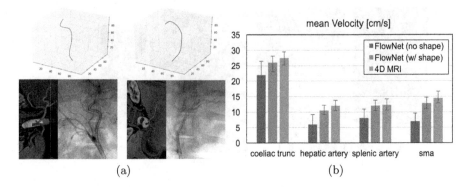

(a)   (b)

Fig. 3. (a) Tracking results on C-arm images with corresponding 3D reconstructed shapes. (b) Validation of blood flow velocities collected from the pig 's arterial tree model with the sensing optical fiber, in comparison to 4D flow MRI measurements.

Table 1. Accuracy in 2D tip and 3D shape tracking in all 5 porcine models, comparing electromagnetic (EM) tracking to standard fibers (SMF-28) and UV enhanced sensors. Blood flow measure reports the mean velocity in 4 points of measures in the arteries using FlowNet with and without input of the shape, using 4D flow as reference.

Model #	2D tip position (mm)			3D RMS shape (mm)			Blood flow (cm/s)	
	EM	SMF-28	**UVE-28**	EM	SMF-28	**UVE-28**	w/o S_f	w/ S_f
1	3.2 ± 1.5	2.7 ± 0.9	1.9 ± 0.6	4.1 ± 1.3	3.5 ± 1.5	2.6 ± 1.1	4.9 ± 2.6	1.9 ± 1.1
2	2.7 ± 1.4	2.2 ± 1.0	1.4 ± 0.5	3.8 ± 1.1	3.1 ± 1.0	2.2 ± 0.8	5.2 ± 3.7	2.0 ± 0.9
3	2.8 ± 1.3	2.3 ± 0.7	1.4 ± 0.4	3.7 ± 1.0	3.0 ± 0.9	2.1 ± 0.7	4.6 ± 2.3	1.7 ± 0.7
4	2.3 ± 1.2	1.6 ± 0.5	1.1 ± 0.3	3.2 ± 1.2	2.9 ± 0.8	2.0 ± 0.6	4.3 ± 2.6	1.6 ± 0.6
5	5.1 ± 2.2	3.3 ± 1.6	2.0 ± 0.7	5.9 ± 2.5	4.0 ± 1.9	3.2 ± 1.0	6.5 ± 3.4	2.1 ± 1.0
Overall	3.2 ± 1.5	2.4 ± 0.9	1.6 ± 0.5	4.1 ± 1.3	3.3 ± 1.3	2.4 ± 0.9	5.1 ± 2.8	1.8 ± 0.8

similar conditions. The experiments yielded overall 3D errors of 0.9 ± 0.4 mm and flow errors of 0.6 ± 0.3 cm/s compared to ground-truth flow values.

The following experiments consisted in testing the framework in 5 separate in-vivo porcine models, evaluating the sensing guidewire's performance. MRA to C-arm registration was obtained with a commercial software (Syngo, Siemens). Table 1 presents the 2D/3D shape tracking accuracy comparing the position of radio-opaque markers with the sensed position, as well as the 3D shape compared to data on CBCT acquisitions, and blood flow error measurements. Figure 3 presents shape sensing results with 2D overlay, as well as flow measure validation in different arterial branches, compared to original 4D MR flow.

4   Conclusion

We presented a new framework for image-guided intravascular procedures, exploiting random UV exposed grating sensors, where backscattered optical wavelength signals are used to recover the three-dimensional shape and velocity information from a guidewire using a dual-branch spatio-temporal neural network, integrating the angiographic and intra-procedural images. These new sensors are based on principles of Rayleigh scattering which are several orders of magnitude superior than typical SMF-28 optical fibers, leading to improved 3D shape and flow sensing capabilities. In-vivo experiments demonstrate the ability of the dual-branch network to process the backscattered wavelength data generating close to theoretical estimations of 3D shape, and offering realistic physiologic representations of the arterial vessels. Because the 3D-2D projection requires no user input, the technique can be used in clinical practice for PTA of arterial occlusions. Future work will validate the technique's robustness towards different anatomical variants and tests within occluded vessels.

References

1. Badoual, A., Gerard, M., De Leener, B., Abi-Jaoudeh, N., Kadoury, S.: 3D vascular path planning of chemo-embolizations using segmented hepatic arteries from MR angiography. In: IEEE 13th International Symposium on Biomedical Imaging (ISBI 2016), pp. 225–228 (2016)
2. Clarke, S.E., Hammond, R.R., Mitchell, J.R., Rutt, B.K.: Quantitative assessment of carotid plaque composition using multicontrast mri and registered histology. Mag. Reson. Med. Official J. Int. Soc. Mag. Reson. Med. 50(6), 1199–1208 (2003)
3. Gao, C., Liu, X., Peven, M., Unberath, M., Reiter, A.: Learning to see forces: surgical force prediction with RGB-point cloud temporal convolutional networks. In: Stoyanov, D., et al. (eds.) CARE/CLIP/OR 2.0/ISIC -2018. LNCS, vol. 11041, pp. 118–127. Springer, Cham (2018). https://doi.org/10.1007/978-3-030-01201-4_14
4. Gehrig, D., Rebecq, H., Gallego, G., Scaramuzza, D.: Asynchronous, photometric feature tracking using events and frames. In: Proceedings of the European Conference on Computer Vision (ECCV), pp. 750–765 (2018)
5. van Herwaarden, J.A., et al.: First in human clinical feasibility study of endovascular navigation with fiber optic realshape (fors) technology. European Journal of Vascular and Endovascular Surgery (2020)
6. Horehledova, B., et al.: Ct angiography in the lower extremity peripheral artery disease feasibility of an ultra-low volume contrast media protocol. Cardiovasc. Int. Radiol. 41(11), 1751–1764 (2018)
7. Jansen, M., Khandige, A., Kobeiter, H., Vonken, E.J., Hazenberg, C., van Herwaarden, J.: Three dimensional visualisation of endovascular guidewires and catheters based on laser light instead of fluoroscopy with fiber optic realshape technology: preclinical results. Eur. J. Vasc. Endovasc. Surg. 60(1), 135–143 (2020)
8. Lea, C., Vidal, R., Reiter, A., Hager, G.D.: Temporal convolutional networks: a unified approach to action segmentation. In: Hua, G., Jégou, H. (eds.) ECCV 2016. LNCS, vol. 9915, pp. 47–54. Springer, Cham (2016). https://doi.org/10.1007/978-3-319-49409-8_7
9. Lessard, S., Plantefève, R., Michaud, F., Huet, C., Soulez, G., Kadoury, S.: Blood-flow estimation in the hepatic arteries based on 3D/2D angiography registration. In: Stoyanov, D., et al. (eds.) LABELS/CVII/STENT -2018. LNCS, vol. 11043, pp. 3–10. Springer, Cham (2018). https://doi.org/10.1007/978-3-030-01364-6_1

10. Loranger, S., Gagné, M., Lambin-Iezzi, V., Kashyap, R.: Rayleigh scatter based order of magnitude increase in distributed temperature and strain sensing by simple uv exposure of optical fibre. Sci. Rep. **5**, 11177 (2014)
11. Monet, F., Loranger, S., Lambin-Iezzi, V., Drouin, A., Kadoury, S., Kashyap, R.: The rogue: a novel, noise-generated random grating. Optics Express **27**(10), 13895–13909 (2019)
12. Monet, F., et al.: High-resolution optical fiber shape sensing of continuum robots: a comparative study. In: 2020 IEEE International Conference on Robotics and Automation (ICRA), pp. 8877–8883. IEEE (2020)
13. Noothout, J.M., de Vos, B.D., Wolterink, J.M., Leiner, T., Išgum, I.: Cnn-based landmark detection in cardiac cta scans. arXiv preprint arXiv:1804.04963 (2018)
14. Parent, F., Gerard, M., Kashyap, R., Kadoury, S.: Uv exposed optical fibers with frequency domain reflectometry for device tracking in intra-arterial procedures. In: International Conference on Medical Image Computing and Computer-Assisted Intervention, pp. 594–601 (2017)
15. Parent, F., et al.: Intra-arterial image guidance with optical frequency domain reflectometry shape sensing. IEEE Trans. Med. Imaging **38**(2), 482–492 (2018)
16. Rutt, B.K., Clarke, S.E., Fayad, Z.A.: Atherosclerotic plaque characterization by mr imaging. Current Drug Targets-Cardiovasc. Hematol. Disord. **4**(2), 147–159 (2004)
17. Sefati, S., Gao, C., Iordachita, I., Taylor, R.H., Armand, M.: Data-driven shape sensing of a surgical continuum manipulator using an uncalibrated fiber bragg grating sensor. IEEE Sensors Journal (2020)
18. Song, J., Li, W., Lu, P., et al.: Long-range high spatial resolution distributed temperature and strain sensing based on optical frequency-domain reflectometry. IEEE Photonics **6**, 1–8 (2014)
19. Tomoi, Y., et al.: Spot stenting versus full coverage stenting after endovascular therapy for femoropopliteal artery lesions. J. Vasc. Surg. **70**(4), 1166–1176 (2019)
20. Xu, L., Xu, W., Golyanik, V., Habermann, M., Fang, L., Theobalt, C.: Event-cap: monocular 3D capture of high-speed human motions using an event camera. In: Proceedings of the IEEE/CVF Conference on Computer Vision and Pattern Recognition, pp. 4968–4978 (2020)

Image-to-Graph Convolutional Network for Deformable Shape Reconstruction from a Single Projection Image

Megumi Nakao[1]([✉])[iD], Fei Tong[1], Mitsuhiro Nakamura[2][iD],
and Tetsuya Matsuda[1][iD]

[1] Graduate School of Informatics, Kyoto University, Kyoto, Japan
megumi@i.kyoto-u.ac.jp
[2] Graduate School of Medicine, Human Health Sciences, Kyoto University,
Kyoto, Japan

Abstract. Shape reconstruction of deformable organs from two-dimensional X-ray images is a key technology for image-guided intervention. In this paper, we propose an image-to-graph convolutional network (IGCN) for deformable shape reconstruction from a single-viewpoint projection image. The IGCN learns relationship between shape/deformation variability and the deep image features based on a deformation mapping scheme. In experiments targeted to the respiratory motion of abdominal organs, we confirmed the proposed framework with a regularized loss function can reconstruct liver shapes from a single digitally reconstructed radiograph with a mean distance error of 3.6 mm.

Keywords: Graph convolutional network · Shape reconstruction · Respiratory motion · X-ray image

1 Introduction

Three-dimensional (3D) medical imaging is widely used for diagnosis and pre-treatment planning. However, organs can move or deform during surgery or radiotherapy, preventing accurate tumor localization and making precise treatment difficult. Treatment is performed using only two-dimensional (2D) images, such as endoscopic images or X-ray images, because of the limitations of available imaging devices. In this paper, we focus on 3D shape reconstruction from a single projection image, for image-guided interventions.

In the last decade, shape reconstruction for image-guided interventions has been investigated in two areas of research: 2D/3D registration for rigid bodies [1–3] and deformable modeling of soft organs [4,5]. In contrast to rigid registration, deformable registration has to handle point-to-point correspondence between large numbers of voxels, making the application of deep learning difficult. For

Electronic supplementary material The online version of this chapter (https://doi.org/10.1007/978-3-030-87202-1_25) contains supplementary material, which is available to authorized users.

© Springer Nature Switzerland AG 2021
M. de Bruijne et al. (Eds.): MICCAI 2021, LNCS 12904, pp. 259–268, 2021.
https://doi.org/10.1007/978-3-030-87202-1_25

this issue, statistical modeling [3,6,7] based on curved manifolds or point-sampled mesh representations is a clinically important, computationally efficient approach for estimating the shapes of deformable organs. Wu et al. proposed a 3D shape reconstruction method based on a convolutional neural network (CNN) [5], and showed that the 3D shape of the lungs during a pneumothorax deformation can be reconstructed from only a single-viewpoint image. However, because the shape was represented as point clouds, surface information and topological information about the relationships between vertices, which are important for deformation field computation, were lost. Wang et al. proposed a CNN-based framework to calculate lung respiratory deformation from a digitally reconstructed radiograph (DRR) [8]. However, 3D shapes were artificially generated from multiple initial 3D templates with free-form deformation. Hence, the CNN-based reconstruction of organ shape for real patients has not yet been achieved or investigated except for our previous study [9].

This paper introduces an image-to-graph convolutional network (IGCN) for deformable shape reconstruction from a single-viewpoint projection image. Specifically, the abdominal organs are hard to detect in low-contrast X-ray images, which contain considerable shape variability between patients. A 2D deformation mapping, an improved feature learning scheme is introduced to learn the relationship between 3D shape/deformation and the deep image features, while preserving the global shape using a regularized loss function.

This study is the first to demonstrate the prediction performance of 2D/3D shape reconstruction targeted to the abdominal organs of real patients. We applied the IGCN to this task by using organ meshes, with point-to-point local correspondence obtained by deformable mesh registration (DMR) [10]. Because the IGCN can generate the shape of a 3D organ in real time, for example, in radiotherapy, it can be used to estimate the area of organs at risk from only X-ray images or perform tumor localization despite respiratory deformation.

2 Methods

2.1 Dataset and Problem Definition

3D-CT volumes of 124 cases and 4D-CT volumes of 35 cases, acquired from different patients who underwent intensity-modulated radiotherapy in Kyoto University Hospital, were collected. Each 4D-CT volume used in this study consists of two time phases (the end-inhalation and end-exhalation phases) of 3D-CT volumes. Each 3D-CT volume consists of 512×512 pixels and 88–152 slices (voxel resolution: 1.0 mm $\times$ 1.0 mm $\times$ 2.5 mm). During routine clinical procedures, the regions of the entire body, stomach, liver, duodenum, left and right kidneys, and the clinical target volume (CTV) were labeled by board-certified radiation oncologists, as shown in Fig. 1(a).

We generated the surface meshes (400–500 vertices and 796–996 triangles for an organ) from the region labels, and obtained organ mesh models with point-to-point correspondence using DMR. The DMR algorithm and the registration performance for the abdominal organ shapes were published in a previous paper [10],

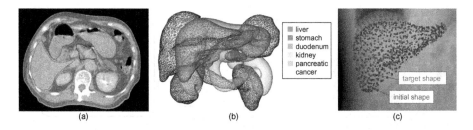

Fig. 1. Abdominal CT dataset and preprocessing. (a) CT slice image with 3D contours of the liver, stomach, duodenum, and kidneys. (b) Statistically reconstructed mean shapes (translucent) and patient-specific models (mesh). (c) DRR of the target state combined with the projected vertices of the initial and target state.

and it was confirmed that a template mesh was registered to patient-specific organ shapes with a 0.2 mm mean distance error, and 1.1 mm Hausdorff distance error, on average. Because the obtained models have point-to-point correspondence, the average shape can be obtained by calculating the average of each coordinate. Figure 1(b) shows one example of the registered organ models of one patient (mesh) and mean shapes (translucent) computed from all registered models.

Figure 1(c) is a part of DRR with 640 × 640 pixels generated from a 3D-CT volume of the target state, overlaid with the projected vertices of the initial and target shape. The patient's body is fixed to the reference position so that the CTV in the initial state corresponds to the focus of radiation beams. Therefore, we assume that the camera parameters (i.e., the projection matrix) for generating the DRR are given, and the organ models can be projected to the DRR using the CTV center in the end-inhalation state as the origin. The shape of the diaphragm visualized in the DRR does not match the projected initial shape because the two states are the most distant in terms of the respiratory phase. The respiratory motion contains nonlinear deformation with local rotation and sliding motion [10,11], and simple linear transformation is not sufficient to register these two states.

The IGCN is designed as a generalized, organ-independent framework. Although the reconstruction performance for all organs could be investigated, we first targeted liver shapes with partly detectable features (i.e., the diaphragm), to discuss how reconstruction error occurs locally in the cases with and without visual cues. The shape and location of the liver in the end-exhalation phase is the prediction target, and the shape in the end-inhalation phase is the initial patient anatomy used as the input data. This problem definition is important for investigating the prediction performance in 2D/3D deformable organ reconstruction because nonlinear deformation between the two states [10] can cause the maximum prediction error in one respiratory cycle.

Because 4D-CT volumes of 35 cases are not sufficient for learning the relationship between shape/deformation variability and the 2D projection images, we created an augmented training dataset from the 3D-CT images of another 124 cases. The 4D-CT volume dataset was divided into a training set (20 cases) and

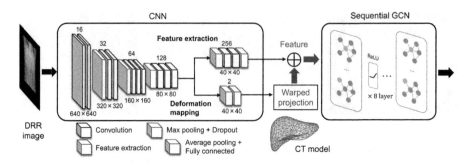

Fig. 2. The full image-to-graph convolutional network. The CNN is used for extracting image features from a single projection image, and the GCN is for learning 3D deformation. $\oplus$ represents the concatenation of image features and vertex coordinates.

test set (15 cases). The mean and standard deviation of the vertex displacement were obtained from the training data, and a similar global translation, generated using Gaussian noise, was applied to all the 3D-CT liver models. A total of 144 datasets (20 4D-CT volumes and 124 volumes from the augmented 3D-CT data) were used to train the IGCN network.

2.2   Image-to-Graph Convolutional Network

Figure 2 shows the IGCN, which consists of a CNN that extracts perceptual features from the input image and a graph convolutional network (GCN) that learns mesh deformation according to the extracted image features. In this framework, the initial model is projected onto the input DRR image. The image features corresponding to each vertex v_i can be extracted. The image features and vertex coordinates are concatenated and incorporated into the GCN for learning deformation. Both networks are optimized simultaneously using a loss function. We used an extended VGG-16 model [14] without pretraining. The width (number of channels) of each CNN layer is marked above the layers, and the size of each layer is marked below.

Pixel2Mesh (P2M) employs a similar network architecture with hierarchical extension to fit an ellipsoid template to a variety of 3D objects [13]. However, it concentrates on mesh deformation and does not consider movement of the target object, meaning that image features distant from the initial template are not learned. This is inadequate for our application because our prediction target contains both local deformation and global translation. In addition, the X-ray images have no clear edges in most parts of the organs. To address these complications, we introduce an improved feature learning scheme using 2D deformation mapping and use a regularized loss function to predict respiratory-associated deformation accurately.

Because the patient's initial liver shape M_I is determined from 3D-CT images in our framework, M_I is used as the initial template for the IGCN. With the projection matrix, reflecting the camera parameters, each vertex v_i is projected

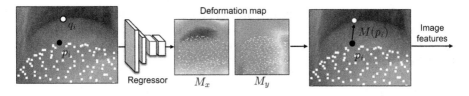

Fig. 3. The image feature learning scheme using 2D deformation mapping for distant target. A part of CNN layer is used to estimate the deformation map over all vertices.

to the corresponding 2D pixel coordinate p_i in the DRR image space (see Fig. 3). P2M uses the projected point p_i of the initial mesh template to capture image features; however, the corresponding features are distant from p_i because of the displacement of the target organ. P2M relies on convolution to capture distant image features but convolution is not effective for capturing high-resolution features at the ideal, corresponding position q_i.

We introduce a deformation mapping, a new concept to overcome the above limitation. The projection point is mapped to a new position $M(p_i)$ (called warped projection in Fig. 2), where a higher probability of obtaining effective image features is expected. The deformation map is a learnable spatial mapping function determined by the 2D vector field (M_x, M_y). The color map in Fig. 3 represents the mapping function M, describing the learned displacement in the x and y directions in 2D-projection image coordinates. This scheme was implemented as an extension to the feature extraction scheme of the CNN part.

For the GCN layers that generate the predicted 3D organ shapes, graph convolution is applied to obtain hierarchical topological features in non-Euclidean space [12]. The mesh is a type of graph $G(\mathcal{V}, \mathcal{E})$, where $\mathcal{V}$ is the set of vertices and $\mathcal{E}$ is the set of edges. Per-vertex features are shared with neighbor vertices. The GCN in our study consists of eight sequential graph convolutional layers, each of which is defined in Eq. (1).

$$X^{(l+1)} = \sigma(\hat{D}^{-\frac{1}{2}}\hat{A}\hat{D}^{-\frac{1}{2}}X^{(l)}W^{(l)}), \tag{1}$$

where $X^{(l)}$ and $X^{(l+1)}$ denote the feature matrix before and after convolution. $A \in \mathbb{R}^{n \times n}$ (where n is the number of vertices) is the adjacency matrix: a symmetric matrix with binary values, in which element A_{ij} is 1 if there is an edge between v_i and v_j, or 0 if the two vertices are not connected. $D \in \mathbb{R}^{n \times n}$ is the degree matrix: a diagonal matrix, in which each element A_{ii} represents the number of edges connected to v_i. W is the learnable parameter matrix and the feature $X^{(l)}$ is the concatenation of 2D image features from the CNN and 3D vertex coordinates. The initial shape is deformed by updating $X^{(l)}$.

2.3   Loss Functions

We introduce three loss functions to constrain mesh deformation and projection point registration. Unlike the P2M framework [13], which is intended for generating target surfaces without corresponding vertices, in our framework, the ground-truth positions of the target models are obtained from the deformable registration process. For a strict evaluation of point-to-point correspondence, we define the mean distance loss $\mathcal{L}_{pos}$ of vertex positions between the estimated shape and the ground truth. $\mathcal{L}_{pos}$ is defined by

$$\mathcal{L}_{pos} = \frac{1}{n} \sum_{i=1}^{n} \|v_i - \hat{v}_i\|_2^2, \tag{2}$$

where n is the number of vertices, $v_i \in \mathcal{V}(i = 1, 2, ..., n)$ is the target 3D position, and $\hat{v}_i$ is the predicted position. This loss function induces the convergence of the estimated vertex to the correct position.

In addition to evaluating the similarity of 3D surfaces, we found that matching the projected 2D points improves feature learning and accuracy of 2D/3D reconstruction results. Specifically, stable learning of the deformation map is important when the target contains both translation and local deformation. We introduce the mapping loss $\mathcal{L}_{map}$, based on the projected points.

$$\mathcal{L}_{map} = \frac{1}{n} \sum_{i=1}^{n} \|q_i - M(p_i)\|_2^2, \tag{3}$$

where p_i is the projected point of the initial shape, and q_i is the projected point that corresponds to the target vertex v_i. M is the mapping function trained by the CNN. The organ deformation is expected to remain within a limited range in our problem setting. To preserve the curvature and smoothness of the initial surface, we use a regularization term that evaluates a discrete Laplacian of the mesh. The Laplacian loss $\mathcal{L}_{laplacian}$ is defined as follows.

$$\mathcal{L}_{laplacian} = \frac{1}{n} \sum_{i=0}^{n} \|L(v_i) - L(\hat{v}_i)\|_2^2, \tag{4}$$

where $L(\cdot)$ is the Laplace–Beltrami operator and $L(v_i)$ is the discrete Laplacian of the vertex v_i defined by $L(v_i) = \sum_{j \in N(v_i)} (v_i - v_j)/N(v_i)$. $N(v_i)$ is the number of adjacent vertices v_j of the 1-ring connected by the vertex v_i. This loss constrains the shape changes from the initial state and avoids the generation of unexpected surface noise and low-quality meshes.

The values of loss functions are normalized using the maximum values in each coordinate. The total loss is the weighted sum of three loss functions:

$$\mathcal{L}_{total} = \mathcal{L}_{pos} + \lambda_{map}\mathcal{L}_{map} + \lambda_{laplacian}\mathcal{L}_{laplacian}. \tag{5}$$

To facilitate feature learning using deformation mapping, we used 10.0 for λ_{map} and 1.0 for $\lambda_{laplacian}$ after examination of several parameter sets.

Table 1. Quantitative comparison of the shape reconstruction performance. The mean±standard deviation of MD, RSME, and DSC for predicted and target shapes.

	Initial	P2M	IGCN (no mapping)	IGCN
MD [mm]	5.7 ± 2.9	5.1 ± 1.5	3.9 ± 0.7	3.6 ± 1.2
RMSE [mm]	12.1 ± 5.3	9.9 ± 5.1	9.3 ± 1.8	8.4 ± 2.2
DSC [%]	84.2 ± 7.6	86.8 ± 4.0	89.4 ± 2.0	91.5 ± 3.4

3 Experiments

In the experiments, the performance of 2D/3D shape reconstruction of the liver was confirmed while comparing it with the results from the existing end-to-end deep learning framework. The whole network was implemented with Python 3.6.8, TFLearn, and the TensorFlow GPU library. The network was trained using an Adam optimizer with a learning rate of 1×10^{-4}. The batch size was 1, and the total number of training epochs was 1000. 0.5 was used for the dropout rate. The training took 4.5 h on a single NVIDIA GeForce RTX 2070.

The implemented IGCN framework can provide the 3D meshes from the initial shape and the DRR in the target state. In this study, the mean distance (MD) between surfaces [6], the root mean square error (RMSE) between corresponding vertices, and the Dice similarity coefficient (DSC) were used as the shape similarity metrics. MD is the mean value of the shortest bidirectional point-to-surface distance, and DSC measures the volume overlap between the deformed meshes and the ground-truth meshes. We compared the performance of the proposed IGCN framework, with and without 2D deformation mapping, and P2M [13]. It should be noted that the correct position of each vertex was obtained from registered models. Hence, in this comparison, we replaced the Chamfer loss (used by P2M) with the mean distance loss defined by Eq. (2), leaving the remaining losses in P2M unchanged. Hierarchical learning was not used to match the end-to-end prediction process of the methods.

The evaluation results for the 15 test cases are listed in Table 1. The results show that MD, RMSE, and DSC for the IGCN framework are significantly smaller than those obtained with the P2M (oneway analysis of variance, ANOVA; $p < 0.05$ significance level). Because these two metrics do not reflect the smoothness or estimated shape quality, we visualize the estimated shape in Fig. 4, to better assess these aspects of performance. The 3D shape was projected onto the DRR, with the pixels corresponding to each 3D vertex colored in red, green, and blue, corresponding to initial, target, and predicted, respectively. Although the projection of the initial shape contains considerable errors for radiotherapy, the estimated 3D shape has been improved, particularly the diaphragm part, where an obvious contour can be observed. The overlap between the predicted shape and the target shape is shown in Fig. 4(d) and (e). Despite that only very low-contrast textures are confirmed in most parts of the liver, the deformation can be spatially reconstructed. The graph convolutions embed per-vertex features

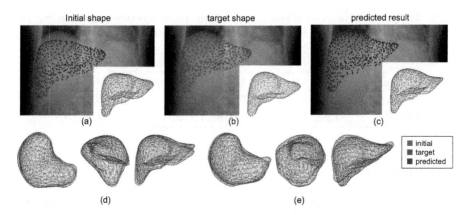

Fig. 4. 3D shape reconstruction results. (a) Initial liver shape and its projection to target DRR image. (b) Target shape. (c) Predicted liver shape with the average deformation error. (d) Deformation error between target and estimation result. (e) Estimation result with the largest deformation error.

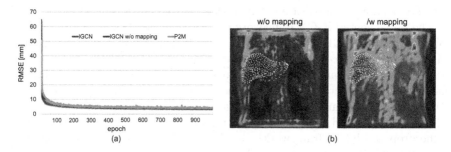

Fig. 5. Training results. (a) Learning curves showing RMSE trained in 1000 iterations. (b) Averaged output features of the 128-channel layer learned with and without deformation mapping. The projected points of the initial shape are overlaid for reference.

with connected neighbors, which results in better estimation performance of the regions with no visual cues. We measured the computation time for the whole shape reconstruction process performed in the CNN and GCN layers. The average computation time was 35.4 ms (28 frames per second), demonstrating the real-time performance of the IGCN.

Figure 5(a) demonstrates the training curves of the three methods for RMSE of training datasets. Each model closely followed and converged before 1000 iterations, with IGCN converging fastest and P2M showing a slightly unstable curve. On the other hand, the deep features learned from DRR are different, and it was confirmed that low-contrast textures such as the spine and vessel structures were extracted in addition to the diaphragm, as shown in Fig. 5(b). We tried other conditions and settings, but the performance of P2M was not improved. This is because P2M does not use dropout, and there is a possibility of overfitting by the number of training data against the number of layers of CNN.

4   Conclusion

This paper proposed IGCN that combines a GCN with a CNN, to reconstruct the 3D shape of organs from low-contrast, 2D projection images. To achieve stable and accurate shape reconstruction, we introduced an improved feature learning scheme using deformation mapping and a newly designed loss function. Our future work includes the performance analysis of our method on other abdominal organs.

Acknowledgments. This research was supported by a JSPS Grant-in-Aid for Scientific Research (B) (Grant number 18H02766a and 19H04484). We thank Edanz Group (https://en-author-services.edanz.com/ac) for editing a draft of this manuscript.

References

1. Miao, S., Wang, Z.J., Liao, R.: A CNN regression approach for real-time 2D/3D registration. IEEE Trans. Med. Imaging **35**(5), 1352–1363 (2016)
2. Markelj, P., Tomaževič, D., Likar, B., Pernuš, F.: A review of 3D/2D registration methods for image-guided interventions. Med. Image Anal. **16**(3), 642–661 (2012)
3. Reyneke, C.J.F., Lüthi, M., Burdin, V., Douglas, T., Vetter, T., Mutsvangwa, T.: Review of 2-D/3-D reconstruction using statistical shape and intensity models and X-Ray image synthesis: toward a unified framework. IEEE Rev. Biomed. Eng. **12**, 269–286 (2019)
4. Koo, B., Özgür, E., Le Roy, B., Buc, E., Bartoli, A.: Deformable registration of a preoperative 3d liver volume to a laparoscopy image using contour and shading cues. In: Descoteaux, M., Maier-Hein, L., Franz, A., Jannin, P., Collins, D.L., Duchesne, S. (eds.) MICCAI 2017. LNCS, vol. 10433, pp. 326–334. Springer, Cham (2017). https://doi.org/10.1007/978-3-319-66182-7_38
5. Wu, S., Nakao, M., Tokuno, J., Chen-Yoshikawa, T., Matsuda, T.: Reconstructing 3D lung shape from a single 2D image during the deaeration deformation process using model-based data augmentation. In: IEEE International Conferences on Biomedical and Health Informatics (BHI), pp. 1–4 (2019)
6. Rigaud, B., Simon, A., Gobeli, M., Leseur, J., Duverge, L., Williaume, D., et al.: Statistical shape model to generate a planning library for cervical adaptive radiotherapy. IEEE Trans. Med. Imag. **38**(2), 406–416 (2019)
7. Nakamura, M., Nakao, M., Mukumoto, N., Ashida, R., Hirashima, H., Yoshimura, M., et al.: Statistical shape model-based planning organ-at-risk volume: application to pancreatic cancer patients. Phys. Med. Biol. **66**, 014001 (2021)
8. Wang, Y., Zhong, Z., Hua, J.: DeepOrganNet: on-the-fly reconstruction and visualization of 3D / 4D lung models from single-view projections by deep deformation network. IEEE Trans. Visual. Comput. Graph. **26**(1), 960–970 (2020)
9. Tong, F., Nakao, M., Wu, S., Nakamura, M., Matsuda, T.: X-ray2Shape: reconstruction of 3D liver shape from a single 2D projection image. In: Proceedings of the IEEE Engineering in Medicine & Biology Society (EMBC), pp. 1608–1611 (2020)
10. Nakao, M., Nakamura, M., Mizowaki, T., Matsuda, T.: Statistical deformation reconstruction using multi-organ shape features for pancreatic cancer localization. Med. Image Anal. **67**, 101829 (2021)

11. Jud, C., Giger, A., Sandkühler, R., Cattin. P.C.: A localized statistical motion model as a reproducing kernel for non-rigid image registration. In: Medical Image Computing and Computer-Assisted Intervention (MICCAI), pp. 261–269 (2017)

12. Kipf, T.N., Welling, M.: Semi-supervised classification with graph convolutional networks. In: Proceedings of the 5th International Conference on Learning Representations (ICLR) (2017)

13. Wang, N., Zhang, Y., Li, Z., Fu, Y., Yu, H., Liu, W., et al.: Pixel2Mesh: 3D mesh model generation via image guided deformation. In: IEEE Transactions on Pattern Analysis and Machine Intelligence (2020)

14. Simonyan, K., Zisserman, A.: Very deep convolutional networks for large-scale image recognition. arXiv, Art no. 1409.1556 (2014)

Class-Incremental Domain Adaptation
with Smoothing and Calibration
for Surgical Report Generation

Mengya Xu[1,2,5], Mobarakol Islam[3], Chwee Ming Lim[4],
and Hongliang Ren[1,2,5(✉)]

[1] Department of Biomedical Engineering, National University of Singapore,
Singapore, Singapore
`mengya@u.nus.edu, hlren@ieee.org`
[2] NUSRI Suzhou, Suzhou, China
[3] Department of Computing, Imperial College London, London, UK
`mobarakol@u.nus.edu`
[4] Department of Otolaryngology-Head and Neck Surgery, Singapore General
Hospital, Singapore, Singapore
[5] Department of Electronic Engineering, The Chinese University of Hong Kong,
Shatin, Hong Kong

Abstract. Generating surgical reports aimed at surgical scene understanding in robot-assisted surgery can contribute to documenting entry
tasks and post-operative analysis. Despite the impressive outcome, the
deep learning model degrades the performance when applied to different
domains encountering domain shifts. In addition, there are new instruments and variations in surgical tissues appeared in robotic surgery.
In this work, we propose class-incremental domain adaptation (CIDA)
with a multi-layer transformer-based model to tackle the new classes
and domain shift in the target domain to generate surgical reports during robotic surgery. To adapt incremental classes and extract domain
invariant features, a class-incremental (CI) learning method with supervised contrastive (SupCon) loss is incorporated with a feature extractor. To generate caption from the extracted feature, curriculum by one-
dimensional gaussian smoothing (CBS) is integrated with a multi-layer
transformer-based caption prediction model. CBS smoothes the features
embedding using anti-aliasing and helps the model to learn domain
invariant features. We also adopt label smoothing (LS) to calibrate prediction probability and obtain better feature representation with both
feature extractor and captioning model. The proposed techniques are
empirically evaluated by using the datasets of two surgical domains, such
as nephrectomy operations and transoral robotic surgery. We observe
that domain invariant feature learning and the well-calibrated network
improves the surgical report generation performance in both source and
target domain under domain shift and unseen classes in the manners of
one-shot and few-shot learning. The code is publicly available at https://
github.com/XuMengyaAmy/CIDACaptioning.

M. Xu and M. Islam—Equal contributions.

M. de Bruijne et al. (Eds.): MICCAI 2021, LNCS 12904, pp. 269–278, 2021.
https://doi.org/10.1007/978-3-030-87202-1_26

1   Introduction

Automatically generating the description for a surgical procedure can free the surgeons from the low-value document entry task, allow them to devote their time to patient-centric tasks, and do post-operative analysis. Image captioning model cannot generalize well to target domain (TD) since existing domain adaptation (DA) approaches such as consistency learning [21], hard-soft DA [25] are utilized to solve the domain shift problem assuming that the source domain (SD) and the TD share the same class set. However, this assumption is impractical in a surgical environment where the TD often includes novel instrument classes and surgical regions that do not appear in the SD.

Class-Incremental (CI) learning methods can learn new instruments absent from SD but will fail if there is a domain shift in robotic surgery [4,14]. Cross-Entropy (CE) loss is sensitive to adversarial samples and leads to poor results if the inputs differ from the training data even a bit [9]. To overcome these issues, supervised contrastive (SupCon) learning [13] applies extensive augmentation and maximizes the mutual information for different views. In this work, we incorporate SupCon with CI learning for novel TD instrument classes under surgical domain shift.

Most recently, transformer based models are showing state-of-the-art performance in the task of classification [7], medical image segmentation [5] and caption generation [6]. The mesh-memory transformer [6] (M^2 transformer) forms of multi-layer encoder-decoder and learn to describe object interaction using extracted features from the image. Though the model shows excellent performance in caption prediction, it is unable to deal with domain shift. There is evidence that curriculum by smoothing (CBS) [22] can learn domain invariant features by applying anti-aliasing. Studies have also shown that feature representation can be improved by a well-calibrated model [12] and label smoothing (LS) enhances the model calibration by limiting network from over-confidence prediction [16]. In this work, we design class-incremental domain adaption (CIDA) with CI learning and SupCon for novel class adaptation and domain invariant feature extraction. To deal with domain shift and network calibration in the caption generation model, we develop a one-dimensional (1D) CBS and incorporate it with LS for M^2 transformer.

Our contributions can be summarized as the following points: (1) Propose CIDA in feature extractor to tackle novel TD classes under domain shift; (2) Design 1D CBS and integrate it with transformer based captioning model to learn domain invariant features and adapt to the new domain for generating the surgical report in robotic surgery; (3) Investigate model calibration with LS for both feature extraction and captioning and observe the effect of well-calibrated model into feature representation and the one/few shot DA; (4) Annotate robot-assisted surgical datasets with proper captions to generate the surgical report for MICCAI robotic instrument segmentation challenge and Transoral robotic surgery (TORS) dataset.

2   Proposed Method

Overall caption generation pipeline consists of two stages of networks, as shown in Fig. 1. The first stage contains a feature extraction model with CIDA and a transformer-based caption model with 1D CBS and LS is designed in the second stage.

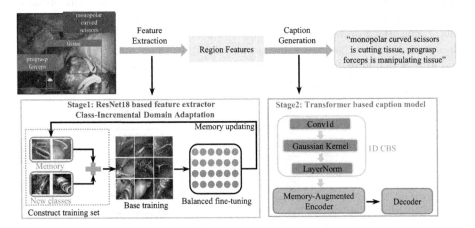

Fig. 1. Overall workflow. The input image is sent into the ResNet18 based feature extractor augmented with CIDA, and output region features. Inside the transformer-based caption model with 1D CBS, the encoder takes in the region features and understands the relationship between regions. The decoder receives the encoder's output and generates the medical report.

2.1   Preliminaries

Label Smoothing (LS), where a model is trained on a smoothened version of the true label, $T_{LS} = T(1 - \epsilon) + \epsilon/K$, with CE loss, shows great effectiveness in improving model calibration [16]. The CE loss with LS can be formulate as $CE_{LS} = -\sum_{k=1}^{K} T_{LS} log(P)$ where T is true label, ϵ is smoothing factor, K is the number of all classes and P is predicted probability.

2.2   Feature Extraction

ResNet18 [10] is employed as the feature extractor. CI learning proposed in [4] aims to handle continually added new classes. However, it is unsuitable to deal with domain shift [14] due to the sensitivity of CE loss to training data.

Class-Incremental Learning. The feature representation and classifier in the feature extractor are updated jointly by minimizing the weighted sum of two loss functions: CE loss $L_{CE} = -\sum_{k=1}^{K} T log(P)$ to learn the new classes and the distillation loss $L_{DT} = -\sum_{k=1}^{K} T_{dist} log(P_{dist})$ to preserve the learned knowledge from the old classes. T_{dist} and P_{dist} are got by dividing T and P by the distillation parameter D [11]. D is set to 3 for our experiments.

Class-Incremental Learning with SupCon. We decouple the classifier and the representation learning in the feature extraction model and propose a novel CIDA with SupCon loss [13] to enable the identification of both shared and new classes in the presence of domain shift. The CIDA framework is composed of four steps, as shown in Fig. 1. The first step is to construct the training data contains the data from new classes and data from old classes saved in the memory. The second step is the training process which aims to fit a model based on the training data. In the third step, the model is fine-tuned with the balanced subset, consisting of data from memory and partial data from new classes. Each class in the subset has the same number of data. The final step is memory updating which aims to add some data from the new classes into the memory.

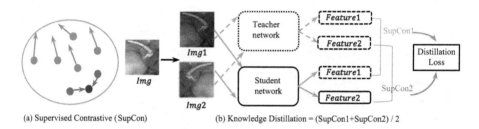

(a) Supervised Contrastive (SupCon)   (b) Knowledge Distillation = (SupCon1+SupCon2) / 2

Fig. 2. Class-Incremental learning with SupCon. (a) explains the idea of SupCon. The red dot represents the original image of one class, blue dots are augmented versions of the same image (Positive images), and the green dots are all the other images in the dataset (Negative images). Positive images are pulled close together and negative images are pushed apart. (b) shows the designed knowledge distillation loss which further upgrades CI method to CIDA method. The student network is the initial copy of teacher network.

The feature extractor is trained by minimizing the CI loss consists of SupCon loss [13] and a novel distillation loss, as shown in Fig. 2. The SupCon loss function can be formulated as $\mathcal{L}^{sup} = \sum_{i=1}^{2N} \mathcal{L}_i^{sup} = \sum_{i=1}^{2N} (\frac{-1}{2N_{\tilde{y}_i}-1} \sum_{j=1}^{2N} \mathbb{1}_{i\neq j} \cdot \mathbb{1}_{\tilde{y}_i=\tilde{y}_j} \cdot \log \frac{\exp(z_i \cdot z_j/\tau)}{\sum_{k=1}^{2N} \mathbb{1}_{i\neq k} \cdot \exp(z_i \cdot z_k/\tau)})$ which means samples z_i and z_j which have the same label $\tilde{y}_i = \tilde{y}_j$ should be maximized in the inner product and pulled together while everything else should be pushed apart in feature representation space [13]. Thus the network learns about these random transformations and possesses the potential to handle domain shift.

We design a novel distillation loss for CIDA, which can be constructed using the equation below:

$$\mathcal{L}_{DT} = \frac{L^{sup}(Timg1 - Simg1) + L^{sup}(Timg2 - Simg2)}{2} \tag{1}$$

where $Timg1$ represents the feature of Image1 output from the teacher network and $Simg1$ represents the feature of Image1 come from the student network.

Domain Adaptation with CBS. The σ, which is the standard deviation of the gaussian kernel, controls the degree of output being blurred after a convolution operation, and increasing σ will lead to a greater amount of blur. We implement CBS by annealing σ, which gradually reduces the amount of blur and allows the model to learn from the incrementally increased information in the feature map. It is also difficult for the model to learn the good representation from the feature maps generated by untrained parameters since these feature maps contain a high amount of aliasing information. Such information can be smoothed out by using a gaussian kernel.

2.3 Captioning Model

A transformer-based multi-layer encoder-decoder [6] network is used for captioning surgical images. The encoder builds of memory augmented self-attention layer and fully-connected layer. The decoder consists of self-attention on words and cross attention over the outputs of all the encoder layers, similar to [6]. There are three encoder and decoder blocks stack to encode input features and predict the word class label. The encoder module takes features of the regions from images as input and understands relationships between regions. The decoder reads each encoding layer's output to model each word's probability in the vocabulary and generate the output sentence.

Feature Representation with 1D CBS. A 1D Gaussian Kernel can be formulated as $g(x) = \frac{1}{\sqrt{2\pi}\sigma}e^{-\frac{x^2}{2\sigma^2}}$, where $g(x)$ represents the x spatial dimensions in the kernel. We propose to augment M^2 Transformer [6] using our designed 1D Gaussian Kernel layer equipped with curriculum learning by annealing the σ values as training progresses.

$$y_i = \text{ReLU}\left(\text{LayerNorm}\left(\theta_{G_\sigma} \circledast (\theta_w \circledast x_i)\right)\right) \tag{2}$$

where x_i is the input region features, θ_w are the learned weights of the 1D convolutional kernel, θ_{G_σ} is a Gaussian Kernel whose standard deviation is σ, ReLU is a non-linearity [17], y_i is the output of the layer. A single 1D CBS layer is added at the beginning of the encoder (as shown in Fig. 1). By applying the blur to the output of a 1D convolutional layer, output features are smoothed, and high-frequency information is reduced.

3 Experiments

3.1 Dataset

Instrument Segmentation Challenge. The SD dataset is from the MICCAI robotic instrument segmentation dataset of endoscopic vision challenge 2018 [1]. The training set includes 15 robotic nephrectomy operations obtained by the da Vinci X or Xi system. We used 14 sequences out of 15 by ignoring the 13th sequence due to the less interaction. 9 objects (instruments and surgical region) appear in the dataset. These instruments have 11 kinds of interaction with the tissue. The captions are annotated by an experienced surgeon in robotic surgery (as shown in Fig. 3). The 1st, 5th, 16th sequences are chosen for validation, and the 11 remaining sequences are selected for training following the work [12,24]. The splitting strategy ensures that most interactions are presented in both sets.

TORS Dataset. The TD dataset used for DA is from 11 patients' surgical videos about transoral robotic surgery (TORS) provided by hospitals. The average time length of the video with rich interactions is the 50s. A total of 5 objects and 6 kinds of semantic relationships appear in the dataset. TORS is used for TD experiments where it appears new instruments with different tissues and different surgical backgrounds (Fig. 3). The TD dataset is further expanded with frames extracted from one youtube surgical procedure video[1] which is about robotic nephroureterectomy with DaVinci Xi.

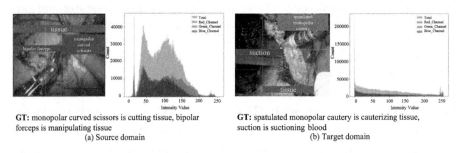

GT: monopolar curved scissors is cutting tissue, bipolar forceps is manipulating tissue
(a) Source domain

GT: spatulated monopolar cautery is cauterizing tissue, suction is suctioning blood
(b) Target domain

Fig. 3. Dataset visualization. The frames, captions, and histograms of SD and TD are shown. The histogram can prove the existence of the domain shift.

3.2   Implementation Details

The feature extractor is trained using stochastic gradient descent with a batch size of 20, weight decay of 0.0001, and momentum of 0.6. We perform 50 and 15 epochs for training and balanced fine-tuning in every incremental step. The learning rate (lr) starts at 0.001 and decays with the factor of 0.8 every 5 epochs. The same lr reduction is also used for fine-tuning except that the initial value is 0.0001. The captioning model is trained using adam optimizer with a batch size of 50 and a beam size of 5. The training epochs are set to 50. For all experiments involved CBS, we use an initial σ of 1 and decay the σ with a factor of 0.9 every 2 epochs. The whole network is implemented by PyTorch and trained in the NVIDIA RTX 2080 Ti GPU.

4   Results and Evaluation

We evaluate the model using four metrics for image captioning, namely BLEU-n [20], ROUGE [15], METEOR [3], CIDEr [23] and miscalibration with Expected

---

[1] https://youtu.be/bwpEul4KCSc

Calibration Error (ECE), Static Calibration Error (SCE), Thresholded Adaptive Calibration Error (TACE) [where, threshold $= 10^{-3}$], Brier Score (BS) [2,18].

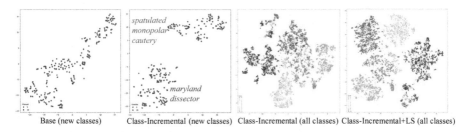

Base (new classes)   Class-Incremental (new classes)   Class-Incremental (all classes)   Class-Incremental+LS (all classes)

Fig. 4. tSNEs for two novel classes and all classes. CI learning forms good clusters for the 2 novel instruments. LS leads to tighter clusters.

We plot tSNEs for two novel classes and all classes in Fig. 4. CI learning can extract better features for novel instruments. LS can improve the feature representation in the penultimate layer of the feature extractor. In TD experiments, the first baseline is to first train on the SD and then fine-tune on the TD, and the second baseline is to train directly on the TD. All the proposed methods outperform the baseline in both SD and TD, as shown in Table 1. Class-Incremental learning with SupCon and CBS (CISC) almost obtains the best performance and performs slightly better than Class-Incremental learning with CBS and LS (CICL) in terms of caption metrics and calibration error since SupCon learning can handle the domain shift. The CI method can only deal with novel instruments and DANN can only handle domain shift. SupCon upgrade the CI method to CIDA method which can handle domain shift and novel classes simultaneously, as shown in Table 2. Meanwhile, CBS plays an auxiliary role since it has been proven to achieve better feature extraction [22]. Benefit from CBS, even though CICL and CI learning all use CE loss, CICL still performs better than the CI learning method.

GT: monopolar curved scissors is cutting tissue, prograsp forceps is manipulating tissue
Base: bipolar forceps and prograsp forceps and monopolar curved scissors are idle
Ours: monopolar curved scissors is cutting tissue, bipolar forceps is retracting tissue

Source domain

GT: suction is suctioning blood, spatulated monopolar cautery and maryland dissector are manipulating tissue
Base: maryland dissector is grasping tissue
Ours: spatulated monopolar cautery is cauterizing tissue, suction is suctioning blood

Target domain

Fig. 5. The predicted caption of our CIDA method for source and target domain

Table 3 investigates the results of applying 1D CBS on the caption prediction model. We observe that 1D CBS enhances the performance in the TD while no

Table 1. Evaluation metrics of the proposed models in SD and TD. The meaning of the abbreviations are: Class-Incremental (CI), Class-Incremental+CBS+LS (CICL), Class-Incremental+SupCon+CBS (CISC), CBS+LS (CL), ResNet18 (Res), M^2 transformer (M2T), Domain Adversarial Neural Network (DANN)

Domain	Stage 1 feature extractor	Stage 2 caption model	Metric						
			BLEU-1↑	BLEU-2↑	BLEU-3↑	BLEU-4↑	METEOR↑	ROUGE↑	CIDEr↑
SD	Res [10]	X-LAN [19]	0.5733	0.5053	0.4413	0.3885	0.3484	0.5642	2.0599
	Res [10]	M2T [6]	0.5703	0.5097	0.4572	0.4156	0.3817	0.599	2.5385
	Res [10]	M2T [6]+DANN [8]	0.5995	0.5318	0.4748	0.4301	**0.5995**	0.5994	2.4672
	Res [10]	Xu et al. [24]	0.5875	0.5190	0.4599	0.4123	0.3621	0.5982	2.5930
	Res+CI	M2T	0.5571	0.4947	0.4395	0.3932	0.3609	0.5791	2.319
	Res+CICL	M2T+CL	0.6204	0.5498	0.4923	0.4452	0.3532	0.6017	2.6524
	Res+CISC	M2T+CL	**0.6246**	**0.5624**	**0.5117**	**0.472**	0.38	**0.6294**	**2.8548**
TD one shot	Res [10]	M2T [6](direct)	0.2408	0.098	0.0319	0.	0.1051	0.2407	0.1348
	Res [10]	M2T [6](fine-tune)	**0.5678**	0.4534	0.3891	0.3305	0.2759	0.5006	1.936
	Res+CI	M2T	0.5439	0.4568	0.403	0.352	**0.2886**	**0.5279**	2.2741
	Res+CICL	M2T+CL	0.5088	0.4277	0.3758	0.3286	0.2687	0.5071	2.3617
	Res+CISC	M2T+CL	0.5626	**0.472**	**0.417**	**0.3648**	0.2857	0.5147	**2.4641**
TD few shot	Res [10]	M2T [6](direct)	0.5331	0.4567	0.4114	0.3712	0.2738	0.5348	2.7496
	Res [10]	M2T [6](fine-tune)	0.5677	0.4807	0.4285	0.3836	0.289	0.5669	2.9209
	Res [10]	M2T [6]+DANN [8]	0.6338	0.5367	0.4819	0.4321	0.3173	0.5794	3.0407
	Res [10]	Xu et al. [24]	0.6286	0.5422	0.4919	**0.4457**	0.3235	0.5921	3.3620
	Res+CI	M2T	0.6156	0.534	0.4859	0.4400	0.3189	0.5975	3.2223
	Res+CICL	M2T+CL	0.6314	0.5434	0.4912	0.4444	0.3262	0.6003	**3.3930**
	Res+CISC	M2T+CL	**0.6455**	**0.5518**	**0.4935**	0.4387	**0.328**	**0.6021**	3.3913

Table 2. Ablation study of the proposed methods

Method	SD			TD few shot		
	BLEU1↑	METEOR↑	ROUGE↑	BLEU1↑	METEOR↑	ROUGE↑
CI	0.5571	0.3609	0.5791	0.6156	0.3189	0.5973
CI+CBS	0.5704	0.3528	0.5856	0.6185	0.3119	0.5722
DANN [8]	0.5995	0.5995	0.5994	0.6338	0.3173	0.5794
CI+SupCon(our CIDA)	0.6009	0.3963	0.6317	0.6309	0.3205	0.6046

Table 3. 1D CBS. The model trained with 1D CBS is able to learn domain invariant features. Despite some sacrifices in performance on SD, the model achieves excellent performance on TD

Stage 1 feature extractor	Stage 2 caption model	SD				TD few shot			
	1D CBS	BLEU-1↑	BLEU-2↑	ROUGE↑	CIDEr↑	BLEU-1↑	BLEU-2↑	ROUGE↑	CIDEr↑
Res	✗	0.5703	0.5097	0.599	2.5385	0.5677	0.4807	0.5669	2.9209
Res+CICL	✓	0.5972	0.5334	0.5803	2.6499	0.6375	0.551	0.6071	3.4956
	✗	0.5873	0.5269	0.6067	2.9042	0.6156	0.5261	0.591	3.2187
Res+CISC	✓	0.5837	0.5264	0.6069	2.7454	0.6478	0.559	0.6072	3.2902
	✗	0.626	0.5651	0.6077	3.2374	0.5669	0.4861	0.5488	2.8877

improvements in SD. Therefore, 1D CBS helps to learn domain-invariant features. The effects of LS on model calibration and DA are shown in Table 4. LS can improve the model calibration to fix overconfident prediction and enhances the model robustness and uncertainty. The model trained with LS obtains performance improvement on SD for all the metrics but few metrics for TD. Predicted

Table 4. Model calibration. LS improves model calibration, boost the SD performance for both two approaches, and improves TD performance for the CISC approach

Stage 1 feature extractor	Stage 2 caption model		SD		TD few shot		Calibration error			
	1D CBS	LS	BLEU-1↑	CIDEr↑	BLEU-1↑	CIDEr↑	ECE↓	SCE↓	TACE↓	BS↓
Res+**CICL**	✓	✓	0.6204	2.6524	0.6314	3.393	0.1194	0.0618	0.0613	0.6299
		✗	0.5972	2.6499	0.6375	3.4956	0.1367	0.0627	0.0624	0.9773
Res+**CISC**	✓	✓	0.6246	2.8548	0.6455	3.3913	0.1385	0.0597	0.0584	0.4346
		✗	0.5837	2.7454	0.6478	3.2902	0.1431	0.0593	0.0592	0.9537

captions of the CISC approach for SD and few shot TD are shown in Fig. 5. Our model can recognize the instruments and interactions more accurately.

5   Discussion and Conclusion

We presented the class-incremental domain adaptation, which aims to handle novel target domain classes under domain shift without the need to re-train all datasets. The feature extractor and transformer-like model trained with CBS can extract the domain-invariant features, and generate the surgical report which describes the instruments-tissue interaction. We also improve the model calibration by using label smoothing. In future work, we will investigate surgical report generation by incorporating temporal information from the surgical video.

Acknowledgements. This work was supported by the Shun Hing Institute of Advanced Engineering (SHIAE project, 8115064#BME-p1-21) at the Chinese University of Hong Kong (CUHK) and Singapore Academic Research Fund under Grant R397000353114. We would like to express sincere thanks to Lalithkumar Seenivasan for his help on incremental learning of our work.

References

1. Allan, M., et al.: 2018 robotic scene segmentation challenge. arXiv preprint arXiv:2001.11190 (2020)
2. Ashukha, A., Lyzhov, A., Molchanov, D., Vetrov, D.: Pitfalls of in-domain uncertainty estimation and ensembling in deep learning. arXiv preprint arXiv:2002.06470 (2020)
3. Banerjee, S., Lavie, A.: Meteor: an automatic metric for MT evaluation with improved correlation with human judgments. In: Proceedings of the ACL Workshop on Intrinsic and Extrinsic Evaluation Measures for Machine Translation and/or Summarization, pp. 65–72 (2005)
4. Castro, F.M., Marín-Jiménez, M.J., Guil, N., Schmid, C., Alahari, K.: End-to-end incremental learning. In: Proceedings of the European Conference on Computer Vision (ECCV), pp. 233–248 (2018)
5. Chen, J., et al.: Transunet: transformers make strong encoders for medical image segmentation. arXiv preprint arXiv:2102.04306 (2021)
6. Cornia, M., Stefanini, M., Baraldi, L., Cucchiara, R.: Meshed-memory transformer for image captioning. In: Proceedings of the IEEE/CVF Conference on Computer Vision and Pattern Recognition, pp. 10578–10587 (2020)

7. Dosovitskiy, A., et al.: An image is worth 16x16 words: transformers for image recognition at scale. arXiv preprint arXiv:2010.11929 (2020)
8. Ganin, Y., et al.: Domain-adversarial training of neural networks. J. Mach. Learn. Res. **17**(1), 2096–2030 (2016)
9. Gunel, B., Du, J., Conneau, A., Stoyanov, V.: Supervised contrastive learning for pre-trained language model fine-tuning. arXiv preprint arXiv:2011.01403 (2020)
10. He, K., Zhang, X., Ren, S., Sun, J.: Deep residual learning for image recognition. In: Proceedings of the IEEE Conference on Computer Vision and Pattern Recognition, pp. 770–778 (2016)
11. Hinton, G., Vinyals, O., Dean, J.: Distilling the knowledge in a neural network. arXiv preprint arXiv:1503.02531 (2015)
12. Islam, M., Seenivasan, L., Ming, L.C., Ren, H.: Learning and reasoning with the graph structure representation in robotic surgery. In: Martel, A.L., et al. (eds.) MICCAI 2020. LNCS, vol. 12263, pp. 627–636. Springer, Cham (2020). https://doi.org/10.1007/978-3-030-59716-0_60
13. Khosla, P., et al.: Supervised contrastive learning. arXiv preprint arXiv:2004.11362 (2020)
14. Kundu, J.N., Venkatesh, R.M., Venkat, N., Revanur, A., Babu, R.V.: Class-incremental domain adaptation. arXiv preprint arXiv:2008.01389 (2020)
15. Lin, C.Y.: Rouge: a package for automatic evaluation of summaries. In: Text Summarization Branches Out, pp. 74–81 (2004)
16. Müller, R., Kornblith, S., Hinton, G.E.: When does label smoothing help? In: Advances in Neural Information Processing Systems, pp. 4694–4703 (2019)
17. Nair, V., Hinton, G.E.: Rectified linear units improve restricted Boltzmann machines. In: ICML (2010)
18. Nixon, J., Dusenberry, M.W., Zhang, L., Jerfel, G., Tran, D.: Measuring calibration in deep learning. In: CVPR Workshops, pp. 38–41 (2019)
19. Pan, Y., Yao, T., Li, Y., Mei, T.: X-linear attention networks for image captioning. In: Proceedings of the IEEE/CVF Conference on Computer Vision and Pattern Recognition, pp. 10971–10980 (2020)
20. Papineni, K., Roukos, S., Ward, T., Zhu, W.J.: Bleu: a method for automatic evaluation of machine translation. In: Proceedings of the 40th Annual Meeting of the Association for Computational Linguistics, pp. 311–318 (2002)
21. Sahu, M., Strömsdörfer, R., Mukhopadhyay, A., Zachow, S.: Endo-Sim2Real: consistency learning-based domain adaptation for instrument segmentation. In: Martel, A.L., et al. (eds.) MICCAI 2020. LNCS, vol. 12263, pp. 784–794. Springer, Cham (2020). https://doi.org/10.1007/978-3-030-59716-0_75
22. Sinha, S., Garg, A., Larochelle, H.: Curriculum by smoothing. arXiv e-prints pp. arXiv-2003 (2020)
23. Vedantam, R., Lawrence Zitnick, C., Parikh, D.: Cider: consensus-based image description evaluation. In: Proceedings of the IEEE Conference on Computer Vision and Pattern Recognition, pp. 4566–4575 (2015)
24. Xu, M., Islam, M., Lim, C.M., Ren, H.: Learning domain adaptation with model calibration for surgical report generation in robotic surgery. arXiv preprint arXiv:2103.17120 (2021)
25. Zia, A., et al.: Surgical visual domain adaptation: results from the MICCAI 2020 SurgVisDom challenge (2021)

Real-Time Rotated Convolutional Descriptor for Surgical Environments

Adam Schmidt$^{(\boxtimes)}$ and Septimiu E. Salcudean

Robotics and Control Lab, Electrical and Computer Engineering,
University of British Columbia, 2329 West Mall, Vancouver, BC V6T 1Z4, Canada
adamschmidt@ece.ubc.ca

Abstract. Many descriptors exist that are usable in real-time and tailored for indoor and outdoor tracking and mapping, with a small subset of these being learned descriptors. In order to enable the same in deformable surgical environments without ground truth data, we propose a Real-Time Rotated descriptor, ReTRo, that can be trained in a weakly-supervised manner using stereo images. We propose a novel network that creates these fast, high-quality descriptors that have the option to be binary-valued. ReTRo is the first convolutional feature descriptor to learn a sampling pattern as part of the network, in addition to being the first real-time learned descriptor for surgery. ReTRo runs on multiple scales and has a large receptive field while only requiring small patches for input, affording it great speed. We quantify ReTRo by using it for pose estimation and tissue tracking, demonstrating its efficacy and real-time speed. ReTRo outperforms classical descriptors used in surgery and it will enable surgical tracking and mapping frameworks.

Keywords: Feature descriptor · Real-time · Endoscopic surgery · Weak supervision

1 Introduction

Tissue tracking in surgery can be used for mapping in arthroscopy [21], tracking heart motion for surgical motion compensation [23], screening of colorectal cancer [9], and many other applications [15,28]. Prior work in tissue tracking depends either directly on image intensity or classical descriptors. Surgical environments have specularity, smoke, blood, and other artifacts. We propose ReTRo, the first surgical descriptor that runs in real-time and learns in non-rigid environments. ReTRo, is fast, low memory, and uses a novel network structure that is adaptive to rotation. First, a sampling pattern centred around each keypoint, like that of ORB's [26] is learned without the need for greedy optimization and ground truth. Second, similarly to classical descriptors that estimate orientation using intensity centroid, our sampling pattern is rotated by an angle, with the difference being that our network estimates this rotation angle. This network that samples, rotates, and outputs a scale-dependent descriptor is replicated at multiple scales using different weights. Our final descriptor is the concatenation

© Springer Nature Switzerland AG 2021
M. de Bruijne et al. (Eds.): MICCAI 2021, LNCS 12904, pp. 279–289, 2021.
https://doi.org/10.1007/978-3-030-87202-1_27

of these scale-dependent descriptors. ReTRo (Real-Time, and Rotated) brings classical descriptor ideas into the modern convolutional space, hence its name. We do not require ground truth and train using pose supervision, cycle consistency, and Pearson correlation loss.

ReTRo is the first feature descriptor to learn a sparse sampling pattern as a part of its network, in addition to being the first learned real-time convolutional feature descriptor for surgical environments. Finally, ReTRo is free of rigid assumptions about the environment.

We will begin with a brief survey of prior work in surgical scene tracking, and relevant work from the wider field of image descriptors, matching, and tracking. Afterwards, we introduce and quantify our novel descriptor. We conclude with timing benchmarks.

2   Related Works

In surgical applications, Richa et al. [23] track heart motion using thin-plate splines and image intensity for matching. More recently, [9] track regions using a Haar-like descriptor. In [36], Yip et al. create a tracking method for image-guided surgery using classical descriptors. SuPer [17] and the following SuPerDeep [20] build a framework to track and reconstruct a surgical scene. SuPerDeep uses learned depth maps and instrument features, but still uses classical descriptors to track tissue motion. MIS-SLAM [28] and DefSLAM [15], extend SLAM to surgical environments using ORB for image descriptors. SD-DefSLAM [24] extends DefSLAM with an illumination invariant Lucas-Kanade optical flow. Liu et al. [18] create a feature descriptor for these environments, introducing a high-quality dense offline descriptor that requires a rigid environment and SIFT keypoints for supervision.

Prior work in image descriptors can be separated into learned descriptors [22,35] and classical descriptors such as ORB [26] and SIFT [19], with more recent extensions, e.g. [16,29,32]. These require ground truth matches or data augmentation for training. Only those learned with pose as supervision or policy gradient [30,31] are able to learn without having ground truth. This is especially important in our case as the surgical environment is non-rigid. Taking inspiration from the weakly-supervised regime, we use the epipolar and cycle loss introduced in CAPS [31].

There are multiple convolutional descriptors that work in real-time. CDBin [33] trains a fast binary descriptor with a small network in a supervised environment. Superpoint and UnsuperPoint detect points and calculate descriptors in parallel [6,8], predicting one descriptor for each 8×8 patch. TFeat [2], and PN-Net [1] both learn shallow and efficient real descriptors given ground truth patches.

After the process of descriptor creation, robust matching is extremely important. GMSMatch [5] introduces a fast heuristic that achieves great results. On the learning side, SuperGlue [27] introduce a graph neural network matcher.

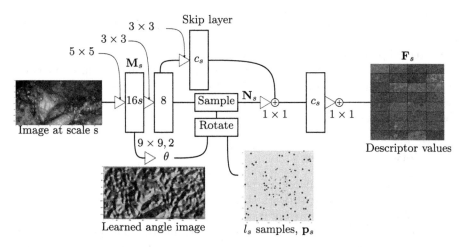

Fig. 1. Our network takes in an image and outputs a descriptor value for detected points. We train in a multi-scale manner and learn weights for three different scales. This figure shows the network running on the full scale image. This network is replicated with different weights for the half and quarter scales. The final descriptor is the concatenation of each of these scales.matching to and from the other image

Motivating ReTRo, Spatial Transformer Networks [12], and Deformable Convolutional Networks [7] use bilinear interpolation to transform images or kernels in a differentiable manner. For segmentation of images, OBELISK [11] learns a deformable sample pattern, and BRIEFNet [10] uses a fixed random pattern. For image descriptors, LIFT [35] learns an orientation, but not a sampling pattern.

3   Method

We design ReTRo with quality and speed in mind, designing a neural network that runs in a multi-scale manner, with few operations before and after sampling. ReTRo can run in either a sparse or dense manner, extracting descriptors for points detected using a keypoint detector, or for the entire image. We enable ReTRo to train in a binary or real manner as well.

ReTRo takes in an image, I, and downsamples it by each factor of $s \in (1, 2, 4)$ resulting in the downsampled I_s. Then, taking inspiration from ORB [26] and deformable neural networks [7], we learn a sparse sampling pattern that samples l_s points per scale. ORB uses an intensity centroid to estimate the rotation at which to sample its sampling pattern. We instead opt to learn the angle to sample at, directly learning to steer to the input data around a keypoint. By sparsely sampling we let the network reach far away regions with one operation, rather than through multiple filters.

To train ReTRo, we detect FAST [25] keypoints which are based on threshold comparison tests. We use these as the points at which we evaluate our model.

The ORB descriptor uses FAST as well, so we choose to do the same for comparison and speed. We train in a sparse manner, not requiring the full image after the detected points have been selected. For loss, we use epipolar loss and a cycle consistency loss between the left and right stereo images as introduced in Wang et al.'s CAPS paper [31] on camera pose supervision. CAPS introduces the concept of using pose and cycle consistency as weak signals to train a ResNet-style convolutional neural network.

3.1   Network Structure

All convolutions in our network are paired with ReLUs, except those for output and angle estimation. Our network starts with a convolutional layer to convert the image into a feature representation of channel size 16s, denoted as $\mathbf{M}_s$. This feature image is then passed through an angle estimator to estimate θ which is an angle image that is used to rotate the sampling pattern in a data-dependent way for each feature. We use a 9×9 convolution to estimate the relative centroid position, and then use $atan2$ to estimate the angle, denoted $angleconv$:

$$\theta = angleconv_s(\mathbf{M}_s). \tag{1}$$

The sampling pattern $\mathbf{n}_s$ is a $l_s \times 2$ set of offset locations learned in the network in an unnormalized form. The normalized form is:

$$\mathbf{p}_s = tanh(\mathbf{n}_s) \cdot r_{\max}, \mathbf{p}_s \in [-r_{\max}, r_{\max}]. \tag{2}$$

This allows us to limit the data requirement for a descriptor. To sample closely for the sake of speed and keeping input patch size small, we set the maximum sampling radius, r_{max}, to 8. Then, we sample using the rotated offsets of $\mathbf{p}_s$:

$$\mathbf{N}_s(z) = sample(conv_s(\mathbf{M}_s), z, rotate(\mathbf{p}_s, \theta)) \in \mathbb{R}^{8 \times l_s} \tag{3}$$

For each position z, this results in l_s samples of size 8 each. We use grid sampling to implement this in a differentiable manner: all sampled points are evaluated by bilinear interpolating the feature image. We pass these through a pointwise convolution with a ReLU to obtain a c_s-sized feature for each keypoint. We add in a skip layer that allows the network to maintain a central sample, and follow it with a final pointwise convolution as shown in Fig. 1. The final descriptor is then the concatenation of the descriptor $\mathbf{F}_s$ from each scale s. We set the channel size c_s for coarse, medium, and fine as $128, 64, 64$ respectively, resulting in a 256-dimensional descriptor.

3.2   Matching

Once we have a descriptor for each point, we have to match it with those in another image. We do this in a similar manner to how it is done in dense images: the match position is the softmax probability weighted position of all possible candidate matches. Soft-argmax is used for finding the point location of a point $\mathbf{x}$ in image I':

$$loc(\mathbf{x}, I') = \sum_{\mathbf{y} \in \mathcal{P}'} \mathbf{y} C(\mathbf{y}, \mathbf{x}) \tag{4}$$

with C as the softmax correlation between point features, and $\mathcal{P}'$ as the keypoints in I'.

3.3 Model Training

For each image we sample a mixture of 70% keypoints selected by FAST, and the rest as randomly sampled points. We use the Hamlyn dataset for stereo depth estimation introduced in [34]. We train our model with a combination of epipolar loss and cycle consistency loss for stereo pairs. Since we train only on the sparse set of keypoints we are able to avoid the cropping done in CAPS [31]. The epipolar loss, $\mathcal{L}_{epipolar}$ is based on the distance between the matched point and its ground truth epipolar line given by the fundamental matrix, F. The cycle loss, $\mathcal{L}_{cycle}$ is the distance between the original point, x, and its estimated position after matching to and from the other image.

$$\mathcal{L}_{epipolar} = \text{dist}\left(loc\left(\mathbf{x}, I_2\right), F\mathbf{x}\right) \tag{5}$$
$$\mathcal{L}_{cycle} = \|loc\left(loc\left(\mathbf{x}, I_2\right), I_1\right) - \mathbf{x}\|_2 \tag{6}$$

Finally, to reduce correlation of descriptor channels with one another in order to make the network more discriminative, we add a Pearson correlation loss on the descriptors for each image batch. Given a set of descriptors $D \in \mathbb{R}^{n \times c}$ for a batch, we define the Pearson loss $\mathcal{L}_{Pearson}$ as the sum of the correlation $Corr$ between every channel in this batch. Denoting $\hat{D}$ as the normalized D, the correlation is:

$$Corr = \hat{D}^{\top} \hat{D} \in \mathbb{R}^{c \times c} \tag{7}$$

$$\mathcal{L}_{Pearson} = \frac{1}{2c(c-1)} \sum_{i \neq j} (Corr_{ij})^2 \tag{8}$$

4 Experiments

In order to quantify ReTRo's performance we use it for the tasks of relative pose estimation and point tracking. Then, we benchmark its speed in both a dense and a per-patch manner. We train our model on the Hamlyn stereo dataset introduced in [34], using the same test/train split of 20,000 stereo pairs, with 3000 in test. We test our model using the test split of the Hamlyn dataset as well as the SuPer [17] dataset which provides annotated tracked points over time. We use PyTorch with an Adam optimizer and a learning rate of 10^{-4}. We train for 100,000 steps, which takes less than a day on a Nvidia 1080 Ti. We use FAST keypoints for our model and the ORB model, and use SIFT's keypoints for the SIFT descriptor. Finally, for comparison, we also use the CAPS model with their implementation as a benchmark for how well a larger weakly-supervised descriptor performs. We additionally test ReTRo with different scales, using just $(\mathbf{F}_2, \mathbf{F}_4)$ for ReTRoM (medium), and $(\mathbf{F}_4)$ for ReTRoC (coarse). We call the binary valued model ReTRoB (binary), which we train using the Straight-Through Estimator as introduced in [4].

4.1 Model Accuracy

We use the 3000 test pairs of the Hamlyn stereo dataset [34] for quantifying our descriptor for the task of estimating relative pose. We use MAGSAC++ for our robust estimator [3] as RANSAC and other sample consensus methods are sensitive to parameterization [13]. We report the mean Average Accuracy (mAA) at 10°, which is the area under the Average Accuracy curve. The curves and their mean Average Accuracy are shown in Fig. 2. This metric can be used as a way to determine the quality and precision of a descriptor as it demonstrates how well the descriptor works for matching between multiple views [13].

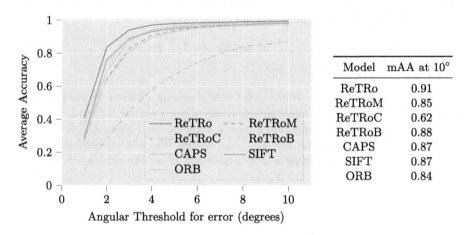

Model	mAA at 10°
ReTRo	0.91
ReTRoM	0.85
ReTRoC	0.62
ReTRoB	0.88
CAPS	0.87
SIFT	0.87
ORB	0.84

Fig. 2. Average Accuracy of relative stereo pose estimation on the Hamlyn dataset.

We then test ReTRo for the task of tracking points over time. For our tracking scheme, we brute-force match descriptors (using the ratio test for initial outlier rejection in SIFT as recommended), and add GMS [5] as a outlier rejection method on top of all descriptors. Without GMSMatch, SIFT and ORB have large amounts of outliers and poor results for tracking. See Fig. 4 for an example of what temporal matches look like without outlier rejection. For a simple, but effective tracking scheme, we track a point by moving it by the average motion of its four nearest neighbor keypoints. Using this tracking scheme, we test on the dataset introduced in SuPeR [17,20], starting with their initial 20 point locations and updating the points with each frame. In Fig. 3 we report the average error

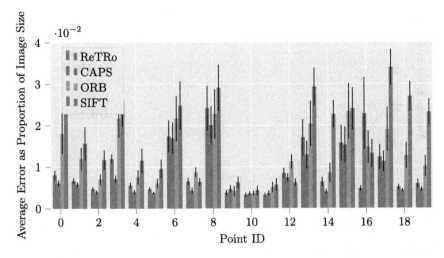

Fig. 3. Our tracking scheme evaluated on the SuPeR dataset; we use a simple nearest-neighbor average as the motion estimate for a point along with GMSMatch for outlier filtering. Error bars for standard error are in black

as a fraction of the image size for each of the labelled points. Both ReTRo and the CAPS descriptor beat ORB and SIFT on average in this dataset.

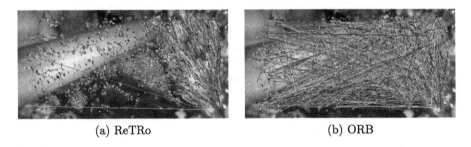

(a) ReTRo   (b) ORB

Fig. 4. Brute-force matching without outlier rejection in a temporal sequence. Points are colored randomly, with lines joining matches from prior frame.

We use the same tracking scheme to estimate forward-backward tracking error [14]. This error is the difference in position between an initial tracked point in a sequence and where the point ends up after tracking forward and backward in time. We pick sixteen random n-frame subsequences from the Hamlyn dataset test split, and report their mean error in Fig. 5. This helps serve as a metric for how much a tracked point can drift over time given each descriptor, and both ReTRo and CAPS outperform ORB and SIFT. CAPS performs well in this case at the cost of speed and memory. Our coarse and medium scale descriptors also perform well in this case, likely due to the smoothness introduced by running on coarser scales.

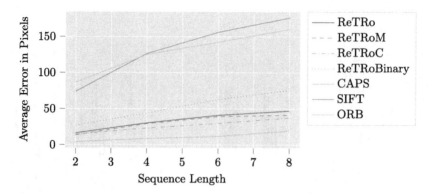

Fig. 5. Forward-backward error of sequences tracked using 4-nearest neighbor and GMSMatch for outlier rejection.

Table 1. Evaluation speed per image on the Hamlyn $(384, 192)$ size images. CAPS outputs at quarter resolution, while ours is full resolution.

Batch Size	ReTRoC	ReTRoM	ReTRo	CAPS
1	1.2 ms	2.3 ms	6.4 ms	9.5 ms
2	0.75 ms	2.1 ms	5.5 ms	6.6 ms
4	0.58 ms	1.6 ms	4.78 ms	5.88 ms
8	0.49 ms	1.4 ms	4.26 ms	5.68 ms

In conclusion, for relative pose esimation, ReTRo outperforms CAPS, ORB, and SIFT. ReTRo performs better than SuPeR, ORB, and SIFT on the SuPeR tissue tracking dataset, while being extremely close to the larger and slower CAPS. Finally, for forward-backward error, ReTRo lies just behind CAPS with ORB and SIFT being further off.

4.2 Model Speed

We first benchmark our descriptor by evaluating densely on full images in order to compare to CAPS. These results are shown in Table 1. Our method performs faster than CAPS, while also outputting 16× more data. Our network only requires 31×31 image patches at each scale, while CAPS has a receptive field of 217×217. By running parallel networks at each resolution, we gain receptive field while also keeping input size small.

We then test a version of our network that takes in patches to fairly estimate single-descriptor evaluation time for keypoints. This model runs in 15.9 μs per descriptor, faster than a recent fast patch based descriptor CDBin which is reported as 53.2 μs (also on an Nvidia 1080). Finally in terms of model size, our network only has $168,638$ parameters, compared to the $20,508,992$ of CAPS. Through making the network small we enable real-time descriptor calculation while also leaving room for matching, tracking, and mapping applications to run on the same device.

5   Conclusion

We introduce ReTRo, a Real-Time, Rotated descriptor that can be learned using just stereo data for supervision. We compare ReTRo to other descriptors with favorable results for the tasks of pose estimation and tissue tracking. ReTRo can be trained directly on surgical environments without labelling, and thus learns a descriptor tuned directly for these environments. We plan to use ReTRo in real-time surgical tracking and mapping applications.

References

1. Balntas, V., Johns, E., Tang, L., Mikolajczyk, K.: PN-Net: Conjoined Triple Deep Network for Learning Local Image Descriptors. ArXiv160105030 Cs (2016)
2. Balntas, V., Riba, E., Ponsa, D., Mikolajczyk, K.: Learning local feature descriptors with triplets and shallow convolutional neural networks. In: British Machine Vision Conference 2016 (2016)
3. Barath, D., Noskova, J., Ivashechkin, M., Matas, J.: MAGSAC++, a fast, reliable and accurate robust estimator. In: 2020 IEEE/CVF Conference on Computer Vision and Pattern Recognition (CVPR), June 2020
4. Bengio, Y., Léonard, N., Courville, A.: Estimating or Propagating Gradients Through Stochastic Neurons for Conditional Computation. ArXiv13083432 Cs, Auguest 2013
5. Bian, J.W., et al.: GMS: grid-based motion statistics for fast, ultra-robust feature correspondence. Int. J. Comput. Vis. **128**(6), 1580–1593 (2019). https://doi.org/10.1007/s11263-019-01280-3
6. Christiansen, P.H., Kragh, M.F., Brodskiy, Y., Karstoft, H.: UnsuperPoint: End-to-end Unsupervised Interest Point Detector and Descriptor. ArXiv190704011 Cs, July 2019
7. Dai, J., et al.: Deformable convolutional networks. In: Proceedings of the IEEE International Conference on Computer Vision (2017)
8. DeTone, D., Malisiewicz, T., Rabinovich, A.: SuperPoint: self-supervised interest point detection and description. In: Proceedings of the IEEE Conference on Computer Vision and Pattern Recognition Workshops (2018)
9. Gong, H., Chen, L., Li, C., Zeng, J., Tao, X., Wang, Y.: Online tracking and relocation based on a new rotation-invariant haar-like statistical descriptor in endoscopic examination. IEEE Access **8**, 101867–101883 (2020)
10. Heinrich, M.P., Oktay, O.: BRIEFnet: deep pancreas segmentation using binary sparse convolutions. In: Descoteaux, M., Maier-Hein, L., Franz, A., Jannin, P., Collins, D.L., Duchesne, S. (eds.) MICCAI 2017. LNCS, vol. 10435, pp. 329–337. Springer, Cham (2017). https://doi.org/10.1007/978-3-319-66179-7_38
11. Heinrich, M.P., Oktay, O., Bouteldja, N.: OBELISK-Net: fewer layers to solve 3D multi-organ segmentation with sparse deformable convolutions. In: Medical Image Analysis, vol. 54, May 2019
12. Jaderberg, M., Simonyan, K., Zisserman, A., Kavukcuoglu, K.: Spatial Transformer Networks. ArXiv150602025 Cs, February 2016

13. Jin, Y., et al.: Image matching across wide baselines: from paper to practice. Int. J. Comput. Vis. **6**, 1–31 (2020). https://doi.org/10.1007/s11263-020-01385-0

14. Kalal, Z., Mikolajczyk, K., Matas, J.: Forward-backward error: automatic detection of tracking failures. In: 2010 20th International Conference on Pattern Recognition (ICPR), Auguest 2010

15. Lamarca, J., Parashar, S., Bartoli, A., Montiel, J.M.M.: DefSLAM: tracking and mapping of deforming scenes from monocular sequences. IEEE Trans. Robot. **37**(1), 291–303 (2021)

16. Levi, G., Hassner, T.: LATCH: learned arrangements of three patch codes. In: 2016 IEEE Winter Conference on Applications of Computer Vision (2016)

17. Li, Y., et al.: SuPer: a surgical perception framework for endoscopic tissue manipulation with surgical robotics. IEEE Robot. Autom. Lett. **5**(2), 2294–2301 (2020)

18. Liu, X., et al.: Extremely dense point correspondences using a learned feature descriptor. In: 2020 IEEE/CVF Conference on Computer Vision and Pattern Recognition (2020)

19. Lowe, D.G.: Object recognition from local scale-invariant features. In: Proceedings of the Seventh IEEE International Conference on Computer Vision, vol. 2, pp. 1150–1157, September 1999

20. Lu, J., Jayakumari, A., Richter, F., Li, Y., Yip, M.C.: SuPer Deep: A Surgical Perception Framework for Robotic Tissue Manipulation using Deep Learning for Feature Extraction. ArXiv200303472 Cs, September 2020

21. Marmol, A., Banach, A., Peynot, T.: Dense-ArthroSLAM: dense intra-articular 3-D reconstruction with robust localization prior for arthroscopy. IEEE Robot. Autom. Lett. **4**(2), 918–925 (2019)

22. Mishchuk, A., Mishkin, D., Radenovic, F., Matas, J.: Working hard to know your neighbor's margins: Local descriptor learning loss. ArXiv170510872 Cs, January 2018

23. Richa, R., Bó, A.P., Poignet, P.: Towards robust 3D visual tracking for motion compensation in beating heart surgery. Med. Image Anal. **15**(3), 302–315 (2011)

24. Rodríguez, J.J.G., Lamarca, J., Morlana, J., Tardós, J.D., Montiel, J.M.M.: SD-DefSLAM: Semi-Direct Monocular SLAM for Deformable and Intracorporeal Scenes. ArXiv201009409 Cs, October 2020

25. Leonardis, A., Bischof, H., Pinz, A. (eds.): ECCV 2006. LNCS, vol. 3951. Springer, Heidelberg (2006). https://doi.org/10.1007/11744023

26. Rublee, E., Rabaud, V., Konolige, K., Bradski, G.: ORB: an efficient alternative to SIFT or SURF. In: 2011 International Conference on Computer Vision (2011)

27. Sarlin, P.E., DeTone, D., Malisiewicz, T., Rabinovich, A.: SuperGlue: learning feature matching with graph neural networks. In: 2020 IEEE/CVF Conference on Computer Vision and Pattern Recognition (CVPR), June 2020

28. Song, J., Wang, J., Zhao, L., Huang, S., Dissanayake, G.: MIS-SLAM: real-time large-scale dense deformable slam system in minimal invasive surgery based on heterogeneous computing. IEEE Robot. Autom. Lett. **3**(4), 4068–4075 (2018)

29. Suárez, I., Sfeir, G., Buenaposada, J.M., Baumela, L.: BEBLID: boosted efficient binary local image descriptor. Pattern Recogn. Lett. **133**, 366–372 (2020)

30. Tyszkiewicz, M.J., Fua, P., Trulls, E.: DISK: learning local features with policy gradient. ArXiv200613566 Cs, June 2020

31. Vedaldi, A., Bischof, H., Brox, T., Frahm, J.-M. (eds.): ECCV 2020. LNCS, vol. 12375. Springer, Cham (2020). https://doi.org/10.1007/978-3-030-58577-8

32. Xompero, A., Lanz, O., Cavallaro, A.: MORB: a multi-scale binary descriptor. In: 2018 25th IEEE International Conference on Image Processing (2018)

33. Ye, J., Zhang, S., Huang, T., Rui, Y.: CDbin: compact discriminative binary descriptor learned with efficient neural network. IEEE Trans. Circ. Syst. Video Technol. **30**(3), 862–874 (2020)
34. Ye, M., Johns, E., Handa, A., Zhang, L., Pratt, P., Yang, G.Z.: Self-Supervised Siamese Learning on Stereo Image Pairs for Depth Estimation in Robotic Surgery. ArXiv170508260 Cs, May 2017
35. Leibe, B., Matas, J., Sebe, N., Welling, M. (eds.): ECCV 2016. LNCS, vol. 9909. Springer, Cham (2016). https://doi.org/10.1007/978-3-319-46454-1
36. Yip, M.C., Lowe, D.G., Salcudean, S.E., Rohling, R.N., Nguan, C.Y.: Tissue tracking and registration for image-guided surgery. IEEE Trans. Med. Imaging **31**(11), 2169–2182 (2012)

Surgical Instruction Generation with Transformers

Jinglu Zhang[1], Yinyu Nie[2(✉)], Jian Chang[1], and Jian Jun Zhang[1]

[1] National Centre for Computer Animation (NCCA),
Bournemouth University, Poole, UK
[2] Technical University of Munich, Munich, Germany
`yinyu.nie@tum.de`

Abstract. Automatic surgical instruction generation is a prerequisite towards intra-operative context-aware surgical assistance. However, generating instructions from surgical scenes is challenging, as it requires jointly understanding the surgical activity of current view and modelling relationships between visual information and textual description. Inspired by the neural machine translation and imaging captioning tasks in open domain, we introduce a transformer-backboned encoder-decoder network with self-critical reinforcement learning to generate instructions from surgical images. We evaluate the effectiveness of our method on DAISI dataset, which includes 290 procedures from various medical disciplines. Our approach outperforms the existing baseline over all caption evaluation metrics. The results demonstrate the benefits of the encoder-decoder structure backboned by transformer in handling multimodal context.

Keywords: Surgical instruction generation · Transformer · Image captioning · Reinforcement learning

1 Introduction

Surgical instruction generation is a task of automatically generating a natural language sentence to guide surgeons of how to perform the operation based on the current surgical view. It is an essential component towards building context-aware surgical system, which aims to utilize available information inside the operation room to provide clinicians with contextual support at appropriate time. Moreover, when on-site mentoring is unavailable or a rare case is detected, providing intra-operative surgical instructions by expert surgeons is imperative. However, surgical data has high heterogeneity even for the same type of surgery due to different surgical skill level, medical condition, and patient specific situation. Accordingly, understanding surgical content and generating a natural language sentence to guide the procedure is challenging.

Previously, telementoring [5], which exchanges medical information through video and audio in real time, has been proved as an efficient solution for intra-operative guidance, including pointing out target anatomical structure from the monitor, controlling the camera or the robotic arm, etc. Nonetheless, telementoring is limited by the cost of specific equipment and software, the high demand of

© Springer Nature Switzerland AG 2021
M. de Bruijne et al. (Eds.): MICCAI 2021, LNCS 12904, pp. 290–299, 2021.
https://doi.org/10.1007/978-3-030-87202-1_28

transport speed, and legal and ethic issues [5,9]. With the huge development of related techniques of context-aware surgical assistance, understanding and analyzing the surgical activities inside the operation room opens up the possibility of providing intra-operative assistance for surgeons. Most of the existing researches focus on surgical phases and fine-grained gestures recognition [10,22,27]. However, these methods can be regarded as the segmentation and classification problems based on pre-defined phases and gestures, thus have no ability of generating the unseen instructions.

The most related research topic to us is medical report generation [6,7,12], which describes the *impression*, *findings*, *tags*, etc. of a patient in reference to the radiology or pathology. One of the earliest medical report generation works based on natural language is [12], which jointly predicts tags and generates long paragraphs with co-attention and hierarchical LSTM. More recently, [7] improves the transformer model [23] by designing a relational memory to record key information of the generation process and provides a memory-driven layer normalization for transformer decoder. Despite the challenges, medical reports also have their own discriminating characters. They often share predefined topics and follow similar writing templates, while surgical instruction generation with natural language has no template to follow.

To our best knowledge, [20] is the only prior work for surgical instruction generation. In their work, the authors create the Database for AI Surgical Instruction dataset (DAISI) and use a bidirectional recurrent neural network (RNN) to generate the description for a surgical image. However their work has two limitations. For one, although RNNs are designed for sequence generation with arbitrary length, they suffer from the essential vanishing gradient problem [17]. For another, they apply the BLEU [16] score as the only evaluation metric, which is insufficient for natural language evaluation.

In this paper, inspired by the great performance of transformer model in machine translation [23] and image captioning [8] from the open domain, we build our network with an encoder-decoder fully backboned transformer to generate surgical instructions. Taking an surgical image as the input, we first extract its visual attention features by a fine-tuned ResNet-101 module. Then the encoder attention blocks, decoder attention blocks, and encoder-decoder attention blocks model the dependencies for visual features, textual features, and visual-textural relational features, respectively. On the other hand, sequence generation models are often trained using the cross-entropy (XE) loss and evaluated using non-differential metrics such as BLEU, CIDEr [24], etc. In order to alleviate the mismatch between training and testing and improve the evaluation performance, we apply the reinforcement learning based self-critical approach [19] to directly optimize the CIDEr score. Experimentally, we extensively explore the performance of different baselines (LSTM-based fully connected and soft-attention models) on DAISI dataset [20]. The experiments demonstrate that our transformer-backboned architecture outperforms the existing methods as well as our other proposed baselines. The promising instructions generated from the network bring potential value in clinical practice.

2 Methodology

In this section, we introduce our framework in details. It involves two sub-modules: 1) the transformer-backboned encoder-decoder structure for surgical instruction generation (see Sect. 2.1); 2) the self-critical reinforcement learning for optimizing the CIDEr score (see Sect. 2.2).

2.1 Encoder-Decoder with Transformer Backbone

The whole encoder-decoder structure can be seen in Fig. 1. Following the modern learning paradigm, we design this kind of structure to encode latent features from images and decode them into natural languages. Before our network, a ResNet-101 [11] is adopted to output $14 \times 14 \times 2048$ image features, which are afterwards embeded by a linear embedding layer to reduce the dimension to $14 \times 14 \times 512$ followed by a ReLU and a dropout layer. Subsequently, our encoder firstly processes the flattened spatial features (196×512) and produces non-local relationships between image regions. Then the decoder takes hidden attentive representation from the encoder outputs and generates the corresponding instruction with natural language. The essential attention mechanism behind the transformer is called scaled dot-product attention [23], which is defined as:

$$\text{Attention}(Q, K, V) = \text{Softmax}(\frac{QK^T}{\sqrt{d}})V \tag{1}$$

where Q is the packed query matrix, K and V are packed key-value pairs, and d is a scaling factor (equals to the dimension of K). It calculates a weighted sum of the values based on the similarity distribution between the query with all the keys.

The whole encoder is a stack of 6 attention blocks with identical structure. Specifically, each block consists of a **multi-head self-attention** layer (8 heads) and a position-wise **feed-forward network**. The multi-head self-attention layer is represented as:

$$MultiHead(Q, K, V) = Concat(head_1, ..., head_h)W^O$$
$$head_i = \text{Attention}(QW_i^Q, KW_i^K, VW_i^V) \tag{2}$$

where matrices W_i^Q, W_i^Q, W_i^Q and W^O are projection parameters to be learned during the training phase. Different linear transformations are applied to the queries, values, and keys for each attention head. A simple position-wise fully connected feed-forward network is then applied to each attention layer:

$$\text{FFN(x)} = \max(0, xW_1 + b_1)W_2 + b_2, \tag{3}$$

where W_1, W_2 and b_1, b_2 are corresponding weights and biases for two fully connected layers.

The input of the decoder is the hidden representation exported from the last encoder layer. The decoder also consists of six identical blocks, where each has two multi-head attention layers (a decoder self-attention layer and an encoder-decoder attention layer) and one fully connected feed-forward network. Every decoder self-attention layer is masked to prevent from attending to future locations. For further explanation of the decoder, please refer to the original transformer paper [23].

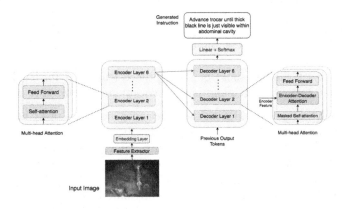

Fig. 1. The transformer-backboned surgical instruction generation architecture

2.2 Reinforcement Learning

Sequence generation models are often trained in "Teacher-Forcing" [4] mode, which inputs the ground-truth to maximize the likelihood of next prediction during training and uses previously generated words from the model distribution to predict the next word during test time. In order to bridge this gap, we apply the self-critical reinforcement learning as proposed in [19]. After pre-training the model with standard word-level XE loss, the CIDEr score [24] is directly optimized as the reward. All the detailed formula derivation can be found in [19].

3 Evaluation

3.1 Experimental Settings

Dataset Description. We evaluate our approach on DAISI dataset [20], which contains 17,255 color images from 290 medical procedures, including external fetal monitoring, laparoscopic sleeve gastrectomy, laparoscopic ventral hernia repair, etc. The availability of the dataset is upon request[1]. Every procedure consists of few images with their corresponding instruction texts. We further clean the dataset by deleting noisy and irrelevant images and descriptions. Finally, there are 16,413 images in total with one caption for each image. While some surgical procedures have only one sample due to the limited dataset size, we

---

[1] https://engineering.purdue.edu/starproj/.

split the data in per image manner. We assign 13,094 images for training, 1,646 for validation, and 1,673 for testing.

Text Preprocessing. Text preprocessing is a significant step to transform the text into a more analyzable and predictable format for the deep learning model. Raw text instructions need to be preprocessed to learn meaningful features and not overfit on irrelevant noise. We follow these steps to clean the text instruction: 1) Converting all words to lower case; 2) Expanding abbreviations, including medical abbreviations (e.g. 'a.' to 'artery') and English contractions (e.g. i've to 'i have'); 3) Removing numbers, punctuation, and whitespace; 4) Tokenizing the sentence into words.

We further set the threshold of the sentence length to 16, label any word count less than five as 'UNK', and build a vocabulary of size 2212 words.

Evaluation Metrics. Besides the instruction generation task, how to automatically evaluate the generated sentences has become increasingly important. The key idea is to measure the correlation of generated captions with human judgments. Following most of the image captioning methods, we apply 1–4 g BLEU [16], Rouge-L [14], METEOR [3], CIDEr [24], and SPICE [1] to evaluate our model, while the first three metrics are originated from machine translation and the last two metrics are specifically designed for image captioning.

3.2   Implementation and Training Details

All the models are implemented in PyTorch and trained on a single NVIDIA GeForce GTX 1080 graphics card. We first train our model with a word-level cross-entropy (XE) loss, then optimize the model using reinforcement learning. It takes around 30 h for the training process (30 epochs for general XE loss, and 30 epochs for reinforcement learning). During the XE training process, the model is trained to predict the next word given previous ground-truth word, while the reinforcement learning process is trained to predict next word based on the previous prediction. It takes around 30 h for the training process (30 epochs for general XE loss, and 30 epochs for reinforcement learning)

Transformer Encoder-Decoder. We use ResNet-101 [11] pre-trained on ImageNet classification task to extract image features. A spatially adaptive max-pooling layer is applied after the final convolution layer. It ends up with a fixed size of $14 \times 14 \times 2048$-d (196 image regions in total) output. For the XE training, we initialize the learning rate to 3×10^{-4} and follow the learning rate scheduling strategy with 20000 warm-up steps for 30 epochs. During the self-critical evaluation, we use a fixed learning rate of 1×10^{-5} for another 30 epochs. Both models are optimized using ADAM optimizer [13] with a batch size of 5.

LSTM-based Models. In order to comprehensively evaluate the surgical instruction generation task, we implement two additional models for comparison and discussion, namely LSTM model and LSTM-based soft-attention model similarly to [25,26]. For LSTM model, images are encoded to 2048 dimension vectors with the final convolution layer of ResNet-101 followed by an average pooling layer. The LSTM-based soft-attention model shares the same image feature maps with transformer model. For both models, the image embedding, words embedding dimension and LSTM hidden state size are set to 512.

4 Results and Discussion

4.1 Comparison with the State-of-the-Art

We clean the original dataset [20] by removing noisy and wrong image-text pairs. Thus a new benchmark is required. As the code in [20] is not publicly available, we re-implement their Bi-RNN model. The 4096 dimensional image features are extracted using the last convolutional layer from a pre-trained VGG16 [21]. The Bi-RNN model is trained with 50 epochs by the initial learning rate at 5×10^{-4} and the batch size at 10. Table 1 compares our proposed models with [20], which shows that Bi-RNN has relatively lower performance, especially for the 3-gram and 4-gram BLEU scores (11.3% and 9.3%) compared with ours (46.4% and 44.9%). In BLEU score evaluation, long $n - gram$ score measures the fluency of the instruction. It can be concluded that Bi-RNN is not capable of generating "human-like" instructions.

LSTM model achieves slightly better performance than LSTM-based soft-attention approach, and the transformer model outperforms all the others in all metrics. This indicates that the conventional RNN-based methods have limited ability of catching the dependencies between image features and text information. While transformer-backboned encoder-decoder layers can encode the dependencies for image pixels, the self-attention layers in decoder are able to model dependencies for textual information, and the encoder-decoder attention builds the relationship between image and textual features. Figure 2 shows some visualization samples using the proposed transformer-backboned framework.

Table 1. Comparison with the state-of-the-art [20] for surgical instruction generation task. B1, B2, B3, B4, C, M, R and S stands for 1–4 g BLEU, CIDEr, METEOR, ROUGE-L and SPICE score respectively.

Surgical Instruction	B1	B2	B3	B4	C	M	R	S
DAISI (Bi-RNN)	21.0	14.4	11.3	9.3	8.32	10.3	22.0	12.1
LSTM	43.7	39.4	37.3	36.2	34.0	24.9	44.6	40.2
LSTM + soft-attn	43.2	38.7	36.3	34.9	32.4	24.3	43.7	38.0
Transformer + rl	**52.8**	**48.7**	**46.4**	**44.9**	**42.7**	**30.7**	**53.1**	**48.4**

4.2 Effects of Reinforcement Learning

To further explore the functionality of each design in our network, we decouple three networks and design an ablative experiment in six settings: (1) LSTM only; (2) LSTM + reinforcement learning; (3) LSTM + soft-attention; (4) LSTM +

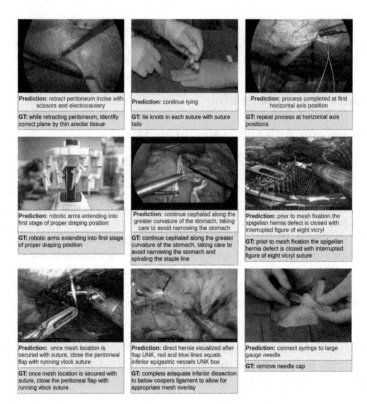

Fig. 2. Qualitative results with transformer-backboned encoder-decoder framework.

soft-attention + reinforcement learning; (5) Transformer only; (6) Transformer + reinforcement learning. The experiment results are shown in Table 2.

(1) v.s. (3): We add the soft-attention module on the top of LSTM to sequentially attend to different parts of image and aggregate information, but it performs slightly worse (around 1% for each evaluation standard) than the baseline model. This indicates that simple soft-attention mechanism cannot build the correlation between salient pixels and the next word prediction.

(1) v.s. (3) v.s. (5): Without using any recurrent neural units as LSTM-based models, transformer-backboned model only use the attention mechanism to encode the image information and decode its corresponding text instruction. Transformer-backboned model achieves better performance than two LSTM models, which demonstrate its ability in handling multi-modal contexts.

(1) v.s. (2), (3) v.s. (4), and (5) v.s. (6): During the training procedure, we first train each model with standard XE loss, then we add the reinforcement learning block to optimize the CIDEr score directly. From the results, it can be seen that not only the CIDEr score, but also the performance of other evaluation metrics has been lifted. Specifically, we observe a significant increasing when using reinforcement training after the transformer-backboned model.

Table 2. Ablative study to explore the influence of reinforcement learning. B1, B2, B3, B4, C, M, R and S stands for 1–4 g BLEU, CIDEr, METEOR, ROUGE-L and SPICE score respectively.

Surgical Instruction	B1	B2	B3	B4	C	M	R	S
LSTM	43.7	39.4	37.3	36.2	34.0	24.9	44.6	40.2
LSTM + rl	44.6	40.3	38.3	37.1	35.1	25.4	45.3	41.1
LSTM + attn	43.2	38.7	36.3	34.9	32.4	24.3	43.7	38.0
LSTM + attn + rl	43.4	38.8	36.4	34.8	33.1	24.8	44.1	38.5
Transformer	45.5	41	38.7	37.2	34	25.6	44.3	39.7
Transformer + rl	**52.8**	**48.7**	**46.4**	**44.9**	**42.7**	**30.7**	**53.1**	**48.4**

4.3 Limitations and Challenges

In this part, we discuss the current challenges and limitations for automatic surgical instruction generation.

1. **Small dataset size.** Deep learning algorithms often require huge amount of data to tune the parameters and prevent overfitting, e.g., COCO dataset [15] has more than 120K samples. Surgical instruction generation is a multi-modal problem, which relates visual, text, and the relationship between them. Therefore, the solution space is much larger than other tasks (e.g. classification and segmentation). However excluding the noisy and irrelevant images, DAISI dataset only contains 16,413 images.
2. **No fine-grained supervisions.** In feature extraction, some image captioning algorithms use Faster R-CNN algorithm [18] to detect object bounding boxes and identify attribute features with Visual-Genome data [2]. However, obtaining semantic and attributive annotations in medical science is quite challenging since it requires expert annotators.
3. **One caption per image.** In real life, an image can be described in different ways. For example, COCO captioning task has equipped with 5 different reference translations for each image. Nonetheless, we have only one annotation per image. It is possible that the evaluation metrics grade an adequate caption a low score only because it does not fit the ground truth.

5 Conclusion

In this paper, we propose an encoder-decoder architecture fully backboned by transformer to generate surgical instructions from various medical disciplines. The experiment results demonstrate that transformer architecture is capable of creating the pixel-wise patterns from self-attention encoder, developing text relationships for masked self-attention decoder, and devising the image-text dependencies from encoder-decoder attention. In order to solve the mismatching between the training and testing procedure, we optimize the model with self-critical reinforcement learning, which takes the CIDEr score as the reward after the general cross-entropy training.

Understanding surgical activity and generating instruction is still at its early stage. Future works include collecting the large training dataset, building the specialized pre-trained model for medical images, regularizing and annotating more reference captions for surgical images.

References

1. Anderson, P., Fernando, B., Johnson, M., Gould, S.: SPICE: semantic propositional image caption evaluation. In: Leibe, B., Matas, J., Sebe, N., Welling, M. (eds.) ECCV 2016. LNCS, vol. 9909, pp. 382–398. Springer, Cham (2016). https://doi.org/10.1007/978-3-319-46454-1_24

2. Anderson, P., et al.: Bottom-up and top-down attention for image captioning and visual question answering. In: Proceedings of the IEEE Conference on Computer Vision and Pattern Recognition, pp. 6077–6086 (2018)

3. Banerjee, S., Lavie, A.: Meteor: an automatic metric for mt evaluation with improved correlation with human judgments. In: Proceedings of the acl Workshop on Intrinsic and Extrinsic Evaluation Measures for Machine Translation and/or Summarization, pp. 65–72 (2005)

4. Bengio, S., Vinyals, O., Jaitly, N., Shazeer, N.: Scheduled sampling for sequence prediction with recurrent neural networks. In: Advances in Neural Information Processing Systems, pp. 1171–1179 (2015)

5. Bilgic, E., et al.: Effectiveness of telementoring in surgery compared with on-site mentoring: a systematic review. Surg. Innov. **24**(4), 379–385 (2017)

6. Bustos, A., Pertusa, A., Salinas, J.M., de la Iglesia-Vayá, M.: Padchest: a large chest x-ray image dataset with multi-label annotated reports. Med. Image Anal. **66**, 101797 (2020)

7. Chen, Z., Song, Y., Chang, T.H., Wan, X.: Generating radiology reports via memory-driven transformer. arXiv preprint arXiv:2010.16056 (2020)

8. Cornia, M., Stefanini, M., Baraldi, L., Cucchiara, R.: Meshed-memory transformer for image captioning. In: Proceedings of the IEEE/CVF Conference on Computer Vision and Pattern Recognition, pp. 10578–10587 (2020)

9. Erridge, S., Yeung, D.K., Patel, H.R., Purkayastha, S.: Telementoring of surgeons: a systematic review. Surg. Innov. **26**(1), 95–111 (2019)

10. Funke, I., Bodenstedt, S., Oehme, F., von Bechtolsheim, F., Weitz, J., Speidel, S.: Using 3d convolutional neural networks to learn spatiotemporal features for automatic surgical gesture recognition in video. In: International Conference on Medical Image Computing and Computer-Assisted Intervention, pp. 467–475. Springer (2019)

11. He, K., Zhang, X., Ren, S., Sun, J.: Deep residual learning for image recognition. In: Proceedings of the IEEE Conference on Computer Vision and Pattern Recognition, pp. 770–778 (2016)

12. Jing, B., Xie, P., Xing, E.: On the automatic generation of medical imaging reports. arXiv preprint arXiv:1711.08195 (2017)

13. Kingma, D.P., Ba, J.: Adam: A method for stochastic optimization. arXiv preprint arXiv:1412.6980 (2014)

14. Lin, C.Y.: Rouge: a package for automatic evaluation of summaries. In: Text Summarization Branches Out, pp. 74–81 (2004)
15. Lin, T.Y., et al.: Microsoft COCO: common objects in context. In: Fleet, D., Pajdla, T., Schiele, B., Tuytelaars, T. (eds.) ECCV 2014. LNCS, vol. 8693, pp. 740–755. Springer, Cham (2014). https://doi.org/10.1007/978-3-319-10602-1_48
16. Papineni, K., Roukos, S., Ward, T., Zhu, W.J.: Bleu: a method for automatic evaluation of machine translation. In: Proceedings of the 40th Annual Meeting of the Association For Computational Linguistics, pp. 311–318 (2002)
17. Pascanu, R., Mikolov, T., Bengio, Y.: On the difficulty of training recurrent neural networks. In: International Conference on Machine Learning, pp. 1310–1318 (2013)
18. Ren, S., He, K., Girshick, R., Sun, J.: Faster r-cnn: towards real-time object detection with region proposal networks. IEEE Trans. Pattern Anal. Mach. Intell. 39(6), 1137–1149 (2016)
19. Rennie, S.J., Marcheret, E., Mroueh, Y., Ross, J., Goel, V.: Self-critical sequence training for image captioning. In: Proceedings of the IEEE Conference on Computer Vision and Pattern Recognition, pp. 7008–7024 (2017)
20. Rojas-Muñoz, E., Couperus, K., Wachs, J.: Daisi: Database for ai surgical instruction. arXiv preprint arXiv:2004.02809 (2020)
21. Simonyan, K., Zisserman, A.: Very deep convolutional networks for large-scale image recognition. arXiv preprint arXiv:1409.1556 (2014)
22. Twinanda, A.P., Shehata, S., Mutter, D., Marescaux, J., De Mathelin, M., Padoy, N.: Endonet: a deep architecture for recognition tasks on laparoscopic videos. IEEE Trans. Med. Imaging 36(1), 86–97 (2016)
23. Vaswani, A., et al.: Attention is all you need. In: Advances in Neural Information Processing Systems, pp. 5998–6008 (2017)
24. Vedantam, R., Lawrence Zitnick, C., Parikh, D.: Cider: consensus-based image description evaluation. In: Proceedings of the IEEE Conference on Computer Vision and Pattern Recognition, pp. 4566–4575 (2015)
25. Vinyals, O., Toshev, A., Bengio, S., Erhan, D.: Show and tell: a neural image caption generator. In: Proceedings of the IEEE Conference on Computer Vision and Pattern Recognition, pp. 3156–3164 (2015)
26. Xu, K., et al.: Show, attend and tell: neural image caption generation with visual attention. In: International Conference on Machine Learning, pp. 2048–2057 (2015)
27. Zhang, J., et al.: Symmetric dilated convolution for surgical gesture recognition. In: Martel, A.L., et al. (eds.) MICCAI 2020. LNCS, vol. 12263, pp. 409–418. Springer, Cham (2020). https://doi.org/10.1007/978-3-030-59716-0_39

Adversarial Domain Feature Adaptation for Bronchoscopic Depth Estimation

Mert Asim Karaoglu[1,2(✉)], Nikolas Brasch[2], Marijn Stollenga[1],
Wolfgang Wein[1], Nassir Navab[2,3], Federico Tombari[2,4], and Alexander Ladikos[1]

[1] ImFusion GmbH, Munich, Germany
karaoglu@imfusion.com
[2] Computer Aided Medical Procedures, Technische Universität München,
Munich, Germany
[3] Computer Aided Medical Procedures, Johns Hopkins University,
Baltimore, MD, USA
[4] Google, Zurich, Switzerland

Abstract. Depth estimation from monocular images is an important task in localization and 3D reconstruction pipelines for bronchoscopic navigation. Various supervised and self-supervised deep learning-based approaches have proven themselves on this task for natural images. However, the lack of labeled data and the bronchial tissue's feature-scarce texture make the utilization of these methods ineffective on bronchoscopic scenes. In this work, we propose an alternative domain-adaptive approach. Our novel two-step structure first trains a depth estimation network with labeled synthetic images in a supervised manner; then adopts an unsupervised adversarial domain feature adaptation scheme to improve the performance on real images. The results of our experiments show that the proposed method improves the network's performance on real images by a considerable margin and can be employed in 3D reconstruction pipelines.

Keywords: Bronchoscopy · Depth estimation · Domain adaptation

1 Introduction

Lung cancer is the leading cause of all cancer-related deaths, accounting for 24% of them in the US in 2017 [26]. The statistics highlight that the patient's survival profoundly depends on the disease stage, intensifying the importance of early diagnosis [26]. Diminishing the potential risks of more invasive techniques, transbronchial needle aspiration (TBNA) is a modern approach for pulmonary specimen retrieval for a decisive diagnosis [15]. Conducted by operating a bronchoscope to reach the suspected lesion site, segmented on a pre-operative CT scan, navigation is an existing challenge for TBNA procedures since it requires registration to the pre-operative CT plan.

Electronic supplementary material The online version of this chapter (https://doi.org/10.1007/978-3-030-87202-1_29) contains supplementary material, which is available to authorized users.

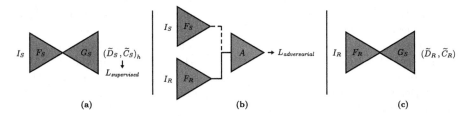

Fig. 1. Outline of the proposed domain-adaptive pipeline. (a) Supervised training of the encoder-decoder structure using the synthetic images (I_S), predicted depth images ($\tilde{D}_S$) and confidence maps ($\tilde{C}_S$). (b) Adversarial training scheme to train a new encoder (F_R) for the images from the real domain (I_R). During the optimization, the weights are updated only on the flow drawn with the solid line. (c) For inference on the real domain, F_R is connected to the decoder trained in the first step (G_S).

One way to achieve this is to register electromagnetic (EM) tracking data captured from the bronchoscope to the segmented airway tree of the pre-operative CT-scan [10,14,24]. In addition to the sensory errors caused by electromagnetic distortion, anatomical deformations are a principal challenge for EM-based approaches.

Motivated by the higher resilience of the camera data against such deformations, researchers have focused on vision-based approaches in recent years. Inspired by their success in natural scenes, direct and feature-based video-CT registration techniques [27] and simultaneous localization and mapping (SLAM) pipelines [1,30] have been investigated in various studies. Even though these approaches have shown a certain level of success, the feature-scarce texture and the photometric-inconsistencies caused by specular reflections are found to be common challenges.

The shortcomings of the direct and feature-based methods have led researchers to focus on adopting depth information to exploit the direct relationship with the scene geometry. Following the advancements in learning-based techniques, supervised learning has become a well-proven method for monocular depth estimation applied to natural scenes. However, it is challenging to employ for endoscopy tasks due to the difficulty of obtaining ground-truth data. An alternative way is to train the network on synthetic images with their rendered depth ground-truths. But due to the domain gap between the real and synthetic images, these models tend to suffer from a performance drop at inference time. To address this [2] utilizes Siemens VRT [5] to render realistic-looking synthetic data for supervised training. However, the renderer is not publicly available and it has only been demonstrated on colonoscopy data. [19] employs a separate network for an image-level transfer before utilizing the depth-estimation model. Aside from the second network's additional runtime cost, executed at the more complex image-level it is not easy to ensure the preservation of the task-related features during the domain transfer. Adopting a fully unsupervised approach, [25] employs a CycleGAN [31] based method for bronchoscopic depth estimation from unpaired images. However, CycleGANs focus on the image appearance and might not preserve the task-relevant features. In [17,18] a structure-from-motion approach is used to obtain a sparse set of points to supervise a dense depth estimation network for sinus endoscopy scenes. While this works well for nasal and sinus passages with their feature rich texture, it is much more challenging to apply to feature-scarce bronchoscopic data.

In this work, we therefore propose a novel, two-step domain-adaptive approach, for monocular depth estimation of bronchoscopic scenes, as outlined in Fig. 1. Overcoming the lack of labeled data from the real domain, the first step trains a network utilizing synthetic images with perfect ground-truth depths. In the second step, inspired by [29], we employ an unsupervised adversarial training scheme to accommodate the network for the real images, tackling the domain adaptation problem at the task-specific feature-level. Our method requires neither the ground-truth depth nor synthetic pairs for the real monocular bronchoscopic images for training. Moreover, unlike some of the explicit domain transfer methods, it does not need a secondary network at inference, improving the overall runtime efficiency. To evaluate our method, we develop a CycleGAN-based explicit domain transfer pipeline connected to the supervised network trained only on synthetic data in the first step. We conduct a quantitative test on a phantom and a qualitative test on human data. Furthermore, we employ our method in a 3D reconstruction framework to assess its usefulness for SLAM-based navigation.

2   Method

2.1   Supervised Depth Image and Confidence Map Estimation

Targeting optimal information compression at their code, encoder-decoder models, such as U-net [23], are a favored choice for various pixel-wise regression tasks. Aiming at an optimal point between accuracy and runtime performance, we utilize a U-net variant [7] with a ResNet-18 [9] encoder. On the decoder side, a series of nearest neighbor upsampling and convolutional layers are configured to regain the input's original size. After each upsampling operation, the corresponding features from the encoder level are concatenated to complete the skip-connection structure.

Motivated by [6], we alter this architecture with the addition of coordinate convolution layers [16]. Coordinate convolution layers introduce an additional signal that describes a connection between the spatial extent and the values of the extracted features, resulting in a faster convergence for supervised training [16]. In total, we employ five coordinate convolution layers at the skip connections and the bottleneck, right before connecting to the decoder.

The decoder's last four levels' outputs form the scaled versions of the estimated depth images and the confidence maps. We utilize the output set in a multi-scale loss consisting of three components: depth estimation loss, scale-invariant gradient loss, and confidence loss.

For depth regression, our objective is to estimate the depth values in the original scale of the ground-truth (D). As the pixel-wise error, we employ the BerHu loss ($\mathcal{B}$) as defined in [13]:

$$L_{depth}(D, \widetilde{D}) = \sum_{i,j} \mathcal{B}(|\, D(i,j) - \widetilde{D}(i,j)\,|, c) \tag{1}$$

where $D(i,j)$ and $\widetilde{D}(i,j)$ are the ground-truth and the predicted depth values at the pixel index (i,j). The threshold c is computed over a batch as:

$$c = k \max_{t,i,j} \left(|D^t(i,j) - \widetilde{D}^t(i,j)| \right) \tag{2}$$

where k is set to 0.2 as in [13] and t is an instance of the depth images inside the given batch.

To ensure smooth output depth images, we employ the scale-invariant gradient loss as introduced in [28] as:

$$L_{gradient}(D, \widetilde{D}, h) = \sum_{i,j} \left\| g(D(i,j), h) - g(\widetilde{D}(i,j), h) \right\|_2 \tag{3}$$

where g is the discrete scale-invariant finite differences operator [28] and h is the step size.

Inspired by [6], we use a supervised confidence loss. To provide the supervision signal, the ground-truth confidence map is calculated as:

$$C(i,j) = e^{-|D(i,j) - \widetilde{D}(i,j)|} \tag{4}$$

Based on this, we define the confidence loss to be the $\mathcal{L}_1$ norm between the ground truth and the prediction as:

$$L_{confidence}(C, \widetilde{C}) = \sum_{i,j} |C(i,j) - \widetilde{C}(i,j)| \tag{5}$$

To form the total loss, we combine the three factors with a span over the four different output scales as:

$$
\begin{aligned}
L_{supervised}(D, \widetilde{D}_h) = \sum_{h \in \{1,2,4,8\}} & (\lambda_{depth} L_{depth}(D, u_h(\widetilde{D}_h)) \\
& + \lambda_{gradient} L_{gradient}(D, u_h(\widetilde{D}_h), h) \\
& + \lambda_{confidence} L_{confidence}(C, u_h(\widetilde{C}_h)))
\end{aligned} \tag{6}
$$

λs are the hyper-parameters to weight each factor, h is the size ratio of the ground-truth to the predicted depth images, and u_h is the bilinear upsampling operator that upsamples an image by a scale of h.

2.2   Unsupervised Adversarial Domain Feature Adaptation

In the second step of the training pipeline, we incorporate the pre-trained encoder F_S to adversarially train a new encoder F_R for the images from the real domain.

Rendered color	Rendered depth	Phantom	Human

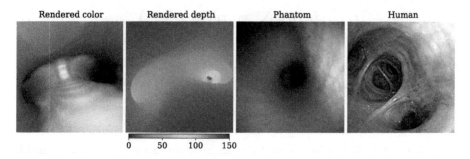

0   50   100   150

Fig. 2. Sample images from our renderings, phantom recordings, and human dataset acquired from [4]. The depth values are in *mm*.

For this task, we empirically decided to utilize three separate PatchGAN discriminators [11] (A^i, where $i \in \{1, 2, 3\}$) to be employed at the encoder's last two skip-connections and the bottleneck, after the coordinate convolution layers, to reduce the domain gap at the task-specific feature levels. At inference, the new encoder, F_R, is connected to the previously trained decoder G_S for depth image and confidence map estimation. Like other neural network models, GANs have limited learning capacity [8]. Trained with a lack of direct supervision, it is oftentimes inevitable for GANs to fall into local minima that are not an optimal hypothesis, referred to as a mode collapse. To increase the model's resilience against this phenomena, we initialize the new encoder F_R with the same weights as the pre-trained encoder F_S. Furthermore, in addition to its improvement for convergence time in supervised learning, we use coordinate convolution layers to lower the chance of a possible mode collapse in adversarial training [16].

For each discriminator, we employ the original GAN loss [8]. The total adversarial loss is the sum of discriminator and encoder losses across all i, where $i \in \{1, 2, 3\}$ is the index of the encoder's feature level (F^i) and its corresponding discriminator (A^i):

$$L_{adversarial}(A, F_S, F_R, I_S, I_R) = \sum_{i \in \{1,2,3\}} (L_{discriminator}(A^i, F_S^i, F_R^i, I_S, I_R) \\ + L_{encoder}(A^i, F_R^i, I_R)) \quad (7)$$

3   Experiments

Data. For our experiments, we use three different datasets: synthetic renderings, recordings inside a pulmonary phantom, and human recordings (Fig. 2). The synthetic dataset consists of 43,758 rendered color (I_S) and depth (D_S) image pairs. For the supervised training, we randomly split twenty percent of it to be utilized as the validation set. The airway tree used for data generation is segmented from the CT scan of a static pulmonary phantom and rendered using a ray-casting volume renderer. We model the virtual camera after our bronchoscope's intrinsic properties.

The training split of the pulmonary phantom dataset consists of 7 sequences, in total 12,720 undistorted frames, recorded inside the same phantom. For the quantitative evaluation, we use a separate set with known 3D tracking information acquired with the bronchoscope's EM sensor. Registered to the phantom's segmented airway tree, we manually verify and pick 62 correctly registered frames and render their synthetic depth images as the ground-truth.

Our human recordings consist of two sets. For training, we use our in-house dataset of three sequences with a total of 16,502 frames. For qualitative testing, a set of 240 frames are acquired from a publicly available video [4]. All of our input color images and outputs are of size 256 by 256.

Experiment Setup. The networks are implemented on PyTorch 1.5 [22]. For the supervised training, we employ random vertical and horizontal flips and color jitters as the data augmentations. For the adversarial training, we only apply the random flips to the real domain but keep the rest for the synthetic images. In the first step, we train the supervised model for 30 epochs using the synthetic dataset. At the output, we employ a ReLU activation [20] for the depth image, as in [13], and a sigmoid activation for the confidence map. We set the batch size to 64 and utilize the Adam optimizer [12] with a learning rate of 10^{-3}. In the second step, the new encoder is domain-adapted to the real images with a training of 12,000 iterations. For the adversarial training, we use the Adam optimizer with the learning rate set to $5(10)^{-6}$, and it is halved at three-fifth and four-fifth of the total iterations. In both cases, we set the β_1 and β_2 values of the Adam optimizers to their defaults, 0.9 and 0.999.

To have a baseline comparison against an explicit domain transfer approach, we train two standard CycleGANs [31] (one for the phantom and the other for the human data) and transfer the real images to the synthetic domain. We use these outputs as the input to our model trained only on synthetic data to complete the pipeline for depth and confidence estimation.

We employ our model in a 3D reconstruction framework based on [3,21]. The method essentially employs multiple pose graphs to approach the problem from local and global perspectives using feature-based tracking information and colored ICP [21] for a multi-scale point cloud alignment. Unlike [3,21], we do not consider loop closures in our settings because our test sequences are small in size and do not contain loops.

Quantitative Results. In Table 1, we evaluate the proposed method against the vanilla model (only trained on synthetic images in a supervised manner) and the explicit domain transfer approach. Moreover, in this test, we compare these models against their versions without coordinate convolution layers. The results reveal that our proposed method performs better across all metrics than the others. Additionally, we confirm that the use of the coordinate convolution layers significantly improves the adversarial training step of our approach.

306 M. A. Karaoglu et al.

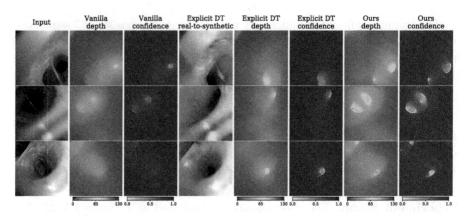

Fig. 3. Qualitative analysis on the real images from the human dataset [4]. The depth values are in *mm*.

Qualitative Results. In Fig. 3, we display a qualitative comparison of our method against the other approaches on the human pulmonary dataset [4]. Only trained on sharper, rendered synthetic data, we observe that the vanilla network shows difficulties generalizing to the bronchoscope's relatively smoother imaging characteristics. Additionally, the differences in illumination properties and anatomical discrepancies across the two domains further challenge the vanilla model causing it to fail to assess the deeper regions. Even though the depth perception of the vanilla model improves using the images that are domain-transferred, the structural inconsistencies generated by CycleGAN become the primary source of error. Results of our proposed method show that adversarial domain feature adaptation readjusts the encoder to accommodate the characteristics of the real scenes of the human anatomy without the shortcomings of the explicit domain transfer pipelines. Furthermore, adversarially trained on various patient data, our model shows generalizability to unseen scenes.

Table 1. Quantitative analysis on our pulmonary phantom dataset. The depth values are estimated in *mm*. The ground-truth depth values for the test are in the range of 1.74 mm and 142.37 mm. The best value for each metric is shown in **bold** characters.

Model	Mean abs. rel. diff.	RMSE	Accuracy		
			$\sigma = 1.25$	$\sigma = 1.25^2$	$\sigma = 1.25^3$
Without coordinate convolution					
Vanilla	0.659	8.407	0.315	0.570	0.733
Explicit DT	0.524	8.731	0.293	0.619	0.810
Domain-adapted	0.543	8.505	0.324	0.695	0.835
With coordinate convolution					
Vanilla	0.699	8.145	0.348	0.587	0.747
Explicit DT	0.580	8.566	0.280	0.579	0.807
Domain-adapted (ours)	**0.379**	**7.532**	**0.458**	**0.735**	**0.856**

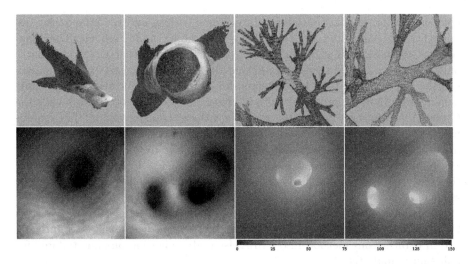

Fig. 4. Top-left pair: Point-cloud representation with color information from different perspectives. Top-right pair: Manual scaling and registration of the reconstruction (orange) to the pulmonary volume (purple) from different perspectives. Bottom-left pair: The first and the last frames of the input color sequence. Bottom-right pair: Corresponding depth estimation results using our approach in *mm* (Color figure online).

During the first-step, supervised training, the vanilla model learns to assign lower confidence to deeper locations and the edges. We interpret that the first property is caused by the ambiguities in the darker regions, while the latter is supported by virtue of the scale-invariant gradient loss, introduced in Eq. 3. Analyzing confidence estimates in Fig. 3, we see that our model obliges to these characteristics more strictly than the others. This affirms that our adversarial training step correctly adapts the vanilla model's encoder for more reliable feature extraction at the new domain, ultimately enhancing its depth perception.

3D Reconstruction Results. For this experiment, we use a sequence of 300 frames recorded inside the pulmonary phantom starting from a midpoint of a bronchus and pivot forward towards a bifurcation point. We estimate the depth images using our model to create a set of RGB-D frames. Figure 4 displays the point cloud output of the framework, manually scaled and registered to the point cloud of the phantom extracted from its CT scan. The results reveal that our method adheres well to the scene's structure, with a certain amount of outliers.

4   Conclusion

This paper approaches the problem of depth estimation in bronchoscopic scenes from an adversarial domain feature adaptation perspective. Our novel pipeline compensates for the lack of labeled bronchoscopy data and the pulmonary system's feature-scarce anatomy by utilizing a two-step training scheme. The quantitative results show that our method's accuracy is higher than the model that is only trained on the synthetic dataset and the CycleGAN based explicit domain transfer approach we have implemented for comparison. Our qualitative results demonstrate that the proposed method preserves the scene's structure while the CycleGAN often fails during the real to synthetic transformation. The experiments reveal that the adopted adversarial training approach is capable of improving over the base model, accomodating for various sources of domain gaps such as illumination, sensory, and anatomical discrepancies. Furthermore, we show that our approach is capable of correctly reconstructing chunks of bronchoscopic sequences.

In the future, our primary objective will be to increase the quantitative tests for a more comprehensive evaluation. Ultimately, we will focus on improving and integrating our method into a SLAM-based navigation pipeline tailored for pulmonary anatomy.

References

1. Chen, L., Tang, W., John, N.W., Wan, T.R., Zhang, J.J.: Slam-based dense surface reconstruction in monocular minimally invasive surgery and its application to augmented reality. Comput. Methods Prog. Biomed. **158**, 135–146 (2018)
2. Chen, R.J., Bobrow, T.L., Athey, T., Mahmood, F., Durr, N.J.: Slam endoscopy enhanced by adversarial depth prediction. arXiv preprint arXiv:1907.00283 (2019)
3. Choi, S., Zhou, Q.Y., Koltun, V.: Robust reconstruction of indoor scenes. In: Proceedings of the IEEE Conference on Computer Vision and Pattern Recognition, pp. 5556–5565 (2015)
4. Deutsche Gesellschaft für Internistische Intensivmedizin und Notfallmedizin (DGIIN): "bronchoskopie anatomie der unteren atemwege". Accessed 30 Feb 2021. [YouTube video]. https://www.youtube.com/watch?v=xPE4V8bU-Lk
5. Eid, M., et al.: Cinematic rendering in ct: a novel, lifelike 3d visualization technique. Am. J. Roentgenol. **209**(2), 370–379 (2017)
6. Facil, J.M., Ummenhofer, B., Zhou, H., Montesano, L., Brox, T., Civera, J.: Camconvs: camera-aware multi-scale convolutions for single-view depth. In: Proceedings of the IEEE/CVF Conference on Computer Vision and Pattern Recognition, pp. 11826–11835 (2019)
7. Godard, C., Mac Aodha, O., Firman, M., Brostow, G.J.: Digging into self-supervised monocular depth estimation. In: Proceedings of the IEEE/CVF International Conference on Computer Vision, pp. 3828–3838 (2019)
8. Goodfellow, I., et al.: Generative adversarial nets. In: Advances in Neural Information Processing Systems, pp. 2672–2680 (2014)
9. He, K., Zhang, X., Ren, S., Sun, J.: Deep residual learning for image recognition. In: Proceedings of the IEEE Conference on Computer Vision and Pattern Recognition, pp. 770–778 (2016)
10. Hofstad, E.F., et al.: Intraoperative localized constrained registration in navigated bronchoscopy. Med. Phys. **44**(8), 4204–4212 (2017)
11. Isola, P., Zhu, J.Y., Zhou, T., Efros, A.A.: Image-to-image translation with conditional adversarial networks. In: Proceedings of the IEEE Conference on Computer Vision and Pattern Recognition, pp. 1125–1134 (2017)

12. Kingma, D.P., Ba, J.: Adam: A method for stochastic optimization. arXiv preprint arXiv:1412.6980 (2014)
13. Laina, I., Rupprecht, C., Belagiannis, V., Tombari, F., Navab, N.: Deeper depth prediction with fully convolutional residual networks. In: 2016 Fourth International Conference on 3D Vision (3DV), pp. 239–248. IEEE (2016)
14. Lavasani, S.N., et al.: Bronchoscope motion tracking using centerline-guided gaussian mixture model in navigated bronchoscopy. Phys. Med. Biol. **66**(2), 025001 (2021)
15. Liu, Q.H., Ben, S.Q., Xia, Y., Wang, K.P., Huang, H.D.: Evolution of transbronchial needle aspiration technique. J. Thorac. Dis. **7**(Suppl 4), S224 (2015)
16. Liu, R., Lehman, J., Molino, P., Such, F.P., Frank, E., Sergeev, A., Yosinski, J.: An intriguing failing of convolutional neural networks and the coordconv solution. In: Advances in Neural Information Processing Systems, pp. 9605–9616 (2018)
17. Liu, X., et al.: Dense depth estimation in monocular endoscopy with self-supervised learning methods. IEEE Trans. Med. Imaging **39**(5), 1438–1447 (2019)
18. Liu, X., et al.: Reconstructing sinus anatomy from endoscopic video – towards a radiation-free approach for quantitative longitudinal assessment. In: Martel, A.L., et al. (eds.) MICCAI 2020. LNCS, vol. 12263, pp. 3–13. Springer, Cham (2020). https://doi.org/10.1007/978-3-030-59716-0_1
19. Mahmood, F., Chen, R., Durr, N.J.: Unsupervised reverse domain adaptation for synthetic medical images via adversarial training. IEEE Trans. Med. Imaging **37**(12), 2572–2581 (2018)
20. Nair, V., Hinton, G.E.: Rectified linear units improve restricted boltzmann machines. In: ICML (2010)
21. Park, J., Zhou, Q.Y., Koltun, V.: Colored point cloud registration revisited. In: Proceedings of the IEEE International Conference on Computer Vision, pp. 143–152 (2017)
22. Paszke, A., et al.: Automatic differentiation in pytorch (2017)
23. Ronneberger, O., Fischer, P., Brox, T.: U-Net: convolutional networks for biomedical image segmentation. In: Navab, N., Hornegger, J., Wells, W.M., Frangi, A.F. (eds.) MICCAI 2015. LNCS, vol. 9351, pp. 234–241. Springer, Cham (2015). https://doi.org/10.1007/978-3-319-24574-4_28
24. Schwarz, Y., Greif, J., Becker, H.D., Ernst, A., Mehta, A.: Real-time electromagnetic navigation bronchoscopy to peripheral lung lesions using overlaid ct images: the first human study. Chest **129**(4), 988–994 (2006)
25. Shen, M., Gu, Y., Liu, N., Yang, G.Z.: Context-aware depth and pose estimation for bronchoscopic navigation. IEEE Robot. Autom. Lett. **4**(2), 732–739 (2019)
26. Siegel, R.L., Miller, K.D., Jemal, A.: Cancer statistics, 2020. CA: Cancer J. Clin. **70**(1), 7–30 (2020)
27. Sinha, A., Liu, X., Reiter, A., Ishii, M., Hager, G.D., Taylor, R.H.: Endoscopic navigation in the absence of CT imaging. In: Frangi, A.F., Schnabel, J.A., Davatzikos, C., Alberola-López, C., Fichtinger, G. (eds.) MICCAI 2018. LNCS, vol. 11073, pp. 64–71. Springer, Cham (2018). https://doi.org/10.1007/978-3-030-00937-3_8
28. Ummenhofer, B., et al.: Demon: depth and motion network for learning monocular stereo. In: Proceedings of the IEEE Conference on Computer Vision and Pattern Recognition, pp. 5038–5047 (2017)
29. Vankadari, M., Garg, S., Majumder, A., Kumar, S., Behera, A.: Unsupervised monocular depth estimation for night-time images using adversarial domain feature adaptation. In: Vedaldi, A., Bischof, H., Brox, T., Frahm, J.-M. (eds.) ECCV 2020. LNCS, vol. 12373, pp. 443–459. Springer, Cham (2020). https://doi.org/10.1007/978-3-030-58604-1_27

30. Visentini-Scarzanella, M., Sugiura, T., Kaneko, T., Koto, S.: Deep monocular 3D reconstruction for assisted navigation in bronchoscopy. Int. J. Comput. Assist. Radiol. Surg. **12**(7), 1089–1099 (2017)

31. Zhu, J.Y., Park, T., Isola, P., Efros, A.A.: Unpaired image-to-image translation using cycle-consistent adversarial networks. In: Proceedings of the IEEE International Conference on Computer Vision, pp. 2223–2232 (2017)

2.5D Thermometry Maps
for MRI-Guided Tumor Ablation

Julian Alpers[1]([✉]), Daniel L. Reimert[1,3], Maximilian Rötzer[1],
Thomas Gerlach[2], Marcel Gutberlet[3], Frank Wacker[3], Bennet Hensen[3],
and Christian Hansen[1]

[1] Faculty of Computer Science, University of Magdeburg, Magdeburg, Germany
julian.alpers@ovgu.de
[2] Faculty of Electrical Engineering and Information Technologies,
Institute of Medical Technologies, University of Magdeburg, Magdeburg, Germany
[3] Institute for Diagnostic and Interventional Radiology, Medical School Hanover,
Hanover, Germany

Abstract. Fast and reliable monitoring of volumetric heat distribution
during MRI-guided tumor ablation is an urgent clinical need. In this
work, we introduce a method for generating 2.5D thermometry maps
from uniformly distributed 2D MRI phase images rotated around the
applicator's main axis. The images canbe fetched directly from the MR
device, reducing the delay between image acquisition and visualization.
For reconstruction, we use a weighted interpolation on a cylindric coordinate representation to calculate the heat value of voxels in a region of
interest. A pilot study on 13 ex vivo bio protein phantoms with flexible tubes to simulate a heat sink effect was conducted to evaluate our
method. After thermal ablation, we compared the measured coagulation
zone extracted from the post-treatment MR data set with the output
of the 2.5D thermometry map. The results show a mean Dice score of
0.75 ± 0.07, a sensitivity of 0.77 ± 0.03, and a reconstruction time within
$18.02\,\text{ms} \pm 5.91\,\text{ms}$. Future steps should address improving temporal resolution and accuracy, e.g., incorporating advanced bioheat transfer simulations.

Keywords: Image-guided interventions · Image reconstruction ·
Simulation · 2.5D thermometry

1 Introduction

A wide range of minimally invasive therapies have been developed for cancer
treatment, additionally to open surgery [1,11,19]. One of these methods is the

The work of this paper is funded by the Federal Ministry of Education and Research
within the Forschungscampus STIMULATE under grant numbers '13GW0473A' and
'13GW0473B'. This work was also supported by PRACTIS - Clinician Scientist Program, funded by the German Research Foundation (DFG, ME 3696/3- 1).
J. Alpers and D. Reimert—Joint first authorship
B. Hensen and C. Hansen—Joint senior authorship.

M. de Bruijne et al. (Eds.): MICCAI 2021, LNCS 12904, pp. 311–320, 2021.
https://doi.org/10.1007/978-3-030-87202-1_30

use of microwave ablation (MWA). Especially for smaller tumors, MWA shows promising results for treatment [18]. As the minimal ablative margin (MAM) is crucial for the local tumor progression (LTP), it is of greatest importance to assess if the malignancy has been adequately and completely treated, regardless of the etiology. For each millimeter increase of the MAM, a 30% reduction of the relative risk for LTP was found. The MAM is especially important as the only significant independent predictor of LTP ($p = 0.036$) [8]. During the intervention, magnetic resonance (MR) imaging offers several advantages like a good soft-tissue contrast without the need of contrast agent, the free orientation and positioning of single slice scans and the possibility to accurately track changes in the temperature inside the tissue [5,7,14,16].

Contribution. In this work, we propose a novel approach for the creation of a volumetric thermometry map without the development of a fully 3D sequence. The introduced 2.5D thermometry method utilizes any common 2D gradient-echo (GRE) sequences. Therefore, possible temporal limitations are less restricting than for the 3D sequences and images with higher resolution may be acquired offering standard thermometry accuracy of around 1 °C deviation while being more robust towards MR inhomogeneities [5]. We will show that our method is well-suited to reconstruct the actual coagulation zone after thermal ablation.

Related Work. Zhang et al. [20] propose a golden-angle-ordered 3D stack-of-radial multi-echo spoiled gradient-echo sequence with a variable flip angle. The image reconstruction is performed offline offering a temporal resolution between 2s-5s. Jiang et al. [6] use an accelerated 3D echo-shifted sequence and the Gadgetron framework for image reconstruction. Temporal resolution lies at around 3s with a temperature error of less than 0.65 °C. Quah et al. [13] are aiming at an increased volume coverage for thermometry without multiple receive coils. An extended k-space hybrid reconstruction was used, yielding an error of < 1 °C and an acquisition time of 3.5s for each image. Fielden et al. [3] present a comparison study between cartesian, spiral-out and retraced spiral-in/out (RIO) trajectories. Using the 3D RIO sequence, they achieved a true temporal resolution of 5.8s with a temporal standard deviation of 1.32 °C. Marx et al. [10] introduced the MASTER sequence for volumetric MR thermometry acquisition, acquiring six slices in around 5s. In a later work [9] they use optimized multiple-echo spiral thermometry sequences, which yield a better precision than the usual 2D Fourier transform thermometry. Image acquisition takes between 7s-11s. Svedin et al. [17] make use of a multi-echo pseudo-golden angle stack-of-stars sequence and offline image reconstruction using MATLAB. They achieved a temporal resolution of around 2s and a spatial average of the standard deviation through a time of 0.3-1.0 °C. Odéen et al. [12] propose the use of a 3D gradient recalled echo pulse sequence with segmented EPI readout. To estimate the temperature change, they also integrate a bioheat equation. They achieved a temperature root mean square error of 1.1 °C. Golkar et al. [4] introduce a fast GPU based simulation approach for cryoablation monitoring. The reconstruction takes 110s and the final result shows a Dice coefficient of 0.82. A summarize of the related work in comparison to our method is shown in Table 1.

Table 1. Overview about the related work in comparison to this work. Every work has been observed according to the following: 1) The kind of image sequence used. 2) The online or offline capability of the reconstruction framework. 3) The temporal resolution of the whole image acquisition in seconds. 4) The temperature accuracy in °C. 5) The resulting Dice Score similarity measurement if available.

	Image sequence	Reconstruction framework	Temporal resolution [s]	Temperature accuracy [°C]	Dice score
Zhang et al. [20]	3D	Offline	2–5	—	—
Jiang et al. [6]	3D	Online (Gadgetron)	3	<0.65	—
Quah et al. [13]	Stack of 2D	Hybrid	3.5	<1	—
Fielden et al. [3]	3D	Online	5.8	1.32	—
Marx et al. [10]	Stack of 2D	—	5	1.3	—
Marx et al. [9]	Stack of 2D	Online	7–11	0.29–0.65	—
Svedin et al. [17]	3D	Offline (MATLAB)	2	0.3–1.0	—
Odeen et al. [12]	Stack of 2D	Offline (MATLAB)	2.4–4.8	<1.1	—
Golkar et al. [4]	3D	—	110	—	0.82
This work	Single 2D	Online	Variable	1	0.75

2 Material and Method

2.1 Image Acquisition

The proposed 2.5D thermometry relies on sampling the volume of interest (VOI) using a common 2D GRE sequence and [1, ..., n] different orientations. The GRE sequence can directly reconstruct magnitude and phase images simultaneously. To ensure a proper sampling of the VOI in our setup the GRE sequence is rotated by 22.5° around the applicator's main axis. This results in an evenly distributed sample of eight different orientations. To increase the spatial resolution, the angles between the acquired scans should be as high as possible, resulting in the following acquisition order: 0°, 90°, 45°, 135°, 22.5°, 112.5°, 67.5°, and 157.5°. To reduce the delay between image acquisition and visualization of the volumetric thermometry map, the SIEMENS Healthineers Access-I Framework was integrated. The framework allows for fetching the image data directly from SIEMENS MR devices without an intermediate imaging archive system.

2.2 2.5D Thermometry Reconstruction

Before treatment starts, reference phase images are acquired for each of the eight orientations. Each newly acquired phase image will start computing the up-to-date 2D thermometry map for the current orientation during the treatment. To do so, the proton resonance frequency shift (PRFS) method is used as described by Rieke et al. [14]. The temperature T based on the PRFS is computed using the following Equation

$$T = \frac{\phi(t) - \phi(t_0)}{\gamma \alpha B_0 TE} + T_0 \tag{1}$$

with $\phi(t) - \phi(t_0)$ defining the phase difference between the current time point $\phi(i)$ and the reference timepoint $\phi(i_0)$, $\gamma = 42,576 \frac{MHz}{T}$ representing the gyromagnetic ratio of hydrogen protons, $\alpha = 0.01 \frac{ppm}{\Delta T}$ representing the proton resonance frequency change coefficient, B_0 representing the used magnetic field strength and TE representing the used echo time. The constant T_0 needs to be added to the temperature since Eq. 1 otherwise only computes the temperature change, neglecting the tissue's base temperature. The Access-I integration and 2D thermometry computation were implemented as modules using MeVisLab 3.4.1 [15]. The 2.5D thermometry reconstruction itself was implemented using C++. A schematic overview of the method can be seen in Fig. 1. To handle the voxel values during slice rotation every cartesian coordinate was mapped to the corresponding cylindrical coordinate representation using Eq. 2,

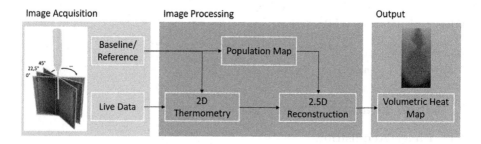

Fig. 1. Schematic overview of the proposed method.

$$P_r(x, y, z) = P_c(r, \theta, z) \tag{2}$$

$$r = \sqrt{(x - x_c)^2 + (y - y_c)^2}$$

$$\theta = atan2\left(\frac{x - x_c}{y - y_c}\right)$$

where x, y represents the Cartesian coordinates of the current voxel and x_c, y_c represents the Cartesian coordinates of the centerline corresponding to the applicator's axis for every slice z in the reconstructed volume. Upon acquisition of the reference images, a multi-dimensional population map is created. For each voxel (x_i, y_i, z_i) in the reconstructed volume, this population map holds information about the radius r and angle θ of the cylindrical coordinates, the general interpolation weight I_w, the adjacent interpolation partner coordinates $IP_{left}(x, y)$ and $IP_{right}(x, y)$ in the 2D live data as Cartesian representation and the weights

w_1 and w_2 of those interpolation partners. The weights may be acquired using Eq. 3,

$$w_1 = \left| \frac{\theta_{IP_{left}} - \theta_i}{\theta_{IP_{left}} - \theta_{IP_{right}}} \right| \tag{3}$$

$$w_2 = 1 - w_1$$

with θ_i representing the cylindric angle of the current Voxel i and $\theta_{IP_{left}}$, $\theta_{IP_{right}}$ representing the orientation angles of the left and right interpolation partners, respectively. The 2D population map can be applied to every slice of the final 3D output volume, reducing the computational power needed. During the intervention, every acquired live image triggers the reconstruction of the up-to-date 2.5D thermometry map. Here, the heat value for each voxel is reconstructed using Eq. 4,

$$T_i = I_w \cdot (w_1 \cdot T_{IP_{left}} + w_2 \cdot T_{IP_{right}}) \tag{4}$$

with T_i representing the temperature of the current voxel i and $T_{IP_{left}}$, $T_{IP_{right}}$ representing the temperature of the adjacent interpolation partners. Occurring vessels or other structures, which cause a heat sink effect are segmented during the intervention planning. Subsequently, the segmented structure is saved as an additional Look-Up Volume. Here, each voxel can be checked if it is part of a heat sink structure. Using this knowledge, the interpolation weight I_w, which ranges between $[0,1]$, may be adjusted. Figure 2 shows a single dimension of the population map for parameter weighting, a reconstructed heat map, a coagulation estimation based on an empirically defined threshold and the corresponding ground truth segmentation. The source code is available for download at https://github.com/jalpers/2.5DThermometryReconstruction.

2.3   Evaluation

Phantom Design. To create a first proof of concept, a pilot study was conducted to evaluate the 2.5D thermometry reconstruction using 13 bio protein phantoms as described by Bu Lin et al. [2]. The coagulation zone's visibility in the post-treatment data sets increased by adding a contrast agent $(0, 5\,\mu\text{mol/L}$ Dotarem) to the phantoms. For six phantoms, additional polyvinyl chloride (PVC) tubes with a diameter of $5\,\text{mm}$ and a wall thickness of $1\,\text{mm}$ were integrated into the phantoms (three single-tubes, three double-tubes) to simulate a possible heat sink (HS) effect.

Experimental Setup. The applicator of the permittivity feedback control MWA system (MedWaves Avecure, Medwaves, San Diego, CA, USA, 14G) was placed inside the phantom by sight and secured in position. Subsequently, the phantoms were placed inside a $1,5T$ MR scanner (Siemens Avanto, Siemens Healthineers, Germany). The coaxial cables connected to the applicator and MW generator were led through a waveguide. Chokes and electrical grounding measures were added as described by Gorny et al. [5] to reduce radio frequency interference. In the case of the perfusion phantoms, the PVC tubes were led through the wave guide. They

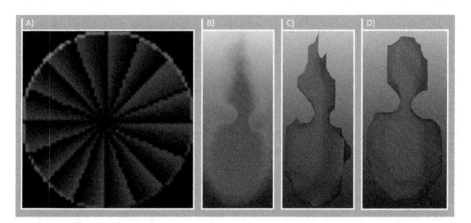

Fig. 2. A) Example population map for output weights color coded on gray scale. B) Reconstructed volumetric heat map. C) Estimated coagulation necrosis based on a threshold of 57 °C. D) Manually segmented ground truth.

were connected to a diaphragm pump and a water reservoir outside the scanning room. A flow meter (SM6000, ifm electronic, Essen, Germany) was interposed between the reservoir and the pump, providing a flow rate of 800 mL/min. Observations showed a moderate HS effect using this setup with a maximum antenna power of 36W. Additionally, temperature sensors were inserted in two phantoms to experimentally verify the temperature accuracy of 1 °C. Right before treatment, ten reference phase images were acquired and averaged for each orientation to compensate for static noise. The MWA duration was set to 15 min with a temperature limit of 90 °C. The GRE sequence offers a slice thickness of 5 mm, a field of view (FOV) of 256 mm * 256 mm, a matrix of 256 * 256, and a bandwidth of 260 Hz/Px. Image acquisition took around 1.1 s with a 5 s break to simulate the temporal resolution for a breathing patient. The TE was 3.69 ms, the TR 7.5 ms, and the flip angle 7°. For post-treatment observation a 3D turbo spin echo (TSE) sequence (TE = 156 ms, TR = 11780 ms, flip angle = 180°, matrix = 256 * 256, FOV = 256 mm * 256 mm, bandwidth = 40 Hz/Px, slice thickness = 1 mm) was used. The 3D TSE allows for proper visualization of the real coagulation zone due to a very high tissue contrast. Extraction of the coagulation ground truth was done manually by a clinical expert using MEVIS draw (Fraunhofer MEVIS, Bremen, Germany). All data sets used are available for download at http://open-science.ub.ovgu.de/xmlui/handle/684882692/89.

Statistical Evaluation. Final evaluation of the acquired data was performed using the dice similarity coefficient (DSC) as explained in Eq. 5

$$DSC = \frac{2 * TP}{2 * TP + FP + FN} \tag{5}$$

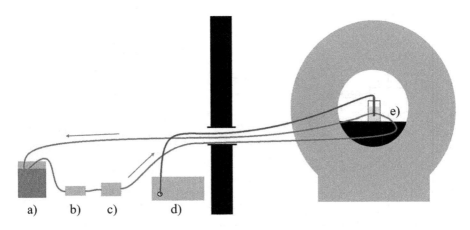

Fig. 3. Experimental evaluation setup. Flexible tubes (blue) lead the water (a) through a flow meter (b), a diaphragm pump (c) and the bio protein phantom (e). The coaxial cables (red) connect the applicator with the MW generator d). (Color figure online)

with TP representing the true positives, FP the false positives and FN the false negatives. Additionally, the standard error of the mean (SEM) was computed at a confidence level of 95% (p = 0.05) using Eq. 6

$$\sigma = \sqrt{\frac{\sum (x_i - \bar{x})^2}{N - 1}} \qquad (6)$$

$$SEM = \frac{\sigma}{\sqrt{N}} * 1.96$$

with σ representing the standard deviation, x_i the current sample, $\bar{x}$ the mean value and N the sample size. To compute the SEM at a confidence level of 95% it has to be multiplied by 1.96, which is the approximated value of the 97.5 percentile of the standard normal distribution.

3 Results

Summarized evaluation results can be seen in Fig. 4. Empirically determined coagulation thresholds were set between 51 °C and 61 °C depending on each phantom's pH value. It is noticeable that the DSCs for HS phantoms show a very high SEM with 0.70 ± 0.15(±21.25%) and 0.74 ± 0.06(±8.49%) regarding the sensitivity. The high range results from a corrupted dataset due to heavy artifacts within the image data. Leaving the corrupted dataset out of the evaluation, the SEM shows a significantly lower deviation of 0.76 ± 0.062(±8.07%) and 0.77 ± 0.048(±6.25%) for the DSC and sensitivity, respectively. Observations show a slightly higher DSC and sensitivity for phantoms without any HS effect. Here, the values range from 0.79 ± 0.04(±4.53%) and 0.79 ± 0.04(±5.55%), respectively. Evaluation showed an overall SEM for the DSC of 0.75 ± 0.07(±9.76%)

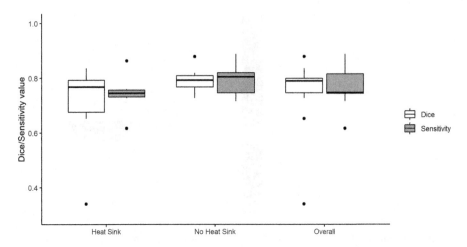

Fig. 4. Summarized evaluation results for phantoms without HS effect, phantoms with HS effect and the overall results. Note that the data range [0, 0.3] was left out because no data points are present in that range.

and a SEM for sensitivity of $0.77 \pm 0.04(\pm 4.99\%)$. To evaluate the computational effort, every major step was performed 100 times. The creation of the population map and the heat sink look up volume took $25.53\,ms \pm 3.33\,ms$ and $3.91\,s \pm 0.59\,s$, respectively. These two steps need to be done just once before start of the treatment. The reconstruction of the 2.5D thermometry map was performed in $18.02\,ms \pm 5.91\,ms$ on a customary workstation (Intel(R) Core(TM) i5-6200U CPU, double-core 2.30 GHz, 8 GB RAM, Intel(R) HD Graphics 520). This reconstruction will be performed every time a new image is acquired during treatment.

4   Discussion and Conclusion

The aim of this work is to develop a volumetric thermometry map, which can be applied to a wide variety of clinical setups. Therefore, our work heavily relies on the up-to-date standard 2D GRE sequence for image acquisition. This allows for the standard accuracy of the thermometry up to $1.0\,°C$. Nonetheless, the sampling of the 3D volume also results in some disadvantages, which need to be addressed in the future. First, the diffusion of the heat inside the tissue is not linear over time. Therefore, it would be necessary to include an adaptive temporal and spatial resolution depending on the current intervention time. A new study should be conducted to identify an optimal sequence protocol for this 2.5D thermometry approach. Second, we found that the reconstruction sometimes shows stair-case artifacts. Because only one image is acquired every few seconds, the time difference between adjacent orientations may be very high. The temperature difference for each voxel dependent on the applicator's radius may be computed and applied to the corresponding voxel on every other out-of-date data to compensate for this error. This transfer of the heat gradient may improve the reconstruction accuracy. Another approach may be the use of a model-based reconstruction to take different tissue characteristics into account. To pseudo-increase the temporal resolution, bio heat transfer simulations may also be included during reconstruction. The acquired live data may be able to adjust the simulation parameters to increase the simulation accuracy. Finally, our study only performs on bio protein phantoms. Results show a proof of concept for the proposed method, but it still has to be evaluated in real tissue and a more realistic clinical environment. Therefore, perfused ex vivo livers may be a way to go in the future. Additionally, we currently assume a breath-holding state or at least a breath-triggered image acquisition. Research shows that a wide range of interventional registration methods is available, but further investigations in this area still need to be done to create an applicable method. The last issues arise because of the MR inhomogeneity during image acquisition. The slightest disturbances may result in heavy image artifacts. Proper shielding of the MW generator is needed to reduce the SNR loss over time thus increasing the thermometry and reconstruction accuracy.

In conclusion, we proposed a novel method for 2.5D thermometry map reconstruction based on common GRE sequences rotated around the applicator's main axis. A pilot study was conducted using bio protein phantoms to simulate cases with possible heat sink effects and without. The evaluation shows promising results regarding the DSC of the reconstructed 2.5D thermometry map and a manually defined ground truth. Future work should address the reconstruction method's improvement by integrating further apriori knowledge like the estimated shape of the heat distribution. Furthermore, a more realistic study should be conducted with bigger sample size and real tissue. In sum, the method shows a high potential to improve the clinical success rate of minimally invasive ablation procedures without necessarily hampering the standard clinical workflow of the individual clinician.

References

1. Ahmed, M., et al.: Image-guided tumor ablation: standardization of terminology and reporting criteria–a 10-year update. Radiology **273**(1), 241–260 (2014)
2. Bu-Lin, Z., Bing, H., Sheng-Li, K., Huang, Y., Rong, W., Jia, L.: A polyacrylamide gel phantom for radiofrequency ablation. Int. J. Hyperth. **24**(7), 568–576 (2008)
3. Fielden, S.W., et al.: A spiral-based volumetric acquisition for MR temperature imaging. Magn. Reson. Med. **79**(6), 3122–3127 (2018)
4. Golkar, E., Rao, P.P., Joskowicz, L., Gangi, A., Essert, C.: Fast GPU computation of 3D isothermal volumes in the vicinity of major blood vessels for multiprobe cryoablation simulation. In: Frangi, A.F., Schnabel, J.A., Davatzikos, C., Alberola-López, C., Fichtinger, G. (eds.) MICCAI 2018. LNCS, vol. 11073, pp. 230–237. Springer, Cham (2018). https://doi.org/10.1007/978-3-030-00937-3_27

5. Gorny, K.R., et al.: Practical implementation of robust MR-thermometry during clinical MR-guided microwave ablations in the liver at 1.5 T. Physica Medica **67**, 91–99 (2019)
6. Jiang, R., et al.: Real-time volumetric MR thermometry using 3D echo-shifted sequence under an open source reconstruction platform. Magn. Reson. Imaging **70**, 22–28 (2020)
7. Kägebein, U., Speck, O., Wacker, F., Hensen, B.: Motion correction in proton resonance frequency-based thermometry in the liver. Top. Magn. Reson. Imaging **27**(1), 53–61 (2018)
8. Laimer, G., et al.: Minimal ablative margin (mam) assessment with image fusion: an independent predictor for local tumor progression in hepatocellular carcinoma after stereotactic radiofrequency ablation. Eur. Radiol. **30**(5), 2463–2472 (2020)
9. Marx, M., Ghanouni, P., Butts Pauly, K.: Specialized volumetric thermometry for improved guidance of MR g FUS in brain. Magn. Reson. Med. **78**(2), 508–517 (2017)
10. Marx, M., Plata, J., Pauly, K.B.: Toward volumetric MR thermometry with the MASTER sequence. IEEE Trans. Med. Imaging **34**(1), 148–155 (2014)
11. Mauri, G., et al.: Technical success, technique efficacy and complications of minimally-invasive imaging-guided percutaneous ablation procedures of breast cancer: a systematic review and meta-analysis. Eur. Radiol. **27**(8), 3199–3210 (2017)
12. Odéen, H., Almquist, S., de Bever, J., Christensen, D.A., Parker, D.L.: MR thermometry for focused ultrasound monitoring utilizing model predictive filtering and ultrasound beam modeling. J. Ther. Ultrasound **4**(1), 1–13 (2016)
13. Quah, K., Poorman, M.E., Allen, S.P., Grissom, W.A.: Simultaneous multislice MRI thermometry with a single coil using incoherent blipped-controlled aliasing. Magn. Reson. Med. **83**(2), 479–491 (2020)
14. Rieke, V., Butts Pauly, K.: MR thermometry. J. Magn. Reson. Imaging Official J. Int. Soc. Magn. Resonance Med. **27**(2), 376–390 (2008)
15. Ritter, F., et al.: Medical image analysis. IEEE Pulse **2**(6), 60–70 (2011)
16. de Senneville, B.D., Mougenot, C., Quesson, B., Dragonu, I., Grenier, N., Moonen, C.T.W.: MR thermometry for monitoring tumor ablation. Eur. Radiol. **17**(9), 2401–2410 (2007)
17. Svedin, B.T., Payne, A., Bolster, B.D., Jr., Parker, D.L.: Multiecho pseudo-golden angle stack of stars thermometry with high spatial and temporal resolution using k-space weighted image contrast. Magn. Reson. Med. **79**(3), 1407–1419 (2018)
18. Tehrani, M.H., Soltani, M., Kashkooli, F.M., Raahemifar, K.: Use of microwave ablation for thermal treatment of solid tumors with different shapes and sizes-a computational approach. Plos One **15**(6), e0233219 (2020)
19. Tomasian, A., Gangi, A., Wallace, A.N., Jennings, J.W.: Percutaneous thermal ablation of spinal metastases: recent advances and review. Am. J. Roentgenol. **210**(1), 142–152 (2018)
20. Zhang, L., Armstrong, T., Li, X., Wu, H.H.: A variable flip angle golden-angle-ordered 3d stack-of-radial MRI technique for simultaneous proton resonant frequency shift and t1-based thermometry. Magn. Reson. Med. **82**(6), 2062–2076 (2019)

Detection of Critical Structures in Laparoscopic Cholecystectomy Using Label Relaxation and Self-supervision

David Owen[1]([✉]), Maria Grammatikopoulou[1], Imanol Luengo[1],
and Danail Stoyanov[1,2]

[1] Digital Surgery, A Medtronic Company, London, UK
david.owen@medtronic.com
[2] Wellcome/EPSRC Centre for Interventional and Surgical Sciences,
University College London, London, UK

Abstract. Laparoscopic cholecystectomy can be subject to complications such as bile duct injury, which can seriously harm the patient or even result in death. Computer-assisted interventions have the potential to prevent such complications by highlighting the critical structures (cystic duct and cystic artery) during surgery, helping the surgeon establish the Critical View of Safety and avoid structure misidentification.

A method is presented to detect the critical structures, using state of the art computer vision techniques. The proposed label relaxation dramatically improves performance for segmenting critical structures, which have ambiguous extent and highly variable ground truth labels. We also demonstrate how pseudo-label self-supervision allows further detection improvement using unlabelled data.

The system was trained using a dataset of 3,050 labelled and 3,682 unlabelled laparoscopic cholecystectomy frames. We achieved an IoU of .65 and presence detection F1 score of .75. The model's outputs were further evaluated qualitatively by three expert surgeons, providing preliminary confirmation of our method's benefits.

This work is among the first to perform detection of critical anatomy during laparoscopic cholecystectomy, and demonstrates the great promise of computer-assisted intervention to improve surgical safety and workflow.

Keywords: Surgical video · Anatomy detection · Self-supervised learning

1 Introduction

Laparoscopic cholecystectomy is a common surgery in which the gallbladder is removed. This involves exposing the critical structures (cystic duct and artery),

Electronic supplementary material The online version of this chapter (https://doi.org/10.1007/978-3-030-87202-1_31) contains supplementary material, which is available to authorized users.

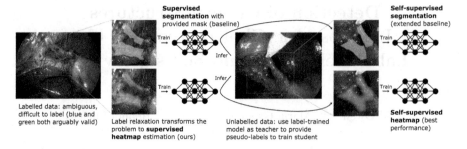

Fig. 1. Overview of our methods. Segmentation is challenged by ground truth structures with ambiguous extent (left). Label relaxation transforms the problem to heatmap estimation with down-weighting of ambiguous regions, which improves detection performance. Self-supervision feeds in unlabelled data via pseudo-labels, allowing further improvement. (Color figure online)

clipping and dividing them, then extracting the gallbladder [7]. Complications can occur when the structures are misidentified or confused with the common bile duct, particularly as they may be difficult to distinguish without thorough dissection. Official guidance has encouraged that surgeons establish "Critical View of Safety" (CVS) before clipping and division [10]. In CVS, both structures can clearly and separately be identified, and traced as they enter the gallbladder.

Computer assistance in achieving CVS has great potential to improve surgical safety and workflow, but has only recently become possible due to advances in computer vision [7]. Namazi *et al.* demonstrated a proof of principle approach using binary CVS classification [9]. Tokuyasu *et al.* developed a bounding box detection system, focused on anatomical landmarks that included the common bile duct and cystic duct but not the cystic artery [12]. Most recently, Mascagni *et al.* used joint segmentation of the hepatobiliary anatomy and classification of CVS [8], arguably combining the best aspects of prior work. Our work differs by focusing on the critical structures directly, as these are the structures that surgeons must identify and divide. This may be beneficial for guiding surgical workflow, providing visual cues that can help achieve CVS.

We present a novel method for detecting critical structures that outperforms conventional segmentation, using label relaxation (Sect. 2.1) to better handle challenging ground truth labels in images where structures are ambiguous. Subsequently, we incorporated pseudo-label self-supervision (Sect. 2.2), using unlabelled data to further improve performance. We trained and evaluated these methods using 3,050 labelled images from 75 videos and 3,682 unlabelled images from 90 videos for self-supervision. Finally, we gathered feedback from three experienced surgeons (Sect. 3.4), comparing different methods and confirming our method can improve clinical significance when detecting critical structures.

2   Methods

2.1   Critical Structures Identification via Label Relaxation

Our objective was to label the critical structures, here treated as a single foreground class, with the rest of the image considered as background. This is naturally posed as a binary segmentation problem. Standard segmentation approaches struggled to perform well in this task, because of the ambiguous and subjective nature of critical structures annotation (see Fig. 1). This problem was exacerbated by the use of conventional one-hot encoding: a given pixel is assigned as either 100% structure or 100% background class. This impairs generalisation, and led the model to struggle with false negatives.

To overcome this, we developed a technique inspired by related work in surgical tool detection [4]. Rather than segmentation, we trained a network for heatmap regression, where the ground truth heatmap is derived from the original annotations' Euclidean distance transforms.

Given a binary segmentation ground truth, x_k for structure k, we defined the relaxed label as $x'_k = 1 - \exp \frac{-\,\text{edt}(x_k \oplus t)}{d}$, where $\text{edt}(\cdot)$ is the Euclidean distance transform, $\oplus t$ represents dilation with a square of t pixels and d is a parameter to control the relaxation. Each x'_k is then normalised by its maximum value to allow use as a probability heatmap. Where heatmaps overlap for different structures within an image, the maximum value was used.

Consequently, central pixels are assigned high confidence, and more distant pixels are assigned low confidence as shown in Fig. 1. This label relaxation better reflects the ambiguity of the structure boundaries, and copes better with variation in annotations. This contrasts with pre-existing work, which largely focuses on improving segmentation results near object edges, and assumes unambiguously correct edge labels in ground truth data [14,15].

2.2   Pseudo-label Self-supervision

Labelling medical imagery is widely recognised as a bottleneck due to its difficulty, high time cost and compliance challenges [11]. This is particularly true for surgical video, which generates large amounts of unstructured data. In this work, we further improved our model by using unlabelled data via self-supervision [1]. Unlike previous work on self-supervision in endoscopic surgery [11], which uses generative models and consistency-based losses, we propose a simple pseudo-label approach that requires minimal computational overhead [1].

After training an initial model on labelled data, we used its predictions to provide pseudo-labels in unlabelled data [1]. This serves as teacher in a teacher-student architecture, where a newly initialised student model is trained on both pseudo-labelled images and the original labelled images. Previous work explored similar methods in segmentation for vehicle imagery [1] and demonstrated that the student learns a superior distillation of feature space compared to its teacher, leading to improved segmentation performance. Here we adapted the approach for heatmap regression by using softmax outputs as the pseudo-labels, rather than hard segmentation outputs. All models used the same architecture, regularisation and hyperparameters (Sect. 2.3).

2.3 Implementation

We used convolutional neural networks throughout, with FCN segmentation architecture [6] – a common baseline for segmentation. All networks used ResNet101 as a backbone [5]. For the segmentation and self-supervised segmentation models we trained the FCN with cross-entropy loss. We initially considered using class frequency weighted cross-entropy loss, in case class imbalance was the cause for segmentation model under-performance, but results were similar to equally weighted cross-entropy loss. To assist with comparison, the proposed heatmap model was kept similar to the segmentation model, simply using softmax to convert raw logits to a heatmap. For our proposed heatmap methods, we used soft cross-entropy loss and relaxed the ground truth label as discussed in Sect. 2.1.

All models were implemented in PyTorch 1.5 and optimised using Adam with learning rate 1e-4 and a "poly" learning rate schedule [1], trained until convergence. During training, models used random image augmentations (padding, cropping, flipping, blurring, rotation, noising) and model regularisation via dropout. We did not perform extensive hyperparameter tuning for augmentation, nor for label relaxation parameters t and d (Sect. 2.1). Performance did not seem sensitive, and we used $t = 15$ and $d = 10$ throughout. For evaluation in each case, the model with lowest validation loss was used. A supervised model training takes approximately 50 epochs (10 h) using four 16GB NVIDIA GPUs, with teacher-student self-supervision requiring approximately twice this time. For self-supervision experiments, the student model is pre-trained on teacher-generated pseudo-labels for 10 epochs, then fine tuned on ground truth labels for 50 epochs [13], again using "poly" schedule this time across the combined 60 epochs. Validation performance was not improved by pretraining for any longer, perhaps due to the relatively small size of the dataset. Similarly, we did not iterate the self-supervision as we saw no further benefit [1].

3 Experiments and Results

3.1 Data and Training

We used 3,050 labelled images from 75 separate laparoscopic cholecystectomy videos, frames chosen near where CVS is achieved. Frames are sampled at 1fps in a window of approximately 40s for each video. Most images contain cystic duct (90%) and/or cystic artery (87%). Labelling was performed by surgical data annotators under supervision of an anatomy specialist. Guidelines and tutorials for annotation were validated by surgeons. Labelled images were separated by video into train/val/test (60/20/20%) – with the test set held out for final evaluation of performance. We additionally used 3,682 unlabelled images derived from 90 videos for the self-supervision experiments – all used as training data.

We trained four models: a baseline segmentation method as described in Sect. 2.3, a heatmap method as described in Sect. 2.1, and variants of both methods using self-supervision to exploit the unlabelled data. We assessed performance on the validation and test sets, and then provided example model outputs from the test set for evaluation by surgeons.

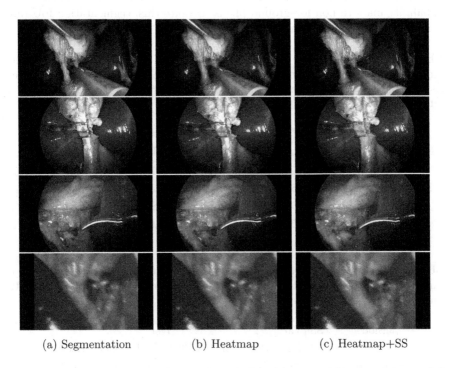

(a) Segmentation   (b) Heatmap   (c) Heatmap+SS

Fig. 2. Example frames from the test set, with softmax model outputs imposed in blue-green. Heatmaps generally reduced false negatives e.g. in the bottom two rows, and show some reduction in false positives (top row). Self-supervision led to further improvements, e.g. the false positive in the top row and false negatives in bottom three rows. (Color figure online)

3.2 Ablation Study

Table 1 shows pixel-level metrics, ordered by method (segmentation versus our proposed heatmap method) and whether self-supervision was used. To accommodate edge ambiguity, the evaluation uses VOC-style metrics, in which a 10 pixel margin around each structure is assigned as "ignore" [3] (not used during training). Heatmap detection consistently performs better than segmentation in IoU, regardless of whether self-supervision is used. Comparing without self-supervision, the IoU is higher for our proposed heatmap approach by 9.7pp/11.7pp (val/test). For both segmentation and heatmaps, our proposed self-supervision seems beneficial: it increased segmentation results by IoU by 1.5pp/0.9pp (val/test); and the IoU of heatmap methods by 3.7pp/3.1pp (val/test).

Notably, although the performance is generally best for our proposed heatmap method with self-supervision and second-best for heatmaps without self-supervision, segmentation achieved a higher pixel precision on the test set. This makes sense in light of the label relaxation, which inevitably assigns some probability mass to non-foreground pixels. Despite this, segmentation IoU (and overall performance) remains worse due to its much lower recall.

Table 1. Pixel-level accuracy metrics, by method. Heatmaps typically outperformed segmentation, as shown in improvements in IoU and other metrics in val and test sets. Self-supervision (SS) generally improves models, with the possible exception of segmentation precision in the test set.

Method	SS	Val			Test		
		IoU	Precision	Recall	IoU	Precision	Recall
Segmentation	✗	.547	.764	.658	.501	**.869**	.542
Segmentation	✓	.562	.807	.649	.512	.849	.563
Heatmap	✗	.644	.836	.750	.618	.811	.721
Heatmap	✓	**.681**	**.867**	**.761**	**.649**	.823	**.755**

Table 2. Higher-level presence detection metrics, evaluated with IoU threshold 0.5 to count as a true positive detection. In every metric, heatmaps outperformed segmentation. Self-supervision (SS) generally improved results.

Method	SS	Val			Test		
		F1	Precision	Recall	F1	Precision	Recall
Segmentation	✗	.597	.599	.594	.615	.626	.606
Segmentation	✓	.640	.640	.641	.616	.616	.617
Heatmap	✗	.716	.721	.711	.694	.703	.685
Heatmap	✓	**.811**	**.833**	**.790**	**.749**	**.750**	**.749**

Table 2 shows metrics for frame-level presence detection, where artery and duct detections must exceed an IoU threshold 0.5 to count as true positives in a given frame. This means that low IoU detections count as false positives. Such statistics are conservative, as a lower IoU overlap may nonetheless be fairly accurate given the ambiguity of ground truth annotation extent (see Fig. 1). Nevertheless, results show a similar pattern to the pixel-level performance metrics, with our proposed heatmap method outperforming segmentation, and self-supervision improving models' performance. Notably, the increased pixel-level precision of segmentation methods does not translate to structure detection, where our heatmap method performs better by every metric.

3.3   Qualitative Performance Across Surgery

Figure 3 shows example frames and model outputs from an 11 min excerpt of a laparoscopic cholecystectomy video in the test set. The full example video is included in Supplementary Material. Performance is generally strong. The model typically does not suffer false detections before structures are visible (Fig. 3a), although it can be fooled by similar shapes near a tool tip, particularly if such shapes are visible near the gallbladder. Even when the structures are heavily coated by fat, the model tends to recognise them at least partially (Fig. 3b, 3c). Manipulation does not usually prevent detection (Fig. 3d).

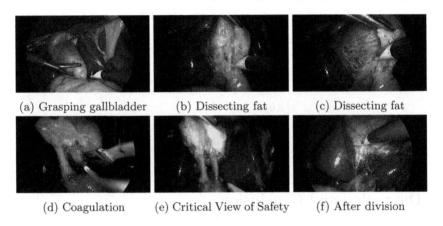

(a) Grasping gallbladder   (b) Dissecting fat   (c) Dissecting fat

(d) Coagulation   (e) Critical View of Safety   (f) After division

Fig. 3. Example frames taken across an 11 min excerpt from a test set video, using our self-supervised heatmap model. Model detections in blue-green. (Color figure online)

Structures often remain detectable after division (Fig. 3e). Although we were impressed by the model's generalisation, this might be undesirable in a practical implementation. This would be fixable by using surgical phase recognition [16], which could deactivate detection after division of structures.

3.4   Surgeon Preference

Table 3 shows how surgeons ranked the different methods in frames taken from the test set. Example frames were intentionally chosen to show differences between the methods, as model outputs were often similar between methods. Set 1 was randomly chosen from frames with greatest differences between segmentation and heatmap outputs ($|IoU_{seg} - IoU_{heat}|$). Set 2 was randomly chosen from frames with greatest differences between supervised and self-supervised heatmap outputs ($|IoU_{sup} - IoU_{ss}|$). Each set used frames from eight different videos.

Surgeons preferred our heatmap method to segmentation, and preferred heatmaps with self-supervision to vanilla heatmaps. These preferences failed to show significance in Set 1, but did show statistical significance in Set 2 ($p < 0.05$).

We discuss this further in Sect. 4.1. Free text feedback was generally positive where provided ("all pretty good"), although one participant did note that in one frame common bile duct was detected.

Table 3. Surgeon preferences (blind) by surgeon and model. Average ranking (1–3, lower is better) for each method and set, with post hoc Wilcoxon signed-rank p value, bolded for significance ($p < 0.05$) after multiple comparison adjustment. p_{heat} is for difference between heatmap and segmentation, p_{SS} is for difference between self-supervised and supervised heatmaps.

Set	Participant	Segm	Heat	Heat+SS	p_{heat}	p_{SS}
1	1	2.50	2.50	1.75	.340	.425
	2	2.13	2.00	2.00		
	3	2.50	1.88	2.13		
	Avg	2.22	2.06	1.97		
2	1	2.50	2.25	1.50	**5.57e-3**	**0.0221**
	2	2.63	1.75	1.88		
	3	2.63	2.00	1.75		
	Avg	2.38	1.97	1.63		

4   Discussion and Conclusion

4.1   Heatmaps Improve Accuracy, but Can Impair Visualisation

Our heatmap models are more accurate than segmentation, as shown in low-level pixel metrics such as IoU and higher-level presence detection such as F1 score. This is supported by blind ratings from surgeons, where they favoured our heatmaps over segmentations in Set 2. Counterintuitively, surgeons did not show a statistically significant preference in Set 1 – despite this set being selected for maximum differences between segmentation and heatmap outputs.

We believe this discrepancy was due to visualisation preferences, based on free text feedback ("I like narrow overlays not zones"). Our heatmap models, by design, tend to detect larger areas than the segmentation models. When we selected Set 1 to maximise differences between segmentation and heatmaps, this selected several frames where the difference is due to the heatmap highlighting a larger area (see Fig. 2, row 1). Conversely, Set 2 was chosen to maximise differences between supervised and self-supervised models, and did not show this effect to the same extent. This emphasises the importance of visualisation, and suggests an important direction for future work.

4.2   Self-supervision Particularly Helps in Difficult Videos

Self-supervision improves accuracy in general, but is particularly beneficial for a few difficult cases. This can be seen in Fig. 2. In rows 1, 3 and 4, self-supervision slightly improved the accuracy, but the overall detection was not changed significantly, and hence the IoU remains similar. In row 2, however, the accuracy improvement was much larger as the supervised model entirely misses the cystic artery, whereas the self-supervised model detected it. This finding is borne out by per-video IoU: for most videos the IoU difference between methods is on the order of 0–5pp, but for three videos it is 10pp or greater.

4.3   Conclusion

Our work is among the first to detect the critical structures during laparoscopic cholecystectomy. When trying to detect structures with ambiguous extent and challenging annotations, a novel heatmap-based approach based on label relaxation significantly outperformed a segmentation baseline. Self-supervision provided further improvement by using unlabelled data for additional training. Our method was validated on held-out test data and surgeon evaluations supported these findings. We hope to develop the method further by using a greater variety of anatomic classes [8,12], such as considering the cystic artery and duct separately, and possibly annotating the common bile duct. Modelling temporal consistency across frames might also be beneficial [2]. Finally, another important advance would be to further validate the method in a large dataset covering the full range of variability. Automatic detection of critical structures in surgery has tremendous potential to improve surgical safety, training and workflow and ultimately patient outcomes. Our work will contribute towards this goal.

Acknowledgements. This work was supported by the Wellcome/EPSRC Centre for Interventional and Surgical Sciences (WEISS) at UCL (203145Z/16/Z), EPSRC (EP/P012841/1, EP/P027938/1, EP/R004080/1) and the H2020 FET (GA 863146). Danail Stoyanov is supported by a Royal Academy of Engineering Chair in Emerging Technologies (CiET18196) and an EPSRC Early Career Research Fellowship (EP/P012841/1).

References

1. Chen, L.-C., et al.: Naive-Student: leveraging semi-supervised learning in video sequences for urban scene segmentation. In: Vedaldi, A., Bischof, H., Brox, T., Frahm, J.-M. (eds.) ECCV 2020. LNCS, vol. 12354, pp. 695–714. Springer, Cham (2020). https://doi.org/10.1007/978-3-030-58545-7_40
2. Colleoni, E., Moccia, S., Du, X., De Momi, E., Stoyanov, D.: Deep learning based robotic tool detection and articulation estimation with spatio-temporal layers. IEEE Rob. Autom. Lett. 4(3), 2714–2721 (2019)
3. Everingham, M., Van Gool, L., Williams, C.K., Winn, J., Zisserman, A.: The PASCAL visual object classes (VOC) challenge. Int. J. Comput. Vision 88(2), 303–338 (2010)

4. Fuentes-Hurtado, F., Kadkhodamohammadi, A., Flouty, E., Barbarisi, S., Luengo, I., Stoyanov, D.: EasyLabels: weak labels for scene segmentation in laparoscopic videos. Int. J. Comput. Assist. Radiol. Surg. **14**(7), 1247–1257 (2019). https://doi.org/10.1007/s11548-019-02003-2

5. He, K., Zhang, X., Ren, S., Sun, J.: Deep residual learning for image recognition. In: Proceedings of the IEEE Conference on Computer Vision and Pattern Recognition, pp. 770–778 (2016)

6. Long, J., Shelhamer, E., Darrell, T.: Fully convolutional networks for semantic segmentation. In: Proceedings of the IEEE Conference on Computer Vision and Pattern Recognition, pp. 3431–3440 (2015)

7. Mascagni, P., et al.: Formalizing video documentation of the Critical View of Safety in laparoscopic cholecystectomy: a step towards artificial intelligence assistance to improve surgical safety. Surg. Endosc., 1–6 (2019)

8. Mascagni, P., et al.: Artificial intelligence for surgical safety: automatic assessment of the Critical View of Safety in laparoscopic cholecystectomy using deep learning. Ann. Surg. (2021)

9. Namazi, B., et al.: AI for automated detection of the establishment of Critical View of Safety in laparoscopic cholecystectomy videos. J. Am. Coll. Surg. **231**(4), e48 (2020)

10. Pucher, P.H., Brunt, L.M., Fanelli, R.D., Asbun, H.J., Aggarwal, R.: SAGES expert Delphi consensus: critical factors for safe surgical practice in laparoscopic cholecystectomy. Surg. Endosc. **29**(11), 3074–3085 (2015). https://doi.org/10.1007/s00464-015-4079-z

11. Ross, T., et al.: Exploiting the potential of unlabeled endoscopic video data with self-supervised learning. Int. J. Comput. Assist. Radiol. Surg. **13**(6), 925–933 (2018). https://doi.org/10.1007/s11548-018-1772-0

12. Tokuyasu, T., et al.: Development of an artificial intelligence system using deep learning to indicate anatomical landmarks during laparoscopic cholecystectomy. Surg. Endosc., 1–8 (2020)

13. Yalniz, I.Z., Jégou, H., Chen, K., Paluri, M., Mahajan, D.: Billion-scale semi-supervised learning for image classification. arXiv preprint arXiv:1905.00546 (2019)

14. Yuan, Y., Xie, J., Chen, X., Wang, J.: SegFix: model-agnostic boundary refinement for segmentation. In: Vedaldi, A., Bischof, H., Brox, T., Frahm, J.-M. (eds.) ECCV 2020. LNCS, vol. 12357, pp. 489–506. Springer, Cham (2020). https://doi.org/10.1007/978-3-030-58610-2_29

15. Zhu, Y., Sapra, K., Reda, F.A., et al.: Improving semantic segmentation via video propagation and label relaxation. In: Proceedings of the IEEE/CVF Conference on Computer Vision and Pattern Recognition, pp. 8856–8865 (2019)

16. Zisimopoulos, O., et al.: DeepPhase: surgical phase recognition in CATARACTS videos. In: Frangi, A.F., Schnabel, J.A., Davatzikos, C., Alberola-López, C., Fichtinger, G. (eds.) MICCAI 2018. LNCS, vol. 11073, pp. 265–272. Springer, Cham (2018). https://doi.org/10.1007/978-3-030-00937-3_31

EMDQ-SLAM: Real-Time High-Resolution Reconstruction of Soft Tissue Surface from Stereo Laparoscopy Videos

Haoyin Zhou and Jagadeesan Jayender[✉]

Surgical Planning Laboratory, Brigham and Women's Hospital,
Harvard Medical School, Boston, USA
jayender@bwh.harvard.edu

Abstract. We propose a novel stereo laparoscopy video-based non-rigid SLAM method called EMDQ-SLAM, which can incrementally reconstruct thee-dimensional (3D) models of soft tissue surfaces in real-time and preserve high-resolution color textures. EMDQ-SLAM uses the expectation maximization and dual quaternion (EMDQ) algorithm combined with SURF features to track the camera motion and estimate tissue deformation between video frames. To overcome the problem of accumulative errors over time, we have integrated a g2o-based graph optimization method that combines the EMDQ mismatch removal and as-rigid-as-possible (ARAP) smoothing methods. Finally, the multi-band blending (MBB) algorithm has been used to obtain high resolution color textures with real-time performance. Experimental results demonstrate that our method outperforms two state-of-the-art non-rigid SLAM methods: MISSLAM and DefSLAM. Quantitative evaluation shows an average error in the range of 0.8–2.2 mm for different cases.

Keywords: Non-rigid SLAM · EMDQ · g2o-based graph optimization · High resolution texture · Multi-band blending · GPU parallel computation

1 Background

Three-dimensional (3D) reconstruction of tissue surfaces from intraoperative laparoscopy videos has found applications in surgical navigation [1,2] and planning [3]. One of the most important video-based 3D reconstruction methods is simultaneous localization and mapping (SLAM). Most existing SLAM methods assume that the environment is static, and the data at different time steps are aligned rigidly according to the estimated 6-DoF camera motion. This assumption is invalid for deformable soft tissues, which require much higher degrees of

Electronic supplementary material The online version of this chapter (https://doi.org/10.1007/978-3-030-87202-1_32) contains supplementary material, which is available to authorized users.

© Springer Nature Switzerland AG 2021
M. de Bruijne et al. (Eds.): MICCAI 2021, LNCS 12904, pp. 331–340, 2021.
https://doi.org/10.1007/978-3-030-87202-1_32

freedom to represent the non-rigid deformation and cannot be accurately recovered by the rigid SLAM methods.

Non-rigid SLAM is an emerging topic and the pioneering work is Dynamic-Fusion [4], which first addressed the problem of incrementally building the 3D model of deformable objects in real-time. Following DynamicFusion, many non-rigid SLAM works have been proposed in the computer vision field [5,6]. However, these methods mainly focused on small regions of interest that require the object to be placed directly in front of the camera. This may not be appropriate for medical applications since the laparoscope may need to scan large areas of the tissue surface. This motivates the need for the development of large-scale non-rigid SLAM methods, which have attracted significant attention in recent years. For example, Mahmoud et al. considered the tissue deformation to be negligible and applied rigid SLAM for dense reconstruction [8,9]. Mountney et al. analyzed and compensated for the periodic motion of soft tissue caused by respiration using the extended Kalman filter [10]. These methods worked for specific situations with assumptions of the underlying tissue deformation, but cannot handle general tissue deformation. To date, only a few non-rigid SLAM works without any assumptions of the tissue deformation have been reported. For example, Collins et al. proposed to use features and tissue boundaries to track tissue deformation [11]. Song et al. proposed to combine ORB-SLAM [20] and a deformation model for tracking the motion of the laparoscope and estimating the deformation of soft tissues [15]. Recently, Lamarca et al. proposed a monocular non-rigid SLAM method called as DefSLAM for large-scale non-rigid environments, which combines the shape-from-template (SfT) and non-rigid structure-from-motion (NRSfM) methods and has obtained impressive results on laparoscopy videos [16].

In this paper, we propose a novel stereo video-based non-rigid SLAM method called as EMDQ-SLAM, which can track large camera motion and significant tissue deformation in real-time. The key algorithm of EMDQ-SLAM is the expectation maximization and dual quaternion (EMDQ) algorithm [17], which can generate dense deformation field from sparse and noisy SURF feature matches in real-time. Hence, EMDQ can efficiently track the camera motion and estimate the non-rigid tissue deformation simultaneously. However, EMDQ tracking suffers from a problem that the estimated tissue deformation may have accumulative errors. To solve this problem, we have developed a graph optimization method based on the g2o library [18], which uses the results of EMDQ as the initial values for further refinement. To preserve the high resolution color textures, we have adapted the multi-band blending (MBB) method [19] for real-time incremental applications, and have implemented GPU-based parallel computation for real-time performance.

2   Method

We first perform a GPU-based stereo matching method to estimate depths of image pixels, which was developed following Ref. [21]. Then, EMDQ-SLAM mosaics the stereo matching results at different time steps by estimating the camera motion and tissue deformation.

Without loss of generality, EMDQ-SLAM considers time $t = 0$ as the canonical frame, and estimates the camera motion and surface deformation at time steps $t = 1, 2, ...$ with respect to time 0 for non-rigid mosaicking. The 6-DoF camera motion at time t is represented using the rotation matrix $\mathbf{R}_t \in SO(3)$ and translational vector $\mathbf{t}_t \in \mathbb{R}^3$. We denote the world coordinate of a template point p at time 0 as $\mathbf{x}_{p,0}$. The tissue deformation at time t is represented by displacements of each point p, which is denoted as $\varDelta \mathbf{x}_{p,t}$. And the world coordinate of point p at time t is

$$\mathbf{x}_{p,t} = \mathbf{x}_{p,0} + \varDelta \mathbf{x}_{p,t}. \tag{1}$$

Hence, tissue deformation recovery is equivalent to the estimation of $\varDelta \mathbf{x}_{p,t}$ for all template points p. Since the template may have millions of points, we have adapted the method from DynamicFusion [4] that uses sparse control points, or deformation nodes, to reduce the computational burden. For each template point p, its displacement $\varDelta \mathbf{x}_{p,t}$ is represented by the weighted average of its neighboring deformation nodes i, which is

$$\varDelta \mathbf{x}_{p,t} = \sum_i^N \left(w_i^p \varDelta \mathbf{x}_{i,t} \right), \tag{2}$$

where $w_i^p = \exp(-\alpha \left\| \mathbf{x}_{i,0} - \mathbf{x}_{p,0} \right\|^2)$, $\sum_i^N w_i^p = 1$ is the normalized weight between node i and point p, and node i is omitted if w_i^p is too small.

In summary, the parameters that need to be estimated include the 6-DoF camera motion $\mathbf{R}_t$ and $\mathbf{t}_t$, and the nodes displacements $\varDelta \mathbf{x}_{i,t}$, $i = 1, 2, ..., N$. We use a two-step framework to solve this problem, which includes EMDQ tracking and g2o-based graph optimization.

2.1   EMDQ Tracking

At time t, the coordinate of node i in the camera frame is $\mathbf{x}_{i,t}^c = \mathbf{R}_t(\mathbf{x}_{i,0} + \varDelta \mathbf{x}_{i,t}) + \mathbf{t}_t$. At time $t + 1$, we first perform SURF [23] matching between video frame t and $t + 1$, and obtain the related 3D coordinates of the SURF matches according to the pixel depths obtained by stereo matching. The number of SURF octave layers is set to one to avoid building the image pyramid and reduce the computational burden, which is reasonable because the change of image scale is small between adjacent image frames. Then, using the 3D SURF matches as the input, the EMDQ algorithm (1) obtains the displaced coordinates of nodes $\mathbf{x}_{i,t+1}^c$ from $\mathbf{x}_{i,t}^c$, and (2) removes SURF mismatches. The first output is directly used for updating the camera motion and nodes displacements at time $t + 1$ in this EMDQ tracking method, and the second output will be used in the subsequent g2o-based graph optimization.

The estimated displacements of nodes between time t and $t+1$, $\mathbf{x}_{i,t+1}^c - \mathbf{x}_{i,t}^c$, include both the rigid and non-rigid components, which are caused by camera motion and tissue deformation respectively. To decompose the rigid and non-rigid components, we follow the method in Ref. [22] to estimate the rigid transformation, $\mathbf{R}_{t \to t+1}$ and $\mathbf{t}_{t \to t+1}$, between two 3D point clouds $\left\{ \mathbf{x}_{i,t}^c \right\}$ and

$\{\mathbf{x}_{i,t+1}^c\}$, $i = 1, 2, ...N$. This method minimizes the sum of squared residuals and we consider the residuals as the non-rigid component, which is also the nodes displacements and we denote it as $\Delta\mathbf{x}_{i,t\to t+1} = \mathbf{x}_{i,t+1}^c - (\mathbf{R}_{t\to t+1}\mathbf{x}_{i,t}^c + \mathbf{t}_{t\to t+1})$. Finally, we update the camera motion in the world frame at time $t + 1$ by

$$\mathbf{R}_{t+1} = \mathbf{R}_{t\to t+1}\mathbf{R}_t, \mathbf{t}_{t+1} = \mathbf{R}_{t\to t+1}\mathbf{t}_t + \mathbf{t}_{t\to t+1}. \tag{3}$$

The node displacements at time $t + 1$ are updated by

$$\Delta\mathbf{x}_{i,t+1} = \mathbf{R}_t^T \Delta\mathbf{x}_{i,t\to t+1} + \Delta\mathbf{x}_{i,t}, i = 1, 2, ...N. \tag{4}$$

2.2 g2o-based Graph Optimization

The proposed EMDQ tracking method suffers from a problem that accumulative errors may exist in the estimated shape deformation. Specifically, Eq. (4) shows that the error of $\Delta\mathbf{x}_{i,t\to t+1}$ will result in the accumulative error of $\Delta\mathbf{x}_{i,t+1}$. Hence, in practice we found that EMDQ tracking works well for short video sequences, but may not be robust for long video sequences. The errors of $\Delta\mathbf{x}_{i,t+1}, i = 1, 2, ...N$ are mainly reflected as the differences among the neighboring deformation nodes. Hence, the deformation recovery errors can be reduced by using the as-rigid-as-possible (ARAP) [24] method, which aims to maintain the shape of deformable templates. We have developed a graph optimization method based on the g2o library [18] as the refinement step for EMDQ tracking, which integrates the EMDQ mismatch removal results and the ARAP costs. The vertices and edges of the graph are introduced in the following section.

Vertices: The graph has a total of $N + 1$ vertices, which include the camera motion ($\mathbf{R}_{t+1}$ and $\mathbf{t}_{t+1}$), and displacements of N nodes ($\Delta\mathbf{x}_{i,t+1}, i = 1, 2, ...N$).

SURF Matching Edges: We denote the 3D camera coordinates of a SURF feature m at time t as $\mathbf{x}_{m,t}^c$, which can be directly obtained by the stereo matching results. Then, its world coordinate at time 0, $\mathbf{x}_{m,0}$, can be obtained according to the estimated camera motion and nodes displacements at time t. Ideally, the estimated $\mathbf{x}_{m,0}$ obtained from time t and $t + 1$ should be the same, which are denoted as $\mathbf{x}_{m,0}^t$ and $\mathbf{x}_{m,0}^{t+1}$ respectively. We use the differences between $\mathbf{x}_{m,0}^t$ and $\mathbf{x}_{m,0}^{t+1}$ as the cost, that is

$$f_{\text{SURF}}(\mathbf{R}_{t+1}, \mathbf{t}_{t+1}, \Delta\mathbf{x}_{i,t+1}) = \sum_m w_m \left\| \mathbf{x}_{m,0}^t - \mathbf{x}_{m,0}^{t+1} \right\|^2, \tag{5}$$

where $w_m = 1/(e_m + 1)$ is the weight of match m, and e_m is the related error of match m in the EMDQ algorithm to distinguish inliers and outliers. We use this soft weight w_m to handle situations when EMDQ does not distinguish inliers and outliers correctly.

ARAP Edges: The basic idea of the standard ARAP method is to minimize non-rigid component after removing the rigid components [24]. Since in our method, the rigid and non-rigid components have already been decomposed, the ARAP term in EMDQ-SLAM is simplified to

$$f_{\mathrm{ARAP}}(\Delta\mathbf{x}_{i,t+1}) = \sum_{i1,i2} w_{i1}^{i2} \left\| \Delta\mathbf{x}_{i1,t+1} - \Delta\mathbf{x}_{i2,t+1} \right\|^2, \tag{6}$$

where w_{i1}^{i2} is the weight between node $i1$ and $i2$. It is worth noting that this is a simplified ARAP method since it uses the same rigid component for all points.

2.3   GPU-Based Dense Mosaicking and MBB Texture Blending

We propose a planar TSDF [25] method to merge the template and stereo matching results at each time step. Our method reprojects each template point p to the image pixel I according to the estimated camera motion and nodes displacements, and merges the stereo matching depth of pixel I following the standard TSDF method. New template points are inserted from the stereo matching results if no existing point is reprojected to the related image pixel.

Since small misalignments are unavoidable, template point p and pixel I may have different RGB colors. Linear blending of the RGB colors may lead to blurry textures, as shown in Fig. 1(d). Hence, we employ the multi-band blending (MBB) method [19]. Traditional MBB method is offline and we refer to Ref. [19] for more details. In this paper, we mainly introduce our modifications for incremental and real-time blending. We use same notations as Ref. [19] for readers to follow easily. Our method reserves the previous information with the template points, and performs RGB color blending between the template and current image. The information at each template point p includes three spatial frequencies B_σ^p and the weight W^p. When blending with the current image, B_σ^p equals to the weighted average of that of p and I, i.e.,

$$B_\sigma^p = (1 - W_\sigma^I) B_\sigma^p + W_\sigma^I B_\sigma^I. \tag{7}$$

The weight of pixel I is obtained by Gaussian blurring $W_\sigma^I = W_{\mathrm{max}}^I * g_\sigma$. $W_{\mathrm{max}}^I = 1$ if $W^I > W^p$, otherwise $W_{\mathrm{max}}^I = 0$. W^p is updated by taking the max value of W^p and W^I. W^I is obtained in the same way as in Ref. [19].

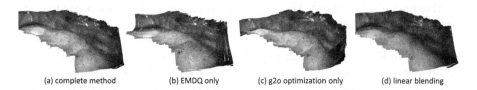

(a) complete method   (b) EMDQ only   (c) g2o optimization only   (d) linear blending

Fig. 1. Comparative study. For results in (c), g2o used estimations at the previous time step as the initial values, and Huber kernels were applied to handle outliers.

Comparative Study: As shown in Fig. 1, we conducted a comparative study to intuitively demonstrate the effects of the methods introduced in this paper. Both EMDQ tracking and g2o-based graph optimization contribute to the accuracy of EMDQ-SLAM. EMDQ tracking cannot handle long sequences robustly due to accumulative errors (Fig. 1(b)). g2o-based graph optimization suffers from the local minima problem and may be affected by SURF mismatches, which requires EMDQ results as the initial values (Fig. 1(c)). MBB blending can obtain better color textures than traditional linear blending (Fig. 1(d)).

3   Results

The source code was implemented in CUDA C++ on a desktop with an Intel Core i9 3.0 GHz CPU and NIVIDA Titan RTX GPU.

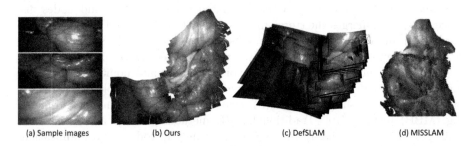

(a) Sample images   (b) Ours   (c) DefSLAM   (d) MISSLAM

Fig. 2. Experiments on the *in vivo* porcine abdomen video from the Hamlyn dataset. (a) Ours. (b) DefSLAM result, which includes multiple overlapped local templates. (c) MISSLAM results, which is the screenshot reported in their paper.

We introduced MISSLAM [15] and DefSLAM [16] for comparison, which are recent video-based non-rigid SLAM algorithms. MISSLAM is a stereo video-based method, which is not open source software hence we used the same data reported in their paper for comparison. It may not be fair for comparing our method with DefSLAM since DefSLAM uses monocular video. Further, DefS-LAM does not fully address the mosaicking problem and its mapping results are sets of overlapped local templates, rather than complete 3D point clouds or mesh models. The stereo laparoscopy videos used for validation were obtained from the Hamlyn online dataset[1]. We also captured intraoperative stereo laparoscopy videos on soft tissues at our hospital.

As shown in Fig. 2, we first conducted experiments on the Hamlyn dataset, which was obtained by scanning a porcine abdomen. The tissue had small deformation but the laparoscope motion was large. EMDQ-SLAM is able to generate a large 3D mosaic with clear textures corresponding to the area covered by the laparoscope. DefSLAM was able to track the camera motion from the monocular video, and provided multiple local templates. We also include a screenshot of the result from the MISSLAM paper for comparison. Figure 3(a) shows the results on another Hamlyn dataset, which scanned large areas of the liver surface. Due to low texture, the DefSLAM reported loss of tracking. Although it is difficult to provide quantitative comparisons since no ground truth is available, qualitative

---

[1] http://hamlyn.doc.ic.ac.uk/vision/.

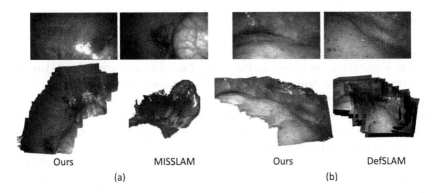

Ours   MISSLAM   Ours   DefSLAM

(a)   (b)

Fig. 3. (a) Experiments on the *in vivo* porcine liver video from the Hamlyn dataset. DefSLAM failed on this data. (b) Experiments on a stereo laparoscopy video captured during a minimally invasive sublobar lung surgery at our hospital.

comparisons show that our result is visually more accurate and can preserve high resolution texture.

For the experiments shown in Fig. 3(b), the video was captured during a minimally invasive sublobar lung surgery at our hospital. The surgeon was asked to move the stereo laparoscope within the patient's thoracic cavity. Due to heartbeat and respiratory motion caused by the adjacent lung, the deflated lung had significant and fast deformation. This experiment demonstrates that our method can handle highly deformable surfaces.

For the experiments shown in Fig. 4, the tissues had significant deformation due to heartbeat although the camera motion was small. Our method was able to track the deformation robustly and monitor the tissue deformation. Videos showing the results of the EMDQ-SLAM algorithm are provided as supplemental material for the paper.

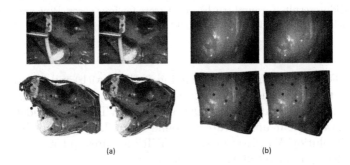

(a)   (b)

Fig. 4. Experiments on the Hamlyn dataset. (a) *In vivo* heart. (b) Heart phantom. The tissues had significant deformation and the camera motion was small.

Quantitative Experiments: The quantitative experiments were conduced using *ex vivo* porcine lungs and livers, which were placed on an Ambu bag and an anesthesia machine inflated/deflated the Ambu bag periodically to simulate the respiration, as shown in Fig. 5(a). Two electromagnetic (EM) trackers were attached to the laparoscope and on the tissue surface respectively. The laparoscope was gradually moved along the surface to create a 3D mosaic while estimating the tissue deformation, which was also measured using the EM sensor on the tissue surface. The EM coordinate frame was registered to the laparoscope coordinate frame, and our results were compared with the EM tracking results, as shown in Fig. 5(b). The errors for four cases are reported in Fig. 5(c), which were small when the EM tracker was in the field of view (FOV), but increased as the laparoscope moved far away, which is expected since the deformation of areas outside FoV cannot be monitored directly. The mean/standard deviation of errors when the EM tracker was in the FoV are 0.87/0.40, 2.1/0.58, 1.7/0.70 and 2.2/0.70 mm respectively for the four cases. The problem that areas outside FoV cannot be accurately estimated is an intrinsic drawback for large-scale non-rigid SLAM methods, since the deformation of invisible areas is obtained by extrapolation of the visible areas. However, this problem does not affect the mosaicking process because mosaicking is only performed at areas in FoV.

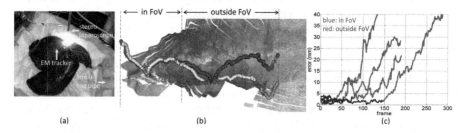

Fig. 5. Quantitative experiments. (a) Configuration. (b) Red dots are the estimated trajectory of the EM tracker by EMDQ-SLAM in the camera frame, yellows dots are the EM tracking results (ground truth). (c). Errors for four cases. Blue and red suggest the EM tracker was in and outside the field of view (FoV) respectively. (Color figure online)

Runtime: For the experiments shown in Fig. 2, Fig. 3(a)(b) and 4(a)(b), the average runtime to process one video frame was 92, 101, 97, 38 and 53 ms respectively, which included stereo matching, EMDQ-SLAM and VTK-based rendering. Hence, our method works at an update rate of around 10–26 Hz, which mostly depends on the image resolution. We use at most 1500 SURF features and 500 SURF matches at each time step to maintain the real-time performance.

4   Conclusion

The problem of large-scale non-rigid SLAM for medical applications is still an open problem. In this paper, we propose a novel non-rigid SLAM method called

EMDQ-SLAM, which uses a two-step framework to track the camera motion and estimate the tissue deformation. Although it is difficult to provide quantitative comparisons with other methods due to the lack of ground truth, qualitative comparisons shows our method can obtain visually more accurate mosaic with clear color textures. Quantitative experiments show that our method has an average error of 0.8–2.2 mm when estimating areas in the field of view.

Acknowledgments. This project was supported by the National Institute of Biomedical Imaging and Bioengineering of the National Institutes of Health through Grant Numbers K99EB027177, R01EB025964 and P41EB015898. Unrelated to this publication, Jagadeesan Jayender owns equity in Navigation Sciences, Inc. He is a co-inventor of a navigation device to assist surgeons in tumor excision that is licensed to Navigation Sciences. Dr. Jayender interests were reviewed and are managed by BWH and Partners HealthCare in accordance with their conflict of interest policies.

References

1. Maier-Hein, L., et al.: Optical techniques for 3D surface reconstruction in computer-assisted laparoscopic surgery. Med. Image Anal. **17**(8), 974–996 (2013)
2. Totz, J., Mountney, P., Stoyanov, D., Yang, G.-Z.: Dense surface reconstruction for enhanced navigation in MIS. In: Fichtinger, G., Martel, A., Peters, T. (eds.) MICCAI 2011. LNCS, vol. 6891, pp. 89–96. Springer, Heidelberg (2011). https://doi.org/10.1007/978-3-642-23623-5_12
3. Lacher, R.M., et al.: Nonrigid reconstruction of 3D breast surfaces with a low-cost RGBD camera for surgical planning and aesthetic evaluation. Med. Image Anal. **53**, 11–25 (2019)
4. Newcombe, R.A., Fox, D., Seitz, S.M.: DynamicFusion: reconstruction and tracking of non-rigid scenes in real-time. In: CVPR, pp. 343–352 (2015)
5. Miroslava, S., Baust, M., Ilic, S.: Variational level set evolution for non-rigid 3D reconstruction from a single depth camera. IEEE TPAMI (2020)
6. Miroslava, S., Baust, M., Ilic, S.: SobolevFusion: 3D reconstruction of scenes undergoing free non-rigid motion. In: CVPR, pp. 2646–2655 (2018)
7. Cadena, C., et al.: Past, present, and future of simultaneous localization and mapping: toward the robust-perception age. IEEE Trans. Rob., 1309–1332 (2016)
8. Mahmoud, N., Hostettler, A., Collins, T., Soler, L., Doignon, C., Montiel, J.M.: SLAM based quasi dense reconstruction for minimally invasive surgery scenes. arXiv preprint arXiv:1705.09107 (2017)
9. Mahmoud, N., Collins, T., Hostettler, A., Soler, L., Doignon, C., Montiel, J.M.: Live tracking and dense reconstruction for handheld monocular endoscopy. IEEE Trans. Med. Imaging **13**, 38(1), 79–89 (2018)
10. Mountney, P., Yang, G.-Z.: Motion compensated SLAM for image guided surgery. In: Jiang, T., Navab, N., Pluim, J.P.W., Viergever, M.A. (eds.) MICCAI 2010. LNCS, vol. 6362, pp. 496–504. Springer, Heidelberg (2010). https://doi.org/10.1007/978-3-642-15745-5_61
11. Collins, T., Bartoli, A., Bourdel, N., Canis, M.: Robust, real-time, dense and deformable 3D organ tracking in laparoscopic videos. In: Ourselin, S., Joskowicz, L., Sabuncu, M.R., Unal, G., Wells, W. (eds.) MICCAI 2016. LNCS, vol. 9900, pp. 404–412. Springer, Cham (2016). https://doi.org/10.1007/978-3-319-46720-7_47

12. Schoob, A., Kundrat, D., Kahrs, L.A., Ortmaier, T.: Stereo vision-based tracking of soft tissue motion with application to online ablation control in laser microsurgery. Med. Image Anal., 80–95 (2017)
13. Modrzejewski, R., Collins, T., Bartoli, A., Hostettler, A., Marescaux, J.: Soft-body registration of pre-operative 3D models to intra-operative RGBD partial body scans. In: Frangi, A.F., Schnabel, J.A., Davatzikos, C., Alberola-López, C., Fichtinger, G. (eds.) MICCAI 2018. LNCS, vol. 11073, pp. 39–46. Springer, Cham (2018). https://doi.org/10.1007/978-3-030-00937-3_5
14. Petit, A., Lippiello, V., Siciliano, B.: Real-time Tracking of 3D Elastic Objects with an RGB-D Sensor. In: IROS (2015)
15. Song, J., Wang, J., Zhao, L., Huang, S., Dissanayake, G.: MIS-SLAM: real-time large-scale dense deformable SLAM system in minimal invasive surgery based on heterogeneous computing. IEEE Rob. Autom. Lett. 3(4), 4068–4075 (2018)
16. Lamarca, J., Parashar, S., Bartoli, A., Montiel, J.M.: DefSLAM: tracking and mapping of deforming scenes from monocular sequences. IEEE Trans. Rob. (2020)
17. Zhou, H., Jayender, J.: Smooth deformation field-based mismatch removal in real-time. arXiv preprint arXiv:2007.08553 (2020)
18. Kmmerle, R., Grisetti, G., Strasdat, H., Konolige, K., Burgard, W.: G2o: a general framework for graph optimization. In: ICRA, pp. 3607–3613 (2011)
19. Brown, M., Lowe, D.G.: Automatic panoramic image stitching using invariant features. Int. J. Comput. Vision 74(1), 59–73 (2007)
20. Mur-Artal, R., Tard, J.D.: ORB-SLAM2: an open-source SLAM system for monocular, stereo, and RGB-D cameras. IEEE Trans. Rob., 1255–1262 (2017)
21. Zhou, H., Jayender, J.: Real-time dense reconstruction of tissue surface from stereo optical video. IEEE Trans. Med. Imaging 39(2), 400–412 (2019)
22. Arun, K.S., Huang, T.S., Blostein, S.D.: Least-squares fitting of two 3-D point sets. IEEE TPAM I, 698–700 (1987)
23. Bay, H., Tuytelaars, T., Van Gool, L.: SURF: speeded up robust features. In: Leonardis, A., Bischof, H., Pinz, A. (eds.) ECCV 2006. LNCS, vol. 3951, pp. 404–417. Springer, Heidelberg (2006). https://doi.org/10.1007/11744023_32
24. Sorkine, O., Alexa, M.: As-Rigid-As-Possible Surface Modeling. In: Symposium on Geometry Processing, vol. 4, pp. 109–116 (2007)
25. Osher, S., Fedkiw, R.: Level Set Methods and Dynamic Implicit Surfaces. AMS, vol. 153. Springer, New York (2003). https://doi.org/10.1007/b98879

Efficient Global-Local Memory for Real-Time Instrument Segmentation of Robotic Surgical Video

Jiacheng Wang[1], Yueming Jin[2], Liansheng Wang[1(✉)], Shuntian Cai[3], Pheng-Ann Heng[2], and Jing Qin[4]

[1] Department of Computer Science at School of Informatics, Xiamen University, Xiamen, China
`jiachengw@stu.xmu.edu.cn`, `lswang@xmu.edu.cn`
[2] Department of Computer Science and Engineering, The Chinese University of Hong Kong, Hong Kong, China
{`ymjin,pheng`}`@cse.cuhk.edu.hk`
[3] Department of Gastroenterology, Zhongshan Hospital affiliated to Xiamen University, Xiamen, China
`xuhongzhi@xmu.edu.cn`
[4] Center for Smart Health, School of Nursing, The Hong Kong Polytechnic University, Hong Kong, China
`harry.qin@polyu.edu.hk`

Abstract. Performing a real-time and accurate instrument segmentation from videos is of great significance for improving the performance of robotic-assisted surgery. We identify two important clues for surgical instrument perception, including local temporal dependency from adjacent frames and global semantic correlation in long-range duration. However, most existing works perform segmentation purely using visual cues in a single frame. Optical flow is just used to model the motion between only two frames and brings heavy computational cost. We propose a novel dual-memory network (DMNet) to wisely relate both global and local spatio-temporal knowledge to augment the current features, boosting the segmentation performance and retaining the real-time prediction capability. We propose, on the one hand, an efficient local memory by taking the complementary advantages of convolutional LSTM and non-local mechanisms towards the relating reception field. On the other hand, we develop an active global memory to gather the global semantic correlation in long temporal range to current one, in which we gather the most informative frames derived from model uncertainty and frame similarity. We have extensively validated our method on two public benchmark surgical video datasets. Experimental results demonstrate that our method largely outperforms the state-of-the-art works on segmentation accuracy while maintaining a real-time speed.

J. Wang and Y. Jin—Contributed equally.
Code is available at https://github.com/jcwang123/DMNet.

© Springer Nature Switzerland AG 2021
M. de Bruijne et al. (Eds.): MICCAI 2021, LNCS 12904, pp. 341–351, 2021.
https://doi.org/10.1007/978-3-030-87202-1_33

1   Introduction

Robotic-assisted surgery has greatly improved the surgeon performance and patient safety. Semantic segmentation of instrument segmentation, aiming to separate instrument and identify its sub-type and parts, serves as an essential prerequisite in various applications in assisted surgery. Achieving high segmentation accuracy while with low latency for real-time prediction is vital in the real-world deployment. For example, fast and accurate instrument segmentation can advance the context-awareness of surgeons when performing surgery, providing timely decision-making support, and generating real-time warning of potential deviations and anomalies. The real-time analysis of robotic surgical videostream can facilitate coordination and communication among the surgical team members. However, fast and accurate instrument segmentation from surgical video is very challenging, due to the complicated surgical scene, various lighting conditions, incomplete and distorted instrument structure caused by small FoV of endoscopic camera and inevitable visual occlusion by blood, smoke or instrument overlap.

Most existing methods [8,17,18,22] on surgical instrument segmentation treat sequential video as static image, and ignore the valuable clues in temporal dimension. For example, the ToolNet [8] uses a holistically-nested fully convolutional network to impose multi-scale constraint of predictions. Shvets *et al.* introduce a skip-connection model trained with transfer learning, winning the 2017 EndoVis Challenge [2]. The LWANet [17] and PAANet [18] apply attention-based mechanism into instrument segmentation to encode the semantic dependencies between channels and global space. Different from these methods that perform segmentation solely relied on visual clues from a single frame, Jin *et al.* [11] utilize optical flow to incorporate temporal prior into the network, which is hence able to capture inherent temporal clues from the instrument motion to boost results. Zhao *et al.* [28] use motion flow to largely improve the performance of semi-supervised instrument segmentation, with low-frequency annotation available in the surgical video. However, leveraging optical flow incurs significant computational cost in the calculation process, also can only capture the temporal information in the local range of two frames.

Recently, many works in surgical video analysis have shown successes of using long-range temporal information for improving results, mainly focus on recognition tasks for surgical workflow, gesture and tool presence [3,5,12,13,25,27]. For example, Czempiel et al. [5] proposed to use multi-stage TCN [7] to leverage temporal relations in all previous frame for workflow recognition. Meanwhile, plenty of studies in natural computer vision domain verify that aggregating the global information that is distant in time to augment the current features can boost the performance. This series of methods achieves the promising performance on various video analysis tasks, such as action recognition [24], video super resolution [26], video object detection [4], and semi video object segmentation [19,23]. Although these methods are hardly employed in our instrument segmentation task given different problem settings, and generally do not take the low computational latency into consideration which is the core consideration for robotic surgery, this stream inspires us to raise a question that how to wisely incorporate longer-range cues in temporal dimension for improving both robustness and efficiency to instrument segmentation.

In this paper, we propose a novel Dual-Memory Network (DMNet) for achieving accurate and real-time instrument segmentation from surgical videos, by holistically and efficiently aggregating spatio-temporal knowledge. The dual-memory framework are based on two important intuitions for humans to perceive instruments in videos, i.e., local temporal dependence and global semantic information, therefore more temporal knowledge can be transferred to current semantic representation. More importantly, we are the first trials that enable such holistic aggregation in real-time setting by carefully considering the properties of these two-level horizons. Concretely, we first design a local memory to take the complementary advantages of RNN and self-attention mechanisms. Considering that the local-range frame segments demonstrate the similar spatial information and highly temporal continuity, we exploit RNN to collect temporal cues with small reception field, following a self-attended non-local block to relate the current frame in larger spatial space with only single collected frame. We further develop a global memory to enable the current frame to efficiently get access to much longer-span content, by incorporating active learning strategy. Most informative and representative frames diffusing in global-range segments are selected based on two sample selection criteria. Relating such frames to current features can achieve sufficient clue enhancement while remaining the fast prediction. We extensively validate our method on two public benchmark surgical video dataset for instrument segmentation. Our approach attains a remarkable segmentation improvement over the state-of-the-art methods, with retaining the fast and real-time speed.

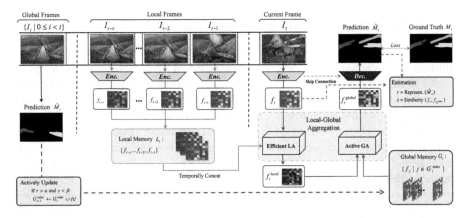

Fig. 1. Illustration of our proposed Dual-Memory Network. Wisely leveraging both local temporal dependence and global semantic information, our method achieves real-time surgical instrument segmentation with high accuracy.

2 Methodology

Figure 1 illustrates the overall architecture of the proposed dual-memory network (DMNet) for instrument segmentation from surgical videos. Two memories are storing and updating the previous frame knowledge from both local-range and global-range. Efficient local aggregation and active global aggregation modules are designed for relating the two-level memories to enhance the current frame features.

2.1 Dual-Memory Architecture

The goal of surgical instrument segmentation is to yield a segmentation map for each frame of the video in real-time to facilitate robot or human manipulation. Given the current time step as t, our dual-memory framework contains a local memory and a global memory for the current frame I_t, respectively defined as $\mathbf{L}_t$ and $\mathbf{G}_t$. They are composed of the extracted feature maps f obtained from previous frames, illustrated as the green stream and red stream in Fig. 1, respectively. The local memory $\mathbf{L}_t$ contains τ feature maps extracted from previous τ frames: $\{f_{t-\tau}, ..., f_{t-2}, f_{t-1}\}$, which is used for local temporal aggregation to avoid redundant computation. The global memory $\mathbf{G}_t$ is composed of a set of feature maps which are selected from the feature maps of all previous frames. For global temporal aggregation, we randomly select a certain number of frames from global memory to offer the dependencies, thus will be able to reduce computation cost significantly. Leveraging the principle of active learning, we further introduce an actively updating strategy to update memory with the most valuable samples to build a more terse global memory. With local and global aggregation, current frame feature is augmented for accurate instrument segmentation.

2.2 Efficient Local Temporal Aggregation

Considering inherent temporal continuity within a video, modeling temporal information, especially from the relevant adjacent frames in a local-range, is crucial to improve segmentation for each frame. Previous works leverage optical flow to explicitly incorporate the motion information of surgical instrument, yet its calculation is time-consuming. Convolutional LSTM (*ConvLSTM*) is another powerful approach generally used for temporal modeling. However, the receptive field captured from each previous frame is limited given the small size of convolutional kernel. Stacking several layers to increase the receptive field is a relatively common strategy to tackle this problem, yet inevitably brings the large computational cost. On the other hand, non-local (*NL*) attention mechanism shows its success in image segmentation recently, by enhancing each pixels through weighted consideration on other pixels of all positions. It is computationally feasible for single image analysis, however, relating several adjacent frames in the video clip shall largely harm the model efficiency. We propose an efficient local aggregation (ELA) module to take the complementary advantages of *ConvLSTM* operation and *NL* mechanism. Temporal dependency in the local memory can be incorporated using ELA to enhance each frame features with light-weight computation.

Formally, for segmenting the current frame I_t, we first form the local feature map clip from the local memory $\{f_{t-\tau}, ..., f_{t-2}, f_{t-1}\}$. *ConvLSTM* operation is then utilized to aggregate the information along the temporal dimension with small convolution kernel. Notably, to enable the maximum efficiency, we employ *BottleneckLSTM* [15] instead of the standard *ConvLSTM*. It benefits from replacing the standard convolution with depth-wise separable convolution and setting a bottlenecked layer after input to reduce the feature dimension. We can obtain the temporal enriched feature from the local memory $\tilde{f}_t = \mathcal{F}_{LSTM}(f_{t-\tau}, ..., f_{t-2}, f_{t-1}, f_t)$. However, each pixel in $\tilde{f}_t$ only considers small region information searched within the kernel size.

Instead of stacking the LSTM layers to increase the receptive field, our ELA leverages *NL* mechanism to enlarge the reference region. Concretely, based on the encoded temporal feature $\tilde{f}_t$ for the current frame, we generate two kinds of feature maps (**key**, **value**) of $\tilde{f}_t$, denoted as $(\tilde{\mathbf{k}}_t, \tilde{\mathbf{v}}_t)$. Each with its own simple convolution layer that preserves their spatial size, while reducing the dimensionality. We set the channel number of **key** feature map as a low value, as we aim to use it to efficiently calculate the similarity matching scores. While the channel number of **value** is set as a relatively high value, in order to store more detailed information for producing the mask estimation for surgical instruments. Next, we compare every spatial location in $\tilde{\mathbf{k}}_t$ with other locations to perform the similarity matching, and the similarity function $\mathcal{F}_{sim}$ is defined as: $\mathcal{F}_{sim}(\mathbf{x}, \mathbf{y}) = \exp(\mathbf{x} \circ \mathbf{y})$, where $\circ$ denotes the dot product. The value in $\tilde{\mathbf{v}}_t$ is then retrieved by a weighted summation with the soft weights and concatenated with the original $\tilde{\mathbf{v}}_t$. To this end, output feature calculated by our ELA module is as follow:

$$f_t^{local_i} = \mathcal{F}_{NL}(\tilde{f}_t^i) = [\tilde{\mathbf{v}}_t^i, \frac{1}{\mathcal{Z}} \sum_{\forall j} \mathcal{F}_{sim}(\tilde{\mathbf{k}}_t^i, \tilde{\mathbf{k}}_t^j) \tilde{\mathbf{v}}_t^j], \tag{1}$$

where $\mathcal{Z} = \sum_{\forall j} \mathcal{F}_{sim}(\tilde{\mathbf{k}}_t^i, \tilde{\mathbf{k}}_t^j)$ is the normalizing factor; $[\cdot, \cdot]$ denotes the concatenation; i is the index of an output position whose response is to be computed and j is the index that enumerates all possible positions in the temporal enriched feature $\tilde{f}_t$. In this regard, f_t is enhanced to f_t^{local} by the efficient feature aggregation along both temporal and visual space from our built local memory.

2.3 Active Global Temporal Aggregation

Apart from augmenting the current frame feature using the local temporal dependency, we propose to introduce the cues that is more distant in time from the global memory. The standard approach for global temporal aggregation is to put feature maps of all the previous frames into the global memory and randomly select some samples when activating it. However, this approach has significant memory-cost, particularly when the video is long and contains plenty of frames. More importantly, different frames indeed count for different values for global temporal aggregation. For example, frames with the motion blur and lighting reflection contain little beneficial information. Including them into the global memory shall contribute less to augment the current frame features. Additionally, adjacent frames commonly have similar visual appearance, leading to redundant features which help nothing but cost more. We propose to selectively recommend frames to form a more supportive global memory G_t for aggregation, named as active global aggregation (AGA). We introduce two selection criteria complementarily considering both representativeness and similarity when forming the global memory, where the representativeness denotes the general semantic context and the similarity is used to evaluate whether the coming frame is supplementary or redundant.

Specifically, given the segmentation map $\hat{M}_t$ of the t-th frame, we first compute the prediction confidences $\mathbf{r}$ through the entropy on behalf of the representativeness. As the samples with low confidence are more likely to be abnormal, causing the prediction at decision boundary, we only queue the frame into G_t if $\mathbf{r} > \alpha$. The predicted entropy is defined as: $\mathbf{r} = \frac{1}{N} \sum_i \sum_c p_i^c \log p_i^c$, where p_i^c denotes probability of the c-th class at location i and N is the number of pixels in the map.

We also aim at the sample diversity in the global memory. The straightforward way is to make a comparison with all feature maps when adding a new one. However, the full comparison is computation-intensive. Instead, we only compare with the latest queued feature map $f_{\mathbf{G}_t^{latest}}$ in the global memory G_t. If dissimilar, the newly coming one is less likely to resemble further features. We employ a negative euclidean distance to measure the multivariate similarity of a pair of feature maps as $\mathbf{s} = -\sqrt{\sum (f_t - f_{\mathbf{G}_t^{latest}})^2}$. We include the feature map into the global memory if $\mathbf{s} < \beta$.

After we build the global memory G_t in which each element is generally informative, we randomly select n frame features $\{f_g\}_{g=1}^n$, from G_t for aggregation. The AGA module is also developed based on the NL mechanism. However, considering that the n frame features are independent without holding temporal continuity, AGA module separately relates each of them to enhance current frame feature. Therefore, different from ELA module to conduct self-attention of feature itself, AGA module input two kinds of features including $\{f_g\}$ and the current frame features augmented by ELA f_t^{local}. Two pairs of (**key**, **value**) are calculated for different features, while other operations such as similarity matching remain the same as ELA module. We denote the globally aggregated feature as $f^{global} = \mathcal{F}_{NL}(f_t^{local}, \{f_g\}_{g=1}^n)$. To this end, the f^{global} can capture the semantic content of the frames that is distant in time. Note that, to ensure the online prediction, only previous frame features are accessible in global memory for augmenting each frame feature. The f^{global} are then decoded to produce the final mask of surgical instrument. We employ **Dice** loss between ground-truth and predicted mask for model optimization.

3   Experiments

Datasets. We extensively evaluate the proposed model on two popular public datasets: 2017 MICCAI EndoVis Instrument Challenge (EndoVis17) and 2018 MICCAI EndoVis Scene Segmentation Challenge (EndoVis18). EndoVis17 [2] records different porcine procedures using da Vinci Xi surgical system. For fair comparison, we follow the same evaluation manner in [11,22], by using the released 8× 225-frame videos for 4-fold cross-validation, also with the same fold division. We perform the most challenging problem, type segmentation, on this dataset to validate our method. EndoVis18 [1] made up of 19 more complex surgery sequences with 15 training videos and 4 test videos. We randomly split three videos out of training sets for model validation. Complete annotation in this challenge contains 12 classes which also includes anatomical objects. Among them, surgical instruments are labeled by different parts (shaft, wrist, and jaws). In this work we focus on these three classes regarding instruments and ignore other segmentation classes. Two commonly used evaluation metrics Mean intersection-over-union (mIoU) and mean Dice coefficient (mDice) are adopted for validation. To evaluate model complexity and time performance, we calculate the number of parameters (Param.), FLOPS, inference time and FPS. We average the inference time on a single frame by 100 times for fair comparison.

Implementation Details. We adopt the lightweight *RefineNet* [16] as the backbone and replace its encoder with *MobileNetv2* [21]. All the training images are resized to 512×640 and augmented by vertical flip, horizontal flip and random scale change (limited 0.9–1.1). The network is pre-trained on ImageNet [6], and optimized by Adam [14], with weight decay as 0.0001. We take a mini-batch size of 16 on 4 TITAN RTX GPUs and apply synchronous Batch Normalization to adapt multi-GPU environment [20]. Learning rate is initialized as 0.0001 and reduced by one-tenth in the 70-th epoch with training 100 epochs in total. We empirically set $\tau = 4$ and $n = 4$ considering the trade-off between accuracy and efficiency. We set $\alpha = -0.08$ and $\beta = -4.65$, by averaging $\mathbf{r}, \mathbf{s}$ of all frames, to represent the general distribution. All experiments are repeated 5 times and we report average values to account for the stochastic nature of training.

Table 1. Type and part segmentation results of instrument on the EndoVis17 and EndoVis18 datasets, respectively. Note that underline denotes the methods in real-time.

Methods	EndoVis17		EndoVis18		Param. (M)	FLOPS (G)	Time (ms)	FPS
	mDice (%)	mIOU (%)	mDice (%)	mIOU (%)				
TernausNet [10]	44.95	33.78	61.78	50.25	36.92	275.45	58.32	17
MF-TAPNet [11]	48.01	36.62	–	–	–	–	–	–
PAANet [18]	56.43	49.64	75.01	64.88	21.86	60.34	38.20	26
LWANet [17]	49.79	43.23	71.73	61.06	**2.25**	**2.77**	13.21	<u>76</u>
TDNet [9]	54.64	49.24	76.22	66.30	21.23	47.60	22.23	<u>45</u>
DMNet	**61.03**	**53.89**	**77.53**	**67.50**	4.38	11.53	26.37	<u>38</u>

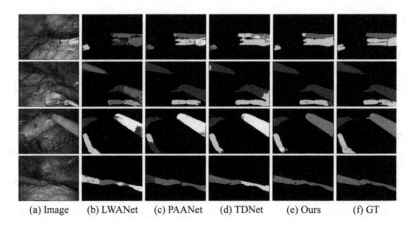

(a) Image   (b) LWANet   (c) PAANet   (d) TDNet   (e) Ours   (f) GT

Fig. 2. Visual comparison of type segmentation on EndoVis17 produced by different methods.

Comparison with State-of-the-Arts. We compare the proposed DMNet with state-of-the-art approaches, including: (1) Ternausnet [10]: the winner method of EndoVis17, (2) MF-TAPNet [11]: the latest instrument segmentation model that applies temporal information by optical flow, (3) PAANet [18]: a 2D segmentation model achieving best performance in instrument segmentation, (4) LWANet [17]: the latest instrument segmentation model with real-time performance. (5) TDNet [9]: a NL-based model processing local-range temporal context in real-time for natural video analysis. For the first two methods, we refer the results from their papers. We re-implement PAANet, LWANet, TDNet and perform 4-fold cross-validation for fair comparison.

As shown in Table 1, DMNet achieves 61.03% mDice and 53.89% mIoU on instrument type segmentation, outperforming the other methods in terms of both metrics by a large margin. Without using temporal aggregation, the real-time approach LWANet only achieves the 49.79% on mDice and 43.23% on mIoU, even if it's the fastest model, demonstrating the importance of temporal information in boosting the segmentation performance. In particular, compared with TDNet, which also harnessed local temporal context to improve performance, the proposed DMNet yields better segmentation accuracy with a nearly closed speed, demonstrating the effectiveness of the proposed active global aggregation scheme. What's more, our results achieve the consistent improvement in part segmentation on EndoVis18 dataset, outperforming our rivals on the test set. It proves that our local-global temporal context is not limited to type segmentation, but can provide general feature enhancement. We further present visual comparison results of some challenging cases in Fig. 2. It is observed that without temporal information, models can hardly identify different types (Fig. 2 (b, c)). Solely using local (Fig. 2 (d)) temporal information, TDNet still performs unsatisfactory segmentations with structural incompleteness caused by pose variation (3rd row) and large occlusion caused by tool overlapping (4th row). Our

DMNet achieves segmentation results closest to the ground truth, demonstrating the effectiveness of the proposed temporal information aggregation schemes in tackling challenging cases.

Table 2. In-depth analysis of our method.

| (a) Ablation for key components. | | | | (b) Different fashions for local aggregation. | | | |
ELA	AGA	mIoU(%)	FLOPS(G)	Type	mIoU(%)	Param.(M)	FLOPS(G)
		46.51	10.21	NL	48.03	6.40	11.82
✓		52.25	10.93	CLSTM	53.07	7.93	17.68
	✓	51.64	10.77	BLSTM	50.14	3.65	10.84
✓	✓	53.89	11.53	ELA(Ours)	52.25	4.05	10.93

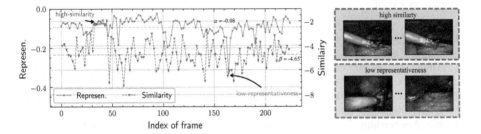

Fig. 3. Similarity and representativeness of a video sequence in global aggregation.

In-depth Analysis. We analyze effectiveness of each key component proposed in our method on EndoVis17. As shown in Table 2(a), applying ELA individually increases 5.74% of mIoU, while adopting AGA individually improves 5.13% of mIoU. With both ELA and AGA, mIoU further boosts by 7.38%, indicating the complementary advantages of local and global information aggregation. More importantly, we achieve these improvements with comparable FLOPS. Without introducing much computational cost, our method maintain real-time inference capability.

We conduct an analysis on our ELA module design, by comparing it with other aggregation manners, including (1) a *NL* operator, which is a basic aggregation module attempting to leveraging more spatial information and (2) two *ConvLSTM* operators, a standard convolutional LSTM (CLSTM) and a Bottleneck LSTM (BLSTM). As shown in Table 2(b), our ELA module outperforms NL and BLSTM in mIoU by 4.18% and 2.11%, respectively, with little extra computational cost. While CLSTM achieves slightly better performance than ELA, its computational costs are almost twice more than ELA. The experimental results demonstrate the ELA achieves a good balance between accuracy and efficiency.

We present the calculated **r** and **s** of a video sequence and visualize some typical frame samples in Fig. 3, to verify the motivation of active selection in proposed AGA module. We see from the green box that the video sequence with close similarity scores shows high-similar appearance. In the orange box, we see that the sample with low confidence presents the abnormal appearance due to motion blur. Using the global information of data, i.e., average values of α and β as thresholds, our AGA module bypasses to utilize such information.

4 Conclusion

This paper presents a novel real-time surgical instruments segmentation model by efficiently and holistically considering the spatio-temporal knowledge in videos. We develop an efficient local cache for leveraging the most favorable region per frame for local-range aggregation, and an active global cache to select the most informative frames to cover the global cues using only few frames. Experimental results on two public datasets shows that our method outperforms state-of-the-arts by a large margin in accuracy while maintaining the fast prediction speed.

References

1. Allan, M., et al.: 2018 robotic scene segmentation challenge. arXiv preprint arXiv:2001.11190 (2020)
2. Allan, M., et al.: 2017 robotic instrument segmentation challenge. arXiv preprint arXiv:1902.06426 (2019)
3. van Amsterdam, B., Clarkson, M.J., Stoyanov, D.: Multi-task recurrent neural network for surgical gesture recognition and progress prediction. In: 2020 IEEE International Conference on Robotics and Automation (ICRA), pp. 1380–1386. IEEE (2020)
4. Chen, Y., Cao, Y., Hu, H., Wang, L.: Memory enhanced global-local aggregation for video object detection. In: Proceedings of the IEEE Conference on Computer Vision and Pattern Recognition (CVPR), pp. 10337–10346 (2020)
5. Czempiel, T., et al.: TeCNO: surgical phase recognition with multi-stage temporal convolutional networks. In: Martel, A.L., et al. (eds.) MICCAI 2020. LNCS, vol. 12263, pp. 343–352. Springer, Cham (2020). https://doi.org/10.1007/978-3-030-59716-0_33
6. Deng, J., Dong, W., Socher, R., Li, L.J., Li, K., Fei-Fei, L.: ImageNet: a Large-Scale Hierarchical Image Database. In: CVPR 2009 (2009)
7. Farha, Y.A., Gall, J.: MS-TCN: multi-stage temporal convolutional network for action segmentation. In: Proceedings of the IEEE/CVF Conference on Computer Vision and Pattern Recognition, pp. 3575–3584 (2019)
8. Garcia-Peraza-Herrera, L.C., et al.: ToolNet: holistically-nested real-time segmentation of robotic surgical tools. In: 2017 IEEE/RSJ International Conference on Intelligent Robots and Systems (IROS), pp. 5717–5722. IEEE (2017)
9. Hu, P., Caba, F., Wang, O., Lin, Z., Sclaroff, S., Perazzi, F.: Temporally distributed networks for fast video semantic segmentation. In: Proceedings of the IEEE Conference on Computer Vision and Pattern Recognition (CVPR), pp. 8818–8827 (2020)
10. Iglovikov, V., Shvets, A.: TernausNet: U-net with vgg11 encoder pre-trained on ImageNet for image segmentation. arXiv preprint arXiv:1801.05746 (2018)
11. Jin, Y., Cheng, K., Dou, Q., Heng, P.-A.: Incorporating temporal prior from motion flow for instrument segmentation in minimally invasive surgery video. In: Shen, D., et al. (eds.) MICCAI 2019. LNCS, vol. 11768, pp. 440–448. Springer, Cham (2019). https://doi.org/10.1007/978-3-030-32254-0_49

12. Jin, Y., et al.: SV-RCNet: workflow recognition from surgical videos using recurrent convolutional network. IEEE Trans. Med. Imaging **37**(5), 1114–1126 (2018)
13. Jin, Y., Long, Y., Chen, C., Zhao, Z., Dou, Q., Heng, P.A.: Temporal memory relation network for workflow recognition from surgical video. IEEE Trans. Med. Imaging (2021)
14. Kingma, D.P., Ba, J.: Adam: a method for stochastic optimization (2017)
15. Liu, M., Zhu, M.: Mobile video object detection with temporally-aware feature maps. In: Proceedings of the IEEE Conference on Computer Vision and Pattern Recognition (CVPR), pp. 5686–5695 (2018)
16. Nekrasov, V., Shen, C., Reid, I.: Light-weight RefineNet for real-time semantic segmentation. arXiv preprint arXiv:1810.03272 (2018)
17. Ni, Z.L., Bian, G.B., Hou, Z.G., Zhou, X.H., Xie, X.L., Li, Z.: Attention-guided lightweight network for real-time segmentation of robotic surgical instruments. arXiv preprint arXiv:1910.11109 (2019)
18. Ni, Z.L., et al.: Pyramid attention aggregation network for semantic segmentation of surgical instruments. In: AAAI, pp. 11782–11790 (2020)
19. Oh, S.W., Lee, J.Y., Xu, N., Kim, S.J.: Video object segmentation using space-time memory networks. In: Proceedings of the IEEE International Conference on Computer Vision (ICCV), pp. 9226–9235 (2019)
20. Peng, C., et al.: MegDet: a large mini-batch object detector. In: Proceedings of the IEEE Conference on Computer Vision and Pattern Recognition (CVPR), June 2018
21. Sandler, M., Howard, A., Zhu, M., Zhmoginov, A., Chen, L.C.: MobileNetv 2: inverted residuals and linear bottlenecks. In: Proceedings of the IEEE Conference on Computer Vision and Pattern Recognition (CVPR), June 2018
22. Shvets, A.A., Rakhlin, A., Kalinin, A.A., Iglovikov, V.I.: Automatic instrument segmentation in robot-assisted surgery using deep learning. In: 2018 17th IEEE International Conference on Machine Learning and Applications (ICMLA), pp. 624–628. IEEE (2018)
23. Voigtlaender, P., Chai, Y., Schroff, F., Adam, H., Leibe, B., Chen, L.C.: FEELVOS: fast end-to-end embedding learning for video object segmentation. In: Proceedings of the IEEE Conference on Computer Vision and Pattern Recognition, pp. 9481–9490 (2019)
24. Wu, C.Y., Feichtenhofer, C., Fan, H., He, K., Krahenbuhl, P., Girshick, R.: Long-term feature banks for detailed video understanding. In: Proceedings of the IEEE Conference on Computer Vision and Pattern Recognition (CVPR), pp. 284–293 (2019)
25. Yi, F., Jiang, T.: Hard frame detection and online mapping for surgical phase recognition. In: Shen, D., et al. (eds.) MICCAI 2019. LNCS, vol. 11768, pp. 449–457. Springer, Cham (2019). https://doi.org/10.1007/978-3-030-32254-0_50
26. Yi, P., Wang, Z., Jiang, K., Jiang, J., Ma, J.: Progressive fusion video super-resolution network via exploiting non-local spatio-temporal correlations. In: Proceedings of the IEEE International Conference on Computer Vision, pp. 3106–3115 (2019)
27. Zhang, J., et al.: Symmetric dilated convolution for surgical gesture recognition. In: Martel, A.L., et al. (eds.) MICCAI 2020. LNCS, vol. 12263, pp. 409–418. Springer, Cham (2020). https://doi.org/10.1007/978-3-030-59716-0_39
28. Zhao, Z., Jin, Y., Gao, X., Dou, Q., Heng, P.-A.: Learning motion flows for semi-supervised instrument segmentation from robotic surgical video. In: Martel, A.L., et al. (eds.) MICCAI 2020. LNCS, vol. 12263, pp. 679–689. Springer, Cham (2020). https://doi.org/10.1007/978-3-030-59716-0_65

C-Arm Positioning for Spinal Standard Projections in Different Intra-operative Settings

Lisa Kausch[1,2(✉)], Sarina Thomas[1], Holger Kunze[3], Tobias Norajitra[1],
André Klein[1,2], Jan Siad El Barbari[4], Maxim Privalov[4], Sven Vetter[4],
Andreas Mahnken[5], Lena Maier-Hein[6], and Klaus H. Maier-Hein[1]

[1] Division of Medical Image Computing, German Cancer Research Center,
Heidelberg, Germany
l.kausch@dkfz-heidelberg.de
[2] Medical Faculty, University, Heidelberg, Germany
[3] Advanced Therapy Systems Division, Siemens Healthineers, Erlangen, Germany
[4] MINTOS Research Group, Trauma Surgery Clinic,
Ludwigshafen, Germany
[5] Division of Diagnostic and Interventional Radiology, University Hospital, Marburg,
Germany
[6] Division of Computer-Assisted Medical Interventions,
German Cancer Research Center, Heidelberg, Germany

Abstract. Trauma and orthopedic surgeries that involve fluoroscopic guidance crucially depend on the acquisition of correct anatomy-specific standard projections for monitoring and evaluating the surgical result. This implies repeated acquisitions or even continuous fluoroscopy. To reduce radiation exposure and time, we propose to automate this procedure and estimate the C-arm pose update directly from a first X-ray without the need for a pre-operative computed tomography scan (CT) or additional technical equipment. Our method is trained on digitally reconstructed radiographs (DRRs) which uniquely provide ground truth labels for arbitrary many training examples. The simulated images are complemented with automatically generated segmentations, landmarks, as well as a k-wire and screw simulation. To successfully achieve a transfer from simulated to real X-rays, and also to increase the interpretability of results, the pipeline was designed by closely reflecting on the actual clinical decision-making of spinal neurosurgeons. It explicitly incorporates steps like region-of-interest (ROI) localization, detection of relevant and view-independent landmarks, and subsequent pose regression. To validate the method on real X-rays, we performed a large specimen study with and without implants (i.e. k-wires and screws). The proposed procedure obtained superior C-arm positioning accuracy ($p_{wilcoxon} \ll 0.01$), robustness, and generalization capabilities compared to the state-of-the-art direct pose regression framework.

Electronic supplementary material The online version of this chapter (https://doi.org/10.1007/978-3-030-87202-1_34) contains supplementary material, which is available to authorized users.

M. de Bruijne et al. (Eds.): MICCAI 2021, LNCS 12904, pp. 352–362, 2021.
https://doi.org/10.1007/978-3-030-87202-1_34

Keywords: Pose estimation · C-arm positioning · Standard projections

1   Introduction

Guidance and quality control during interventions increasingly rely on fluoroscopy using a mobile C-arm. For surgical validation, it is important to obtain anatomy-specific accurate standard projections during surgical instrumentation. A standard projection corresponds to a specific C-arm pose relative to the patient's position such that fracture reduction and implant positioning can be assessed. C-arm positioning is usually performed manually using a trial-and-error approach of *fluoro hunting* at a cost of radiation exposure and intervention time. Due to the complexity of the procedure and spatial configuration, the results are often error-prone and highly surgeon dependent increasing the risk of overlooked errors [22].

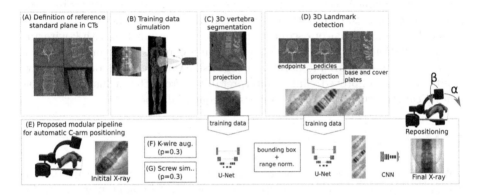

Fig. 1. Overview of the proposed sequential pipeline for automatic C-arm pose estimation for standard projections. The pipeline is trained on simulated data. Techniques for spinal implant simulation are proposed and integrated into the training pipeline.

Many state-of-the-art approaches require either external tracking hardware [6,20], preoperative CTs [3,5,7,9,21] or manual landmark selection [2]. [20] propose an approach independent of all mentioned requirements employing augmented reality. Their method does not estimate an optimal pose but addresses the problem of re-aligning the scanner with recorded C-arm poses. Deep learning-based automatic pose regression approaches give promising results and indicate that neural networks outperform other regression approaches [4,10]. Recent work [13] tackle this inherently ill-posed problem without patient-individual prior information or additional technical equipment and present a fully automatic approach that learns the necessary hints for C-arm positioning directly from an initial X-ray just as the operator. The proposed convolutional neural network (CNN) for pose regression is solely trained on simulated images since they uniquely provide ground truth labels. The state-of-the-art direct pose regression does not address two major challenges arising in an intra-operative setting: (1) Intra-operative X-rays are likely to contain background artifacts which can easily fool the pose regressor and lead to a wrong prediction. (2) Learning from simulations leads to the challenge of transferring to real-world applications. Intra-operative X-rays may contain implants or surgical tools that partly overlay the anatomy. Different approaches for tool augmentation have been presented based on transfer learning [27], realistic tool modeling [18, 26], and domain randomization [17, 23].

Our contribution is three-fold: (1) Instead of a direct pose regression, we propose a sequential deep learning-based pipeline for automatic C-arm pose estimation (Fig. 1(E)). Building upon ideas of [1, 8], we leverage domain knowledge to guide robust decision-making. Proceeding U-Net-based modules locate the ROI and detect relevant view-independent landmarks for subsequent CNN-based pose regression. This restricts the focus to anatomical regions and prevents irrelevant regions, such as background, from distorting the pose estimation. The landmark detection module mimics human decision-making to identify correct standard projections. (2) We tackle the domain gap between DRRs and intraoperative X-rays by integrating a k-wire and screw simulation into the training pipeline. (3) A large specimen study was performed, simulating different stages in a real clinical procedure, to demonstrate generalizability of the method.

We demonstrate that our sequential pose regression pipeline with k-wire and screw simulation significantly boosts the accuracy, robustness, and generalization capabilities compared to naive pose regression. Furthermore, our qualitative results show an improved clinical interpretability from the inclusion of visible anatomical structures within the proposed sequential pipeline.

2 Methods

Accurate C-arm positioning towards anatomy-specific standard projections is essential to properly evaluate fracture reduction and implant placement. We focused our analysis exemplary on the 4th lumbar vertebra (L4) and the anterior-posterior (AP) projection. The method can also be applied to other standard projections or vertebra levels.

2.1 Mobile C-arm Device

The C-arm has 6 degrees of freedom, including three rotational and three translational parameters. The primary orbital rotation, denoted by α, describes the rotation within the range of the C-arm gantry. The secondary angular rotation, denoted by β, describes the rotation perpendicular to the plane of the C-arm gantry (Fig. 1). The in-plane rotation in the detector plane is denoted by γ.

2.2 Training Data Simulation

Due to the lack of a constant reference frame in interventional fluoroscopy, X-rays with annotated pose labels do not exist in clinical routine. Therefore, training data was synthetically generated from 47 full-body CTs acquired at different institutions and scanners. A clinical expert defined the standard projection planes for the L4 vertebra in each 3D volume (Fig. 1(A)). The CTs were randomly divided into 60% training, 20% validation, 20% test, assuring that images of one patient are not shared between folds. Realistic DRRs were computed with the DeepDRR method for a reduced resolution of 256^2 pixel (Fig. 1(B)) [25]. The source and detector poses were varied around the C-arm pose, defined by the reference standard. During training, angular and orbital rotation were uniformly sampled from α, $\beta \in [-40°, 40°]$, while validation was limited to $[-30°, 30°]$ to assure equal coverage of all poses. For simulation, the system parameters were defined according to a Siemens Cios Spin® with 300 mm detector, 1952^2 pixel and 1164 mm source-detector distance. The simulations were converted from intensity to line integral domain by applying the negative Log transform, decreasing the dynamic range. The in-plane rotation γ was sampled from a uniform distribution $\gamma \in \mathcal{U}(-180°, 180°)$ and augmented during training, thus reducing simulation and computation time. Translation in the detector plane was sampled from a uniform distribution $t \in \mathcal{U}(-50\,mm, 50\,mm)^2$. Translation along the beam direction was modeled by scaling. Moreover, contrast transform, random additive Gaussian noise and Gaussian blur were applied [12]. Ground truth pose labels were established by labeling DRRs according to their distance to the optimal view $(d\alpha, d\beta, d\gamma, t)$. The intensity range was normalized to $[-1, 1]$.

2.3   Generation of Ground Truth Segmentations and Landmarks

2D ground truth segmentations and landmark locations were established by projecting automatically generated 3D annotations (Fig. 1(C)). Annotating the landmarks in 3D has the advantage of 2D consistency and reduces the time consumption compared to manual 2D annotations. For 3D vertebra segmentation, we employed an nnU-Net [11] trained on the VerSe challenge dataset [19]. Clinically meaningful landmarks were selected in consultation with a spinal neurosurgeon: (1) endpoints of the transverse and spinous processes, (2) pedicles, (3) vertebra base and cover plates. Landmarks were manually annotated for vertebra levels L2-L5 in 10 CT volumes. To automatically predict the selected landmarks on the remaining CTs, a vertebra-instance based nnU-Net was trained on the annotated CTs. The 3D landmarks were projected and one projection image was generated for endpoints, pedicles and plates respectively (Fig. 1(D)). In the projection domain, landmarks were represented as heatmaps by convolution with a Gaussian kernel with $\sigma = 3$. This reflects uncertainty since 3D manual annotation can differ between several experts.

2.4   K-Wire and Screw Simulation

Different modeling techniques were proposed to address the domain gap between training DRRs and intra-operative X-rays with metal implants. For k-wire augmentation, we used quadratic Bézier curves. For screw simulation, we developed a realistic screw modeling consisting of the following steps: (1) Screw trajectories were automatically planned in CT volumes for vertebra levels L2-S1 [14]. (2) Screw CAD models were aligned with the derived screw parameters. (3) CAD models were converted to screw segmentation masks. Derived masks were projected in the detector. The number of implants was selected randomly in the interval $[1, 10]$.

2.5   Pose Regression Framework

The proposed method predicts the pose update for acquiring the desired standard projection from one initial X-ray. We propose a sequential pose regression that is designed to reflect and automate the approaches used by spinal neurosurgeons for identifying correct standard projections. Proceeding U-Net-based modules were trained for ROI localization and detection of relevant view-independent landmarks (Fig. 1(E)).

In developing the proposed pipeline, we followed good surgical practice:

ROI Localization Module: Surgeons adjust the standard projection only based on the bone structures. We reflect this by constraining the input only to the bone region employing a vertebra segmentation mask. Therefore, a U-Net model [16] for view-independent vertebra segmentation of spinal X-rays was trained on simulated data. The model weights were optimized using a combination of binary cross-entropy and dice loss. A rotating bounding box was computed from the predicted vertebra segmentation and the X-ray was masked by the estimated rotated bounding box followed by range normalization to $[-1, 1]$. That reflects the use of collimators during surgery to increase image contrast. Besides, it guides the focus on the anatomy and prevents irrelevant regions from distorting the prediction.

Landmark Detection Module: In the clinic, correct standard projections are typically evaluated based on 2D anatomy-specific landmarks because this provides implicit additional 3D information based on anatomical prior knowledge. To reflect this, we trained a U-Net model [16] for view-independent landmark detection in spinal X-rays on simulated data after ROI localization. The model weights were optimized using L2 loss on the difference between target and predicted heatmaps.

Pose Regression Module: The predicted heatmaps were used as input for the pose regression network. Landmark prediction uncertainties were covered by training the pose regressor on the predicted landmarks instead of the ground truth.

We compared the proposed approach to the direct intensity-based pose regression [13]. The pose regression CNNs share the same architecture. The models were implemented using PyTorch v1.6.0 and trained with an 11 GB GeForce RTX 2080 Ti. The networks were optimized using the Adam optimizer [15] with a base learning rate of $\eta = 10^{-4}$ and batchsize 64 until convergence. During inference, test time ensembling was employed: The initial X-ray was augmented with different in-plane rotations and the median of all derived pose predictions was computed to increase robustness.

2.6   Validation Data

To validate the proposed sequential pose regression and implant simulation strategies, a specimen study was performed. The ethics committee Rheinland-Pfalz approved the specimen study (No. 2020-15423). During acquisition, different stages of a real clinical procedure were simulated. Therefore, k-wires and screws were inserted in selected pedicles. The datasets were acquired with a Siemens Cios Spin® mobile C-arm. To acquire data with ground truth poses, the torso specimens were initially aligned with the C-arm such that the C-arms' isocenter aligns with L4 and its base position at $\alpha = 0°$, $\beta = 0°$ corresponds to the AP standard pose. A dataset consists of orbital projection sequences acquired at different angulations ($\beta \in [-15°, 15°]$) and corresponding 3D reconstructions. The Cios Spin® stores the α/β angles in the metadata of every single acquisition. These served as the ground truth poses. For validation, all projections with $\alpha \in [-30°, 30°]$ were sampled from the orbital sequence, resulting in 1364 projection images per dataset. 16 datasets were acquired in total: 7 specimens without metal, 5 specimens with k-wires, 4 specimens with screws.

3   Experiments and Results

The following research questions were investigated:

- Can the 2-stage intensity-based pose regression [13] be replaced by a sequential 1-stage approach without decreasing robustness and accuracy (Sect. 3.1).
- How does the proposed method solely trained on simulated DRRs generalize to real X-rays without metal acquired in a specimen study (Sect. 3.2).
- Does the proposed k-wire and screw simulation techniques improve generalization towards surgical instrumentation (Sect. 3.3).

The proposed sequential C-arm pose regression (Sect. 2.5) and the baseline approach [13] were trained with the proposed k-wire augmentation and screw simulation integrated into the training pipeline (Sect. 2.4, see suppl. material for a visualization). The pose accuracy was evaluated based on the angle $\theta = \arccos(\langle v_{pred}, v_{gt} \rangle)$ between the principal rays of the ground truth v_{gt} and predicted pose v_{pred}. In addition, we report the mean absolute error (MAE) and standard deviation (SD) of rotation along individual axes denoted by $d\alpha$, $d\beta$, $d\gamma$. The average runtime per X-ray is 0.12 s.

3.1   Accuracy and Robustness Analysis

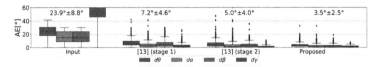

Fig. 2. Quantitative performance comparison of the proposed sequential method to the 2-stage direct pose regression [13] evaluated on 3969 DRRs. Our method outperforms [13] already in a single prediction step. Left: Initial offset distribution.

Validation was performed on the test DRRs simulated from 9 CTs excluded from training, covering an initial pose offset distribution of $d\theta = 23.8° \pm 8.8°$ around the AP standard pose (3969 DRRs). The proposed sequential approach ($d\theta = 3.5° \pm 2.5°$) is more accurate and robust than the 2-stage direct intensity-based approach ($d\theta_1 = 7.2° \pm 4.6°$, $d\theta_2 = 5.0° \pm 4.0°$) [13] while only requiring one prediction step (Fig. 2). Significant accuracy improvement in beam direction was confirmed by a Wilcoxon test on $d\theta$, $d\alpha$, $d\beta$, $d\gamma$ with $p \ll 0.01$.

3.2   Generalization Analysis: DRRs to Real X-Rays Without Metal

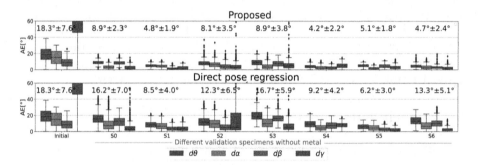

Fig. 3. Evaluation of the proposed sequential method on real X-rays without metal with $MAE_{d\theta} \pm SD_{d\theta}$ indicated for each specimen S0-S6. The proposed method shows improved generalization capabilities compared to direct intensity-based regression.

To assess the generalization capabilities from simulation-based training to real X-rays, validation was performed on 7 specimens without metal. For each specimen 1364 X-rays with an initial pose offset distribution of $d\theta = 18.3° \pm 7.6°$ were acquired. Figure 3 shows that the proposed method outperforms direct pose regression [13] across all validation specimens. On 4 specimens (S1, S4, S5, S6) the proposed method shows similar performance as on the test DRRs indicating that there is no intrinsic domain gap between simulations and real X-rays, while specimens S0, S2, S3 show lower performance due to low bone-tissue contrast resulting in blurred bone boundaries (see suppl. material for a visualization).

3.3   Analysis of Domain Adaptation: DRRs to Real X-Ray with Spinal Implants

Furthermore, validation was conducted on $9 \cdot 1364$ real X-rays containing k-wires (5 specimens) and screws (4 specimens) to investigate if the proposed k-wire augmentation and realistic screw simulation address the domain gap. Applying the proposed approach trained without implant simulation does not improve upon the initial offset distribution. Direct intensity-based pose regression [13] combined with the proposed implant simulation is outperformed in all specimens

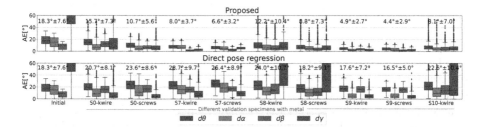

Fig. 4. Evaluation of the proposed sequential method on real X-rays with k-wires and screws with $MAE_{d\theta} \pm SD_{d\theta}$ indicated for each specimen S0, S7-S10. The proposed implant simulation strategies combined with the sequential pose regression shows to improve generalization towards spinal instrumentation.

by the proposed sequential method with simulated metal augmentation (Fig. 4). Evaluation on specimens S7, S9, S10 with k-wires and with screws indicates that the proposed implant simulation strategies combined with the sequential pose regression successfully address the domain gap. Specimens S0, S8 show limited accuracy, which can be explained by projection artifacts resulting from the body bag and edge overlays resulting from corpse decay air pockets (see suppl. material for a visualization).

The predicted pose updates are verified by forward projection for exemplary cases (with good, mean, bad performance) in Fig. 5. Even for challenging anatomical deformations (specimen S4), the method shows robust performance. Bad performance is observed for low contrast X-rays (specimen S3, S8) leading to inaccurate ROI localization and landmark detection.

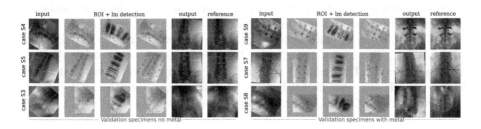

Fig. 5. The initial input is shown along with the predicted ROI and landmark detection for cases with good, mean, and bad performance (top-to-bottom). The predicted pose update is verified by projection and shown side-by-side with the reference standard.

4   Discussion and Conclusion

We proposed a sequential framework for C-arm pose estimation for standard projections, where the differential pose change is obtained from one initial X-ray. The framework is solely trained on simulated data since they uniquely provide ground truth pose annotations. To address the domain gap to real X-rays containing superimposed metal implants we proposed a k-wire augmentation and realistic screw simulation technique. On the example of the AP standard projection for the spine anatomy, we demonstrated that the proposed sequential pose regression significantly increases accuracy, robustness, and generalization properties compared to direct intensity-based pose regression. This was validated on a large specimen study covering different intra-operative settings, thus validating its ability to handle intra-class anatomical variation and its capabilities to generalize from in silico data to ex vivo human cadaver data. To derive a clinical tolerance of targeted positioning, we evaluated the variability of the manually acquired AP standard views relative to a reference standard plane defined in the reconstructed volumes. Across the validation specimens, we computed a mean deviation of $d\theta = 6.1° \pm 4.2°$. This was reached by the proposed method for all specimens without metal and for 6 out of 9 with metal. Failure cases can be attributed to low tissue-to-bone contrast (resulting from low bone density or obese patients) and to projection artifacts (resulting from corpse bags or corpse decay air pockets) not present during training. The proposed method has the potential to reduce the number of necessary acquisitions and improve standard projection accuracy in clinical practice. The radiation dose is drastically reduced by replacing the iterative manual positioning procedure with the presented approach requiring only a single X-ray. Furthermore, our method explicitly promotes clinical interpretability of results, by incorporating visible anatomical structures into the decision-making process as part of the proposed sequential pipeline (Fig. 5). Incorporating a robotic C-arm for automatic positioning would require additional handling of environmental obstacles. Currently our pose regression framework is trained specifically for the standard projection of the 4th lumbar vertebra. In future work, we also seek to adapt our approach to other vertebra levels by utilizing the vertebra-specific masks as an initialization source for a vertebra-specific standard projection. Furthermore, we plan to extend the implant simulation techniques to represent a broader variability of surgical tools. The approach can generalize to other spinal standard projections or vertebra levels. This requires the definition of the reference standard planes in the CTs. It can also be applied to other anatomical regions after defining anatomy-specific ROIs and clincally meaningful landmarks.

References

1. Bier, B., et al.: Learning to detect anatomical landmarks of the pelvis in X-rays from arbitrary views. Int. J. Comput. Assist. Radiol. Surg. **14**(9), 1463–1473 (2019). https://doi.org/10.1007/s11548-019-01975-5
2. Binder, N., Bodensteiner, C., Matthäus, L., Burgkart, R., Schweikard, A.: Image guided positioning for an interactive C-arm fluoroscope. Int. J. Comput. Assist. Radiol. Surg., 5–7 (2006)
3. Bott, O., Dresing, K., Wagner, M., Raab, B., Teistler, M.: Informatics in radiology: use of a C-arm fluoroscopy simulator to support training in intraoperative radiography. Radiographics **31**(3), E65–E75 (2011). https://doi.org/10.1148/rg.313105125

4. Bui, M., Albarqouni, S., Schrapp, M., Navab, N., Ilic, S.: X-ray PoseNet: 6 DoF pose estimation for mobile X-ray devices. In: 2017 IEEE Winter Conference on Applications of Computer Vision, pp. 1036–1044 (2017). https://doi.org/10.1109/WACV.2017.120

5. De Silva, T., et al.: C-arm positioning using virtual fluoroscopy for image-guided surgery. In: Medical Imaging: Image-Guided Procedures, Robotic Interventions, and Modeling 10135, p. 101352K (2017)

6. Fotouhi, J., et al.: Interactive flying frustums (IFFs): spatially aware surgical data visualization. Int. J. Comput. Assist. Radiol. Surg., 913–922 (2019)

7. Gong, R., Jenkins, B., Sze, R., Yaniv, Z.: A cost effective and high fidelity fluoroscopy simulator using the image-guided surgery toolkit (IGSTK). In: Medical Imaging: Image-Guided Procedures, Robotic Interventions, and Modeling 9036, p. 903618 (2014)

8. Grupp, R., et al.: Automatic annotation of hip anatomy in fluoroscopy for robust and efficient 2D/3D registration. Int. J. Comput. Assist. Radiol. Surg., 1–11 (2020). http://dx.doi.org/10.1007/s11548-020-02162-7

9. Haiderbhai, M., Turrubiates, J., Gutta, V., Fallavollita, P.: Automatic C-arm positioning using multi-functional user interface. CMBES Proc. **42** (2019)

10. Hou, B., et al.: Predicting slice-to-volume transformation in presence of arbitrary subject motion. In: Descoteaux, M., Maier-Hein, L., Franz, A., Jannin, P., Collins, D.L., Duchesne, S. (eds.) MICCAI 2017. LNCS, vol. 10434, pp. 296–304. Springer, Cham (2017). https://doi.org/10.1007/978-3-319-66185-8_34

11. Isensee, F., et al.: nnU-Net: self-adapting framework for u-net-based medical image segmentation. arXiv preprint arXiv:1809.10486 (2018)

12. Isensee, F., et al.: batchgenerators - a python framework for data augmentation. (2020). https://doi.org/10.5281/zenodo.3632567

13. Kausch, L., et al.: Toward automatic C-arm positioning for standard projections in orthopedic surgery. Int. J. Comput. Assist. Radiol. Surg., 1–11 (2020). https://doi.org/10.1007/s11548-020-02204-0

14. Kausch, L., Scherer, M., Thomas, S., Klein, A., Isensee, F., Maier-Hein, K.: Automatic image-based pedicle screw planning. In: Medical Imaging 2021: Image-Guided Procedures, Robotic Interventions, and Modeling 11598, pp. 115981I (2021). https://doi.org/10.1117/12.2582571

15. Kingma, D., Ba, J.: Adam: a method for stochastic optimization. arXiv:1412.6980 (2014)

16. Klein, A., Wasserthal, J., Greiner, M., Zimmerer, D., Maier-Hein, K.: MIC-DKFZ/basic_unet_example: Release (v2019.01) (2019). Zenodo. https://doi.org/10.5281/zenodo.2549509

17. Kordon, F., Maier, A., Swartman, B., Kunze, H.: Font augmentation: implant and surgical tool simulation for X-ray image processing. Bildverarbeitung für die Medizin, 176–182 (2020). http://dx.doi.org/10.1007/978-3-658-29267-6_36

18. Kügler, D., et al.: i3PosNet: instrument pose estimation from X-ray in temporal bone surgery. Int. J. Comput. Assist. Radiol. Surg. **15**(7), 1137–1145 (2020). https://doi.org/10.1007/s11548-020-02157-4

19. Löffler, M., et al.: A vertebral segmentation dataset with fracture grading. Radiol. Artif. Intell. **2**(4), e190138 (2020). http://dx.doi.org/10.1148/ryai.2020190138

20. Matthews, F., et al.: Navigating the fluoroscope's C-arm back into position: an accurate and practicable solution to cut radiation and optimize intraoperative workflow. J. Orthopaedic Trauma **21**(10), 687–692 (2007)

21. Miao, S., Wang, Z., Liao, R.: A CNN regression approach for real-time 2D/3D registration. IEEE Trans. Med. Imaging **35**(5), 1352–1363 (2016). https://doi.org/10.1109/TMI.2016.2521800

22. Rikli, D., et al.: Optimizing intraoperative imaging during proximal femoral fracture fixation – a performance improvement program for surgeons. Injury **104**, 19–19 (2018). https://doi.org/10.1016/j.injury.2017.11.024

23. Toth, D., Cimen, S., Ceccaldi, P., Kurzendorfer, T., Rhode, K., Mountney, P.: Training deep networks on domain randomized synthetic X-ray data for cardiac interventions. In: International Conference on Medical Imaging with Deep Learning, pp. 468–482 (2019)

24. Unberath, M., et al.: Augmented reality-based feedback for technician-in-the-loop C-arm repositioning. Healthcare Technol. Lett., 143–147 (2018). http://dx.doi.org/10.1049/htl.2018.5066

25. Unberath, M., et al.: DeepDRR – a catalyst for machine learning in fluoroscopy-guided procedures. In: Frangi, A.F., Schnabel, J.A., Davatzikos, C., Alberola-López, C., Fichtinger, G. (eds.) MICCAI 2018. LNCS, vol. 11073, pp. 98–106. Springer, Cham (2018). https://doi.org/10.1007/978-3-030-00937-3_12

26. Unberath, M., et al.: Enabling machine learning in x-ray-based procedures via realistic simulation of image formation. Int. J. Comput. Assist. Radiol. Surg. **14**(9), 1517–1528 (2019). https://doi.org/10.1007/s11548-019-02011-2

27. Zhang, Y., Miao, S., Mansi, T., Liao, R.: Task driven generative modeling for unsupervised domain adaptation: application to X-ray image segmentation. In: Frangi, A.F., Schnabel, J.A., Davatzikos, C., Alberola-López, C., Fichtinger, G. (eds.) MICCAI 2018. LNCS, vol. 11071, pp. 599–607. Springer, Cham (2018). https://doi.org/10.1007/978-3-030-00934-2_67

Quantitative Assessments for Ultrasound Probe Calibration

Elvis C. S. Chen[1,2,3](✉) (iD), Burton Ma[4](iD), and Terry M. Peters[1,2,3](iD)

[1] Robarts Research Institute, Western University, London, ON, Canada
chene@robarts.ca
[2] School of Biomedical Engineering, Western University, London, ON, Canada
[3] Department of Medical Biophysics, Western University, London, ON, Canada
[4] School of Computing, Queen's University, Kingston, ON, Canada

Abstract. Ultrasound probe calibration remains an area of active research but the science of validation has not received proportional attention in current literature. In this paper, we propose a framework to improve, assess, and visualize the quality of probe calibration. The basis of our framework is a heteroscedastic fiducial localization error (FLE) model that is physically quantifiable, used to i) derive an optimal calibration transform in the presence of heteroscedastic FLE, ii) assess the quality of a particular instance of probe calibration using a registration circuit, and iii) visualize the distribution of target registration error (TRE). The novelty of our work is the extension of the registration circuit to Procrustean point-line registration, and a demonstration that it produces a quantitative metric that correlates with true TRE. By treating ultrasound calibration as a heteroscedastic errors-in-variables regression instead of a least-squares regression, a more accurate calibration can be consistently obtained. Our framework has direct implication to many calibration techniques using point- and line-based calibration phantoms.

Keywords: Ultrasound probe calibration · Target registration error · Fiducial localization error · Heteroscedasticity · Registration circuits · Procrustes analysis · Non-perspective-n-point · Error-in-variable

1 Introduction

Probe calibration for freehand 3D ultrasound systems has been an area of active research for the past three decades [15,22,23]. In a freehand ultrasound system, a spatial tracking sensor is integrated into the ultrasound transducer and the coordinate system of the ultrasound image must be related to that of the tracking sensor through a spatial calibration process. Tracked and calibrated ultrasound effectively transforms a standard 2D ultrasound system into a 3D image modality, with vast clinical utilities including 3D visualization of the entire structure of an organ, direct measurements of quantitative attributes in 3D, and fusion with other 3D image modalities [23].

© Springer Nature Switzerland AG 2021
M. de Bruijne et al. (Eds.): MICCAI 2021, LNCS 12904, pp. 363–372, 2021.
https://doi.org/10.1007/978-3-030-87202-1_35

The quality of 3D ultrasound reconstruction from tracked 2D ultrasound ultimately depends on accurate probe calibration. Many *indirect* quantitative metrics used to assess probe calibration quality were comprehensively discussed in two review papers [15, 22]. To briefly summarize, *precision* of a calibration technique can be assessed via repeated calibrations, but together they do not provide a quantitative measurement for a specific instance of the calibration. *Trueness* of an instance of the probe calibration can be measured against a "known" ground truth but, as is the case for every registration problem, such a ground truth calibration may never be obtained. Alternatively, fitness of a probe calibration can be indirectly measured via methods including point-reconstruction [18, 24, 28], distance reconstruction [18, 25], and volume measurements [26]. These validation methods often require a separate *validation phantom*, and the validation process is time- and labour-intensive. An instance of the probe calibration can be validated against a known calibration obtained via an alternative means, but it is difficult to separate contribution of errors between the calibration to be validated from the calibration against which it is being compared. Ultimately, the validation phantom and technique themselves need to be validated leading to a vicious cycle that is difficult to complete practically. What is clinically relevant is the target registration error [9] (TRE), thus calibration accuracy should be assessed using TRE or a metric that correlates with TRE.

An ultrasound probe calibration technique has two components: i) a calibration phantom from which image features are captured and, ii) a numerical solution from which a calibration transform is derived. Many of the calibration techniques in current literature can be categorized into those using point-based and line-based phantoms [15, 22], and among those, a solution is often derived using Procrustean methods [12]. Several authors have employed the tip of a tracked stylus as the calibration phantom [24, 28], with the calibration transform derived using a least-squares solution for fiducial registration [1, 14]. Many calibration techniques are based on the construct of a "z-fiducial" [5]: while the z-fiducial is a line-based phantom, the calibration is still solved using fiducial registration. It should be noted that a popular, open-source, implementation of the z-fiducial calibration [17] reports a quantitative metric as the calibration error, but it is cautioned that, to avoid bias, validation should be performed using a different phantom [18, 27]. Khamene and Sauer were the first to introduce the use of the orientation of the stylus shaft as a form of line-fiducial [16], it is now understood that this calibration approach is equivalent to the non-perspective-n-point problem (NPnP) where many Procrustean solutions exist [4, 11]. However, none of these existing ultrasound calibration techniques account for heteroscedasticity of fiducial localization error (FLE) distribution in their solutions.

By treating ultrasound calibration as a Procrustean registration, a rich set of error estimation techniques can be utilized to provide quantitative assessments for an instance of the calibration. In this paper, we propose a unified assessment framework for Procrustean ultrasound probe calibration with consideration for heteroscedastic FLE. As the ground truth calibration can never be obtained, we demonstrate the efficacy of our framework using Monte Carlo simulation while using real data for illustration. Simulation results show that correctly accounting for FLE heteroscedasticity in the registration problem leads to lower TRE, and that our extended registration circuit analysis correlates with TRE. An open-source implementation of our framework can be accessed at https://github.com/chene/ARQOPUS.

2   Methods

We model the process of probe calibration as a Procrustean registration between corresponding points (ultrasound fiducial measurements in image space) and lines (tracked stylus in surgical space), where the ultrasound image space and surgical space are related by the calibration transform. Following [16], a tracked needle is used as the calibration phantom. A hyperechoic reflection is generated when the needle shaft is placed in the ultrasound image plane; the image depends on the ultrasound beam profile and angle of insonation and is heteroscedastic in nature (Fig. 1). For each recorded tracked ultrasound image, a bounding box is manually placed at the brightest hyperechoic region, the center of which denotes one ultrasound image fiducial for probe calibration.

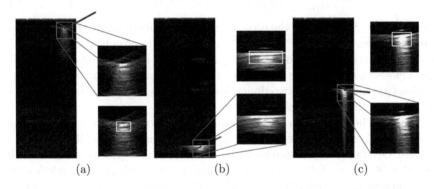

(a)   (b)   (c)

Fig. 1. Visual illustration of heteroscedasticity of needle reflection in ultrasound. The yellow bounding box depicts where the strongest needle reflection occurs. The geometry of the needle reflection depends on the ultrasound beam profile and angle of insonation: (a) at the proximal region of the ultrasound image, the needle reflections tends to be clear, leading to small bounding box, (b) at the distal portion of the image, the needle reflections tend to be large, possibly due to the wide beam profile thickness, and (c) a comet-tail image artifact often occurs when the needle is perpendicular to ultrasound image plane. Images taken from a Canon I18LX5 linear probe at 10 cm imaging depth.

As the hyperechoic signal is generated only when the ultrasound wave is reflected by an object, this bounding box constrains where the needle phantom **must be** after probe calibration. That is, we hypothesize and denote the centroid of the brightest reflection to be the ultrasound point fiducial where the needle intersects with the ultrasound image plane, and is contained within the image of the reflection, but whose true location is unknown. We further hypothesize that the geometry of the needle reflection, and hence the size of the bounding box, is proportional to FLE. For each calibration measurement, we record a homologous tuple (X_i, S_i, O_i, D_i), where (X_i, S_i) denotes the centroid and the size of the ith bounding box in the ultrasound image, and (O_i, D_i) denotes the ith measured needle position and orientation obtained from the tracking system. We take the size of the bounding box to be proportional to the in-plane component of the FLE covariance for the ith point measurement. The out-of-plane component is taken to be a small value to avoid having a singular covariance matrix.

The set of homologous tuples represents the Procrustean point-line registration, where the correspondence between ultrasound reflection and needle pose is explicitly known. As this is equivalent to the NPnP problem [11], many solutions exist [4]. Conceptually, these algorithms often decompose the difficult case of Procrustean point-line registration (where there is no known closed-form solution) to the simpler case of Procrustean point-point registration [1,14], by iteratively searching for the closest point on the ith line fiducial for each point fiducial X_i in an ICP-like framework [2]. We use heteroscedastic error-in-variable (HEIV) regression [21] in place of Horn's method in an ICP-like algorithm to incorporate the observed FLE into deriving the calibration transform:

$$^{SURG}T_{IMG} = HEIV(X^{IMG}, S^{IMG}, O^{SURG}, D^{SURG}) \qquad (1)$$

where $(X_{3\times n}^{IMG}, S_{3\times 3\times n}^{IMG})$ are the ultrasound fiducial locations and FLE covariance in image space, and $(O_{3\times n}^{SURG}, D_{3\times n}^{SURG})$ are the needle pose in surgical space. In Eq. 1, the corresponding closest point was taken to be the point on the ith line that is closest to X_i in terms of the Mahalanobis distance. The ultrasound fiducial locations are image-based (two-dimensional); we append a 0 for the z-component when solving the HEIV problem.

To assess the quality of a particular instance of probe calibration, we extended the concept of the registration circuit [7] to the Procrustean point-line registration problem. Originally formulated for Procrustean fiducial registration, the AQUIRC algorithm [8] forms a fully connected graph of m nodes, where each node is a fiducial configuration and the edge represents a registration between nodes. The image fiducial is perturbed with a known FLE into $m-1$ configurations, each of which is registered among themselves as well as to the fiducial configuration in the surgical space, producing a quantitative metric ϵ that was shown to correlate with the true TRE. In our extension of the registration circuit to point-line registration, each perturbed ultrasound fiducial configuration would be registered to a *different* set of fiducial configurations in surgical space, hence a fully-connected graph could no longer be maintained. For each red link in Fig. 2, we calculate the ground truth TRE and ϵ-values.

To visualize the TRE distribution across the entire ultrasound image as a result of fiducial configuration, we employed a line-based TRE modeling technique that accounts for heteroscedastic FLE [3]. Based on a spatial stiffness model of Procrustean point-line registration, this model estimates the *expected* target registration error magnitude for a target r, given the point fiducial locations X, FLE covariances S and line orientations D after registration:

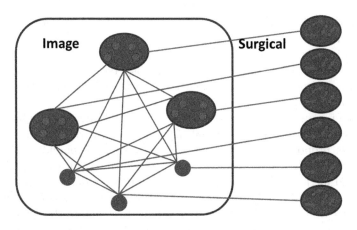

Fig. 2. Conceptualization of the registration circuit for Procrustean point-line registration. Image space contains $m-1$ ultrasound fiducial configurations that are created by adding random FLE values, drawn from the FLE covariance S_i, to the original fiducial location. Surgical space contains the nearest corresponding points on the registered line fiducials. Blue edges represents fiducial registration, whereas red edges represent point-line registration. Our circuit differs from the point-to-point circuit in that there are $m-1$ instances in our surgical space as opposed to only one instance in the point-to-point circuit.

$$eTRE = lineTRE(X^{IMG}, S^{IMG}, D^{IMG}, r^{IMG}) \qquad (2)$$

where $r_{3\times 1}^{IMG}$ is a target location in the image space. We note that inputs to Eq. 2 must reside in the same coordinate system. The efficacy of this line-based TRE estimation thus depends on the quality of the calibration registration.

3   Experiments and Results

To evaluate the predictive power of the registration circuit and line-based TRE model, we designed a Monte Carlo simulation based on real ultrasound calibration data, where the ground truth TRE is known.

The simulation was bootstrapped using real ultrasound probe calibration data: a 6-degrees-of-freedom (6-DoF) magnetic pose sensor (Aurora, NDI, Canada) was rigidly attached to the exterior chasing of a Canon i18LX5 linear transducer (width ≈ 46 mm), with a 17 gauge 5-DoF tracked needle used as a calibration phantom. Ultrasound images were acquired using a USB frame grabber (Epiphan Systems, USA) at a size of 312×714 pixels at an imaging depth of 10 cm. A total of 18 tracked ultrasound images were acquired, with the needle placed at varying locations and angulation across the ultrasound image. For each recorded image, a bounding box was manually placed at the brightest hyperechoic region (Fig. 1). The centroid of the bounding box, along with the tracked needle pose, was used to derive a probe calibration transform using an iterative least-squares solution [4]. Using this estimated calibration as the

368   E. C. S. Chen et al.

ground truth calibration, the *idealized, noise-free*, ultrasound fiducial locations were calculated as the intersection between the $x - y$ plane (*i.e.* $z = 0$ where the ultrasound images lies) and needle poses inverted by the calibration transform. Figure 3b depicts the spatial extent and line dispersion of the fiducial configuration. The centroid of the ultrasound image was designated as the target for TRE evaluation.

We then perturbed the locations of these idealized ultrasound fiducials using a FLE drawn from a random normal distribution with zero mean and a variance that is proportional to the size of their respective bounding boxes (*e.g.* heteroscedastic FLE). This was performed 10 000 times: for each configuration of perturbed ultrasound fiducials, two calibrations were derived using the least-squares [4] and HEIV [21] solutions. The *noise-free* fiducial configuration was used in our extension of the registration circuit to generate the quantitative ϵ-value, using $m = 25$ to generate a connected graph of m nodes. To visualize the TRE distribution for the entire ultrasound image, the line-based TRE model [3] was used to generate an *expected* TRE for every pixel, as depicted in Fig. 3b.

The efficacy of the Procrustean point-line registration solutions were evaluated. As the ground truth calibration is known, the accuracy of the calibration transform using both the least-squares and HEIV solutions can be assessed using TRE. Over 10 000 perturbations of the ultrasound image fiducials, the HEIV solution [21] consistently achieved a more accurate calibration than the least-squares solution [1]. After the normality of their TRE distributions were confirmed by the One-sample Kolmogorov-Smirnov test, the paired t-test and Wilcoxon signed rank test rejected the null hypothesis that their difference comes from a distribution with zero mean/median (both with $p < 0.0001$). Since the HEIV algorithm achieved a lower mean TRE and smaller standard deviation, we conclude that HEIV indeed achieves a more accurate calibration with statistical significance. The statistics of their TRE distributions is shown in Table 1.

Table 1. Statistics of the TRE distribution based on the least-squares over 10000 simulated calibrations. The skewness and kurtosis confirm the normality of their TRE distributions.

TRE distribution	Mean (mm)	Std (mm)	Skewness	Kurtosis
Least-squares [1]	1.73	1.02	1.11	4.45
HEIV [21]	0.90	0.42	0.72	3.72

To assess the quality of the calibration, we ran our point-line registration circuit 10000 times. At each iteration, the *noise-free* ultrasound fiducial was used, but perturbed into $m = 24$ noisy configurations and produced $m = 24$ ϵ-values. Following [7], we only considered the non-traditional registration circuit (NTRC) of size 3, as we also found no correlation between the result of the traditional registration circuit (TRC) and the true TRE. We found significant correlations between the ϵ-value and true TRE ($r = 0.1860, p < 0.0001$, Fig. 3a).

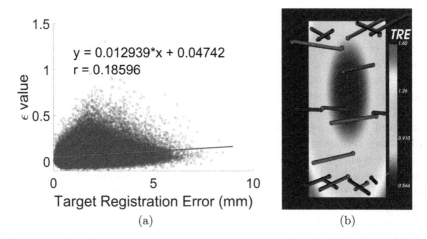

Fig. 3. (a) Scatter plot of the TRE and ϵ-value using line-based registration circuit, and (b) TRE visualization of the calibration heatmap: cold colors have lower eTRE. (Color figure online)

An *expected* TRE value can be assigned to a pixel location using Eq. 2 for a given fiducial configuration and known calibration. As shown in Fig. 3b, the expected TRE has an elliptical distribution, lower in the centroid of the fiducial configuration (blue) and gradually increasing outwards. The distal part of the ultrasound image has greater FLE in our model which is reflected in the greater estimated TRE in that part of the image.

4   Discussion and Conclusion

In this paper, we presented a framework for improving, assessing, and visualizing probe calibration accuracy for freehand ultrasound. Our framework is based on a heteroscedastic FLE model that is physically quantifiable. By treating probe calibration as a Procrustean registration, we improved calibration accuracy by solving the calibration transform as a heteroscedastic-error-in-variable regression [21]. To assess the accuracy of a particular instance of calibration, we extended the concept of a registration circuit [7] to Procrustean point-line registration, and demonstrated that the resulting ϵ-value correlates with the true TRE. To visualize the distribution of TRE, we employed the line-based TRE model [3] to estimate the expected TRE value for the ultrasound image. Our framework accounts for zero mean normally distributed FLE.

In our extension of the registration circuit for point-line Procrustean analysis, a fully-connected graph can no longer be maintained. This deviates from the original formulation [8], and is because each FLE perturbation introduced in image space creates an ultrasound fiducial configuration that is registered to a different set of point fiducials in surgical space (Fig. 2). The inability to maintain a complete graph implies there is additional information (*i.e.* edges) in surgical space that is not utilized, and is one of the possible reason why our ϵ-value has a lower correlation ($r = 0.1860$) with the true TRE when compared to [8]. Extension of our registration circuit for line-based registration using a *fully-connected* and *directed* graph is one of our future goals.

In our simulation with known ground truth, we demonstrated that HEIV [21] achieved more accurate calibrations than the standard least-squares solution [1,14]. The heteroscedasticity of ultrasound fiducials is due to the finite thickness of the ultrasonic beam and angle of insonation with the calibration needle, which is quantifiable by the image appearance of the ultrasound fiducial. We estimated the magnitude of individual FLE by manually placing a bounding box surrounding the ultrasound fiducial, but this task can be automated by training a Mask R-CNN [13] for ultrasound fiducial segmentation and bounding box recognition.

Our framework can be applied to other calibration techniques involving Procrustean fiducial registration [24,28]. In this scenario, the heteroscedastic FLE would incorporate the thickness of the ultrasonic beam profile in the $z-$dimension. The original formulation of the HEIV algorithm [20,21], registration circuit [7,8], and point-based TRE modeling technique [6,19] apply directly. Our framework can also be applied to $z-$fiducial calibration [5,17]: instead of using the principle of similar triangles to infer the 3D location of the slanted wire or rod, the entire z-fiducial should be modeled as 3 separate line-fiducials instead.

The main limitation of our work is the lack of variation in fiducial configurations in our validation. As evident from Fig. 3, accuracy of line-based registration depends not only on the spread of point fiducial [10], but also on the dispersion of line fiducial orientation. Consider the trivial scenario: if all needle poses are parallel to each other, such a line fiducial configuration cannot constrain the calibration to a unique solution. To fully validate the efficacy of the registration circuit framework, a large number of (> 5000) different fiducial configurations should be tested instead of just one configuration as it is done in our work and in [8]. Creating such a vast variable, and yet representative, set of calibration fiducial configurations is practically challenging. One mitigation strategy is to formulate a video-based probe calibration technique and draw from it samples to form different fiducial configurations.

The TRE visualization (Fig. 3) can be used to visually guide the calibration process. The TRE distribution for a given calibration is ellipsoidal, lowest at the centroid of the fiducial configuration but higher at the periphery of the ultrasound image. In our example, the fiducial measurements are roughly symmetrical across the ultrasound image, and yet the expected TRE is higher at the bottom of the image. In our calibration framework where each calibration fiducial is measured incrementally, additional calibration fiducial measurements should be taken from regions with high expected TRE.

The key to our ultrasound calibration assessment framework is the incorporation of the heteroscedastic FLE which is quantifiable by the size of the ultrasound image fiducial. For all ultrasound calibration techniques based on

Procrustean registration, we advocate the use of the HEIV [21] solution, as it consistently provides a better estimate for probe calibration in the presence of heteroscedastic FLE. When multiple calibrations are available, the registration circuit can be used to compare the relative accuracy of these calibrations as the output of the registration circuit is shown to correlate with TRE. Lastly, TRE visualization can be used to incrementally improve the calibration quality by suggesting regions of the ultrasound image from which additional calibration fiducials should be measured.

Conflict of Interest. The authors declare that they have no conflict of interest.

References

1. Arun, K.S., Huang, T.S., Blostein, S.D.: Least-squares fitting of two 3-D point sets. IEEE Trans. Pattern Anal. Mach. Intell. **PAMI-9**(5), 698–700 (1987)
2. Besl, P.J., McKay, N.D.: A method for registration of 3-D shapes. IEEE Trans. Pattern Anal. Mach. Intell. **14**(2), 239–256 (1992)
3. Chen, E.C.S., Peters, T.M., Ma, B.: Guided ultrasound calibration: where, how, and how many calibration fiducials. Int. J. Comput. Assist. Radiol. Surg. **11**(6), 889–898 (2016)
4. Chen, E.C.S., Peters, T.M., Ma, B.: Which point-line registration? In: Webster, R.J., III., Fei, B. (eds.) Medical Imaging 2017: Image-Guided Procedures, Robotic Interventions, and Modeling, vol. 10135, pp. 70–82. International Society for Optics and Photonics, SPIE (2017)
5. Comeau, R.M., Fenster, A., Peters, T.M.: Integrated MR and ultrasound imaging for improved image guidance in neurosurgery. In: Hanson, K.M. (ed.) Medical Imaging 1998: Image Processing, vol. 3338, pp. 747–754. International Society for Optics and Photonics, SPIE (1998)
6. Danilchenko, A., Fitzpatrick, J.M.: General approach to first-order error prediction in rigid point registration. IEEE Trans. Med. Imaging **30**(3), 679–693 (2011)
7. Datteri, R.D.: Assessing registration quality via registration circuits. Ph.D. thesis, Vanderbilt University, Nashville, Tennessee, USA (2014)
8. Datteri, R.D., Dawant, B.M.: Estimation and reduction of target registration error. In: Ayache, N., Delingette, H., Golland, P., Mori, K. (eds.) MICCAI 2012. LNCS, vol. 7512, pp. 139–146. Springer, Heidelberg (2012). https://doi.org/10.1007/978-3-642-33454-2_18
9. Fitzpatrick, J.M.: Fiducial registration error and target registration error are uncorrelated. In: Miga, M.I., Wong, K.H. (eds.) Medical Imaging 2009: Visualization, Image-Guided Procedures, and Modeling, vol. 7261, pp. 21–32. International Society for Optics and Photonics, SPIE (2009)
10. Fitzpatrick, J.M., West, J.B., Maurer, C.R.: Predicting error in rigid-body point-based registration. IEEE Trans. Med. Imaging **17**(5), 694–702 (1998)
11. Fusiello, A., Crosilla, F., Malapelle, F.: Procrustean point-line registration and the NPnP problem. In: 2015 International Conference on 3D Vision, pp. 250–255 (2015)
12. Gower, J.C., Dijksterhuis, G.B.: Procrustes Problems. Oxford University Press, Oxford (2004)
13. He, K., Gkioxari, G., Dollár, P., Girshick, R.: Mask R-CNN. In: 2017 IEEE International Conference on Computer Vision (ICCV), pp. 2980–2988 (2017)

14. Horn, B.K.P.: Closed-form solution of absolute orientation using unit quaternions. J. Opt. Soc. Am. A **4**(4), 629–642 (1987)
15. Hsu, P.W., Prager, R.W., Gee, A.H., Treece, G.M.: Freehand 3D ultrasound calibration: a review. In: Sensen, C.W., Hallgrímsson, B. (eds.) Advanced Imaging in Biology and Medicine: Technology, Software Environments, Applications, pp. 47–84. Springer, Heidelberg (2009). https://doi.org/10.1007/978-3-540-68993-5_3
16. Khamene, A., Sauer, F.: A novel phantom-less spatial and temporal ultrasound calibration method. In: Duncan, J.S., Gerig, G. (eds.) MICCAI 2005. LNCS, vol. 3750, pp. 65–72. Springer, Heidelberg (2005). https://doi.org/10.1007/11566489_9
17. Lasso, A., Heffter, T., Rankin, A., Pinter, C., Ungi, T., Fichtinger, G.: PLUS: open-source toolkit for ultrasound-guided intervention systems. IEEE Trans. Biomed. Eng. **61**(10), 2527–2537 (2014)
18. Lindseth, F., Tangen, G.A., Langø, T., Bang, J.: Probe calibration for freehand 3-D ultrasound. Ultrasound Med. Biol. **29**, 1607–1623 (2003)
19. Ma, B., Moghari, M.H., Ellis, R.E., Abolmaesumi, P.: Estimation of optimal fiducial target registration error in the presence of heteroscedastic noise. IEEE Trans. Med. Imaging **29**(3), 708–723 (2010)
20. Matei, B.: Heteroscedastic errors-in-variables models in computer vision. Ph.D. thesis, Rutgers University (2001)
21. Matei, B., Meer, P.: Optimal rigid motion estimation and performance evaluation with bootstrap. In: Proceedings of 1999 IEEE Computer Society Conference on Computer Vision and Pattern Recognition, vol. 1, pp. 339–345 (1999)
22. Mercier, L., Langø, T., Lindseth, F., Collins, D.L.: A review of calibration techniques for freehand 3-D ultrasound systems. Ultrasound Med. Biol. **31**(4), 449–471 (2005)
23. Mozaffari, M.H., Lee, W.S.: Freehand 3-D ultrasound imaging: a systematic review. Ultrasound Med. Biol. **43**, 2099–2124 (2017)
24. Muratore, D.M., Galloway, R.L., Jr.: Beam calibration without a phantom for creating a 3-D freehand ultrasound system. Ultrasound Med. Biol. **27**(11), 1557–1566 (2001)
25. Prager, R.W., Rohling, R.N., Gee, A.H., Berman, L.H.: Rapid calibration for 3-D freehand ultrasound. Ultrasound Med. Biol. **24**(6), 855–869 (1998)
26. Rousseau, F., Hellier, P., Barillot, C.: Confhusius: a robust and fully automatic calibration method for 3D freehand ultrasound. Med. Image Anal. **9**(1), 25–38 (2005)
27. Treece, G.M., Gee, A.H., Prager, R.W., Cash, C.J.C., Berman, L.H.: High-definition freehand 3-D ultrasound. Ultrasound Med. Biol. **29**, 529–546 (2003)
28. Zhang, H., Banovac, F., White, A., Cleary, K.: Freehand 3D ultrasound calibration using an electromagnetically tracked needle. In: Cleary, K.R., Galloway, R.L., Jr. (eds.) Medical Imaging 2006: Visualization, Image-Guided Procedures, and Display, vol. 6141, pp. 775–783. International Society for Optics and Photonics, SPIE (2006)

Intra-operative Update of Boundary Conditions for Patient-Specific Surgical Simulation

Eleonora Tagliabue[1]([✉]), Marco Piccinelli[1], Diego Dall'Alba[1], Juan Verde[2],
Micha Pfeiffer[3], Riccardo Marin[4], Stefanie Speidel[3], Paolo Fiorini[1],
and Stéphane Cotin[5]

[1] University of Verona, Verona, Italy
`eleonora.tagliabue@univr.it`
[2] Institut de Chirurgie Guidée par l'Image, Strasbourg, France
[3] National Center for Tumor Diseases, Dresden, Germany
[4] Sapienza University of Rome, Rome, Italy
[5] INRIA, Strasbourg, France

Abstract. Patient-specific Biomechanical Models (PBMs) can enhance computer assisted surgical procedures with critical information. Although pre-operative data allow to parametrize such PBMs based on each patient's properties, they are not able to fully characterize them. In particular, simulation boundary conditions cannot be determined from pre-operative modalities, but their correct definition is essential to improve the PBM predictive capability. In this work, we introduce a pipeline that provides an up-to-date estimate of boundary conditions, starting from the pre-operative model of patient anatomy and the displacement undergone by points visible from an intra-operative vision sensor. The presented pipeline is experimentally validated in realistic conditions on an ex vivo pararenal fat tissue manipulation. We demonstrate its capability to update a PBM reaching clinically acceptable performances, both in terms of accuracy and intra-operative time constraints.

Keywords: Intra-operative model update · Boundary conditions · Biomechanical modeling

1 Introduction

An up-to-date Patient-specific Biomechanical Model (PBM) of the surgical scenario can bring benefits to the surgical practice in several ways. In computer-assisted interventions, such PBM can guide surgeons towards the structures of interest, which might be hidden from the partial view available intra-operatively [11]. A further application is in the field of autonomous robotic surgery, where a PBM of the current surgical condition is required for verification of the robotic actions in a controlled environment, before execution [5]. Moreover, PBMs can play the role of virtual sensors to estimate interaction forces between instruments and tissues when direct force measurement systems are not available, and provide

© Springer Nature Switzerland AG 2021
M. de Bruijne et al. (Eds.): MICCAI 2021, LNCS 12904, pp. 373–382, 2021.
https://doi.org/10.1007/978-3-030-87202-1_36

a feedback to the surgeon [8]. The predictive power of such PBMs highly relies on their accurate parametrization, which has to be tailored to each new patient. Patient-specific geometry and tissues' elastic properties can be extracted from pre-operative anatomical images or using ad-hoc modalities such as elastographic techniques, allowing to build a PBM with personalized properties [7]. However, information available before the intervention is often insufficient to fully characterize PBMs to the extent required to achieve clinically accepted accuracy. In particular, there is usually no way to delineate the adhesions between neighboring organs from pre-operative data. Such adhesions define simulation Boundary Conditions (BCs), thus they are key to obtain an accurate model [12,13,15]. As a consequence, reliable BCs can be estimated only from data that are collected intra-operatively [11,13,15].

The problem of intra-operative estimation of BCs has been tackled by only few works. In [13,18], the position of BCs is initialized based on statistical atlases, leading to a method which is not robust to inter-patient variations. Other approaches [14] propose to update BCs exploiting additional intra-operative sensors, undermining their direct applicability to the standard clinical practice. In [13,15], authors propose to use stochastic filters to estimate the elasticity of the hepatic ligaments, exploiting intra-operative observations of the tissue state. However, the filters' inference time strongly depends on their parameters initialization, which is highly sensitive to each patient's properties, possibly introducing a degradation in the performances from case to case.

A recent research trend has focused on the usage of Deep Neural Networks (DNNs) to update a biomechanical model based on intra-operative data [4,10,16,17,19]. These works have shown that DNNs can learn biomechanical models even when trained with synthetic data only, while guaranteeing very low inference time. Furthermore, Pfeiffer et al. [16,17] demonstrated that DNNs can also learn a surface representation, thus being able to handle any input geometric model. However, all these works have focused on the estimation of the full 3D displacement field to accomplish the task of intra-operative registration, either without taking advantage of any PBM parametrization inferred from pre-operative data or assuming that BCs are fixed and a-priori known.

In this work, we present a complete pipeline that allows to continuously update an existing PBM by estimating model BCs, starting directly from the raw intra-operative point cloud of the deforming anatomy provided by a vision sensor and without relying on any a-priori assumption about their location. We conduct an experimental validation of the complete framework on an ex vivo human model, including the anterior renal fascia (Gerota's fascia), the pararenal adipose tissue and the kidney. The manipulation of these tissues is a key step during most of surgical kidney procedures (most importantly during partial nephrectomy). Obtained results demonstrate that the proposed pipeline is able to update a real PBM, respecting clinically acceptable requirements both in terms of accuracy and timing, thus allowing to account for possible intra-operative changes of the BCs caused by surgical manipulations.

2   Method

In order to provide an up-to-date PBM that continuously follows the current surgical scenario, we rely on a framework involving two independent processes which run concurrently. The first process is entirely dedicated to a physics-based simulation of the surgical environment capable of real-time performances [24]. Such simulation leverages the PBM created from pre-operative data, characterized by both the undeformed 3D geometry of the anatomy and its known mechanical properties. The second process is devoted to a strategy for updating PBM parametrization during the intervention, starting from intra-operative sensor data. In this work, we focus on this second task. In particular, we present a pipeline to update simulation BCs from the 3D point cloud of the surgical scene with a very short latency[1]. This allows the simulation to continuously reflect the changes introduced in the environment by surgical manipulations (Fig. 1).

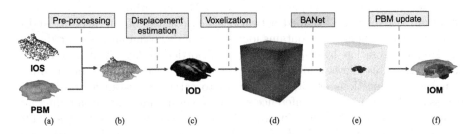

Fig. 1. Overview of the pipeline to update PBM boundary conditions. (a) Initial data: IOS and PBM; (b) rigidly aligned data; (c) estimated IOD, where brighter color is associated with highest displacement; (d) voxelized representation, where grid cells are colored based on the signed distance field from PBM surface; (e) estimated BCs in grid space; (f) PBM annotated with estimated BCs, giving the intra-operative model IOM.

Pre-processing. The acquired raw 3D point cloud passes through an initial pre-processing step. First, both color and spatial segmentation are performed to extract the current view of the deformed Intra-Operative Surface (IOS) from the full anatomical point cloud. The extracted IOS is then rigidly aligned to its corresponding portion of the PBM, based on geometric features and known spatial relations estimated at the beginning of surgery.

Displacement Estimation. We calculate the Intra-Operative Displacement field (IOD) which maps each point in the IOS to its corresponding one on the undeformed PBM. To achieve this, we estimate a correspondence with a nearest-neighbor pairing between the point cloud and the PBM, and we refine it using the ZoomOut method [9]. This non-rigid approach is entirely intrinsic, promotes isometric solutions (i.e., correspondences that preserve the surface

---

[1] Project available at https://gitlab.com/altairLab/banet.

distances between the points), and has approximately linear complexity. Thus, ZoomOut guarantees a trade-off between accuracy and timing, making it ideal for real-time precision surgical operations, where we assume the folding and bending of the surface preserves surface geometry (i.e., boundaries and metric) without introducing dramatic stretching.

Voxelization. In order to exploit convolutional filters, input information is converted into a grid-like volume of dimension $64 \times 64 \times 64$ and side length 30 cm. The PBM is encoded into the grid using its signed distance field and the IOD through a Gaussian interpolation.

BANet. The boundary condition update is performed by BANet, a DNN estimating at which points a given deformable tissue is attached to the surrounding environment [22]. BANet has been validated on phantom data with simple geometry, but has never been applied to a real PBM and within a realistic clinical situation. BANet is a U-shaped network, which consists of an encoding and a decoding path. The former contains four stages of downsampling, each reducing the spatial resolution by half, resulting in a bottleneck layer which is 4^3 grid cells in size. This allows information to travel across the spatial domain. Additional skip-connections enable the network to carry fine detail forward where necessary. To ensure a high inference speed, downsampling and upsampling are done via simple MaxPool and interpolation functions. The network is trained to approximate the function $f(PBM, IOD) = AP$, where AP is the binary voxel field representation of the attachment points. The training dataset is composed of synthetic samples representing adipose tissues PBMs (with different random geometries and mechanical parameters) and annotated with randomly extracted BCs. In this work, we use the publicly available implementation of BANet with the provided pre-trained weights [22].

PBM Update. Finally, BANet-estimated BCs are mapped from grid space to the original PBM space, giving the Intra-Operative Model (IOM). This step completes the proposed pipeline, and the obtained IOM is used to update the intra-operative simulation running in the synchronous process.

3   Experiments and Results

Validation of the presented method is carried out on ex vivo pararenal fat tissue manipulation (Fig. 2a). Tissue's PBM is initialized with the specimen 3D geometry, generated from manual segmentation of its CT scan, and discretized with 65,538 tetrahedral elements. Its biomechanical properties are selected to be aligned with those observed for adipose tissues (i.e., St Venant Kirchhoff material with Young's modulus 3 kPa and Poisson ratio 0.45) [1]. Leveraging on the constructed PBM (Fig. 2b), it has been also possible to generate a synthetic dataset based on a real anatomical model. Such dataset allows us to evaluate method

(a)   (b)

Fig. 2. (a) An example tissue manipulation. The surgeon manipulates the pararenal fat and anterior fascia with a laparoscopic instrument, grasping from point A. Point clouds of the tissue state are acquired with an RGB-D camera capturing the scene from the same perspective of the picture. (b) The PBM of the ex-vivo perinephric tissue. The position of the grasping points is marked with a letter and a green circle. (Color figure online)

performances within a scenario which is influenced neither by sensor noise nor by inaccuracies introduced by ZoomOut, since the IOD can be estimated by directly matching corresponding points in the deformed and the undeformed configurations.

The capability of the method to update the PBM such that the virtual environment resembles the current tissue condition is assessed by comparing the deformed state in the simulated environment with the available ground truth deformed configuration. The deformed state in the virtual environment is obtained by performing a finite element (FE) simulation where model BCs are defined by the proposed approach and computing the Root Mean Squared Error (RMSE) between the simulated and the ground truth configurations. A state-of-the-art direct solver [20] is used together with an iterative Newton-Raphson method to solve the non-linear system of equations in the static domain, within the open-source SOFA framework [6]. The obtained simulation result is weakly sensitive to possible inaccuracies in the selected mechanical parameters, since the driving input to the simulation is represented by a displacement (i.e., the same displacement which is applied to the corresponding ground truth configuration) [12].

3.1 Synthetic Adipose Tissue Manipulation

A synthetic dataset of 600 samples of adipose tissue manipulation is generated following the same training data generation strategy used in [22], but keeping the considered PBM fixed (Fig. 2b). The median RMSE (interquartile range IQR) calculated on all the 3D model points between the deformed IOM and the corresponding reference sample is 1.4 $(0.7 - 3.2)$ mm, indicating an overall good matching (Fig. 3a). Since ground truth BCs are available for the synthetic dataset, we further assess prediction accuracy by computing the Dice coefficient (DSC)[23], which measures the overlapping area between ground truth and predicted BCs. Median (IQR) value for DSC is 0.51 (0.40−0.60). Figure 3b and c show that the network is challenged by the complex geometry and fails to accurately identify BCs when they are distributed along the sharp edges of the mesh, especially when the amount of visible surface is very small and does not capture the region undergoing the highest deformation. However, when the visible surface captures the region with greatest deformation (Fig. 3b), BANet is able to provide a plausible prediction, which leads to a precise matching between the simulated and the reference configuration.

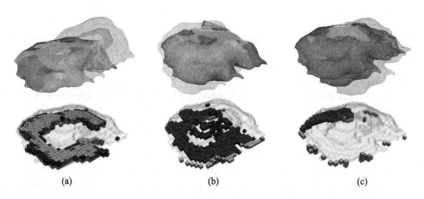

(a)   (b)   (c)

Fig. 3. Results on the synthetic dataset considering input displacements above 25 mm. Upper row: ground truth deformed configuration (green) and simulated deformed configuration obtained when using predicted BCs (yellow), with the considered visible surface (light blue). Pink mesh represents the deformed configuration when BCs are unknown, thus undefined. Lower row: ground truth BCs (green) and predicted BCs (red) in grid space. (a) Good overlap between simulated and ground truth configurations (RMSE = 1.6 mm), due to good prediction accuracy (DSC = 0.68). (b) Good overlap (RMSE = 1.7 mm) is possible even if prediction is not the same as the ground truth one (DSC = 0.09). (c) High RMSE (14.6 mm) is due to an inaccurate prediction (DSC = 0.02), when visible surface does not provide enough information about tissue state. (Color figure online)

3.2   Real Adipose Tissue Manipulation

The presented pipeline is employed to update the PBM during ex vivo pararenal tissue manipulation. An expert surgeon is asked to grasp the human tissue with a laparoscopic tool from four pre-defined points, whose position is known in the 3D model space thanks to CT-visible markers, and lift it to the maximum reachable extent that prevents tissue tearing (Fig. 2a). After pulling the tissue from all the points, the surgeon introduces a change in BCs by dissecting some tissue adhesions, and repeats the acquisitions. Acquired data are relative to three different initial states (i.e., two dissection stages); however, it has not been always possible to lift from all the grasping points due to excessive tissue damage introduced by dissection. Intra-operative point clouds capturing deformed tissue states are acquired through an Intel RealSense D435 RGB-D camera. Collected images allow to extract the displacement applied to the tissue by the surgeon, by tracking a STag marker attached to the instrument (Fig. 2a) [3]. To evaluate the presented method, point clouds at regular lifting steps of 10 mm are extracted and passed through the complete pipeline described in Sect. 2. As soon as a new estimate of BCs is available, we perform a FE simulation to the obtained IOM applying the same input displacement applied by the surgeon. We then calculate the RMSE between each point in the acquired point cloud and its corresponding one, as estimated by ZoomOut, in the deformed IOM.

Table 1 reports the average RMSE over 10 runs of the whole pipeline, relative to the error at rest (average 3.62 mm, which includes contributions of segmentation and registration errors) at increasing pulling levels. The reason why we report results over multiple runs is that we rely on an approximate nearest neighbour for ZoomOut to minimize the computational overhead, which might however introduce some differences between different runs. The average time required to update the synchronous PBM simulation is 1.44 ± 0.14 s (with an average of 0.039 s dedicated to pre-processing, 1.39 s to displacement estimation, and 0.013 s to BANet), tested on a workstation with an AMD Ryzen7 3700X CPU and NVIDIA RTX 2070 SUPER graphics card. This update rate allows to provide feedback to the operating surgeon and/or the autonomous system based on a model reflecting the changes in clinical settings with a latency compatible with standard surgical workflows. Average RMSEs remain below 5 mm in almost all the cases, which is aligned with the accuracy levels required for model-guided intra-operative applications in the context of minimally-invasive surgery [4,11,16]. Furthermore, obtained RMSEs are significantly lower than the ones achieved when BCs are unknown, thus they cannot be defined and the simulation remains unconstrained (last column in the table). In general, RMSE increases with increasing deformation. This might be due to the fact that the high deformations cause poor surface estimation from [21], which assumes that the point cloud already represents the desired surface without noise or topological artifacts. In such case, ZoomOut promotes a disturbed correspondence (i.e., isometric to a wrong surface). To improve this step, investigating the point cloud denoising techniques seems a promising future direction. This further motivates the worse RMSE values obtained in correspondence to grasping point B, whose point clouds are partially occluded by the surgical tool.

380   E. Tagliabue et al.

Table 1. RMSE between ground truth point clouds and their corresponding points in the virtual environment when using predicted BCs, at increasing deformation levels [mm]. Acquisitions are grouped by grasping point (A, B, C, D) and initial tissue state (1, 2, 3). Reported errors are relative to the error at rest and represent the average over 10 runs of the entire pipeline. Missing values are due to failures in instrument tracking. Last column reports the RMSE obtained when no BCs are defined in the virtual environment ($Mean_{UC}$).

Grasp	State	$10\,mm$	$20\,mm$	$30\,mm$	$40\,mm$	$50\,mm$	Mean	Std	$Mean_{UC}$
A	1	-	0.94	1.06	1.28	6.81	2.52	2.48	24.93
	2	-	-	1.42	2.42	2.75	2.20	0.65	31.99
B	1	2.96	6.75	-	-	-	4.86	1.90	13.02
	2	5.05	9.91	11.47	-	-	8.81	2.84	27.06
	3	3.27	11.71	9.12	3.92	-	7.01	3.60	31.53
C	1	-	1.65	2.49	-	-	2.07	0.42	18.66
	2	5.30	2.54	3.84	-	-	3.89	1.42	14.23
	3	2.59	4.61	6.93	10.76	-	6.22	3.04	18.79
D	1	1.07	-	-	-	-	1.07	0.00	7.72
	2	-	1.32	1.57	3.98	5.25	3.03	1.71	19.76
	3	1.51	2.28	3.57	-	-	2.45	0.85	7.29
Mean		3.11	4.64	4.61	4.47	4.94			
Std		1.50	3.90	3.52	3.32	1.80			

4   Discussion and Conclusion

In this work, we have presented a complete pipeline that allows to update a patient-specific pre-operative model for surgical assistance, based on data acquired during the intervention. Validation experiments have shown that the presented pipeline can be used to successfully update a PBM exploiting data coming directly from intra-operative sensors, while respecting both accuracy and time constraints compatible with standard minimally-invasive surgical applications. The quality of the final result is influenced by the different sources of errors that are introduced throughout the various stages of the pipeline, from an imprecise initial rigid alignment, to the presence of sensor noise and inaccurate computation of corresponding points. In future works, we plan to improve the pre-processing stage, for instance by reconstructing the point cloud from the stereo-endoscope view [2]. Inaccurate surface matching can be addressed by either letting the network implicitly solve the surface correspondence problem as in [16], by providing salient points extracted from camera view to ZoomOut, or by improving surface estimation [25]. In particular, relying on DNN to directly solve for surface correspondences seems promising to further improve the time performances of the current implementation, where ZoomOut is responsible of the main computational overhead. By providing an update of model BCs with a very short delay, our method could handle situations with dynamically changing

BCs, for example involving dissection, sutures removal or topological changes. However, due to the fact that the surgical environment is intrinsically evolving in time and the network has shown to benefit from the availability of more informative input data, we expect that the robustness of the prediction will be improved by considering time dynamics and we will tackle this in future works.

Acknowledgements. Authors would like to thank the preclinical research staff at IHU Strasbourg for their assistance and support during the experiments. This project has received funding from the European Research Council (ERC) under the European Union's Horizon 2020 research and innovation programme (grant agreement No. 742671 "ARS"), from French state funds managed within the "Plan Investissements d'Avenir" and from the ANR (reference ANR-10-IAHU-02).

References

1. Alkhouli, N., et al.: The mechanical properties of human adipose tissues and their relationships to the structure and composition of the extracellular matrix. Am. J. Physiol. Endocrinol. Metab. **305**(12), E1427–E1435 (2013)
2. Allan, M., et al.: Stereo correspondence and reconstruction of endoscopic data challenge. arXiv preprint arXiv:2101.01133 (2021)
3. Benligiray, B., Topal, C., Akinlar, C.: Stag: a stable fiducial marker system. Image Vis. Comput. **89**, 158–169 (2019)
4. Brunet, J.-N., Mendizabal, A., Petit, A., Golse, N., Vibert, E., Cotin, S.: Physics-based deep neural network for augmented reality during liver surgery. In: Shen, D., et al. (eds.) MICCAI 2019. LNCS, vol. 11768, pp. 137–145. Springer, Cham (2019). https://doi.org/10.1007/978-3-030-32254-0_16
5. Choi, H., et al.: On the use of simulation in robotics: opportunities, challenges, and suggestions for moving forward. Proc. Natl. Acad. Sci. **118**(1) (2021)
6. Faure, F., et al.: Sofa: a multi-model framework for interactive physical simulation. In: Payan, Y. (ed.) Soft Tissue Biomechanical Modeling for Computer Assisted Surgery, pp. 283–321. Springer, Heidelberg (2012). https://doi.org/10.1007/8415_2012_125
7. Galbusera, F., Cina, A., Panico, M., Albano, D., Messina, C.: Image-based biomechanical models of the musculoskeletal system. Eur. Radiol. Exp. **4**(1), 1–13 (2020)
8. Haouchine, N., Kuang, W., Cotin, S., Yip, M.: Vision-based force feedback estimation for robot-assisted surgery using instrument-constrained biomechanical three-dimensional maps. IEEE Robot. Autom. Lett. **3**(3), 2160–2165 (2018). https://doi.org/10.1109/LRA.2018.2810948
9. Melzi, S., Ren, J., Rodolà, E., Sharma, A., Wonka, P., Ovsjanikov, M.: Zoomout: spectral upsampling for efficient shape correspondence. ACM Trans. Graph. (TOG) **38**(6), 155 (2019)
10. Mendizabal, A., Tagliabue, E., Brunet, J.-N., Dall'Alba, D., Fiorini, P., Cotin, S.: Physics-based deep neural network for real-time lesion tracking in ultrasound-guided breast biopsy. In: Miller, K., Wittek, A., Joldes, G., Nash, M.P., Nielsen, P.M.F. (eds.) MICCAI 2018-2019, pp. 33–45. Springer, Cham (2020). https://doi.org/10.1007/978-3-030-42428-2_4
11. Mendizabal, A., Tagliabue, E., Hoellinger, T., Brunet, J.-N., Nikolaev, S., Cotin, S.: Data-driven simulation for augmented surgery. In: Developments and Novel Approaches in Biomechanics and Metamaterials. ASM, vol. 132, pp. 71–96. Springer, Cham (2020). https://doi.org/10.1007/978-3-030-50464-9_5

12. Miller, K., Lu, J.: On the prospect of patient-specific biomechanics without patient-specific properties of tissues. J. Mech. Behav. Biomed. Mater. **27**, 154–166 (2013)
13. Nikolaev, S., Cotin, S.: Estimation of boundary conditions for patient-specific liver simulation during augmented surgery. Int. J. Comput. Assist. Radiol. Surg. **15**, 1107–1115 (2020)
14. Peterlik, I., Courtecuisse, H., Duriez, C., Cotin, S.: Model-based identification of anatomical boundary conditions in living tissues. In: Stoyanov, D., Collins, D.L., Sakuma, I., Abolmaesumi, P., Jannin, P. (eds.) IPCAI 2014. LNCS, vol. 8498, pp. 196–205. Springer, Cham (2014). https://doi.org/10.1007/978-3-319-07521-1_21
15. Peterlik, I., Haouchine, N., Ručka, L., Cotin, S.: Image-driven stochastic identification of boundary conditions for predictive simulation. In: Descoteaux, M., Maier-Hein, L., Franz, A., Jannin, P., Collins, D.L., Duchesne, S. (eds.) MICCAI 2017. LNCS, vol. 10434, pp. 548–556. Springer, Cham (2017). https://doi.org/10.1007/978-3-319-66185-8_62
16. Peterlik, I., Haouchine, N., Ručka, L., Cotin, S.: Image-driven stochastic identification of boundary conditions for predictive simulation. In: Descoteaux, M., Maier-Hein, L., Franz, A., Jannin, P., Collins, D.L., Duchesne, S. (eds.) MICCAI 2017. LNCS, vol. 10434, pp. 548–556. Springer, Cham (2017). https://doi.org/10.1007/978-3-319-66185-8_62
17. Pfeiffer, M., Riediger, C., Weitz, J., Speidel, S.: Learning soft tissue behavior of organs for surgical navigation with convolutional neural networks. Int. J. Comput. Assist. Radiol. Surg. **14**(7), 1147–1155 (2019)
18. Plantefève, R., Peterlik, I., Haouchine, N., Cotin, S.: Patient-specific biomechanical modeling for guidance during minimally-invasive hepatic surgery. Ann. Biomed. Eng. **44**(1), 139–153 (2016)
19. Saeed, S.U., Taylor, Z.A., Pinnock, M.A., Emberton, M., Barratt, D.C., Hu, Y.: Prostate motion modelling using biomechanically-trained deep neural networks on unstructured nodes. In: Martel, A.L., et al. (eds.) MICCAI 2020. LNCS, vol. 12264, pp. 650–659. Springer, Cham (2020). https://doi.org/10.1007/978-3-030-59719-1_63
20. Schenk, O., Gärtner, K.: Solving unsymmetric sparse systems of linear equations with Pardiso. Futur. Gener. Comput. Syst. **20**(3), 475–487 (2004)
21. Sharp, N., Crane, K.: A laplacian for nonmanifold triangle meshes. In: Computer Graphics Forum, vol. 39, pp. 69–80. Wiley Online Library (2020)
22. Tagliabue, E., et al.: Data-driven intra-operative estimation of anatomical attachments for autonomous tissue dissection. IEEE Robot. Autom. Lett. **6**(2), 1856–1863 (2021)
23. Taha, A.A., Hanbury, A.: Metrics for evaluating 3D medical image segmentation: analysis, selection, and tool. BMC Med. Imaging **15**(1), 29 (2015)
24. Taylor, Z.A., Cheng, M., Ourselin, S.: High-speed nonlinear finite element analysis for surgical simulation using graphics processing units. IEEE Trans. Med. Imaging **27**(5), 650–663 (2008)
25. Williams, F., Schneider, T., Silva, C., Zorin, D., Bruna, J., Panozzo, D.: Deep geometric prior for surface reconstruction. In: Proceedings of the IEEE/CVF Conference on Computer Vision and Pattern Recognition (CVPR), June 2019

Deep Iterative 2D/3D Registration

Srikrishna Jaganathan[1,2(✉)], Jian Wang[2], Anja Borsdorf[2], Karthik Shetty[1], and Andreas Maier[1]

[1] Pattern Recognition Lab, FAU Erlangen-Nürnberg, Erlangen, Germany
srikrishna.jaganathan@fau.de
[2] Siemens Healthineers AG, Forchheim, Germany

Abstract. Deep Learning-based 2D/3D registration methods are highly robust but often lack the necessary registration accuracy for clinical application. A refinement step using the classical optimization-based 2D/3D registration method applied in combination with Deep Learning-based techniques can provide the required accuracy. However, it also increases the runtime. In this work, we propose a novel Deep Learning driven 2D/3D registration framework that can be used end-to-end for iterative registration tasks without relying on any further refinement step. We accomplish this by learning the update step of the 2D/3D registration framework using Point-to-Plane Correspondences. The update step is learned using iterative residual refinement-based optical flow estimation, in combination with the Point-to-Plane correspondence solver embedded as a known operator. Our proposed method achieves an average runtime of around 8s, a mean re-projection distance error of 0.60 ± 0.40 mm with a success ratio of 97% and a capture range of 60 mm. The combination of high registration accuracy, high robustness, and fast runtime makes our solution ideal for clinical applications.

Keywords: Deep learning · Image fusion · 2D/3D registration

1 Introduction

In X-ray-based image-guided interventions, 2D X-ray fluoroscopy is preferable for providing image guidance. However, due to the projective nature of the 2D images acquired, there is an inherent loss of information. Overlaying the preoperative 3D volume onto the 2D images can provide the necessary additional information during the intervention. To obtain an accurate overlay, one needs to register the preoperative 3D volume with the current patient position, which is accomplished using 2D/3D registration. The goal of 2D/3D registration is to find the optimal transformation of the preoperative 3D volume to the current patient position.

The problem of 2D/3D registration for medical images has already been explored for decades and has well-established techniques for specific use cases [9]. In many of these classical techniques, the registration problem is often formulated as an optimization problem and solved with iterative schemes. If the initial misalignment is large, the optimization problem is often non-convex, thus requiring

© Springer Nature Switzerland AG 2021
M. de Bruijne et al. (Eds.): MICCAI 2021, LNCS 12904, pp. 383–392, 2021.
https://doi.org/10.1007/978-3-030-87202-1_37

global optimizers to avoid the problem of getting stuck in local minima [9]. Such global optimizers are computationally expensive and are not sufficiently fast for application during the intervention if the inital registration error is large.

Deep Learning (DL) based techniques for 2D/3D registration problem have shown promising results, by improving the computational efficiency [11,17] and robustness [8,10,19]. Recent works have also shown that the robustness can be increased to a much greater extent using DL-based methods and even propose fully automatic 2D/3D registration solution [2,3]. However, most of these techniques rely on a final refinement step based on the classical methods to achieve the necessary registration accuracy for interventional application, which limits the computational efficiency of DL-based registration.

The recent trend has clearly shown that DL-based techniques can provide good initialization and, when used in combination with a refinement step, can provide a hybrid 2D/3D registration method. A DL-based registration method that doesn't rely on refinement for registration accuracy is highly desirable. We propose a DL-based solution to accomplish this goal, where we learn the update step of the iterative 2D/3D registration framework based on Point-to-Plane Correspondence (PPC) constraint [22]. Learning such an update step prediction, proposed in [5], showed significant improvement (around two times) for single-step update prediction. However, the iterative application using the learned update step to the actual 2D/3D registration problem was lacking. We retain the structure of the previously proposed update step prediction [5] but make significant architectural changes to incorporate another domain-prior, which approximately models the iterative nature of the problem. For this, we use iterative residual refinement [4] based optical flow estimation. In combination with the PPC solver embedded as a known operator, we can learn the update step end-to-end directly on the registration loss.

2   Methods

We use the PPC-based iterative registration framework [22] along with our DL-based update step prediction. The schematic of the proposed registration framework is depicted in Fig. 1. The proposed DL-based solution acts as a drop-in replacement for the update step prediction in the original framework. We provide a brief background on the original framework [22], what constitutes an update step, and the PPC constraint in the following section.

2.1   Background

The input to the framework is the fluoroscopic X-ray image $\mathbf{I}_{\mathrm{flr}}$, the preoperative CT volume $\mathbf{V}$ along with a rough initial registration estimate $\mathbf{T}_{\mathrm{init}}$. We assume the intrinsic camera parameters $\mathbf{K}$ is known. The framework consists of an initialization step and an update step. During the initialization, the surface points along with its gradients are extracted from $\mathbf{V}$. Digitally Reconstructed Radiograph (DRR) $\mathbf{I}_{\mathrm{drr}} = \mathcal{R}(\mathbf{V}, \mathbf{T}_i)$ is rendered from $\mathbf{V}$ and the current pose

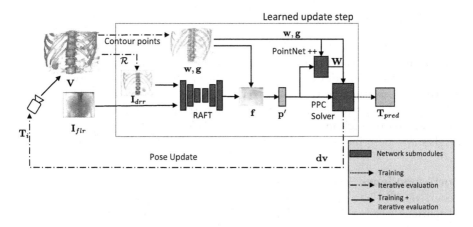

Fig. 1. Overview of the proposed registration framework with the DL-driven learned update step prediction $\phi((\mathbf{I}_{flr}, \mathbf{I}_{drr}, \mathbf{w}, \mathbf{g}), \mathcal{W})$. The RAFT architecture ϕ_f is used for correspondence estimation and the PointNet++ architecture ϕ_w is used to estimate per correspondence weight matrix $\mathbf{W}$. The differentiable PPC solver $\mathcal{K}_{ppc}$ computes the 3D motion update.

$\mathbf{T}_i$ at iteration i using the rendering operator $\mathcal{R}$. Pose dependent apparent contour points $\mathbf{w}$, with the corresponding gradients $\mathbf{g}$ are selected from the surface points [22] before the start of each update step.

The update step $\mathcal{U}(\mathbf{I}_{flr}, \mathbf{I}_{drr}, \mathbf{w}, \mathbf{g})$ predicts the 3D motion update $\mathbf{dv}$ given $(\mathbf{I}_{flr}, \mathbf{I}_{drr}, \mathbf{w}, \mathbf{g})$ as inputs to the update step. The update step has the following operations. Initially, 2D correspondences search is performed between $\mathbf{I}_{flr}$ and $\mathbf{I}_{drr}$ at the projected contour points $\mathbf{p} = \mathcal{P}(\mathbf{w}, \mathbf{K})$ in $\mathbf{I}_{drr}$, where $\mathcal{P}$ is the projection operation. This results in a set of correspondences $(\mathbf{p}, \mathbf{p}')$, where $\mathbf{p}'$ is the corresponding points of $\mathbf{p}$ found in $\mathbf{I}_{flr}$. The 2D correspondences can only reveal the observable 2D misalignment. However, to recover the true 3D motion error and thus finding the optimal transformation, the unobservable 3D motion components should also be considered. The total 3D motion is effectively constrained using the PPC model. The PPC model constraints the 3D motion to a plane spanning between $(\mathbf{w} \times \mathbf{g})$ and $\mathbf{p}'$ [22]. We use the weighted variant of the PPC constraint to reduce the effects of noisy 2D correspondences. The weighted PPC constraint is as follows:

$$\mathbf{W} \, \mathbf{A} \mathbf{dv} = \mathrm{diag}(\mathbf{W}) \, b \; , \tag{1}$$

where $\mathbf{A} = ((\mathbf{n} \times \mathbf{w}^{\mathsf{T}}) - \mathbf{n}^{\mathsf{T}})$ and $\mathbf{b} = \mathbf{n}^T \mathbf{w}$. The matrix $\mathbf{A} \in \mathbb{R}^{N \times 6}$ and vector $\mathbf{b} \in \mathbb{R}^N$ are computed from the N_{cp} point correspondences $(\mathbf{p}, \mathbf{p}')$, $\mathbf{n}$ is the plane normal. The weight matrix $\mathbf{W}$ is a diagonal matrix providing individual weights for each estimated correspondence. The 3D motion vector $\mathbf{dv}$ can be computed from the PPC constraint using closed-form solution with pseudo-inverse of $\mathbf{A}$ and converted to a transformation matrix $\mathbf{T}_{i+1}$ which generates the input to the next update step. The process is repeated until convergence criteria is reached [22].

2.2   DL-Based Update Step Prediction

We parameterize $\mathcal{U}$ with $\phi(\mathcal{U}, \mathcal{W})$, where ϕ is the Deep Neural Network (DNN) and $\mathcal{W}$ are the network parameters. DNN-based parameterization for $\mathcal{U}$ was proposed earlier [5], however we have significant changes in the network architecture to model the iterative nature of the problem more precisely. The network ϕ is composed of two sub networks, ϕ_f for correspondence estimation, ϕ_w for correspondence weighting and a PPC solver layer $\mathcal{K}_{ppc}$ which has no learnable weights. We describe the different submodules in the following paragraphs.

Correspondence Estimation Network ϕ_f. We use the recently proposed Recurrent All-Pairs Field Transforms (RAFT) [20] architecture for optical flow estimation between $\mathbf{I}_{drr}$ and $\mathbf{I}_{flr}$. The network uses iterative residual refinement [4] for estimating the optical flow. The network consists of a recurrent component (referred as update operator in [20]) which unrolls the flow prediction and enables the use of residual refinement. The flow prediction is iteratively refined $\mathbf{f}_{j+1} = \mathbf{f}_j + \Delta\mathbf{f}$ by repeated application of the update operator until it converges to a fixed point (mimicking the behavior of optimization methods). At each update, the network only needs to predict the small residual $\Delta\mathbf{f}$. During training, the update operator is unrolled for a fixed number of iterations N_{FL}. We use a version of the RAFT architecture with shared weights between the updates, as it is both memory efficient and gives the best performance [20]. Since pixel level ground truth flow is not available, we use sparse ground truth flow labels at the projected contour points $\mathbf{p}$ which is directly computed from the ground truth registration matrix $\mathbf{T}_{gt}$. We use a modified optical flow loss function compared to [20] to counter the sparsity of the labels [19]. The network outputs dense correspondence for all pixels, but we sample the predicted flow $\mathbf{f}$ at the last flow update step N_{FL} of the network, to compute flow $\mathbf{dp}$ at the projected contour points $\mathbf{p}$. We find the corresponding points $\mathbf{p}' = \mathbf{p} + \mathbf{dp}$ in $\mathbf{I}_{flr}$ and use it as input to the next layers.

Correspondence Weighing ϕ_w. Our correspondence weighting network is based on the PointNet++ architecture [16], which takes in as input feature vector $\mathbf{f}_{\phi_w} = \{\mathbf{w}, \mathbf{g}, \mathbf{nw}, \mathbf{p}'\} \in \mathbb{R}^{N_{cp} \times 9}$. The PointNet++ architecture performs hierarchical feature learning on 3D point cloud data and incorporates local context information, which was missing in the original PointNet [15]. It can handle additional features along with the 3D Euclidean coordinate features thus we don't need any modification for our input feature vector. We use single scale grouping variant of the PointNet++ segmentation architecture, which can already provide per point label. We only modify the final activation to sigmoid activation to predict correspondence weights. The network predicts per correspondence weight similar to the attention model [17]. The weight matrix $\mathbf{W}$ obtained from a set of N_{cp} contour points is used to solve the weighted PPC model in Eq. (1).

PPC Solver $\mathcal{K}_{ppc}$. The differentiable PPC layer takes in the correspondence set $\{\mathbf{p}, \mathbf{p}'\}$, the contour points with its gradients $\{\mathbf{w}, \mathbf{g}\}$ and the estimated per correspondence weights $\mathbf{W}$ and solves for $\mathbf{dv}$ from the PPC constraint (Eq. (1)) using closed-form solution. The motion vector $\mathbf{dv}$ is converted to a transformation matrix $\mathbf{T}_{pred}$ which serves as the predicted output from our proposed network ϕ.

2.3 Loss Function

The registration loss is computed between the contour point positions at $\mathbf{T}_{gt}$ and $\mathbf{T}_{pred}$ using $\mathcal{L}_{reg} = ||\mathbf{T}_{pred}(\mathbf{w}) - \mathbf{T}_{gt}(\mathbf{w})||_1$. We use a modified average End Point Error (EPE) to compute optical flow (Eq. (2)) loss at the projected contour points. A binary mask image $\mathbf{M}_\mathbf{p}$ of the projected contour points is used to zero out the loss at other pixels [19]. Since the RAFT architecture has an unrolled update operator, which produces flow at N_{FL} as $\{\mathbf{f}_j\}$, we use the same ground truth flow for all iteration with a discount factor $\gamma^{N_{\mathrm{FL}}-j}$ [20]. The optical loss is computed as follows,

$$\mathcal{L}_{flow} = \sum_{j=1}^{N_{\mathrm{FL}}} \gamma^{N_{\mathrm{FL}}-j} \frac{1}{N_{\mathrm{cp}}} \mathbf{M}_\mathbf{p} ||\mathbf{f}^j - \mathbf{f}_{gt}||_1 \ . \tag{2}$$

The combined training loss function with the regularizers is as follows,

$$\mathcal{L} = \alpha \, \mathcal{L}_{\mathrm{flow}} + \beta \, ||\mathbf{T}_{\mathrm{pred}}(\mathbf{w}) - \mathbf{T}_{\mathrm{gt}}(\mathbf{w})||_1 + \lambda \, ||\mathbf{dv}||_2 + \frac{\zeta}{2} \, ||\mathcal{W}||^2 \ , \tag{3}$$

where we use a regularization on the estimated $\mathbf{dv}$ computed from the PPC solver and weight decay regularizer on the network weights $\mathcal{W}$. The hyper-parameters α, β are used to control the strength of optical flow and registration loss respectively. The hyper-parameters λ, ζ are used to control the motion and weight decay regularizer strength respectively.

3 Experiments and Results

We validate our method using clinical cone beam CT (CBCT) reconstruction data set for a single view scenario, since it is harder for the registration methods and largely remains unsolved compared to multi-view scenario which was the focus of many of the previous works [8,10,18].

3.1 Data

The data set consists of reconstructed CBCT volumes and the corresponding X-ray images used for reconstruction. The ground truth registration between the X-ray images and the CBCT volume is available. The data set is from vertebra body region, consisting of 55 patient volumes (includes both thoracic and lumbar regions). The number of X-ray images per CBCT volume varies depending on the slice thickness, between 190 to 390. The slice resolution also varies between 256×256 to 512×512 and the voxel spacing between 0.49 mm to 0.99 mm for all the dimension. The X-ray images has a resolution of 616×480 with a pixel spacing of 0.616 mm. Random initial transformation is used to create training samples with initial registration error measured in mean Target Registration Error (mTRE) [7] is in the range of $[0, 30]$ mm, with translations in range of $[0, \pm 30]$ mm and rotation in range of $[0, \pm 20]$ degrees for all the three axes. Each training sample consists of the initial transformation matrix $\mathbf{T}_{init}$, the ground truth registration matrix $\mathbf{T}_{gt}$, $(\mathbf{w}, \mathbf{g})$, $\mathbf{I}_{drr}$ rendered based on $\mathbf{T}_{init}$ and $\mathbf{I}_{flr}$. From each patient we generate around 1200 to 1800 such samples using the random initial transformations depending on the number of fluoroscopic images available. The data set is split into training, validation and test sets with 43 patients for training, 6 patients for validation, 6 patients for testing. We have combined total of around 80,000 samples for training and validation.

3.2   Training

We pretrain ϕ_f for 50 epochs on our training data using Eq. (2). This is to avoid the pseudo-inverse computation used in PPC solver from failing when the correspondence estimation is bad, which can be the case when ϕ_f is randomly initialized. Following this we train our proposed network ϕ with the loss function described in Eq. (3) for 100 epochs with early stopping criteria used on validation data set. The hyper-parameters of the loss function (Eq. (3)) used are set to $\alpha = 1$, $\beta = 0.5$, $\lambda = 1e - 3$ and $\zeta = 1e - 5$ and γ in Eq. (2) set to 0.8 [20]. We unroll ϕ_f with $N_{FL} = 6$ iterations for both training and evaluation. ADAM [6] optimizer is used with a cyclical learning rate varying from $1e - 4$ to $1e - 6$ and a batch size of 16. We implemented the network using the PyTorch framework [13]. We compute the gradients of all the layers by back-propagation directly using autograd [12] in PyTorch.

Data augmentation plays a crucial role for us to overcome the limited amount of training samples available considering the number of network parameters and the complex nature of the problem. Online data augmentation is used with color space transforms (where we adjust brightness and contrast of both I_{drr} and I_{flr}), geometric transforms (affine 2D rotation with translation, horizontal and vertical flips) and random erasing [23]. The corresponding modification in ground truth data is performed to account for the augmentation.

3.3   Evaluation

We use the standardized evaluation measures [7] for 2D/3D registration. The initial error range is reported in mTRE and the final registration error is measured in mean Re-Projection Distance (mRPD) [7] as it is standard practice for the evaluation of single view registration [7,17,19]. Along with the registration error, we also indicate the Success Ratio (SR) and the Capture Range (CR) to quantify the robustness. We define the success threshold as mRPD ≤ 2.0 mm final registration error [17]. The capture range measures the highest sub-interval of initial error for which we can achieve SR $\geq 95\%$. Due to the large initial error range considered here, we only report capture range at the intervals of 5mm. We run all the evaluations using the Intel Core i7-6850K CPU and Nvidia GeForce Gtx Titan X GPU with 12 GB graphics memory and report the average run time for the registration. Our proposed network is evaluated for iterative 2D/3D registration on the test data set (6 Patients). For each patient, 600 test samples are created from ground truth registration using random initial transformations of the AP and LAT views. The initial registration error measured in mTRE varies from $[0,60]$ mm with translations in the range of $[0, \pm60]$ mm and rotation in the range of $[0, \pm40]$ degrees for all three axes. We run our proposed method for 10 iterations.

We compare our proposed method with the existing state-of-the-art techniques. The considered models, as well as their method used to compute the update step, are described below. We consider DPPC [21], the classical depth-aware PPC model which relies on patch matching for correspondence estimation and heuristics for correspondence weighting, DPPC Attention [17] uses patch matching for correspondence estimation and PointNet based learned correspondence weighting, DPPC CL Refinement (DPPC-CL-A in [19]) which uses a FlowNet [1] architecture to estimate the correspondence, however, a refinement step is required using patch-matching and heuristics are used for correspondence weighting [19]. PPC Flow Attention [5] which uses FlowNetC [1] architecture for correspondence estimation and PointNet based attention [17] for correspondence weighting. The baseline methods were run for the proposed evaluation configurations in the respective work. We use pretrained weights obtained from the authors of the respective baseline methods, trained on the same data set for the learnable modules in baseline methods. In this way, we ensure that the most optimized version of the baseline methods are used.

3.4 Results

Table 1. Comparison of our proposed method with the existing state-of-the-art techniques. The initial error range measured in mTRE varies between $[0, 60]$ mm. We report final registration error in mRPD [mm], Success Ratio (SR) (mRPD ≤ 2.0 mm), Capture Range (CR) and the average runtime [s] for solving one registration problem. $\uparrow$ indicates that higher values are better and $\downarrow$ indicates that lower values are better.

	mRPD [mm] $\downarrow \mu \pm \sigma$	SR [%] $\uparrow$	CR [mm] $\uparrow$	Runtime [s] $\downarrow \mu \pm \sigma$
DPPC [21]	0.58 ± 0.218	61.9	10–15	36.13 ± 26.20
DPPC CL Refinement [19]	$\mathbf{0.40 \pm 0.09}$	94.58	20–25	37.53 ± 11.40
DPPC Attention [17]	0.47 ± 0.23	95.5	35–40	16.42 ± 23.43
PPC Flow Attention [5]	1.60 ± 0.34	35.4	0	12.82 ± 0.2
Proposed method	0.60 ± 0.40	$\mathbf{97.0}$	$\mathbf{55 - 60}$	$\mathbf{8.05 \pm 0.2}$

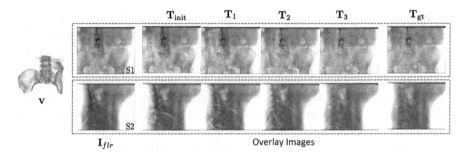

Fig. 2. Overview of the qualitative results shown on two example cases S1 (AP) and S2 (LAT) on a test data set patient using our proposed technique. $\mathbf{T}_{\text{init}}$ shows a large inital misalignment for both the examples. The ground truth overlay is also provided for comparison. We show the first three iterations which demonstrates the speed of convergence from large initial registration error.

We evaluate our proposed methods and the baseline methods as described in Sect. 3.3. The evaluation results of all the considered methods are presented in Table 1. DPPC CL Refinement [19] achieves the best registration accuracy of 0.47 ± 0.23 mm. Our proposed methods has the best SR with 97%, CR with 55–60 mm and average runtime for a registration problem at 8.05 ± 0.2 s. Qualitative results of our proposed registration method is shown in Fig. 2, evaluated for two different views on one test data set patient.

4 Discussion and Conclusion

The results presented in Table 1 show that our proposed method improves the SR by 1.5% and increases the capture range by 20 mm (50% improvement) compared to the other state-of-the-art methods without sacrificing registration accuracy, as we achieve sub-millimeter registration error. Our technique is robust for a comparatively higher range of initial error. We are also twice faster compared to other methods, which is a crucial factor for the interventional application. Our proposed technique performs significantly better, compared to [5] which fails for the iterative registration task using a similar learned update step. This shows modeling the iterative nature of the problem is essential for such a learned update step, as this allows the network to learn intermediate flow steps, thus allowing the network to perform well for both large and small displacements.

In summary, we proposed a DL-driven iterative 2D/3D registration framework which is fast, robust and provides highly accurate registration. To the best of our knowledge, we are one of the first DL-driven method to retain high robustness and also achieve highly accurate registration without any further refinement. Future research direction can be extending the proposed method to fully automatic registration and multi-view scenario. One challenge for using the proposed method, is that, it requires a large number of annotated training data.

Use of simulated data in combination with domain randomization [3] or adversarial data augmentation [14] strategies can be explored to reduce the burden of annotated training data requirements.

Disclaimer. The concepts and information presented in this paper are based on research and are not commercially available.

References

1. Dosovitskiy, A., et al.: Flownet: learning optical flow with convolutional networks. In: Proceedings of the IEEE International Conference on Computer Vision, pp. 2758–2766 (2015)
2. Esteban, J., Grimm, M., Unberath, M., Zahnd, G., Navab, N.: Towards fully automatic X-ray to CT registration. In: Shen, D., et al. (eds.) MICCAI 2019. LNCS, vol. 11769, pp. 631–639. Springer, Cham (2019). https://doi.org/10.1007/978-3-030-32226-7_70
3. Grimm, M., Esteban, J., Unberath, M., Navab, N.: Pose-dependent weights and domain randomization for fully automatic X-ray to CT registration. arXiv preprint arXiv:2011.07294 (2020)
4. Hur, J., Roth, S.: Iterative residual refinement for joint optical flow and occlusion estimation. In: Proceedings of the IEEE/CVF Conference on Computer Vision and Pattern Recognition, pp. 5754–5763 (2019)
5. Jaganathan, S., Wang, J., Borsdorf, A., Maier, A.: Learning the update operator for 2D/3D image registration. In: Palm, C., Deserno, T.M., Handels, H., Maier, A., Maier-Hein, K., Tolxdorff, T. (eds.) Bildverarbeitung für die Medizin 2021. I, pp. 117–122. Springer, Wiesbaden (2021). https://doi.org/10.1007/978-3-658-33198-6_27
6. Kingma, D.P., Ba, J.: Adam: a method for stochastic optimization. arXiv preprint arXiv:1412.6980 (2014)
7. van de Kraats, E.B., Penney, G.P., Tomaževič, D., van Walsum, T., Niessen, W.J.: Standardized evaluation of 2D-3D registration. In: Barillot, C., Haynor, D.R., Hellier, P. (eds.) MICCAI 2004. LNCS, vol. 3216, pp. 574–581. Springer, Heidelberg (2004). https://doi.org/10.1007/978-3-540-30135-6_70
8. Liao, H., Lin, W.A., Zhang, J., Zhang, J., Luo, J., Zhou, S.K.: Multiview 2D/3D rigid registration via a point-of-interest network for tracking and triangulation. In: Proceedings of the IEEE/CVF Conference on Computer Vision and Pattern Recognition, pp. 12638–12647 (2019)
9. Markelj, P., Tomaževič, D., Likar, B., Pernuš, F.: A review of 3D/2D registration methods for image-guided interventions. Med. Image Anal. **16**(3), 642–661 (2012)
10. Miao, S., et al.: Dilated FCN for multi-agent 2D/3D medical image registration. In: Proceedings of the AAAI Conference on Artificial Intelligence, vol. 32 (2018)
11. Miao, S., Wang, Z.J., Zheng, Y., Liao, R.: Real-time 2D/3D registration via CNN regression. In: 2016 IEEE 13th International Symposium on Biomedical Imaging (ISBI), pp. 1430–1434. IEEE (2016)
12. Paszke, A., et al.: Automatic differentiation in pytorch (2017)
13. Paszke, A., et al.: Pytorch: an imperative style, high-performance deep learning library. arXiv preprint arXiv:1912.01703 (2019)

14. Peng, X., Tang, Z., Yang, F., Feris, R.S., Metaxas, D.: Jointly optimize data augmentation and network training: adversarial data augmentation in human pose estimation. In: Proceedings of the IEEE Conference on Computer Vision and Pattern Recognition, pp. 2226–2234 (2018)

15. Qi, C.R., Su, H., Mo, K., Guibas, L.J.: Pointnet: deep learning on point sets for 3D classification and segmentation. In: Proceedings of the IEEE Conference on Computer Vision and Pattern Recognition, pp. 652–660 (2017)

16. Qi, C.R., Yi, L., Su, H., Guibas, L.J.: Pointnet++: deep hierarchical feature learning on point sets in a metric space. arXiv preprint arXiv:1706.02413 (2017)

17. Schaffert, R., Wang, J., Fischer, P., Borsdorf, A., Maier, A.: Learning an attention model for robust 2-D/3-D registration using point-to-plane correspondences. IEEE Trans. Med. Imaging 39(10), 3159–3174 (2020)

18. Schaffert, R., Wang, J., Fischer, P., Maier, A., Borsdorf, A.: Robust multi-view 2-D/3-D registration using point-to-plane correspondence model. IEEE Trans. Med. Imaging 39(1), 161–174 (2019)

19. Schaffert, R., Weiß, M., Wang, J., Borsdorf, A., Maier, A.: Learning-based correspondence estimation for 2-D/3-D registration. In: Bildverarbeitung für die Medizin 2020. I, pp. 222–228. Springer, Wiesbaden (2020). https://doi.org/10.1007/978-3-658-29267-6_50

20. Teed, Z., Deng, J.: RAFT: recurrent all-pairs field transforms for optical flow. In: Vedaldi, A., Bischof, H., Brox, T., Frahm, J.-M. (eds.) ECCV 2020. LNCS, vol. 12347, pp. 402–419. Springer, Cham (2020). https://doi.org/10.1007/978-3-030-58536-5_24

21. Wang, J.: Robust 2-D/3-D Registration for Real-time Patient Motion Compensation: Robuste 2-D/3-D Registrierung Zur Echtzeitfähigen, Dynamischen Bewegungskompensation. Verlag Dr, Hut (2020)

22. Wang, J., et al.: Dynamic 2-D/3-D rigid registration framework using point-to-plane correspondence model. IEEE Trans. Med. Imaging 36(9), 1939–1954 (2017)

23. Zhong, Z., Zheng, L., Kang, G., Li, S., Yang, Y.: Random erasing data augmentation. In: Proceedings of the AAAI Conference on Artificial Intelligence, vol. 34, pp. 13001–13008 (2020)

hSDB-instrument: Instrument Localization Database for Laparoscopic and Robotic Surgeries

Jihun Yoon[1], Jiwon Lee[1], Sunghwan Heo[1], Hayeong Yu[1], Jayeon Lim[1], Chi Hyun Song[1], SeulGi Hong[1], Seungbum Hong[1], Bokyung Park[1], SungHyun Park[2], Woo Jin Hyung[1,2], and Min-Kook Choi[1(✉)]

[1] hutom, Seoul, Republic of Korea
mkchoi@hutom.io
[2] Department of Surgery, Yonsei University College of Medicine, Seoul, Republic of Korea

Abstract. Automated surgical instrument localization is an important technology to understand the surgical process and in order to analyze them to provide meaningful guidance during surgery or surgical index after surgery to the surgeon. We introduce a new dataset that reflects the kinematic characteristics of surgical instruments for automated surgical instrument localization of surgical videos. The hSDB(hutom Surgery DataBase)-instrument dataset consists of instrument localization information from 24 cases of laparoscopic cholecystecomy and 24 cases of robotic gastrectomy. Localization information for all instruments is provided in the form of a bounding box for object detection. To handle class imbalance problem between instruments, synthesized instruments modeled in Unity for 3D models are included as training data. Besides, for 3D instrument data, a polygon annotation is provided to enable instance segmentation of the tool. To reflect the kinematic characteristics of all instruments, they are annotated with head and body parts for laparoscopic instruments, and with head, wrist, and body parts for robotic instruments separately. Annotation data of assistive tools (specimen bag, needle, etc.) that are frequently used for surgery are also included. Moreover, we provide statistical information on the hSDB-instrument dataset and the baseline localization performances of the object detection networks trained by the MMDetection library and resulting analyses (The dataset, additional dataset statistics and several trained models are publicly available at https://hsdb-instrument.github.io/).

Keywords: Surgical instrument localization · Object detection · Class imbalance · Domain randomization

1 Introduction

In the last 30 years, the number of open surgeries has decreased significantly, and the number of minimally invasive surgeries that can speed up recovery by minimizing patient complications has increased significantly. In particular, minimally invasive surgeries using robots are increasing more rapidly in terms of the

© Springer Nature Switzerland AG 2021
M. de Bruijne et al. (Eds.): MICCAI 2021, LNCS 12904, pp. 393–402, 2021.
https://doi.org/10.1007/978-3-030-87202-1_38

Table 1. Summary of datasets for surgical instruments recognition Surgery type, recognition type, the number of surgeries, the number of instruments, the number of frames and if the synthetic data was included were investigated. The additional frames of synthetic and domain randomization data are highlighted in bold.

	Surgery type	Recognition type	# of Surgeries	# of Instruments	# of Frames	Synthetic data
[4]	Laparoscopic colorectal surgeries	Instance segmentation	30 cases	7 laparoscopic	10,040	X
[5]	Laparoscopic cholecystectomy	Tool presence	80 cases	7 laparoscopic	176,020	X
[6]	Robotic abdominal porcine procedures	Binary(parts) segmentation	10 cases	7 robotic	600	X
[7]	Robotic surgery(skill training)	Object detection	99 cases for 6 surgical tasks	1 robotic	22,467	X
hSDB-instrument	Laparoscopic cholecystectomy and robotic gasterctomy	Object detection	24 cases for cholecystectomy and 24 cases for gastrectomy	10 laparoscopic, 6 robotic and 4 assistive	26,919(**+12,428**) for cholecystectomy and 42,891(**+13,247**) for gastrectomy	O

convenience of the surgical procedure and the minimization of deviations of surgical performances among specialists [1,2]. Owing to the nature of laparoscopic surgery, which is performed by a specialist while looking at the inside of the patient, minimally invasive surgery is recorded in the form of a video that covers all the operations taking place inside the patient. If the operation process in the vidoe is recognized, the operation can be automatically evaluated and analyzed [3]. Furthermore, it may be possible to automate certain surgical procedures, like autonomous driving, as well as real-time guidance during operations, like ADAS (Advanced Driver Assisted System) in vehicle driving (Table 1).

To analyze and evaluate the surgical procedures by using surgical videos in an automated manner, localization, and motion estimation of the surgical tools are essential. To this end, several research teams have released datasets for instrument localization of laparoscopic instruments for colorectal surgeries [4], cholecystectomy [5] and of robotic instruments for abdominal porcine procedures [6] and robotic surgery skill training [7]. However these datasets did not provide sufficient annotations for pixel-level instrument localization and various surgical instruments and did not handle the class imbalance problem. In the computer vision community, with the help of publicly available datasets such as MS-COCO [8] and KITTI [9], visual recognition techniques for object detection have been developed to a very high level [10–12]. To achieve a similar improvement in surgical instrument localization, large-scale datasets that contain a wider range of localization information about surgical tools are necessary.

To address this need, we present a hSDB(hutom Surgery DataBase)-instrument dataset for instrument localization in laparoscopic and robotic surgery. The hSDB-instrument dataset is produced from real surgical situation videos, and it contains the localization information of tools used in the surgical procedure. The hSDB-instrument contains instrument information for a total of 24 cases of laparoscopic cholecystectomy and 24 cases of robotic gastrectomy for gastric cancer. Localization information of the surgical instruments is provided in the form of a bounding box to train object detection networks, and each instrument's bounding box is divided into head, wrist, and body parts according to the kinematic characteristics of each tool. The information also includes needles, specimen bags, and surgical tubes that are often used in the surgical process, even if they are not laparoscopic or robotic surgical instruments. Besides, the robotic surgery includes several assistive laparoscopic instruments that are frequently used for gastrectomy.

Moreover, the hSDB-instrument dataset provides localization information on virtual surgical instruments made by Unity and real tools from real minimally invasive surgery videos. There is a severe class imbalance between the surgical tools in the real surgeries. To resolve the class imbalance, we provide localization annotations in two-dimensional (2D) projection images of three-dimensional (3D) models of several laparoscopic and robotic surgical instruments. In the case of synthetic tools, the annotations are free and include both pixel-level information and bounding box. The synthetic data can be used for training the recognition models by itself, and can also be used for training by applying techniques such as image-to-image translation using generative adversarial neural networks that were recently proposed [13–16]. We also provide baseline performances for object detection models for the hSDB-instrument dataset, which is expected to help develop new models and algorithms for instrument localization. The models were trained by the MMDetection library [17] based on PyTorch [18].

The technical contributions of the hSDB-instrument dataset are:

- The hSDB-instrument dataset was created based on the instrument's localization information of real surgical procedures. The datasets were split into training and validation sets by each case of surgeries, not just the number of annotations. This means that the generalization error of the recognition models according to different patients' condition is considered.
- We provide synthetic datasets to handle class imbalance that may occur during data acquisition. Two-dimensional projected pixel-based annotations of 3D synthetic models for several instruments used in the surgical procedure are provided. Several approaches can be employed to address class imbalance using the annotations of the synthetic data.
- The hSDB-instrument dataset provides baseline performances in instrument localization using MMDetection-based model evaluations. We expect that the baseline models can be used as a guideline in developing new surgical tool localization models or algorithms.

2   Data Collection

Figure 1 shows the data acquisition platform and storage structure of the hSDB-instrument dataset. In both cholecystectomy and gastrectomy, the format of the annotation database follows the basic structure of the MS-COCO dataset [8]. This section describes the details of the annotation generation tool and method used in the data acquisition process shown on the left side of Fig. 1.

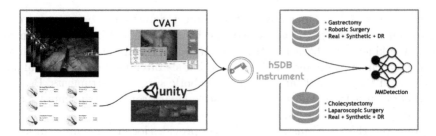

Fig. 1. Schematic representation of data acquisition and storage of the hSDB-instrument dataset. The box on the left shows the data acquisition process, and the box on the right shows the type of stored data. The hSDB-instrument dataset provides a comparative evaluation of the baseline models trained by using the MMDetection library from the training set.

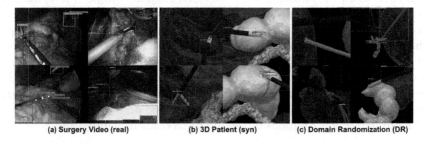

(a) Surgery Video (real)   (b) 3D Patient (syn)   (c) Domain Randomization (DR)

Fig. 2. Annotation visualization using hSDB-instrument API. (a) shows bounding boxes of tools with respect to a frame extracted from a video. (b) shows the localization information of the tools using the synthetic data from patient CT (Computer Tomography), and (c) shows the localization information of the data generated by the domain randomization method.

2.1 Laparoscopic Cholecystectomy

The original videos of cholecystectomy included in the hSDB-instrument dataset were a subset of the Cholec80 dataset [5]. A web-based computer vision annotation tool called CVAT [19] was used to annotate surgical instruments in the form of a bounding box. All the tools were divided into the head and body structures according to the kinematic characteristics of laparoscopic tools. Annotations for all tools were sampled at an average of 1 frame per second. All the annotations were marked by two trained annotators, cross-validated with each other, and only the annotations finally approved by a supervisor, medical expert were stored in the database.

The synthetic dataset was generated by two types of 3D modeling. The first is a synthetic dataset with a 3D environment assuming the real surgical procedure, near the gallbladder modeled with the CT (Computer Tomography) data of a specific patient and the laparoscopic tool. Similar to the robotic surgery console interface, the user inputs commands for the operation of the tool to generate the data while the user interacts with the synthetic environment inside the patient. The second is a synthetic dataset generated by a method so-called domain randomization [20]. Domain randomization-based data were randomly generated within a set of constraints of all the tools and organs in the 3D synthetic environment, including location, camera type, lighting, viewpoint, and color of objects. All the 3D modeling data was produced by Unity, and the 3D model information was projected as 2D information corresponding to a specific camera viewpoint and stored as the final training data.

2.2 Robotic Gastrectomy for Gastric Cancer

The original videos of gastrectomy for gastric cancer, included in the hSDB-instrument dataset, consist of 24 robotic surgery recordings by experienced specialists. The da Vinci Si system was used to record 14 videos and the da Vinci Xi system was used for the rest 10 of videos. Gastrectomy was performed by an experienced medical specialist, and each time of the recorded videos was between 1.5 and 4 h, depending on types of surgery. As in the case of cholecystectomy, 1 f/s sampling was applied to all the frames in which the tool appeared to obtain annotated information.

Unlike the laparoscopic instruments, the robotic instruments in gastrectomy were annotated in three parts according to their kinematic characteristics. In addition to robotic surgical instruments, it includes several hand-assisted laparoscopic instruments and assistive tools. In the case of the 3D modeling data, the CT data of the cholecystectomy patient were used, and the two types of 3D training data (surgical environment and domain randomization) for robotic surgery instruments were generated in the same manner as that for the laparoscopic surgery. Figure 2 shows the visualization of our dataset.

3   Statistics of hSDB-instrument Dataset

The annotations for each of the 24 surgery cases for both cholecystectomy and gastrectomy have a different distribution of instrument quantities. Figure 3 shows a distribution of annotations for the surgical instruments in the hSDB-instrument dataset. 18 of the 24 surgery cases were used to generate training data, 3 cases for validation data, and the last 3 cases for test data assuming unseen surgical situations. 10 types of laparoscopic tools are used for cholecystectomy-atraumatic grasper, clip applier (ham-o-lock), clip applier (metal), curved atraumatic grasper, electrichook, ligasure, overholt, scissors, specimen bag, and suction-irrigation. In the case of gastrectomy, a total of 17 surgical instruments are used, including robotic tools, laparoscopic tools, and

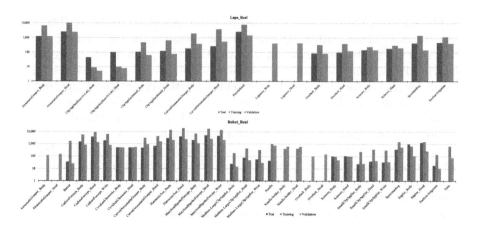

Fig. 3. Distribution of annotations obtained from laparoscopic cholecystectomy and gastrectomy for gastric cancer videos. The number of annotations is adjusted on a logarithmic scale. Both cholecystectomy and gastrectomy have severe class imbalance problems with several instruments.

auxiliary tools. Atraumatic grasper, baxter, cadiere forceps, covidien ultrasonic, curved atraumatic grasper, harmonic ace, maryland bipolar forceps, medium-large clip applier, needle, needle holder, overholt, scissors, small clip applier, stapler, specimen bag, suction-irrigation, and tube.

In the case of the synthetic data, it can be randomly generated regardless of data distribution. Therefore, we generated a uniform amount of synthetic data for some of the tools available for 3D models. Next, the available 3D models include 11 tools for gastrectomy (atraumatic grasper, cadiere forceps, curved atraumatic grasper, harmonic ace, maryland bipolar forceps, medium-large clip applier, overholt, scissors, small clip applier, stapler, and suction-irrigation) and 9 laparoscopic tools (atraumatic grasper, clip applier (hem-o-lok), clip applier (metal), curved atraumatic grasper, electrichook, ligasure, overholt, scissors, and suction-irrigation). Two different types of training data were generated with different distributions of synthetic data. First type is training data including the real data and synthetic data with the same amount of each instrument, regardless of a class distribution of the real data to identify the effect of the synthetic data on training. Second type is training data in which synthetic data is added to the real data in order to make up for the lack of annotations for each instrument and verify if the virtual data can solve class imbalance problem. Figure 4 shows the distribution of the real and synthetic training data for the two types of data[1].

---

[1] Additional statistics is described at https://hsdb-instrument.github.io.

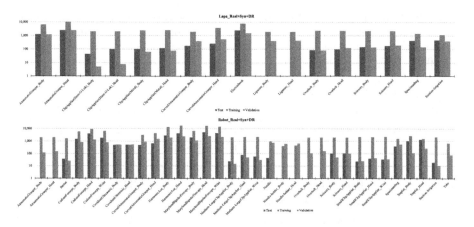

Fig. 4. Distribution of annotations obtained from laparoscopic cholecystectomy and gastrectomy for gastric cancer videos and 3D patient model. Compared to the distribution of annotations obtained from only the surgery videos (real), the class imbalance problem is alleviated in both surgeries in the real+synthetic+domain randomization data distribution.

4   Baseline Localization Performances

We trained CNN-based object detection models with MMDetection library [17] to provide the baseline performances of the hSDB-instrument dataset. We evaluated the mAP averaged for IoU $\in [0:5 : 0:05 : 0:95]$ (COCO's standard metric) where APs were computed at 10 IoU thresholds. The AP is the average of precision $P(\beta)$ over the different levels of recall $R(\beta)$ achieved by varying a confidence threshold β. The mAP is the mean of APs over all object categories. We trained representative two-stage models: Faster R-CNN [10], Cascade R-CNN [21] and one-stage models: RetinaNet [22], FCOS [23], FoveaBox [24]. The trained models also included feature pyramid structure [12], Libra module [25], double head module [26], and several backbones [27–29] with each model. Table 2 summarizes the performance changes depending on each detector and its associated submodules[2]. And Table 3 shows how our synthetic data helps models to be improved in both types of surgeries (Fig. 5).

---

[2] Additional experiment details and results are described at https://hsdb-instrument. github.io/.

Table 2. Performance evaluation result for validation set using hSDB-instrument dataset. Each object detection model combines with a submodule and backbone to show different performances. We evaluated the mAP averaged for IoU ∈ [0:5 : 0:05 : 0:95]

Laparoscopic cholecystectomy (real)			
Model	Backbone	Submodule	mAP [0.5 : 0.05 : 0.95]
Faster R-CNN	ResNet50	FPN	20.2
		FPN+OHEM	20.1
		FPN+Double head	22.9
	ResNet101	FPN	22.8
	ResNeXt101-32x4d	FPN	24.1
	ResNeXt101-64x4d	FPN	24.9
	HRNetV2p-W18	–	23.4
	HRNetV2p-W40	–	24.1
Cascade R-CNN	ResNet101	FPN	24.7
	ResNeXt101-64x4d	FPN	24.5
	HRNetV2p-W32	–	**25.7**
FCOS	HRNetV2p-W32-GN	FPN	11.8
	ResNeXt101-64x4d	FPN+mstrain	22.0
FoveaBox	ResNet50	FPN	22.9
	ResNeXt101	FPN+align-gn-ms	23.7
Robotic gastrectomy for gastric cancer (real)			
Model	Backbone	Submodule	mAP [0.5 : 0.05 : 0.95]
Faster R-CNN	ResNet50	FPN+Double head	37.9
	HRNetV2p-W40	–	37.8
Cascade R-CNN	ResNet101	FPN	38.2
	ResNeXt101-64x4d	FPN	**39.9**
	HRNetV2p-W32	–	38.8
FoveaBox	ResNeXt101	FPN+align-gn-ms	37.5

Table 3. Performance evaluation result for validation set using the second type of training set. We evaluated the mAP averaged for IoU ∈ [0:5 : 0:05 : 0:95].

Laparoscopic cholecystectomy			mAP [0.5 : 0.05 : 0.95]			
Model	Backbone	Submodule	real	real+syn	real+DR	real+syn+DR
Cascade R-CNN	HRNetV2p-W32	–	25.7	25.2	25.8	**27.1**
FoveaBox	ResNeXt101	FPN+align-gn-ms	23.7	–	–	**26.4**
Robotic gastrectomy for gastric cancer			mAP [0.5 : 0.05 : 0.95]			
Model	Backbone	Submodule	real	real+syn	real+DR	real+syn+DR
Cascade R-CNN	HRNetV2p-W32	–	38.8	**39.8**	39.7	39.6
FoveaBox	ResNeXt101	FPN+align-gn-ms	37.5	–	–	**40.1**

Fig. 5. Visualization of inference results for laparoscopic and robotic surgical tools. The left two images are the results of Cascade R-CNN for the laparoscopic tools, and the right two images are the results of FoveaBox for the robotic tools.

5 Conclusion

We have published the hSDB-instrument dataset with localization information for the analysis and evaluation of surgical procedures in laparoscopic and robotic surgery. The hSDB-instrument dataset provides bounding box annotations for laparoscopic and robotic surgery, while providing 3D synthetic data to handle class imbalance problems. Besides, the hSDB-instrument dataset is part-specific annotations of the instruments in order to enable the networks to recognize the desired level of subparts of the tools. We also provides the baseline performances of object detection networks for the dataset. We analyzed the hSDB-instrument dataset to provide guidelines for the localization of surgical instruments in surgical videos, while also making the dataset open to the public to contribute to relevant research. We expect that the hSDB-instrument dataset will be of great help in the development of new algorithms related to applications of surgical instrument recognition.

Acknowledgement. This work was supported by the Korea Medical Device Development Fund grant funded by the Korea government (the Ministry of Science and ICT, the Ministry of Trade, Industry and Energy, the Ministry of Health & Welfare, the Ministry of Food and Drug Safety) (Project Number: 202012A02-02).

References

1. Hughes-Hallett, A., Mayer, E.K., Pratt, P.J., Vale, J.A., Darzi, A.W.: Quantitative analysis of technological innovation in minimally invasive surgery. Br. J. Surg. **102**(2), 151–157 (2015)
2. Perez, R.E., Schwaitzberg, S.D.: Robotic surgery: finding value in 2019 and beyond. Ann. Laparosc. Endosc. Surg. **4**(51) (2019)
3. Jin, A., et al.: Tool detection and operative skill assessment in surgical videos using region-based convolutional neural networks. In: Proceedings of WACV (2018)
4. Maier-Hein, L., Wagner, M., Ross, T., et al.: Heidelberg colorectal data set for surgical data science in the sensor operating room. Sci. Data **8**, 101 (2021)
5. Twinanda, A.P., Shehata, S., Mutter, D., Marescaux, J., de Mathelin, M., Padoy, N.: EndoNet: a deep architecture for recognition tasks on laparoscopic videos. IEEE Trans. Med. Imaging **36**(1), 86–97 (2017)

6. Allan, M., et al.: 2017 Robotic Instrument Segmentation Challenge. arXiv: 1902.06426 (2019)
7. Sarikaya, D., Corso, J.J., Guru, K.A.: Detection and localization of robotic tools in robot-assisted surgery videos using deep neural networks for region proposal and detection. IEEE Trans. Med. Imaging **36**(7), 1542–1549 (2017)
8. Lin, T.-Y., et al.: Microsoft COCO: common objects in context. In: Fleet, D., Pajdla, T., Schiele, B., Tuytelaars, T. (eds.) ECCV 2014. LNCS, vol. 8693, pp. 740–755. Springer, Cham (2014). https://doi.org/10.1007/978-3-319-10602-1_48
9. Geiger, A., Lenz, P., Urtasun, R.: Are we ready for autonomous driving? The KITTI vision benchmark suite. In: Proceedings of CVPR (2012)
10. Ren, S., He, K., Girshick, R., Sun, J.: Faster R-CNN: towards real-time object detection with region proposal networks. In: Proceedings of NIPS (2015)
11. Liu, W., et al.: SSD: single shot MultiBox detector. In: Leibe, B., Matas, J., Sebe, N., Welling, M. (eds.) ECCV 2016. LNCS, vol. 9905, pp. 21–37. Springer, Cham (2016). https://doi.org/10.1007/978-3-319-46448-0_2
12. Lin, T.-Y., Dolláir, P., Girshick, R., He, K., Hariharan, B., Belongie, S.: Feature pyramid networks for object detection. In: Proceedings of CVPR (2017)
13. Huang, X., Liu, M.-Y., Belongie, S., Kautz J.: Multimodal unsupervised image-to-image translation. In: Proceedings of ECCV (2018)
14. Lee, K., Choi, M. -K., Jung, H.: DavinciGAN: unpaired surgical instrument translation for data augmentation. In: Proceedings of MIDL (2019)
15. Park, T., Liu, M.-Y., Wang, T.-C., Zhu, J.-Y.: Semantic image synthesis with spatially-adaptive normalization. In: Proceedings of CVPR (2019)
16. Pfeiffer, M., et al.: Generating large labeled data sets for laparoscopic image processing tasks using unpaired image-to-image translation. In: Shen, D., et al. (eds.) MICCAI 2019. LNCS, vol. 11768, pp. 119–127. Springer, Cham (2019). https://doi.org/10.1007/978-3-030-32254-0_14
17. Chen, K., et al.: MMDetection: Open MMLab Detection Toolbox and Benchmark. arXiv:1906.07155 (2019)
18. Paszke, A., et al.: PyTorch: an imperative style, high-performance deep learning library. In: Proceedings of NeurIPS (2019)
19. Computer Vision Annotation Tool (CVAT). https://github.com/opencv/cvat
20. Tremblay, J., et al.: Training deep networks with synthetic data: bridging the reality gap by domain randomization. In: Proceedings of CVPRW (2018)
21. Cai Z., Vasconcelos, N.: Cascade R-CNN: delving into high quality object detection. In: Proceedings of CVPR (2018)
22. Lin, T.-Y., Goyal, P., Girshick, R., He, K., Dollar, P.: Focal loss for dense object detection. In: Proceedings of ICCV (2017)
23. Tian, Z., Shen, C., Chen, H., He, T.: FCOS: fully convolutional one-stage object detection. In: Proceedings of ICCV (2019)
24. Kong, T., Sun, F., Liu, H., Jiang, Y., Shi J.: FoveaBox: Beyond Anchor-based Object Detector. arXiv:1904.03797 (2019)
25. Pang, J., Chen, K., Shi, J., Feng, H., Ouyang, W., Lin, D.: Libra R-CNN: towards balanced learning for object detection. In: Proceedings of CVPR (2019)
26. Li, A., Yang, X., Zhang, C.: Rethinking classification and localization for object detection. In: Proceedings of BMVC (2019)
27. He, K., Zhang, X., Ren, S., Sun, J.: Deep residual learning for image recognition. In: Proceedings of CVPR (2016)
28. Xie, S., Girshick, R., Dolláir, P., Tu, Z., He K.: Aggregated residual transformations for deep neural networks. In: Proceedings of CVPR (2017)
29. Wang, J., et al.: Deep High-Resolution Representation Learning for Visual Recognition. arXiv:1908.07919 (2019)

Co-generation and Segmentation for Generalized Surgical Instrument Segmentation on Unlabelled Data

Megha Kalia[1]($\boxtimes$)(iD), Tajwar Abrar Aleef[2](iD), Nassir Navab[3], Peter Black[4],
and Septimiu E. Salcudean[1]

[1] Electrical and Computer Engineering, University of British Columbia,
Vancouver, Canada
{mkalia,tims}@ece.ubc.ca
[2] School of Biomedical Engineering, University of British Columbia,
Vancouver, Canada
tajwaraleef@ece.ubc.ca
[3] Computer Aided Medical Procedures, Technical University of Munich,
Munich, Germany
nassir.navab@tum.de
[4] Vancouver Prostate Centre, Vancouver, Canada
peter.black@ubc.ca

Abstract. Surgical instrument segmentation for robot-assisted surgery is needed for accurate instrument tracking and augmented reality overlays. Therefore, the topic has been the subject of a number of recent papers in the CAI community. Deep learning-based methods have shown state-of-the-art performance for surgical instrument segmentation, but their results depend on labelled data. However, labelled surgical data is of limited availability and is a bottleneck in surgical translation of these methods. In this paper, we demonstrate the limited generalizability of these methods on different datasets, including robot-assisted surgeries on human subjects. We then propose a novel joint generation and segmentation strategy to learn a segmentation model with better generalization capability to domains that have no labelled data. The method leverages the availability of labelled data in a different domain. The generator does the domain translation from the labelled domain to the unlabelled domain and simultaneously, the segmentation model learns using the generated data while regularizing the generative model. We compared our method with state-of-the-art methods and showed its generalizability on publicly available datasets and on our own recorded video frames from robot-assisted prostatectomies. Our method shows consistently high mean Dice scores on both labelled and unlabelled domains when data is available only for one of the domains.

Keywords: Surgical instrument segmentation · Unpaired image to image translation · Generative adversarial learning

M. Kalia and T. A. Aleef—Contributed equally to the manuscript

© Springer Nature Switzerland AG 2021
M. de Bruijne et al. (Eds.): MICCAI 2021, LNCS 12904, pp. 403–412, 2021.
https://doi.org/10.1007/978-3-030-87202-1_39

1   Introduction

Surgical instrument segmentation is fundamental to Augmented Reality (AR) in image-guided robot-assisted surgery (RAS) [11] and has been an active topic of research, with convolutional neural network (CNN)-based methods surpassing prior methods by a significant margin [2,5,10]. CNN-based methods depend on the availability of annotated surgical data, which may be difficult to obtain [9]. Their performance has been reported for publicly available *ex-vivo* and porcine *in-vivo* RAS surgeries, but not in human RAS.

Recently, many generative approaches have been proposed to mitigate the problem of limited clinical labelled data [4,7,13]. For laparoscopic instrument segmentation, [13] proposed a generative adversarial network (GAN)-based method to use a small amount of labelled data. In [7], labelled data from cadaver surgery was transferred to *in-vivo* surgery. Then a separate segmentation model was trained using either the translated cadaver data or translated *in-vivo* data to the cadaver domain. In [4], an image-to-image (I2I) mapping of simulated to real surgical instruments was proposed, with blending into the camera background. In the above methods, the translated data was used to train a segmentation model. Finding validated quantitative metrics for the quality of translated data is difficult and is the topic of on-going research [6,15]. In many cases, the generative models change the surgical instruments' shape and introduce artefacts while the overall accuracy decreases [Fig. 1]; this is undesirable for clinical application. Hence, a segmentation strategy leveraging the power of generative models to alleviate the problem of unlabelled clinical data while addressing the predominant current challenges of generative models is imperative.

Therefore, in the current paper, we present a joint unpaired I2I mapping and segmentation strategy for better generalizability of a surgical instrument segmentation model to a domain with no labelled data. The generative and segmentation models are trained together and reach convergence in a synergistic manner. The generative model maps from a source domain with labelled data to a target domain with unlabelled data with constant feedback from the segmentation model. The segmentation model trains in parallel on the generated target images and on the labelled source images. The convergence criterion of this joint-system is the segmentation quality. The segmentation model also regularizes the generative model that can otherwise change the shape of the surgical instruments during the I2I mapping. We call our method coSegGAN. The closest method to it is presented in [7]. However, unlike in [7], our segmentation model is not pre-trained. It provides feedback to the generators as it learns using the generated data, thus seeing much more varied data. Unlike prior work, we provide an explicit shape constraint on latent space to provide intermediate supervision during generative training. Through evaluation on real surgical sequences and publicly available datasets, we show that coSegGAN has better generalizability than existing methods. The main contribution of the paper is presenting a joint generation and segmentation framework that provides state of the art (SOTA) results to segment surgical instruments on unlabelled data. The method performs better than using a generative model for data augmentation as a separate

Method	Test / Train	Endovis	UCL	Surgery
Ternaus	Endovis	94.9	58.2	86.0
	UCL	63.6	93.2	43.5
Rasnet	Endovis	89.2	84.2	80.5
	UCL	56.1	92.7	36.8
U-Net	Endovis	95.0	33.5	58.8
	UCL	72.2	87.7	45.1

Fig. 1. (Left) Table showing the limited generalizability of state of the art (SOTA) methods across domains. Mean Dice scores are shown for different methods on datasets *Endovis, UCL,* and *Surgery*. (Right) Figures showing the problem with cycleGAN where the intermediate generated output can be unrealistic while overall cycle consistency loss is low. Figures A, B, and C show the original, translated, and reconstructed domain.

step. To the best of our knowledge, this is the first method that segments surgical instruments with no labelled data by jointly training the generative and segmentation model as a joint feedback system to perform an I2I mapping between the labelled and the unlabelled domain.

2 Methods

2.1 Network Details

The generative part of coSegGAN uses a cycleGAN like architecture with two generators and two discriminators [16]. Let x_{ai} and x_{bi} denote the two i^{th} images in the domains ψ_A and ψ_B, respectively, and x_a and x_b denote the set of all images in domains a and b, respectively. y_{ai} denotes the corresponding individual label for the ith x_{ai} image and y_a is the set of all such labels. G_A and G_B are the two generators estimating the mappings, with $G_A : x_{b \to a}$ and $G_B : x_{a \to b}$, respectively. The discriminator D_A is responsible to discriminate between given true images in domain ψ_A and generated images $G_A(x_b)$. Similarly, D_B is responsible for discriminating between the true domain ψ_B and generated images $G_B(x_a)$. Both G_A and G_B have a U-Net-like architecture [12] with a contracting and expanding path. The contracting path consists of four 4×4 convolutional layer with stride 2 + Leaky ReLu + Instance normalization [14] blocks where in each subsequent block the output is halved and the channel numbers are doubled. The expanding path consists of three blocks with each block having an up-sampling layer + 4×4 convolution with a stride of 1 + ReLu activation + Instance normalization. The output of each block was concatenated with the low-level features from the contracting path by skip connections and then passed as an input to the next block. The output of the final block was passed though a convolutional layer followed by a tanh activation. For the discriminator, we used a patchGAN similar to [16]. For the segmentation model (S) in coSegGAN we used the original U-Net architecture but with 16 base filters to prevent over-fitting and to reduce computation. This did not decrease the performance of segmentation when compared to the original U-Net, as determined empirically.

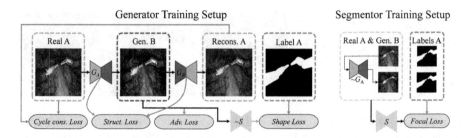

Fig. 2. Overview of the training setup for the generation and segmentation side. The diagram on the left shows the A to B mapping side of the cycleGANs that is modified to incorporate shape loss and structural loss during training. The grey blocks indicate the different losses used for updating the weights of the generator. On the right side, the input and loss used for the segmentation model training are shown. Any network indicated by '$\sim$' in this figure indicates it has frozen weights.

2.2   Training Strategy

We trained the generators, discriminators & the segmentation model in an alternative fashion. In the first run, the weights through the generators, G_A and G_B were back-propagated while freezing the weights of the discriminators and the segmentation model. Then in the next run, the discriminators as well as the segmentation model, D_A, D_B, & S were trained and updated. For training S, both x_a and $G_B(x_a)$ were fed as the input. Since the generated images are translated versions of the real image, the corresponding labels for $G_B(x_a)$ are the same as x_a. Note that S is seeing different variations of the generated target domain images in every epoch because the generators and S are learning in parallel. While the quality of the I2I mapping from the generators increases, the quality of the images seen by S also increases. Details can be seen in Fig. 2.

2.3   Loss Functions

Segmentation Model. In order not to overwhelm the loss with the higher number of background pixels, we used an α-balanced variant of focal loss, $\mathcal{L}_{foc}$ [8], a modification of cross-entropy, where the γ factor controls the contribution of high-probability samples in the loss calculation. We used the hyper-parameters γ and α as 2.0 and 0.25, respectively. The total segmentation loss, $\mathcal{L}_{seg}$, is

$$\mathcal{L}_{seg} = \mathcal{L}_{foc}\left(x_a, y_a\right) + \mathcal{L}_{foc}\left(G_B\left(x_a\right), y_a\right) \tag{1}$$

Generative Model. For cycleGAN we used an adversarial loss, $\mathcal{L}_{GAN}$, and a pixel-level cycle consistency loss $\mathcal{L}_{cyc}$ proposed in [16]. Although $\mathcal{L}_{cyc}$ reduces the number of possibilities when mapping across domains and regularizes the cycleGAN, it does not suffice to preserve the higher-level semantics in the image. This can change the shapes of surgical instruments during the translation, which is not desirable. Therefore we included feedback from the segmentation model in the total generative loss. This penalizes the generation of unrealistic surgical instrument shapes in $G_B(x_a)$. Since we are interested in the mapping from x_a to x_b, which later is fed as an input to the segmentation model, we included this constraint only on the generator G_B. This shape preservation loss, $\mathcal{L}_{shape}$, is

$$\mathcal{L}_{shape} = \mathcal{L}_{foc}\left(G_B\left(x_a,\right)y_a\right) \tag{2}$$

In cycleGAN models, $\mathcal{L}_{cycTotal}$ is the sum of two cycle consistency losses such that, $\mathcal{L}_{cycTotal} = \mathcal{L}_{cyc}(x_a, G_A(G_B(x_a))) + \mathcal{L}_{cyc}(x_b, G_B(G_A(x_b)))$. These losses enforce pixel level constraints between the original inputs x_a and x_b and reconstructed outputs $G_A(G_B(x_a))$ and $G_A(G_B(x_a))$, where the two GANs are optimized together. There is no intermediate supervision after each generative step $G_A : x_{b \to a}$ and $G_B : x_{a \to b}$. Thus G_A and G_B can produce unrealistic images while the total $\mathcal{L}_{cyc}$ is reduced (Shown in Fig. 1, (right)). In particular, the mapping across domains should change only the 'appearance' of the scene while retaining the domain-invariant structural elements. To preserve the structural properties of the scene across domains, we introduce an explicit, intermediate, feature level, latent space loss. This latent space loss, and the total generated loss, are:

$$\mathcal{L}_{structure} = \mathbb{E}\left[\|e_A(x_a) - e_B(G_B(x_a))\|_1\right] + \mathbb{E}\left[\|e_B(x_b) - e_A(G_A(x_b))\|_1\right] \tag{3}$$
$$\mathcal{L}_{generator} = \lambda_1 \mathcal{L}_{GANTotal} + \lambda_2 \mathcal{L}_{cycTotal} + \lambda_3 \mathcal{L}_{shape} + \lambda_4 \mathcal{L}_{structure} + \lambda_5 \mathcal{L}_I \tag{4}$$

where, e_A and e_B are encoders in G_B and G_A, respectively, $\mathcal{L}_{GANTotal} = \mathcal{L}_{GAN}(G_B, D_B, x_a, x_b) + \mathcal{L}_{GAN}(G_A, D_A, x_b, x_a)$ and $\mathcal{L}_I$ is the identity mapping loss as given in [16]. Values of λ_1, λ_2, λ_3, λ_4, and λ_5 are 1, 10, 1, 5, and 1, respectively. These values were tuned during the hyper-parameter tuning phase.

Training Details and Hyper-Parameters. For training and testing our models, we use Tensorflow & Keras API on a NVIDIA Tesla V100 GPU (16 GB). For training the proposed models, we used a batch size of 8, and Adam optimizer with β_1 and β_2 of 0.9 and 0.999, respectively, with a learning rate of 10^{-3}. We trained our models for 100 epochs (approximately 12 h) and saved weights of the segmentation model with the highest validation Dice score [17]. Code is available at: https://github.com/tajwarabraraleef/coSegGAN.

3   Experiments

Datasets: Endovis Challenge, 2017, *in-vivo* Dataset [1]: It is a porcine surgery procedure with a training set consisting of 8 videos of 225 frames each and a test set consisting of 8 videos of 75 frames and 2 videos of 300 frames each. We used 6 videos for training and 2 videos for validation from the training set. We used 8 videos from the test set for testing; these were not used for validation. In the paper, we refer to this dataset as *Endovis*. In Table 1 *Endovis* is abbreviated as *Endo*.

UCL *ex-vivo* Dataset [3]: The dataset consists of 14 videos with different animal tissues as background. Similar to [3], we used 8, 2 and 4 videos for training, validation and testing, respectively.

Prostatectomy Dataset. We prepared the training dataset from 5 videos of robot-assisted radical prostatectomy procedures with the da Vinci Si surgical system from Vancouver General Hospital, Vancouver, Canada. We manually selected 1327 frames to isolate surgical instruments from other visible objects in the surgical field of view. These frames do not have corresponding labels. To evaluate the performance of the various methods on actual surgical data, we prepared a test set of 182 frames taken from 4 different surgeries independent from the training set. The test data represents approximately 12% of the entire surgical data used. We manually labelled surgical instruments in these frames only for the purpose of testing coSegGAN and existing methods. All the frames were center cropped to give a final size of 721×503 pixels. We will refer to this dataset as *Surgery* in the rest of the paper. Ethics to collect data was obtained from the Institutional Clinical Research Ethics Board. For all three datasets, we resized the frames to 256×256 to accelerate the computation.

Evaluation. We compared coSegGAN with Ternausnet, the best performing method in the Endovis Challenge [1] for binary segmentation and RASnet, reporting a mean 94.65% Dice coefficient on Endovis. For a fair comparison to coSegGAN, we performed data augmentation with the cycleGAN architecture given in Sect. 2. The cycleGAN model was run for 50 epochs in all cases as it converged in 50 epochs. After cycleGAN I2I translation from source (with labels) to target domain, the SOTA segmentation models were trained with both the translated and original domain data. We also performed an ablation experiment comparing coSegGAN with and without the proposed $\mathbf{L}_{structure}$ loss. We refer to RASnet, Ternausnet, and our U-Net variant with focal loss, trained using the augmented data generated from a separate cycleGAN (unlike our joint strategy) as $RASnet+$, $Ternausnet+$ and $U\text{-}Net_{FL}+$ respectively. The coSeg-GAN network without $\mathbf{L}_{structure}$ is called $coSegGAN-$. We performed evaluation of four combinations of datasets for labelled and unlabelled domains. For ease of reporting, we refer to *Endovis* (labelled) + *Surgery* (Unlabelled), *UCL* (labelled) + *Surgery* (Unlabelled), *Endovis* (labelled) + *UCL* (Unlabelled), and *UCL* (labelled) + *Endovis* (Unlabelled) data combinations as case 1, case 2, case 3, and case 4, respectively. Since, we want to quantify the generalizability of our method across labelled and unlabelled domains, for a particular dataset combination, we also calculated an absolute difference in the Dice scores, Δ *Dice*, and absolute difference in Intersection over Union (IoU), Δ *IoU*, between labelled domain A and unlabelled domain B. The lower the Δ *Dice* and Δ *IoU*, the higher is the generalizability between domains (refer to Table 1).

Table 1. Comparison of Mean Dice and IoU scores of coSegGAN with SOTA methods

Method		Dom A	DomB (Unlabelled)	Domain A		Domain B (Unlabelled)		Δ Values	
				Dice	IoU	Dice	IoU	Dice	IoU
coSegGAN	Case 1	Endo	Surgery	**93.7%**	88.4%	**92.8%**	84.7%	**0.9%**	3.7%
	Case 2	UCL	Surgery	91.1%	84.2%	74.3%	59.8%	16.8%	24.4%
	Case 3	Endo	UCL	**93.2%**	88.3%	**90.0%**	82.2%	**3.2%**	6.1%
	Case 4	UCL	Endo	93.5%	91.1%	79.4%	66.8%	14.1%	24.3%
RASnet+	Case 1	Endo	Surgery	88.3%	79.9%	78.1%	64.7%	10.2%	15.2%
	Case 2	UCL	Surgery	92.3%	85.8%	47.8%	33.0%	44.5%	52.8%
	Case 3	Endo	UCL	88.4%	80.0%	83.3%	71.9%	5.1%	8.1%
	Case 4	UCL	Endo	92.4%	85.9%	66.8%	52.9%	25.6%	33.0%
Ternaus+	Case 1	Endo	Surgery	94.2%	89.9%	88.7%	80.4%	5.5%	9.5%
	Case 2	UCL	Surgery	95.8%	92.1%	46.0%	31.3%	49.8%	60.8%
	Case 3	Endo	UCL	93.3%	89.2%	41.7%	29.0%	51.6%	60.2%
	Case 4	UCL	Endo	93.4%	87.8%	55.0%	41.2%	38.4%	46.6%
$U\text{-}Net_{FL}+$	Case 1	Endo	Surgery	91.8%	85.7%	58.0%	42.5%	33.8%	43.5%
	Case 2	UCL	Surgery	93.2%	87.4%	36.0%	22.9%	57.2%	64.5%
	Case 3	Endo	UCL	83.9%	73.5%	23.3%	13.6%	60.7%	31.5%
	Case 4	UCL	Endo	74.6%	61.2%	56.5%	42.0%	18.1%	19.2%
coSegGAN–	case 1	Endo	Surgery	94.1%	89.6%	92.3%	86.0%	**1.8%**	3.6%
	Case 2	UCL	Surgery	93.5%	93.0%	74.5%	69.7%	19.0%	23.3%
	Case 3	Endo	UCL	93.3%	88.2%	90.8%	83.0%	**2.5%**	5.2%
	Case 4	UCL	Endo	94.2%	89.2%	74.4%	64.8%	19.8%	24.4%

4   Results and Discussion

For case 1, the proposed coSegGAN network gave significantly higher Dice
(92.8%) and IoU scores (84.7%) on unlabelled domain B (*Surgery*) when compared to *RASnet+*, *Ternausnet+* and *U-Net$_{FL}$+* which have Dice scores
of 78.1% (IoU = 64.7%), 88.7% (IoU = 80.4%), and 84.1% (IoU = 42.5%),
respectively. For case 2 as well, the Dice score for coSegGAN on unlabelled
domain (*Surgery*) is 74.3% (IoU = 59.8%) while *RASnet+*, *Ternausnet+*, and
U-Net$_{FL}$+ have lower Dice scores of 47.8% (IoU =33.0%), 46.0% (IoU = 31.3%),
and 45.6% (IoU = 22.9%), respectively. Similarly, for case 3, the Dice (IoU =
82.2%) score for coSegGAN on unlabelled data (*UCL*) is 90%, which is higher
than *RASnet+*, *Ternausnet+* and *U-Net$_{FL}$+* with Dice scores of 83.3% (IoU
= 71.9%), 41.7% (IoU = 29.0%), and 81.8% (IoU = 13.6%), respectively. For
case 4, the Dice score for coSegGAN on unlabelled *Endovis* data is 79.4% (IoU
= 66.8%), which, similar to other cases, is higher than the rest of the methods;
Dice scores of *RASnet+*, *Ternausnet+* and *U-Net$_{FL}$+* being 66.8% (IoU =
52.9%), 55.0% (IoU = 52.9%) and 56.5% (IoU = 42.0%), respectively.

The Δ *Dice*, for coSegGAN for case 1 is much lower 0.9% (IoU = 3.7%) while for *RASnet+*, *Ternausnet+*, and *U-Net$_{FL}$+* it is 10.2% (IoU = 15.2%), 5.5% (IoU = 9.5%), and 33.8% (IoU = 43.5%), respectively. For case 2, Δ *Dice* for coSegGAN is 16.8% (IoU = 24.4%), while for *RASnet+*, *Ternausnet+* and *U-Net$_{FL}$+* it is 44.5% (IoU = 52.8%), 49.8% (IoU = 60.8%) and 57.2% (IoU = 64.5%), respectively. For case 3, Δ *Dice*, for coSegGAN is 3.2% (IoU = 6.1%), which is much lower than *RASnet+*, *Ternausnet+* and *U-Net$_{FL}$+* with Δ *Dice* of 5.1% (IoU = 8.1%), 51.6% (IoU = 60.2%), and 60.7% (IoU = 31.5%), respectively. For case 4, similarly, the Δ *Dice* for coSegGAN is 14.1% (IoU = 24.3%) when compared to *RASnet+*, *Ternausnet+*, and *U-Net$_{FL}$+* with Δ *Dice* of 25.6% (IoU = 33.0%), 38.4% (IoU = 46.6%), and 18.1% (IoU = 19.2%), respectively. Consistently higher Dice and IoU on unlabelled data and significantly lower Δ *Dice* and Δ *IoU* of coSegGAN show its generalizability when compared to all other methods for all the cases.

For coSegGAN, in cases 2 and 4, when the mapping is from *UCL* (labelled) to either *Surgery* or *Endovis*, the Δ *Dice* is higher than cases 1 and 3, showing comparatively less generalizability. This could be because the *UCL* data is an *ex-vivo* dataset where data distribution potentially differs from a real surgery, with remarkably different lighting and background. Also, there is only one type of surgical instrument visible in the *UCL* dataset, which might have hindered the mapping to multiple types of instruments.

In the ablation experiment, coSegGAN–, i.e., coSegGAN without the $\mathbf{L}_{structure}$, showed comparable performance with coSegGAN, except case 4, where the performance of coSegGAN is significantly higher (approximately 5%) on the unlabelled *Endovis* dataset. coSegGAN– has higher Δ *Dice* for all cases except case 4, showing that with the $\mathbf{L}_{structure}$ loss coSegGAN generalizes better to both labelled and unlabelled datasets.

A qualitative comparison of coSegGAN with other methods for different surgeries can be seen in Fig. 3. As can be seen (column 1), coSegGAN performs better in preserving overall tool structure, with finer details, when compared to other methods. In comparison to *Ternausnet+* and *RASnet+*, the method also produces fewer false positives [Fig. 3 (column 2)]. Although coSegGAN performs better than SOTA methods in identifying tools, it occasionally fails to identify the tool in the presence of blood where surgical instrument blends in with the background. Usually this happens at the image periphery where the region is relatively dark compared to the well-lit image center. Figure 3 (column 4) shows one such failure case.

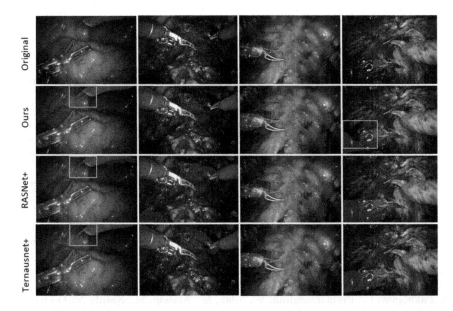

Fig. 3. Figure showing a qualitative comparison of our method with other methods. It can be seen that overall, our method preserves the shape of the instruments better with fewer false positives. (column 1) Inset showing preservation of instrument shape in our method. (column 4) Inset showing a failure case of our method.

5   Conclusion

We presented a joint generative and segmentation strategy, coSegGAN, that outperforms SOTA methods in its generalization capability to unlabelled domain data. The evaluated SOTA methods use separate I2I mapped data augmentation and segmentation steps. The proposed losses helped to preserve finer tool structure. The method is easy to adapt to other deep learning segmentation methods and thus can significantly improve the existing methods. The method aims to utilize unlabelled surgical data, which is much easier to acquire than labelled data, to improve any instrument segmentation model in a simple yet effective manner. Therefore, coSegGAN has the potential to significantly facilitate surgical translation of current and future surgical tool segmentation methods because it effectively alleviates the problem of unlabelled data. Current testing of coSeg-GAN has been limited to footage from prostatectomy procedures. A thorough performance analysis for different types of RAS surgeries is part of future work.

References

1. Allan, M., et al.: 2017 robotic instrument segmentation challenge. arXiv preprint arXiv:1902.06426 (2019)
2. Attia, M., Hossny, M., Nahavandi, S., Asadi, H.: Surgical tool segmentation using a hybrid deep CNN-RNN auto encoder-decoder. In: 2017 IEEE International Conference on Systems, Man, and Cybernetics (SMC), pp. 3373–3378. IEEE (2017)

3. Colleoni, E., Edwards, P., Stoyanov, D.: Synthetic and real inputs for tool segmentation in robotic surgery. In: Martel, A.L., et al. (eds.) MICCAI 2020. LNCS, vol. 12263, pp. 700–710. Springer, Cham (2020). https://doi.org/10.1007/978-3-030-59716-0_67

4. Colleoni, E., Stoyanov, D.: Robotic instrument segmentation with image-to-image translation. IEEE Robot. Autom. Lett. **6**(2), 935–942 (2021)

5. Iglovikov, V., Shvets, A.: TernausNet: U-net with VGG11 encoder pre-trained on imagenet for image segmentation. arXiv preprint arXiv:1801.05746 (2018)

6. Kynkäänniemi, T., Karras, T., Laine, S., Lehtinen, J., Aila, T.: Improved precision and recall metric for assessing generative models. arXiv preprint arXiv:1904.06991 (2019)

7. Lin, S., Qin, F., Li, Y., Bly, R.A., Moe, K.S., Hannaford, B.: LC-GAN: image-to-image translation based on generative adversarial network for endoscopic images. arXiv preprint arXiv:2003.04949 (2020)

8. Lin, T.Y., Goyal, P., Girshick, R., He, K., Dollár, P.: Focal loss for dense object detection. In: Proceedings of the IEEE International Conference on Computer Vision, pp. 2980–2988 (2017)

9. Maier-Hein, L., et al.: Surgical data science for next-generation interventions. Nature Biomed. Eng. **1**(9), 691–696 (2017)

10. Pakhomov, D., Premachandran, V., Allan, M., Azizian, M., Navab, N.: Deep residual learning for instrument segmentation in robotic surgery. In: Suk, H.-I., Liu, M., Yan, P., Lian, C. (eds.) MLMI 2019. LNCS, vol. 11861, pp. 566–573. Springer, Cham (2019). https://doi.org/10.1007/978-3-030-32692-0_65

11. Pauly, O., Diotte, B., Habert, S., Weidert, S., Euler, E., Fallavollita, P., Navab, N.: Relevance-based visualization to improve surgeon perception. In: Stoyanov, D., Collins, D.L., Sakuma, I., Abolmaesumi, P., Jannin, P. (eds.) IPCAI 2014. LNCS, vol. 8498, pp. 178–185. Springer, Cham (2014). https://doi.org/10.1007/978-3-319-07521-1_19

12. Ronneberger, O., Fischer, P., Brox, T.: U-Net: convolutional networks for biomedical image segmentation. In: Navab, N., Hornegger, J., Wells, W.M., Frangi, A.F. (eds.) MICCAI 2015. LNCS, vol. 9351, pp. 234–241. Springer, Cham (2015). https://doi.org/10.1007/978-3-319-24574-4_28

13. Ross, T., et al.: Exploiting the potential of unlabeled endoscopic video data with self-supervised learning. Int. J. Comput. Assist. Radiol. Surg. **13**(6), 925–933 (2018). https://doi.org/10.1007/s11548-018-1772-0

14. Ulyanov, D., Vedaldi, A., Lempitsky, V.: Instance normalization: the missing ingredient for fast stylization. arXiv preprint arXiv:1607.08022 (2016)

15. Zhou, S., Gordon, M.L., Krishna, R., Narcomey, A., Fei-Fei, L., Bernstein, M.S.: HYPE: a benchmark for human eye perceptual evaluation of generative models. arXiv preprint arXiv:1904.01121 (2019)

16. Zhu, J.Y., Park, T., Isola, P., Efros, A.A.: Unpaired image-to-image translation using cycle-consistent adversarial networks. In: Proceedings of the IEEE International Conference on Computer Vision, pp. 2223–2232 (2017)

17. Zou, K.H., et al.: Statistical validation of image segmentation quality based on a spatial overlap index1: scientific reports. Acad. Radiol. **11**(2), 178–189 (2004)

Surgical Data Science

E-DSSR: Efficient Dynamic Surgical Scene Reconstruction with Transformer-Based Stereoscopic Depth Perception

Yonghao Long[1]([✉]), Zhaoshuo Li[2], Chi Hang Yee[3], Chi Fai Ng[3], Russell H. Taylor[2], Mathias Unberath[2], and Qi Dou[1,4]

[1] Deptment of Computer Science and Engineering,
The Chinese University of Hong Kong, Shatin, Hong Kong
`yhlong@cse.cuhk.edu.hk`
[2] Department of Computer Science, Johns Hopkins University, Baltimore, USA
[3] SH Ho Urology Centre, Department of Surgery,
The Chinese University of Hong Kong, Shatin, Hong Kong
[4] T Stone Robotics Institute, The Chinese University of Hong Kong,
Shatin, Hong Kong

Abstract. Reconstructing the scene of robotic surgery from the stereo endoscopic video is an important and promising topic in surgical data science, which potentially supports many applications such as surgical visual perception, robotic surgery education and intra-operative context awareness. However, current methods are mostly restricted to reconstructing static anatomy assuming no tissue deformation, tool occlusion and de-occlusion, and camera movement. However, these assumptions are not always satisfied in minimal invasive robotic surgeries. In this work, we present an efficient reconstruction pipeline for highly dynamic surgical scenes that runs at 28 fps. Specifically, we design a transformer-based stereoscopic depth perception for efficient depth estimation and a lightweight tool segmentor to handle tool occlusion. After that, a dynamic reconstruction algorithm which can estimate the tissue deformation and camera movement, and aggregate the information over time is proposed for surgical scene reconstruction. We evaluate the proposed pipeline on two datasets, the public Hamlyn Centre Endoscopic Video Dataset and our in-house DaVinci robotic surgery dataset. The results demonstrate that our method can recover the scene obstructed by the surgical tool and handle the movement of camera in realistic surgical scenarios effectively at real-time speed.

Keywords: Dynamic surgical scene reconstruction ·
Transformer-based depth estimation · Stereo image perception

Y. Long and Z. Li—Authors contributed equally to this work.

Electronic supplementary material The online version of this chapter (https:// doi.org/10.1007/978-3-030-87202-1_40) contains supplementary material, which is available to authorized users.

M. de Bruijne et al. (Eds.): MICCAI 2021, LNCS 12904, pp. 415–425, 2021.
https://doi.org/10.1007/978-3-030-87202-1_40

1   Introduction

Reconstructing the surgical scene from stereo endoscopic video in robotic-assisted minimally invasive surgeries (MIS) is an important topic as it is central to down-stream tasks. For example, during surgical training, it is desirable to expose the trainees to the complete soft-tissue even if surgical tools block the view partially [22,23] in order to provide enriched context for understanding the surgical manipulation. As illustrated in Fig. 1(a), given the reconstruction from recorded surgical videos and the current video frame with instrument blocking the view, a transparent overlay can be generated for AI-augmented demonstration which brings new possibilities for robotic surgery education. A similar method potentially may be used to provide additional useful context intraoperatively.

However, reconstruction of the surgical scene in laparoscopy is challenging for three reasons. First, the soft-tissue is constantly deforming. This gives rise to challenges to soft-tissue localization. Secondly, it presents heavy occlusion and dynamic movement of the surgical instruments. Identification of occlusion and proper handling of de-occlusion require spatial and temporal coherence and consistency. Lastly, the changes of camera view points compound the aforementioned difficulties into an ego motion task with dynamic objects. Even though there exists prior works in 3D reconstruction in surgical scene, they are generally limited by assuming a static scene [12] or no presence of surgical tools [20].

Closest to our work are [10,13], which jointly handle tissue deformation and surgical tool occlusion but using kinematics information. We improve upon prior work with an *image-only* reconstruction pipeline shown in Fig. 1(b) that improves the quality and run-time speed. We implement an efficient transformer-based depth perception module and a light-weight tool segmentor to reconstruct the surgical scenes with only stereo endoscopic image frames as inputs. The two modules run in parallel to output a masked depth estimation without surgical instruments. The masked depth map is later used to produce a temporally and spatially coherent reconstruction of the soft-tissue. We also demonstrate the effectiveness of our pipeline with camera motion and smoke, which is missing from [10,13].

Our main contributions are summarized as follows: We propose a novel online reconstruction pipeline called E-DSSR which can reconstruct the surgical scene with *only* stereo endoscopic videos as input, and handle the cases of tissue deformation, tool occlusion and the camera movement simultaneously. We qualitatively and quantitatively evaluate our proposed pipeline on both the public Hamlyn Centre Endoscopic Video Dataset (Hamlyn Dataset) [27] and our in-house DaVinci robotic surgery data to validate the effectiveness of our method.

2   Related Work

Recent works of [7,21] have explored SLAM system to handle the non-rigidity of tissues. However, they assume a simplified environment where surgical tools are not present. Later, Li et al. [10] proposes a reconstruction framework to simultaneously reconstruct the soft tissue and track surgical instruments using kinematics. However, instrument location is prone to noise and error given the long kinematics

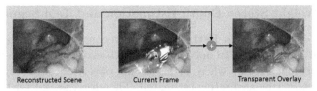

(a) An illustrative demonstration of instrument occlusion in robotic surgery.

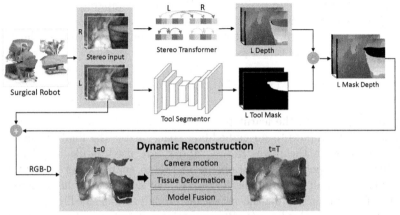

(b) An overview of our E-DSSR pipeline

Fig. 1. Illustration of the insight (a) and the pipeline of our proposed E-DSSR (b).

chain as demonstrated in [2]. Lu et al. [13] further improves the previous framework with deep learning methods to estimate depth, and a key-point+kinematics hybrid approach to localize the surgical instrument. However, the result is only demonstrated in an *ex vivo* environment with a fixed camera pose, without camera motion and realistic artifacts such as blood. Furthermore, the surgical instrument is removed by computing the pose of the surgical tool and then rendering a binary mask from a 3D model, which is slower as we will show in Sect. 4.3. In this work, we present an efficient reconstruction pipeline that only uses a stream of stereo images as input, as well as being capable of handling camera motion and surgical effects such as smoke in addition to the previous challenges.

3   Method

Figure 1(b) shows the overview of our proposed dynamic surgical scene reconstruction pipeline E-DSSR. We denote left (L) and right (R) RGB image at time stamp $t \in [0, T]$ as l_t and r_t. A transformer-based stereo depth estimation estimates the depth image d_t using l_t and r_t. A tool segmenting network predicts the mask of the tool l_t^m concurrently. Note that both networks are designed to be light-weight to enable real-time performance. The masked depth image d_t^m and left frame l_t from $t = 0$ to $t = T$ are input to a dynamic reconstruction algorithm which can dynamically recover the 3D information of the surgical scene.

3.1 Light-Weight Stereo Depth Estimation and Tool Segmentation

Transformer-Based Depth Estimation. To reconstruct the soft-tissue with high quality, it is important to only include depth estimations with high confidence. In the occluded region where pixels are not observed commonly by both l_t and r_t, depth cannot be accurately estimated. Therefore, we opt for the recently proposed Stereo Transformer (STTR) [11] as the depth estimation module as it explicitly identifies these regions. STTR densely compares pixels in l_t and r_t along epipolar lines to find the best matches to estimate depth. STTR uses the attention mechanism [24], which computes the attention (feature similarities α) between source p_{src} and target p_{tgt} and outputs the updated features p_{out} as:

$$p_{out} = \mathrm{softmax}(\frac{\alpha}{\sqrt{C_{attn}}})W_v p_{tgt}, \text{ with } \alpha = p_{src}^T W_q^T W_k p_{tgt}. \tag{1}$$

where W_q, W_k, W_v are the learnt projection weights of dimension $\mathbb{R}^{C_{attn} \times C_{attn}}$ to transform the inputs to an embedding space. After the final attention, the similarities between the pixels from l_t and r_t are used as likelihood for pixel matching. We propose a new variant of STTR, to enable faster computational speed while without harming performance. We note that a large amount of FLOPS of STTR is within the attention module. Since STTR is comparing the pixels along epipolar lines (of W^2 potential matches) for H lines, its number of parameters and FLOPS are in the order of

$$\text{Number of parameters} = \mathcal{O}(C_{attn}^2 N_{attn}), \text{ FLOPS} = \mathcal{O}(HW^2 N_{attn}). \tag{2}$$

where H, W are the height and width of the input image, and N_{attn} is the number of attentions computed. To avoid significant deterioration of the performance, we keep number of parameters by making $C_{attn} N_{attn}$ constant while quartering the number of attentions. Following prior work, we train STTR on the large-scale synthetic Scene Flow dataset [14], which is a commonly used pre-training dataset for stereo networks. We show that transformer-based depth module improves the reconstruction quality in surgical scene quantitatively in Sect. 4.3 and our light-weight STTR is faster without much performance sacrifice. This is also the first time transformer-based module is applied to surgical scene depth estimation.

Efficient Surgical Tool Segmentation. Given depth estimation of the scene, it is desirable to isolate the depth of soft-tissue from surgical tools since surgical tools are considered as "outlier" in soft-tissue reconstruction. For this purpose, we predict a binary mask indicating which pixels belong to the tools. U-Net [17] has been proven to be a light-weight yet accurate model for segmentation tasks [1,6]. In our pipeline, we design a light-weight U-Net with VGG11 [19] as backbone and with 5 scales of down-sample layers to maintain a run-time of around 12ms per frame (input resolution of 640 × 512). We train the model on public dataset of Robotic Instrument Segmentation from the 2017 MICCAI EndoVis Challenge [1] and directly use it to predict binary instrument masks on our robotic surgery datasets. To mitigate the performance variation of the trained

U-Net due to slight domain gap between the training data and the data we evaluate our pipeline on, we perform morphological operation on the segmentation mask to refine tool segmentation boundaries.

3.2 Dynamic Reconstruction

Surfel Representation. Different from most existing methods that adopt a volumetric model [15], we rely on a memory-efficient data representation surfel [3], which is suitable to the varying environment in surgical scenes. Surfels are tuples of variables including 3D position v, surface normal n, confidence c and timestamp of last observation t stored as an unordered list S. In our pipeline, position v is computed using inverse camera intrinsic matrix and estimated masked depth d_t^m. Confidence c is computed using radial distance from the camera center, with the intuition that more oblique points will exhibit larger uncertainty. The canonical surfels S_{ref} is stored in the coordinate defined by the first frame observed, which is then continuously updated by incorporating the surfels S_{obs} of newly observed frames.

Camera Pose. The stereoscope will move during the surgery and the movement would not be continuous nor fast to avoid harming tissue based on consultation with clinical collaborators. As a result, we can assume that the motion between adjacent frames is relatively small, S_{ref} is transformed to the current view given the most recent camera pose and projected to the camera plane. Between S_{ref} and S_{obs}, if the surfels' normals and depth values d are closer than a threshold, a correspondence is found. Given all correspondences P, the camera pose T_{cam} is solved by minimizing the energy function defined as follows:

$$E_{depth}(T_{cam}) = \sum_{(S_{ref}, S_{obs}) \in P} (n_{obs}^T(d_{ref} - d_{obs}))^2. \tag{3}$$

Tissue Deformation Field. As the surgical scene and tissue is non-rigid and dynamic, to efficiently model the deformation, a sparse node graph $\mathcal{W}$ [8] is built with each node represented by its position v similar to surfels, but also a local 6 DoF deformation T. Given this set of sparse nodes, any points can interpolate its deformation using a weighted sum of node deformation based on the distance and radius. For a given query position x and node j in $\mathcal{W}$, the weight can be found as $w_j(x) = \exp(-\frac{||v_j - x||^2}{2r_j^2})$. After applying the updated rigid motion of the camera from previous section, the deformations in node graph $\mathcal{W}$ is solved using following equation with as-rigid-as-possible regularization:

$$E_{deform}(\mathcal{W}) = E_{depth}(\mathcal{W}) + \sum_{j \in \mathcal{W}} \sum_{i \in N_j} ||T_j v_j - T_i v_j||^2. \tag{4}$$

where N_j is the neighboring nodes of node j, with the intuition that motion within a neighborhood should be as small as possible.

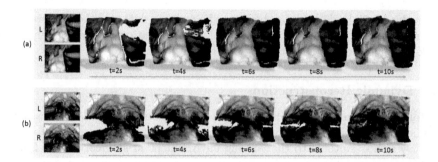

Fig. 2. Qualitative result of the in-house DaVinci dataset. The first column shows the stereo RGB frames when $t = 0$, while the other columns show the result of the reconstructed result along time axis.

Model Fusion. Lastly, to recover and reconstruct the whole surgical scene, the observation will get integrated to the canonical model S_{ref}. For surfels with correspondences, they are fused as one where confidence values get accumulated and the timestamp gets updated. The normal and position are updated as the confidence-weighted sum of observed points and model points. For the fused surfel to be added to the canonical model, the sum of confidence in local neighborhood needs to be higher than a pre-defined threshold and local motion needs to be consistent. At the same time, a surfel in the canonical model will be removed if it is not observed for a long time.

4   Experiments

4.1   Experimental Setting

We evaluate the effectiveness of our proposed dynamic reconstruction framework on two datasets: (1) the public Hamlyn Dataset [27], and (2) our in-house DaVinci robotic surgery dataset of prostatectomy procedure. The Hamlyn Dataset consists of rectified stereo images with resolution of 384×192 collected in partial nephrectomy and without camera calibration information. Our in-house dataset contains 6 cases of high-resolution stereo videos, each records the whole procedure of robotic prostatectomy. We employ the method of Zhang *et al.* [28] to calculate the camera calibration information of our surgical stereoscope.

We collected 5 video clips (with 1200 pairs of rectified stereo frames) from our in-house dataset and 2 video clips (600 pairs) from Hamlyn Dataset. Each clip lasts for around 10 s, including scenarios of surgical tool occlusion, camera movement and tissue deformation. All the clips are used for testing, as all components in our method do not require the clips for training. In the experiment, we downsample the video frames of our in-house dataset from the original resolution 1280 × 1024 to 640 × 512. We implement the tool segmentation network and the transformer-based depth estimation with PyTorch [16]. The dynamic reconstruction is implemented using C++ with CUDA [18] to accelerate the

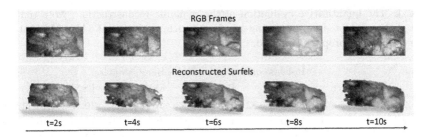

Fig. 3. Qualitative results on the Hamlyn Dataset. The first row shows the RGB frames while the second row shows the reconstructed result along time axis.

running speed. Our experiment is conduced on a PC with one Nvidia TITAN RTX GPU and Intel Xeon(R) W-2123 CPU (3.60 GHz × 8).

4.2   Qualitative Result

To demonstrate the effectiveness of the reconstruction pipeline, we show two examples in Fig. 2 from our in-house dataset. In Fig. 2, the scenes are reconstructed with a fixed camera view, moving surgical tools, and tissue deformations. It can be shown from Fig. 2 that our pipeline can recover the blocked surgical scene, and continuously improve and complete the canonical model S_{ref} given subsequent frames which expose the obstructed tissue.

We show another example in Fig. 3 from the Hamlyn Dataset where the recorded video is more challenging with camera movement, surgical tool movement, tissue deformation and smoke. As shown in Fig. 3, even with camera movement, the dynamic reconstruction algorithm can still track the soft-tissue and complete the surgical scene of the canonical model S_{ref}. Furthermore, the reconstruction pipeline is also robust against smoke.

4.3   Quantitative Evaluation and Analysis

Since acquiring the 3D model of the tissue or ground-truth depth with additional sensors in an *in vivo* environment is currently impractical due to clinical regulations, we use image-based metrics to evaluate the dynamic reconstruction results. Structural Similarity Index Measure (SSIM) [25] has been widely used for computing consistencies between two images, with a higher value indicating a higher similarity. Peak-Signal-to-Noise Ratio (PSNR) [5] is a metric used to estimate the distortion between target image and synthesized (or noisy) image with a higher value indicating a smaller distortion. Both metrics have been used as evaluation metrics to assess the similarities of the predicted (re-projected) image and original image in absence of ground truth depth information [4,9]. Following prior works, we adopt them to evaluate the performance of our dynamic reconstruction results.

Table 1. The quantitative evaluation of the dynamic reconstruction. Note that the reported time is tested on processing one frame with resolution of 640 × 512. Asterisk sign indicates our proposed method.

Method	In-house dataset		Hamlyn dataset		Speed
	$SSIM_a(\%)$	$PSNR_a$	$SSIM_a(\%)$	$PSNR_a$	
HSM [26] + DR w/o mask	58.48 ± 6.82	11.14 ± 1.83	37.98 ± 7.32	10.27 ± 1.70	18 Hz
HSM [26] + DR w/ mask	59.10 ± 6.71	11.76 ± 1.67	38.39 ± 7.10	10.55 ± 1.86	15 Hz
E-DSSR w/o mask	64.17 ± 4.67	13.59 ± 1.72	40.83 ± 8.45	12.01 ± 2.05	36 Hz
DSSR w/o mask	65.09 ± 5.64	13.00 ± 1.61	41.83 ± 7.20	13.04 ± 2.07	18 Hz
E-DSSR* (efficient)	66.65 ± 4.59	**13.68 ± 1.65**	41.97 ± 7.32	12.85 ± 2.03	28 Hz
DSSR* (high-quality)	**66.81 ± 4.90**	13.64 ± 1.81	**42.41 ± 7.12**	**13.09 ± 2.14**	15 Hz

Specifically, we compare the similarities between the observed frame l_t and the re-projected frame $\hat{l}_t$ using the reconstructed canonical surfel. Since our goal is to reconstruct soft-tissue, the mask of the surgical tool is applied to the predicted frame to obtain the masked frames l_t^m and $\hat{l}_t^m$ with only tissue information. We computed the average SSIM and PSNR across all the frames and for all video clips as the final evaluation metrics, which is shown below:

$$\text{SSIM}_a = \frac{1}{T}\sum_{t\in T} SSIM(l_t^m, \hat{l}_t^m); \quad \text{PSNR}_a = \frac{1}{T}\sum_{t\in T} PSNR(l_t^m, \hat{l}_t^m). \quad (5)$$

To demonstrate the effectiveness of our method and assess the contribution of each component in our pipeline, we conduct a set of experiments with our proposed E-DSSR pipeline. We compare E-DSSR with a similar pipeline DSSR with our light-weight STTR replaced by STTR. We further compare with a reconstruction pipeline using a different depth estimation network HSM [26], which is a state-of-the-art fully convolutional stereo depth estimation method. The full list of experiments are shown in Table 1, including: (1) HSM + Dynamic Reconstruction (DR) without tool mask, (2) HSM + Dynamic Reconstruction (DR) with tool masked, (3) DSSR without tool mask, (4) E-DSSR without tool mask, and our proposed methods (5) DSSR and (6) E-DSSR.

We evaluate the advantages of using transformer-based depth module by comparing E-DSSR with HSM + DR w/ mask, we can see that our method outperforms by 7.55% $SSIM_a$ and 1.92 $PSNR_a$ for the in-house dataset and by 3.58% $SSIM_a$ and 2.30 $PSNR_a$ for Hamlyn Dataset.

We also evaluate the contribution of tool segmentation module in the reconstruction result. By comparing E-DSSR and E-DSSR w/o mask, we show that without explicit tool identification, the $SSIM_a$ drops from 66.65% to 64.17% and $PSNR_a$ drops from 13.68 to 13.59 for in-house dataset. As for the Hamlyn Dataset, the $SSIM_a$ drops from 41.97% to 40.83% and $PSNR_a$ drops from 12.85 to 12.01. The same observation can be found by comparing DSSR and DSSR w/o mask, HSM + DR w/ mask and HSM + DR w/o mask. All results demonstrate the benefit of the proposed tool segmentation module.

Finally, while DSSR achieves the best result of $SSIM_a$ on both dataset and $PSNR_a$ on Hamlyn Dataset, the E-DSSR is nearly two times faster than the DSSR (28Hz v.s. 15Hz) and achieves best result of $PSNR_a$ on in-house dataset while with little performance compromise (0.16% $SSIM_a$ on in-house dataset, 0.44% $SSIM_a$ and 0.24 $PSNR_a$ on Hamlyn Dataset).

We also compare our approach with the previous proposed work [10]. Based on their reported result, our light-weight pipeline is 14× faster (28Hz v.s. 2Hz) given the same image resolution. Comparing our image-based tool segmentation module with their kinematics-driven 3D model rendering approach, our module is more than 5 times faster (83Hz v.s. 15Hz).

5   Conclusion

In conclusion, we propose an efficient, image-only reconstruction pipeline for surgical scenes. Our light-weight modules enable real-time reconstruction result in the presence of camera motion, surgical tools, and tissue deformation. We evaluate our pipeline qualitatively and quantitatively to demonstrate its effectiveness. We did not evaluate on longer duration, because the camera views may be completely different within a long video clip, which could be considered as several short clips to be reconstructed. Future works include acquiring larger amounts of data with more significant variations to expand evaluation of our approach, and investigating the effectiveness of our approach in downstream applications such as AI-augmented robotic surgery education.

Acknowledgement. This project was supported by CUHK Shun Hing Institute of Advanced Engineering (project MMT-p5-20), CUHK T Stone Robotics Institute, Hong Kong RGC TRS Project No. T42-409/18-R, and Multi-Scale Medical Robotics Center InnoHK under grant 8312051.

References

1. Allan, M., et al.: 2017 robotic instrument segmentation challenge. arXiv preprint arXiv:1902.06426 (2019)
2. Ferguson, J.M., et al.: Comparing the accuracy of the da Vinci xi and da Vinci si for image guidance and automation. Int. J. Med. Robot. Comput. Assist. Surgery **16**(6), 1–10 (2020)
3. Gao, W., Tedrake, R.: SurfelWarp: efficient non-volumetric single view dynamic reconstruction. arXiv preprint arXiv:1904.13073 (2019)
4. Godard, C., Mac Aodha, O., Brostow, G.J.: Unsupervised monocular depth estimation with left-right consistency. In: Proceedings of the IEEE Conference on Computer Vision and Pattern Recognition, pp. 270–279 (2017)
5. Hore, A., Ziou, D.: Image quality metrics: PSNR vs. SSIM. In: 2010 20th International Conference on Pattern Recognition, pp. 2366–2369. IEEE (2010)
6. Jin, Yueming, Cheng, Keyun, Dou, Qi., Heng, Pheng-Ann.: Incorporating temporal prior from motion flow for instrument segmentation in minimally invasive surgery video. In: Shen, D., et al. (eds.) MICCAI 2019. LNCS, vol. 11768, pp. 440–448. Springer, Cham (2019). https://doi.org/10.1007/978-3-030-32254-0_49

7. Lamarca, J., Parashar, S., Bartoli, A., Montiel, J.: DefSLAM: tracking and mapping of deforming scenes from monocular sequences. IEEE Trans. Robot. **37**, 291–303 (2020)

8. Li, H., Adams, B., Guibas, L.J., Pauly, M.: Robust single-view geometry and motion reconstruction. ACM Trans. Graph. (ToG) **28**(5), 1–10 (2009)

9. Li, L., Li, X., Yang, S., Ding, S., Jolfaei, A., Zheng, X.: Unsupervised learning-based continuous depth and motion estimation with monocular endoscopy for virtual reality minimally invasive surgery. IEEE Trans. Ind. Inform. **17**, 3920–3928 (2020)

10. Li, Y., et al.: SuPer: a surgical perception framework for endoscopic tissue manipulation with surgical robotics. IEEE Robot. Autom. Lett. **5**(2), 2294–2301 (2020)

11. Li, Z., et al.: Revisiting stereo depth estimation from a sequence-to-sequence perspective with transformers. arXiv preprint arXiv:2011.02910 (2020)

12. Liu, X., et al.: Reconstructing sinus anatomy from endoscopic video – towards a radiation-free approach for quantitative longitudinal assessment. In: Martel, A.L., et al. (eds.) MICCAI 2020. LNCS, vol. 12263, pp. 3–13. Springer, Cham (2020). https://doi.org/10.1007/978-3-030-59716-0_1

13. Lu, J., Jayakumari, A., Richter, F., Li, Y., Yip, M.C.: Super deep: a surgical perception framework for robotic tissue manipulation using deep learning for feature extraction. arXiv preprint arXiv:2003.03472 (2020)

14. Mayer, N., et al.: A large dataset to train convolutional networks for disparity, optical flow, and scene flow estimation. In: Proceedings of the IEEE Conference on Computer Vision and Pattern Recognition, pp. 4040–4048 (2016)

15. Newcombe, R.A., Fox, D., Seitz, S.M.: Dynamicfusion: reconstruction and tracking of non-rigid scenes in real-time. In: Proceedings of the IEEE Conference on Computer Vision and Pattern Recognition, pp. 343–352 (2015)

16. Paszke, A., et al.: Pytorch: an imperative style, high-performance deep learning library. arXiv preprint arXiv:1912.01703 (2019)

17. Ronneberger, Olaf, Fischer, Philipp, Brox, Thomas: U-Net: convolutional networks for biomedical image segmentation. In: Navab, Nassir, Hornegger, Joachim, Wells, William M.., Frangi, Alejandro F.. (eds.) MICCAI 2015. LNCS, vol. 9351, pp. 234–241. Springer, Cham (2015). https://doi.org/10.1007/978-3-319-24574-4_28

18. Sanders, J., Kandrot, E.: CUDA by Example: An Introduction to General-Purpose GPU Programming. Addison-Wesley Professional (2010)

19. Simonyan, K., Zisserman, A.: Very deep convolutional networks for large-scale image recognition. arXiv preprint arXiv:1409.1556 (2014)

20. Song, J.: 3D non-rigid SLAM in minimally invasive surgery. Ph.D. thesis (2020)

21. Song, J., Wang, J., Zhao, L., Huang, S., Dissanayake, G.: MIS-SLAM: real-time large-scale dense deformable slam system in minimal invasive surgery based on heterogeneous computing. IEEE Robot. Autom. Lett. **3**(4), 4068–4075 (2018)

22. Stoyanov, D., Mylonas, G.P., Lerotic, M., Chung, A.J., Yang, G.Z.: Intra-operative visualizations: perceptual fidelity and human factors. J. Display Technol. **4**(4), 491–501 (2008)

23. Taylor, Russell H.., Menciassi, Arianna, Fichtinger, Gabor, Fiorini, Paolo, Dario, Paolo: Medical robotics and computer-integrated surgery. In: Siciliano, Bruno, Khatib, Oussama (eds.) Springer Handbook of Robotics, pp. 1657–1684. Springer, Cham (2016). https://doi.org/10.1007/978-3-319-32552-1_63

24. Vaswani, A., et al.: Attention is all you need. arXiv preprint arXiv:1706.03762 (2017)

25. Wang, Z., Bovik, A.C., Sheikh, H.R., Simoncelli, E.P.: Image quality assessment: from error visibility to structural similarity. IEEE Trans. Image Process. **13**(4), 600–612 (2004)

26. Yang, G., Manela, J., Happold, M., Ramanan, D.: Hierarchical deep stereo matching on high-resolution images. In: Proceedings of the IEEE/CVF Conference on Computer Vision and Pattern Recognition, pp. 5515–5524 (2019)
27. Ye, M., Johns, E., Handa, A., Zhang, L., Pratt, P., Yang, G.Z.: Self-supervised siamese learning on stereo image pairs for depth estimation in robotic surgery. arXiv preprint arXiv:1705.08260 (2017)
28. Zhang, Z.: A flexible new technique for camera calibration. IEEE Trans. Pattern Anal. Mach. Intell. **22**(11), 1330–1334 (2000)

CataNet: Predicting Remaining Cataract Surgery Duration

Andrés Marafioti[1(✉)], Michel Hayoz[1], Mathias Gallardo[1],
Pablo Márquez Neila[1], Sebastian Wolf[2], Martin Zinkernagel[2],
and Raphael Sznitman[1]

[1] AIMI, ARTORG Center, University of Bern, Bern, Switzerland
andres.marafioti@artorg.unibe.ch
[2] Department for Ophthalmology, Inselspital, University Hospital,
University of Bern, Bern, Switzerland

Abstract. Cataract surgery is a sight saving surgery that is performed over 10 million times each year around the world. With such a large demand, the ability to organize surgical wards and operating rooms efficiently is critical to delivery this therapy in routine clinical care. In this context, estimating the remaining surgical duration (RSD) during procedures is one way to help streamline patient throughput and workflows. To this end, we propose CataNet, a method for cataract surgeries that predicts in real time the RSD jointly with two influential elements: the surgeon's experience, and the current phase of the surgery. We compare CataNet to state-of-the-art RSD estimation methods, showing that it outperforms them even when phase and experience are not considered. We investigate this improvement and show that a significant contributor is the way we integrate the elapsed time into CataNet's feature extractor.

1 Introduction

Cataract surgery is one of the most common surgeries in the world, with over 10 million procedures conducted each year. Worldwide, 100 million people suffer from cataract-induced vision impairments and with the aging world population growing, the number of patients at risk of complete blindness is sharply increasing [1]. Yet, even though cataracts can easily be treated, the shear number of surgeries needed poses an organizational challenge of unprecedented scale.

At its core, cataract surgery involves using a surgical microscope to help replace a patient's eye lens, that has become opaque, with a synthetic clear lens. Depending on the risk of the patient [2,3] and the experience of the operating surgeon [4,5], the procedure can be performed in under 20 min, whereby the majority of delicate surgical phases last 6–15 min. In major outpatient cataract clinics, a single surgeon can operate over 50 patients in a given day. As such, the

Electronic supplementary material The online version of this chapter (https://doi.org/10.1007/978-3-030-87202-1_41) contains supplementary material, which is available to authorized users.

M. de Bruijne et al. (Eds.): MICCAI 2021, LNCS 12904, pp. 426–435, 2021.
https://doi.org/10.1007/978-3-030-87202-1_41

ability to streamline patients and prepare them for surgery plays an important role in surgical workflow and the organization around the operating room. In this context, the ability to appropriately estimate remaining surgical duration (RSD) is imperative to prepare the stream of upcoming patients and doing so as early as possible is critical.

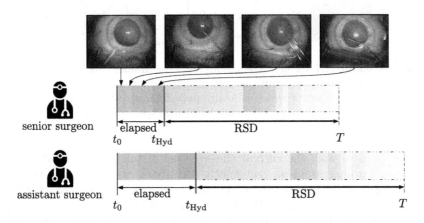

Fig. 1. Surgeon experience and surgical phases play important roles in estimating remaining surgical duration (RSD) in cataract surgery. RSD predictions at t_{Hyd} allows for optimal operating room patient management.

To date, considerable efforts have been put into designing automated methods to predict RSD [6–12]. Namely, [13] presented the TimeLSTM network, which combined a CNN and an RNN to perform RSD prediction. This method, which achieves good results for cholecystectomy surgeries, pre-trained its CNN for phase recognition thus requiring phase annotations. In an attempt to avoid this requirement, [14] introduced RSDNet which only used unlabeled surgical videos to predict the RSD. Relying on the implicit *progress* label of the videos, the authors showed that either the surgical phase or progress labels could be effectively utilized for RSD prediction on laparoscopic surgeries. In contrast to laparoscopic procedures however, no RSD methods have focused on cataract surgery. However, there is related research such as that of Neumuth et al. [15], which proposed a surgical workflow management system potentially applicable to RSD estimation. Similarly, [16,17] detected the current phase in cataract sequences from which RSD could be estimated. Yet these methods overlook important aspects: (1) surgeon experience plays a major factor in cataract surgery duration [4,5] and (2) assessing the risk of the patient by inspecting the initial eye anatomy plays a key role in determining the difficulty and length of the procedure [2,3].

In this work, we thus present a novel approach for online RSD prediction in cataract surgery. Our approach is to explicitly incorporate information from observed surgical phases, the operating surgeon's experience and the elapsed time at any given point to infer RSD prediction. We do this by embedding

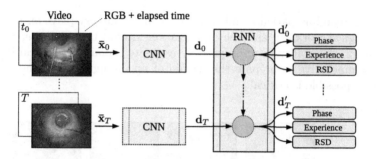

Fig. 2. The end-to-end system. The inputs are video frames concatenated with the elapsed time of the surgery. Inputs are individually fed into the CNN and aggregated by the RNN, the output of which is finally passed through three independent fully connected layers to predict surgical phase, surgeon's experience, and RSD.

the video frames with the current elapsed time of the surgery, establishing a multi-task learning problem, and jointly identifying the surgeon's experience and the surgical phase, whereby overcoming a number of important limitations from recent methods (*i.e.*, RSDNet and TimeLSTM). By doing so, our approach avoids introducing additional complexities and yet considerably outperforms competing methods on both average RSD measures and RSD estimates at early stages of the surgery. In addition, we present an ablation study to identify the components of our method that give rise to the performance reported[1].

2   Approach

2.1   Model

Following [5], we identify three key factors that influence the RSD: the surgeon's experience, the current surgical phase, and the elapsed time of the surgery (Fig. 1). For accurate RSD estimation, it is thus critical that the predictive model is aware of these factors when processing the input video. To that end, we incorporate the factors into the model in a number of ways. The elapsed time, readily available at both training and inference time, is appended as an additional channel to the input video frames. On the other hand, surgeon's experience and surgical phase are unknown at inference time. Instead, we train the model to estimate them from the input data.

Figure 2 depicts our model and how these three predictive factors are incorporated into it. Formally, our model consists of a CNN $f : [0,1]^{3+1} \to \mathcal{D}$ that maps the input tensor $\bar{\mathbf{x}}_t$ to a frame descriptor vector $\mathbf{d}_t \in \mathcal{D}$, followed by a RNN [18] $g : \mathcal{D} \to \mathcal{D}'$ that incorporates temporal information to produce a video descriptor vector $\mathbf{d}'_t \in \mathcal{D}'$. We pass the input tensor $\bar{\mathbf{x}}_t = \left[\mathbf{x}_t, 1\frac{t}{T_{\max}}\right]$, which contains the input frame $\mathbf{x}_t$ at time t and the elapsed time t as an additional channel,

---

[1] Code and instructive examples are available at github.com/aimi-lab/catanet.

to the CNN. The elapsed time is scaled to the range $[0, 1]$ by dividing t by the expected maximum video length $T_{\max}$ that we set to 20 min. Passing the elapsed time at the image level enables the CNN to learn its embedding.

Every video descriptor vector $\mathbf{d}'_t$ produced by the LSTM is finally processed with three independent fully connected layers ($h^{\exp}$, h^{phase}, h^{rsd}) to estimate the surgeon's experience $\hat{\mathbf{y}}_t^{\exp}$, the surgical phase $\hat{\mathbf{y}}_t^{\text{phase}}$, and the RSD $\hat{y}_t^{\text{rsd}}$. A softmax non-linearity is applied to obtain the probabilities $\hat{\mathbf{y}}_t^{\exp}$ and $\hat{\mathbf{y}}_t^{\text{phase}}$.

2.2 Training Objectives

Our training dataset is a collection of tuples $\left(\{\mathbf{x}_t\}_t, \{y_t^{\text{rsd}}\}_t, \{y_t^{\text{phase}}\}_t, y^{\exp} \right)$ consisting of a video sequence $\mathbf{x}_t$, the corresponding remaining surgical duration y_t^{rsd} per frame, surgical phases y_t^{phase} per frame, the surgeon's experience label $y^{\exp}$ per sequence. The index t is the elapsed time of the sequence.

We use the labeled data to train our model by minimizing two different loss functions. First, the CNN loss ℓ_{cnn} is used to train the standalone CNN, without the RNN, to classify the phase and experience of individual frames. To this end, we append two temporary linear layers, akin to h^{phase} and $h^{\exp}$ above, acting on the output of the CNN $\mathbf{d}_t$ to produce frame-level predictions $\hat{\mathbf{y}}_{\text{cnn},t}^{\text{phase}}$ and $\hat{\mathbf{y}}_{\text{cnn},t}^{\exp}$. The CNN loss minimizes the cross-entropies of both predictions,

$$\ell_{\text{cnn}} = \text{H}(\hat{\mathbf{y}}_{\text{cnn},t}^{\text{phase}}, y_t^{\text{phase}}) + \text{H}(\hat{\mathbf{y}}_{\text{cnn},t}^{\exp}, y_t^{\exp}). \tag{1}$$

The RNN loss ℓ_{rnn}, on the other hand, is used with video sequences to train the RNN and to fine-tune the entire model end-to-end. It is a combination of the cross-entropies on phase and experience predictions, and the L1-norm of RSD predictions,

$$\ell_{\text{rnn}} = \alpha \left| \hat{y}_t^{\text{rsd}} - y_t^{\text{rsd}} \right| + \text{H}(\hat{\mathbf{y}}_t^{\text{phase}}, y_t^{\text{phase}}) + \text{H}(\hat{\mathbf{y}}_t^{\exp}, y_t^{\exp}), \tag{2}$$

where the hyperparameter α weights the relative contribution of the L1-norm.

3 Experiments

3.1 Training and Test Data

We used the cataract-101 dataset [5] containing 101 videos (1'263'116 frames) with a resolution of 720×540 pixels acquired at 25 fps. We did not choose a minimum video length, but used every video in the dataset. Each video is annotated with 10 surgical phases and the experience of the operating surgeon. Surgeries were performed by four different surgeons, divided in two senior surgeons (56 surgeries) and two assistant surgeons (45 surgeries). In addition, we manually labelled the start and end of each surgery, respectively, as the start of the first incision and the last tool interaction with the patient's eye.

The dataset was randomly split into 81 training and 20 test videos, so that 5 videos per surgeon remained in the test set. In the following experiments, we perform 6-fold cross-validation on the training split for model selection and hyper-parameter tuning. For inference, the output of all models is averaged.

3.2   Implementation and Baseline Methods

Our CNN uses a DenseNet-169 [19] architecture pre-trained on ImageNet. Input images are reshaped and cropped to 224×224, and the network produces descriptor vectors $\mathbf{d}_t$ of 1664 dimensions. We implement our RNN as a LSTM [18] with two layers of 128 cells, producing 128-dimensional video descriptor vectors $\mathbf{d}'_t$.

Training is performed in four stages: (1) First, to tackle class imbalance in surgical phases, we apply stratified sampling over the whole training dataset and sample 8000 frames per phase. We train using the Adam optimizer with early stopping in all training stages. The CNN is trained to minimize ℓ_{cnn} for 3 epochs with a learning rate of 10^{-4}, batch size of 100 and early stopping on sub-epoch validation loss. (2) We minimize ℓ_{rnn} to train the RNN on full video sequences, temporally downsampled to 2.5 fps, for 50 epochs and a learning rate 10^{-3}. The weights of the CNN are frozen during this stage. (3) The entire model is trained end-to-end minimizing ℓ_{rnn}. We apply truncated back-propagation on sub-sequences of 48 frames and setting the learning rate to $5 \cdot 10^{-4}$ for 10 epochs. (4) Finally, we fine-tune the RNN minimizing ℓ_{rnn} for another 20 epochs while keeping the learning rate at $5 \cdot 10^{-4}$. The weights of the CNN are frozen during this stage. For the ℓ_2 loss, we set $\alpha = 1$. We implemented our method with PyTorch 1.6 and trained models using two Nvidia GeForce GTX 1080 Ti GPUs.

Given that no method for cataract RSD estimation exists, we compare our approach to two methods originally designed for laparoscopic surgery:

TimeLSTM [13]: A ResNet CNN trained for phase recognition, followed by a LSTM trained for RSD prediction.

RSDNet [14]: A modified version of [13], where the CNN is trained for progress prediction and the elapsed time is concatenated to the LSTM's output.

Both methods were originally proposed for cholecystectomy surgeries and did not provide implementations. Therefore, we use our own implementations for both baselines, following the respective publications.

We measure the quality of RSD predictions with the mean absolute error (MAE) per video, $\text{MAE} = \frac{1}{T} \sum_{t=0}^{T-1} \left| \hat{y}_t^{rsd} - y_t^{rsd} \right|$. Similarly, we also provide MAE averaged over the last two (MAE-2) and five (MAE-5) minutes, as well as at the end of *Hydrodissection* phase (MAE@Hyd). The latter metric is of clinical relevance in cataract surgery, as it highlights an appropriate time to prepare the following patient for surgery. In addition, we compute frame-wise accuracy (ACC) and F1-score per video to quantify surgical phase classification.

3.3   Results

Table 1 shows the RSD prediction performance for all methods grouped by surgeon experience level. We group results by surgeon's experience level as both MAE and MAE@Hyd indirectly depend on the duration of the surgery and these take 5.6 and 11.8 min on average for senior and assistant surgeons, respectively. CataNet outperforms both RSDNet and TimeLSTM in all but one metric. At

Table 1. RSD prediction results. The MAE (mean±std in minutes) is shown for entire videos, the last two and five minutes, and at the end of *Hydrodissection*.

	Exp	CataNet	RSDNet	TimeLSTM
MAE@Hyd	All	**1.66 ± 1.35**	2.32 ± 1.27	2.34 ± 1.54
	Senior	**1.22 ± 0.97**	2.86 ± 1.31	3.30 ± 1.06
	Assistant	2.10 ± 1.56	1.78 ± 1.02	**1.39 ± 1.37**
MAE-5	All	**0.64 ± 0.56**	1.37 ± 0.83	1.47 ± 0.78
	Senior	**0.78 ± 0.60**	1.98 ± 0.73	2.06 ± 0.70
	Assistant	**0.51 ± 0.23**	0.76 ± 0.28	0.88 ± 0.14
MAE-2	All	**0.35 ± 0.20**	1.23 ± 0.53	1.22 ± 0.32
	Senior	**0.37 ± 0.22**	1.42 ± 0.45	1.43 ± 0.32
	Assistant	**0.34 ± 0.18**	1.04 ± 0.56	1.03 ± 0.13
MAE	All	**0.99 ± 0.65**	1.59 ± 0.69	1.66 ± 0.79
	Senior	**0.83 ± 0.64**	1.97 ± 0.73	2.11 ± 0.70
	Assistant	**1.15 ± 0.65**	1.19 ± 0.36	1.20 ± 0.59

the end of the critical *Hydrodissection* phase, for all experiences CataNet performs 0.66 min better than RSDNet and 0.68 min better than TimeLSTM. At the end of this phase, CataNet is considerably better than the baselines for senior surgeons, but worse for assistant surgeons. Considering the prediction over the whole video, CataNet performs on average 0.6 min better than RSD-Net and 0.67 min better than TimeLSTM. This can be explained by the fact that CataNet achieves comparable results for both senior and assistant surgeons. Overall, detection of the surgeon's experience is achieved with 0.92 ± 0.16 accuracy and can thus exploit the fact that senior surgeons show low variance in surgery duration, however we do not claim that this accuracy would translate to new surgeons. The competing methods, on the other hand, tend to overestimate the duration of surgeries performed by senior surgeons.

We visualize CataNet's results for individual videos shown in Fig. 3 (see Supplementary material for more examples). Here, we see that predicting the surgeon experience on every frame can be beneficial in determining the confidence in RSD predictions. That is, given that the experience of the surgeon is known by the operating staff, an incorrect classification in experience can serve as an easy and interpretable indicator when the system is performing poorly (*i.e.* overestimating the RSD for false *assistant* predictions, or underestimating it for false *senior* predictions). Additionally, considering that *experience* is not a binary label, but a multi factored and scaled concept, our approach could be used to help assistant surgeons detect which phases of the surgery they could improve on. Finally, in two test set sequences, the surgeons fails to correctly perform the *lens implantation* phase, leading to unexpected extensions of the surgeries by 2–3 minutes and consequently underestimate of RSD before the mistake. However, our approach corrects the RSD predictions shortly thereafter. Details of these two sequences can be found in the supplementary material.

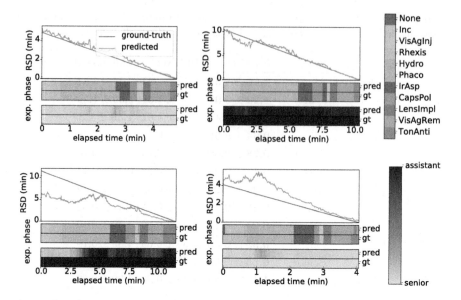

Fig. 3. Four examples of our method's outputs. For each plot, we show (**top**) the concordance between ground-truth and predicted RSD, (**middle**) the ground-truth and predicted surgical phases over time, and (**bottom**) the ground-truth and predicted probability of the surgeon's experience level.

Ablation Study: CataNet is trained to classify the experience of the surgeon, the surgical phase, and the RSD, while its input is the video frames concatenated with the elapsed surgical time. To characterize the effects on performance of these different components, we show the performance of the following different approaches in Table 2: (i) train the CNN to only predict surgical phases and the RNN to predict both phases and RSD; (ii) train the CNN to only predict the surgeons experience and the RNN to predict both experience and RSD; (iii) train the CNN and the RNN to estimate only the RSD; (iv) same as (iii) but concatenate the elapsed time to the output of the LSTM (*i.e.*, as in RSDNet) instead of to the video frames.

From these experiments, we can see that (i) generally performs as well as CataNet, even outperforming it for some metrics. However, CataNet generally achieves a better performance for senior surgeons, who conduct the bulk of actual cataract surgeries [20]. In addition, we notice that (iii) performs better than (iv), showing that using the elapsed time as an input for the model considerably outperforms having it after the LSTM layer. Last, even when training without any labels, our approach (iii) performs better than that of RSDNet.

Results on Surgical Phase Classification: Table 3 shows CataNet's performance for phase classification. Compared to the state-of-the-art by Qui et al. [21], CataNet achieves an increase of 12% in accuracy from 0.84 to 0.95. Furthermore, CataNet reliably detects the *Hydrodissection* phase, which is critical

Table 2. Ablation evaluation for RSD prediction.

	Exp	CataNet	(i) phase	(ii) exp	(iii) RSD	(iv) elapsed
MAE@Hyd	All	1.66 ± 1.35	$\mathbf{1.43 \pm 1.19}$	1.82 ± 1.63	1.99 ± 1.38	2.28 ± 1.34
	Senior	$\mathbf{1.22 \pm 0.97}$	1.42 ± 1.38	1.46 ± 1.3	1.71 ± 1.24	1.45 ± 0.59
	Assistant	2.10 ± 1.56	$\mathbf{1.43 \pm 1.05}$	2.18 ± 1.91	2.26 ± 1.52	3.12 ± 1.37
MAE-5	All	$\mathbf{0.64 \pm 0.46}$	0.74 ± 0.56	0.87 ± 0.62	0.76 ± 0.41	0.75 ± 0.34
	Senior	$\mathbf{0.78 \pm 0.60}$	0.88 ± 0.73	0.99 ± 0.77	0.98 ± 0.44	0.85 ± 0.31
	Assistant	$\mathbf{0.51 \pm 0.23}$	0.59 ± 0.27	0.76 ± 0.43	0.55 ± 0.24	0.64 ± 0.34
MAE-2	All	0.35 ± 0.20	$\mathbf{0.35 \pm 0.23}$	0.51 ± 0.27	0.39 ± 0.28	0.44 ± 0.20
	Senior	0.37 ± 0.22	$\mathbf{0.35 \pm 0.26}$	0.52 ± 0.36	0.45 ± 0.38	0.51 ± 0.22
	Assistant	0.34 ± 0.18	0.36 ± 0.21	0.50 ± 0.18	$\mathbf{0.33 \pm 0.10}$	0.36 ± 0.15
MAE	All	0.99 ± 0.65	$\mathbf{0.98 \pm 0.58}$	1.22 ± 0.92	1.11 ± 0.62	1.34 ± 0.73
	Senior	$\mathbf{0.83 \pm 0.64}$	0.91 ± 0.77	1.03 ± 0.80	1.03 ± 0.46	$\mathbf{0.83 \pm 0.30}$
	Assistant	1.15 ± 0.65	$\mathbf{1.04 \pm 0.31}$	1.41 ± 1.03	1.20 ± 0.76	1.85 ± 0.67

in the clinical context. Indeed, knowing the RSD at the end of this phase will improve the OR management since it corresponds to the moment where the next patient could be prepared for surgery.

Inference Speed: RSD estimation is intended to be performed on real-time. We measured the execution time using a GeForce MX250 and avoided any overhead produced by other components of the system. We first run 100 frames through the GPU after which we measured the inference time on the next 1000 frames. The average time per frame was 34.3 ± 1.9 ms, which corresponds to 29.09 fps. Considering that we sample the videos at 2.5 fps, we conclude that CataNet can easily be applied at 10 times real-time speed.

Table 3. Macro F1-score and micro accuracy averaged over the 6-fold models.

	CataNet	Qui et al. [21]	TimeLSTM-CNN [13]
F1	$\mathbf{0.93 \pm 0.06}$	–	0.80 ± 0.07
F1-Hyd	$\mathbf{0.94 \pm 0.08}$	–	0.84 ± 0.17
ACC	$\mathbf{0.95 \pm 0.05}$	0.84 ± 0.06	0.84 ± 0.07

4   Conclusion

We have proposed a novel real-time method for estimating RSD for cataract surgeries from video feeds. Our approach jointly predicts the RSD, the surgeon's experience and the surgical phase, as these three elements are interconnected. Even when training our method without any labels, it outperforms the previous state-of-the-art RSD estimation models. We investigated the sources of this improvement and attribute these to (1) concatenating the video frames with the elapsed time and (2) including the phase and experience labels. Predicting the experience on every frame additionally increases the clinical applicability of our method by identifying low method confidence by observing predicted and real experience levels. Moving forward, a major challenge is in establishing large datasets to evaluate generalization capabilities and major clinical impact [22], for which assuring data consistency will be critical [23]. In the future, we plan to investigate this and how pre-operative data can be used to further improve RSD predictions for cataract surgery.

Acknowlegements. This work was partially supported by the Haag-Streit Foundation and the University of Bern.

References

1. Wang, W., Yan, W., Müller, A., He, M.: A global view on output and outcomes of cataract surgery with national indices of socioeconomic development. Invest. Ophthalmol. Vis. Sci. **58**, 3669–3676 (2017)
2. Achiron, A., Haddad, F., Gerra, M., Bartov, E., Burgansky-Eliash, Z.: Predicting cataract surgery time based on preoperative risk assessment. Eur. J. Ophthalmol. **26**(3), 226–229 (2016)
3. Lanza, M.: Application of artificial intelligence in the analysis of features affecting cataract surgery complications in a teaching hospital. Front. Med. **7**, 607870 (2020)
4. Devi, S.P., Rao, K.S., Sangeetha, S.S.: Prediction of surgery times and scheduling of operation theaters in ophthalmology department. J. Med. Syst. **36**(2), 415–430 (2012)
5. Schoeffmann, K., Taschwer, M., Sarny, S., Münzer, B., Primus, M.J., Putzgruber, D.: Cataract-101 - Video dataset of 101 cataract surgeries. In: Proceedings of the 9th ACM Multimedia Systems Conference, MMSys 2018, pp. 421–425 (2018)
6. Padoy, N., Blum, T., Feussner, H., Berger, M.O., Navab, N.: On-line recognition of surgical activity for monitoring in the operating room. In: Proceedings of the National Conference on Artificial Intelligence, vol. 3 (2008)
7. Franke, S., Meixensberger, J., Neumuth, T.: Intervention time prediction from surgical low-level tasks. J. Biomed. Inform. **46**(1), 152–159 (2013)
8. Guédon, A.C., et al.: 'It is time to prepare the next patient' real-time prediction of procedure duration in laparoscopic cholecystectomies. J. Med. Syst. **40**(12), 271 (2016)
9. Spangenberg, N., Wilke, M., Franczyk, B.: A big data architecture for intra-surgical remaining time predictions. Procedia Comput. Sci. **113**, 310–317 (2017)
10. Maktabi, M., Neumuth, T.: Online time and resource management based on surgical workflow time series analysis. Int. J. Comput. Assist. Radiol. Surg. **12**(2), 325–338 (2017)
11. Bodenstedt, S., et al.: Prediction of laparoscopic procedure duration using unlabeled, multimodal sensor data. Int. J. Comput. Assist. Radiol. Surg. **14**(6), 1089–1095 (2019)

12. Rivoir, D., Bodenstedt, S., von Bechtolsheim, F., Distler, M., Weitz, J., Speidel, S.: Unsupervised temporal video segmentation as an auxiliary task for predicting the remaining surgery duration. In: Zhou, L., et al. (eds.) OR 2.0/MLCN -2019. LNCS, vol. 11796, pp. 29–37. Springer, Cham (2019). https://doi.org/10.1007/978-3-030-32695-1_4

13. Aksamentov, I., Twinanda, A.P., Mutter, D., Marescaux, J., Padoy, N.: Deep neural networks predict remaining surgery duration from cholecystectomy videos. In: Descoteaux, M., Maier-Hein, L., Franz, A., Jannin, P., Collins, D.L., Duchesne, S. (eds.) MICCAI 2017. LNCS, vol. 10434, pp. 586–593. Springer, Cham (2017). https://doi.org/10.1007/978-3-319-66185-8_66

14. Twinanda, A.P., Yengera, G., Mutter, D., Marescaux, J., Padoy, N.: RSDNet: learning to predict remaining surgery duration from laparoscopic videos without manual annotations. IEEE Trans. Med. Imag. **38**(4), 1069–1078 (2019)

15. Neumuth, T., Liebmann, P., Wiedemann, P., Meixensberger, J.: Surgical workflow management schemata for cataract procedures process model-based design and validation of workflow schemata. Meth. Inform. Med. **51**(5), 371–382 (2012)

16. Zisimopoulos, O., et al.: DeepPhase: surgical phase recognition in CATARACTS videos. In: Frangi, A.F., Schnabel, J.A., Davatzikos, C., Alberola-López, C., Fichtinger, G. (eds.) MICCAI 2018. LNCS, vol. 11073, pp. 265–272. Springer, Cham (2018). https://doi.org/10.1007/978-3-030-00937-3_31

17. Primus, M.J., et al.: Frame-based classification of operation phases in cataract surgery videos. In: Schoeffmann, K., et al. (eds.) MMM 2018. LNCS, vol. 10704, pp. 241–253. Springer, Cham (2018). https://doi.org/10.1007/978-3-319-73603-7_20

18. Hochreiter, S., Schmidhuber, J.: Long short-term memory. Neural Comput. **9**(8), 1735–1780 (1997)

19. Huang, G., Liu, Z., Van Der Maaten, L., Weinberger, K.Q.: Densely connected convolutional networks. In: Proceedings of the IEEE Conference on Computer Vision and Pattern Recognition, pp. 4700–4708 (2017)

20. Campbell, R.J., et al.: Association of cataract surgical outcomes with late surgeon career stages: a population-based cohort study. JAMA Ophthalmol. **137**(1), 58–64 (2019)

21. Qi, B., Qin, X., Liu, J., Xu, Y., Chen, Y.: A deep architecture for surgical workflow recognition with edge information. In: Proceedings - 2019 IEEE International Conference on Bioinformatics and Biomedicine, BIBM 2019, pp. 1358–1364 (2019)

22. Bar, O., et al.: Impact of data on generalization of AI for surgical intelligence applications. Sci. Rep. **10**(1), 1–12 (2020)

23. Ghamsarian, N., Taschwer, M., Schoeffmann, K.: Deblurring cataract surgery videos using a multi-scale deconvolutional neural network. In: 2020 IEEE 17th International Symposium on Biomedical Imaging (ISBI), pp. 872–876 (2020)

Task Fingerprinting for Meta Learning in Biomedical Image Analysis

Patrick Godau[1,2](✉) and Lena Maier-Hein[1,2]

[1] German Cancer Research Center, Heidelberg, Germany
patrick.scholz@dkfz-heidelberg.de
[2] Ruprecht-Karls University of Heidelberg, Heidelberg, Germany

Abstract. Shortage of annotated data is one of the greatest bottlenecks in biomedical image analysis. Meta learning studies how learning systems can increase in efficiency through experience and could thus evolve as an important concept to overcome data sparsity. However, the core capability of meta learning-based approaches is the identification of similar previous tasks given a new task - a challenge largely unexplored in the biomedical imaging domain. In this paper, we address the problem of quantifying task similarity with a concept that we refer to as *task fingerprinting*. The concept involves converting a given task, represented by imaging data and corresponding labels, to a fixed-length vector representation. In fingerprint space, different tasks can be directly compared irrespective of their data set sizes, types of labels or specific resolutions. An initial feasibility study in the field of surgical data science (SDS) with 26 classification tasks from various medical and non-medical domains suggests that task fingerprinting could be leveraged for both (1) selecting appropriate data sets for pre-training and (2) selecting appropriate architectures for a new task. Task fingerprinting could thus become an important tool for meta learning in SDS and other fields of biomedical image analysis.

Keywords: Meta learning · Knowledge transfer · Medical image analysis · Surgical data science · Task similarity

1 Introduction

Shortage of annotated data is one of the greatest bottlenecks related to biomedical image analysis in general, and surgical data science (SDS) in particular [30]. Methods proposed to address this issue include transfer learning [9,35], crowdsourcing [29] and self-supervised learning [36]. More recently, first attempts to leverage the concept of meta learning have been made [44,45]. Meta learning studies how learning systems can increase in efficiency through experience [41], where experience can be represented by solutions to tasks connected to previously acquired data, for

Electronic supplementary material The online version of this chapter (https://doi.org/10.1007/978-3-030-87202-1_42) contains supplementary material, which is available to authorized users.

M. de Bruijne et al. (Eds.): MICCAI 2021, LNCS 12904, pp. 436–446, 2021.
https://doi.org/10.1007/978-3-030-87202-1_42

example. A core capability of meta learning-based approaches is the identification of similar previous tasks given a new task [20]. Previous work has pioneered the idea of quantifying similarity between tasks in the domain of biomedical image analysis [8]. However, their proposed distance between any two given tasks depends on the set of tasks currently available. In consequence, re-computation of all distances is required every time a new task is added, rendering the approach unscalable. Further, validation was performed for a relatively small set of tasks.

We address this gap in the literature from both a methodological and a validation perspective. As illustrated in Fig. 1, we present the concept of biomedical *task fingerprinting*, which involves representing a task (comprising images and labels), by a vector of fixed length irrespective of data set size, types of labels or specific resolutions. For the task embedding, we investigate several complementary approaches, which can be assigned to one of the two following categories: (1) those leveraging images *and labels* for the embedding and (2) those directly comparing the distributions of the images with sample-based and optimal transport-based methods. Although all following methods could be transferred to a wide range of task types, we restrict ourselves to visual classification tasks as the primary scope. Visual classification tasks are relevant for a multitude of medical applications, such as the classification of colonoscopic or dermoscopic data for cancer screening or the classification of medical instruments for action recognition during surgery.

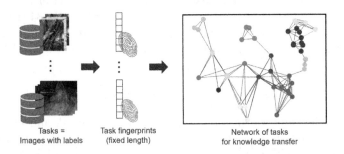

| Tasks = | Task fingerprints | Network of tasks |
| Images with labels | (fixed length) | for knowledge transfer |

Fig. 1. Concept of biomedical *task fingerprinting*. A given task, represented by imaging data and labels, is converted to a fixed-length vector representation such that similarity of fingerprints indicates potential knowledge transfer.

Our validation was performed on a total of 26 tasks from various domains with a focus on endoscopic vision (laparoscopy, colonoscopy, laryngoscopy). Specifically, we investigated the following hypotheses in the context of medical image analysis:

H1: Task fingerprinting can capture semantic relationships between tasks.
H2: Task fingerprinting can be used to select appropriate pretraining tasks for a given new task.
H3: Task fingerprinting can be used for model selection for a given new task.

2   Methods

This section presents the proposed concept of task fingerprinting along with the experimental conditions.

2.1   Task Fingerprinting

We represent a **task** by $\mathfrak{T} = \{(x_i, y_i)\}_{i=1}^{N}$ comprising N samples, each consisting of an image x_i and a label y_i, as proposed by Achille et al. [2]. We assume that the images corresponding to different tasks are of the same dimension after a homogenization process, detailed in Sect. 2.2. Our aim is to find a corresponding embedding $\mathbf{e}_\mathfrak{T}$ together with a computable distance function d, such that for any two tasks $\mathfrak{T}, \mathfrak{T}'$: $d(\mathfrak{T}, \mathfrak{T}') := d(\mathbf{e}_\mathfrak{T}, \mathbf{e}_{\mathfrak{T}'}) \geq 0$ and $d(\mathfrak{T}, \mathfrak{T}')$ is an indicator of *how related* the two tasks are (defined in Sect. 2.3). For defining $\mathbf{e}_\mathfrak{T}$ and d we investigate a total of four complementary approaches: an embedding-based approach based on the Fisher Information Matrix (FIM) that leverages both images and labels and three approaches based on comparing only the image distributions:

Fisher Embedding Distance (FED). Inspired by Achille et al. [1], we generate a task embedding that encodes both, data and labels, in one joint feature vector of fixed length by making use of the FIM. More precisely, for each task $\mathfrak{T}$, we use a pretrained probe-network (here: a standard ResNet34 [19], that has been trained on ImageNet [13]) and retrain the final layer for the current task. For this tuned model, we then calculate the diagonal elements of the FIM, which is defined as:

$$F := \mathbb{E}_{x,y \sim \hat{p}(x), p_w(y|x)} [\nabla_w \log p_w(y|x) \nabla_w \log p_w(y|x)^T] \tag{1}$$

with $\hat{p}$ being the empirical distribution of images in the task and p_w being the prediction distribution of the trained model with respect to the weights w. The resulting vector $\mathbf{e}$, with $\mathbf{e}_i := F_{i,i}$, represents the embedding of $\mathfrak{T}$. Note that we simplify the FIM, specifically, we partly summarize parameters of a convolutional kernel by taking the mean value and ignore bias parameters as well as some of the early network layers. Furthermore, we disregard the parameters of the final layer, as their shapes may differ from task to task. In contrast to Achille et al. [1], we calculate the FIM directly and chose a different training setup based on the data augmentation and sampling strategy detailed in Sect. 2.2. Generally speaking, the FIM is a measure for how much information the predictions of a model carry about its parameters: If a parameter of the probe-network is highly decisive for the predictions made by it, then the corresponding entry in $\mathbf{e}$ will be large. To compare any two tasks $\mathfrak{T}, \mathfrak{T}'$, we compute the Fisher embedding distance $d_{\text{FED}}(\mathfrak{T}, \mathfrak{T}')$ as the cosine distance of the embeddings. This approach is the only one incorporating the labels into the distance function.

Sample-Based Embeddings. For the second kind of embedding we proceed as follows: For a given task $\mathfrak{T}$, we use a fixed pretrained model (the same configuration as in the previous method) to extract deep feature vectors $\mathbf{v} \in \mathbb{R}^n$ (n = 512 in our case) from sampled images of that task, where $\mathbf{v}$ represents the activations prior to the classification layer. To ensure fair comparison, we use exactly as many samples m as during the classifier training for the FIM computation ($m = 10,000$). The sample-feature matrix M may then be further processed to generate the embeddings.

Maximum Mean Discrepancy (MMD) is the largest difference in expectations over functions in the unit ball of a reproducing kernel Hilbert space (RKHS) [16]. It is calculated on a sample base, hence using the full matrix M, which serves itself as embedding in this case. Given two sample-based distributions p, p' on $\mathbb{R}^n$, the MMD is an estimator for the maximal discrepancy between expectation operators on them (which is zero if $p = p'$) and can be computed as

$$d_{\mathrm{MMD}}(\mathfrak{T}, \mathfrak{T}') = \mathbb{E}_{x_1, x_2 \sim p} k(x_1, x_2) - 2\mathbb{E}_{x \sim p, y \sim p'} k(x, y) + \mathbb{E}_{y_1, y_2 \sim p'} k(y_1, y_2), \quad (2)$$

where for $k(\cdot, \cdot)$ we choose the so called Cauchy kernel $k(x, y) = (1 + \|x - y\|^2 \sigma^{-2})^{-1}$ with hyperparameter σ.

Kullback-Leibler Divergence (KLD) is a concept closely connected to Fisher Information [12]. To leverage the KL-divergence for sample-based comparison, we base our work on an approach by Bhattacharjee et al. [5]: Computing the arithmetic mean across all samples on M and normalizing the outcome generates a discrete probability distribution p over the finite space $\{1, ..., n\}$, which serves as a low dimensional embedding. Applying the entropy-based KL-divergence upon two of these distributions p, p' defines the KL-distance

$$d_{\mathrm{KLD}}(\mathfrak{T}, \mathfrak{T}') := \mathrm{KL}(p, p') = \sum_{1 \le j \le n} p(j) \log \frac{p(j)}{p'(j)} \quad (3)$$

of their corresponding tasks $\mathfrak{T}$ and $\mathfrak{T}'$. The KL-distance is the only asymmetric distance function we investigate.

Earth Mover's Distance (EMD) is another approach for measuring distances between distributions and originates from the theory of optimal transport [38]. It also uses the full matrix M as embedding. The rationale behind this approach is to quantify the minimal energy necessary to move the mass of one distribution to form another. More formally, given two distributions p and p' on $\mathbb{R}$, the first Wasserstein distance is defined as

$$W(p, p') := \inf_{\gamma \in \Gamma(p, p')} \mathbb{E}_{(x, y) \sim \gamma} |x - y|, \quad (4)$$

where $\Gamma(p, p')$ represents the set of distributions whose marginals are p and p' on the first and second factors respectively. Since the high-dimensional computation of W is complex, we restrict ourselves to the 1D case and average these distances across all feature dimensions of M.[1] More precisely, if p_j, p'_j represent

---

[1] We use the SciPy [42] implementation of the Wasserstein distance.

the distributions of j-th feature dimension ($1 \leq j \leq n$) in the samples from tasks $\mathfrak{T}, \mathfrak{T}'$, then

$$d_{\text{EMD}}(\mathfrak{T}, \mathfrak{T}') := \frac{1}{n} \sum_{1 \leq j \leq n} W(p_j, p_j'). \tag{5}$$

According to our experience, the most important hyperparameters are the optimizer (e.g. learning rate) and selection of FIM elements to form the embedding for FED and kernel choice as well as kernel parameter σ for MMD. KLD and EMD are solely dependent on the sampling strategy and feature extractor, which also holds true for FED and MMD. While we did not perform a systematic grid search to optimize all of them in parallel, we individually refined them in the course of method development using tasks not in the test set of tasks applied in this work.

2.2   Data

We assessed the performance of our method in the specific context of 2D medical image classification with a focus on SDS data. As we assume that medical image processing tasks can potentially benefit from data of other domains [7], we included both medical and non-medical data in this study, as summarized in Tab. 2 within the Suppl. Materials. Overall, we used a total of 26 publicly available tasks (including 18 from biomedical imaging) with highly varying sample sizes (ranging from 434 to 184,498) and number of classes (ranging from 2 to 256) [3,4,6,10,11,14,15,17,18,22,23,25–28,31–34,37,39,40]. To obtain semantic meta labels for the tasks, we assigned them to their respective *domains* representing the application context. The medical tasks were further assigned an additional *task type* meta label based on the target structure that the image processing task focuses on (e.g. medical instrument, pathology, artefact). The labels are also given in Tab. 2. All data sets were split into training and validation set, either according to their standard splits, or (if not available) according to a random 80:20 split per class. We used the same data preprocessing (resizing to 224×224 pixels), data augmentation (random rotation and flipping, except for the digit classification tasks) and sampling strategies (class-balanced sampling of 10,000 images per epoch) across all tasks.

To avoid the danger of 'model selection' based on test data, we left approx. 45% (n = 8) of the medical imaging tasks untouched until we fully completed the processing pipeline, including the fixation of all hyperparameters. The quantitative validation was then performed exclusively on this untouched test set of tasks.

2.3   Experimental Design

All of the following experiments were performed five times and averaged results are reported.

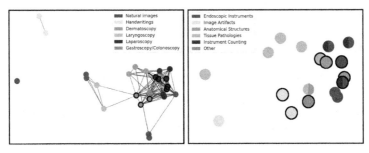

(a) Colour coding of domain   (b) Colour coding of task type

Fig. 2. 2D visualizations of task fingerprints computed with the FED method. Tasks (dots) are coloured by domain ((a); all 26 tasks) and task type ((b); only medical tasks). Encircled dots represent tasks of the untouched 'test set' of tasks. Distances below a threshold of 65% mean distance are represented by edges. Dimension reduction was achieved with the Kamada-Kawai algorithm [21].

H1 Semantic Relationships. To investigate whether task fingerprinting captures semantic relationships between tasks, we converted all 26 tasks into the *fingerprint space* and analyzed whether our method results in a grouping of tasks corresponding to (1) similar domains (e.g. laparoscopic images *vs.* dermatological images *vs.* natural images) and (2) similar tasks (e.g. artefact detection *vs.* medical instrument classification). Specifically, we paired all test tasks with every other task and computed their distance with the four approaches described in Sect. 2. We then determined intra- and inter-domain similarity based on the provided meta labels.

H2 Benefit for Pretraining. The purpose of the second experiment was to assess whether the quantification of task similarity can be leveraged to select appropriate pretraining tasks. To this end, we proceeded as follows for each of the 8 test tasks: We sampled a (class-balanced) random subset of size 300 (train: 240, validation: 60) from the training split to simulate a data scarcity scenario. Given any of the other data sets as source task, we used a pretrained ResNet50 [19] model and tuned the model for 15 epochs (each 10,000 samples) on the source task. After replacing the final classification layer, we tuned for 15 more epochs on the target task. Based on the best validation performance we evaluated the resulting model on the unseen original validation split of the target task. As a baseline, we also omitted the first step and directly tuned on the target task. Based on these results we defined the *relative transfer learning (TL) success* as the tuned cross-entropy loss divided by the baseline cross-entropy loss. This makes the results for different target tasks of varying complexity and number of classes comparable.

H3 Benefit for Architecture Selection. The purpose of the third experiment was to assess whether task fingerprints can be leveraged to select an appropriate

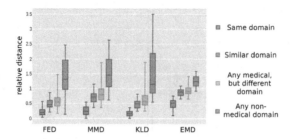

Fig. 3. Task fingerprinting captures semantic similarities. The distance between tasks in fingerprint space increases with the semantic dissimilarity of domains for all four methods.

model for a new given task. To this end, we applied the following strategy. Initially, we trained a set of $n = 11$ different neural network architectures (inclusion criteria: ImageNet performance, parameter count, variety of models and implementation availability)[2] for all of our 18 medical imaging tasks. Training was done for 50 epochs (each 10,000 samples again) with Adam [24]. For each task, we then ranked the different models according to their performance, resulting in an n-dimensional *performance trace* for each task. To quantify the quality of model transfer for a given pair of two tasks, we computed the Spearman's rank correlation coefficient on the performance traces and compared the result to the distance between the tasks in fingerprint space.

3   Results

According to our results, both the label-based method (FED) and all three sample-based methods are well-suited for capturing semantic relationships between tasks, thus confirming our hypothesis H1. Figure 2 shows a 2D visualization of the task fingerprint space for the FED method, demonstrating a meaningful structuring with respect to both domain and target structure. For example, the dermatoscopic, laparoscopic and gastroscopic tasks are clustered and related task pairs, such as (CIFAR-10, CIFAR-100), (MNIST, EMNIST) and (Caltech101, Caltech256) are close to each other in Fig. 2a. Similarly, an analysis of the medical tasks reveals that similar tasks according to algorithm targets (e.g. pathology *vs.* artefact) cluster as well (Fig. 2b). The rather qualitative results were confirmed by our quantitative assessment which shows that the distance between tasks in fingerprint space increases with the semantic dissimilarity of domains for all four methods (Fig. 3). Similarly, tasks with similar targets feature

---

[2] We used the implementations from [43] and the models CSPNet (cspdarknet53, cspresnext50), ECA-Net (ecaresnet50d), EfficientNet (tf_efficientnet_b2_ns), DPN (dpn68b), MixNet (mixnet_xl), RegNetY (regnety_032), ResNeXt (swsl_resnext50_32x4d), ReXNet (rexnet_200), VovNet2 (ese_vovnet39b) and Xception (xception). Please refer to the documentation of [43] for full references of these.

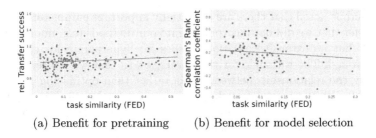

(a) Benefit for pretraining   (b) Benefit for model selection

Fig. 4. Task fingerprinting benefits knowledge transfer for both pretraining and model selection. **(a)** Relative transfer learning success and **(b)** Spearman's rank correlation coefficient (see Sect. 3) both plotted against FED distance. Trendline has been created using ordinary least squares regression.

a comparatively low task distance with the FED and KLD methods, as shown in Tab. 1 in the Suppl. Materials.

A further important result of our study is that task fingerprinting with the FED method can be used for selecting appropriate tasks for pretraining or model selection, thus confirming hypotheses H2 and H3 (Fig. 4). To compare the four methods with respect to these aspects, we computed the Spearman's Rank correlation coefficients ρ and performed significance tests for all of them. The three sample-based methods perform similarly for the pretraining task (H2) with $\rho_{MMD} = 0.04$ ($p = 0.53$), $\rho_{KLD} = 0.06$ ($p = 0.36$) and $\rho_{EMD} = 0.07$ ($p = 0.27$) (see Suppl. Materials Fig. 5) as well as for the performance rankings of neural network architectures (H3) with $\rho_{MMD} = -0.06$ ($p = 0.45$), $\rho_{KLD} = -0.07$ ($p = 0.36$) and $\rho_{EMD} = -0.06$ ($p = 0.42$) (see Suppl. Materials Fig. 6). In contrast, the FED method outperforms the other methods in both experiments with $\rho_{FED} = 0.12$ ($p = 0.07$) for pretraining and $\rho_{FED} = -0.16$ ($p = 0.05$) for model selection.

4   Discussion

To our knowledge, this paper is the first to address the challenge of task similarity quantification in the field of SDS. According to our results, our approach is well-suited for capturing semantic relationships between tasks and can be leveraged for selecting appropriate tasks for pretraining and model selection. Incorporation of labels in the fingerprints turned out to be particularly useful.

The closest work to ours in the field of biomedical image analysis was presented by Cheplygina et al. [8] (see Sect. 1). We built upon this general idea by exploring approaches that produce fingerprints invariant to the (other) available data sets and that are thus inherently scalable. Outside of the field of biomedical imaging, task similarity estimation is an increasingly active field of research, and our research has been inspired by previous work in the general machine learning community [1,5]. However, the research is still in its infancy, and no widely accepted approaches have been established.

It should be noted that there are potentially important parameters for knowledge transfer that we did not incorporate into our methods (e.g. number of samples) and that no extensive ablation studies for hyperparameter optimization were performed. The rationale for this strategy was that we are seeking a way to quantify the relationship between tasks rather than optimizing a model for a given task. While our initial work is encouraging, further work is required (1) to enable drawing broad conclusions, (2) to transfer the methodology to other types of tasks (e.g. semantic segmentation) and (3) to implement the approach in a complete meta learning framework.

In the long-term, the ability to quantify similarities between tasks could evolve as an enabling technique to leverage the full potential of meta learning for the field of biomedical imaging, and thus pave the way for translating biomedical image analysis research into clinical practice.

Acknowledgements. We would like to thank our colleagues Minu Dietlinde Tizabi, Tim Adler, Thuy Nuong Tran, Tobias Ross and Lucas-Raphael Müller for their valuable feedback on the drafts of this work. This project has been funded by the Surgical Oncology Program of the National Center for Tumor Diseases (NCT) Heidelberg. The present contribution is also supported by the Helmholtz Imaging Platform (HIP), a platform of the Helmholtz Incubator on Information and Data Science.

References

1. Achille, A., et al.: Task2Vec: Task Embedding for Meta-Learning. ArXiv (2019)
2. Achille, A., et al.: The Information Complexity of Learning Tasks, their Structure and their Distance. ArXiv (2019)
3. Ali, S., et al.: Endoscopy artifact detection (EAD 2019) challenge dataset. ArXiv (2019)
4. Allan, M., et al.: 2017 robotic instrument segmentation challenge. ArXiv (2019)
5. Bhattacharjee, B., et al.: P2L: predicting transfer learning for images and semantic relations. In: CVPR Workshops (2020)
6. Borgli, H., et al.: Hyperkvasir, a comprehensive multi-class image and video dataset for gastrointestinal endoscopy. Sci. Data **7**(1), 1–14 (2020)
7. Cheplygina, V.: Cats or cat scans: transfer learning from natural or medical image source datasets? ArXiv (2018)
8. Cheplygina, V., et al.: Exploring the similarity of medical imaging classification problems. In: CVII-STENT/LABELS@MICCAI (2017)
9. Cheplygina, V., et al.: Not-so-supervised: a survey of semi-supervised, multi-instance, and transfer learning in medical image analysis. Med. Image Anal. **54**, 280–296 (2019)
10. Cohen, G., et al.: EMNIST: an extension of MNIST to handwritten letters. ArXiv (2017)
11. Combalia, M., et al.: BCN20000: dermoscopic lesions in the wild. ArXiv (2019)
12. Dabak, A., et al.: Relations between kullback-leibler distance and fisher information (2002)
13. Deng, J., et al.: Imagenet: a large-scale image database. In: 2009 IEEE Conference on Computer Vision and Pattern Recognition (2009)

14. Faria, S.M., et al.: Light field image dataset of skin lesions. In: 2019 41st Annual International Conference of the IEEE Engineering in Medicine and Biology Society (EMBC) (2019)
15. Fei-Fei, L., et al.: Learning generative visual models from few training examples: an incremental Bayesian approach tested on 101 object categories. In: CVPR Workshops (2004)
16. Gretton, A., et al.: A kernel two-sample test. J. Mach. Learn. Res. **13**(1), 723–773 (2012)
17. Griffin, G., et al.: Caltech-256 object category dataset (2007)
18. Gutman, D., et al.: Skin lesion analysis toward melanoma detection: a challenge at the 2017 international symposium on biomedical imaging (ISBI), hosted by the international skin imaging collaboration (ISIC). In: 2018 IEEE 15th International Symposium on Biomedical Imaging (ISBI 2018) (2018)
19. He, K., et al.: Deep residual learning for image recognition. In: 2016 IEEE Conference on Computer Vision and Pattern Recognition (CVPR) (2016)
20. Hospedales, T.M., et al.: Meta-learning in neural networks: a survey. ArXiv (2020)
21. Kamada, T., et al.: An algorithm for drawing general undirected graphs. Inf. Process. Lett. **31**(1), 7–15 (1989)
22. Kawahara, J., et al.: Seven-point checklist and skin lesion classification using multitask multimodal neural nets. IEEE J. Biomed. Health Inform. **23**(2), 538–546 (2019)
23. Khosla, A., et al.: Novel dataset for fine-grained image categorization. In: First Workshop on Fine-Grained Visual Categorization, IEEE Conference on Computer Vision and Pattern Recognition. Colorado Springs, CO (2011)
24. Kingma, D.P., et al.: Adam: a method for stochastic optimization. CoRR (2015)
25. Krizhevsky, A.: Learning multiple layers of features from tiny images (2009)
26. LeCun, Y., et al.: Gradient-based learning applied to document recognition (1998)
27. Leibetseder, A., et al.: LAPGYN4: a dataset for 4 automatic content analysis problems in the domain of laparoscopic gynecology. In: Proceedings of the 9th ACM Multimedia Systems Conference (2018)
28. Leibetseder, A., et al.: Glenda: gynecologic laparoscopy endometriosis dataset. In: MMM (2020)
29. Maier-Hein, L., et al.: Can masses of non-experts train highly accurate image classifiers? - a crowdsourcing approach to instrument segmentation in laparoscopic images. In: MICCAI International Conference on Medical Image Computing and Computer-Assisted Intervention (2014)
30. Maier-Hein, L., et al.: Surgical Data Science – from Concepts to Clinical Translation. ArXiv, October 2020
31. Moccia, S., et al.: Confident texture-based laryngeal tissue classification for early stage diagnosis support. J. Med. Imaging **4**(3), 034502 (2017)
32. Moccia, S., et al.: Learning-based classification of informative laryngoscopic frames. Comput. Methods Programs Biomed. **158**, 21–30 (2018)
33. Netzer, Y., et al.: Reading digits in natural images with unsupervised feature learning (2011)
34. Pogorelov, K., et al.: Nerthus: a bowel preparation quality video dataset. In: Proceedings of the 8th ACM on Multimedia Systems Conference, MMSys 2017. ACM, New York (2017)
35. Raghu, M., et al.: Transfusion: understanding transfer learning for medical imaging. In: NeurIPS (2019)

36. Ross, T., et al.: Exploiting the potential of unlabeled endoscopic video data with self-supervised learning. Int. J. Comput. Assist. Radiol. Surg. **13**(6), 925–933 (2018). https://doi.org/10.1007/s11548-018-1772-0

37. Ross, T., et al.: Robust medical instrument segmentation challenge 2019 (2020)

38. Rubner, Y., et al.: A metric for distributions with applications to image databases. In: Sixth International Conference on Computer Vision (1998)

39. Tschandl, P., et al.: The HAM10000 dataset, a large collection of multi-source dermatoscopic images of common pigmented skin lesions. Sci. Data **5**(1), 1–9 (2018)

40. Twinanda, A.P., et al.: Endonet: a deep architecture for recognition tasks on laparoscopic videos. IEEE Trans. Med. Imaging **36**(1), 86–97 (2017)

41. Vilalta, R., et al.: A perspective view and survey of meta-learning. Artif. Intell. Rev. **18**(2), 77–95 (2005)

42. Virtanen, P., et al.: SciPy 1.0: fundamental algorithms for scientific computing in python. Nat. Methods **17**(3), 261–272 (2020)

43. Wightman, R.: Pytorch image models (2019). https://doi.org/10.5281/zenodo.4414861

44. Yuan, P., et al.: Few is enough: task-augmented active meta-learning for brain cell classification. In: Marte, A.L., et al. (eds.) MICCAI 2020. LNCS, vol. 12261, pp. 367–377. Springer, Cham (2020). https://doi.org/10.1007/978-3-030-59710-8_36

45. Zhang, L., et al.: Generalizing deep learning for medical image segmentation to unseen domains via deep stacked transformation. IEEE Trans. Med. Imaging **39**(7), 2531–2540 (2020)

Acoustic-Based Spatio-Temporal Learning for Press-Fit Evaluation of Femoral Stem Implants

Matthias Seibold[1,2]([✉]), Armando Hoch[3], Daniel Suter[3], Mazda Farshad[3], Patrick O. Zingg[3], Nassir Navab[1], and Philipp Fürnstahl[2,3]

[1] Computer Aided Medical Procedures (CAMP), Technical University of Munich, Munich 85748, Germany
matthias.seibold@tum.de
[2] Research in Orthopedic Computer Science (ROCS), University Hospital Balgrist, University of Zurich, Zurich 8008, Switzerland
[3] Balgrist University Hospital, University of Zurich, Zurich 8008, Switzerland

Abstract. In this work, we propose a method utilizing tool-integrated vibroacoustic measurements and a spatio-temporal learning-based framework for the detection of the insertion endpoint during femoral stem implantation in cementless Total Hip Arthroplasty (THA). In current practice, the optimal insertion endpoint is intraoperatively identified based on surgical experience and dependent on a subjective decision. Leveraging spectogram features and time-variant sequences of acoustic hammer blow events, our proposed solution can give real-time feedback to the surgeon during the insertion procedure and prevent adverse events in clinical practice. To validate our method on real data, we built a realistic experimental human cadaveric setup and acquired acoustic signals of hammer blows during broaching the femoral stem cavity with a novel inserter tool which was enhanced by contact microphones. The optimal insertion endpoint was determined by a standardized preoperative plan following clinical guidelines and executed by a board-certified surgeon. We train and evaluate a Long-Term Recurrent Convolutional Neural Network (LRCN) on sequences of spectrograms to detect a reached target press fit corresponding to a seated implant. The proposed method achieves an overall per-class recall of $93.82 \pm 5.11\%$ for detecting an ongoing insertion and $70.88 \pm 11.83\%$ for identifying a reached target press fit for five independent test specimens. The obtained results open the path for the development of automated systems for intra-operative decision support, error prevention and robotic applications in hip surgery.

Keywords: Spatio-temporal learning · Acoustic sensing · Total hip arthroplasty · Femoral stem insertion

1 Introduction

For the preparation of the femur for implant insertion in cementless THA, the femoral head is resected and broaches of increasing size are driven into the femoral

© Springer Nature Switzerland AG 2021
M. de Bruijne et al. (Eds.): MICCAI 2021, LNCS 12904, pp. 447–456, 2021.
https://doi.org/10.1007/978-3-030-87202-1_43

stem cavity with hammer blows, before the final femoral stem implant is inserted. A frequent intraoperative complication during this procedure is periprosthetic femoral fracture which has been reported to occur with rates of 3.5% to 5.4% [2,3,19]. Hereby, the majority of periprosthetic fractures (46.5%) happen during the preparation of the femur for stem insertion [1] and are mainly caused by excessive broaching beyond the optimal insertion endpoint. Figure 1 shows a radiograph of a periprosthetic fracture. A relevant part of these fractures, so-called occult periprosthetic fractures, cannot even be discovered intraoperatively [23]. The standard clinical procedure to assess the seating of the femoral stem implant intraoperatively and determine the optimal insertion endpoint is based on preoperative planning, surgical experience, simple distance measurements, and radiologic verification of the implant seating in radiographs [10]. Therefore, a system which is able to assess the seating of the implant in the femoral cavity, inform the surgeon when the target press-fit is reached, and reduce the risk of intraoperative periprosthetic fracture would be highly desirable.

An ideal system would detect the optimal insertion endpoint during broaching and inform the surgeon to stop the insertion procedure. Conventional navigation systems can provide support in finding the correct implant position [18], but they cannot guide the surgeon in finding the insertion endpoint. As the underlying problem is not geometric, we chose to focus on a different data modality. Acoustic signals have been shown to be a rich source of information in medical scenarios, for example for the applications in bone drilling [20], arthroscopy [21], or needle injection [9]. Also for the assessment of the insertion endpoint in femoral stem insertion, acoustic signals which are generated by the impact of the hammer onto the broach inserter have been analyzed in prior work. For femoral stem insertion, hammer blow sounds have been identified to be correlated to complications by Morohashi et al. [14]. A cadaveric study was performed by Oberst et al. [15] and a nonlinear time-series analysis of the impulse response function showed relations to the process of femoral broaching. However, no clear insertion endpoint or stopping criterion was identified in this work.

Goossens et al. identified a number of hand-crafted features which correlate with the implant seating in *in-vitro* [5] and *in-vivo* [6] experiments using air-borne microphones and defined a stopping criterion based on the convergence of these features. However, air-borne sensors have the disadvantage of capturing environmental noise of the operating room in contrast to contact-based vibroacoustic measurements. Furthermore, they did not implement a system for the automated assessment of the implant seating. Additionally, learned features have been shown to have advantages over handcrafted features, such as better generalization and better performance with an increasing amount of training data [11].

A sensorized instrument for the monitoring of cementless femoral stem insertion has been proposed by Tijou et al. [22] employing a force sensor attached to the impacting face of the hammer to measure the impact force. This preliminary study using artificial bone models was later validated in a cadaveric experiment [4]. They showed that a time-domain and peak-based feature shows correlations to the displacement of the implant, measured using optical markers. Also in this work, no automated system was developed and features were handcrafted and not learned.

In this work, we propose a method which enables smart instruments for orthopedic surgery through tool-integrated vibroacoustic measurements and developed a deep learning framework for the automated analysis of hammer blow sounds in THA. By not attaching the piezo element to the hammer, but to the inserter tool itself, we capture the full vibration response of the broach-inserter structure instead of only the impact force. We furthermore advance the state-of-the-art in vibroacoustic analysis of surgical procedures by introducing a spatio-temporal model for sequence-based press fit evaluation and show the benefits of including time-domain data in our framework. We thoroughly evaluate the performance of our model in cadavers using 5-fold cross validation. In the following sections we introduce our method and describe the experimental setup and data generation process.

2 Materials and Method

2.1 Data Pre-processing and Spatio-Temporal Model for Press-Fit Evaluation

To capture acoustic signals of the surgical procedure, we used a custom piezo-based contact microphone and attached it to the inserter tool to capture structure-borne sounds of the hammer blows and the resonance of the tool during the insertion process. The piezo-sensor is electromagnetically shielded and impedance-buffered to minimize noise and optimize the frequency transmission. We attach the sensor to the tool using electrical tape for firm fixation. The inserter with attached contact sensor is illustrated in Fig. 1.

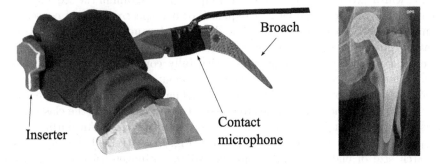

Fig. 1. Left: The inserter-broach structure with attached piezo-based contact microphone. Right: A radiograph of a periprosthetic fracture.

From the recorded soundclips, we extract spectrogram features which are lower-dimensional compared to raw audio and have been shown to yield promising performance for general audio signal processing [17], but also for automated acoustic analysis in surgical applications [16,20]. Log-mel-spectrograms, a sparse and high resolution spectrogram variant, with dimensions $256 \times 344 \times 1$ were computed from the individual sound clips of hammer blows using the python library *librosa 0.7.2* [13]. Log-mel spectrograms are two-dimensional matrices with time windows as columns, frequency mel-bins as rows, and amplitude as scalar matrix values. First, Short-Time Fourier Transformation (STFT) was applied to each sound clip of length N using:

$$X(m,k) = \sum_{n=0}^{N-1} x(n+mH)w(n)exp(\frac{-2\pi ikn}{N}) \qquad (1)$$

In our implementation, the term $w(n)$ corresponds to the *Hann* window function to avoid *spectral leakage* [12]. The resulting STFT spectrogram X contains the kth Fourier coefficient for the mth time frame as matrix values. We used a hop length of $H = 16$ for all spectrograms. We computed the power spectrogram from X and mapped it to the logarithmic decibel scale using:

$$X_{pow}(m,k) = 10\, log_{10}(X(m,k)^2) \qquad (2)$$

To finally convert this log STFT spectrogram to the mel scale, a total number of 256 triangular filters which are spaced evenly on the Mel scale (Eq. 3) were applied to X_{pow}. The mel scale can be computed from frequency by:

$$f_{mel} = 2595\, log_{10}(1 + \frac{f}{700}) \qquad (3)$$

We normalized all spectrograms $X_{norm,mel} = (X_{mel} - \mu)/\sigma$, where (μ) is overall mean and (σ) is the overall standard deviation.

The network architecture consists of a ResNet-50 backbone [7], an architecture which has been shown to perform especially well for audio classification tasks [8,20], which extracts features from each input spectrogram in the sequence. The input of the proposed model is a sequence of five spectrograms of consecutive hammer blows. The output of the time-distributed, randomly initialized ResNet is a sequence of five feature vectors of size 1×2048 which is passed to an LSTM layer and subsequently to two consecutive fully connected layers. In between the two fully connected layers, we apply a dropout of 0.5 to reduce the model's tendency towards overfitting. The entire model has a total of 26,050,945 parameters and outputs the probabilities to classify the sequences into the classes $C :=$ {insertion, press fit}, where c_i denotes the respective class. An outline of the proposed detection pipeline is illustrated in Fig. 2. We trained the model using a batch size of eight sequences and the Adam optimizer with an empirically determined optimal learning rate of $1 * 10^{-5}$, minimizing a binary crossentropy loss. Early stopping is used for additional regularization. Data augmentation was applied during training using time-stretching by a factor of [0.5, 0.7, 1.2, 1.5] and pitch-shifting by [$-3, -1, +1, +3$] semitones on the audio file level. All experiments were performed using Tensorflow/Keras 2.2 on a NVIDIA GeForce RTX 2080 SUPER GPU. The implementation is available under https://caspa. visualstudio.com/CARD%20public/_git/AudioFemoralStem.

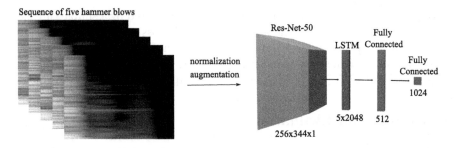

Fig. 2. The outline of the proposed spatio-temporal model. The input sizes of each layer of the pipeline is stated at the bottom. For visualization purposes, the spectrograms are displayed as color images, however the implementation uses features of size $256 \times 344 \times 1$.

2.2   Experimental Setup and Data Generation

To build an experimental setup, which is as close as possible to the clinical scenario, we conducted experiments with five thawed fresh-frozen human cadaveric hip specimens including soft tissue. An ethical approval and informed consent from all study subjects was obtained. One orthopedic surgeon prepared the femur using a standard anterior approach. To follow the clinical procedure for femoral stem insertion, we broached the femoral stem cavity using the Medacta *QUADRA* system (Medacta International SA, Castel San Pietro, Switzerland), consisting of broaches of increasing size. The target broach size was planned using the clinical planning software *mediCAD* (mediCAD Hectec GmbH, Altdorf/Landshut, Germany) and the surgical plan was executed by a trained surgeon for each specimen. The outline of the preoperative plan and the cadaveric experiments is illustrated in the last column of Table 1. The insertion endpoint was classified as the preoperatively planned broach size fully seated in the anatomy and confirmed by the surgeon.

Structure-borne acoustic signals were acquired from the contact sensor attached to the tool during the broaching process. We additionally captured video footage of the whole experiment to facilitate the labelling process. For the generation of the training data set, the recorded sequences of hammer blows were labelled into two classes, $\{c_0, c_1\}$. The class c_0 contains audio samples of hammer blows during insertion of the increasing sizes of broaches until the planned target size. The operating surgeon identified the target press fit as the preoperatively planned broach size fully seated in the anatomy. The class c_1 contains samples of hammer blows after reaching the target press fit.

We use the PreSonus Studio 68 (PreSonus Audio Electronics, Inc., Baton Rouge, LA, USA) audio interface and the Audio Stream Input/Output (ASIO) low-latency driver to capture loss-less audio with a sampling rate of 44.1 kHz and a bit depth of 24 bit. All samples were cut to a length of $N = 5500$ samples, which corresponds to a duration of 125 ms which has been empirically determined as the optimal length to capture the whole duration of a hammer blow sound.

During real surgery, these hammer blows could be detected by a simple threshold in the recorded structure-borne audio. For 5-fold cross validation, we split the data set on the specimen level and use the data from four specimens for training and from one specimen for testing, respectively. The final data set contains a total number of 1795 ($n_{c_0} = 1245$, $n_{c_1} = 550$) sequences of five hammer blows.

3   Results and Evaluation

In the following sections we present the evaluation of the proposed spatio-temporal model including an in-depth inference analysis for 5-fold cross validation and show the benefits of incorporating sequence data in our framework.

3.1   Model Performance

The proposed method achieves an overall per-class recall (mean and standard deviation) of 93.82±5.11% for detecting an ongoing insertion and 70.88±11.83% for identifying a reached target press fit for five independent test specimens. Table 1 illustrates the per-class recall and precision for each fold. Hereby, we consider the per-class recall (bold values) as main metric, as it corresponds to the ratio of correctly identified sequences. Furthermore, we present and in-depth analysis of the network performance throughout the whole insertion procedure for each independent test specimen in Fig. 3.

Table 1. Performance for 5-fold cross validation

		Spatio-temporal model		Non-sequence baseline		Planned broach size	Specimen number
		ongoing insertion	press fit reached	ongoing insertion	press fit reached		
Fold 0	recall	**94.67%**	**49.37%**	**62.35%**	**53.55%**	4	1
	precision	63.96%	90.70%	55.50%	60.49%		
Fold 1	recall	**93.50%**	**69.86%**	**95.91%**	**19.31%**	4	2
	precision	89.47%	79.69%	74.82%	65.38%		
Fold 2	recall	**96.72%**	**71.74%**	**42.86%**	**95.09%**	4	3
	precision	81.94%	94.29%	91.58%	57.20%		
Fold 3	recall	**99.71%**	**81.90%**	**92.08%**	**38.97%**	7	4
	precision	94.23%	98.96%	81.76%	62.35%		
Fold 3	recall	**84.51%**	**81.54%**	**76.67%**	**29.33%**	7	5
	precision	96.28%	48.18%	85.36%	18.97%		

3.2   Comparison with Non-sequence Data

We compared the proposed spatio-temporal model with a single spectrogram based detection method. Therefore, we implemented the model architecture

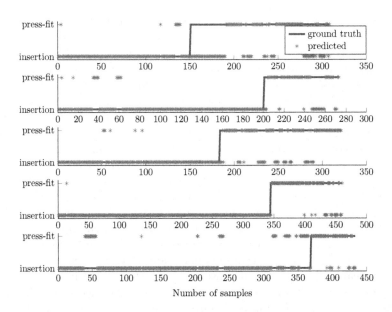

Fig. 3. An in-depth analysis of the results of the spatio-temporal model during 5-fold cross validation. Each plot corresponds to an independent test specimen (Specimen 1–5 in ascending order from top to bottom), the network was trained on the remaining four folds.

proposed by Seibold et al. [20], which yielded promising performance for the application in surgical drill breakthrough detection, and evaluate it on single spectrograms from the data set collected in the cadaveric experiments. For pre-processing, we applied the same augmentation strategy to the non-sequence data set and normalize every sample spectrogram. Without the temporal context of the sequence of hammer blows the detection performance decreases dramatically. The model reaches a per-class recall of $73.97 \pm 19.59\%$ and $47.25 \pm 26.45\%$ for c_0 (ongoing insertion) and c_1 (press fit reached), respectively. The per-class recall and precision for each fold is given in Table 1.

4   Discussion

The proposed automated method for the assessment of the optimal insertion endpoint of femoral stem implants could be an important step towards reducing the risk of periprosthetic femoral fracture during cementless THA, which would consequently improve patient safety and treatment quality. To the best knowledge of the authors, this is the first work to employ a sensorized instrument for capturing vibroacoustic signals directly from the operation area and to develop a state-of-the-art learning based method for the automated assessment of the seating of the femoral stem implant. We furthermore evaluated the proposed model in cadaver experiments in which the real intervention was simulated as closely as possible on thawed fresh-frozen cadavers.

In comparison to previous work, we demonstrated the feasibility of an automated system for the assessment of the optimal insertion endpoint. Even though the full potential of deep learning models and the advantages over handcrafted features is revealed when large amounts of training data are available [11], we show that the proposed model is able to learn useful information for the majority of the presented test cases even though the presented data set is small.

The results of 5-fold cross validation show that the proposed network yields promising overall performance, however the model confuses the samples in the critical region of the optimal insertion endpoint (when changing from "insertion" to "press fit" in Fig. 3) for the first and last fold. These outliers and the resulting relatively large standard deviation in cross validation can be attributed to the fact that the sample size is relatively small with five cadaveric specimens also due to the relevant inter-subject variance in bone density. An increased size of data would improve the model's capability for generalization and improve the performance of the algorithm. Nevertheless, we consider our sample size sufficient for a technical feasiblity study, because human cadaver experiments are associated with significant cost and ethical considerations. More extensive data collection will be addressed in future work together with additional postprocessing steps, such as majority vote or additional convergence criteria, to improve generalization capabilities. Even though the presented dataset is not highly imbalanced, the influence of class imbalance should be investigated in future work.

In the presented work, we showed that the spatio-temporal model clearly outperforms non-sequence data for the application in the assessment of the femoral stem press fit. However, in future work, other temporal modelling approaches, such as Temporal Convolutional Neural Networks (TCNs), could be employed. Furthermore, the influence of additional mechanisms, such as Self-attention, could be investigated. A technical limitation of the presented work is the subjective decision of the surgeon for the definition of the optimal insertion endpoint. However, this definition of the ground truth is in line with all prior work and a quantitative measurement is infeasible, as e.g. a measurement of the pull out force would require an extensive measurement setup and is out of scope of the presented work.

5   Conclusion

To the best knowledge of the authors, we propose the first automated detection approach for assessing the press fit of femoral stem implants in THA. Our method consists of a sensorized smart instrument and a spatio-temporal model capable of inferring the optimal insertion endpoint. The proposed solution shows the general feasibility of such an automated system and is thoroughly evaluated on human cadaveric data with an in-depth analysis of the model performance in 5-fold cross validation using independent test specimens, achieving a per-class sensitivity of $93.82 \pm 5.11\%$ and $70.88 \pm 11.83\%$ for identifying an ongoing insertion or a reached target press fit, respectively.

The presented system could not only be valuable as supplementary system for navigation systems and robotic applications, but also for error prevention in conventional surgery. Additionally, an automated system for the assessment of the optimal femoral stem insertion point could be employed for surgical skill assessment to define the level of surgical expertise.

Acknowledgment. This work is part of the SURGENT project and was funded by University Medicine Zurich/Hochschulmedizin Zürich. Matthias Seibold and Nassir Navab are partly funded by the Balgrist Foundation in form of the guest professorship at Balgrist University Hospital.

We would like to thank Navid Navab from Topological Media Lab at Concordia University, Montreal, Canada, for initial discussions, as well as creative and valuable interactions.

References

1. Abdel, M.P., Houdek, M.T., Watts, C.D., Lewallen, D.G., Berry, D.J.: Epidemiology of periprosthetic femoral fractures in 5417 revision total hip arthroplasties: a 40-year experience. Bone Joint J. **98**, 468–474 (2016)
2. Berend, K.R., Lombardi, A.V., Mallory, T.H., Chonko, D.J., Dodds, K.L., Adams, J.B.: Cerclage wires or cables for the management of intraoperative fracture associated with a cementless, tapered femoral prosthesis: results at 2 to 16 years. J. Arthroplasty **19**, 17–21 (2004)
3. Capello, W.N., Houdek, M.T., Watts, C.D., Lewallen, D.G., Berry, D.J.: Periprosthetic fractures around a cementless hydroxyapatite-coated implant: a new fracture pattern is described. Clin. Orthop. Relat. Res. **472**, 604–610 (2014)
4. Dubory, A., Rosi, G., Tijou, A., Lomami, H.A., Flouzat-Lachaniette, C.H., Haiat, G.: A cadaveric validation of a method based on impact analysis to monitor the femoral stem insertion. J. Mech. Behav. Biomed. Mater. **103**, 103535 (2020)
5. Goossens, Q., Leuridan, S., Roosen, J.: Monitoring of reamer seating using acoustic information. In: Annual meeting of the European Society of Biomechanics (2015)
6. Goossens, Q., et al.: Acoustic analysis to monitor implant seating and early detect fractures in cementless THA: an in vivo study. J. Orthop. Res. **39**, 1164–1173 (2020)
7. He, K., Zhang, X., Ren, S., Sun, J.: Deep residual learning for image recognition. In: 2016 IEEE Conference on Computer Vision and Pattern Recognition (CVPR), pp. 770–778 (2016)
8. Hershey, S., et al.: CNN architectures for large-scale audio classification. In: International Conference on Acoustics, Speech and Signal Processing (ICASSP), pp. 131–135 (2017)
9. Illanes, A., et al.: Novel clinical device tracking and tissue event characterization using proximally placed audio signal acquisition and processing. Sci. Rep. **8**, 12070 (2018)
10. Le Béguec, P., Canovas, F., Roche, O., Goldschild, M., Batard, J.: Uncemented Femoral Stems for Revision Surgery. Springer, Cham (2015). https://doi.org/10.1007/978-3-319-03614-4
11. Lin, W., Hasenstab, K., Cunha, G.M., Schwartzman, A.: Comparison of handcrafted features and convolutional neural networks for liver MR image adequacy assessment. Sci. Rep. **10**, 20336 (2020)
12. Lyon, D.A.: The discrete fourier transform, part 4: spectral leakage. J. Object Technol. **8**(7), 23–34 (2009)

456   M. Seibold et al.

13. McFee, B., et al.: librosa: audio and music signal analysis in python. In: 14th Python in Science Conference, pp. 18–25 (2015)

14. Morohashi, I., et al.: Acoustic pattern evaluation during cementless hip arthroplasty surgery may be a new method for predicting complications. In: SICOT-J 3 (2017)

15. Oberst, S., et al.: Vibro-acoustic and nonlinear analysis of cadavric femoral bone impaction in cavity preparations. Int. J. Mech. Sci. 144, 739–745 (2018)

16. Ostler, D., et al.: Acoustic signal analysis of instrument-tissue interaction for minimally invasive interventions. Int. J. Comput. Assist. Radiol. Surg. 15, 771–779 (2020). https://doi.org/10.1007/s11548-020-02146-7

17. Purwins, H., Li, B., Virtanen, T., Schlüter, J., Chang, S.y., Sainath, T.: Deep learning for audio signal processing. IEEE J. Sel. Top. Signal Process. 14, 206–219 (2019)

18. Renner, L., Janz, V., Perka, C., Wassilew, G.I.: What do we get from navigation in primary THA? EFORT Open Rev. 1, 205–210 (2016)

19. Ricioli, W., Queiroz, M.C., Guimarães, R.P., Honda, E.K., Polesello, G., Fucs, P.M.M.B.: Prevalence and risk factors for intra-operative periprosthetic fractures in one thousand eight hundred and seventy two patients undergoing total hip arthroplasty: a cross-sectional study. Int. Orthopaedics 39(10), 1939–1943 (2015). https://doi.org/10.1007/s00264-015-2961-x

20. Seibold, M., et al.: Real-time acoustic sensing and artificial intelligence for error prevention in orthopedic surgery. Sci. Rep. 11, 3993 (2021)

21. Suehn, T., Pandey, A., Friebe, M., Illanes, A., Boese, A., Lohman, C.: Acoustic sensing of tissue-tool interactions - potential applications in arthroscopic surgery. Curr. Dir. Biomed. Eng. 6, 20203152 (2020)

22. Tijou, A.: Monitoring cementless femoral stem insertion by impact analyses: an in vitro study. J. Mech. Behav. Biomed. Mater. 88, 102–108 (2018)

23. Yun, H.H., Lim, J.T., Yang, S.H., Park, P.S.: Occult periprosthetic femoral fractures occur frequently during a long, trapezoidal, double-tapered cementless femoral stem fixation in primary THA. PLos ONE 19, e0221731 (2019)

Surgical Planning and Simulation

Deep Simulation of Facial Appearance Changes Following Craniomaxillofacial Bony Movements in Orthognathic Surgical Planning

Lei Ma[1], Daeseung Kim[2], Chunfeng Lian[1], Deqiang Xiao[1], Tianshu Kuang[2], Qin Liu[1], Yankun Lang[1], Hannah H. Deng[2], Jaime Gateno[2,3], Ye Wu[1], Erkun Yang[1], Michael A.K. Liebschner[4], James J. Xia[2,3(✉)], and Pew-Thian Yap[1(✉)]

[1] Department of Radiology and Biomedical Research Imaging Center (BRIC), University of North Carolina at Chapel Hill, Chapel Hill, NC 27599, USA
ptyap@med.unc.edu
[2] Department of Oral and Maxillofacial Surgery, Houston Methodist Research Institute, Houston, TX, USA
jxia@houstonmethodist.org
[3] Department of Surgery (Oral and Maxillofacial Surgery), Weill Medical College, Cornell University, Ithaca, NY, USA
[4] Department of Neurosurgery, Baylor College of Medicine, Houston, TX, USA

Abstract. Facial appearance changes with the movements of bony segments in orthognathic surgery of patients with craniomaxillofacial (CMF) deformities. Conventional bio-mechanical methods, such as finite element modeling (FEM), for simulating such changes, are labor intensive and computationally expensive, preventing them from being used in clinical settings. To overcome these limitations, we propose a deep learning framework to predict post-operative facial changes. Specifically, FC-Net, a facial appearance change simulation network, is developed to predict the point displacement vectors associated with a facial point cloud. FC-Net learns the point displacements of a pre-operative facial point cloud from the bony movement vectors between pre-operative and post-operative bony models. FC-Net is a weakly-supervised point displacement network trained using paired data with strict point-to-point correspondence. To preserve the topology of the facial model during point transform, we employ a local-point-transform loss to constrain the local movements of points. Experimental results on real patient data reveal that the proposed framework can predict post-operative facial appearance changes remarkably faster than a state-of-the-art FEM method with comparable prediction accuracy.

Keywords: Facial appearance change · Point transform network · Topology preservation.

M. de Bruijne et al. (Eds.): MICCAI 2021, LNCS 12904, pp. 459–468, 2021.
https://doi.org/10.1007/978-3-030-87202-1_44

1   Introduction

Craniomaxillofacial (CMF) deformities involve congenital and acquired deformities of the skull and face [12]. Patients with CMF deformities suffer from both functional and aesthetic impairments [1]. Orthognathic surgery is a surgical procedure specifically designed to correct skeletal deformities [13]. In orthognathic surgery, the jaws (both maxilla and mandible) are first osteotomized into multiple bony segments, which are then moved to their desired positions [18]. Facial soft tissues, while not directly operated on during the surgical procedure, are passively changed following the underlying bony movements [4]. Due to the complex anatomy and aesthetical concerns in the CMF region, orthognathic surgery requires an extensive and accurate surgical plan with the help of computer-aided surgical simulation (CASS) techniques [17].

Surgeons can accurately and efficiently plan the bony movements using CASS [16,19]. However, they are still unable to practically predict the post-operative facial appearance during the surgical planning process, and just hope for the best that a postoperative face will "automatically" become normal following the normalized skeleton [4]. This is often not true because the relationship between bony and facial soft tissue movements is based on complex and non-trivial physical interactions. Post-operative facial appearance does not directly reflect bony changes [2].

The ultimate orthognathic surgical outcome is judged by not only the normal skeletal anatomy and function, but also facial appearance, which the first thing seen by others in daily life. Therefore, predicting post-operative facial appearance following the movements of underlying bony segments is critical [14]. Conventional simulation methods predict post-operative facial appearance using biomechanical simulation (e.g., finite-element modeling (FEM), mass-spring modeling (MSM), and mass-tensor modeling (MTM)) [4,5,15]. However, these methods require labor-intensive and time-consuming data processing and computationally expensive simulation procedures. These problems prevent clinicians from utilizing the bio-mechanical simulation methods in daily clinical practice. Recently, deep learning has been reported to demonstrate potential in modeling mechanical behavior of soft tissues more efficiently than traditional FEM methods [7,8]. We believe that deep learning based method can mitigate the time-consuming and labor-intense data preparation process in bio-mechanical simulation-based facial appearance prediction.

In this paper, we propose a deep-learning framework to accelerate the simulation of facial appearance changes following bony movements. The core of the proposed framework is a facial appearance change simulation network, FC-Net, that learns to predict a post-operative facial point cloud from a pre-operative facial point cloud based on the movement vectors associated with a pre-operative bony point cloud. FC-Net uses the bony movement vectors to guide the movements of the pre-operative facial point cloud. FC-Net is a weakly-supervised point displacement network trained using paired data of pre-operative facial models and the movement vectors of the bony models. Compared with a state-of-the-art FEM method, our framework can predict facial appearance changes significantly faster with comparable prediction accuracy.

The contribution of our method is two-fold: (1) A weakly-supervised point transform network, the FC-Net, can simulate facial changes by learning point displacement vectors of pre-operative facial point cloud from bony movement vectors. (2) To preserve the topology of the facial model during point transform, a local-point-transform loss is developed for FC-Net to constrain the movements of facial points within a local area. To the best our knowledge, this is the first deep-learning based method developed for predicting post-operative facial appearance following bony movements. FC-Net is efficient and has already shown its value in clinical practice.

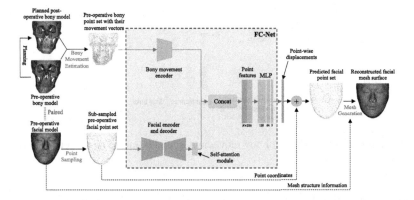

Fig. 1. The proposed framework for predicting facial appearance changes according to CMF bony movements in orthognathic surgical planning.

2   Method

The schematic diagram of the proposed framework is shown in Fig. 1. Bony movement vectors are first estimated from paired pre-operative and planned post-operative bony models. Then, the bony movement vectors and the facial point cloud, which is sampled from pre-operative facial soft tissue model, are fed to the facial appearance change prediction network FC-Net to predict the point movement vectors of the pre-operative facial point cloud. Finally, the pre-operative facial mesh is generated by applying dense movement vectors, interpolated from the predicted point movement vectors, on the pre-operative facial mesh.

2.1   Bony Movement Vector Estimation

For a given pre-operative bony model $\mathbf{M}_{\text{B-pre}}$, a skilled surgeon simulates a normalized post-operative bony model $\mathbf{M}'_{\text{B-post}}$ by cutting the jaw of $\mathbf{M}_{\text{B-pre}}$ into several segments and moving them rigidly to their desired positions. To estimate the movement vectors of the bony segments, we need to determine the point-to-point correspondence between $\mathbf{M}_{\text{B-pre}}$ and $\mathbf{M}'_{\text{B-post}}$. To this end, we first deform $\mathbf{M}_{\text{B-pre}}$ to $\mathbf{M}'_{\text{B-post}}$ using landmark based thin plate splines (TPS), and then

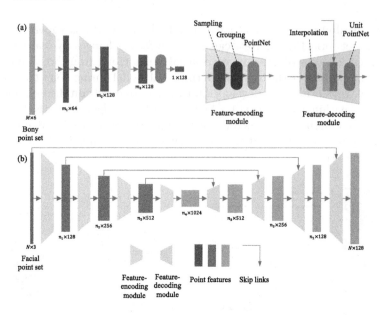

Fig. 2. Architectures of the (a) bony movement encoder and (b) facial encoder-decoder of FC-Net.

further register the deformed bony model to $\mathbf{M}'_{\text{B-post}}$ by non-rigid registration via coherent point drift (CPD) [9], resulting in a registered pre-operative bony model $\mathbf{M}^{R}_{\text{B-pre}}$. We can determine the point-to-point correspondence by finding the closest point of $\mathbf{M}^{R}_{\text{B-pre}}$ to $\mathbf{M}'_{\text{B-post}}$, thus obtaining the point movement vectors from $\mathbf{M}_{\text{B-pre}}$ to $\mathbf{M}'_{\text{B-post}}$.

2.2   FC-Net

Input: FC-Net simulates the facial appearance change by learning the point displacement vectors of the pre-operative facial point cloud from the bony movement vectors of the pre-operative bony point cloud. Therefore, FC-Net needs two inputs to perform the simulation. First, the coordinate of the pre-operative facial point cloud $\mathbf{P}_{\text{F-pre}}$ sampled from $\mathbf{M}_{\text{F-pre}}$ with size of $N \times 3$, where N is the number of points. Second, the combination of the coordinates of the pre-operative bony point cloud $\mathbf{P}_{\text{B-pre}}$ sampled from $\mathbf{M}_{\text{B-pre}}$ and its movement vectors $\mathbf{V}_{\text{B}}$ with size of $M \times 6$, where M is the number of points. In our implementation, N and M are both set to 4096.

FC-Net Architecture: FC-Net consists of three components (Fig. 1), bony movement encoder, facial encoder-decoder, and a multi-layer perceptron (MLP). The bony movement encoder takes the pre-operative bony point cloud $\mathbf{P}_{\text{B-pre}}$ and its movement vectors $\mathbf{V}_{\text{B}}$ as input, and outputs a 1×128 point feature as shown in Fig. 2(a). The bony movement encoder starts from a number of feature-encoding

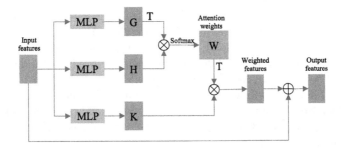

Fig. 3. Architecture of the self-attention module adapted in FC-Net.

modules derived from PointNet++ [11] and ends with a PointNet layer [10]. Each feature-encoding module consists of a sampling layer, a grouping layer, and a PointNet layer [11].

Figure 2(b) presents the facial encoder and decoder. The structure of the facial encoder, similar to the bony movement encoder, consists of a series of the feature-encoding modules. The facial encoder takes point cloud $\mathbf{P}_{\text{F-pre}}$ sub-sampled from the pre-operative facial model, an $N \times 3$ matrix, as input and abstracts the input to a $N' \times 1024$ matrix of point features with N' being the number of points. The facial decoder propagates the abstracted point features from the sub-sampled points to the original points for point-wise displacement prediction, outputting a $N \times 128$ matrix of point features. The facial decoder is composed of a group of feature-decoding modules, which consist of an interpolation layer and a unit PointNet layer.

A self-attention module is further employed to enhance the structural information, e.g., nose and lips in the pre-operative facial point cloud. The self-attention module is implemented following [6] as shown in Fig. 3. The input facial features are first transformed into $\mathbf{G}$, $\mathbf{H}$, and $\mathbf{K}$ features through MLPs. Then, attention weights W are generated as follows:

$$\mathbf{W} = f_{\text{softmax}}(\mathbf{G}^{\top}\mathbf{H}), \tag{1}$$

where $f_{\textbf{softmax}}$ represents the softmax function. The output features of the self-attention module is the sum of the input features and the weighted features (i.e., $\mathbf{W}^{\top}\mathbf{K}$).

We use the bony movement vectors to guide the shape deformation of the pre-operative facial point cloud. To this end, we concatenate the point features abstracted by the bony movement encoder with the outputted point features of the facial decoder point-wisely, and feed the concatenated point features through a MLP to predict the point displacement vectors $\mathbf{V}'_{\text{F}}$ of the pre-operative facial point cloud $\mathbf{P}_{\text{F-pre}}$. The simulated post-operative facial point cloud $\mathbf{P}'_{\text{F-post}}$ can be achieved by adding the predicted point displacement vectors to the pre-operative facial point cloud point-wise following $\mathbf{P}'_{\text{F-post}} = \mathbf{P}_{\text{F-pre}} + \mathbf{V}'_{\text{F}}$.

Loss Functions: The objective of FC-Net training is to make the predicted post-operative facial shape represented by point cloud $\mathbf{P}'_{\text{F-post}}$ as close as possible to its target shape of point cloud $\mathbf{P}_{\text{F-post}}$. To this end, a geometric loss $\mathbf{L}_g$ is used to measure the geometric difference between the predicted and target point clouds [20]. The geometric loss $\mathbf{L}_g$ consists of two terms, *i.e.*, shape loss and point density loss. The shape loss $\mathbf{L}_s$ computes the sum of distance errors between the predicted facial point cloud $\mathbf{P}'_{\text{F-post}}$ and the target facial point cloud $\mathbf{P}_{\text{F-post}}$. In this work, we employ the Chamfer Distance [3] as the measurement of the shape loss as follows:

$$\mathbf{L}_s = \frac{1}{N} \left(\sum_{x \in \mathbf{P}_{\text{F-post}}} \min_{y \in \mathbf{P}'_{\text{F-post}}} \|x - y\|^2 + \sum_{y \in \mathbf{P}'_{\text{F-post}}} \min_{x \in \mathbf{P}_{\text{F-post}}} \|x - y\|^2 \right), \quad (2)$$

where N is the number of points in $\mathbf{P}'_{\text{F-post}}$. $\|x - y\|^2$ denotes the squared Euclidean distance between points x and y. The point density loss $\mathbf{L}_d$ measures the similarity between the density of the predicted facial point cloud $\mathbf{P}'_{\text{F-post}}$ and the density of the target shape $\mathbf{P}_{\text{F-post}}$. It is defined as the summation of distances between density vectors of $\mathbf{P}'_{\text{F-post}}$ and $\mathbf{P}_{\text{F-post}}$ [20]:

$$\mathbf{L}_d = \frac{1}{k} \sum_{x \in \mathbf{P}_{\text{F-post}}} \sum_{i=1}^{k} |d(x, \mathbf{N}_i[\mathbf{P}_{\text{F-post}}, x]) - d(x, \mathbf{N}_i[\mathbf{P}'_{\text{F-post}}, x])|, \quad (3)$$

where $\mathbf{N}_i[\mathbf{P}_{\text{F-post}}, x]$ denotes the i-th point of the k nearest points of x in the target point cloud $\mathbf{P}_{\text{F-post}}$ from the same point cloud, and $\mathbf{N}_i[\mathbf{P}'_{\text{F-post}}, x]$ represents the i-th point of the k nearest points of x in the target point cloud $\mathbf{P}_{\text{F-post}}$ from the predicted point cloud $\mathbf{P}'_{\text{F-post}}$

The geometric loss is a point-to-shape distance measure. However, with only the geometric loss, the intrinsic topology of the pre-operative facial point cloud cannot be preserved in the predicted post-operative facial point cloud, which is critical for mesh surface recovery from the predicted point cloud. To preserve the intrinsic topology in the predicted point cloud, we use a local-point-transform (LPT) loss $\mathbf{L}_{\text{LPT}}$ to constrain relative movements between one point and its neighbors. we define the LPT loss as follows:

$$\mathbf{L}_{\text{LPT}} = \frac{1}{k} \sum_{\substack{p \in \mathbf{P}_{\text{F-pre}} \\ p' \in \mathbf{P}'_{\text{F-post}}}} \sum_{i=1}^{k} |d(p, K_i) - d(p', K'_i)|, \quad (4)$$

where p' in the predicted point cloud is the point transformed from p in the input point cloud. K_i denote the i-th point of the k nearest points of p, and K'_i represents the point transformed from the point K_i. Therefore, the final loss function used to train FC-Net can be summarized as follows:

$$\text{Loss} = \mathbf{L}_s + \mu \mathbf{L}_d + \lambda \mathbf{L}_{\text{LPT}}, \quad (5)$$

where μ and λ are the weights for the density loss and the LPT loss, respectively.

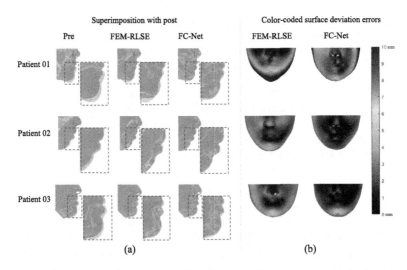

Fig. 4. (a) Comparison of the simulated facial models (green part) with their ground truth(red part). (b) Color-coded surface deviation errors between the simulated models and their ground-truth models. (Color figure online)

Implementation: The bony movement encoder of FC-Net was implemented using 3 feature-encoding modules. The numbers of points in the point clouds sampled in these modules were $\{1024, 512, 256\}$, and the dimensions of the output point features of the 3 feature-encoding modules were $\{128, 128, 128\}$. The facial encoder and decoder were implemented using 4 feature-encoding modules and 4 feature-decoding modules with $\{1024, 512, 256, 64, 256, 512, 1024, 4096\}$ point number for sampling and $\{128, 256, 512, 1024, 512, 256, 128, 128\}$ dimensions of the output point features. The channels of the MLP were $\{128, 64, 3\}$. The model of FC-Net was trained for 200 epochs in total with the batch size 2 using the ADAM optimizer. The learning rate was set to $1e{-}3$ and decayed to $1e{-}4$ at discrete intervals during training. The value of k in (3) (4), μ and λ in (5) were empirically set to 8, 0.3 and 5, respectively. The models were trained on an Nvidia Titan XP graphics processing unit with 12 GB memory.

3   Experimental Results

Dataset and Experimental Setup: From our digital archive, we randomly selected 40 paired pre- and post-operative head CT scans of patients who had undergone orthognathic surgery in treating jaw deformities. All patients had a complete record of pre-operative surgical plan generated with CASS. Patients who had syndromic conditions, previous jaw or facial cosmetic surgeries, or CMF trauma, were excluded. The study was approved by Institutional Review Board (Pro00009723). Experienced oral surgeons segmented bony and facial models from each CT scan and digitized 22 midface landmarks following clinical routine [21]. For effective model training, all the models were rigidly aligned with the

Table 1. The average landmark-based error (mean ± standard deviation) and maximum error between the predicted and ground-truth facial models given by FEM-RLSE and our method. The landmark errors in the jaw region (16 landmarks), midface region (22 landmarks) and whole face (38 landmarks) are shown.

	FEM-RLSE		Our method	
	Landmark distance	Maximum error	Landmark distance	Maximum error
16 landmarks (Jaw)	3.19 ± 0.81 mm	4.69 mm	3.15 ± 0.82mm	4.58 mm
22 landmarks (Midface)	2.52 ± 0.57 mm	3.32 mm	2.40 ± 0.47mm	3.26 mm
38 landmarks (Whole Face)	2.84 ± 0.61 mm	3.90 mm	2.76 ± 0.51mm	3.75 mm

model of the first patient. The segmented bony and facial models were cropped to retain only surgically affected regions. A point cloud of 4096 points was subsampled from the vertices of each cropped model as inputs to FC-Net.

We randomly divided the dataset into 5 groups, 8 patients in each group. 5-fold cross-validation was carried out. For each experiment, we selected 1 group as the testing set and the remaining 4 as the training set. The results of our method were compared with a state-of-the-art FEM method with realistic lip sliding effect (FEM-RLSE) [4]. However, only 23 out of 40 patients were used because FEM-RLSE was very time-consuming and only 23 predictions were available. Quantitative evaluation was first performed by measuring the distance errors between the corresponding landmarks on the predicted and ground-truth models. The same experienced oral surgeons digitized 38 facial landmarks (i.e., 22 landmarks in the midface region and 16 landmarks in the jaw region) on each facial model for use in the quantitative evaluation. Paired t-tests were performed to determine whether there was a statistically significant difference in errors between the two methods. Then, qualitative evaluation was carried out by an experienced oral surgeon by visually comparing the facial models generated by our method and FEM-RLSE to the ground truth (the actual post-operative models). Finally, the time costs for data preparation and computation were also evaluated.

Results and Discussion: Figure 4(a) shows 3 random examples of the facial models predicted by the two methods and their corresponding ground-truth models. To ease visual comparison, we overlay the simulated models (in green) with their corresponding ground truth (in red). Figure 4(b) shows the corresponding color-coded surface deviation errors between the predicted and ground-truth facial models. The results of quantitative evaluation are shown in Table 1. While no statistically significant differences were found between two methods for the whole face, jaw, and midface ($P > 0.05$), errors given by our method were generally smaller, indicating accuracy at least comparable to FEM-RLSE. Qualitative evaluation indicates that our method is better in 7 patients, equally good in 9 and worse in 7 when compared with FEM-RLSE. This suggests that our method is capable of predicting post-operative facial models as accurately as FEM-RLSE.

The time consumed for each simulation using our framework was less than 1.67 min, including the time for point sub-sampling (~10 s), bony vector estimation (~50 s), network prediction (~20 s), and mesh generation (~20 s). In contrast, FEM-RLSE consumed around 30 min for a single simulation. The construction of the global stiffness matrix in FEM consumes approximately 75% of the total time. Surgeons often perform multiple simulations to test different surgical plans before an optimal plan is finalized. Our method is therefore a significant time saver. More importantly, there was an even larger difference in time spent on data preparation. Our method does not require laborious data preparation other than bony and facial segmentation and landmark digitization, which are completed as part of the clinical routine. On the other hand, an additional 5–7 h is required by FEM-RLSE to generate the FE models and to improve the lip mesh [4]. Therefore, our method has the potential to significantly accelerate the surgical planning process.

We compared the performance of FC-Net trained with and without the self-attention module. The average landmark error of the jaw region is reduced from 3.26 mm to 3.15 mm by adding the self-attention module.

4 Conclusions

We have developed a learning-based framework to simulate facial appearance changes following movements of bony segments for orthognathic surgical planning. FC-Net, a weakly supervised point displacement network, predicts facial changes from bony movements. Evaluation results using a real clinical dataset demonstrate that FC-Net can predict facial changes with accuracy comparable to an FEM method but at a remarkably greater speed.

Acknowledgement. This work was supported in part by United States National Institutes of Health (NIH) grants R01 DE022676, R01 DE027251, and R01 DE021863.

References

1. Alanko, O.M., Svedström-Oristo, A.L., Tuomisto, M.T.: Patients' perceptions of orthognathic treatment, well-being, and psychological or psychiatric status: a systematic review. Acta Odontologica Scandinavica **68**(5), 249–260 (2010)
2. Bell, W.H., Ferraro, J.W.: Modern practice in orthognathic and reconstructive surgery. Plastic Reconstruct. Surg. **92**(2), 362 (1993)
3. Fan, H., Su, H., Guibas, L.J.: A point set generation network for 3D object reconstruction from a single image. In: Proceedings of the IEEE Conference on Computer Vision and Pattern Recognition, pp. 605–613 (2017)
4. Kim, D., et al.: A new approach of predicting facial changes following orthognathic surgery using realistic lip sliding effect. In: Shen, D., et al. (eds.) MICCAI 2019. LNCS, vol. 11768, pp. 336–344. Springer, Cham (2019). https://doi.org/10.1007/978-3-030-32254-0_38
5. Knoops, P.G., et al.: A novel soft tissue prediction methodology for orthognathic surgery based on probabilistic finite element modelling. PloS ONE **13**(5), e0197209 (2018)

6. Li, R., Li, X., Fu, C.W., Cohen-Or, D., Heng, P.A.: PU-GAN: a point cloud upsampling adversarial network. In: Proceedings of the IEEE/CVF International Conference on Computer Vision, pp. 7203–7212 (2019)

7. Liang, L., Liu, M., Martin, C., Sun, W.: A deep learning approach to estimate stress distribution: a fast and accurate surrogate of finite-element analysis. J. Roy. Soc. Interface 15(138), 20170844 (2018)

8. Mendizabal, A., Márquez-Neila, P., Cotin, S.: Simulation of hyperelastic materials in real-time using deep learning. Med. Image Anal. 59, 101569 (2020)

9. Myronenko, A., Song, X.: Point set registration: coherent point drift. IEEE Trans. Pattern Anal. Mach. Intell. 32(12), 2262–2275 (2010)

10. Qi, C.R., Su, H., Mo, K., Guibas, L.J.: PointNet: deep learning on point sets for 3D classification and segmentation. In: Proceedings of the IEEE Conference on Computer Vision and Pattern Recognition (CVPR), pp. 652–660 (2017)

11. Qi, C.R., Yi, L., Su, H., Guibas, L.J.: PointNet++: deep hierarchical feature learning on point sets in a metric space. In: Advances in Neural Information Processing Systems, pp. 5099–5108 (2017)

12. Rankin, M., Borah, G.L.: Perceived functional impact of abnormal facial appearance. Plastic Reconstr. Surg. 111(7), 2140–2146 (2003)

13. Shafi, M., Ayoub, A., Ju, X., Khambay, B.: The accuracy of three-dimensional prediction planning for the surgical correction of facial deformities using maxilim. Int. J. Oral Maxillofac. Surg. 42(7), 801–806 (2013)

14. Shahim, K., Jürgens, P., Cattin, P.C., Nolte, L.-P., Reyes, M.: Prediction of craniomaxillofacial surgical planning using an inverse soft tissue modelling approach. In: Mori, K., Sakuma, I., Sato, Y., Barillot, C., Navab, N. (eds.) MICCAI 2013. LNCS, vol. 8149, pp. 18–25. Springer, Heidelberg (2013). https://doi.org/10.1007/978-3-642-40811-3_3

15. Ullah, R., Turner, P., Khambay, B.: Accuracy of three-dimensional soft tissue predictions in orthognathic surgery after le fort i advancement osteotomies. Br. J. Oral Maxillofac. Surg. 53(2), 153–157 (2015)

16. Wang, L., et al.: Estimating patient-specific and anatomically correct reference model for craniomaxillofacial deformity via sparse representation. Med. Phys. 42(10), 5809–5816 (2015)

17. Xia, J.J., Gateno, J., Teichgraeber, J.F.: New clinical protocol to evaluate craniomaxillofacial deformity and plan surgical correction. J. Oral Maxillofac. Surg. 67(10), 2093–2106 (2009)

18. Xia, J., et al.: Algorithm for planning a double-jaw orthognathic surgery using a computer-aided surgical simulation (cass) protocol. part 1: planning sequence. Int. J. Oral Maxillofac. Surg. 44(12), 1431–1440 (2015)

19. Xiao, D., et al.: Estimating reference bony shape models for orthognathic surgical planning using 3D point-cloud deep learning. IEEE J. Biomed. Health Inform. 25(8), 2958–2966 (2021)

20. Yin, K., Huang, H., Cohen-Or, D., Zhang, H.: P2P-NET: bidirectional point displacement net for shape transform. ACM Trans. Graph. (TOG) 37(4), 1–13 (2018)

21. Yuan, P., et al.: Design, development and clinical validation of computer-aided surgical simulation system for streamlined orthognathic surgical planning. Int. J. Comput. Assist. Radiol. Surg. 12(12), 2129–2143 (2017). https://doi.org/10.1007/s11548-017-1585-6

A Self-supervised Deep Framework for Reference Bony Shape Estimation in Orthognathic Surgical Planning

Deqiang Xiao[1], Hannah H. Deng[2], Tianshu Kuang[2], Lei Ma[1], Qin Liu[1],
Xu Chen[1], Chunfeng Lian[1], Yankun Lang[1], Daeseung Kim[2], Jaime Gateno[2,3],
Steve Guofang Shen[4], Dinggang Shen[1], Pew-Thian Yap[1(✉)],
and James J. Xia[2,3(✉)]

[1] Department of Radiology and Biomedical Research Imaging Center (BRIC),
University of North Carolina at Chapel Hill, Chapel Hill, NC 27599, USA
ptyap@med.unc.edu
[2] Department of Oral and Maxillofacial Surgery, Houston Methodist Hospital,
Houston, TX 77030, USA
jxia@houstonmethodist.org
[3] Department of Surgery (Oral and Maxillofacial Surgery), Weill Medical College,
Cornell University, New York, NY 10065, USA
[4] Oral and Craniomaxillofacial Surgery at Shanghai Ninth Hospital,
Shanghai Jiaotong University College of Medicine, Shanghai 200011, China

Abstract. Virtual orthognathic surgical planning involves simulating surgical corrections of jaw deformities on 3D facial bony shape models. Due to the lack of necessary guidance, the planning procedure is highly experience-dependent and the planning results are often suboptimal. A reference facial bony shape model representing normal anatomies can provide an objective guidance to improve planning accuracy. Therefore, we propose a self-supervised deep framework to automatically estimate reference facial bony shape models. Our framework is an end-to-end trainable network, consisting of a simulator and a corrector. In the training stage, the simulator maps jaw deformities of a patient bone to a normal bone to generate a simulated deformed bone. The corrector then restores the simulated deformed bone back to normal. In the inference stage, the trained corrector is applied to generate a patient-specific normal-looking reference bone from a real deformed bone. The proposed framework was evaluated using a clinical dataset and compared with a state-of-the-art method that is based on a supervised point-cloud network. Experimental results show that the estimated shape models given by our approach are clinically acceptable and significantly more accurate than that of the competing method.

Keywords: Orthognathic surgical planning · Shape estimation · Point-cloud network

Electronic supplementary material The online version of this chapter (https://doi.org/10.1007/978-3-030-87202-1_45) contains supplementary material, which is available to authorized users.

M. de Bruijne et al. (Eds.): MICCAI 2021, LNCS 12904, pp. 469–477, 2021.
https://doi.org/10.1007/978-3-030-87202-1_45

1   Introduction

Orthognathic surgery requires a detailed surgical plan to correct jaw deformities. During computer-aided orthognathic surgical planning, a patient's head computed tomography (CT) or cone-beam computed tomography (CBCT) scan is acquired to generate three-dimensional (3D) craniomaxillofacial (CMF) bony shape models [1]. The deformed maxilla and mandible (i.e., "jaw" in Fig. 1(a)) are then osteotomized from the 3D models into multiple bony segments (Fig. 1(b)). Under the guidance of 3D cephalometric analysis [2], a surgeon discretely moves each bony segment to a desired position, thus forming a new normal-looking CMF skeletal anatomy. A group of surgical splints are then designed and fabricated to correct the patient's jaw according to the plan during the surgery (Fig. 1(c)). Cephalometric analysis is based on a group of specific distances and angles calculated from the anatomical landmarks on the 3D models, these measurements provide a limited guidance to the planning process, forming an "ideal" surgical plan that still heavily relies on the surgeon's clinical experience and imagination of what the patient's normal bone should look like. Therefore, from the surgeon's perspective, a reference CMF bony shape model is highly desirable to accurately guide the movement of the bony segments.

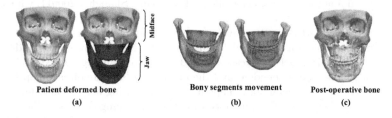

Fig. 1. (a) The bony surface of a patient with jaw deformity, consisting of a normal "midface" and a deformed "jaw" (in red). (b) The jaw is cut into several segment pieces and moved to form a new normal-looking shape model. (c) The patient's facial bony shape model after treatments. (Color figure online)

Wang et al. [3] proposed to estimate patient-specific reference bony shapes based on sparse representation. Specifically, they represented a patient's bony shape with a set of landmarks, and split them into midface landmarks and jaw landmarks. By representing the patient's midface landmarks with a midface dictionary from normal subjects, a set of sparse coefficients were obtained, and applied to a normal jaw dictionary to estimate the patient's normal jaw landmarks. The reference bony shape model was then generated based on the estimated landmarks. This method can work well in certain circumstances, but is heavily dependent on a small group of pre-digitized landmarks, causing the estimation performance to be sensitive to the locations of landmarks. To alleviate this problem, a recent method [4] was introduced to first simulate paired deformed-normal bony surfaces based on sparse representation, and then train a supervised point-cloud network using the simulated data to generate patient-specific reference bony surfaces. Although this network [4] outperforms the method in [3], its generalizability is limited by the extent deformity can be simulated.

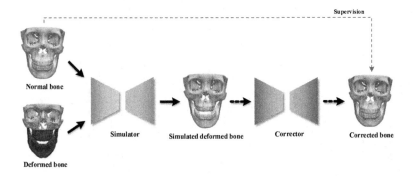

Fig. 2. An overview of our self-supervised deep framework for estimating reference CMF bony shape models.

We hypothesize that a fully end-to-end trainable network is capable of estimating reference CMF bony shapes, and the network can be directly trained using random pairs of deformed-normal bones. We propose a self-supervised deep framework to first map jaw deformities of a patient bone to a normal bone to generate a simulated deformed bone with a simulator network, and then restore the simulated deformed bone back to normal with a corrector network. We train the framework using a series of bony surfaces from patients and normal subjects. The trained corrector is applied to patients to generate patient-specific reference shape models. The proposed framework was evaluated using a clinical dataset and showed superior performance over the method that is based on supervised learning [4].

2   Methods

An overview of our framework is provided in Fig. 2. Given any pair of deformed-normal bony surfaces, the simulator network takes as input the coordinates of N vertices from the two surfaces and outputs the vectors that shift the vertices of the normal bony surface to generate the simulated deformed bony surface with the jaw similar to the input deformed bone and the midface same as the input normal bone. The corrector network learns a set of displacement vectors from the vertex coordinates of the simulated deformed bony surface and generates the corrected bony surface that matches the input normal bony surface.

Simulator Network: The simulator network consists of a series of encoding, fusion, and decoding layers to learn point features from the coordinates of vertices (Fig. 3). There are two encoding branches with shared weights to extract point features from the two input bony surfaces separately. Each encoding layer selects a subset of N_{sub} points from the input points via furthest point sampling [5]. For each sampled point, its neighboring points in a 3D ball region with radius r are gathered from the input points. The point convolution is

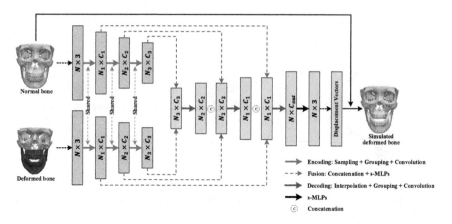

Fig. 3. The architecture of the simulator network, consisting of point-feature encoding, fusion, and decoding layers.

then performed over point features of the gathered points via PointConv [6], which is implemented based on shared multi-layer perceptron (s-MLP) and max-pooling [7]. Following the encoding layers, the outputs of corresponding layers from two encoding branches are fused via concatenation and s-MLPs. The fused features are then decoded by upsampling via point interpolation [5], grouping via 3D ball neighboring, and convolution via PointConv. The decoding layer increases the number N_{sub} of points and reduces the dimension C_{feature} of each feature vector. A set of cascaded fusion and decoding layers are applied to generate a $N \times 3$ output vector $V_{\text{simulator}}$, which updates the positions of the vertices of the input normal bone to obtain the simulated deformed bone.

We train the simulator network using loss function

$$L_{\text{simulator}} = L_{\text{jaw}} + L_{\text{midface}} + \alpha L_{\text{smooth}} + \beta L_{\text{simu\_reg}}, \tag{1}$$

which encourages the jaw of the simulated deformed bone to be similar to the input deformed bone based on the relative coordinates of vertices. Specially, a set of landmarks are defined on each bony surface, the relative coordinate vector $r_{\text{def\_jaw}}(i, j)$ of the i-th jaw vertex relative to the j-th landmark on the deformed bony surface is calculated as

$$r_{\text{def\_jaw}}(i, j) = c_{\text{def\_jaw}}(i) - c_{\text{def\_landmark}}(j), \tag{2}$$

where $c_{\text{def\_jaw}}(i)$ and $c_{\text{def\_landmark}}(j)$ are respectively the coordinate vectors of the i-th vertex and the j-th landmark. Similarly, the relative coordinate vector for the i-th jaw vertex of the simulated deformed bony surface is

$$r_{\text{simu\_jaw}}(i, j) = c_{\text{simu\_jaw}}(i) - c_{\text{simu\_landmark}}(j). \tag{3}$$

We calculate L_{jaw} by

$$L_{\text{jaw}} = \frac{1}{N_{\text{jaw}}} \sum_{i=1}^{N_{\text{jaw}}} \sum_{j=1}^{K} \|r_{\text{def\_jaw}}(i, j) - r_{\text{simu\_jaw}}(i, j)\|_2, \tag{4}$$

where N_{jaw} is the number of vertices that belong to the jaw, K is the number of landmarks, and $\|\cdot\|_2$ is the ℓ_2-norm. L_{midface} forces the network to fix the midface during simulation. To keep the midface unchanged, i.e., $L_{\text{midface}} = 0$, we set the displacement vectors corresponding to midface vertices to zero in $V_{\text{simulator}}$. A smooth deformation field is encouraged by calculating

$$L_{\text{smooth}} = \frac{1}{N} \sum_{i=1}^{N} \frac{1}{\|\mathcal{N}(v_i)\|} \sum_{v_j \in \mathcal{N}(v_i)} \|u(i) - u(j)\|_2, \tag{5}$$

where $u(i) = c_{\text{simu}}(i) - c_{\text{norm}}(i)$ and $u(j) = c_{\text{simu}}(j) - c_{\text{norm}}(j)$ are respectively the displacement vectors of v_i and v_j on the simulated deformed bony surface. $\mathcal{N}(v_i)$ is the set of one-ring neighboring vertices of v_i. N is the total number of vertices on the surface. Finally, a ℓ_2-norm regularization $L_{\text{simu\_reg}}$ of the network parameters is incorporated to avoid overfitting.

Corrector Network: The corrector is an encoding-decoding network (Fig. 4). Like the simulator network, each encoding layer in the corrector performs sampling, grouping, and convolution operations; each decoding layer performs interpolation, grouping, and convolution. The skip connection between two corresponding layers from the encoding and decoding streams is applied to learn point features integrating local and global shape information. Following the decoding stream, the point features are processed by a set of s-MLPs to obtain a $N \times 3$ vector, which corrects the positions of vertices of the simulated deformed bone to get a corrected bone.

The loss function for the corrector network is

$$L_{\text{corrector}} = \frac{1}{N} \sum_{i=1}^{N} \|c_{\text{correct}}(i) - c_{\text{norm}}(i)\|_2 + \lambda L_{\text{correct\_reg}}, \tag{6}$$

where $c_{\text{correct}}(i)$ and $c_{\text{norm}}(i)$ are the coordinate vectors of the i-th vertex on the corrected and input normal bony surface, respectively, $L_{\text{correct\_reg}}$ enforces ℓ_2-norm regularization on the network parameters.

Network Training and Inference: To train the proposed networks, vertex-wise correspondences of surfaces are established by non-rigidly matching a template surface with all bony surfaces in the training set. First, a group of corresponding landmarks localized on the training surfaces are rigidly aligned. From the aligned surfaces, we then choose the surface that is closest to the average landmarks as the template, and warp it to each rest surface via non-rigid coherent point drift (CPD) matching [8]. All warped surface templates are used for network training. We reduce the number of vertices via surface simplification [9] to reduce the computation cost. The vertex coordinates are min-max normalized. The loss functions are minimized with the Adam [10] optimizer to determine the

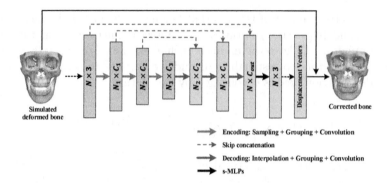

Fig. 4. The architecture of the corrector network with an encoding-decoding structure.

optimal network parameters. The two networks are alternatively trained in turn on each batch of samples. During the inference stage, the trained corrector is directly applied on a patient bony surface to estimate a reference bony shape model. The corrector network is invariant to the number and the order of vertices on the testing surface [7].

3   Experimental Results

Materials and Settings: We used CT scans of 67 normal subjects from a previous study [11] and CT scans of 116 patients with jaw deformities to train our networks. The study was approved by our Institutional Review Board (#Pro00009723). Following the clinical routine [12], all CT data were segmented to reconstruct 3D bony shape models. 51 landmarks were localized on each bony surface by experienced oral surgeons for calculating the loss function during the training only. Following Sect. 2, a bony surface template from the training surfaces was selected based on the 51 landmarks to establish dense vertex correspondences. Ultimately, a total of 7772 (67 × 116) random pairs of deformed-normal bones were used to train the networks.

The simulator network was implemented with 4 encoding, 4 fusion, and 4 decoding layers. The corrector network was configured with 4 layers each for encoding and decoding. The numbers of points for the encoding and decoding layers were $N_{\text{sub}} = \{\frac{N}{4}, \frac{N}{16}, \frac{N}{64}, \frac{N}{128}, \frac{N}{64}, \frac{N}{16}, \frac{N}{4}, N\}$, where $N = 4724$ during training, and $N > 10\,000$ during testing. The radius r of the 3D neighboring ball was set to $\{0.1, 0.2, 0.4, 0.8\}$ and $\{0.8, 0.4, 0.2, 0.1\}$ for the encoding and decoding layers, respectively. The output point features from the 4 encoding and 4 decoding layers had dimensions $C_{\text{feature}} = \{64, 128, 256, 512\}$ and $C_{\text{feature}} = \{512, 256, 128, 128\}$, respectively. Empirically, we set $\alpha = 0.3$, $\beta = 0.1$, and $\lambda = 0.1$ in the loss functions (1) and (6). The number $K = 51$ of landmarks was set for calculating L_{jaw} in (4). The networks were trained for 400 epochs with the learning rate of 0.0001.

Table 1. Statistics for VD (mm), ED (mm), SC, and LD (mm) based on evaluation using 24 patients.

	Method	VD	ED	SC	LD
Jaw	DefNet	4.08 ± 1.02	0.31 ± 0.06	0.69 ± 0.05	4.26 ± 0.95
	Ours	3.49 ± 0.52	0.26 ± 0.05	0.72 ± 0.04	3.62 ± 0.53
Midface	DefNet	1.36 ± 0.36	0.15 ± 0.04	0.96 ± 0.04	1.25 ± 0.31
	Ours	1.39 ± 0.37	0.15 ± 0.04	0.95 ± 0.04	1.32 ± 0.36

Note: A larger SC indicates better estimation performance.

For testing, we acquired the data from another 24 patients with paired pre- and post-operative CT scans. The post-operative bones were used as ground truth for performance evaluation. Following the method in [4], we remeshed each post-operative bony surface according to its pre-operative bony surface to construct the vertex-wise correspondences for calculating estimation accuracy. Four evaluation metrics [4], i.e., vertex distance (VD), edge-length distance (ED), surface coverage (SC), and landmark distance (LD), were employed in the quantitative evaluation. We compared our method with DefNet [4], which was trained using 6834 paired samples simulated by sparse representation, 102 deformed bones were synthesized for each of 67 normal bones from 116 patient bones. Qualitative evaluation was performed by an experienced oral surgeon. Statistical significance of accuracy improvement was evaluated using the paired t-test.

Results: The trained framework was tested on the data of 24 patients. Figure 5 shows the estimated bony surfaces and the vertex-wise distance heatmaps for three randomly selected patients[1]. All results from our method inspected by the expert suggest that the estimated bones are clinically acceptable. For quantitative evaluation, the estimation accuracy for the jaw and the midface were calculated separately. Table 1 shows that our method is statistically significantly more accurate in estimating the normal jaw ($p < 0.05$) than DefNet based on the four metrics. The accuracy in maintaining the midface is comparable between the two methods ($p > 0.05$). Overall, the proposed method outperforms DefNet in the term of accuracy.

Our method takes about 30 min for one training epoch, and 10 s for testing on each patient data, evaluated based on an NVIDIA Titan Xp GPU.

---

[1] A video illustration of the three estimation examples is available as a supporting material.

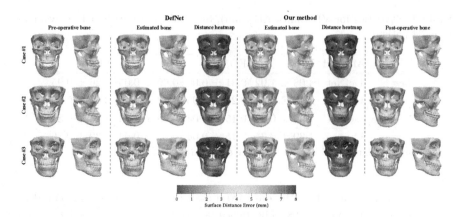

Fig. 5. Corrected bones estimated by the two methods for three randomly selected patients. Heatmaps of surface vertex distance are calculated by comparing the estimated bony surfaces with the corresponding post-operative bony surfaces (ground truth).

4   Discussion and Conclusion

As sufficient paired deformed-normal bones are almost impossible to acquire clinically, simulating paired data is necessary for training supervised deep learning models for our task. Data simulation based on sparse representation as done in DefNet [4] is limited in mimicking deformities, and requires significant efforts from experts to confirm simulation quality. Our self-supervised deep learning framework can be trained using unpaired data. The embedded simulator network has a stronger non-linear representation ability than sparse representation, and is able to accurately synthesize a broader range of jaw deformities based on patient bones. Compared with DefNet, our framework simplifies the training process and improves prediction accuracy.

Acknowledgement. This work was supported in part by United States National Institutes of Health (NIH) grants R01 DE022676, R01 DE027251, and R01 DE021863.

References

1. Xia, J., et al.: Algorithm for planning a double-jaw orthognathic surgery using a computer-aided surgical simulation (CASS) protocol. Part 1: planning sequence. Int. J. Oral Maxillofacial Surg. **44**(12), 1431–1440 (2015)
2. Xia, J., et al.: Algorithm for planning a double-jaw orthognathic surgery using a computer-aided surgical simulation (CASS) protocol. Part 2: three-dimensional cephalometry. Int. J. Oral Maxillofacial Surg. **44**(12), 1441–1450 (2015)
3. Wang, L., et al.: Estimating patient-specific and anatomically correct reference model for craniomaxillofacial deformity via sparse representation. Med. Phys. **42**(10), 5809–5816 (2015)

4. Xiao, D., et al.: Estimating reference bony shape models for orthognathic surgical planning using 3D point-cloud deep learning. IEEE J. Biomed. Health Informat. (2021). https://doi.org/10.1109/JBHI.2021.3054494

5. Qi, C.R., et al.: PointNet++: deep hierarchical feature learning on point sets in a metric space. In: Proceedings of the Neural Information Processing Systems, pp. 5099–5108 (2017)

6. Wu, W., et al.: PointConv: deep convolutional networks on 3D point clouds. In: Proceedings IEEE Conference Computer Vision Pattern Recognition, pp. 9621–9630 (2019)

7. Qi, C.R., et al.: PointNet: deep learning on point sets for 3D classification and segmentation. In: Proceedings IEEE Conference Computer Vision Pattern Recognition, pp. 652–660 (2017)

8. Myronenko, A., Song, X.: Point set registration: coherent point drift. IEEE Trans. Pattern Anal. Mach. Intell. 32(12), 2262–2275 (2010)

9. Garland, M., Heckbert, P.: Surface simplification using quadric error metrics. In: Proceedings SIGGRAPH, vol. 97, pp. 209–216 (1997)

10. Kingma, D., Ba, J.: Adam: a method for stochastic optimization. In: Proceedings International Conference Learning Representations, pp. 1–41 (2015)

11. Yan, J., et al.: Three-dimensional CT measurement for the craniomaxillofacial structure of normal occlusion adults in Jiangsu, Zhejiang and Shanghai Area. China J. Oral Maxillofac. Surg. 8, 2–9 (2010)

12. Yuan, P., et al.: Design, development and clinical validation of computer-aided surgical simulation system for streamlined orthognathic surgical planning. Int. J. Comput. Assist. Radiol. Surg. 12(12), 2129–2143 (2017). https://doi.org/10.1007/s11548-017-1585-6

DLLNet: An Attention-Based Deep Learning Method for Dental Landmark Localization on High-Resolution 3D Digital Dental Models

Yankun Lang[1], Hannah H. Deng[2], Deqiang Xiao[1], Chunfeng Lian[1], Tianshu Kuang[2], Jaime Gateno[2,3], Pew-Thian Yap[1(✉)], and James J. Xia[2,3(✉)]

[1] Department of Radiology and Biomedical Research Imaging Center (BRIC), University of North Carolina at Chapel Hill, Chapel Hill, NC, USA
ptyap@med.unc.edu
[2] Department of Oral and Maxillofacial Surgery, Houston Methodist Hospital, Houston, TX, USA
jxia@houstonmethodist.org
[3] Department of Surgery (Oral and Maxillofacial Surgery), Weill Medical College, Cornell University, NY, USA

Abstract. Dental landmark localization is a fundamental step to analyzing dental models in the planning of orthodontic or orthognathic surgery. However, current clinical practices require clinicians to manually digitize more than 60 landmarks on 3D dental models. Automatic methods to detect landmarks can release clinicians from the tedious labor of manual annotation and improve localization accuracy. Most existing landmark detection methods fail to capture local geometric contexts, causing large errors and misdetections. We propose an end-to-end learning framework to automatically localize 68 landmarks on high-resolution dental surfaces. Our network hierarchically extracts multi-scale local contextual features along two paths: a landmark localization path and a landmark area-of-interest segmentation path. Higher-level features are learned by combining local-to-global features from the two paths by feature fusion to predict the landmark heatmap and the landmark area segmentation map. An attention mechanism is then applied to the two maps to refine the landmark position. We evaluated our framework on a real-patient dataset consisting of 77 high-resolution dental surfaces. Our approach achieves an average localization error of 0.42 mm, significantly outperforming related start-of-the-art methods.

Keywords: 3D dental surface · Landmark localization · Geometric deep learning

1 Introduction

Digitalization (a.k.a. Localization) of dental landmarks is a necessary step in dental model analysis during treatment planning for patients with jaw and teeth deformities. In the modern era of digital dentistry, high-resolution digital dental

© Springer Nature Switzerland AG 2021
M. de Bruijne et al. (Eds.): MICCAI 2021, LNCS 12904, pp. 478–487, 2021.
https://doi.org/10.1007/978-3-030-87202-1_46

surface mesh models are either generated by a three-dimensional (3D) intraoral surface scanner or constructed from cone-beam computed tomography (CBCT) images. In the current standard of care, over 60 commonly used dental landmarks are digitized manually for each patient by orthodontists, surgeons, or trained technicians, which is time-consuming and labor-intense.

Automatic localization of dental landmarks on a 3D surface mesh model is challenging. A high degree of accuracy (less than 0.5 mm error) is required. The shapes of dental landmark areas (cusps and fossa) vary dramatically across patients due to normal wear or tooth restoration. Processing these high-resolution models is computationally intensive since they usually contain more than 100,000 mesh cells. Over the years, deep neural networks have been shown to be effective in the localization of anatomical landmarks [7,11,13,14]. However, these networks are developed mainly for medical images and can not be directly used on 3D mesh models. A potential solution, as described in [4,6], is to map the 3D mesh to a 2D planar flat-torus, which is then fed to a fully convolutional network [5] to annotate the landmarks. This approach is susceptible to transformation artifacts and information loss. More recently, PointNet++ [9] was proposed to learn group-wise geometric features by applying PointNet [8] hierarchically on grouped points. PointConv [12] learns translation-invariant and permutation-invariant convolution kernels via multi-layer perceptrons (MLP). Lian et al. [3] introduced MeshSegNet to hierarchically extract multi-scale local contextual features with dynamic graph-constrained learning modules by using multiple features extracted from each cell.

Although yielding promising results in classification and segmentation tasks, the methods described above suffer from several limitations when applied to detecting dental landmarks. First, they are agnostic to curvature features and are not necessarily catered to learning edge features inside the landmark areas. Second, the high-resolution model is usually significantly down-sampled to meet GPU limitations. Essential structural information is hence lost and localization accuracy might not be able to meet the clinical requirements.

In this paper, we propose an end-to-end deep learning method, DLLNet, to automatically localize 68 commonly used dental landmarks on 3D high-resolution dental models. All landmarks are detected with a coarse-to-fine two-stage strategy (Fig. 1). In the first stage, a segmentation network [3] is applied on a down-sampled mesh model for tooth segmentation. The teeth are grouped into four partitions. The proposed network takes each partition as input, and outputs a coarse localization result of each landmark. In the second stage, DLLNet is applied to mesh patches sampled in the vicinity of the coarse localization results to refine landmark locations.

The main technical contribution of our paper is three-fold. First, DLLNet hierarchically extracts multi-scale local contextual features along two collaborative task-driven paths (i.e., landmark localization and landmark area segmentation). It captures the global context of each tooth and the local contexts of landmark areas. Second, in addition to features described in [3] (i.e., vertex coordinates, cell normal vectors, and cell centroids), curvature features are

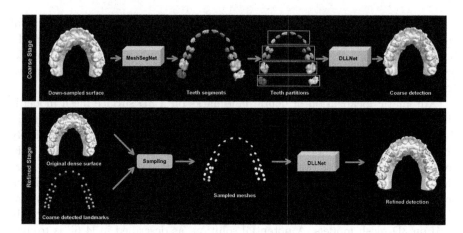

Fig. 1. Coarse-to-fine framework for dental landmark localization on a 3D surface.

included for more comprehensive structural description of landmark areas. Third, an attention mechanism is applied to improve detection accuracy and to reduce misdetections.

2 Methods

As shown in Fig. 2, DLLNet extracts multi-scale local context features along two task-driven paths. The extracted global-to-local features are concatenated to output heatmaps and segmentation probability maps. Additionally, an attention mechanism is adopted for the two outputs, yielding refined heatmaps for landmark localization.

2.1 High-Level Feature Extraction

DLLNet takes a matrix $\mathbf{F}^0 \in R^{N \times 24}$ as input. N is the number of cells in the down-sampled mesh models. Each cell is described by a 24-dimensional feature vector. Following [3], the first 15 elements of the feature vector include the coordinates of the three vertices (9 elements), normal vectors (3 elements) and cell centroid (3 elements). Since dental landmarks are located on the tips or valleys of the tooth surface with typically large curvatures (e.g., cusp landmarks or fossa landmarks), Gaussian curvatures (3 elements), maximum curvatures (3 elements) and minimum curvatures (3 elements) are included to capture edge information.

Given an input feature matrix $\mathbf{F}^0$, the first MLP block consisting of two successive MLP layers is applied to extract high-level geometric features. A feature-transformer module (FTM) [8] follows by aligning the input to a canonical space to learn transformation-invariant features $\mathbf{F}^1 \in R^{N \times 64}$. After FTM, $\mathbf{F}^1$ is fed to two first-level graph-constrained learning modules (GLMs) [3], i.e., GLM_S_1 and

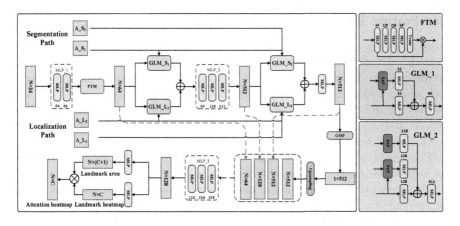

Fig. 2. The architecture of DLLNet and the details of the modules.

GLM_L$_1$ in the segmentation path and the localization path, respectively. Specifically, the segmentation path detects areas where landmarks may exist (landmark RoI). With the same modules but different receptive fields, the localization path detects landmarks from these areas. In each path, symmetric average pooling (SAP) operates on the input feature matrix $\mathbf{F}$ and an $N \times N$ adjacent matrix $\mathbf{A}$ to generate a local contextual feature matrix $\tilde{\mathbf{F}}$, which is calculated by

$$f_{\mathrm{SAP}}(\mathbf{F}|\mathbf{A}) = (\tilde{\mathbf{D}}^{-\frac{1}{2}}(\mathbf{A}+\mathbf{I})\tilde{\mathbf{D}}^{-\frac{1}{2}})\mathbf{F}, \qquad (1)$$

where $\tilde{\mathbf{D}}^{-\frac{1}{2}}$ is the diagonal degree matrix. Adjacent matrix $\mathbf{A}$ controls the receptive field in a sphere with the geodesic radius r. We empirically set $r_{L_1} = 0.25$ and $r_{S_1} = 0.1$ to construct $\mathbf{A}_{L_1}$ and $\mathbf{A}_{S_1}$ since localizing landmark requires a larger receptive field. The output of the first-level GLM is calculated by

$$\hat{\mathbf{F}} = \sigma(\sigma(\tilde{\mathbf{F}}) \oplus \sigma(\mathbf{F})), \qquad (2)$$

where $\sigma(\cdot)$ is the MLP layer and $\oplus$ is a concatenation operator. The outputs of GLM_S$_1$ and GLM_L$_1$, i.e., $\hat{\mathbf{F}}_{S_1}$ and $\hat{\mathbf{F}}_{L_1}$, are concatenated across channels and are then consumed by the second MLP block to generate a feature matrix $\mathbf{F}^2 \in R^{N \times 512}$.

The second-level GLMs (GLM_S$_2$ and GLM_L$_2$) adopt an addition SAP operation on $\mathbf{A}_{L_2}/\mathbf{A}_{S_2}$ with $\mathbf{F}^2$ to output multi-scale contextual features $\hat{\mathbf{F}}_{S_2}$ and $\hat{\mathbf{F}}_{L_2}$. Specifically, $\mathbf{A}_{L_2}$ and $\mathbf{A}_{S_2}$ are constructed by setting $r_{L_1} = 0.3$ and $r_{S_1} = 0.2$ to enlarge the receptive fields. $\hat{\mathbf{F}}_{S_2}$ and $\hat{\mathbf{F}}_{L_2}$ are then concatenated and squeezed by a MLP layer to output $\mathbf{F}^3$. Global max pooling (GMP) is then applied to embed global structural features into a feature vector $\mathbf{F}^4$.

2.2   Feature Fusion and Attention Heatmap

A fusion strategy is employed to concatenate the local-to-global contextual features ($\mathbf{F}^1$, $\mathbf{F}^2$, $\mathbf{F}^3$ and $\mathbf{F}^4$). Followed by the third MLP block, $\mathbf{F}^5 \in R^{N \times 128}$ is obtained as the feature matrix that is shared by two tasks: 1) landmark regression, where a MLP layer is used to predict a Gaussian heatmap matrix $\mathbf{H}$ with size $N \times C$; and 2) landmark area segmentation, where another MLP layer with softmax activation is used to predict a probability map $\mathbf{S}$ with size $N \times (C+1)$. C is the number of landmarks.

The result of landmark localization is sensitive to the accuracy of $\mathbf{H}$ on the foreground mesh cells (mesh cells that are close to the target landmark). Misdetection even happens when background mesh cells are assigned with a high probability due to feature similarity. To eliminate these effects, we use $\mathbf{S}$ as an attention map on $\mathbf{H}$ to generate an attention heatmap $\hat{\mathbf{H}}$:

$$\hat{\mathbf{H}} = \mathbf{H} \odot \hat{\mathbf{S}}, \tag{3}$$

where $\hat{\mathbf{S}} \in R^{N \times C}$ consists of the last C columns of $\mathbf{S}$. $\odot$ is the Hadamard-product. Learning $\hat{\mathbf{H}}$ forces the network to focus on the regression on landmark areas. This procedure also constrains the training of $\mathbf{S}$ and $\mathbf{H}$ by each other. Finally, the landmark localization results are determined by $\hat{\mathbf{H}}$ as the coordinates of the mesh cell with the largest probability value. Additionally, computing $\hat{\mathbf{H}}$ can be regarded as local coarse-to-fine processing since $\mathbf{S}$ can be viewed as a coarse landmark detection result. The total training loss of our network is

$$L = \lambda_h L_H(\mathbf{H}, \mathbf{H}^*) + \lambda_s L_S(\mathbf{S}, \mathbf{S}^*) + \lambda_a L_A(\hat{\mathbf{H}}, \hat{\mathbf{H}}^*), \tag{4}$$

where λ_h, λ_s and λ_a are training weights. $\mathbf{H}^*$, $\mathbf{S}^*$ and $\hat{\mathbf{H}}^*$ are the corresponding ground truths. We employ Adaptive Wing Loss [10] for L_H, MSE loss for L_A, and generalized Dice loss [1] for L_S.

2.3   Implementation and Inference

In the first stage, we use tooth surfaces as the ground truth to train the segmentation network, i.e., MeshSegNet, following the parameter setting in [3]. Each tooth surface is formed by combining all corresponding landmark areas cropped within a non-overlapping geodesic ball ($r = 1.5 \, \mathrm{mm}$). The segmented teeth are grouped into four partitions: anterior teeth (incisors + canines), premolars, first molars and second molars. Before training the DLLNet, we crop teeth partitions that have the same topology as the corresponding segmentation results from the original data, then down-sampled them to 3,000 mesh cells.

DLLNet is trained by ADAM optimizer with an initial rate of 0.01 for 30 epochs (2000 iterations/epoch) in total. The batch size is set to 20. $\mathbf{H}^*$ is created with a Gaussian distribution with variance of $1 \, \mathrm{mm}$ on each landmark. The geodesic radius of landmark areas for $\mathbf{S}^*$ is set to 0.8 mm. $\hat{\mathbf{H}}^*$ is generated by performing Hadamard-product on $\mathbf{H}^*$ with $\mathbf{S}^*$ (last C columns). In the second stage, 150 mesh cells around each predicted landmark ($\leq 0.5 \, \mathrm{mm}$) are sampled

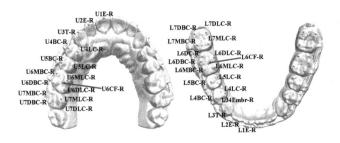

Fig. 3. Names of landmarks annotated on the maxillary (left) and mandibular (right) dental models.

to train another DLLNet to refine the results. We empirically set $\lambda_h = 0.5$, $\lambda_s = 0.5$, and $\lambda_a = 1$.

In the inference phase, all landmarks are localized by directly using the coarse-to-fine strategy with the trained networks. In the second stage, only 150 mesh cells centered at the estimated landmark location are sampled. Our approach takes about 1 min to process a dental model (maxilla or mandible) using an Intel Core i7-8700K CPU with a 12 GB GeForce GTX 1080Ti GPU. All the procedures are implemented by Python based on Keras.

3   Experiments

3.1   Data

Our approach was evaluated quantitatively using 77 sets of high-resolution digital dental models randomly selected from our clinical digital archive, in which 15 sets were partially edentulous (missing tooth/teeth). All personal information were deidentified prior to the study. For each set of the dental models, 32 maxillary and 36 mandibular dental landmarks were digitized by experienced oral surgeons (Fig. 3). Each dental surface has roughly 100,000 ~ 300,000 mesh cells, with a resolution of 0.2 ~ 0.4 mm (the average length of cell edges). Using 5-fold cross-validation, we randomly selected 57 sets for training, 10 sets for validation and the rest for testing. Prior to training, data augmentation (30 times) was performed by random rotation ($[-\frac{\pi}{20}, \frac{\pi}{20}]$), translation ($[-20, 20]$) and re-scaling ($[0.8, 1.2]$) along the three orthogonal direction. The input feature matrix was normalized by Gaussian normalization constant (GNC).

3.2   Comparison Methods

DLLNet was compared with PointNet++ [9], PointConv [12] and the state-of-the-art MeshSegNet [3] with the same network architectures that were described in the original papers. To evaluate the effectiveness of the curvature features, two-task driven paths, and the attention mechanism, which are the main differences between our DLLNet and MeshSegNet, we performed an ablation study by

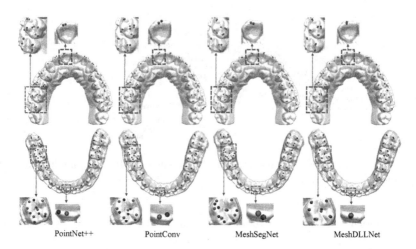

|PointNet++ | PointConv | MeshSegNet | MeshDLLNet|

Fig. 4. Results of maxillary and mandibular landmark localization of a set of randomly selected dental models using the four methods (Red: Algorithm-localized landmarks; Green: Ground truth). (Color figure online)

comparing DLLNet with three variants: 1) DLL-SA with input features identical to MeshSegNet; 2) DLL-C with GLM_S_1, GLM_S_2 and output **S** removed and thus only focuses on heatmap regression; and 3) DLL-CS with the attention module removed. The results of landmark localization were quantitatively evaluated with root mean squared error (RMSE). Finally, the misdetection rate (MDR) was calculated. All compared methods were trained using the same coarse-to-fine strategy, augmented dataset, and training loss for heatmap regression and landmark area segmentation.

3.3   Results

Table 1 summarizes the landmark localization results in RMSE based on anatomical regions, including anterior teeth (AT), central and lateral incisors, canines, premolars (PM), first molars (FM), and second molars (SM). The central dental

Table 1. RMSE (mean ± SD, unit: mm) of landmark localization.

Method	AT	PM	FM	SM	CI	MDR
Point++	0.71 ± 0.49	1.02 ± 0.58	1.40 ± 0.53	1.44 ± 0.64	0.95 ± 0.61	17%
PointConv	0.66 ± 0.41	1.40 ± 0.54	1.39 ± 0.48	1.41 ± 0.58	0.82 ± 0.52	15%
MeshSegNet	0.64 ± 0.49	0.52 ± 0.41	0.59 ± 0.47	0.78 ± 0.43	0.78 ± 0.43	10%
DLL-SA	0.48 ± 0.27	0.51 ± 0.42	0.57 ± 0.48	0.69 ± 0.58	0.49 ± 0.38	3%
DLL-C	0.49 ± 0.29	0.48 ± 0.34	0.56 ± 0.41	0.60 ± 0.42	0.48 ± 0.34	10%
DLL-CS	0.40 ± 0.21	0.45 ± 0.32	0.51 ± 0.36	0.58 ± 0.39	0.42 ± 0.29	8%
DLLNet	**0.30 ± 0.11**	**0.39 ± 0.26**	**0.47 ± 0.28**	**0.49 ± 0.37**	**0.28 ± 0.15**	0%

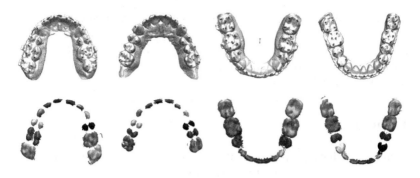

Fig. 5. Localization and pre-segmentation results for partially edentulous patients.

midline points of maxillary and mandibular dental arches are extremely important during planning and treatment, as they are derived from the right and left of the central incisors as the midpoint [2]. Therefore, we single out the accuracy evaluation of the four central incisor (CI) landmarks from the anterior teeth and present them separately in the fifth column. The results show that DLLNet achieves the highest accuracy among the four methods. MeshSegNet outperforms Point++ and PointConv in all 4 dental regions, indicating that using GLM and the combined features is effective in capturing the local-to-global contextual features. However, the errors are still considerably large in all regions because consideration of structural information captured in landmark areas is inadequate. Furthermore, all three competing methods have a large MDR due to surface similarity, hindering them from being used in real clinical applications. Finally, the accuracy in molar regions are slightly lower than the others due to normal wear of the molars. Nonetheless, the accuracy achieved by DLLNet is still within the clinical standard of 0.5 mm. Figure 4 shows the qualitative results of a set of randomly selected dental models, clearly demonstrating that our approach yields in overall better performance than the completing methods. Notably, the localization accuracy on the central and lateral incisors has been significantly improved.

The ablation results compared with the three variants are summarized in Table 1. DLL-C outperforms MeshSegNet and other methods partly due to the curvature features, which are added into the input matrix by considering that landmarks are located on cusps or fossa. However, DLL-C still yields a large MDR (10%). DLL-CS further improves the accuracy by collaboratively performing multi-scale landmark detection via the landmark area segmentation path. The overall performance of DLL-SA is slightly worse than the other two variants because curvature information is not considered. However, DLL-SA still outperforms the other compared methods via local coarse-to-fine processing, where the landmark area segmentation results are used as attention maps, forcing each landmark to be localized within the landmark area, thus reducing misdetections. By integrating all these strategies into our framework, DLLNet ultimately achieves the highest accuracy and the lowest misdetection rate when compared with related methods.

Finally, Fig. 5 shows the localization results on randomly selected partial edentulous subjects, where tooth absence can be directly detected from the pre-segmentation results. All available landmarks are correctly and accurately localized without false detection. This strongly suggests that our approach is capable of handling imperfections.

4   Conclusion

In this paper, we have proposed an attention-based deep learning method, called DLLNet, to accurately localize 68 commonly used tooth landmarks on 3D dental surface models. DLLNet applies curvature features and learns multi-scale local contextual features along two task-driven paths. By using an attention mechanism, the network further refine localization accuracy and reduces misdetections. Experimental results based on a clinical dataset show that DLLNet significantly outperforms related state-of-the-art methods. Future work will focus on validation on more patients with various tooth conditions.

Acknowledgment. This work was supported in part by United States National Institutes of Health (NIH) grants R01 DE022676, R01 DE027251, and R01 DE021863.

References

1. Hong, Y., Kim, J., Chen, G., Lin, W., Yap, P.T., Shen, D.: Longitudinal prediction of infant diffusion MRI data via graph convolutional adversarial networks. IEEE Trans. Med. Imaging **38**(12), 2717–2725 (2019)
2. Hsu, S.S.P., et al.: Accuracy of a computer-aided surgical simulation protocol for orthognathic surgery: a prospective multicenter study. J. Oral Maxillofac. Surg. **71**(1), 128–142 (2013)
3. Lian, C., Wang, L., Wu, T.H., Wang, F., Yap, P.T., Ko, C.C., Shen, D.: Deep multi-scale mesh feature learning for automated labeling of raw dental surfaces from 3d intraoral scanners. IEEE Trans. Med. Imaging **39**(7), 2440–2450 (2020)
4. Liu, S., He, J.L., Liao, S.H.: Automatic detection of anatomical landmarks on geometric mesh data using deep semantic segmentation. In: 2020 IEEE International Conference on Multimedia and Expo (ICME), pp. 1–6. IEEE (2020)
5. Long, J., Shelhamer, E., Darrell, T.: Fully convolutional networks for semantic segmentation. In: Proceedings of the IEEE Conference on Computer Vision and Pattern Recognition, pp. 3431–3440 (2015)
6. Maron, H., et al.: Convolutional neural networks on surfaces via seamless toric covers. ACM Trans. Graph. **36**(4), 71–1 (2017)
7. Payer, C., Štern, D., Bischof, H., Urschler, M.: Regressing heatmaps for multiple landmark localization using CNNs. In: Ourselin, S., Joskowicz, L., Sabuncu, M.R., Unal, G., Wells, W. (eds.) MICCAI 2016. LNCS, vol. 9901, pp. 230–238. Springer, Cham (2016). https://doi.org/10.1007/978-3-319-46723-8_27
8. Qi, C.R., Su, H., Mo, K., Guibas, L.J.: PointNet: deep learning on point sets for 3D classification and segmentation. In: Proceedings of the IEEE Conference on Computer Vision and Pattern Recognition, pp. 652–660 (2017)

9. Qi, C.R., Yi, L., Su, H., Guibas, L.J.: Pointnet++: deep hierarchical feature learning on point sets in a metric space. arXiv preprint arXiv:1706.02413 (2017)
10. Wang, X., Bo, L., Fuxin, L.: Adaptive wing loss for robust face alignment via heatmap regression. In: Proceedings of the IEEE/CVF International Conference on Computer Vision, pp. 6971–6981 (2019)
11. Wang, X., Yang, X., Dou, H., Li, S., Heng, P.A., Ni, D.: Joint segmentation and landmark localization of fetal femur in ultrasound volumes. In: 2019 IEEE EMBS International Conference on Biomedical & Health Informatics (BHI), pp. 1–5. IEEE (2019)
12. Wu, W., Qi, Z., Fuxin, L.: PointConv: deep convolutional networks on 3D point clouds. In: Proceedings of the IEEE/CVF Conference on Computer Vision and Pattern Recognition, pp. 9621–9630 (2019)
13. Zhang, J., Liu, M., Shen, D.: Detecting anatomical landmarks from limited medical imaging data using two-stage task-oriented deep neural networks. IEEE Trans. Image Process. **26**(10), 4753–4764 (2017)
14. Zhang, J., et al.: Joint craniomaxillofacial bone segmentation and landmark digitization by context-guided fully convolutional networks. In: Descoteaux, M., Maier-Hein, L., Franz, A., Jannin, P., Collins, D.L., Duchesne, S. (eds.) MICCAI 2017. LNCS, vol. 10434, pp. 720–728. Springer, Cham (2017). https://doi.org/10.1007/978-3-319-66185-8_81

Personalized CT Organ Dose Estimation from Scout Images

Abdullah-Al-Zubaer Imran[1(✉)], Sen Wang[1], Debashish Pal[2], Sandeep Dutta[2], Bhavik Patel[1], Evan Zucker[1], and Adam Wang[1]

[1] Stanford University, Stanford, CA 94305, USA
aimran@stanford.edu
[2] GE Healthcare, Waukesha, WI 53188, USA

Abstract. With the rapid increase of CT usage, radiation dose across patient populations is also increasing. Therefore, it is desirable to reduce the CT radiation dose. However, the reduction in dose also incurs additional noise and with the degraded image quality, diagnostic performance can be compromised. Existing routine dosimetric quantities are usually based on absorbed dose within cylindrical phantoms and do not appropriately represent the actual patient dose. More comprehensive dose metrics such as effective dose require estimation of patient-specific dose at an organ level. Unfortunately, currently available systems are quite far from achieving this goal as well as limited by a number of manual adjustments, time-consuming and inefficient procedures. To overcome all these challenges in achieving the goal of patient safety through reduced dose without compromising image quality, we devise a fully-automated, end-to-end deep learning-based solution to perform real-time, patient-specific, organ-level dosimetric prediction of CT scans. Leveraging the 2D scout (frontal and lateral) images of the actual patients, which are routinely acquired prior to the CT scan, our proposed *Scout-Net* model estimates the patient-specific mean dose in real-time for six different organs. Our experimental evaluation on real patient data demonstrates the effectiveness of our *Scout-Net* model not only in real-time dose estimation (only 11 ms on average per scan), but also as a potential tool for optimizing CT radiation dose in specific patients.

Keywords: Computed tomography · Ionizing radiation · Organ dose · Scout images · CNN · Segmentation

1 Introduction

Computed tomography (CT) has been one of the most successful imaging modalities and has facilitated countless image-based medical procedures since its invention five decades ago. CT accounts for a large amount of ionizing radiation exposure, especially with the rapid growth in CT examinations [16]. In CT scans, there is always a trade-off between dose and image quality as images are acquired with the two competing objectives of maximizing image quality and minimizing

© Springer Nature Switzerland AG 2021
M. de Bruijne et al. (Eds.): MICCAI 2021, LNCS 12904, pp. 488–498, 2021.
https://doi.org/10.1007/978-3-030-87202-1_47

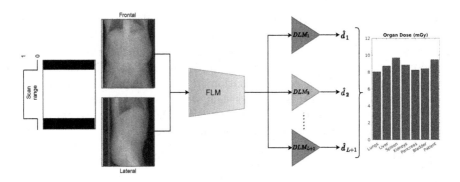

Fig. 1. Schematic of the proposed *Scout-Net* model.

radiation dose [12,13]. CT dose optimization is a multi-step process. Optimization of exam protocols could be performed through establishing baseline dose levels and image quality [5]. Patient-specific dose metrics such as *effective dose* need to account for the different radio-sensitivity of different organs (e.g., using the ICRP tissue weights) [21]. This could facilitate focusing radiation exposure in regions where it is most needed for image quality and the clinical task, while avoiding radiation-sensitive organs.

Over the years, Monte Carlo (MC) simulation has remained the gold-standard for estimating patient dose. The MC simulation can generate and track particles at the voxel level and the deposited energy is calculated for patient-specific estimation of absorbed dose [8]. An accelerated simulation has been introduced through GPU computation, namely MC-GPU, which can accurately model the physics of x-ray photon transport in voxelized geometries [1]. While MC-GPU has made patient-specific dose estimation faster, it is limited by the requirement of expensive computing resources, configuration dependent on a number of heuristics, and 3D patient data. With the advent of deep learning, especially convolutional neural networks (CNNs), data driven automated CT dose estimation has been gaining a lot of attention. Several deep learning-based CT dose calculation approaches have been proposed [7,9,11,14,15,18,25]. Organ-specific and region focused dose estimates would be more practical and potentially a tool to reduce patient effective doses. Maier et al. [18] proposed a 3D U-Net-based dose estimation to reproduce the MC dose distribution from an input CT scan. Other 3D U-Net-based dose calculation methods were proposed by Fan et al. [7] and Kontaxis et al. [14]. Offe et al. [20] developed a rapid and fully automated tool for patient-specific organ dose estimation from pediatric CT scans using a LBTE solver and V-Net segmentation model. However, all of these studies assume a 3D patient representation is available, which is only possible *after* the CT scan is obtained. If we wish to optimize the CT scan itself, these organ dose estimates must be made *before* the CT scan.

CT scout images are preliminary radiographs that are mainly obtained to aid in planning the CT examinations, such as for patient positioning, identifying the scan range, and determining tube current modulation. The scouts make up only about 4% of the total radiation dose and could even be important to diagnosis as well [2]. In contrast to the existing CT dose estimation methods, we propose a simple yet effective, end-to-end, fully-automated CNN-based solution for patient-specific, organ dose estimation in real time from the scouts only. To our best knowledge this is the first such study focused on estimating organ dose from scouts, prospectively even before doing the actual scans. Our specific contributions in the present paper can be summarized as:

- Real-time, patient-specific, CT organ dose estimation prospectively from scout images
- Simple yet effective deep learning-based solution for predicting organ-specific radiation dose from frontal and lateral scout images
- Configuration of CT dose simulation with bowtie filtration and anode heel effect
- Estimation of organ dose leveraging a 3D multi-organ segmentation model

2   Patient-Specific Organ Dose from Scouts

We propose the *Scout-Net* model (see Fig. 1) to predict the CT dose distribution (characterized by mean dose) at organs-of-interest from input frontal x_1 and lateral x_2 scout views. The scouts are almost always longer than the actual scans. Since our goal is to predict the dose in the actual scans, a scan range signal s is passed to inform the model about the focus region in the input scouts. This additional information makes the model informed about where to look at for dose estimation, given the input scout images. The model facilitates jointly learning the dose prediction of the patient body and organs-of-interest. The *Scout-Net* model comprises of two basic modules: a generic feature learning module (FLM) and multiple dose learning modules (DLM). FLM is used to extract the shared features, later utilized through separate organ-specific DLM. The model also estimates the patient's overall body dose through a separate DLM.

The FLM module includes eight 3×3 convolutions with feature maps of 16, 16, 32, 32, 64, 128, 256, and 512 respectively with stride 2 in the second and fourth convolutions; each of the convolutions is followed by an instance normalization and a leaky ReLU activation with negative slope of 0.2. The feature maps are downsized by half after every two convolutions via 2×2 maxpool operations. Through a global average pooling layer, the features are drawn across the channels. The extracted features are then shared across the DLMs. Each DLM consists of two full-connected (FC) layers (512 and 1 neurons respectively), a leakyReLU activation with slope 0.2, and finally a sigmoid to output the mean dose prediction in the normalized scale. From the reference mean doses d_l and the model predicted doses $\hat{d}_l$ at an organ labeled as l ($l \in L$), the loss for the *Scout-Net* model is therefore calculated as:

$$\mathcal{L}^{dose}_{(d,\hat{d})} = \frac{1}{M} \sum_i^M \sum_l^{L+1} \|d_l(i) - \hat{d}_l(i)\|_2, \tag{1}$$

Table 1. Customized MC-GPU parameters to configure the dose calculation engine for our realistic CT system.*

Parameter	Choice(s)
Tube potential	120 kVp
Pitch factor	0.99
Z-axis coverage	80 mm
Source to detector distance	1097.61 mm
Source to rotation axis distance	625.61 mm
Angle between projections	15°
Vertical translation between projections	3.3 mm
Polar and azimuthal apertures	(42.0°, 7.32°)
Voxel spacings (isotropic)	(4, 4, 4) mm^3

*Default MC-GPU parameters are not listed

where M denotes the minibatch size and the patient body dose is denoted at $(L+1)$-th.

3   Organ Dose Estimation with CT Scans

To establish reference organ dose values, we first calculate the patient-specific CT organ dose from CT scans. We make use of the publicly available MC-GPU[1] [1] and configured it to model more realistic CT scanners to compute the 3D dose map (Sect. 3.1). Combined with organ segmentations through a 3D CNN-based model, organ-specific dose statistics are calculated (Sect. 3.2). Although we use a CNN to segment organs rather than by an expert, previous studies have shown that organ dose is robust against small errors in segmentation [20].

In our pipeline for calculating organ dose from CT scans, an input CT scan is interpolated and resampled to 128×128 in each axial slice. A voxelized patient phantom is created by mapping the CT Hounsfield Units (HU) to water-equivalent density, with isotropic voxels of $4 \times 4 \times 4$ mm^3.

$$V = \frac{HU}{1000} + 1. \tag{2}$$

3.1   Monte Carlo Dose Estimation

MC simulations can provide patient-specific estimates of organ dose with the input of 3D patient phantom models. The GPU-based MC tool (MC-GPU) has near real-time performance and was validated independently on an AAPM reference phantom [23].

With the goal of modeling a realistic CT system, the original version of MC-GPU is modified to include the bowtie filtration and anode heel effect (Table 1).

---

[1] https://github.com/DIDSR/MCGPU, version 1.3, accessed on August 10, 2020.

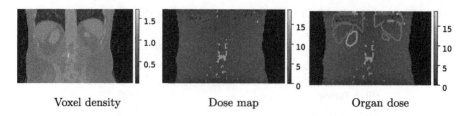

Voxel density   Dose map   Organ dose

Fig. 2. Coronal view: generation of organ-specific dose maps from a CT scan leveraging the MC-GPU dose engine and the multi-organ segmentation.

The anode heel effect leads to photon intensity variation for azimuthal angles that can be modeled as a probability function. The bowtie model is based on basis material decomposition so that the inputs for MC-GPU are the material attenuation coefficients and material thickness combinations at different fan angles. Once the source spectrum is input, we can calculate the filtered spectra and relative photon intensity distribution for all fan angles. Therefore, photon directions and energies can be sampled accordingly.

The modified MC-GPU validated with an air scan enables us to analytically calculate the ground truth of the detected image given the bowtie model, the heel effect model, and the input spectrum. For individualized patient organ dose, the phantoms are generated from patient CT voxel data. The input phantom data for MC simulation should contain the spatial map of both material type and mass density. The density mapping is performed following a piece-wise linear curve which defines the densities of the mixture of water and bone [24].

We simulate a helical scan from the most superior to most inferior slice, with pitch factor 0.99, 80 mm Z-axis coverage, 24 views per rotation, 120 kVp spectrum, and a constant tube current. The MC-GPU dose calculation was repeated with $N = 4$ start angles $(\theta(i))$ uniformly spaced apart and averaged to obtain the dose map.

$$D_{avg} = \frac{1}{n} \sum_{i}^{N} MCGPU(V, \psi)|_{\theta(i)}, \tag{3}$$

where, ψ denotes the set of all the parameters used to configure the MC-GPU simulation. Dose is reported as eV/g/photon and can be scaled to mGy (1 eV/g/photon = 3.79 mGy/100 mAs) using a scanner-specific calibration (e.g., CTDI measurement). A representative dose map can be seen in Fig. 2. We mask out the air using a patient body mask with a threshold $t_{air} = 0.1$ from the voxel geometry V to obtain the body dose map. Therefore, the final dose map is

$$D = D_{avg} \cdot (V > t_{air}). \tag{4}$$

3.2   Multi-organ Segmentation in Chest-Abdomen-Pelvis CT

Our segmentation model leverages a 3D Context Encoder U-Net [4,10] network. The context encoder utilizes atrous convolution at different rates in the encoder network which enables capturing longer range information compared to standard U-Net [3]. The decoder employs residual multi-kernel pooling which performs max-pooling at multiple field-of-views. The network is trained separately for the L different organs. We follow [6] for the pre-processing step.

The encoder and decoder networks utilize 5 convolution blocks followed by downsampling and upsampling respectively. The final convolution is followed by a softmax activation. The model is trained with a focal categorical cross-entropy loss [17].

$$L^{seg}_{(y,\hat{y})} = -\sum_t \sum_i \alpha(1 - \hat{y}_{t,i})^\gamma y_{t,i} \log(\hat{y}_{t,i}), \tag{5}$$

where, i iterates over the number of channels and t iterates over the number of voxels in the image patches. A weight map is calculated to emphasize the voxels near the boundary of the reference and is added to the loss function in (5) [22].

$$w_t = a \cdot \exp(-b \cdot d_t) + 1, \tag{6}$$

where, d is the Euclidean distance from the closest label; weighting factors $a = 2.0$ and $b = 0.5$ are chosen for the experiments.

Organ Dose: In order to calculate the organ-specific doses, the model predicted segmentation mask ($\hat{y}$) is interpolated and resampled to align the voxel coordinates to the CT voxel geometry prepared for MC-GPU. The organ-specific dose map O_l, $l \in L$ is generated from the dose map D in (4).

$$O_l = D \cdot \hat{y}_l; l \in L. \tag{7}$$

Then the organ-specific reference dose distributions characterized by mean dose ($d_l = mean(O_l)$) as well as mean body dose ($d_{L+1} = mean(O_{L+1})$) are determined.

4   Experimental Evaluations

4.1   Materials

We make use of a set of 112 adult body scans, primarily contrast-enhanced, acquired from Revolution CT scanners (GE Healthcare), paired with (frontal and lateral) scout views. Note that the scout images do not contain contrast, and there are small inconsistencies between the scouts and CT scan due to patient motion, as is realistic in clinical use. All the data were collected with IRB approval from the participating institutions. The dataset has representative coverage of patient populations, with water equivalent diameter (DW) [19] ranging from 25 cm–36 cm. The segmentation model for six different organs (liver, lungs, kidneys, spleen, pancreas and bladder) were trained on a combination of

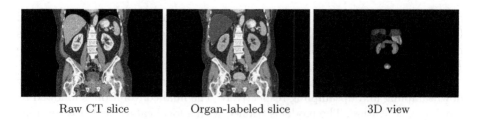

Raw CT slice   Organ-labeled slice   3D view

Fig. 3. Segmentation visualization by the 3D segmentation model in segmenting the six organs: liver, lungs, kidneys, bladder, spleen, and pancreas.

internal and publicly available CT data[234]. Considering the availability of high quality annotated data, our train datasets varied from 317 to 1010 scans.

4.2 Implementations

Scout-Net: The input scout pairs (frontal and lateral) were registered to the scan range at 690×530 pixels and passed to the model. We use the Adam optimizer with adaptive learning rate starting at $1e-4$ and a weight decay of $1e-6$. We train the model for 50 epochs with a minibatch size of 8. For better generalization and enhanced scope of the train data, we perform vertical flips and addition of Gaussian noise to augment the training data (other basic augmentations are not physically consistent between scouts and the CT scan). There is no prior method available that could be used as baseline. Replacing the 2D operations in our Scout-Net by their 3D counterparts, we train a baseline (retrospective) model namely CT-Net which predicts the doses from the entire CT volume. Model performances are evaluated by performing 4-fold cross validation and mean relative error (body/organ-specific) is reported. The model is implemented in Python with PyTorch and run on a *Intel Core i7 64 GiB* machine with a *GeForce RTX 2080-Super* GPU.

MC-GPU: The MC-GPU simulation is modeled after the GE Revolution CT scanner. The modified MC-GPU code is compiled in C++/CUDA, and run on the same machine as the *Scout-Net* model.

Segmentation: The segmentation model is trained for max 50 epochs and the best model is saved based on validation accuracy. Learning Rate Scheduler is used for determining the learning rate. We use a minibatch size of 16 and the Adam optimizer with a initial learning rate of $5e-6$. The best model weights are saved for each of the organ models and later used for inference. The segmentation model implemented in Python and the TensorFlow 1.4 framework, is run on a *Nvidia DGX* machine with a *Nvidia V100* GPUs.

---

[2] https://wiki.cancerimagingarchive.net/display/Public/Pancreas-CT.
[3] https://www.synapse.org/#!Synapse:syn3193805/wiki/89480.
[4] http://medicaldecathlon.com/.

Table 2. Performance comparison (relative error %) of the (prospective) *Scout-Net* against the baseline (retrospective) CT-Net model in estimating the organ doses. Mean ($\pm$ stdev) error rates are reported after cross-validation of the models.

Model	Body	Organ-of-interest					
		Liver	Lungs	Kidneys	Bladder	Spleen	Pancreas
CT-Net	8.88 ± 2.00	10.39 ± 1.64	9.87 ± 0.93	7.97 ± 0.58	11.86 ± 1.75	9.70 ± 2.29	16.55 ± 1.61
Scout-Net	6.26 ± 0.40	9.52 ± 0.35	8.75 ± 0.48	8.78 ± 0.38	13.22 ± 0.43	7.34 ± 0.35	13.86 ± 0.42

Table 3. Performance evaluation (Dice) of the proposed segmentation model in segmenting organs from chest-abdomen-pelvis CT scans.

Q Score	Organ-of-interest					
	Liver	Lungs	Kidneys	Spleen	Pancreas	Bladder
Q1	0.770	0.926	0.889	0.949	0.728	0.664
Q2	0.942	0.945	0.925	0.960	0.807	0.842
Q3	0.955	0.967	0.942	0.968	0.853	0.923

4.3   Results and Discussion

Effectiveness: We employ a Dice-based evaluation for our segmentation model, which achieved an average Dice score of 0.903 across the six organs against our internal test set (on par with other CT multi-organ segmentation models). Table 3 details the segmentation Dice (quartile) scores of organs-of-interest. Table 2 reports the relative error (percentage) for body/organ-specific dose calculation. As can be seen, the *Scout-Net* model obtained better error rates with the proposed (generic) FLM module and DLM modules over the baseline CT-Net. It is relatively harder for the CT-Net to directly learn the retrospective dose prediction from the volumetric data. The model better learns the features when trained jointly through a single FLM over multiple FLMs. Moreover, the model performance is improved after applying the scan range to register the scout pairs with the CT scans (approximately 9% performance gain on average). This justifies the effectiveness of enabling the model to focus on the scan-specific regions while still having the full scout images. Additionally, the careful choice of augmentations physically consistent with the scans further improves the performance—approximately 14% improvement in body dose prediction.

Robustness: For in-depth analysis of the Scout-Net results, we investigated the DWs and body/organ doses. We found strong correlations of the patient DW with doses of each of the organs as well as the patient body ($p < 0.001$). We further found the dose prediction error has no correlation for any of the organs or patient body. Even though we have large DW coverage denoting the variety in patient sizes, this demonstrates the robustness of our Scout-Net model in predicting the body/organ doses.

Efficiency: In addition to accurate and robust estimation of the patient (body) and organ-specific doses, our proposed *Scout-Net* model also facilitates massive gain in the run-time and computing resources. The overall pipeline for CT organ dose calculation from the CT data takes about 200 s including the MCGPU-based dose estimation and the segmentation of the six organs. On the other hand, with the proposed method, the estimation of CT organ dose is only 11 ms, or 18k times faster. In addition to the huge run-time advantage, our *Scout-Net* can be used even with limited computing resources as per scan CPU inference takes only 16 ms.

5 Conclusions

We have proposed an end-to-end, fully-automated, and real-time personalized CT organ dose estimation method (*Scout-Net*) which is capable of calculating organ dose distribution prospectively before doing the CT examinations. We have modified the open source MC-GPU tool to model bowtie filtration and heel effect in order to simulate more realistic CT dose estimation. Leveraging the MC-GPU dose calculation engine and our proposed 3D CNN-based organ segmentation, we have developed an automated pipeline that can provide accurate, reference organ dose distributions from CT scans. Comparing against the retrospective dose estimation, our *Scout-Net* has demonstrated the correct prediction of the prospective dose report at six different organs from the scout views, with improved generalization and execution speed. In ongoing work, we are extending *Scout-Net* to account for tube current modulation and are collecting a much larger dataset for further evaluation. We also envision performing organ dose estimation for additional organs and at other CT scan locations. Ultimately, using *Scout-Net* to plan the CT acquisition can potentially facilitate the reduction of radiation dose by guiding the automatic exposure control to balance CT dose and image quality.

References

1. Badal, A., Badano, A.: Accelerating Monte Carlo simulations of photon transport in a voxelized geometry using a massively parallel graphics processing unit. Med. Phys. **36**(11), 4878–4880 (2009)
2. Brook, O.R., Guralnik, L., Engel, A.: CT scout view as an essential part of CT reading. Australas. Radiol. **51**(3), 211–217 (2007)
3. Chen, L.C., Papandreou, G., Schroff, F., Adam, H.: Rethinking atrous convolution for semantic image segmentation. ArXiv abs/1706.05587 (2017)
4. Çiçek, Ö., Abdulkadir, A., Lienkamp, S.S., Brox, T., Ronneberger, O.: 3D U-Net: learning dense volumetric segmentation from sparse annotation. In: Ourselin, S., Joskowicz, L., Sabuncu, M.R., Unal, G., Wells, W. (eds.) MICCAI 2016. LNCS, vol. 9901, pp. 424–432. Springer, Cham (2016). https://doi.org/10.1007/978-3-319-46723-8_49
5. Damilakis, J.: CT Dosimetry: what has been achieved and what remains to be done. Invest. Radiol. **56**(1), 62–68 (2021)

6. Dutta, S., Das, B., Kaushik, S.: Assessment of optimal deep learning configuration for vertebrae segmentation from CT images. In: Medical Imaging 2019: Imaging Informatics for Healthcare, Research, and Applications. vol. 10954, pp. 298–305. SPIE (2019)

7. Fan, J., Xing, L., Dong, P., Wang, J., Hu, W., Yang, Y.: Data-driven dose calculation algorithm based on deep U-Net. Phys. Med. Biol. **65**(24), 245035 (2020)

8. Furhang, E.E., Chui, C.S., Sgouros, G.: A Monte Carlo approach to patient-specific dosimetry. Med. Phys. **23**(9), 1523–1529 (1996)

9. Götz, T.I., Schmidkonz, C., Chen, S., Al-Baddai, S., Kuwert, T., Lang, E.: A deep learning approach to radiation dose estimation. Phys. Med. Biol. **65**(3), 035007 (2020)

10. Gu, Z., et al.: CE-Net: context encoder network for 2D medical image segmentation. IEEE Trans. Med. Imaging **38**(10), 2281–2292 (2019)

11. Guerreiro, F., et al.: Deep learning prediction of proton and photon dose distributions for paediatric abdominal tumours. Radiother. Oncol. **156**, 36–42 (2021)

12. Imran, A.A.Z., Pal, D., Patel, B., Wang, A.: SSIQA: multi-task learning for non-reference CT image quality assessment with self-supervised noise level prediction. In: 2021 IEEE 18th International Symposium on Biomedical Imaging (ISBI), pp. 1962–1965 (2021)

13. Kachelrieß, M., Rehani, M.M.: Is it possible to kill the radiation risk issue in computed tomography? Physica Medica Eur. J. Med. Phys. **71**, 176–177 (2020)

14. Kontaxis, C., Bol, G., Lagendijk, J., Raaymakers, B.: DeepDose: towards a fast dose calculation engine for radiation therapy using deep learning. Phys. Med. Biol. **65**(7), 075013 (2020)

15. Lee, M.S., Hwang, D., Kim, J.H., Lee, J.S.: Deep-dose: a voxel dose estimation method using deep convolutional neural network for personalized internal dosimetry. Sci. Rep. **9**(1), 1–9 (2019)

16. Lell, M.M., Kachelrieß, M.: Recent and upcoming technological developments in computed tomography: high speed, low dose, deep learning, multienergy. Invest. Radiol. **55**(1), 8–19 (2020)

17. Lin, T.Y., Goyal, P., Girshick, R., He, K., Dollar, P.: Focal loss for dense object detection. In: Proceedings of the IEEE International Conference on Computer Vision (ICCV), October 2017

18. Maier, J., Eulig, E., Dorn, S., Sawall, S., Kachelrieß, M.: Real-time patient-specific CT dose estimation using a deep convolutional neural network. In: 2018 IEEE Nuclear Science Symposium and Medical Imaging Conference Proceedings (NSS/MIC), pp. 1–3. IEEE (2018)

19. McCollough, C., et al.: Use of water equivalent diameter for calculating patient size and size-specific dose estimates (SSDE) in CT: the report of AAPM task group 220. AAPM Rep. **2014**, 6 (2014)

20. Offe, M., et al.: Evaluation of deep learning segmentation for rapid, patient-specific CT organ dose estimation using an LBTE solver. In: Medical Imaging 2020: Physics of Medical Imaging, vol. 11312, p. 113124O. International Society for Optics and Photonics (2020)

21. Protection, R.: ICRP publication 103. Ann. ICRP **37**(2–4), 1–332 (2007)

22. Ronneberger, O., Fischer, P., Brox, T.: U-Net: convolutional networks for biomedical image segmentation. In: Navab, N., Hornegger, J., Wells, W.M., Frangi, A.F. (eds.) Medical Image Computing and Computer-Assisted Intervention - MICCAI 2015, pp. 234–241. Springer, Cham (2015)

23. Sharma, S., Kapadia, A., Fu, W., Abadi, E., Segars, W.P., Samei, E.: A real-time Monte Carlo tool for individualized dose estimations in clinical CT. Phys. Med. Biol. **64**(21), 215020 (2019)
24. Wang, A., et al.: Acuros CTS: a fast, linear Boltzmann transport equation solver for computed tomography scatter-part ii: system modeling, scatter correction, and optimization. Med. Phys. **45**(5), 1914–1925 (2018)
25. Zhu, J., Liu, X., Chen, L.: A preliminary study of a photon dose calculation algorithm using a convolutional neural network. Phys. Med. Biol. **65**(20), 20NT02 (2020)

High-Particle Simulation of Monte-Carlo Dose Distribution with 3D ConvLSTMs

Sonia Martinot[1,2,3($\boxtimes$)], Norbert Bus[1], Maria Vakalopoulou[2],
Charlotte Robert[3], Eric Deutsch[3], and Nikos Paragios[1]

[1] Therapanacea, Paris, France
sonia.martinot@hotmail.fr
[2] CentraleSupélec, 91190 Gif-sur-Yvette, France
[3] Institut Gustave Roussy, 94805 Villejuif, France

Abstract. Monte-Carlo simulation of radiotherapy dose remains an extremely time-consuming task, despite being still the most precise tool for radiation transport calculation. To circumvent this issue, deep learning offers promising avenues. In this paper, we extend ConvLSTM to handle 3D data and introduce a 3D recurrent and fully convolutional neural network architecture. Our model's purpose is to infer a computationally expensive Monte Carlo dose calculation result for VMAT plans with a high number of particles from a sequence of simulations with a low number of particles. We benchmark our framework against other learning methods commonly used for denoising and other medical tasks. Our model outperforms the other methods with regards to several evaluation metrics used to assess the clinical viability of the predictions. Code is available at https://git.io/JcbxD.

Keywords: Deep learning · Radiotherapy · Recurrent neural network · LSTM · Monte-Carlo · Denoising · Convolutional neural network

1 Introduction

In photon radiotherapy, accurate dose modelling is crucial for the treatment planning process in order to target tumours while leaving surrounding healthy tissues unharmed. For that purpose, Monte-Carlo (MC) methods remain unmatched by conventional algorithms such as pencil beam [9] and collapsed cone [1] in terms of precision. MC methods are based on probabilistic simulation of the behaviour of billions of particles in matter. However, due to the full particle transport modeling, MC methods are still extremely computationally expensive which prevents their extensive clinical adoption. Recent research partially addressed this issue by taking advantage of hardware acceleration and using efficient GPU implementations [4,11].

Modern treatment techniques require the calculation of radiation from complex beam configurations. Intensity-modulated radiation therapy (IMRT) allows only a few distinct gantry angles (the direction of the radiation source) while volumetric modulated arc therapy (VMAT) relies on the continuous movement

© Springer Nature Switzerland AG 2021
M. de Bruijne et al. (Eds.): MICCAI 2021, LNCS 12904, pp. 499–508, 2021.
https://doi.org/10.1007/978-3-030-87202-1_48

of the gantry around the center of the tumor. The latter presents less monitor units, more conformity and its delivery time is faster than IMRT [13], improving patient care. On the other hand, the large irradiated area in VMAT requires more simulated particles and therefore more time to reach high quality dose simulations. This is due to the inherent noise present in MC simulations. As the number of simulated particles increase, the noise that corrupts the underlying true dose decreases. Enabling deep neural network architectures to learn the underlying mechanisms of this causal relationship would lead to a flexible dose prediction framework which would be more resilient to various anatomies and allow efficient denoising of MC simulations of VMAT cases.

In this work, we generalize a recurrent neural network structure called a ConvLSTM cell to cope with three dimensional inputs for denoising of dose maps. Our contributions are threefold: *(i)* we present one of the first deep learning based denoising algorithms on VMAT plans, *(ii)* we introduce a 3D, recurrent and fully convolutional deep learning model that infers low uncertainty MC dose distributions from sequences of high uncertainty and low time complexity MC simulations, *(iii)* we release our VMAT dataset (dose distributions) and code. To the best of our knowledge, this is one of the first time convolutional LSTMs are investigated in a fully 3D setting, which could boost a big range of medical imaging applications, such as [2].

2   Related Work

Artificial intelligence and in particular deep learning, offer promising avenues for the clinical integration of MC methods through a colossal gain in terms of simulation time. Deep learning engines for denoising anatomy specific MC simulations have been proposed recently in the literature. In [12] the authors proposed an encoder-decoder architecture to predict high precision simulations from low precision ones in rectal cancer patients treated with IMRT. Neph et al. [10] used combined UNets [14] coupled with additional CT scans as input to solve the same problem in MR-guided beamlet dose for head and neck patients. Vasudevan et al. [18] investigated Generative Adversarial Networks [3] (GANs) to denoise dose simulations in water phantoms reporting promising results.

In the deep learning domain, convolutional architectures which extract relevant features from spatial information differ from recurrent architectures which exploit sequential correlations. ConvLSTM cells [16] were introduced to take advantage of both spatial and sequential information in a two-dimensional setting. This is a generalisation of Long Short-Term Memory (LSTM) [5] and fully convolutional LSTMs [17] architectures. This study aims to demonstrate that this novel recursive framework harnesses its strength from the sequential nature of its input and its ability to derive correlation between the levels of noisiness induced by the different number of particles simulated in the 3D space.

3   Methodology

3.1   Formulation of the Monte-Carlo Progressive Denoising Task

A MC simulation of radiotherapy dose requires inferring the dose deposited by billions of photons in the human body. The method consists in drawing independent random samples from an unknown distribution by means of sequentially sampling empirical measures that describe the dose deposition of individual photons. Let us denote by $M_{N_i} \in \mathbb{R}_+^{d_1 \times d_2 \times d_3}$ the 3D dose volume result of a simulation performed with N_i photons. Since several MC dose simulations for the same patient are independent from each other, the following equation holds:

$$M_{N_i} + M_{N_j} = M_{N_i+N_j}$$

Repeating this addition multiple times allows us to achieve simulations with a high number of samples. We can then assess this cumulative process as a temporal one, where the indices N_i correspond to consecutive time steps. In that case, a dose simulation can be represented by a stochastic variable X_{N_i} that we observe over time, as the number of simulated photons grows. Then, considering a sequence $(X_{N_1}, \ldots, X_{N_T})$ with T observations of that variable, our denoising problem amounts to predicting the most likely observation $X_{N_{T+1}}$ based on the given sequence:

$$X_{N_{T+1}} = \underset{X_{N_{T+1}}}{\mathrm{argmax}}\, p(X_{N_{T+1}}|X_{N_1}, \ldots, X_{N_T})$$

where p denotes an unknown probability. For our denoising task, an observation of X_{N_i} is the radiotherapy dose delivered to a patient at each time step, i.e. M_{N_i}. Hence, there is a need to exploit both spatial and temporal information of the given sequence before inferring the highly sampled dose. This formulation allows us to exploit the temporal and spatial coherence in the process of simulation.

3.2   LSTM and ConvLSTM Cells

LSTMs are a special type of recurrent neural networks (RNNs), able to exploit long-term temporal dependencies. The major asset of LSTM lies in its memory cell which can accumulate information as a cell state C_t. This cell is modified depending on whether the controlling gates are activated. As the input state i_t enters the LSTM, it is processed by an activation function whose final value can activate the forget gate f_t. When the forget gate is on, the past cell status C_{t-1} may be "forgotten". The current cell state is then propagated to the final state h_t in a way that is determined by the output gate o_t. The notation follows [16].

The memory cell allows LSTMs to circumvent the vanishing gradient problem that occurs in the regular RNN model. However, the LSTM only models 1D temporal information and does not make use of potential spatial information. ConvLSTMs overcome the latter limitation of LSTM. They can process 2D data

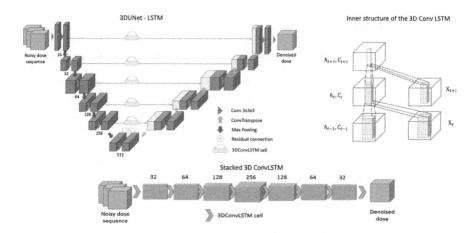

Fig. 1. Proposed architectures and inner structure of the 3D ConvLSTM. The number of output channels after each block appears above or below the layers' output volumes.

such as images by replacing transitional multiplications with convolutions. This innovation allows the model to infer a pixel's next state from its own past status as well as its neighbours.

3.3   Proposed 3DConvLSTM MC Denoiser

3DConvLSTM Cells. As we are considering data that present spatial information in three dimensions, we extended the ConvLSTM framework to deal with temporal sequences of 3D volumes. This can be achieved by using 3D convolutional operators indicated by $*$. In that structure, W_z and b_z in the equations below denote the parameters (filters and bias) of the considered convolutional layers. $\odot$ stands for the Hadamard product and σ for the sigmoid function. The following Eqs. (1–5) describe how gates are activated and states modified:

$$i_t = \sigma(W_{xi} * X_t + W_{hi} * H_{t-1} + W_{ci} \odot C_{t-1} + b_i) \tag{1}$$

$$f_t = \sigma(W_{xf} * X_t + W_{hf} * H_{t-1} + Wcf \odot C_{t-1} + b_f) \tag{2}$$

$$C_t = f_t \odot C_{t-1} + i_t \odot tanh(W_{xc} * X_t + W_{hc} * H_{t-1} + b_c) \tag{3}$$

$$o_t = \sigma(W_{xo} * X_t + W_{co} \odot C_t + b_o) \tag{4}$$

$$H_t = o_t \odot tanh(C_t) \tag{5}$$

Figure 1 presents the structure of the 3D ConvLSTM. All the states of the 3DConvLSTM cell are initialized with zeros which corresponds to ignorance of the future states. To ensure that dimensions match between the input and the various states inside the cell, padding is applied before convolutions. Therefore, the output of the 3DConvLSTM has the same spatial dimensions as the input.

This extension of ConvLSTM allows processing of medical volumetric sequential data in a fully convolutional manner. Each voxel's future state can be seamlessly predicted using contextual information brought by both temporal and spatial features from its own and its neighbours' past states in all dimensions. In the following subsections we present the two different setups that we used to integrate the 3DConvLSTM cells.

Proposed Model with Stacked 3DConvLSTM Cells: The model consists of 7 3DConvLSTM cells stacked on top of each other, without introducing any spatial downsampling. All convolutional layers in the 3DConvLSTM cells contain $3 \times 3 \times 3$ filters. Figure 1 shows the architecture of the Stacked ConvLSTM model. The spatial dimensions remain equal to that of the input through all propagation.

Proposed UNet with 3DConvLSTM Cells as Skip Connections: We also introduce a model based on the 3DUNet [21] architecture and enhanced with 3D ConvLSTM cells in the skip connections to further extract features at each of the 5 down-sampling steps. Down-sampling is performed using max pooling with kernel size and stride of 2 and comprises two identical convolutional layers. All convolutions have $3 \times 3 \times 3$ filters. The bottleneck has two identical convolutions with a residual connection to further exploit the deep features. The up-sampling blocks have a transpose convolution for up-sampling and a regular convolutional layer for further processing. Each convolutional layer is ended with a LeakyReLU [8] activation function, and batch normalization [6] is used for faster convergence. Details regarding the number of channels at each stage are shown in Fig. 1. This model is trained in the same setting as the proposed model with stacked 3D ConvLSTM cells.

To train all these models, we use a hybrid loss function that adds the Structural Similarity Index Measure (SSIM) [20] and the L1 loss. The parameters of the models are optimized by minimizing the following loss function (6):

$$\mathcal{L} = \sum_{i=0}^{N_{samples}} \left(\left\| X_{N_{T+1}}^{(i,estimated)} - X_{N_{T+1}}^{i} \right\|_1 + SSIM \left(X_{N_{T+1}}^{(i,estimated)}, X_{N_{T+1}}^{i} \right) \right)$$

(6)

where $X_{N_{T+1}}^{(i,estimated)}$ is the model's estimation of the i-th denoised dose volume sample. The SSIM metric is renown for giving a quantitative idea of the perceived quality of an image by measuring the similarity between two images. The L1 loss is well known to help keep track of fine grained details while training a model.

4 Dataset Construction

The patient cohort encompasses 50 patients treated with external beam photon therapy using the VMAT technique. Anatomies are diverse, including 22 pelvic and 28 head and neck cases. The dataset was split to 40, 5 and 5 patients for training, validation and test respectively. Anatomies were distributed as evenly as possible between these sets.

A model for a 6 MeV photon beam with standard fluence of a Varian True-beam linear accelerator was constructed using schematics and phase space data available in the IEAE phase space database. Each patient plan comprises two VMAT arcs. For each patient, the multileaf collimator (MLC) shapes and gantry angles were extracted from the original clinical plans. MC simulations of such plans were computed with 5×10^8, 10^9, 5×10^9 particles for the noisy dose volumes and 10^{11} particles for the ground-truth dose. The particle transport was simulated using OpenGate [15] with Geant4. The resolution of the simulation was set to 2 mm^3. For the fully sampled dose, the maximum uncertainty in areas within 20%–100% of the dose maximum remained below the clinically accepted 3% threshold. One complete simulation of a 2 arcs VMAT plan required over 4k hours of computation time on CPU without using any variance reduction technique.

4.1 Implementation Details

A patch-based training was implemented by randomly selecting sub-volumes from the 3D input sequences - ground-truth pairs, in areas within 30%–100% of the dose maximum. The patch size was $64\,mm^3$ subvolumes, i.e. $12.8\,cm^3$. We used Adam optimizer with learning rate, weight decay, beta1, beta2 and epsilon parameters set to 10^{-5}, 10^{-4}, 0.9, 0.999 and 10^{-8} respectively. The learning rate was reduced by half when the validation loss stagnated, i.e. when difference in loss was inferior to $1e^{-2}$ for more than 200 iterations. The batch size was set to 8. All models were trained for $3 \cdot 10^5$ iterations. The final model we kept was the one that performed best on the validation set.

The input sequence comprises 3 decreasingly noisy dose volumes simulated with 5×10^8, 10^9 and 5×10^9 particles of the same patient case. We use random horizontal and vertical flipping as sole augmentation techniques. The ground-truth was the corresponding highly sampled simulation with 10^{11} particles. Each sample was selected and fed to the model along the axial view.

5 Experimental Results

We compare our method with other commonly used learning based denoising methods in the literature. Our first benchmarking model is a 3DUNet [21] with 5 down-sampling blocks. The second one is Pix2Pix [7], a generative adversarial framework. Pix2Pix has been adapted to a 3D setting. Moreover, since we are

Table 1. Evaluation metrics for the performance of the models on the test set.

Method	SSIM	GPR	L1	# parameters
Inputs 5e9 particles	58.1 ± 0.1	59.1 ± 2.1	0.149 ± 0.050	
3DUNet [14]	80.0 ± 2.4	61.2 ± 2.8	0.088 ± 0.007	10 M
Pix2Pix 3D [7]	55.4 ± 8.6	66.6 ± 14.4	0.102 ± 0.009	120 M
3D BiONet [19]	93.0 ± 0.2	90.6 ± 1.2	0.080 ± 0.001	178 M
Proposed 3DUNet ConvLSTM	64.5 ± 6.1	79.1 ± 1.2	0.037 ± 0.004	36 M
Proposed Stacked 2D ConvLSTM	81.6 ± 3.2	74.1 ± 3.1	0.021 ± 0.003	1.5 M
Proposed Stacked 3D ConvLSTM	$\mathbf{97.9 \pm 0.9}$	$\mathbf{94.1 \pm 1.2}$	$\mathbf{0.004 \pm 0.001}$	5 M

considering smaller data in terms of height and width, we remove one down-sampling block and the corresponding decoding block from the generative model. The adapted generator thus consists of 5 down-sampling blocks, giving a fair comparison with the proposed 3DUNet architecture. Since these models don't handle sequential data, the input is the last volume of the sequence fed to the recurrent architecture, i.e. the least noisy simulation of the sequence. Finally, we also compared with the recently proposed BiONet architecture [19] after adapting it to 3D data and also limiting the number of down-sampling blocks.

Extensive quantitative comparison using the L1 error, SSIM and gamma passing rate (GPR) for each model on the test set are presented in Table 1. We evaluate the GPR criteria with a dose to agreement and tolerance on dose values of 3%/3 mm within 30%–100% of maximum dose. Results show that Stacked 3DConvLSTM outperforms all benchmark models in all metrics while having the lowest number of trainable parameters. We also trained the original ConvL-STMs, on slices of dose volumes. Results in Table 1 reveal that the 2D version still performs better than 3DUNet and Pix2Pix3D with regards to all metrics with only 1.5 million parameters indicating the need of sequential data for this task. Nevertheless, it does not outperform its 3D counterpart nor 3D BiONet. This fact highlights that our 3D model as well as 3D BiONet extract volumetric features that greatly improve the quality of the predictions. Moreover, Stacked 3D ConvLSTM achieves the lowest L1 value and displays GPR scores with standard deviations of 1.2. Although BiONet also shows comparative robustness in its predictions, the quality of the denoised dose volumes remain inferior to that of our proposed models. Another remark stemming from these results is that, despite having a higher SSIM than Pix2Pix3D, 3DUNet's GPR is lower. This might indicate that 3DUNet is able to infer structural coherence in the dose volumes but lacks in precision at a voxel level. In contrast, the proposed recurrent 3DUNet outperforms 3DUNet on the GPR by 18% even though its SSIM score fails to match the 3DUNet.

Figure 2 shows the predictions of the best performing models, namely BiONet and the Stacked 3DConvLSTM, on a test case. Both models reproduce high dose regions well. To further assess the denoising ability of the models, dose profiles are provided in Fig. 3. Both models succeed in smoothing the noise of the

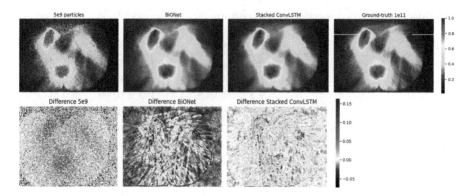

Fig. 2. On the first row from left to right: a single slice of the $5 \cdot 10^9$ dose volume, BiONet's, Stacked 3D ConvLSTM's predictions and ground-truth $1 \cdot 10^{11}$ dose volume. On the second row from right to left error maps for the three different representations.

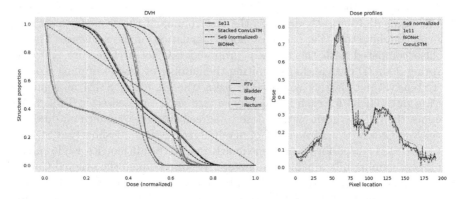

Fig. 3. a) DVH curves, showing that the dose gradients are reproduced faithfully. b) Dose profile along the line indicated on the ground truth image of Fig. 2.

low-simulation input. Error maps between predictions and ground-truth dose associated with BiONet point out that BiONet globally overestimates the dose in low dose gradient regions. Stacked 3DConvLSTM performs better in those regions but underestimates dose in high dose gradient regions where denoising is expected to be more challenging. However, we can notice that both models unfortunately tend to smooth fine details of the Monte-Carlo ground-truth simulation. Figure 3 plots the dose volume histogram (DVH) corresponding to the patient studied in Fig. 2. Both models substantially improve the DVHs towards the ground-truth DVHs. Nevertheless, the DVHs of BiONet indicate that the model still slightly overestimates the dose in voxels, in contrary to the Stacked 3DConvLSTMs.

6   Conclusions

Independently of GPU-accelerated computation, MC simulation time can be further decreased using deep learning based frameworks. The goal of this work is to highlight how considering the MC simulation task as a spatiotemporal problem can be an asset to reach accurate and fast computation of dose. Extensive experiments and comparisons with other state of the art methods highlight the potential of our method. However, the fact that our model does not perform any spatial down-sampling implies that the required GPU memory usage could still be reduced. Achieving high GPR scores while decreasing the computational load could enable real-time Monte-Carlo dose simulation. Future work aims to reduce the number of simulated particles, or in other words increase the level of noise of the input dose volumes.

Acknowledgments. We thank the CEA-TGCC and the TGCC support team for their guidance and the computational resources of the Joliot-Curie supercomputer through a GENCI Grand Challenge. This project has received funding from the European Union's Horizon 2020 research and innovation programme under grant agreement No. 880314.

References

1. Ahnesjö, A.: Collapsed cone convolution of radiant energy for photon dose calculation in heterogeneous media. Med. Phys. **16**(4), 577–592 (1989)
2. Gao, Y., Phillips, J.M., Zheng, Y., Min, R., Fletcher, P.T., Gerig, G.: Fully convolutional structured LSTM networks for joint 4D medical image segmentation. In: 2018 IEEE 15th International Symposium on Biomedical Imaging (ISBI 2018), pp. 1104–1108 (2018). https://doi.org/10.1109/ISBI.2018.8363764
3. Goodfellow, I.J., et al.: Generative adversarial networks (2014)
4. Hissoiny, S., Raaijmakers, A., Ozell, B., Després, P., Raaymakers, B.: Fast dose calculation in magnetic fields with GPUMCD. Phys. Med. Biol. **56**(16), 5119 (2011)
5. Hochreiter, S., Schmidhuber, J.: Long short-term memory. Neural Comput. **9**(8), 1735–1780 (1997)
6. Ioffe, S., Szegedy, C.: Batch normalization: accelerating deep network training by reducing internal covariate shift (2015)
7. Isola, P., Zhu, J.Y., Zhou, T., Efros, A.A.: Image-to-image translation with conditional adversarial networks. In: CVPR (2017)
8. Maas, A.L., Hannun, A.Y., Ng, A.Y.: Rectifier nonlinearities improve neural network acoustic models (2013)
9. Mohan, R., Chui, C., Lidofsky, L.: Differential pencil beam dose computation model for photons. Med. Phys. **13**(1), 64–73 (1986)
10. Neph, R., Huang, Y., Yang, Y., Sheng, K.: DeepMCDose: a deep learning method for efficient Monte Carlo Beamlet dose calculation by predictive denoising in MR-guided radiotherapy. In: Nguyen, D., Xing, L., Jiang, S. (eds.) AIRT 2019. LNCS, vol. 11850, pp. 137–145. Springer, Cham (2019). https://doi.org/10.1007/978-3-030-32486-5_17
11. Neph, R., Ouyang, C., Neylon, J., Yang, Y., Sheng, K.: Parallel beamlet dose calculation via beamlet contexts in a distributed multi-GPU framework. Med. Phys. **46**(8), 3719–3733 (2019)

12. Peng, Z., et al.: Deep learning for accelerating Monte Carlo radiation transport simulation in intensity-modulated radiation therapy. arXiv preprint arXiv:1910.07735 (2019)
13. Quan, E., et al.: A comprehensive comparison of IMRT and VMAT plan quality for prostate cancer treatment. Int. J. Radiat. Oncol. Biol. Phys. **83**, 1169–78 (2012). https://doi.org/10.1016/j.ijrobp.2011.09.015
14. Ronneberger, O., Fischer, P., Brox, T.: U-Net: convolutional networks for biomedical image segmentation. In: Navab, N., Hornegger, J., Wells, W.M., Frangi, A.F. (eds.) MICCAI 2015. LNCS, vol. 9351, pp. 234–241. Springer, Cham (2015). https://doi.org/10.1007/978-3-319-24574-4_28
15. Sarrut, D., et al.: A review of the use and potential of the gate Monte Carlo simulation code for radiation therapy and dosimetry applications. Med. Phys. **41**(6Part1), 064301 (2014)
16. Shi, X., Chen, Z., Wang, H., Yeung, D.Y., Wong, W.K., Woo, W.C.: Convolutional LSTM network: a machine learning approach for precipitation nowcasting. arXiv preprint arXiv:1506.04214 (2015)
17. Sutskever, I., Vinyals, O., Le, Q.V.: Sequence to sequence learning with neural networks. arXiv preprint arXiv:1409.3215 (2014)
18. Vasudevan, V., Huang, C., Simiele, E., Yu, L., Xing, L., Schuler, E.: Combining Monte Carlo with deep learning: Predicting high-resolution, low-noise dose distributions using a generative adversarial network for fast and precise Monte Carlo simulations. Int. J. Radiat. Oncol. Biol. Phys. **108**(3), S44–S45 (2020)
19. Xiang, T., Zhang, C., Liu, D., Song, Y., Huang, H., Cai, W.: BiO-Net: learning recurrent bi-directional connections for encoder-decoder architecture. In: Martel, A.L., et al. (eds.) MICCAI 2020. LNCS, vol. 12261, pp. 74–84. Springer, Cham (2020). https://doi.org/10.1007/978-3-030-59710-8_8
20. Zhou Wang, Bovik, A.C., Sheikh, H.R., Simoncelli, E.P.: Image quality assessment: from error visibility to structural similarity. IEEE Trans. Image Process. **13**(4), 600–612 (2004). https://doi.org/10.1109/TIP.2003.819861
21. Özgün Çiçek, Abdulkadir, A., Lienkamp, S.S., Brox, T., Ronneberger, O.: 3D U-Net: learning dense volumetric segmentation from sparse annotation (2016)

Effective Semantic Segmentation in Cataract Surgery: What Matters Most?

Theodoros Pissas[1,2]($\boxtimes$), Claudio S. Ravasio[1,2], Lyndon Da Cruz[3,4], and Christos Bergeles[2]

[1] Wellcome/EPSRC Centre for Interventional and Surgical Sciences, University College London (UCL), London, UK
rmaptpi@ucl.ac.uk
[2] School of Biomedical Engineering and Imaging Sciences, King's College London (KCL), London, UK
[3] Moorfields Eye Hospital, London, UK
[4] Institute of Ophthalmology, University College London, London, UK

Abstract. Our work proposes neural network design choices that set the state-of-the-art on a challenging public benchmark on cataract surgery, CaDIS. Our methodology achieves strong performance across three semantic segmentation tasks with increasingly granular surgical tool class sets by effectively handling class imbalance, an inherent challenge in any surgical video. We consider and evaluate two conceptually simple data oversampling methods as well as different loss functions. We show significant performance gains across network architectures and tasks especially on the rarest tool classes, thereby presenting an approach for achieving high performance when imbalanced granular datasets are considered. Our code and trained models are available at https://github.com/RViMLab/MICCAI2021_Cataract_semantic_segmentation and qualitative results on unseen surgical video can be found at https://youtu.be/twVIPUj1WZM.

Keywords: Semantic segmentation · Cataract surgery · Oversampling

1 Introduction

Cataract is the leading cause of blindness [1] while also being predominantly preventable through surgery [21]. Surgical data science envisions the deployment of data-driven systems to enhance clinical practice planning, delivery and quality assessment [13]. Intraoperative scene understanding constitutes a necessary building block towards utilising such systems in the context of computer assisted interventions, as it encapsulates recognition problems ranging from image-level

---

T. Pissas and C. S. Ravasio—Contributed equally.
L. Da Cruz and C. Bergeles—Contributed equally.

Electronic supplementary material The online version of this chapter (https://doi.org/10.1007/978-3-030-87202-1_49) contains supplementary material, which is available to authorized users.

© Springer Nature Switzerland AG 2021
M. de Bruijne et al. (Eds.): MICCAI 2021, LNCS 12904, pp. 509–518, 2021.
https://doi.org/10.1007/978-3-030-87202-1_49

classification of the procedure undertaken at a specific time step to the semantic labelling of every pixel in the scene.

Within the domain of cataract surgery, various works have focused on phase and tool presence detection [2,16,18,24,29], estimating only global video-level or frame-level information. Furthermore, some research has addressed the task of tool recognition and localisation via bounding box estimation [26], segmentation [17], or both [7], extracting information about tools against an all-encompassing background class. Recent advancements in computer vision [5,22,25] demonstrate deep convolutional nets to be powerful semantic segmentation models for natural scenes if provided with large, pixel-wise annotated datasets. Our work therefore goes further and aims to obtain a semantic segmentation of both tools and all anatomical regions in the scene. Importantly, different from [17,26], we report results on a publicly available dataset, CaDIS [8], thus enabling reproducibility.

A common characteristic of cataract surgery video datasets is class imbalance caused by the small size of tools relative to anatomical regions, and the overall sparse appearance of certain tools in the duration of a surgical procedure. Coupled with this is the second major challenge of inter-class resemblance: using a more granular set of tool classes further reduces the perceived variation in tool appearance along with the number of samples per class, thus increasing imbalance. Consequently, and as demonstrated in [8], fine-grained tool recognition and segmentation is a particularly challenging task, with increases in class granularity leading to significant performance drops.

Our paper addresses these challenges for pixel-level semantic segmentation of cataract surgery videos. Our contributions are the following:

- We demonstrate that oversampling in the form of repeat factor sampling, previously only proposed for long-tail instance segmentation [9], or a custom adaptive sampling algorithm lead to significant gains in performance, especially for the rarest classes and in tasks where the class distribution is most imbalanced.
- We conduct detailed ablation studies involving different network architectures, data sampling strategies, loss functions, and training policies.
- From this, we determine a training policy consisting of a data oversampling method, a loss function and a training schedule that achieves the highest performance across benchmark sub-tasks and network architectures tested.

Ultimately, our top-performing models significantly outperform the results reported in [8] on all sub-tasks of the CaDIS dataset, setting the state-of-the-art on this challenging benchmark.

2   Materials and Methods

We first present the evaluated design choices spanning network architectures, loss functions and two data oversampling strategies.

2.1   Data

Experiments were conducted using the public CaDIS dataset [8] consisting of 25 surgical videos that contain segments of cataract surgery and comprise 4671 pixel-level annotated frames with labels for *anatomies, instruments* and *miscellaneous objects*. We follow [8] that defined 3, increasingly granular, semantic segmentation tasks. Task 1 that entails 8 classes: 4 anatomies, 1 all-encompassing instrument class and 3 classes for miscellaneous objects. Task 2 increases the number of classes to 17 by splitting the all-encompassing instrument class of task 1 to 9 distinct classes allowing the separate recognition of different tools. Finally, task 3 pushes instrument recognition granularity even further by considering the handles of certain instruments as separate classes, and has 25 classes overall. In the remainder of the manuscript we refer to different tasks as t_1, t_2, and t_3 for brevity. An example image can be seen in Fig. 1.

Data inspection [11] revealed 179 at least partially mislabelled frames, most frequently affecting tool classes, which may obfuscate conclusions. We corrected the tool labels on 40 of these frames, and excluded the rest from consideration. This reduced the overall dataset from 4670 to 4491 records. We use the same training set as Grammatikopoulou et al. [8], and report results on their original unfiltered train-val-test split to ensure reproducibility. However, the original test set [8] is overly limited in size and no robust conclusions can be drawn: some of the rarest classes are present in just one of the three videos in the test set, with as few as seven highly correlated sequential instances in the 587 frames overall. Therefore, we merge their validation and test sets for our more extensive evaluation. This ultimately yields 3550 frames from 19 subjects for training and 1120 frames from 6 different subjects for validation, or 3490 training records and 1001 validation records after correcting for mislabelled frames.

2.2   Network Architectures

We focus on networks with the classic encoder-decoder structure, using ResNets [10] as the standard encoder or network backbone. We made use of the implementations in the Torchvision library [14], removing the final fully connected layers and passing the features to the decoder. As decoders or network heads we

Fig. 1. From left to right: An image from the CaDIS dataset (Video 22, frame 5040), its ground truth segmentation, and a sample prediction from our network for t_3. The labels show the pupil (light blue), iris (red), cornea (white), skin (magenta), surgical tape (orange), an eye retractor (purple), and the surgeon's hand (pink). Two tools are present: a micromanipulator (orange red) and a phacoemulsifier handpiece handle (green). (Color figure online)

selected current state-of-the-art options which instantiate three different mechanisms for enhancing and decoding the encoder's feature maps into dense semantic labelling, namely a Feature Pyramid Network (FPN), an attention based contextual module and a dilated convolution context module.

The **UPerNet** (UPN) head [22] consists of a FPN which takes the features from ResNet layers 2–5 as input and combines them with a Pyramid Pooling Module (PPM) at the lowest resolution level. The outputs are fused into a single feature map, which is connected to a classifier that predicts the semantic segmentation at a quarter scale relative to the original input image. This prediction is then upsampled to the original resolution. We based our code on the implementation in [27,28], with FPN and PPM channel lengths of 512.

The **OCRNet** (OCR) head [25] obtains a coarse intermediate prediction using a convolutional output head, operating on top of features of layer 4 of the encoder. Together with layer 5, this results in an aggregate feature representation for each object in the scene, fed to the Object Contextual Representation module to enhance it with contextual information based on object-pixel similarities. Subsequently, a convolutional layer maps them to the final prediction upsampled to the original resolution. We based our code on the official implementation [25], using 512 channels for all convolutional layers. OCRNet uses a dilated ResNet backbone, replacing the stride 2 convolutions in layers 4 and 5 ([10]) with a dilated convolution with rate 2. The spatial dimensions of the encoder's output features result as 1/8 of the input resolution, rather than the default 1/32.

The **DeepLabv3+** (DLv3+) head [5] utilises an Atrous Spatial Pyramid Pooling (ASSP) module which encodes multi-scale context through 5 parallel dilated convolutions with dilation rates $\{1, 6, 12, 18\}$. This is followed by a coarse-to-fine decoder which fuses layer 2 of the encoder with the ASPP features to produce a prediction at a quarter of the input resolution which is then upsampled. We adapted the official Tensorflow implementation [5] for Pytorch, using 256 channels for all convolutions. DLv3+uses the same dilated ResNet as OCRNet.

2.3 Loss Functions

Cross-entropy as the standard first choice of loss function for semantic segmentation struggles to deal with strong class imbalances such as the ones encountered in t_2 and t_3, and minimising the cross-entropy often correlates poorly with maximising the actual performance metric, i.e. mean intersection over union (mIoU). Therefore, we explored a variation that uses online hard example mining (OHEM) [19] by ignoring the loss in all pixels where the correct label is predicted with a probability above a threshold value, chosen as 0.7. This follows the work on detection in [19], and we used an implementation from [20]. While directly training on the mIoU metric is intractable due to its non-differentiability, surrogates such as the Lovász-Softmax have been developed. It makes use of the Lovász extension of the Jaccard index to create a differentiable loss function whose optimisation corresponds more closely to minimising the intersection over union metric [4]. We used the PyTorch implementation provided by the authors.

2.4 Addressing Class Imbalance

The number of class instances in the dataset vary by three orders of magnitude, as does the average number of pixels per class [8]. This extreme imbalance affecting the labels for t_2 and especially t_3 is the main hurdle to achieving a high mIoU on this dataset. We evaluated two different approaches to overcome this, aimed at increasing the frequency with which the network is presented with the rarer and therefore harder to learn labels. Repeat factor sampling (RF) relies on repeating specific records within the dataset, based on pre-computed image-level scores, increasing the number of records per epoch. Adaptive sampling on the other hand moves away from the concept of an 'epoch' as iterating through the dataset once, and instead follows an online stochastic selection process for each batch, no longer guaranteeing each record will be seen in every epoch.

Repeat Factor Sampling: We tailored repeat factor sampling, introduced in [9] for instance segmentation under long-tailed class distributions, to the task of semantic segmentation. First, for each class c we compute a class-level repeat factor as $r_c = \max(1, \sqrt{t/f_c})$ where f_c is the frequency of image-level occurrence of this class and t is a frequency threshold hyper-parameter. Then, for each image I in the dataset, an image-level repeat factor is computed as $r_I = \max_{c \in I} r_c$, which specifies the number of times I should be repeated during an epoch of training and in general is not an integer. The value of r_I is stochastically rounded before each epoch resulting in a randomly varying number of oversampled images per epoch, while over the course of training $\mathbb{E}[r_I^{rounded}] \approx r_I$. We select $t = 0.15$ to oversample images with classes such that $f_c < 0.15$.

Adaptive Sampling: Adaptive sampling is based on the principle of using the current per-class performance, calculated as a moving average of IoU values achieved during training with an exponential smoothing factor of 0.1, initialised at 0.5 for numerical stability. Therefore, our method is biased towards the selection of whole images with a high incidence of globally underperforming classes, as opposed to Berger et al. [3] who compile training batches selecting patches with high local error values within the image data. The IoU values, which measure class performance, are converted into percentages expressing how many records in the batch should be selected while prioritising each class: $p = softmax(1 - IoU^2)$. Each of these records is chosen as the one with the highest number of pixels labelled with the class of interest out of 10 random samples from the dataset. Thus, the selection is automatically biased in favour of a higher incidence of underperforming, often rarer classes while not restricting it to any specific subset of the data.

2.5 Training Schedule

As default settings, the networks were trained using the Adam optimiser [12] for a maximum of 50 epochs, with each epoch including repeated data if repeat factor or adaptive sampling was used. We chose a batch size of 8, images at their original resolution of 540×960, and online data augmentation. The latter consists of random horizontal flips with probability 0.5, Gaussian blurring with a kernel size randomly chosen between 3 and 7 occurring with a probability of 0.05, and the Torchvision 'colorjitter' augmentation with brightness, contrast and saturation adjustments in $[2/3, 3/2]$, and hue adjustment in $[-0.05, 0.05]$.

Learning Rate: Two learning rate decay functions were tested: the exponential formula $lr = lr_0 * \alpha^i$ and the polynomial formula $lr = lr_0 * (1 - i/n)^p$ [15], where lr is the current learning rate, lr_0 the initial learning rate, α and p hyperparameters to be chosen, i the current epoch number, and n the total number of epochs. We also used restarts at 65 % of the previous lr_0, following the proposal in [15]. Ultimately, we chose the exponential decay with $lr_0 = 10^{-4}$ and $\alpha = 0.98$ as standard for all experiments in this work, as the other options explored yielded no significant benefits.

Initialisation: Decoders were randomly initialised, while we used Imagenet-pretrained weights for backbones. We also experimented with using MoCov2 weights from self-supervised pre-training on Imagenet [6] for a ResNet50.

2.6   Implementation Details

Our implementation uses PyTorch 1.6. Experiments were run on an NVIDIA Quadro P6000 GPU with 24 GB memory for UPN models (around 20 GB of memory, 20 h training time), and an NVIDIA Quadro RTX 8000 GPU with 48 GB memory for OCRNet (around 23 GB of memory, 17 h training time) and DLv3+ models (around 24 GB of memory, 25 h training time).

3   Results and Discussion

The main results are collated in Table 1. The metric reported is the best mean Intersection over Union (mIoU) value achieved on the validation dataset over the course of the 50 training epochs. As shown in Table 1, combining any of the architectures with the Lovász loss and RF (Sect. 2.4) outperforms all other variants consistently across all three tasks t_1, t_2, t_3, providing an effective training policy for effective semantic segmentation across different levels of class granularity. We also observed that a mean ensemble of the best models per head further improves performance on all tasks, as reported in Table 1. Following are our key conclusions for what matters for high performance in this benchmark.

The Superiority of the Lovász-Softmax Loss: A consistent finding was that the Lovász-Softmax loss outperforms cross-entropy based losses as the default option in the literature [5, 22, 25]. Notably, adding the Lovász in t_1 and t_2 already leads to performance that is on par with models using a combination of cross-entropy and oversampling, the latter increasing training time significantly in the case of RF. We also empirically observed that the validation Lovász-Softmax loss closely correlated with the validation mean IoU: the epoch for which the validation loss reached its minimum was consistently close to that with the maximum mIoU, a desirable property not apparent for cross-entropy-based losses.

Table 1. Results for t_1, t_2, and t_3. **Bold** numbers denote the best results per task and model head, † indicates MoCov2 initialisation, Ensemble uses models in bold per task.

Model		Loss			Sampling		Mean IoU		
ResNet	Head	CE	OHEM	Lovász	RF	Adapt.	t_1	t_2	t_3
50	OCR	✓			✓		0.8950	0.8102	0.7624
50	OCR		✓		✓		0.8967	0.8102	0.7742
50	OCR			✓			0.8979	0.8109	0.7345
50	OCR			✓	✓		0.8999	0.8236	**0.7777**
50†	OCR			✓	✓		**0.9013**	**0.8282**	0.7512
50	OCR			✓		✓	0.8997	0.8220	0.7632
34	UPN	✓			✓		0.8957	0.8013	0.7534
34	UPN		✓		✓		0.8967	0.8039	0.7491
34	UPN			✓			0.8990	0.8151	0.7374
34	UPN			✓	✓		**0.8992**	**0.8298**	**0.7735**
34	UPN			✓		✓	0.8988	0.8198	0.7457
50	DLv3+			✓	✓		**0.8996**	**0.8250**	**0.7763**
50†	DLv3+			✓	✓		0.8983	0.8224	0.7578
Ensemble				✓	✓		0.9020	0.8360	0.7870

Oversampling: To demonstrate the effect of oversampling in the presence of extreme class imbalance we separately report the mean IoU over anatomical, tool and rare classes, the latter being tool classes occurring in less than 10% of records. As shown in Table 2, training with either RF or Adaptive sampling significantly boosts performance on rare and tool classes, regardless of architecture.

Backbone Effect: Against a baseline of 0.8109 mIoU at t_2 (UPN head, ResNet34 backbone, Lovász-Softmax loss, no augmentations or oversampling), we found most deeper backbones such as ResNet50, WideResNet50, WideResNet101 to yield no returns. Consistent with [23], only ResNeXt50 and ResNeXt101 were promising with respective mIoU values of 0.8132 and 0.8195, at the cost of higher GPU memory usage. Furthermore, initialising OCRNet's ResNet50 backbone with MoCov2 [6] instead of Imagenet weights slightly boosted performance as shown in Table 1, hinting that transfer learning from natural to surgical scenes from a self-supervised initialisation can be more effective.

Comparison with Existing Work: Our results outperform the state-of-the-art on the CaDIS benchmark established in [8]. For our previous experiments we utilised a filtered version of the dataset as described in Sect. 2.1. For a direct comparison, we report results using the train-validation-test split of [8], without filtering out mislabelled frames. As shown in Table 3 our methodology outperforms results of [8] across all tasks and network architectures by a large margin. Notably, the gain in performance by our methodology increases for t_2 and t_3 that present high imbalance in the class distributions, which can be attributed to the fact that we employ oversampling methods that explicitly address it.

Table 2. Effect of oversampling in t_2 and t_3 (all use Lovász loss)

Model			Oversampling		Mean IoU			
Task	Backbone	Decoder	Adaptive	RF	Anatomies	Tools	Rare	Overall
t_2	ResNet50	OCR			0.9038	0.7502	0.7572	0.8109
t_2	ResNet50	OCR	✓		0.9117	0.7666	0.7614	0.8220
t_2	ResNet50	OCR		✓	0.9063	0.7689	0.7752	0.8236
t_2	ResNet34	UPN			0.9061	0.7546	0.7374	0.8151
t_2	ResNet34	UPN	✓		0.9045	0.7620	0.7665	0.8198
t_2	ResNet34	UPN		✓	0.9045	0.7787	0.7836	0.8298
t_3	ResNet50	OCR			0.9062	0.6695	0.6202	0.7345
t_3	ResNet50	OCR	✓		0.9020	0.7112	0.6913	0.7622
t_3	ResNet50	OCR		✓	0.9059	0.7296	0.7144	0.7777
t_3	ResNet34	UPN			0.9031	0.6744	0.6274	0.7374
t_3	ResNet34	UPN	✓		0.9010	0.6860	0.6571	0.7457
t_3	ResNet34	UPN		✓	0.9024	0.7238	0.7076	0.7735

Table 3. Results on train-val-test dataset split of [8], **bold** denotes the best result per model, red and **blue** the best overall in val and test set of each task respectively.

Model	Training		t_1		t_2		t_3	
	Loss	Sampling	val	test	val	test	val	test
DLv3+ [8]	CE	-	0.8530	0.8262	0.7450	0.7226	0.6860	0.6323
DLv3+	Lovász	RF	**0.8848**	**0.8565**	**0.7914**	**0.7517**	**0.7744**	**0.7051**
UPN [8]	CE	-	0.8790	0.8396	0.7950	0.7376	0.7420	0.6676
UPN	Lovász	RF	**0.8885**	**0.8632**	**0.8154**	**0.7688**	**0.7575**	**0.7044**
HRNetv2 [8]	CE	-	0.8810	0.8491	0.8180	0.7611	0.7240	0.6664
OCR	Lovász	RF	0.8897	**0.8640**	0.8325	**0.7909**	0.7940	**0.7194**

4 Conclusion

The CaDIS dataset is a new benchmark in the domain of cataract surgery. Its main challenge lies in the extreme class imbalance encountered in the highly granular semantic segmentation labels provided, which we meet with two different data oversampling strategies. We set a new state-of-the-art and provide extensive ablation studies on the effect of different design choices. These show that the effect of varying encoder or decoder designs is minor and generally difficult to predict, while the choice of the loss function and data sampling strategy are paramount. Specifically, we recommend the use of the Lovász-Softmax loss as a differentiable surrogate for the Jaccard index [4], and an adaptation of repeat factor sampling [9] to increase the frequency of hard records in the training data. Our findings can be applied to other datasets with a similar class imbalance, and may guide efforts of pushing the state-of-the-art further on CaDIS.

Acknowledgements. The authors would like to thank Martin Huber, Jeremy Birch and Joan M. Nunez Do Rio for their contributions in the EndoVIS challenge participation. This work was supported by the National Institute for Health Research NIHR (Invention for Innovation, i4i; II-LB-0716-20002). The views expressed are those of the authors and not necessarily those of the NHS, the NIHR, or the Department of Health and Social Care.

References

1. Blindness and vision impairment. https://www.who.int/news-room/fact-sheets/detail/blindness-and-visual-impairment. Accessed 1 Mar 2021
2. Al Hajj, H., et al.: Cataracts: challenge on automatic tool annotation for cataract surgery. Med. Image Anal. **52**, 24–41 (2019)
3. Berger, L., Eoin, H., Cardoso, M.J., Ourselin, S.: An adaptive sampling scheme to efficiently train fully convolutional networks for semantic segmentation. In: Nixon, M., Mahmoodi, S., Zwiggelaar, R. (eds.) MIUA 2018. CCIS, vol. 894, pp. 277–286. Springer, Cham (2018). https://doi.org/10.1007/978-3-319-95921-4_26
4. Berman, M., Triki, A.R., Blaschko, M.B.: The lovász-softmax loss: a tractable surrogate for the optimization of the intersection-over-union measure in neural networks. In: Proceedings of the IEEE Conference on Computer Vision and Pattern Recognition, pp. 4413–4421 (2018)
5. Chen, L.C., Zhu, Y., Papandreou, G., Schroff, F., Adam, H.: Encoder-decoder with atrous separable convolution for semantic image segmentation. In: Proceedings of the European Conference on Computer Vision (ECCV), pp. 801–818 (2018)
6. Chen, X., Fan, H., Girshick, R., He, K.: Improved baselines with momentum contrastive learning. arXiv preprint arXiv:2003.04297 (2020)
7. Fox, M., Taschwer, M., Schoeffmann, K.: Pixel-based tool segmentation in cataract surgery videos with mask R-CNN. In: 2020 IEEE 33rd International Symposium on Computer-Based Medical Systems (CBMS), pp. 565–568. IEEE (2020)
8. Grammatikopoulou, M., et al.: Cadis: cataract dataset for surgical RGB-image segmentation. Med. Image Anal. **71**,(2021). https://doi.org/10.1016/j.media.2021.102053
9. Gupta, A., Dollar, P., Girshick, R.: LVIS: a dataset for large vocabulary instance segmentation. In: Proceedings of the IEEE Conference on Computer Vision and Pattern Recognition, pp. 5356–5364 (2019)
10. He, K., Zhang, X., Ren, S., Sun, J.: Deep residual learning for image recognition. In: Proceedings of the IEEE Conference on Computer Vision and Pattern Recognition, pp. 770–778 (2016)
11. Karpathy, A.: A recipe for training neural networks (2019). http://karpathy.github.io/2019/04/25/recipe/
12. Kingma, D.P., Ba, J.: Adam: a method for stochastic optimization. In: Bengio, Y., LeCun, Y. (eds.) 3rd International Conference on Learning Representations, ICLR 2015 (2015)
13. Maier-Hein, L., et al.: Surgical data science-from concepts to clinical translation. arXiv preprint arXiv:2011.02284 (2020)
14. Marcel, S., Rodriguez, Y.: Torchvision the machine-vision package of torch. In: Proceedings of the 18th ACM international conference on Multimedia, pp. 1485–1488 (2010)

15. Mishra, P., Sarawadekar, K.: Polynomial learning rate policy with warm restart for deep neural network. In: TENCON 2019–2019 IEEE Region 10 Conference (TENCON), pp. 2087–2092. IEEE (2019)
16. Morita, S., Tabuchi, H., Masumoto, H., Yamauchi, T., Kamiura, N.: Real-time extraction of important surgical phases in cataract surgery videos. Sci. Rep. **9**(1), 1–8 (2019)
17. Ni, Z.-L., et al.: RAUNet: residual attention U-Net for semantic segmentation of cataract surgical instruments. In: Gedeon, T., Wong, K.W., Lee, M. (eds.) ICONIP 2019. LNCS, vol. 11954, pp. 139–149. Springer, Cham (2019). https://doi.org/10.1007/978-3-030-36711-4_13
18. Padoy, N., Blum, T., Ahmadi, S.A., Feussner, H., Berger, M.O., Navab, N.: Statistical modeling and recognition of surgical workflow. Med. Image Anal. **16**(3), 632–641 (2012)
19. Shrivastava, A., Gupta, A., Girshick, R.: Training region-based object detectors with online hard example mining. In: Proceedings of the IEEE Conference on Computer Vision and Pattern Recognition (CVPR), June 2016
20. Sun, K., Xiao, B., Liu, D., Wang, J.: Deep high-resolution representation learning for human pose estimation. In: CVPR (2019)
21. Wang, W., Yan, W., Fotis, K., Prasad, N.M., Lansingh, V.C., Taylor, H.R., Finger, R.P., Facciolo, D., He, M.: Cataract surgical rate and socioeconomics: a global study. Invest. Ophthalmol. Vis. Sci. **57**(14), 5872–5881 (2016)
22. Xiao, T., Liu, Y., Zhou, B., Jiang, Y., Sun, J.: Unified perceptual parsing for scene understanding. In: Proceedings of the European Conference on Computer Vision (ECCV), pp. 418–434 (2018)
23. Xie, S., Girshick, R., Dollár, P., Tu, Z., He, K.: Aggregated residual transformations for deep neural networks. In: Proceedings of the IEEE Conference on Computer Vision and Pattern Recognition, pp. 1492–1500 (2017)
24. Yu, F., et al.: Assessment of automated identification of phases in videos of cataract surgery using machine learning and deep learning techniques. JAMA Netw. Open **2**(4), e191860–e191860 (2019)
25. Yuan, Y., Chen, X., Wang, J.: Object-contextual representations for semantic segmentation. arXiv preprint arXiv:1909.11065 (2019)
26. Zang, D., Bian, G.-B., Wang, Y., Li, Z.: An extremely fast and precise convolutional neural network for recognition and localization of cataract surgical tools. In: Shen, D., et al. (eds.) MICCAI 2019. LNCS, vol. 11768, pp. 56–64. Springer, Cham (2019). https://doi.org/10.1007/978-3-030-32254-0_7
27. Zhou, B., Zhao, H., Puig, X., Fidler, S., Barriuso, A., Torralba, A.: Scene parsing through ade20k dataset. In: Proceedings of the IEEE Conference on Computer Vision and Pattern Recognition (2017)
28. Zhou, B., et al.: Semantic understanding of scenes through the ade20k dataset. Int. J. Comput. Vis. **127**, 302–321 (2018)
29. Zisimopoulos, O., et al.: DeepPhase: surgical phase recognition in CATARACTS videos. In: Frangi, A.F., Schnabel, J.A., Davatzikos, C., Alberola-López, C., Fichtinger, G. (eds.) MICCAI 2018. LNCS, vol. 11073, pp. 265–272. Springer, Cham (2018). https://doi.org/10.1007/978-3-030-00937-3_31

Facial and Cochlear Nerves Characterization Using Deep Reinforcement Learning for Landmark Detection

Paula López Diez[1,2(✉)], Josefine Vilsbøll Sundgaard[1], François Patou[2], Jan Margeta[2,3], and Rasmus Reinhold Paulsen[1]

[1] DTU Compute, Technical University of Denmark, Kongens Lyngby, Denmark
plodi@dtu.dk
[2] Oticon Medical, Research and Technology Group, Smørum, Denmark
[3] KardioMe, Research and Development, Nova Dubnica, Slovakia

Abstract. We propose a pipeline for the characterization of facial and cochlear nerves in CT scans, a task specifically relevant for cochlear implant surgery planning. These structures are hard to locate in clinical CT scans due to their small size relative to the image resolution, the lack of contrast, and the proximity to other similar structures in this region. We define key landmarks around the facial and cochlear nerves and locate them using deep reinforcement learning with communicative multi-agents based on the C-MARL model. These landmarks are used as initialization for customized characterization methods. These include the automated direct measurement of the diameter of the cochlear nerve canal and extraction of the cochlear nerve cross-section followed by its segmentation using active contours. We also derive a path selection algorithm for optimal geodesic pathfinding selection based on Dijkstra's algorithm for the characterization of the facial nerve. A total of 119 clinical CT images from preoperative patients have been used to develop this pipeline that produces accurate characterizations of these nerves in the cochlear region and provides reliable measurements for computer-aided diagnosis and surgery planning.

Keywords: Cochlear implant · Deep reinforcement learning · Cochlear nerve · Facial nerve · Surgery planning

1 Introduction

Locating the facial nerve (FN) is crucial for trajectory planning to cochlear implant (CI) surgery. It is also believed that proximity between the FN and the basal turn of the cochlea is associated with an increase incidence in cases of FN stimulation [10] because of the FN's proximity to some of the CI electrodes located in the cochlear structure. FN stimulation can lead to severe involuntary motion and pain, and knowing the FN location when passing by the cochlea structure could help predict cases of FN stimulation and/or adjust cochlear implant stimulation parameters to

© Springer Nature Switzerland AG 2021
M. de Bruijne et al. (Eds.): MICCAI 2021, LNCS 12904, pp. 519–528, 2021.
https://doi.org/10.1007/978-3-030-87202-1_50

mitigate FN stimulation after implantation. The cochlear nerve (CN) is primarily responsible for transmitting the electrical impulses generated in the cochlea to the brain. These impulses will be integrated for hearing and localization of sound. Its diameter is a critical measurement for the diagnosis of inner ear conditions as well as a prerequisite for the successful outcome of a CI surgery [23]. We propose an automated pipeline for the characterization of the CN and FN in CT scans.

Locating thin neural structures is a very challenging task, especially in CT scans. To the best of our knowledge, there is no previous approach to automate the characterization of the CN in CT scans. There are different approaches for FN characterization in CTs most of which focus on surgery trajectory planning, which sometimes dismisses the labyrinth segment shown in Fig. 1. Most of the literature in this field discusses atlas-based segmentation [6,19,20], the use of deformable models [22], or a combination of both [15]. Some of these methods are semi-automatic, based on the manual location of landmarks in precise locations of the FN as seen in [6,22]. Recent approaches use state of the art deep learning methods to segment the FN, as seen in [5,14]. These studies are focused on segmentation for trajectory planning of the CI surgery, and thus focusing on other related structures of the region. However, we can in both studies observe how the performance of the method drops significantly when evaluating the FN.

The first stage of our pipeline is automatic location of landmarks which are designed to provide crucial information for the initialization of the subsequent characterization methods. Annotating 3D landmarks in an unlabelled dataset (as ours) is a much faster process than the full 3D segmentation annotation which is time-consuming and, for these structures, challenging even for experts. Landmark location in medical images is a very active field of research. Initially machine learning approaches were based on regression and classification of hand-crafted features but with the rise of deep learning, various neural networks have been used to automate the location of landmarks as in [17,27].

Analyzing the landmark location problem from an object search prespective, the exhaustive scanning or mapping approach can result in a loss of accuracy or very high computational times. Deep reinforcement learning finds search strategies for locating different structures based on the image information at multiple scales which benefits from the different representations of the image. This resembles the human approach to locating a certain landmark and in this rather challenging scenario, it was found to be the best approach. Ghesu et al. in 2016 [8] first used a reinforcement learning agent to navigate through a 3D image. In 2017, Xu et al. [25] developed a supervised method to classify actions using image partitioning. The computational cost of these initial approaches was reduced thanks to the patch-based iterative convolutional neural network (CNN) approach proposed in 2018 [12,16] which can be adapted for single or multiple landmark detection. To exploit multi-scale image representation, Ghesu et al. extended their reinforcement-learning-based landmark detection in [7,9]. In 2019, Vlontzos A et al. [21] proposed the MARL model where multiple agents implicitly communicate by sharing the CNN weights. In 2020, Leroy et al. proposed a C-MARL model [11] by adding explicit communication based on sharing the average weights of the fully connected layers to the MARL model. Communication between agents benefits the system's ability to locate landmarks especially when those present spatial correlation, as in our case of study.

2 Data

The dataset available for this study consists of 119 clinical CT scans from diverse imaging equipment. The CT scans cover the patient's inner ear and were performed before surgery. A region of interest is cropped from each CT scan (32.1^3 mm^3), with voxel resolution average of 0.3 mm and range of [0.13,0.45]. The dataset is therefore representative of routine clinical scans in terms of anatomy, imaging modality, and image quality. The dataset presents a large variation regarding contrast, intensity and noise levels, which is convinient for the development of this application. All the scans were manually labeled by the main author with the designed landmarks using the *ITK-SNAP* software [26].

The location of seven landmarks has been designed to make each landmark as unique as possible within the structure, so it can be easily differentiated from other anatomical locations. This provides more robustness to the manual annotation which influences the robustness of the model. Furthermore, the placement of these landmarks is directly relevant for assessing various clinical metrics. Because these structures are very small as well as their surrounding structures, this task is very challenging, and the resolution does not allow us to locate very specific features of the nerves. Seven landmarks have been selected, two for the CN and five for the FN. The distribution of the landmarks can be seen in Fig. 1.

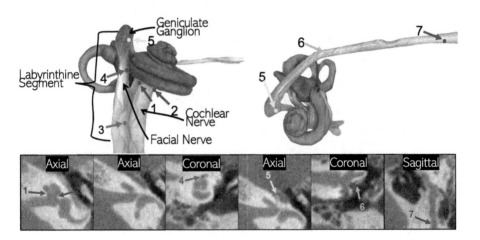

Fig. 1. Top: 3D representation of landmarks in the neural structure. Edited from [18]. Bottom: Landmarks' location in a CT scan (zoomed-in).

3   Methods

Deep-Q-Networks [13] are used to find the optimal strategy for agents to reach their goal. For landmark location, the network architecture resembles the architecture of a typical image classification network, in which convolutional layers extract the relevant features of the state (centered 3D patch in the position of the agent) followed by fully connected layers that map those features to the probability distribution of the Q value of each of the 3D actions (up, down, left, right, forward and backward). Diagonal movements are not considered. The architecture used in this application is shown in Fig. 2 for the case of two agents. The C-MARL model [11] uses multiple agents with implicit (same CNN weights) and explicit communication (average weight sharing of the fully connected layers). We use 21 agents, 3 per landmark, resulting in 21 fully connected networks implemented in the architecture. The dataset containing 119 samples has been randomly split into: 11 samples test set $\approx 10\%$, 12 samples validation set $\approx 10\%$, and 96 training set $\approx 80\%$. The neural networks are trained end-to-end on a 12GB GPU, with a total training time of 6 d. In the training process, the final state is reached when the distance to the landmark is ≤ 1 voxel using the ϵ-greedy search strategy [24]. The forgetting factor γ is set to 0.9 as this has been empirically found to be the best performing option. We use three isotropic resolutions for the multi-scale: 0.9, 0.6, and finally 0.3 mm per voxel dimension. The state of the agent becomes spatially smaller once the agent is oscillating in the current resolution. All the agents must reach the oscillation state in order to either move to the next resolution or to finalize the search. The final model is chosen based on the performance on the validation set.

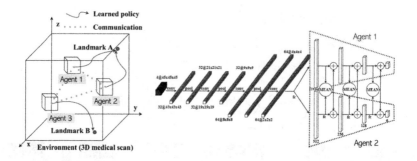

Fig. 2. Left: Diagram of the MARL model Right: architecture of the C-MARL model. Edited from [11,21].

The initial approach for the CN characterization is to measure the distance between landmarks 1 and 2 which are placed on opposite sides of the bony cochlear nerve canal (BCNC) in the axial plane of the CT. To give more robustness to this measurement, the median distance of all the possible distances between the three candidates (one per agent) of each of the landmarks has been chosen as the final measurement. To characterize this structure further than just the diameter, the area of the cross-section was computed. The cross-sectional slice was thus extracted based on the median location of landmarks 1 and 2 and the known parallelism of the CN and this plane.

The Chan-Vese approach [2] for active contours without edge (ACWE) has been used to segment the cross-section of the CN. The algorithm is based on the mean intensity inside and outside of the curve instead of the gradient, which is convenient for a structure with no clear edges, such as the CN. The region of interest of the cross-section is cropped according to the medial location of the landmarks, and the curve is initialized with a 3-pixel radius circle centered in the centroid of both landmarks. Empirically 15 iterations were found to be sufficient for the algorithm to converge. To fine-tune the regularization parameters of the ACWE algorithm is a tedious task, especially for the big variability of the images intensities and contrast present in clinical CTs. To avoid this task and generate a parameterization of this structure, a least-square ellipse is fitted to the ACWE contour. The CN presents an elliptical shape at this point due to its connection with the spiral shape of the cochlea structure. An example of the processing steps of this method are shown in Fig. 3. To corroborate that the parameterization is robust and reliable, the procedure is extended to multiple slices. This allows us to analyze the continuity in the third dimension and generate a full characterization of the canal. Currently, we select 8 slices to estimate the canal. We use the center of the ellipse fitted in the adjacent slice as the initialization seed for the following one.

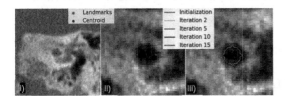 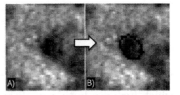

Fig. 3. Left: ACWE segmentations steps (cropping and iterations). Right: Contour of fitted ellipse (B) in black obtained from the ACWE contour (A) in red. (Color figure online)

To characterize the labyrinth segment of the FN, we use Dijkstra's algorithm [3] for geodesic path finding where the edge values are the intensity value of the voxel where the directed graph is pointing. There are three landmarks in the region of interest: number 3 which locates the exit of the FN from the internal acoustic canal, number 4 which locates the closest point of the FN to the cochlea, and number 5 which locates the first geniculate ganglion. To increase the robustness of this approach we design an optimal path selection algorithm. In this algorithm: **I)** Three candidate paths are computed $\text{path}_c^l = \text{path}_6^{3,5}, \text{path}_{18}^{3,5}, \text{path}_{26}^{3,4,5}$, where c represents the connectivity or number of neighbours of each edge and l represents the landmarks used, as illustrated in Fig. 4. **II)** Compute all paths' square derivatives, P_c^l, to have a measurement of the total path intensity variability from adjacent points.

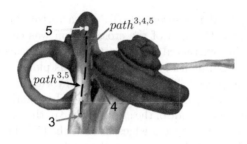

Fig. 4. FN selection algorithm path illustration. Both possibilities of landmarks either 2 (black) or 3 (red) landmarks involved are shown. Edited from [18]. (Color figure online)

$$P_c^l = \sum_{n=1}^{n=N} (\text{path}_c^l[n] - \text{path}_c^l[n-1])^2 \qquad (1)$$

III) The path with the minimal square derivative or minimal weight, P_c^l, is chosen as the optimal path.

A 3D representation of the nerve was generated by convolving the binary image containing the binary trace of the FN with a Gaussian kernel with $\sigma = 0.5$ and thresholding the outcome. This produces a normally distributed width of the nerve in the radial direction of the neural structure related to the normal thickness of the nerve in the labyrinth segment (1.23 ± 0.22 mm according to [1]).

4   Results

The results of the described C-MARL model on our test set are shown in Table 1 which includes both a qualitative and quantitative analysis of the performance where the landmark prediction is classified as correct, very close (1 or 2 voxels from optimal position) or wrong.

Figure 5 shows the table with the evaluation of the direct measurement extracted from the landmarks compared to the manual annotation, the signed difference and the absolute error are shown. The average automated BCNC diameters obtained is 2.4mm ($\sigma = 0.31$) which is a value of similar magnitude as the 2.13 mm ($\sigma = 0.44$) found in the clinical study by Fatterpekar *et al.* [4]. The figure also includes an example of the 3D results in a test image.

Figure 6 shows the performance of the path selection algorithm used for the FN tracking as well as an example of the results obtained with the posterior splatting technique and isosurface generation. For all the samples the method selects the best trace of the nerve. Validation and test sets are used to evaluate this method to provide a more significant performance overview as no difference in landmark location performance was found between both data splits.

The results are finally combined in a 3D representation generated with *ITK-SNAP* based on the different segmentations as shown in Fig. 7. The cochlea struc-

Table 1. Mean distance error in mm, μ_{Error}, and the standard deviation, σ_{Error}, for each landmark based on the selection of median location of the 3 candidates generated with the described C-MARL model. Qualitative analysis in percentage of correctly located, very close to optimal location and wrong location (color-labeled).

Landmark	1	2	3	4	5	6	7	Overall
μ_{Error} [mm]	0.791	0.561	0.889	0.599	0.988	1.897	2.797	1.218
σ_{Error}	0.321	0.224	0.347	0.435	0.409	0.972	2.140	1.185
Correct [%]	100	90.91	100	100	100	100	81.82	96.1
Very close [%]	0	9.09	0	0	0	0	0	1.3
Wrong [%]	0	0	0	0	0	0	18.18	2.6

Direct measurement		μ [mm]	σ	IQR [mm]	min/max [mm]
	Diff	-0.066	0.413	0.733	-1.027/0.625
	||Diff||	0.333	0.252	0.208	0.010/1.027

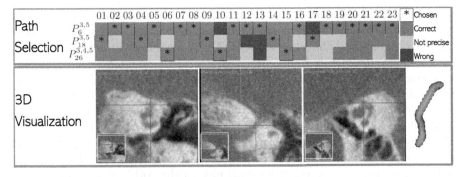

Fig. 5. CN results. Direct measurement: Table showing the performance statistics of the absolute and relative difference between our method and the manual annotations. 3D visualization: CT scan from test set with blue overlay of the 3D ACWE and ellipse fitting approach (left), together with the isosurface 3D representation (right). (Color figure online)

Fig. 6. FN results. Path Selection: Table showing the performance of the selection algorithm in test (1–11) and validation (12–23) scans, all the chosen paths (*) with minimal weight and color-coded the evaluation of the traces. 3D visualization: CT scan from test set with light blue overlay of the splatted selected path (left), together with the isosurface 3D representation (right). (Color figure online)

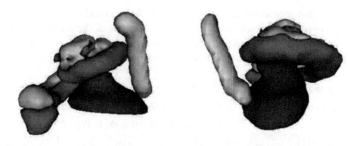

Fig. 7. 3D representation of the segmentation of the anatomical structures. Scala tympani (red), scala vestibuli (green), CN (dark blue), FN (light blue). (Color figure online)

ture has also been included for a better representation of both neural structures together and their relative positions.

5   Discussion and Conclusion

The characterization methods based on the estimated location of the landmarks produce accurate characterization of the neural structures in this region. There is an overall error of 1.218 mm in landmark location within the neural structures, a mean difference of 0.333 mm between the manual and automated BCNC measurement (similar magnitude as the dimension of the voxel), and a correct trace of the FN obtained in all the samples. The results show that both the landmark estimation and the further characterization methods are robust enough to characterize these challenging structures on clinical data. To provide further qualitative analysis of the full characterization with clinical annotations should be obtained.

The main contribution of this study is a novel pipeline using reinforcement learning for the extraction of clinically relevant metrics related to the CN health and the FN anatomy. The automation of the CN measurement in CT has not, to our knowledge, previously been performed. Providing the cross-sectional characterization allows clinicians to study the condition of the CN based on both of its elliptical axes and not only on the axial plane view. This cross-sectional information is rather difficult for non-experts to extract due to the non-orthogonal location relative to the Cartesian axis of the CT image. The correct tracking of the FN without human interaction along the labyrinth segment is a big advantage, as this region is known for the difficulty of its characterization. This pipeline allows for a characterization of its proximity to the cochlear structure in this critical region and helps clinicians prevent the FN stimulation derived from this proximity.

The result of this work is a complete pipeline that automatically segments and parameterizes both neural structures in the close-by area of the cochlea. We have developed an advanced tool for CI surgical planning based on deep learning and computer vision algorithms which may pave the way to support clinicians in the routine assessment of FN anatomy, related risk assessment of FN stimulation, and the evaluation of CN health.

References

1. Celik, O., Eskiizmir, G., Pabuscu, Y., Ulkumen, B., Toker, G.T.: The role of facial canal diameter in the pathogenesis and grade of bell's palsy: a study by high resolution computed tomography. Brazilian J. Otorhinol. **83**, 261–268 (2017)
2. Chan, T.F., Vese, L.A.: Active contours without edges. IEEE Trans. Image Process. **10**(2), 266–277 (2001)
3. Dijkstra, E.W.: A note on two problems in connexion with graphs. Numerische mathematik **1**(1), 269–271 (1959)
4. Fatterpekar, G.M., Mukherji, S.K., Lin, Y., Alley, J.G., Stone, J.A., Castillo, M.: Normal canals at the fundus of the internal auditory canal: CT evaluation. J. Comput. Assist. Tomogr. **23**, 776–780 (1999)
5. Fauser, J., et al.: Toward an automatic preoperative pipeline for image-guided temporal bone surgery. Int. J. Comput. Assist. Radiol. Surg. **14**(6), 967–976 (2019)
6. Gare, B.M., Hudson, T., Rohani, S.A., Allen, D.G., Agrawal, S.K., Ladak, H.M.: Multi-atlas segmentation of the facial nerve from clinical CT for virtual reality simulators. Int. J. Comput. Assist. Radiol. Surg. **15**, 259–267 (2020)
7. Ghesu, F.C., Georgescu, B., Grbic, S., Maier, A.K., Hornegger, J., Comaniciu, D.: Robust multi-scale anatomical landmark detection in incomplete 3D-CT data. Proc. MICCAI **2017**, 194–202 (2017)
8. Ghesu, F.C., Georgescu, B., Mansi, T., Neumann, D., Hornegger, J., Comaniciu, D.: An artificial agent for anatomical landmark detection in medical images. Proc. MICCAI **2016**, 229–237 (2016)
9. Ghesu, F.C., et al.: Multi-scale deep reinforcement learning for real-time 3D-landmark detection in CT scans. IEEE Trans. Pattern Anal. Mach. Intell. **41**, 176–189 (2019)
10. Hatch, J.L., et al.: Can preoperative CT scans be used to predict facial nerve stimulation following CI? Otol. Neurotol. **38**, 1112–1117 (2017)
11. Leroy, G., Rueckert, D., Alansary, A.: Communicative reinforcement learning agents for landmark detection in brain images. In: Kia, S.M., et al. (eds.) MLCN/RNO-AI -2020. LNCS, vol. 12449, pp. 177–186. Springer, Cham (2020). https://doi.org/10.1007/978-3-030-66843-3_18
12. Li, Y., et al.: Fast multiple landmark localisation using a patch-based iterative network. In: Frangi, A.F., Schnabel, J.A., Davatzikos, C., Alberola-López, C., Fichtinger, G. (eds.) MICCAI 2018. LNCS, vol. 11070, pp. 563–571. Springer, Cham (2018). https://doi.org/10.1007/978-3-030-00928-1_64
13. Mnih, V., et al.: Human-level control through deep reinforcement learning. Nature **518**, 529–533 (2015)
14. Nikan, S., Osch, K.V., Bartling, M., Allen, D.G., Rohani, S.A., Connors, B., Agrawal, S.K., Ladak, H.M.: PWD-3DNet: a deep learning-based fully-automated segmentation of multiple structures on temporal bone CT scans. IEEE Trans. Image Process. **30**, 739–753 (2021)
15. Noble, J.H., Warren, F.M., Labadie, R.F., Dawant, B.M.: Automatic segmentation of the facial nerve and chorda tympani in CT images using spatially dependent feature values. Med. Phys. **35**, 5375–5384 (2008)

16. Noothout, J.M.H., de Vos, B.D., Wolterink, J.M., Leiner, T., Isgum, I.: CNN-based landmark detection in cardiac CTA scans. CoRR abs/1804.04963 (2018). http://arxiv.org/abs/1804.04963
17. Oktay, O., et al.: Stratified decision forests for accurate anatomical landmark localization in cardiac images. IEEE Trans. Med. Imaging **36**, 332–342 (2017)
18. Trier, P., Karsten Noe, M.S.S.: The visible ear simulator (2020). https://ves.alexandra.dk/
19. Powell, K.A., Kashikar, T., Hittle, B., Stredney, D., Kerwin, T., Wiet, G.J.: Atlas-based segmentation of temporal bone surface structures. Int. J. Comput. Assist. Radiol. Surg. **14**, 1267–1273 (2019)
20. Powell, K.A., Liang, T., Hittle, B., Stredney, D., Kerwin, T., Wiet, G.J.: Atlas-based segmentation of temporal bone anatomy. Int. J. Comput. Assist. Radiol. Surg. **12**, 1937–1944 (2017)
21. Vlontzos, A., Alansary, A., Kamnitsas, K., Rueckert, D., Kainz, B.: Multiple landmark detection using multi-agent reinforcement learning. In: Shen, D., et al. (eds.) MICCAI 2019. LNCS, vol. 11767, pp. 262–270. Springer, Cham (2019). https://doi.org/10.1007/978-3-030-32251-9_29
22. Voormolen, E.H., et al.: Determination of a facial nerve safety zone for navigated temporal bone surgery. Neurosurgery **70**, 50–60 (2012)
23. Waldman, S.D.: Chapter 9 - the vestibulocochlear nerve—cranial nerve viii. In: Waldman, S.D. (ed.) Pain Review, pp. 22–25. W.B. Saunders, Philadelphia (2009). http://www.sciencedirect.com/science/article/pii/B9781416058939000095
24. Watkins, C.J.C.H.: Learning from Delayed Rewards. Ph.D. thesis, King's College, Cambridge, UK, May 1989
25. Xu, Z., et al.: Supervised action classifier: approaching landmark detection as image partitioning. In: Descoteaux, M., Maier-Hein, L., Franz, A., Jannin, P., Collins, D.L., Duchesne, S. (eds.) MICCAI 2017. LNCS, vol. 10435, pp. 338–346. Springer, Cham (2017). https://doi.org/10.1007/978-3-319-66179-7_39
26. Yushkevich, P.A., Piven, J., Cody Hazlett, H., Gimpel Smith, R., Ho, S., Gee, J.C., Gerig, G.: User-guided 3D active contour segmentation of anatomical structures: significantly improved efficiency and reliability. Neuroimage **31**(3), 1116–1128 (2006)
27. Zhang, D., Liu, Y., Noble, J.H., Dawant, B.M.: Automatic localization of landmark sets in head CT images with regression forests for image registration initialization. Med. Imaging 2016 Image Process. **9784**, 97841M (2016)

Patient-Specific Virtual Spine Straightening and Vertebra Inpainting: An Automatic Framework for Osteoplasty Planning

Christina Bukas[1,2,3(✉)], Bailiang Jian[2,3], Luis Francisco Rodríguez Venegas[2,3], Francesca De Benetti[2,3], Sebastian Rühling[4], Anjany Sekuboyina[4,5], Jens Gempt[6], Jan Stefan Kirschke[4], Marie Piraud[1], Johannes Oberreuter[2,7], Nassir Navab[3,8], and Thomas Wendler[2,3]

[1] Helmholtz AI, Helmholtz Zentrum München, Munich, Germany
christina.bukas@helmholtz-muenchen.de
[2] ScintHealth GmbH, Munich, Germany
[3] Chair for Computer Aided Medical Procedures and Augmented Reality,
Technische Universität München, Garching, Germany
[4] Department of Neuroradiology, Technische Universität München, Munich, Germany
[5] Working Group for Image-Based Biomedical Modeling,
Technische Universität München, Garching, Germany
[6] Department of Neurosurgery, Technische Universität München, Munich, Germany
[7] Reply SpA, Munich, Germany
[8] Computer Aided Medical Procedures Lab, Laboratory for Computational
Sensing+Robotics, Johns Hopkins University, Baltimore, MD, USA

Abstract. Symptomatic spinal vertebral compression fractures are often treated by osteoplasty where a cement-like material is injected into the bone to stabilize the fracture, restore the vertebral body height and alleviate pain. Leakage is a common complication and may occur due to too much cement being injected. Here, we propose an automated patient-specific framework that can allow physicians to calculate an upper bound of the volume of cement for particular types of VCFs and estimate the optimal outcome of osteoplasty. The framework uses the patient CT scan and the segmentation label of the fractured vertebra to build a virtual healthy spine. Firstly, the fractured spine is segmented with a three-step Convolutional Neural Network architecture. Next, a per-vertebra rigid registration to a healthy reference spine restores its curvature. Finally, a GAN-based inpainting approach replaces the fractured vertebra with an estimation of its original shape, the volume of which we use as an estimate of the original healthy vertebra volume. As a clinical application, we derive an upper bound on the amount of bone cement for the injection. We evaluate

ScintHealth GmbH, Munich, Germany—Partially funded by the EIT Health Headstart Grant "AcryRad".

Electronic supplementary material The online version of this chapter (https://doi.org/10.1007/978-3-030-87202-1_51) contains supplementary material, which is available to authorized users.

M. de Bruijne et al. (Eds.): MICCAI 2021, LNCS 12904, pp. 529–539, 2021.
https://doi.org/10.1007/978-3-030-87202-1_51

our framework by comparing the virtual vertebrae volumes of ten patients to their healthy equivalent and report an error of $3.88 \pm 7.63\%$. The presented pipeline offers a first approach to a personalized automatic high-level framework for planning osteoplasty procedures.

Keywords: Spine osteoplasty · Inpainting · Deformable registration

1   Introduction

This work proposes improvements in osteoplasty planning by image analysis using deep learning. Spinal vertebra compression fractures (VCFs) are a painful and debilitating injury to the skeleton. Osteoplasty, namely kyphoplasty or vertebroplasty, is an operative procedure, during which the body of a fractured vertebra is filled with bone cement, which stabilizes the spine and relieves the patient from the pain [4]. Yet, it is difficult to determine the amount of cement to be injected [7]. If too much cement is injected, a leakage may put pressure on the spinal cord and can even lead to pulmonary cement embolisms [19].

The goal of this work is to provide a computer-assisted intervention method, to generate a virtual 3D model of a healthy vertebra in a personalized manner, taking into account the patient's anatomy and the type of vertebra. The only data readily available comes from CT imaging, which is solely used to derive the shape of a healthy-looking vertebra matching the spine of the patient and an upper bound for the bone cement to be injected. We propose an automated framework following the workflow of Figs. 1 and 2. In addition to osteoplasty planning, our framework may also be used for studying biomechanics before and after a fracture, without the need of a healthy CT.

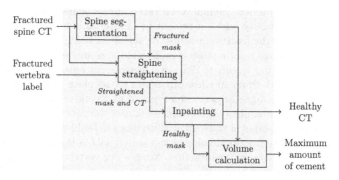

Fig. 1. Proposed framework: The CT image is first used to generate a segmentation mask, which is then fed into the spine straightening block, together with the CT image and the fractured vertebra label (e.g. "L3"). The straightened spine and corresponding segmentation are then sent to the inpainting block to generate the final virtually healthy CT and the upper bound for the cement.

The main contributions in this paper are: 1) To our knowledge, this is the first framework to do an automated end-to-end spine straightening and vertebra generation to estimate the upper bound of bone cement for osteoplasty. 2) We introduce a method for virtually straightening a fractured spine to a healthy-looking

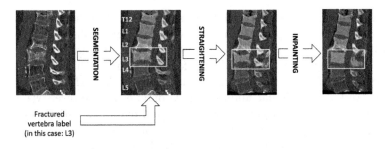

Fig. 2. A visual example of the inputs and outputs of each stage of the framework.

curvature and height solely using a CT image and the label of the damaged vertebra. 3) We propose a patient-tailored 2.5D inpainting method to generate a healthy-looking version of a fractured vertebra and its segmentation mask.

2   Methodology

Each step of the pipeline (Figs. 1 and 2) is detailed below. The models were implemented in PyTorch and their weights, along with code are publicly available[1].

Spine Segmentation. The goal of this step is to automatically generate voxel-level masks of the vertebrae, a task widely analysed in literature (e.g. [16]). Compared to binary segmentation, vertebral multi-class segmentation requires a more complex representation to be learned. Not only are the fields-of-view across CT datasets varying, but also the shapes of adjacent vertebrae are highly correlated. For these reasons, we selected the approach of Sekuboyina et al. splitting the task into three modules: spine detection, vertebrae labeling, and vertebrae segmentation [16,17].

Spine detection enables the pipeline to be applicable to scans of any field of view. For this task we use a variant of the U-Net [14,15], to regress a coarse 3D Gaussian heatmap of the spinal centerline and a 3D bounding box around it to localize the spine. For the second stage, we re-implement the Btrfly-Net [17] architecture, without the energy-based adversarial training stage, since it works on maximum intensity projections (MIPs) thereby reducing complexity. The network predicts 24 refined Gaussian heatmaps centered at each vertebra. Finally, the spine image is cropped to a 3D patch around each vertebral centroid previously detected, along with its corresponding heatmap. Both are then fed into a binary U-Net in order to segment only the vertebra of interest.

Virtual Spine Straightening. Our virtual spine straightening algorithm corrects for the height of the fractured vertebral body and reconstructs the spine's curvature to a healthy shape. The latter is achieved by straightening the spine in all three dimensions on a vertebra by vertebra basis (shown in Fig. 4a).

---

[1] https://github.com/christinab12/bone-cement-injection-planning.

For the task of spine registration, Forsberg et al. proposed registering sub-volumes of a patient's spine to an atlas with non-rigid transformations, which evidently ignored the rigid nature of the vertebrae [5]. Drobny et al. proposed to register the spine with a poly-rigid transformation model thus ensuring a rigid transformation of the vertebrae, but evaluated their approach only on synthetic data [3]. We follow a similar approach as Dronby et al., however, we additionally tackle the issue of varying patient sizes and employ a simpler method for deforming non-vertebra voxels. Our inputs here are the patient's CT, the vertebrae centroids and the spine segmentation mask, as well as centroids and mask of a healthy and straight spine, which we use as a reference. First, the reference spine is scaled to the height of the patient. The scaling factor is computed by comparing the sum of the distances between the centroids of all visible vertebrae on the image, while excluding vertebra(e) with fracture(s). The distances are computed using the first two vectors from the principal component analysis (PCA) of the centroids' coordinates to mitigate the deviation resulting from the patient's positioning during the CT acquisition. Next, a set of displacement fields is generated from a per-vertebra rigid registration of the segmentation masks. Additionally, a distance map of each vertebra is calculated, where the voxel value represents the physical distance to the nearest vertebra voxel. The displacement fields are then compounded with a simplified version of Little et al.'s interpolation approach [10] (see Supplementary Material), based on our biomechanic assumption that the soft tissue is transformed inversely proportional to its distance from vertebrae, similar to linear blend skinning [8]. Finally, we generate the straightened spine by resampling, using the combined displacement field.

Vertebra Inpainting. The next task is to replace the fractured vertebra with its healthy equivalent. We do this by using inpainting, an imaging technique which fills in missing parts of images and is widely used in the field of computer vision. Though it has previously been used in medicine for predicting missing information [1,20], removing lesions [26] or correcting limited-angle acquisitions [9,22,27], none of these works tackles a similar problem to ours. Additionally, in the field of computer vision inpainting has been applied to produce high quality images while large areas can be removed from the original image [6,11,21,24,25]. We therefore chose to transfer state-of-the-art works from this ever-evolving field to the medical imaging domain. Wang et al. propose 3D models to inpaint volumes, but their inputs are point-clouds thus not applicable in our setup [21]. Here, though we are dealing with volumes, when focusing on vertebral bodies the sagittal and coronal views are expected to provide sufficient information for generating a vertebra. Accordingly, we chose to apply two 2D neural networks to a volume, one trained on sagittal slices, the other on coronal, and fuse the outputs. Yeh et al. propose using GANs which perform inpainting on 2D images but with low resolution [24], while Isola et al., and Yu et al. obtain better results at a higher resolution by extending to conditional GANs [6,25]. Liu et al. also achieve impressive results when completing images with irregular holes by introducing partial convolutions in a U-Net [11]. Our approach is nonetheless different: While we are mainly interested in the segmentation mask for our pipeline, we tackle the inpainting as a multi-task learning problem, by training a model to a) generate the inpainted CT image and b) output the segmentation mask. This enforces the model to better learn the training data distribution and offers interpretability to the user. We remove the fractured vertebra from the CT and segmentation mask slices by applying a binary mask, which is generated based on the segmentation and fractured label. These are the inputs to the inpainting network, whose architecture extends that of Yu et al. [25], as shown in Fig. 3. We chose this network as our baseline since i) it outperformed other methods [2,6], or our implementation of them [11], and ii) the contribution of the contextual attention layer is of particular interest in our case. Similarly to the original authors' network, the input first passes through a coarse generator for a first estimate of the inpainted result and next through a refinement network, both of which have a U-like architecture [14]. The last layer of the refinement network has been adjusted to two final parallel layers, which output the inpainted image and the inpainted mask. Four discriminators follow the generator; two, for the image and mask, are local and only take the patch region as input and two are global, for the entire image space. We use the same losses as presented in [25], while extending the generator loss by adding weighted dice losses for the patch region and the background of the mask and adjusting the total discriminator loss to accommodate all four sub-networks. To obtain the final volumes, we apply a simple averaging of the intensities to fuse the two CT volumes, while for the inpainted mask, the output pixel-wise probabilities of the segmentation layer are averaged before acquiring the final predictions. From this we compute the virtually healthy vertebra volume, and by subtracting from it the original fractured vertebra volume, compute an upper bound for the cement.

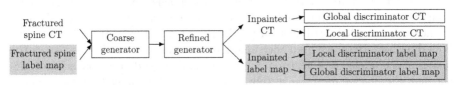

Fig. 3. The proposed inpainting network architecture. The parts highlighted provide our extensions to the architecture of Yu et al. [25].

3   Experimental Setup

Segmentation Training Dataset. We used the VerSe'20 dataset [12,16,18] for the segmentation pipeline, consisting of 100 patients with vertebrae centroid annotations and segmentation masks. We randomly split this dataset into a training set of 70 patients, and a validation and testing set of 15 patients, respectively.

Inpainting Training Dataset. Since our model needs to learn to inpaint a healthy vertebra, we use a healthy dataset as ground truth which includes no fractures. We created two 2D datasets, one of coronal and one of sagittal views, from a total of 110 volumes (95 confirmed healthy spines from our institution and 15 from the CSI challenge[2] which we verified to be healthy) and set aside ten for validation and testing alike. Depending on the size and resolutions of the scans, the number of slices and spine regions taken from each volume varied, making up a total of 3557 sagittal and 1358 coronal images, each consisting of five vertebra.

Dataset for Validation of Virtual Spine Straightening and Overall Pipeline. For the evaluation of the straightening and the overall workflow, we obtained a list of 316 patients who underwent kyphoplasty between 2014 and 2019 at our institution. We filtered these to include only patients who (1) have the pre-fracture, post-fracture and post-operative CT scans, and (2) have CT images with voxel sizes smaller than $1 \times 1 \times 3$ mm^3 (xyz). After applying these criteria, our dataset included ten patients. For the sake of completeness, some of these ten patients have more than one fractured vertebra, which enabled us to evaluate the pipeline on a total of 15 vertebrae. The CT scans included in the dataset have resolutions between 0.28 and 0.97 mm (xy), and between 0.7 and 3.0 mm (z). For clarification the *pre-fracture* scan shows a healthy spine, while the *post-fracture* scan has at least one fracture, the *post-operative* scan is taken after osteoplasty, the *straightened* scan is the post-fracture scan after the virtual spine straightening, and the *inpainted* scan is the straightened scan after the vertebra inpainting.

Segmentation Pipeline Training. We trained each module of the segmentation pipeline independently, following the same dataset split for every task. For the detection stage, we first built the ground-truth spine heatmap by combining the individual heatmaps of the vertebrae landmarks. We then trained the 3D U-Net variant with the L2 loss, using the Adam optimizer, a batch size of two and a learning rate of 1e-3 until convergence (77 epochs). Next, we extracted the bounding box around the predicted spine heatmap. For the labelling stage we trained the modified BtrflyNet on the MIPs extracted from the previous stage bounding box. Here, we used the same hyperparameters and L2 loss, since this task is also modeled as a heatmap regression, and trained until convergence (280 epochs). Finally, we used the vertebral patches and heatmaps from the previous stage to train the 3D U-Net. We used the dice loss and trained for 25 epochs, again with the same hyperparameters.

Inpainting Model Training. We trained the two models on the lateral and coronal datasets, where each slice contains four visible vertebrae and one digitally erased. Thus, each image was sent five times through the networks, which were trained using a batch size of 16, with Adam for the optimization and initial learning rates of 1e-3 and 1e-4 for the generator and discriminator. A grid search was applied to find optimal loss weights and learning rates, in the space around the original authors' choice. We implement early stopping considering the segmentation metrics, since for our pipeline the segmentation result is of higher relevance. The best performing models on the validation sets were chosen.

---

[2] http://csi-workshop.weebly.com/challenges.html.

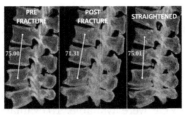

(a) *Straightening*

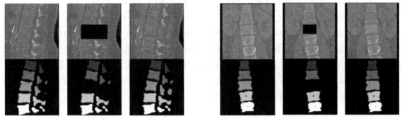

(b) *Inpainting: sagittal view*   (c) *Inpainting: coronal view*

Fig. 4. (a) Visual results of the straightening step. The image includes the distance (in mm) between the vertebrae above and below the fractured one. (b) and (c) Visual results of the inpainting step. From left to right: ground-truth, input and output of the lateral and coronal GANs.

4   Results

Evaluation of Segmentation. For assessing the performance of the vertebrae segmentation framework, we employed the dice coefficient metric. We computed this metric at a vertebra level over all the vertebrae annotated in the ground truth from the VerSe2020 mask labels. The average dice score over all patients in the test set is $87.06 \pm 9.54\%$, a value well in the range of the reported values from the literature (83.06% to 93.01%, e.g., [16]).

Evaluation of Virtual Spine Straightening. We evaluated the spine straightening algorithm by utilizing what we call the *fracture distance*. This is the physical distance between the vertebrae above and below the fracture, which should increase to provide enough space for the inpainting of a healthy vertebra. In order to make the distance comparable, the pre-fracture spine CT was also registered to the reference spine space using the straightening algorithm. We therefore compared the *fracture distance* of straightened pre-fracture, raw post-fracture and straightened post-fracture spine of the patients (Figs. 4a and 5b). The average error on the *fracture distance* yielded 0.23 ± 2.68 mm, or equivalently, $0.50 \pm 3.95\%$.

Evaluation of Inpainting. To evaluate the results of the inpainting model we computed the structural similarity index (SSIM) and peak signal-to-noise ratio (PSNR) of the output CT [23], as well as the intersection over union (IoU) of the segmentation in the region of interest [13], and the mean relative volume error

(MRE) of the vertebra in question. Fusing the coronal and sagittal outputs improves the individual results, with our final model achieving: SSIM = 0.82, PSNR = 26.45 dB, IoU = 0.76 and MRE = 19% (see Figs. 4b, 4c, 5a).

Evaluation of Overall Pipeline. We evaluated the performance of our framework in an end-to-end manner by using ten patients, for whom as mentioned a CT scan before and after the VCF were available. To evaluate the effectiveness of our pipeline, we compared the volumes of the pre-fractured vertebrae with the volumes of the inpainted ones. Overall, our method has an error of 2.61 ± 5.07 mL, which translates to a relative error of 3.08 ± 7.63 %. Figure 5a shows the correlation between the pre-fractured and inpainted vertebra volumes (see also Supplementary Material). As an additional clinical assessment, we segmented the bone cement in the post-operative CTs by thresholding, thus realizing that the actual cement injected was always lower than our upper bound. Additional results are available in the Supplementary Material.

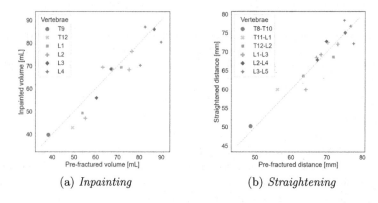

(a) *Inpainting*   (b) *Straightening*

Fig. 5. Quantitative results of the (a) inpainting and (b) straightening steps.

5   Discussion and Future Work

In this work, we have presented an integrated computer-assisted intervention framework to estimate a realistic healthy 3D shape for fractured vertebrae. Using only a patient's post-fracture CT image, a healthy vertebra shape replaces the fractured one, virtually restoring the spine. As a clinical application of our work, we envision that the inpainted vertebra can be used as a guideline to estimate an upper bound of the injected material, thus avoiding leakage, albeit we are not performing this very analysis. Although desired, it is difficult to evaluate the ideal amount of cement to be injected since we do not currently have many cases with leakage in our database, to make a significant statistical analysis. Furthermore, different kinds of VCFs might require a different amount of bone cement. We do not have this information in the dataset to confirm our suspicion.

As the primary output of the workflow is a 2.5D image generated by a GAN trained on healthy spines, it is necessary that the healthy vertebrae in the image to be inpainted are resembling a healthy spine and provide accurate space for the inpainting. The virtual spine straightening step solves this problem. We aim to extend our model to a 3D network, since a current limitation is that the CT outputs are somewhat blurry. Moreover, our framework generates the entire vertebra (body and processes) which may reduce performance. We observe that transverse and spinous processes fail to be correctly segmented and inpainted. It must be pointed out that only the body is filled with cement during osteoplasty such that an automatic removal of the processes may be beneficial. We are therefore currently working on extending our framework to include this.

We have also validated the distance between the vertebrae neighbouring the fracture for those patients, where anecdotally pre-fracture imaging is available - yielding satisfactory results. Additionally, the evaluation of the inpainting showed that it tallies very well with the pre-fractured state at a relative error of below 4%. The number of patients evaluated was low, since finding scans from both healthy and fractured vertebra states is unlikely. We do however, aim to extend our dataset if possible to allow for a more robust statistical analysis. In general, the method is slightly underestimating the volume of the vertebrae, which can be considered conservative given the necessity for an upper bound. In its current state, our framework demonstrates its high potential as an automatic method for generating healthy-looking medical images, requiring minimal diagnostic input. The reported ability to derive quantitative results proves the usefulness of deep learning approaches for planning interventions.

References

1. Armanious, K., Mecky, Y., Gatidis, S., Yang, B.: Adversarial inpainting of medical image modalities. In: ICASSP 2019-2019 IEEE International Conference on Acoustics, Speech and Signal Processing (ICASSP), pp. 3267–3271. IEEE (2019)
2. Barnes, C., Shechtman, E., Finkelstein, A., Goldman, D.B.: Patchmatch: a randomized correspondence algorithm for structural image editing. ACM Trans. Graph. **28**(3), 24 (2009)
3. Drobny, D., et al.: Towards automated spine mobility quantification: a locally rigid CT to X-ray registration framework. In: Špiclin, Ž, McClelland, J., Kybic, J., Goksel, O. (eds.) WBIR 2020. LNCS, vol. 12120, pp. 67–77. Springer, Cham (2020). https://doi.org/10.1007/978-3-030-50120-4_7
4. Filippiadis, D.K., Marcia, S., Masala, S., Deschamps, F., Kelekis, A.: Percutaneous vertebroplasty and kyphoplasty: current status, new developments and old controversies. Cardiovasc. Intervent. Radiol. **40**(12), 1815–1823 (2017)
5. Forsberg, D.: Atlas-based registration for accurate segmentation of thoracic and lumbar vertebrae in CT data. In: Yao, J., Glocker, B., Klinder, T., Li, S. (eds.) Recent Advances in Computational Methods and Clinical Applications for Spine Imaging, pp. 49–59. Springer, Cham (2015). https://doi.org/10.1007/978-3-319-14148-0_5
6. Isola, P., Zhu, J.Y., Zhou, T., Efros, A.A.: Image-to-image translation with conditional adversarial networks. In: Proceedings of the IEEE Conference on Computer Vision and Pattern Recognition, pp. 1125–1134 (2017)

7. Janssen, I., et al.: Risk of cement leakage and pulmonary embolism by bone cement-augmented pedicle screw fixation of the thoracolumbar spine. Spine J. **17**(6), 837–844 (2017)

8. Lewis, J.P., Cordner, M., Fong, N.: Pose space deformation: a unified approach to shape interpolation and skeleton-driven deformation. In: Proceedings of the 27th Annual Conference on Computer Graphics and Interactive Techniques, SIGGRAPH 2000, pp. 165–172. ACM Press/Addison-Wesley Publishing Co., USA (2000). https://doi.org/10.1145/344779.344862

9. Li, Z., et al.: Promising generative adversarial network based sinogram inpainting method for ultra-limited-angle computed tomography imaging. Sensors **19**(18), 3941 (2019)

10. Little, J., Hill, D., Hawkes, D.: Deformations incorporating rigid structures, vol. 66, pp. 223–232 (1997). https://doi.org/10.1006/cviu.1997.0608. https://www.sciencedirect.com/science/article/pii/S1077314297906081

11. Liu, G., Reda, F.A., Shih, K.J., Wang, T.C., Tao, A., Catanzaro, B.: Image inpainting for irregular holes using partial convolutions. In: Proceedings of the European Conference on Computer Vision (ECCV), September 2018

12. Löffler, M.T., et al.: A vertebral segmentation dataset with fracture grading. Radiol. Artif. Intell. **2**(4), e190138 (2020)

13. Rahman, M.A., Wang, Y.: Optimizing intersection-over-union in deep neural networks for image segmentation. In: Bebis, G., et al. (eds.) ISVC 2016. LNCS, vol. 10072, pp. 234–244. Springer, Cham (2016). https://doi.org/10.1007/978-3-319-50835-1_22

14. Ronneberger, O., Fischer, P., Brox, T.: U-Net: convolutional networks for biomedical image segmentation. In: Navab, N., Hornegger, J., Wells, W.M., Frangi, A.F. (eds.) MICCAI 2015. LNCS, vol. 9351, pp. 234–241. Springer, Cham (2015). https://doi.org/10.1007/978-3-319-24574-4_28

15. Roy, A.G., Navab, N., Wachinger, C.: Concurrent spatial and channel 'squeeze & excitation' in fully convolutional networks. In: Frangi, A.F., Schnabel, J.A., Davatzikos, C., Alberola-López, C., Fichtinger, G. (eds.) MICCAI 2018. LNCS, vol. 11070, pp. 421–429. Springer, Cham (2018). https://doi.org/10.1007/978-3-030-00928-1_48

16. Sekuboyina, A., et al.: Verse: a vertebrae labelling and segmentation benchmark. arXiv preprint arXiv:2001.09193 (2020)

17. Sekuboyina, A., et al.: Btrfly net: vertebrae labelling with energy-based adversarial learning of local spine prior. In: Frangi, A.F., Schnabel, J.A., Davatzikos, C., Alberola-López, C., Fichtinger, G. (eds.) MICCAI 2018. LNCS, vol. 11073, pp. 649–657. Springer, Cham (2018). https://doi.org/10.1007/978-3-030-00937-3_74

18. Sekuboyina, A., Rempfler, M., Valentinitsch, A., Menze, B.H., Kirschke, J.S.: Labeling vertebrae with two-dimensional reformations of multidetector CT images: an adversarial approach for incorporating prior knowledge of spine anatomy. Radiol.: Artif. Intell. **2**(2), e190074 (2020)

19. Sørensen, S.T., Kirkegaard, A.O., Carreon, L., Rousing, R., Andersen, M.Ø.: Vertebroplasty or kyphoplasty as palliative treatment for cancer-related vertebral compression fractures: a systematic review. Spine J. **19**(6), 1067–1075 (2019)

20. Torrado-Carvajal, A., et al.: Inpainting as a technique for estimation of missing voxels in brain imaging. Ann. Biomed. Eng. **49**(1), 345–353 (2021)

21. Wang, W., Huang, Q., You, S., Yang, C., Neumann, U.: Shape inpainting using 3D generative adversarial network and recurrent convolutional networks. In: Proceedings of the IEEE International Conference on Computer Vision (ICCV), October 2017

22. Wang, Y., et al.: An effective sinogram inpainting for complementary limited-angle dual-energy computed tomography imaging using generative adversarial networks. J. X-Ray Sci. Technol. 1–25 (2020)
23. Wang, Z., Bovik, A.C., Sheikh, H.R., Simoncelli, E.P.: Image quality assessment: from error visibility to structural similarity. IEEE Trans. Image Process. **13**(4), 600–612 (2004)
24. Yeh, R.A., Chen, C., Yian Lim, T., Schwing, A.G., Hasegawa-Johnson, M., Do, M.N.: Semantic image inpainting with deep generative models. In: Proceedings of the IEEE Conference on Computer Vision and Pattern Recognition (CVPR), July 2017
25. Yu, J., Lin, Z., Yang, J., Shen, X., Lu, X., Huang, T.S.: Generative image inpainting with contextual attention. In: Proceedings of the IEEE Conference on Computer Vision and Pattern Recognition, pp. 5505–5514 (2018)
26. Zhang, H., Bakshi, R., Bagnato, F., Oguz, I.: Robust multiple sclerosis lesion inpainting with edge prior. In: Liu, M., Yan, P., Lian, C., Cao, X. (eds.) MLMI 2020. LNCS, vol. 12436, pp. 120–129. Springer, Cham (2020). https://doi.org/10.1007/978-3-030-59861-7_13
27. Zhao, J., Chen, Z., Zhang, L., Jin, X.: Unsupervised learnable sinogram inpainting network (SIN) for limited angle CT reconstruction. arXiv preprint arXiv:1811.03911 (2018)

A New Approach to Orthopedic Surgery Planning Using Deep Reinforcement Learning and Simulation

Joëlle Ackermann[1,2]([⊠]), Matthias Wieland[1,3], Armando Hoch[1,4],
Reinhold Ganz[5], Jess G. Snedeker[2], Martin R. Oswald[6], Marc Pollefeys[6,7],
Patrick O. Zingg[4], Hooman Esfandiari[1], and Philipp Fürnstahl[1]

[1] ROCS, Balgrist University Hospital, UZH, Zurich, Switzerland
`joelle.ackermann@balgrist.ch`
[2] Laboratory for Orthopedic Biomechanics, ETH Zurich, Zurich, Switzerland
[3] Department of Mechanical and Process Engineering,
ETH Zurich, Zurich, Switzerland
[4] Orthopedic Department, Balgrist University Hospital, UZH, Zurich, Switzerland
[5] Emeritus, University of Berne, Bern, Switzerland
[6] Department of Computer Science, ETH Zurich, Zurich, Switzerland
[7] Microsoft Mixed Reality and AI Zurich Lab, Zurich, Switzerland

Abstract. Computer-assisted orthopedic interventions require surgery planning based on patient-specific three-dimensional anatomical models. The state of the art has addressed the automation of this planning process either through mathematical optimization or supervised learning, the former requiring a handcrafted objective function and the latter sufficient training data. In this paper, we propose a completely model-free and automatic surgery planning approach for femoral osteotomies based on Deep Reinforcement Learning which is capable of generating clinical-grade solutions without needing patient data for training. One of our key contributions is that we solve the real-world task in a simulation environment tailored to orthopedic interventions based on an analytical representation of real patient data, in order to overcome convergence, noise, and dimensionality problems. An agent was trained on simulated anatomy based on Proximal Policy Optimization and inference was performed on real patient data. A qualitative evaluation with expert surgeons and a complementary quantitative analysis demonstrated that our approach was capable of generating clinical-grade planning solutions from unseen data of eleven patient cases. In eight cases, a direct comparison to clinical gold standard (GS) planning solutions was performed, showing our approach to perform equally good or better in 80% (surgeon 1) respectively 100% (surgeon 2) of the cases.

This research was co-funded by Promedica foundation, Switzerland.
J. Ackermann and M. Wieland - Joint first authorship.

Electronic supplementary material The online version of this chapter (https:// doi.org/10.1007/978-3-030-87202-1_52) contains supplementary material, which is available to authorized users.

M. de Bruijne et al. (Eds.): MICCAI 2021, LNCS 12904, pp. 540–549, 2021.
https://doi.org/10.1007/978-3-030-87202-1_52

Keywords: 3D surgery planning · Deep Reinforcement Learning · Femoral Head Reduction Osteotomy

1 Introduction

Legg-Calvé-Perthes (LCP) is one of the most common orthopedic hip disorders in young children [16]. It is caused by insufficient blood supply in the femoral head, leading to necrotic bone tissue and deformation of the femoral head. Impaired hip function, pain and early joint degeneration can be noted as the main conditions associated with LCP [16]. One possible joint-preserving surgical treatment is called Femoral Head Reduction Osteotomy (FHRO), which is, however, rarely performed due to its underlying complexity [33]. Note that even the largest radiological FHRO studies have study sizes of max. 20 patients from multiple centers and over the course of 5 years [19]. FHRO requires to execute two intra-articular osteotomies dividing the femoral head into three parts: a stable part, a central necrotic wedge and a mobile fragment (Fig. 1 C). Afterwards, the central part is resected and the mobile fragment is realigned to its new position and secured using screw implants. The reconstruction is considered successful, if the resulting femoral head is spherically shaped, resembling a healthy joint (Fig. 1 C).

Computer-assisted 3D surgery planning for FHRO is still in its infancy and thus it consists of an extensive manual surgery planning process performed in collaboration between the surgeons and the technical staff. Currently in our clinics, planning is performed on CT-reconstructed 3D bone models using an in-house planning software [9]. The software allows the user to define 3D planes, which can be placed and moved freely in 3D space to define the osteotomies. After simulating the osteotomies, the mobile fragment can be reoriented to its desired position through mouse interaction. During planning, five main clinical criteria are considered [9]. First, the reconstructed femoral head must adhere to a spherical geometry. Second, the deformed central bone area of the femoral head is necrotic, while the area along the widest diameter of the femoral head can be assumed intact. Therefore, the wedge should comprise most of the necrotic bone, while intact bone and cartilage should be preserved. Third, the areas where the blood supply to the bone is introduced and arteries are originating, must be avoided during cutting (no-go zones EN_1, EN_2 in purple, Fig. 1 A). Fourth, the residual articular step-off between the contact surfaces of the fragments must be kept at a minimum (Fig. 1 C). Lastly, to ensure biomechanical stability, a sufficiently large neck pillar should remain (Fig. 1 C).

The creation of a surgery plan is a difficult three-dimensional (3D) geometrical problem, which can take up to several hours per patient [9], since 16 partially interdependent parameters have to be defined to fully describe the location of the two cutting planes and the mobile fragment. This illustrates the necessity of an automatic planning approach that reduces manual workload and treatment costs. Although numerous methods for automating surgery planning of orthopedic interventions have been proposed [11], clinical-grade planning solutions of complex interventions such as FHRO are still obtained manually [29]. One reason is that automating surgical planning by mathematical optimization (as proposed in [4–6,23]) requires handcrafted and fine-tuned multi-objective functions, which means a costly development that is only worthwhile for frequent pathologies. Moreover, optimization-based approaches are computationally expensive to evaluate, have poor generalization and show a tendency to converge to local minima.

Over the past few years, data-driven methods such as deep learning (DL) have become one of the most efficient techniques for solving complex medical problems [7,29,32], as in such methods extraction of handcrafted features and the modeling of objective functions are not required anymore. Only a few approaches have been described for the task of surgery planning of orthopedic interventions [13,28]. Kulyk et al. [13] presented a DL approach for the planning of total shoulder arthroplasty by inferring the position of the cutting plane from CT data. However, the lack of large training data remains a common problem of data-driven methods, particularly in rare interventions such as FHRO. To address this limitation, Deep Reinforcement Learning (DRL) has become an exciting alternative to supervised learning (SL) for solving problems in the medical domain [12,30,31]. Unlike SL, which can only perform as good as the data allows, DRL learns from own experiences and therefore may enable superhuman performance [18]. However, DRL approaches tend to fail when it comes to transfer the trained agent from simulation to the real data and the real-world problem [20].

In this study, we propose an automatic approach for the surgery planning of FHRO based on DRL. A key contribution of our approach is the development of a simulation environment tailored to orthopedic interventions which is based on an analytical representation of real patient data, thereby overcoming convergence, noise, and dimensionality problems. To the authors' knowledge, our approach is the first to successfully use DRL to solve the task of surgery planning of orthopedic interventions on real patient data.

2 Methods

Our key idea to enable the transfer of DRL to the surgery planning task, is the use of an analytical representation of real patient data in a simulation environment tailored to FHRO. Section 2.1 describes how the mapping between real patient data and analytical anatomy was performed. Thereafter, the DRL environment for simulating FHRO is described in Sect. 2.2, whose output were the optimal poses of the two cutting planes P_1 and P_2 as well as the transformation T required for realigning the mobile bone fragment. Details on the training of the agent are given in Sect. 2.3. The implementation is publicly available: https://rocs.balgrist.ch/open-access/.

2.1 Analytical Representation of the Anatomy and Intervention

In contrast to the spherical shape of a healthy femoral head, a typical pathological femoral head resulting from LCP, resembles the shape of a mushroom with an extra large horizontal diameter [33]. To model this kind of pathological bone morphology, we chose an ellipsoid EP with semi-axes wp, dp, hp and center $\vec{cp}$, while a sphere G with radius rg and center $\overrightarrow{cg}$ was used as the reconstruction template (goal sphere) (Fig. 1 A).

The analytical models were derived from 3D triangular surface models of the patients' femur and pelvis bones provided by our clinics from patients who underwent computer-aided surgery (see [9] for information about patient demographics, CT protocol and segmentation). The ellipsoid representation was calculated using least-squares fitting to the convex hull [2] of all surface points of the pathological head. The convex hull was necessary, to avoid fitting to tunnels or holes, which may be present in deformed femoral heads. The areas which must be avoided during cutting, referred to as no-go zones EN_k were modeled as ellipsoids with semi-axes wn_k, dn_k, hn_k and center $\overrightarrow{cn}_k \ \forall \ k \in \{1,2\}$. The no-go zone ellipsoids were manually placed on the pathological bone model on locations where the crucial anatomical soft-tissue structures like blood vessels are present (Fig. 1 A). The goal sphere G was derived as follows. For patients with a healthy contralateral femur, the contralateral bone model was mirrored and registered to the pathological femur model using ICP [22]. Afterwards, G was created by fitting a perfect sphere to the mirrored contralateral femur using least-squares [24] (Fig. 1 A). In case of a pathological contralateral femur, the center of the goal sphere $\overrightarrow{cn}_k$ was estimated by the mechanical joint center of the acetabular surface and its size was determined by adjusting the sphere diameter until it covered the healthy portion of the femoral head [9,21]. Furthermore, we modeled two intact and one necrotic bone area as Gaussian distributions $IZ_1(\overrightarrow{\mu}_1, \Sigma_1)$, $IZ_2(\overrightarrow{\mu}_2, \Sigma_2)$, $NZ(\overrightarrow{\mu}_3, \Sigma_3)$ respectively (Fig. 1 E). The necrotic bone areas were manually selected by the surgeons on the segmented 3D bone surface (recognized by bump-like irregularities), using the original CT data as decision support. The fitted ellipsoids EP, EN_k and the goal sphere G, as well as the Gaussian distributions IZ_1, IZ_2, NZ served as input for the simulation environment described in the following section.

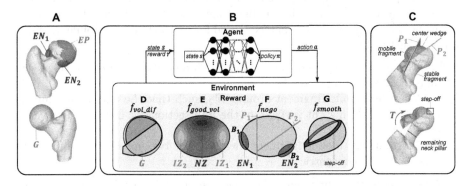

Fig. 1. Overview of our DRL implementation of FHRO. **A** shows the analytical representation derived from patient data, **B** represents the DRL method with the reward function terms (D, E, F, G) and **C** illustrates how the DRL output was applied to the real patient mesh.

2.2   Environment

Mathematically, a reinforcement learning problem is described as a Markov Decision Process, corresponding to a 4-tuple $(\mathcal{S}, \mathcal{A}, P, r)$, where $\mathcal{S} \in \mathbb{R}^N$ denotes the state space, $\mathcal{A} \in \mathbb{R}^M$ represents the action space, $P(s_{t+1}|s_t, a_t)$ represents the state transition probability and $r_t(s_t, a_t)$ is the reward the agent receives as feedback from the environment after taking action a_t in state s_t. To find the optimal policy π_θ, the agent's objective is to maximize the sum of expected rewards $J(\theta)$ throughout an episode $t = [t_0, \ldots, t_n]$. In our simulation environment, states and actions were modeled as continuous parameters. A state $s_t \in \mathcal{S}$ in time step t is defined as

$$s_t = (\alpha_1, \beta_1, \gamma_1, \alpha_2, \beta_2, \gamma_2, dp, hp, rg, \overrightarrow{cn}_1, \overrightarrow{cn}_2, C) \tag{1}$$

The first six values describe the two cutting planes P_1 and P_2 in a Cartesian coordinate system by $\alpha_j = x \cos\beta_j \sin\gamma_j + y \sin\beta_j \sin\gamma_j + z \cos\gamma_j, \forall\, j \in \{1, 2\}$, where α_j, β_j and γ_j denote the radial distance, the azimuthal and polar angle, respectively. Set C represents the patient anatomy and consists of parameters wp, wn_k, dn_k, hn_k, $\overrightarrow{cg}$, IZ_1, IZ_2, NZ as defined in Sect. 2.1. The state vector was normalized such that wp was always equal to one. To keep the state space small, the parameters were described with respect to the center and axes of EP. In the initial state s_{t_0}, the locations of planes P_j were initialized in the area between the no-go zones, such that $((\overrightarrow{cn}_1 - \overrightarrow{p}_j) \cdot \overrightarrow{n}_j)((\overrightarrow{cn}_2 - \overrightarrow{p}_j) \cdot \overrightarrow{n}_j) < 0$, where $\overrightarrow{n}_j$ and $\overrightarrow{p}_j$ are the normal and a point of plane P_j.

At each time step, the agent updates the plane positions by action $a_t = (\Delta\alpha_1, \Delta\beta_1, \Delta\gamma_1, \Delta\alpha_2, \Delta\beta_2, \Delta\gamma_2, \underbrace{0, \ldots, 0}_{no\ update\ of\ dp, hp, rg, \overrightarrow{cn}_1, \overrightarrow{cn}_2, C}) \in \mathcal{A}$, yielding the deterministic state transition $s_{t+1} = s_t + a_t$.

Our reward function is designed to assess both the quality of the osteotomy planes and the reattachment of the mobile fragment. For calculating the latter in each step of the agent, EP is cut by P_1 and P_2 such that the central wedge was removed and the remaining mobile fragment was reattached to the stable fragment. The transformation matrix T used for reattaching the mobile fragment was established by:

1. Rotating the mobile fragment around the intersection line of P_1 and P_2 to align the two polytopes on P_2
2. Rotating the mobile fragment around $\overrightarrow{n}_2$, such that the major (and minor) axes of the resulting cut surfaces of the intersections of EP with P_1 and P_2 are aligned
3. Transforming the mobile fragment such that the two upper points of the two minor axes are concentric

After fragment realignment, the agent received a reward $r_t(s_t, a_t)$ based on the weighted average of four terms quantifying the four clinical objectives described above (Eq. 2).

$$r_t(s_t, a_t) = -w_1 f_{vol\_dif} + w_2 f_{good\_vol} - w_3 f_{nogo} - w_4 f_{smooth} \tag{2}$$

The four terms were modeled as follows.

f_{vol_dif} The degree of head sphericity was designed as a negative reward reflecting the deviation of reconstructed bone shape to the goal sphere G and was modeled as the volume of the Boolean difference (XOR) between G and the post-resection bone shape (Fig. 1 D). To reduce computational complexity, we approximated the Boolean operation by evaluating the distances of closest points from the union of the mobile and the stable fragment (Res) to G at uniformly sampled points with a resolution of approximately 0.12 points/mm^2.

f_{good_vol} This reward reflects the surgical objective of bringing intact bone regions into the weight-bearing zone (Fig. 1 E). Let V be the set of points obtained by sampling equally distributed points with a resolution of 0.05 points/mm^3 from Res. f_{good_vol} was defined as $\gamma_1(V) + \gamma_2(V) - \gamma_3(V)$, where $\gamma_1()$, $\gamma_2()$, $\gamma_3()$ denote the Gaussian measure for IZ_1, IZ_2, NZ respectively.

f_{nogo} If a cut is performed through a nogo-zone, the following large penalty is given (Fig. 1 F):

$$f_{nogo} = \begin{cases} 40 & \text{if } \exists j, k \in \{1, 2\}, \exists \overrightarrow{v}, \overrightarrow{w} \in EN_k, |\overrightarrow{n}_j \cdot (\overrightarrow{v} - \overrightarrow{p}_j) > 0 \\ & \wedge \overrightarrow{n}_j \cdot (\overrightarrow{w} - \overrightarrow{p}_j) < 0 \\ 0 & \text{otherwise} \end{cases} \qquad (3)$$

f_{smooth} This term implements the surgical objective that the intra-articular step-off between the reattached fragments should be minimized. We modeled this relationship as the area calculated by the Boolean subtraction of the cut surfaces of the mobile fragment and the stable fragment after realignment (see red surface in Fig. 1 G).

The weights were empirically set according to their clinical importance, the sphericity f_{vol_dif} being the most important factor ($w_1 = 15$), followed by the no-go zones f_{nogo} ($w_3 = 1$), the smoothness area f_{smooth} ($w_4 = 5$) and the bone quality f_{good_vol} ($w_2 = 1.5$). The agent optimized over a reward range of -40 to 2, where a negative sign did not necessarily represent a penalty.

2.3 DRL Method Training and Evaluation

Due to the encouraging results reported on learning in continuous parameter spaces [3,8,10,15,17,25,26], we chose policy gradient (PG) algorithms for training our agent. PG methods aim to optimize the policy directly by computing the gradient of the objective function (Eq. 4). According to Policy Gradient Theorem [27], the derivative of the expected sum of rewards $J(\theta)$ is the expectation $\mathbb{E}_{\pi_\theta}$ of the product of the sum of rewards $R(\tau)$ and gradient of the log of the policy $\log(\pi_\theta(\tau))$:

$$\nabla_\theta J(\theta) = \mathbb{E}_{\pi_\theta} \left[R(\tau) \nabla_\theta \log\left(\pi_\theta(\tau)\right) \right] = \mathbb{E}_{\pi_\theta} \left[R(\tau) \left(\sum_{t=t_0}^{t_n} \nabla_\theta \log\left(\pi_\theta(a_t|s_t)\right) \right) \right] \quad (4)$$

Among different actor-critic algorithms [8,10,26], we found the Proximal Policy Optimization (PPO) [26] to perform best for our problem in terms of learning progress, achieved rewards and low standard deviation, by defining the objective $J(\theta)$ as a weighted average between an adaptive KL penalty and a clipping mechanism. In our implementation, the actor and critic neural networks each contained two hidden layers, 512 nodes of ReLU activation functions and a learning rate of 5×10^{-5}. Approximately 5'000 episodes (or 500'000 time-steps) were necessary to reach acceptable results and generally after 10'000 episodes the agent peaked.

The training was performed on simulated data only. dp, hp, rg and $\vec{cn}_k$ were reinitialized at the start of each epoch. To ensure an efficient training process, the state parameters were constrained to lie within specific ranges. Every episode lasted 100 time steps which was sufficient to find a global optimum in all our experiments. The agent was trained using Ray's RLlib [14] PPO implementation with a PPO clip parameter of $\epsilon = 0.3$, a GAE lambda parameter of 0.95, an initial coefficient for the KL divergence of $w_0 = 0.6$ and a KL target value of $w_{end} = 0.01$. An asynchronous learning process was used with 8 workers, a SGD minibatch size within each epoch of 1024 and a total size of each SGD epoch 16384.

3   Results and Discussion

The network was evaluated on unseen CT-reconstructed 3D models of 11 LCP patients of which 8 underwent an FHRO surgery between June 2017 and November 2019. For those 8 cases, our method was compared to a gold standard 3D surgery planning (GS), which had been created manually by an expert hip surgeon together with engineers. Two hip surgeons, one world-recognized expert in FHRO (referred to as S_1) and a board-certified orthopedic hip surgeon of our institution (referred to as S_2), performed a qualitative evaluation and classified our surgery planning solutions into four categories: not acceptable, acceptable, equally good (as the GS) and better (than GS). The evaluation was based on the 5 clinical criteria mentioned in the introduction which were: 1) spherical geometry, 2) bringing the intact bone into the load bearing zone, 3) intact no-go zones, 4) minimal residual articular step-off and 5) sufficiently large neck pillar. As shown in Table 1, their evaluations demonstrated that the DRL solutions were equally good or better than GS solutions in 80% (judged by S_1) and 100% (judged by S_2) of the cases. The poorer grading for case 6 and 8 were due to the size of the reconstructed head, which was slightly too large. In addition to these cases, our approach was applied to 3 more cases lacking a GS solution for which it generated clinical-grade solutions classified as 'acceptable' by both experts. A detailed visualization of the DRL solutions of all cases can be found in the supplementary material. Furthermore, a quantitative analysis was conducted by evaluating four metrics: The average volume difference between the goal sphere and the reconstructed bone was superior for GS ($\mu_{gs} = 15.20\%$, $\sigma_{gs} = 7.79\%$) compared to DRL ($\mu_{drl} = 20.86\%$, $\sigma_{drl} = 4.62\%$), which was in-line with the qualitative evaluation, where the size of the reconstructed head was criticized. The mean Hausdorff distance between DRL/GS solutions and the goal sphere

Table 1. Qualitative evaluation by a world-recognized FHRO expert (S_1) and a board-certified orthopedic hip surgeon (S_2) on 11 patients by grading solutions into 'not acceptable' (NA), 'acceptable' (A), 'equally good as GS' (E) or 'better than GS' (B). Cases 1–8 were evaluated in comparison to the GS, whereas for cases 9–11 no GS was available and therefore could only be classified as either 'acceptable' (A) or 'not acceptable' (NA).

case	1	2	3	4	5	6	7	8	9	10	11
	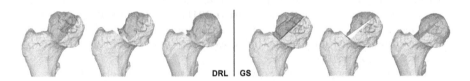										
S_1	E	B	E	B	B	A	B	A	A	A	A
S_2	E	B	B	E	B	B	B	E	A	A	A

was slightly in favor of DRL ($\mu_{gs} = 3.80$ mm, $\sigma_{gs} = 1.33$ mm, $\mu_{drl} = 3.41$ mm, $\sigma_{drl} = 1.01$ mm). According to our goodness measure f_{good_vol}, DRL is similar in preserving intact bone tissue ($\mu_{gs} = 0.72$, $\sigma_{gs} = 0.09$, $\mu_{drl} = 0.71$, $\sigma_{drl} = 0.12$). Moreover, in DRL, the bone contact surfaces of the fragments were on average more similar according to f_{smooth}, which led to a smaller residual articular step-off ($\mu_{gs} = 98.57$ mm^2, $\sigma_{gs} = 41.58$ mm^2, $\mu_{drl} = 78.68$ mm^2, $\sigma_{drl} = 42.37$ mm^2). DRL and GS solutions for case 5 are visualized in Fig. 2. The average error of the ellipsoid fit with the original femoral head shape, was 0.9 mm (closest-point RMSE). We further evaluated the difference between the manually defined and analytically modeled necrotic zone for one example case, yielding a closest-point RMSE of 1.3 mm.

Fig. 2. Visualization of DRL (left) and GS (right) solution for case 5. A visualization of all patient cases can be found in the supplementary material.

Our results indicate that DRL is capable of outperforming human planners in the herein presented FHRO surgery planning task. A clear benefit of our approach over supervised methods is, that no training data is needed. In contrast to standard optimization [6], we achieved clinical-grade results only with a simple reward function thanks to our analytical approach.

Although the analytical representation works well for FHRO, it may be more difficult to generalize it to other anatomies and interventions. Furthermore, even though our reward function was carefully engineered and designed based on well-established clinical criteria where each component was tested individually by capping the remaining rewards, reward hacking may have still occurred [1].

Moreover, our approach simplified the integration of other anatomical structures to no-go zones modeled as ellipsoids. The application of DRL directly on imaging data could be a possible solution in cases were analytical modeling is impossible, but achieving robust solutions in such high-dimensional and noisy data remains a challenge addressed in future work.

4   Conclusion

In this study, we presented the first successful application of DRL to orthopedic surgery planning. A simulation environment tailored to orthopedic interventions was developed, which leverages an analytical representation of patient data in order to overcome convergence, noise and dimensionality problems. We believe that this work will contribute to future planning approaches which will be eventually capable of solving complex surgical problems with super-human performance.

References

1. Amodei, D., Olah, C., Steinhardt, J., Christiano, P., Schulman, J., Mané, D.: Concrete problems in AI safety. arXiv preprint arXiv:1606.06565 (2016)
2. Barber, C.B., Dobkin, D.P., Huhdanpaa, H.: The quickhull algorithm for convex hulls. ACM Trans. Math. Softw. (TOMS) **22**(4), 469–483 (1996)
3. Barth-Maron, G., et al.: Distributed distributional deterministic policy gradients. arXiv preprint arXiv:1804.08617 (2018)
4. Belei, P., Schkommodau, E., Frenkel, A., Mumme, T., Radermacher, K.: Computer-assisted single-or double-cut oblique osteotomies for the correction of lower limb deformities. Proc. Inst. Mech. Eng. Part H **221**(7), 787–800 (2007)
5. Carrillo, F., et al.: An automatic genetic algorithm framework for the optimization of three-dimensional surgical plans of forearm corrective osteotomies. Med. Image Anal. **60**, 101598 (2020)
6. Carrillo, F., Vlachopoulos, L., Schweizer, A., Nagy, L., Snedeker, J., Fürnstahl, P.: A time saver: optimization approach for the fully automatic 3D planning of forearm osteotomies. In: Descoteaux, M., Maier-Hein, L., Franz, A., Jannin, P., Collins, D.L., Duchesne, S. (eds.) MICCAI 2017. LNCS, vol. 10434, pp. 488–496. Springer, Cham (2017). https://doi.org/10.1007/978-3-319-66185-8_55
7. Esfandiari, H., Newell, R., Anglin, C., Street, J., Hodgson, A.J.: A deep learning framework for segmentation and pose estimation of pedicle screw implants based on C-arm fluoroscopy. Int. J. Comput. Assist. Radiol. Surg. **13**(8), 1269–1282 (2018)
8. Fujimoto, S., Hoof, H., Meger, D.: Addressing function approximation error in actor-critic methods, pp. 1587–1596 (2018)
9. Fürnstahl, P., Casari, F.A., Ackermann, J., Marcon, M., Leunig, M., Ganz, R.: Computer-assisted femoral head reduction osteotomies: an approach for anatomic reconstruction of severely deformed legg-calvé-perthes hips. A pilot study of six patients. BMC Musculoskelet. Disord. **21**(1), 1–9 (2020)
10. Haarnoja, T., Zhou, A., Abbeel, P., Levine, S.: Soft actor-critic: off-policy maximum entropy deep reinforcement learning with a stochastic actor, pp. 1861–1870 (2018)

11. Joskowicz, L., Hazan, E.J.: Computer aided orthopaedic surgery: incremental shift or paradigm change? (2016)

12. Kober, J., Bagnell, J.A., Peters, J.: Reinforcement learning in robotics: a survey. Int. J. Robot. Res. **32**(11), 1238–1274 (2013)

13. Kulyk, P., Vlachopoulos, L., Fürnstahl, P., Zheng, G.: Fully automatic planning of total shoulder arthroplasty without segmentation: a deep learning based approach. In: Vrtovec, T., Yao, J., Zheng, G., Pozo, J.M. (eds.) MSKI 2018. LNCS, vol. 11404, pp. 22–34. Springer, Cham (2019). https://doi.org/10.1007/978-3-030-11166-3_3

14. Liang, E., et al.: Ray rllib: a composable and scalable reinforcement learning library. arXiv preprint arXiv:1712.09381, p. 85 (2017)

15. Lillicrap, T.P., et al.: Continuous control with deep reinforcement learning. arXiv preprint arXiv:1509.02971 (2015)

16. Loder, R.T., Skopelja, E.N.: The epidemiology and demographics of Legg-Calvé-Perthes' disease. Int. Sch. Res. Notices **2011** (2011)

17. Mnih, V., et al.: Asynchronous methods for deep reinforcement learning, pp. 1928–1937 (2016)

18. Mnih, V., et al.: Human-level control through deep reinforcement learning. Nature **518**(7540), 529–533 (2015)

19. Paley, D.: The treatment of femoral head deformity and coxa magna by the ganz femoral head reduction osteotomy. Orthop. Clin. North Am. **42**(3), 389–99 (2011)

20. Pan, X., You, Y., Wang, Z., Lu, C.: Virtual to real reinforcement learning for autonomous driving. arXiv preprint arXiv:1704.03952 (2017)

21. Pflugi, S., et al.: A cost-effective surgical navigation solution for periacetabular osteotomy (PAO) surgery. Int. J. Comput. Assist. Radiol. Surg. **11**(2), 271–280 (2016)

22. Rusinkiewicz, S., Levoy, M.: Efficient variants of the ICP algorithm, pp. 145–152 (2001)

23. Schkommodau, E., Frenkel, A., Belei, P., Recknagel, B., Wirtz, D.C., Radermacher, K.: Computer-assisted optimization of correction osteotomies on lower extremities. Comput. Aided Surg. **10**(5–6), 345–350 (2005)

24. Schneider, P., Eberly, D.H.: Geometric tools for computer graphics. Elsevier (2002)

25. Schulman, J., Levine, S., Abbeel, P., Jordan, M., Moritz, P.: Trust region policy optimization, pp. 1889–1897 (2015)

26. Schulman, J., Wolski, F., Dhariwal, P., Radford, A., Klimov, O.: Proximal policy optimization algorithms. arXiv preprint arXiv:1707.06347 (2017)

27. Sutton, R.S., Barto, A.G.: Reinforcement learning: an introduction (2011)

28. Tschannen, M., Vlachopoulos, L., Gerber, C., Székely, G., Fürnstahl, P.: Regression forest-based automatic estimation of the articular margin plane for shoulder prosthesis planning. Med. Image Anal. **31**, 88–97 (2016)

29. Vercauteren, T., Unberath, M., Padoy, N., Navab, N.: CAI4CAI: the rise of contextual artificial intelligence in computer-assisted interventions. Proc. IEEE **108**(1), 198–214 (2019)

30. Yu, C., Liu, J., Nemati, S.: Reinforcement learning in healthcare: a survey. arXiv preprint arXiv:1908.08796 (2019)

31. Zhang, Q., Li, M., Qi, X., Hu, Y., Sun, Y., Yu, G.: 3D path planning for anterior spinal surgery based on CT images and reinforcement learning, pp. 317–321 (2018)

32. Zhou, X.Y., Guo, Y., Shen, M., Yang, G.Z.: Application of artificial intelligence in surgery. Front. Med. 1–14 (2020)

33. Ziebarth, K., Slongo, T., Siebenrock, K.A.: Residual perthes deformity and surgical reduction of the size of the femoral head. Oper. Tech. Orthop. **23**(3), 134–139 (2013)

Whole Heart Mesh Generation for Image-Based Computational Simulations by Learning Free-From Deformations

Fanwei Kong$^{(\boxtimes)}$ ⓘ and Shawn C. Shadden ⓘ

Department of Mechanical Engineering, University of California,
Berkeley, CA 94720, USA
{fanwei_kong,shadden}@berkeley.edu

Abstract. Image-based computer simulation of cardiac function can be used to probe the mechanisms of (patho)physiology, and guide diagnosis and personalized treatment of cardiac diseases. This paradigm requires constructing simulation-ready meshes of cardiac structures from medical image data–a process that has traditionally required significant time and human effort, limiting large-cohort analyses and potential clinical translations. We propose a novel deep learning approach to reconstruct simulation-ready whole heart meshes from volumetric image data. Our approach learns to deform a template mesh to the input image data by predicting displacements of multi-resolution control point grids. We discuss the methods of this approach and demonstrate its application to efficiently create simulation-ready whole heart meshes for computational fluid dynamics simulations of the cardiac flow. Our source code is available at https://github.com/fkong7/HeartFFDNet.

Keywords: Cardiac simulations · Mesh generation · Deep learning

1 Introduction

Patient-specific computer models of the heart derived from medical image data have been developed to simulate a variety of aspects of cardiac function, including electrophysiology [19], hemodynamics [3] and tissue mechanics [12] to explore improvements in cardiovascular diagnoses and treatments [1,25], and the biomechanical underpinnings of diseases [6,16]. However, generating simulation-suitable models of the heart from image data requires significant time and human efforts and is a critical bottleneck limiting clinical applications or large-cohort

This works was supported by the National Science Foundation, Award No. 1663747.

Electronic supplementary material The online version of this chapter (https://doi.org/10.1007/978-3-030-87202-1_53) contains supplementary material, which is available to authorized users.

© Springer Nature Switzerland AG 2021
M. de Bruijne et al. (Eds.): MICCAI 2021, LNCS 12904, pp. 550–559, 2021.
https://doi.org/10.1007/978-3-030-87202-1_53

studies [13,14]. Deforming-domain computational fluid dynamics (CFD) simulations of the intracardiac hemodynamics, in particular, requires both the geometries and the deformation of the heart from a sequence of image snapshots of the heart throughout the cardiac cycle [9,13,21]. Challenges of image-based model construction are related to the entwined nature of the heart, difficulty differentiating individual cardiac structures from each other and the surrounding tissue, the large deformations of these structures over the cardiac cycle, as well as complicated steps to label various surfaces or regions for the assignment of boundary conditions or parameters if the model is to be used to support simulations.

Deep learning methods have demonstrated promising performance in automatic whole heart segmentation [15,26]. However, prior studies have not generally focused on learning to predict surface meshes directly from medical data for the purpose of computational simulations. Prior efforts on accelerating cardiac model construction for simulations have typically adopted a multistage approach whereby 3D segmentations of cardiac structures are first obtained from image volumes using convolutional neural networks (CNN), meshes of the segmented regions are then generated using marching cube algorithms, and finally manual surface post-processing or editing is performed [9,11]. However, CNN based methods may produce segmentations that achieve high average voxel-wise accuracy but contain extraneous regions that are anatomically inconsistent, unphysical and unintelligible for simulation-based analyses. Correcting such artifacts generally requires a number of carefully designed post-processing steps [9] that cannot be easily adapted to generating complicated whole heart models.

Recently, a few studies have sought to directly construct anatomical structures in the form of surface meshes from medical image data [2,5,10,22,23]. Particularly, [10] leverages a graph convolutional neural network to predict deformation on mesh vertices from a pre-defined mesh template to fit multiple anatomical structures in a 3D image volume. [2] uses deep neural networks and patient metadata to directly predict cardiac shape parameters of a statistical shape model of the heart. However, the shape and topology of the predicted meshes are pre-determined by the mesh template and cannot be easily changed to accommodate various mesh requirements for different cardiac simulations. In contrast to learning deformation on a template mesh, [22] proposed to learn the *space* deformation by predicting the displacements of a control point grid to deform template meshes of the lung. However, this approach leveraged memory-intensive, fully-connected neural network layers to predict the displacements with a small number of control points from a 2D X-Ray image and thus cannot be directly applied to model complex whole-heart geometries with large geometric variations from 3D image volumes.

Therefore, we are motivated to automatically and directly generate meshes that are suitable for computational simulations of cardiac function. Our method learns the multi-resolution B-spline free-form deformation of the space to produce detailed whole heart meshes from volumetric CT images by a novel graph convolutional module and feature sampling method. After learning the space deformation, our approach is not limited by a single template and is able to deform mesh templates that have arbitrary resolutions or contains various subsets of cardiac structures. When applied on time-series image data, our approach is able to generate temporally consistent meshes that capture the motion of a beating heart and are suitable for cardiac hemodynamics simulations.

2   Methods

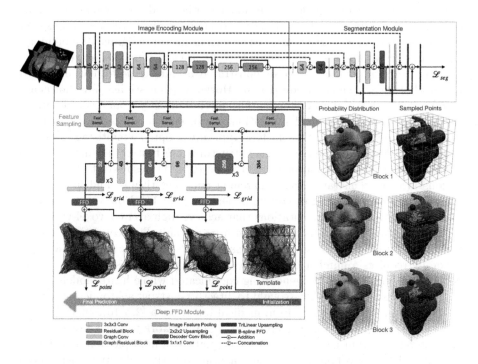

Fig. 1. Diagram of the proposed automatic whole heart reconstruction approach.

Figure 1 shows our proposed whole-heart mesh generation pipeline. Our framework consists of three components to predict the whole-heart meshes from a volumetric input image: (1) an image encoding module, (2) a image feature sampling module, (3) a deep free-form deformation module that predicts control point displacements to deform the template mesh and (4) a segmentation module that predicts a binary segmentation map to allow additional supervision using ground truth annotations.

B-Spline Based FFD. We used a 3D tensor product of the 1D cubic B-spline formulation to deform the space. Namely, for a control point gird of $(l+1) \times (m+1) \times (n+1)$, the relation between a deformed mesh vertex $\mathbf{v}$ and the control points $\mathbf{p}$ is described by $\mathbf{v}(s,t,u) = \sum_{i=0}^{l} \sum_{j=0}^{m} \sum_{k=0}^{n} B_{i,3}(s) B_{j,3}(t) B_{k,3}(u) \mathbf{p}_{i,j,k}$. $B_{i,3}$ is the cubic B-spline basis. Such relation can be expressed in the matrix form $\mathbf{V} = \mathbf{BP}$, $\mathbf{V} \in \mathbb{R}^{N \times 3}$, $\mathbf{B} \in \mathbb{R}^{N \times \psi}$, $\mathbf{P} \in \mathbb{R}^{\psi \times 3}$, where N and ψ are the number of mesh vertices and control points, respectively. $\mathbf{B}$ is the trivariate B-spline tensor and can be pre-computed from the template mesh. $\mathbf{P}$ is the control point coordinates that the network will learn to predict. Compared with the Bernstein deformation tensor implemented in [22], the B-spline-based deformation matrix is sparse since the B-spline basis are defined locally and thus can greatly reduce the computational cost for high-resolution control point grids.

Deep FFD Module. Since the heart involves complicated geometries and significant shape variations during the cardiac cycle and across patients, a dense control point grid is necessary to produce accurate reconstruction of the cardiac structures. Therefore, flattening the image feature vectors of a 3D image and using fully connected layers to predicted control point displacements as proposed in [22] is no longer computationally feasible. We therefore propose to use a graph convolutional network (GCN) to predict the control point displacements based on sampled image feature vectors.

Graph Convolution on Control Grid: We represent the control point grid as a graph $\mathcal{M} = (\mathcal{V}, \mathcal{E})$. Each control point is connected with all its 26 neighbors (7, 11, and 17 neighbors at the corner point, edge point, and surface point, respectively). The graph convolution on a mesh follows [4] and [10]. Briefly, we used a first-order Chebyshev polynomial approximation described as $f_{out} = \sigma(\theta_0 f_{in} + \theta_1 f_{in} \tilde{L})$, where $\theta_0, \theta_1 \in \mathbb{R}^{d_{out} \times d_{in}}$ are trainable weights, $f_{in} \in \mathcal{R}^{d_{in} \times N}$, $f_{out} \in \mathcal{R}^{d_{out} \times N}$ are input and output feature matrices of a graph convolution layer applied on the control point grid, respectively, and $\tilde{L} = 2L_{norm}/\lambda_{max} - I$, $\tilde{L} \in \mathcal{R}^{N \times N}$ is the scaled and normalized Laplacian matrix [4]. N is the number of control points. d_{in} and d_{out} are the input and output graph feature dimensions, respectively. The feature lengths of the intermediate layers match with the numbers in Fig. 1, with 3 for displacements. Compared with conventional convolution with trainable filters, graph convolution requires far fewer parameters as the connection among vertices is encoded in the graph Laplacian matrix.

Deep Multi-Resolution FFD: Our proposed graph decoding module consists of three deformation blocks to progressively deform the template mesh. For the initial mesh deformation blocks, we used lower resolution control point grids conditioned on the more abstracted, high-level image feature maps while using high-resolution control point grids with low-level, high-resolution feature maps for the later mesh deformation blocks. Within each deformation block, we concatenate the sampled image feature vector with the vertex feature vectors on the control points and then use residual graph convolutional blocks to predict the displacements on the control points to deform the template meshes. Between two deformation blocks, we used trilinear interpolation to upsample the features on lower-resolution control point grid to the same grid resolution as in the next deformation block. The numbers of control points along each dimension were 6, 12, and 16, respectively for the 3 deformation blocks.

Probability Sampling of Image Features: Effective sampling of the image features is essential for training a dense volumetric feature encoder. We randomly sample 16 points on the whole heart per control point based on a normal distribution centered at each control points with the covariance determined by the grid resolution (Fig. 1). In each FFD block, we update the coordinates of sampled points based on FFD, sample image features at these coordinates and then compute the expectation of image features over the sampled points for each grid point. These image features are then concatenated with the grid features for displacement prediction using GCN. Control points in the low-resolution grid thus have a larger field of selection than those in the high-resolution grid. Figure 1 visualizes the probability distribution and sampled points correspond to one control point from control grid at different resolutions.

Loss Functions. The training of our networks was supervised by 3D ground truth meshes of the whole heart as well as a binary segmentation indicating occupancy of the heart on the voxel grid that corresponds to the input image volume. The total loss is a weighted combination of a point loss, a grid elasticity loss and a segmentation loss over all deformation blocks B_i. Namely, $\mathcal{L}_{total} = \sum_b^3 \mathcal{L}_{point}(\mathbf{P}^{B_i}, \mathbf{G}^{B_i}) + \alpha_1 \sum_b^3 \mathcal{L}_{grid}(\Delta \mathbf{C}^{B_i}) + \alpha_2 \mathcal{L}_{seg}(I_p, I_g)$. We used $\alpha_1 = 100$ selected from 10, 100, 1000 based on the validation accuracy. α_2 was initially set to 200 (selected from 10, 100 and 200) and decreased by 5% every 5 epochs during training. We use the Chamfer loss as the point loss $\mathcal{L}_{point}(\mathbf{P}_i, \mathbf{G}_i) = \sum_{\mathbf{p} \in \mathbf{P}_i} \min_{\mathbf{g} \in \mathbf{G}_i} ||\mathbf{p} - \mathbf{g}||_2^2 + \sum_{\mathbf{g} \in \mathbf{G}_i} \min_{\mathbf{p} \in \mathbf{P}_i} ||\mathbf{p} - \mathbf{g}||_2^2$, where $\mathbf{p}$ and $\mathbf{g}$ are, respectively, points from vertex sets of the predicted mesh $\mathbf{P}_i$ and the ground truth mesh $\mathbf{G}_i$ of cardiac structure i. Since excessive deformation of the control points especially during early phase of training may introduce undesirable mesh artifacts, we use the grid elasticity loss to regularize the network to predict small displacements of the control points, $\mathcal{L}_{grid}(\Delta \mathbf{C}) = \sum_{\mathbf{c} \in \mathbf{C}} ||\Delta \mathbf{c} - \frac{1}{N} \sum_{\mathbf{c} \in \mathbf{C}} \Delta \mathbf{c}||_2^2$. We used a hybrid loss function $\mathcal{L}_{seg}(I_p, I_g)$ that sums the cross-entropy and the dice losses between the predicted occupancy probability map I_p and the ground truth binary segmentation of the whole heart I_g. The validation loss converged in 36 hrs on a GTX1080Ti GPU.

Cardiac Flow Simulation. We applied the Arbitrary Lagrangian-Eulerian (ALE) formulation of the incompressible Navier-Stokes equations to simulate the intraventricular flow and account for deforming volumetric mesh using the finite element method. The volumetric mesh was created automatically from our FFD predicted surface mesh using TetGen [18]. Blood was assumed to have a viscosity μ of $4.0 \times 10^{-3} Pa \cdot s$ and a density ρ of $1.06 \, \text{g/cm}^3$. The equations were solved with the open-source svFSI solver from the SimVascular project [20].

3   Experiments and Results

Dataset and Preprocessing. We applied our method to public datasets of contrast-enhanced CT images from both normal and abnormal hearts and mostly cover the whole hearts, MMWHS [26], orCalScore [24] and SLAWT [8]. Intensity normalization and resizing as well as data augmentation techniques, random scaling, rotation, shearing and elastic deformation were applied following the procedures in [10]. The training and validation datasets contained 87 and 15 CT images, respectively. The 40 CT images from MMWHS test dataset and 10 sets of times-series CT data [10] were left out for evaluations. The ground truth labels include the 4 heart chambers, aorta, pulmonary artery, parts of the pulmonary veins and venae cavae for the training and validation data.

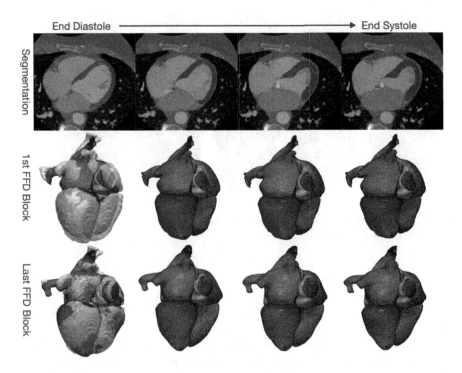

Fig. 2. Whole-heart reconstruction results for time-series CT data. The first row shows predicted segmentation overlaid with CT images. The second and third rows shows mesh predictions from the first and the last deep FFD blocks. The predictions at the first time frame are overlaid with ground truths. Color maps denotes the mesh vertex IDs.

Generation of 4D Meshes for CFD Simulations. We applied our method on time-series CT image data that consisted of images from 10 time frames over the cardiac cycle for each patient. Figure 2 compares the predictions from the first and last FFD blocks. A low-resolution control point grid can capture the general shape and location of the heart in the image stack whereas the high-resolution control point grid can capture further detailed features. From the segmentation results in Fig. 2, our method is able to capture the minor changes between time frames. Furthermore, as denoted by the color maps of vertex IDs, our method consistently deforms the template meshes such that predictions across different time frames have feature correspondence.

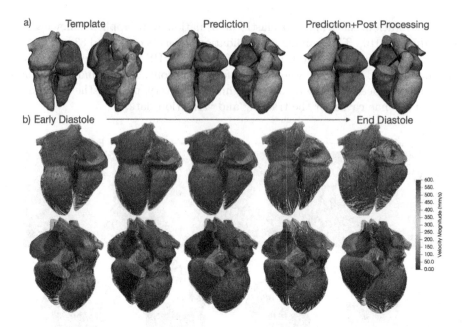

Fig. 3. a) Simulation-ready templates and example predictions. b) CFD simulation results using the predicted 4D meshes.

We also evaluated the potential of our method to generate simulation-ready meshes for deforming-domain CFD simulations of cardiac flow. We used a simulation ready template with trimmed inlet/outlet geometries and tagged face IDs for prescribing boundary conditions. Figure 3a shows the template mesh and our prediction for a representative patient. As our method does not constrain the faces of inlets or outlets to be co-planer, we post-processed our prediction using automatic scripts to project the points on the tagged inlet and outlet faces to fitted planes. We used the resulting meshes to simulate the filling phase of heart after interpolating the 4D meshes to increase the temporal resolution to 0.001s. For the fluid domain, Dirichlet (displacements) boundary conditions were applied on the chamber walls as well as on aorta and pulmonary outlets, while Neumann (pressure) boundaries conditions were applied on pulmonary vein and vena cava inlets. Figure 3b displays the simulation results of the velocity streamlines at multiple time steps during diastole. Videos of the predicted meshes and simulation results of more cases are in our supplementary materials.

Comparison of Different Methods. We compared the whole heart reconstruction performance of different FFD strategies, namely, 1) using the same resolution of control point grid for all deformation blocks as in [22] and uniformly sample the image feature space based the coordinates of the control points (Single-Res + US), 2) using multi-resolution control point grids and uniformly sample the image feature space based on the coordinates of the high resolution grid (Multi-Res + US) and 3) our final model that uses multi-resolution control point grids and probability sampling of the image feature space based on the coordinates of the whole heart template (Multi-Res + WHS). Our supplementary materials additionally include an ablation study of individual loss components and the use of GCN decoder. Furthermore, we compared these FFD-based methods with prior whole-heart reconstruction or segmentation methods, Kong et al. [10], 2DUNet [9,17], residual 3D UNet [7] and Voxel2Mesh[23]. We followed procedures described in [10] to implement those methods.

Table 1. A comparison of prediction accuracy on MMWHS CT test datasets from different deep FFD methods.

		Epi	LA	LV	RA	RV	Ao	PA	WH
Dice	Singe-Res+US	0.72±0.09	0.82±0.08	0.81±0.07	0.79±0.07	0.81±0.06	0.78±0.09	0.69±0.14	0.79±0.05
	Multi-Res+US	0.81±0.06	0.89±0.05	0.88±0.07	0.84±0.07	0.86±0.05	0.87±0.07	0.76±0.13	0.86±0.04
	Multi-Res+WHS	**0.84 ± 0.05**	**0.91 ± 0.04**	**0.89 ± 0.07**	**0.86 ± 0.06**	**0.88 ± 0.04**	**0.91 ± 0.04**	**0.8 ± 0.1**	**0.88 ± 0.03**
ASSD (mm)	Singe-Res+US	2.28±0.76	2.61±0.97	2.72±1.02	2.87±0.93	2.42±0.56	2.41±0.96	3.55±1.62	2.69±0.69
	Multi-Res+US	1.61±0.43	1.5±0.56	1.62±0.66	2.09±0.92	1.59±0.4	1.29±0.52	2.5±1.38	1.75±0.41
	Multi-Res+WHS	**1.41 ± 0.38**	**1.4 ± 0.47**	**1.46 ± 0.68**	**1.87 ± 0.84**	**1.49 ± 0.42**	**1.08 ± 0.35**	**2.19 ± 1.17**	**1.54 ± 0.34**
HD (mm)	Singe-Res+US	15.44±2.42	11.54±3.47	10.45±3.13	15.22±5.35	12.02±2.5	14.9±7.32	26.41±11.49	27.87±10.59
	Multi-Res+US	14.45±2.49	9.25±2.92	8.04±2.25	13.55±5.84	10.86±2.59	12.54±5.53	**23.97 ± 12.62**	**25.67 ± 11.51**
	Multi-Res+WHS	**13.51 ± 2.59**	**8.58 ± 2.87**	**7.66 ± 2.61**	**12.75 ± 5.46**	**10.09 ± 2.48**	**12.24 ± 6.86**	24.79±12.52	26.76±11.17

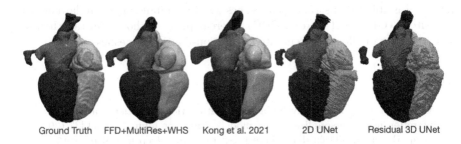

Ground Truth   FFD+MultiRes+WHS   Kong et al. 2021   2D UNet   Residual 3D UNet

Fig. 4. Qualitative comparisons among different methods

Table 1 presents the accuracy scores of different FFD-based methods evaluated on the MMWHS test dataset. Since the ground truths of MMWHS test dataset do not contain the pulmonary veins or venae cavae, we used a template without those structures to generate our predictions. For FFD-based methods, Multi-Res consistently produced more accurate geometries for all cardiac structures (Table 1). Multi-Res + WHS produced more accurate geometries with less surface artifacts than Multi-Res + US, indicating the contribution of our proposed sampling method. Figure 4 displays qualitative whole-heart reconstruction results from different methods. Mesh-deformation based methods, ours and [10], are able to generate smoother and more anatomically consistent geometries compared with segmentation-based approaches, which produced surfaces with staircase artifacts and disconnected regions (Fig. 4). However, Kong, et al., 2021 produced overly smoothed pulmonary veins and vena cava geometries, likely because these are elongated structures and that method deforms spheres rather than a more fitting template of each structure as used in our method here. Quantitatively, our method is generally able to produce similar geometric accuracy compared with prior state-of-the-art methods, while having the additional advantage of directly support various cardiac simulations. Nonetheless, we did observe slightly reduced level of accuracy (see supplementary materials), likely because it is challenging to use a single whole-heart template to fully capture the geometric variations across patients. In ongoing work we plan to add a template retrieval module to automatically select a template that best suits the input.

4   Conclusion

We proposed a novel deep-learning approach to directly construct whole heart meshes from image data. We learn to deform a template mesh to match the input data by predicting displacements of multi-resolution control point grids. To our knowledge, this is the first approach that is able to directly generate whole heart meshes for computational simulations and allows switching template meshes to accommodate different modeling requirements. We demonstrated application of our method on constructing a dynamic whole heart mesh from time-series CT image data to simulate the cardiac flow driven by the cardiac motion. Our method was able to construct such meshes within a minute on a standard desktop computer (3 GHz Intel Core i5 CPU) whereas prior methods can take hours of time and human efforts to generate simulation-ready 4D meshes.

References

1. Arevalo, H.J., et al.: Arrhythmia risk stratification of patients after myocardial infarction using personalized heart models. Nat. Commun. **7**(1), 11437 (2016)
2. Attar, R., et al.: 3D cardiac shape prediction with deep neural networks: simultaneous use of images and patient metadata. In: Shen, D., et al. (eds.) MICCAI 2019. LNCS, vol. 11765, pp. 586–594. Springer, Cham (2019). https://doi.org/10.1007/978-3-030-32245-8_65
3. Bavo, A., et al.: Patient-specific CFD simulation of intraventricular haemodynamics based on 3D ultrasound imaging. Biomed. Eng. Online **15**, 107 (2016)
4. Defferrard, M., Bresson, X., Vandergheynst, P.: Convolutional neural networks on graphs with fast localized spectral filtering. In: Lee, D., Sugiyama, M., Luxburg, U., Guyon, I., Garnett, R. (eds.) Advances in Neural Information Processing Systems, vol. 29, pp. 3844–3852. Curran Associates, Inc. (2016)
5. Ecabert, O., et al.: Automatic model-based segmentation of the heart in CT images. IEEE Trans. Med. Imaging **27**(9), 1189–1201 (2008)
6. Genet, M., Lee, L.C., Baillargeon, B., Guccione, J.M., Kuhl, E.: Modeling pathologies of diastolic and systolic heart failure. Ann. Biomed. Eng. **44**(1), 112–127 (2016)
7. Isensee, F., Maier-Hein, K.: An attempt at beating the 3D U-Net. arXiv abs/1908.02182 (2019)
8. Karim, R., et al.: Algorithms for left atrial wall segmentation and thickness - evaluation on an open-source CT and MRI image database. Med. Image Anal. **50**, 36–53 (2018)

9. Kong, F., Shadden, S.C.: Automating model generation for image-based cardiac flow simulation. J. Biomech. Eng. **142**(11), 111011 (2020)

10. Kong, F., Wilson, N., Shadden, S.: A deep-learning approach for direct whole-heart mesh reconstruction. Med. Image Anal., 102222 (2021). https://doi.org/10.1016/j.media.2021.102222

11. Maher, G., Wilson, N., Marsden, A.: Accelerating cardiovascular model building with convolutional neural networks. Med. Biol. Eng. Comput. **57**, 2319–2335 (2019)

12. Marx, L., et al.: Personalization of electro-mechanical models of the pressure-overloaded left ventricle: fitting of windkessel-type afterload models. Philos. Trans. Royal Soc. A **378**(2173), 20190342 (2020)

13. Mittal, R., et al.: Computational modeling of cardiac hemodynamics: current status and future outlook. J. Comput. Phys. **305**, 1065–1082 (2015)

14. Doost, S.N., Ghista, D., Su, B., Zhong, L., Morsi, Y.S.: Heart blood flow simulation: a perspective review. Biomed. Eng. Online **15**, 1–28 (2016)

15. Payer, C., Štern, D., Bischof, H., Urschler, M.: Multi-label whole heart segmentation using CNNs and anatomical label configurations. In: Pop, M., et al. (eds.) STACOM 2017. LNCS, vol. 10663, pp. 190–198. Springer, Cham (2018). https://doi.org/10.1007/978-3-319-75541-0_20

16. Potse, M., et al.: Patient-specific modelling of cardiac electrophysiology in heart-failure patients. EP Europace **16**, iv56–iv61 (2014)

17. Ronneberger, O., Fischer, P., Brox, T.: U-Net: convolutional networks for biomedical image segmentation. In: Navab, N., Hornegger, J., Wells, W.M., Frangi, A.F. (eds.) MICCAI 2015. LNCS, vol. 9351, pp. 234–241. Springer, Cham (2015). https://doi.org/10.1007/978-3-319-24574-4_28

18. Si, H.: Tetgen, a delaunay-based quality tetrahedral mesh generator. ACM Trans. Math. Softw. **41**(2), 1–36 (2015)

19. Trayanova, N.A., Constantino, J., Gurev, V.: Electromechanical models of the ventricles. Am. J. Physiol. Heart Circ. Physiol. **301**(2), H279–H286 (2011)

20. Updegrove, A., Wilson, N., Merkow, J., Lan, H., Marsden, A., Shadden, S.: Simvascular: an open source pipeline for cardiovascular simulation. Ann. Biomed. Eng. **45**, 525–541 (2016)

21. Vedula, V., Seo, J.H., Lardo, A., Mittal, R.: Effect of trabeculae and papillary muscles on the hemodynamics of the left ventricle. Theor. Comput. Fluid Dyn. **30** (2015)

22. Wang, Y., Zhong, Z., Hua, J.: Deeporgannet: on-the-fly reconstruction and visualization of 3D/4D lung models from single-view projections by deep deformation network. IEEE Trans. Visual Comput. Graphics **26**(1), 960–970 (2020)

23. Wickramasinghe, U., Remelli, E., Knott, G., Fua, P.: Voxel2Mesh: 3D mesh model generation from volumetric data. In: Martel, A.L., et al. (eds.) MICCAI 2020. LNCS, vol. 12264, pp. 299–308. Springer, Cham (2020). https://doi.org/10.1007/978-3-030-59719-1_30

24. Wolterink, J.M., et al.: An evaluation of automatic coronary artery calcium scoring methods with cardiac CT using the orcascore framework. Med. Phys. **43**(5), 2361–2373 (2016)

25. Zahid, S., et al.: Feasibility of using patient-specific models and the "minimum cut" algorithm to predict optimal ablation targets for left atrial flutter. Heart Rhythm **13**(8), 1687–1698 (2016)

26. Zhuang, X., et al.: Evaluation of algorithms for multi-modality whole heart segmentation: an open-access grand challenge. Med. Image Anal. **58**, 101537 (2019)

Automatic Path Planning for Safe Guide Pin Insertion in PCL Reconstruction Surgery

Florian Kordon[1,3,4]([✉]) [ID], Andreas Maier[1,2,3] [ID], Benedict Swartman[5],
Maxim Privalov[5], Jan Siad El Barbari[5], and Holger Kunze[4] [ID]

[1] Pattern Recognition Lab, Friedrich-Alexander-Universität Erlangen-Nürnberg
(FAU), Erlangen, Germany
florian.kordon@fau.de
[2] Machine Intelligence, Friedrich-Alexander-Universität Erlangen-Nürnberg (FAU),
Erlangen, Germany
[3] Erlangen Graduate School in Advanced Optical Technologies (SAOT),
Friedrich-Alexander-Universität Erlangen-Nürnberg (FAU), Erlangen, Germany
[4] Siemens Healthcare GmbH, Forchheim, Germany
[5] Department for Trauma and Orthopaedic Surgery,
BG Trauma Center Ludwigshafen, Ludwigshafen, Germany

Abstract. Reconstruction surgery of torn ligaments typically requires precise and anatomically correct fixation of the graft substitute on the bone surface. Several planning methodologies have been proposed that aim at standardizing the interventional procedure by localizing drill sites or defining the drill tunnel orientation with the help of anatomical landmarks. However, the practical implementation is limited by the often complex and time-consuming nature of the planning steps. For this reason, we propose an automatic solution for safe guide pin path planning based on bone contour extraction, axis detection, anatomical landmark detection, and geometrical construction. We evaluate our approach for the task of double-bundle posterior cruciate ligament reconstruction surgery on the lateral tibia using 38 clinical X-ray images. Our method achieves a median path angulation error of 0.37° and a median localization error of 0.96 mm for the ligament attachment center.

Keywords: Surgical planning · Orthopedics · Drill path guidance · X-ray imaging · Bone segmentation · Landmark localization

1 Introduction

Guide pins are a minimally invasive and flexible tool in today's trauma surgery and orthopedics. They are frequently used in the surgical treatment of intricate

The authors gratefully acknowledge funding of the Erlangen Graduate School in Advanced Optical Technologies (SAOT) by the Bavarian State Ministry for Science and Art.

M. de Bruijne et al. (Eds.): MICCAI 2021, LNCS 12904, pp. 560–570, 2021.
https://doi.org/10.1007/978-3-030-87202-1_54

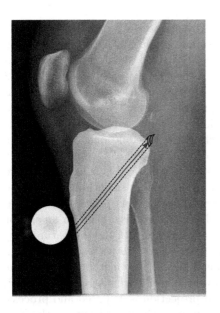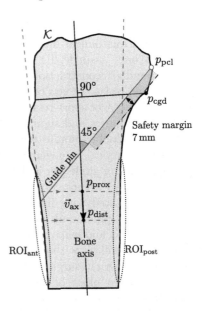

Fig. 1. Drill tunnel for PCL reconstruction surgery on the lateral tibia following [4,7, 10,20]. Identification of anatomic attachment p_{pcl} and the optimal guide pin path for drilling (marked in red) are obtained by geometric construction. (Color figure online)

bone fractures, where they either serve the temporary fixation of bone fragments or enable a definitive fixation, e.g. in the arthrodesis of smaller joints [5]. Moreover, accurate guide pin placement is used to pilot safe tunnel drilling during the anatomical reconstruction of torn ligaments such as the medial patellofemoral ligament (MPFL) [18] or the anterior/posterior cruciate ligaments (ACL/PCL) [4,7,10,20]. A precise drill tunnel is of high clinical relevance to prevent detrimental changes in the reconstructed ligament's biomechanics as well as premature wall breakout with the risk of damaging neurovascular structures [6,12,14,24]. For this reason, several radiographic methodologies for planning and verification were proposed that aim to provide standardized and reliable tools for intraoperative assessment [7,18]. However, these methods usually comprise a number of complex manual planning steps such as the localization of anatomical landmarks, the segmentation of relevant anatomy, or detection of surfaces and lines. Coupled with the need for non-sterile interaction with the planning software, practical implementation is thus cumbersome. Although various (semi-)automatic methods were proposed to aid the surgeon in safe path planning and drilling (e.g. optical navigation and robotics [17,23], 3D registration [3], arthroscopic video [16], or patient-specific drill templates [25]), adapting these for automation of the aforementioned radiographic guidelines is not feasible due to multi-modality constraints or dependencies on pre-operative 3D image data. While such 3D data in general allows for very accurate plannings, a complex 3D-2D registration to fluoroscopic live images complicates real-time assessment.

With these issues in mind, this study investigates a method for automated localization of anatomical insertion points and guide pin path planning which operates on only a single 2D X-ray image. We use tibial tunnel positioning in double-bundle reconstruction surgery of the PCL as an example [4,7,10,20] (Fig. 1). While isolated tears of the PCL are rare injuries with an annual incidence of 0.65%–3% of sport-related knee injuries, they often accompany concurrent ligament injuries and significantly decrease joint stability if left untreated [11]. Recently, double-bundle reconstruction has become one of the primary techniques for surgical management: After identification of the anatomical graft attachment sites, a guide pin is drilled into the tibia and is subsequently over-reamed with direct arthroscopic guidance. After tunnel smoothing, the replacement graft is fixated with inference screws. During this procedure, a correct positioning of the pin and ultimately the drilling tunnel is crucial for an anatomical reconstruction and helps to prevent graft abrasion [22] and premature cortex breakout of the drill bit with potential injury to the neurovascular bundle, which is a common concern in surgical management [2,6,7]. Our method smoothly integrates into this key phase of the surgical workflow. Before insertion of the guide pin, an intra-operative fluoroscopic lateral projection is acquired and processed by the path planning algorithm. The planning is then overlaid on the X-ray image and provides visual guidance for a safe drilling path to the surgeon. Besides path angulation and position, the method also accounts for the drill width and incorporates a safety margin [7,10].

To satisfy such a clinical workflow, our method is designed as a 2-stage approach. First, key anatomic cues are extracted from the X-ray image using a deep learning approach. Then, these cues are forwarded to a geometric pipeline in which a logical partitioning of the bone contour is proposed to ensure a safe distance between the drill tunnel and the bone edge. In contrast to a single-stage algorithm, this allows the user to interactively adjust the location of the inferred anatomical features and enables low-latency modification of the path proposal. We evaluate our approach on clinical, true-lateral radiographs of the knee joint.

2   Methods

The localization of the anatomic ligament attachment point and the calculation of the correct guide pin angulation are based on the extraction of three anatomical structures (Fig. 1): 1) $\mathcal{K}$: Contour of the tibial bone. 2) $\text{ROI}_{\text{ant}}, \text{ROI}_{\text{post}}$: Regions of interest (ROIs) marking the anterior and posterior tangential extension of the bone contour to derive the shaft axis $\vec{v}_{\text{ax}} = (p_{\text{dist}} - p_{\text{prox}})$. 3) Landmark representing the most posterior point of the champagne drop-off ridge.

These structures are first extracted by a deep learning model (Subsect. 2.1) and then geometrically combined in multiple planning steps to form the final surgical planning (Subsect. 2.2). By doing so we can provide a solution which is close to clinical practice, so that the individual planning elements can be easily edited by the executing physician.

2.1   Learning-Based Extraction of Anatomical Structures

We employ the paradigm of multi-task learning to jointly optimize all three local-ization tasks. We use a multi-task Hourglass Module (HG) with three bottleneck prediction heads [9,13]. The anatomical landmark location p_{cgd} is modeled as a Gaussian likelihood distribution with standard deviation $\sigma = 6\,px$ in the 2D image domain to encode positional variance. Similarly, the ROIs are constructed by evaluating the line-orthogonal Gaussian function for every pixel value within a discrete sample region along a descriptive 2D line [8]. This region describes an axisymmetric margin of $37\,px$ ($\sigma = 6\,px$). Both localization tasks are optimized by heatmap matching with a mean squared error loss function. The prediction of the tibial segmentation mask follows a binomial classification, optimized with a binary cross entropy term.

2.2   Geometric Path Planning

A) Tibial Bone Contour Extraction. The tibial bone contour is directly extracted from the predicted segmentation mask. Therefore, a morphological ero-sion with a 3×3 structuring element (ones at $\{(-1,0),(0,-1),(0,0),(0,1),(1,0)\}$) is applied to the mask. A one-pixel border $\mathcal{K}$ is then extracted by logical XOR oper-ation between the original mask and the erosion result [8].

B) Bone Axis Construction. To calculate the longitudinal axis of the bone, the anterior and posterior extension lines of the bone shaft are first determined. To do this, the bone contour extracted in step A) is masked with the learned ROIs and the respective contour points are then smoothed using major axis regression [9,21]. By constructing two parallel intersection lines over both contour sections, the shaft axis is then completely described by the centers p_{dist} and p_{prox} of these two line segments (Fig. 1) [8]. Consequently, the alignment of the bone axis is dependent on the quality of the extracted bone contour and the ROIs, but at the same time implies a precise alignment to the bone shaft center.

C) Initialization of Guide Pin Path. The target guide pin path must be positioned such that both the acute angle of $45°$ to the tibial bone axis and a safety margin between the line and the closest cortex area is maintained. The correct alignment is thus ensured by a sequence of initialization and corrective steps followed by the derivation of the bundle attachment point and optimal guide pin path. First, the orientation of the presented knee is determined. This prerequisite step is necessary as all subsequent planning steps work on oriented vectors which are subject to the global knee orientation. To this end, a two-dimensional matrix with column vectors $\vec{v}_{ax} = (p_{dist} - p_{prox})$ and $\vec{v}_{cgd} = (p_{cgd} - p_{prox})$ is constructed and the determinant of their spanned basis is calculated. The sign of the determinant encodes the relative orientation of the two vectors. Such orientation function f operates on a vector pair $\vec{u}, \vec{v}$ and is calculated by

$$f : \mathbb{R}^2, \mathbb{R}^2 \to \{-1, 0, 1\} \quad \text{where} \quad \vec{u}, \vec{v} \xmapsto{f} \text{sgn}\left(\det \begin{bmatrix} u_x & v_x \\ u_y & v_y \end{bmatrix} \right). \tag{1}$$

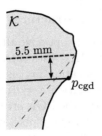
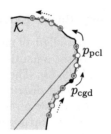

(a) Parallel line construction to approx-   (b) Insertion p_{pcl} obtained by curve in-
imate the anatomic insertion point.   tersection of guide pin and contour.

Fig. 2. Initialization of guide pin path and curve intersection with bone contour.

Second, the guide pin path is initialized. Its orientation vector $\vec{v}_{\text{gw}}$ is obtained by rotating the axis vector $\vec{v}_{\text{ax}}$ by 45° towards the anterior aspect of the tibia. An initial position p_{pcl} is then proposed such that the line intersects the tibial segmentation contour at the anatomic insertion site on a line parallel to the champagne glass drop-off line (line orthogonal to the bone axis passing through p_{cgd}) and at a distance of $d_{\text{pcl}} = 5.5\,\text{mm}$ [7] to p_{cgd} (Fig. 2a). To obtain the respective intersection point p_{pcl} on the segmentation contour that is not affected by false-positive contour-spikes, we employ an iterative search over smoothed contour sections. To this end, the contour points are sub-sampled and sorted by a depth-first-search in a 2-nearest-neighbor graph based on the Euclidean distance norm between points. A Bézier curve is then subsequently fitted to overlapping groups of four contour points and evaluated for curve intersection with the guide pin path via the Bézier clipping algorithm [19] (Fig. 2b).

D) Safety Adjustment of Guide Pin Path. Although an anatomical location of the PCL attachment can be approximated by following this construction, it can not be guaranteed that subsequent drilling won't induce a breakout at the close posterior cortex of the tibia. We tackle this issue by calculating the necessary parallel displacement of the guide pin path accordingly. This displacement can be estimated by extracting the contour point distal to the drilling path which yields the closest orthogonal distance to the initialized guide pin path (Fig. 3b). The potential candidates for this contour point must first be limited to those points which are distal and anterior to p_{cgd} and which do not represent unwanted intersection points of the guide pin path at both the anterior and posterior cortex. Using orientation function f, this subset $\mathcal{K}' \subset \mathcal{K}$ is obtained by evaluating the relative orientation of every point $k \in \mathcal{K}$ with respect to the three characteristic vectors $\vec{v}_{\text{ax}}$, $\vec{v}_{\text{gw}}$, and $\vec{v}_{\text{gw}}^{\perp}$ as

$$
\begin{aligned}
\mathcal{K}' = \{k \in \mathcal{K} \mid \ & f(k - p_{\text{prx}}, \vec{v}_{ax}) = S_1 \\
\wedge \ & f(k - p_{\text{sup}}, \vec{v}_{gw}) = S_2 \\
\wedge \ & f(k - p_{\text{sup}}, \vec{v}_{gw}^{\perp}) = S_3\}
\end{aligned}
\tag{2}
$$

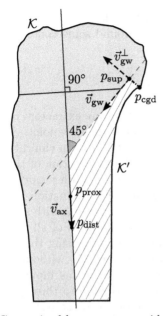

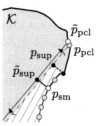

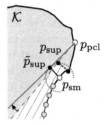

(b) Safety variant (1): Parallel shift of the guide pin path and construction of updated attachment point $\tilde{p}_{pcl}$.

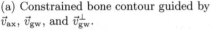

(a) Constrained bone contour guided by $\vec{v}_{ax}$, $\vec{v}_{gw}$, and $\vec{v}_{gw}^{\perp}$.

(c) Safety variant (2): Update of guide pin angulation to maintain the anatomic attachment point p_{pcl}.

Fig. 3. Proposed path correction variants to allow for a safe drilling tunnel.

with sign signature $S = (1, 1, -1)$ or $S = (-1, -1, -1)$ for the right and left knee respectively (Fig. 3a). p_{sup} is an anchor point that is used to calculate the relevant subset of contour points and is initialized such that a safety margin of $d_{sm} = 7\,\text{mm}$ [7] to p_{cgd} (Fig. 3a) can be maintained:

$$\vec{v}_{offset} = \hat{v}_{cgd}^{\perp} * \frac{d_{sm}}{\cos\frac{\pi}{4}} * f\left(\vec{v}_{ax}, \vec{v}_{cgd}\right) \tag{3}$$

$$p_{sup} = p_{cgd} + \vec{v}_{offset} \tag{4}$$

The closest contour point $p_{sm} \in \mathcal{K}'$ which characterizes the safety margin is then obtained by minimizing the orthogonal Euclidean distance to the initial line as

$$p_{sm} = \operatorname*{arg\,min}_{p \in \mathcal{K}'} \|\operatorname{rej}\left(p - p_{sup}, \vec{v}_{gw}\right)\|_2 \tag{5}$$

where *rej* denotes vector rejection. We can subsequently update the position of the anchor point p_{sup} by $\tilde{p}_{sup} = p_{sm} + v_{offset}$ (Fig. 3b) and detect the corrected PCL attachment $\tilde{p}_{pcl}$ with a second search for curve intersection.

If the anatomical insertion point is to be maintained, a safe guide pin path can also be achieved by adjusting the guide pin angulation represented by the updated vector

$$\tilde{v}_{gw} = \operatorname*{arg\,min}_{\vec{v}_{cdt} \in \mathcal{V}} \langle \vec{v}_{ax}, \vec{v}_{cdt} \rangle \tag{6}$$

with $\mathcal{V} = \{h(k) - p_{\text{pcl}} : k \in \mathcal{K}'\}$. h denotes a function that calculates the antero-proximal tangent point through p_{pcl} to a circle around some contour point k with radius $r = d_{\text{sm}}$ (Fig. 3c).

2.3   Dataset and Training

Data for training and validation of the anatomical feature extractor consists of 167/16 (train/val split) conventional X-ray images of the knee joint. Although Johannsen et al. assume a true-lateral image of the tibia [7], in clinical practice this is typically approached using the lateral standard projection of the femur. For this reason, only those images were selected in which both femoral condyle outlines are aligned in the lateral X-ray image [18]. For evaluation, a holdout dataset of 38 radiographs with similar image characteristics but with a standardized reference sphere was used. The ground truth annotation of all anatomical structures was performed by a medically trained engineer using the described geometric pipeline. A single HG with a feature root of 256 without adaptive task balancing is used [9,13]. Given a single X-ray image as input, this HG is optimized to predict 4 spatial output masks distributed to three prediction heads: 1) tibia bone, 2) anterior and posterior contour sections ROI_{ant}, ROI_{post}, and 3) champagne drop-off point p_{cgd} (Subsect. 2.1). An online augmentation sequence was applied to the training data which uses basic affine transformations (rotation, scaling, horizontal flipping, shearing) as well as random margin crops. The data was brought to a common shape by resizing (maintaining the original aspect-ratio) and a subsequent center crop to a spatial resolution of 256 x 256 px.

Optimization was performed with RMSprop in PyTorch v1.6.0 [15] (Python v3.8.5, CUDA v11.0, cuDNN v7.6.5) on a NVIDIA TITAN RTX graphics card. The maximum training time was set to 300 epochs and a batch size of 2 was used. The initial learning rate of 2.5e-4 was halved every 50 epochs and the weights were regularized by L2-norm weight decay with multiplicative strength of 5e-5. Model selection was based on the minimum sum over all three task losses on the validation split where convergence was reached. The subsequent geometric construction was limited to the standard approach with parallel displacement of the guide pin path to comply with clinical practice (Subsect. 2.2, Fig. 3b).

3   Results

Segmentation, Landmark Localization. For the tibial bone segmentation, the trained model consistently achieves a Dice coefficient of 0.99 ± 0.006 (mean±std) and yields an average surface distance of 0.88 ± 1.95 mm. Paired with an average Hausdorff distance of 7.49 ± 11.59 mm over all samples which mostly occurs in regions affected by osteophytic proliferation of the bone, this enables a reliable contour estimate for subsequent geometric planning. In contrast, a significantly larger Euclidean distance error of 1.79 ± 1.025 mm is observed for landmark p_{cgd}. Such larger distances can be explained by overlapping representations of the medial and lateral tibial head on the X-ray image, which leads

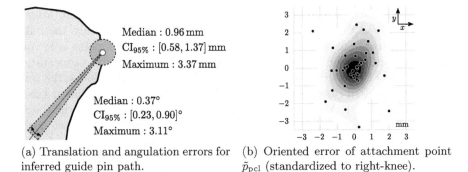

(a) Translation and angulation errors for inferred guide pin path.

(b) Oriented error of attachment point $\tilde{p}_{\mathrm{pcl}}$ (standardized to right-knee).

Fig. 4. Planning accuracy evaluated on 38 test radiographs.

to ambiguities in the exact localization of the champagne drop-off point both in ground truth and during inference. The tibial bone shaft axis is estimated with a mean angular error of $0.66 \pm 0.71°$ and mean positional error of 0.17 ± 0.13 mm which translates into marginal error propagation into the geometric steps.

Guide Pin Path Planning. The proposed planning workflow with safety correction by parallel path shift (Subsect. 2.2) yields consistenly low errors for both the guide pin angulation and the PCL bundle attachment point (Fig. 4). With an anatomical variance of the PCL tibial attachment site of 1.2 mm w.r.t. the CGD reference landmark [1], the mean errors are well within the clinical acceptance range. The maximum angulation error of $3.11°$ is observed in an image with a reduced avulsion fracture of the tibial tuberosity, where a bicortical screw corrupts the ROI detection of the anterior and posterior cortex extension lines.

4 Discussion and Conclusion

This study investigates an automatic approach to assist the trauma surgeon in tibial guide pin placement and tunnel drilling in the technically demanding PCL reconstruction surgery. A deep learning based extraction of anatomical features on X-ray images serves as a prior for geometric construction and subsequent path correction for a safe surgical workflow. Using a two-stage approach to offload most of the planning steps to deterministic post-processing has two major benefits. First, it allows the user to interactively adjust the location of the inferred anatomical features to modify the planning in near real-time (execution time for the geometric steps averages 0.5 s on Intel Core i7-7820HQ). The ability to display all relevant anatomical structures further allows for better interpretability and overall acceptance of the method by the clinician. Second, the segmented bone can be used to constrain the registration of the planning on subsequent live images. In the context of intra-operative imaging, this is especially important since patient-device orientation changes slightly over time, and patient anatomy is typically not fixated.

We see limitations in that an accurate localization of the ligament attachment point along the guide pin path is difficult to achieve if only the outer segmentation contour is used. Since the anatomic attachment point is located in a depression between the two tibial condyles [6], an additional segmentation mask or ROI of this intercondylar region should be integrated. Besides, the complex 3D structure of the tibial head can only be represented to a limited extent by a single lateral 2D X-ray image, which makes it difficult to assess the spatial position of the guide pin. In accordance with medical literature [7], we therefore seek to integrate the antero-posterior view to add further constraints to lateral path planning and to allow for a concurrent evaluation of multi-planar radiographs. This additional information can also be used to estimate the spatial position of implants to resolve ambiguities in the occluded areas. The proposed solution achieves encouraging results and suggests further clinical validation and comparison to the accuracy of manual plannings by trauma surgery experts.

Disclaimer. The methods and information presented here are based on research and are not commercially available.

References

1. Anderson, C.J., Ziegler, C.G., Wijdicks, C.A., Engebretsen, L., LaPrade, R.F.: Arthroscopically pertinent anatomy of the anterolateral and posteromedial bundles of the posterior cruciate ligament. J. Bone Joint Surg. Am. **94**(21), 1936–1945 (2012). https://doi.org/10.2106/JBJS.K.01710
2. Bertollo, N., Walsh, W.R.: Drilling of bone: Practicality, limitations and complications associated with surgical drill-bits. In: Klika, V. (ed.) Biomechanics in Applications, Chap. 3. IntechOpen, Rijeka (2011). https://doi.org/10.5772/20931
3. Caversaccio, M., et al.: Robotic cochlear implantation: surgical procedure and first clinical experience. Acta oto-laryngologica **137**(4), 447–454 (2017). https://doi.org/10.1080/00016489.2017.1278573
4. Chahla, J., Nitri, M., Civitarese, D., Dean, C.S., Moulton, S.G., LaPrade, R.F.: Anatomic double-bundle posterior cruciate ligament reconstruction. Arthroscopy Tech. **5**(1), e149-56 (2016). https://doi.org/10.1016/j.eats.2015.10.014
5. Ewerbeck, V., et al. (eds.): Standardverfahren in der operativen Orthopädie und Unfallchirurgie, 4th edn. Thieme, Stuttgart (2014)
6. Jackson, D.W., Proctor, C.S., Simon, T.M.: Arthroscopic assisted PCL reconstruction: a technical note on potential neurovascular injury related to drill bit configuration. Arthroscopy J. Arthroscopic Related Surg. **9**(2), 224–227 (1993). https://doi.org/10.1016/s0749-8063(05)80381-0
7. Johannsen, A.M., Anderson, C.J., Wijdicks, C.A., Engebretsen, L., LaPrade, R.F.: Radiographic landmarks for tunnel positioning in posterior cruciate ligament reconstructions. Am. J. Sports Med. **41**(1), 35–42 (2013). https://doi.org/10.1177/0363546512465072
8. Kordon, F., Maier, A., Swartman, B., Privalov, M., El Barbari, J.S., Kunze, H.: Contour-based bone axis detection for X-Ray guided surgery on the knee. In: Martel, A.L., et al. (eds.) MICCAI 2020. LNCS, vol. 12266, pp. 671–680. Springer, Cham (2020). https://doi.org/10.1007/978-3-030-59725-2_65

9. Kordon, F., et al.: Multi-task localization and segmentation for X-Ray guided planning in knee surgery. In: Shen, D., et al. (eds.) MICCAI 2019. LNCS, vol. 11769, pp. 622–630. Springer, Cham (2019). https://doi.org/10.1007/978-3-030-32226-7_69

10. LaPrade, R.F., et al.: Double-bundle posterior cruciate ligament reconstruction in 100 patients at a mean 3 years' follow-up: outcomes were comparable to anterior cruciate ligament reconstructions. Am. J. Sports Med. 46(8), 1809–1818 (2018). https://doi.org/10.1177/0363546517750855

11. Longo, U.G., et al.: Epidemiology of posterior cruciate ligament reconstructions in Italy: a 15-year study. J. Clin. Med. 10(3), 499 (2021). https://doi.org/10.3390/jcm10030499

12. Montgomery, S.R., Johnson, J.S., McAllister, D.R., Petrigliano, F.A.: Surgical management of PCL injuries: indications, techniques, and outcomes. Curr. Rev. Musculoskelet. Med. 6(2), 115–123 (2013). https://doi.org/10.1007/s12178-013-9162-2

13. Newell, A., Yang, K., Deng, J.: Stacked hourglass networks for human pose estimation. In: Leibe, B., Matas, J., Sebe, N., Welling, M. (eds.) ECCV 2016. LNCS, vol. 9912, pp. 483–499. Springer, Cham (2016). https://doi.org/10.1007/978-3-319-46484-8_29

14. Nicodeme, J.D., Löcherbach, C., Jolles, B.M.: Tibial tunnel placement in posterior cruciate ligament reconstruction: a systematic review. Knee Surg. Sports Traumatol. Arthroscopy 22(7), 1556–1562 (2014). https://doi.org/10.1007/s00167-013-2563-3

15. Paszke, A., et al.: PyTorch: an imperative style, high-performance deep learning library. In: Wallach, H., Larochelle, H., Beygelzimer, A., Alché-Buc, F.d., Fox, E., Garnett, R. (eds.) Advances in Neural Information Processing System, vol. 32, pp. 8026–8037. Curran Associates, Inc. (2019)

16. Raposo, C., et al.: Video-based computer navigation in knee arthroscopy for patient-specific ACL reconstruction. Int. J. Comput. Assist. Radiol. Surg. 14(9), 1529–1539 (2019). https://doi.org/10.1007/s11548-019-02021-0

17. Rossini, M., Valentini, S., Portaccio, I., Campolo, D., Fasano, A., Accoto, D.: Localization of drilling tool position through bone tissue identification during surgical drilling. Mechatronics 67, 102342 (2020). https://doi.org/10.1016/j.mechatronics.2020.102342

18. Schöttle, P.B., Schmeling, A., Rosenstiel, N., Weiler, A.: Radiographic landmarks for femoral tunnel placement in medial patellofemoral ligament reconstruction. Am. J. Sports Med. 35(5), 801–804 (2007). https://doi.org/10.1177/0363546506296415

19. Sederberg, T.W., Nishita, T.: Curve intersection using Bézier clipping. Comput.-Aided Des. 22(9), 538–549 (1990). https://doi.org/10.1016/0010-4485(90)90039-F

20. Spiridonov, S.I., Slinkard, N.J., LaPrade, R.F.: Isolated and combined grade-III posterior cruciate ligament tears treated with double-bundle reconstruction with use of endoscopically placed femoral tunnels and grafts: operative technique and clinical outcomes. J. Bone Joint Surg. Am. 93(19), 1773–1780 (2011). https://doi.org/10.2106/JBJS.J.01638

21. Warton, D.I., Wright, I.J., Falster, D.S., Westoby, M.: Bivariate line-fitting methods for allometry. Biol. Rev. Cambridge Philos. Soc. 81(2), 259–291 (2006). https://doi.org/10.1017/S1464793106007007

22. Weimann, A., Wolfert, A., Zantop, T., Eggers, A.K., Raschke, M., Petersen, W.: Reducing the "killer turn" in posterior cruciate ligament reconstruction by fixation level and smoothing the tibial aperture. Arthroscopy J. Arthroscopic Related Surg. 23(10), 1104–1111 (2007). https://doi.org/10.1016/j.arthro.2007.04.014

23. Yang, B., Hu, L., Guo, N., Wang, Y., Liu, H., Han, Z.: Anterior cruciate ligament reconstruction surgery navigation and robotic positioning system under X-rays. In: 2018 IEEE International Conference on Robotics and Biomimetics (ROBIO), pp. 156–163 (2018)
24. Yao, J., et al.: Effect of tibial drill-guide angle on the mechanical environment at bone tunnel aperture after anatomic single-bundle anterior cruciate ligament reconstruction. Int. Orthopaedics **38**(5), 973–981 (2014). https://doi.org/10.1007/s00264-014-2290-5
25. Zhu, M., et al.: Tibial tunnel placement in anatomic anterior cruciate ligament reconstruction: a comparison study of outcomes between patient-specific drill template versus conventional arthroscopic techniques. Archiv. Orthopaedic Trauma Surg. **138**(4), 515–525 (2018). https://doi.org/10.1007/s00402-018-2880-6

Improving Hexahedral-FEM-Based Plasticity in Surgery Simulation

Ruiliang Gao and Jörg Peters[(⊠)]

University of Florida, Gainesville, FL 32611, USA

Abstract. Collecting, stretching and tearing soft tissue is common in surgery. These repeated deformations have a plastic component that surgeons take into consideration and that surgical simulation should model. Organs and tissues can often be modeled as curved cylinders or planes, offset orthogonally to form thick shells. A pair of primary directions, e.g., axial and radial for cylinders, then provides a quadrilateral mesh whose offset naturally yields a hexahedral mesh.

To better capture tissue plasticity for such hexahedral meshes, this work compares to and extends existing volumetric finite element models of plasticity. Specifically, we extend the open source simulation framework SOFA in the context of surgical simulation. Based on factored deformation gradients, the extension focuses on the challenge of separating symmetric and asymmetric, elastic and plastic deformation components – while preserving volume and avoiding re-meshing.

Keywords: Laparoscopic simulation · Soft tissue · Real-time simulation · SOFA · Elastoplasticity

1 Motivation

Tearing internal soft tissue is, besides cutting and cauterizing, an important surgical skill – to mobilize vessels and organs held in place by connective and fatty tissue. Realistic tearing requires a plastic deformation of the tissue – that is the tissue does not spring back to its initial position when released.

When surgeons interact with virtual tissues in a real-time training environment, perfect shape memory in the form of perfect elasticity is distracting and some plastic deformation is expected as thick tissues or organ walls stretch. Plastic deformation should also be monitored to penalize over-stretching.

To improve real-time simulation of soft tissue undergoing plastic deformation for training, we build on an existing simulation platform, the Simulation Open Framework Architecture (SOFA). SOFA [1] offers plastic deformations albeit currently only for linear tetrahedral elements. It however efficient and natural to

Electronic supplementary material The online version of this chapter (https://doi.org/10.1007/978-3-030-87202-1_55) contains supplementary material, which is available to authorized users.

M. de Bruijne et al. (Eds.): MICCAI 2021, LNCS 12904, pp. 571–580, 2021.
https://doi.org/10.1007/978-3-030-87202-1_55

Fig. 1. Laparoscopic surgery simulation: stretching fatty tissue. Note the vestigial plastic deformation at ↓ (Color figure online)

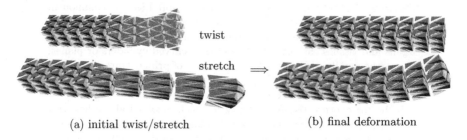

(a) initial twist/stretch   (b) final deformation

Fig. 2. SOFA tet-element plasticity: (*top*) twist, (*bottom*) stretch. (a) initial deformation (b) (*top*) lack of plasticity, (*bottom*) lack of symmetry.

generate meshes for surgical simulation as an offset from a quadrilateral base surface. Fatty tissue can often be presented as an offset of a covering surface sheet Fig. 3b. Thick-walled organs of the gastrointestinal tract have a natural tube structure (and so does the covering tissue, recall Fig. 1). Other organs can be embedded into a hexahedral free-from deformation grid as illustrated in Fig. 3a. Offsetting a quadrilateral mesh yields a hex-mesh, e.g., a partition of a tissue into boxes that is predictable and immediate. When splitting the natural hexahedra into tetrahedra to apply SOFA's plasticity code, we experienced unnatural stiffness Fig. 2, *top*; and asymmetry when choosing an asymmetrical tetrahedralization Fig. 2, *bottom*. We noted also that algorithmically generated tetrahedral partitions can change strongly for small changes the enclosing surface and so lack the symmetry and feature alignment of hex elements.

Currently no *hex*-FEM codes exist that model plasticity in an interactive environment for surgery simulation. This paper reports on an extension of SOFA to allow for plastic deformations of hex-elements. The extension represents a careful trade-off between higher accuracy and simplicity of computation via a 'blended-vertex approach'. The contributions are as follows.

- Extending the linear co-rotational elasto-plastic FEM to hex meshes.
- Extending the third-order accurate blended-vertex deformation [2] to hex meshes. (The increased degrees of freedom enables coarse hex meshes to replace high-resolution tet meshes when modelling large plastic deformations).

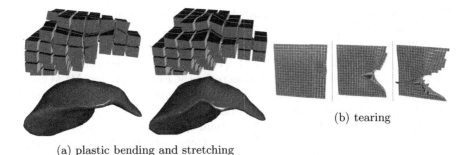

(b) tearing

(a) plastic bending and stretching

Fig. 3. Interactive surgical models (a) Piecewise hexahedral free-form deformation cage enclosing the liver. (b) Tearing a peritoneal sheet.

- Devising a simple plastic decomposition scheme that handles both rotational and stretching plasticity.
- Implementing and comparing the blended vertex approach and the cell-centered piecewise constant approach to plastic rotational deformation. 4 rotational deformation both vertex-centered (The blended vertex approach is more flexible and accurate, the cell-centered approach yields better element quality.)
- Preserving volume over large deformations.

2   Background

The choice of tetrahedral vs hexahedral finite elements to model elasticity has been debated for many years in the context of engineering analysis. [3, 4] observe that contact pressure distribution and contact shear stress distribution predicted by linear tetrahedral mesh are noisy and mesh dependent resulting in patches of locally elevated peak pressures whereas the pressure distribution predicted on hexahedral (and quadratic tetrahedral) meshes was smooth and uniform. More recently [5] argued that total degree quadratic finite elements on tetrahedra provide comparable outcomes to tri-linear elements on hexahedra. The authors later make the case for adding non-polynomial polyhedral elements [6].

In practice, for real-time soft tissue simulation, both tri-linear elements on hex-partitions and linear (total degree) tetrahedral elements are commonly used. Unsurprisingly, compared with linear tetrahedral elements, tri-linear hex elements are more flexible, perform better over viscous regions and have higher accuracy [7,8]. SOFA offers both hex and tet elements. Tessellating organs like the liver into tetrahedra is well automated (e.g., [9]) whereas quick and reliable coarse hex-meshing remains a challenge that has spawned an active research community, see e.g., [10–13]. Luckily, simulated organs and tissues can often be outlined as curved cylinders or planes, that are offset orthogonally to form thick shells: A pair of primary directions, e.g., axial and radial for cylinders, provides a quadrilateral mesh whose offset naturally yields a regular hexahedral mesh.

The flexibility of hex-elements compared to linear tet elements comes at a cost: preservation of volume is tricky, but necessary in surgery simulation so that plastically deformed tissue neither artificially swells nor disappears.

The material point method (MPM, [14–16]) need not be concerned with distortion of mesh elements, a major concern for the FEM approach. MPM excels at modelling plasticity of granular materials like sand or snow. However such materials are not typically relevant to surgical simulation. SPH-type approaches do not take advantage of the available regular quad-offset structure. The well-known linear elastic co-rotational FEM [17] factors out rotational components (the displacement is treated as $\mathbf{R}\hat{\mathbf{x}} - \mathbf{x}$ in the notation developed in the next Section, where $\mathbf{R}$ describes the material rotation). Co-rotational FEM therefore models primarily stretching in a major direction.

3   Methodology

Plastic deformation occurs when a material is subjected to tensile, compressive, bending, or torsion stresses that exceed the material's yield strength. With $\mathbf{x}$ the start position, possibly plastically deformed in a previous iteration, and $\hat{\mathbf{x}}$ the deformed position in world coordinates, displacement is infintesimally character-ized by the Gradient $\hat{\mathbf{J}} := [\frac{\partial \hat{\mathbf{x}}_i}{\partial \mathbf{s}_j}]$, where $\mathbf{s}$ are the domain (reference) coordinates. The relative gradient, called deformation gradient,

$$\mathbf{F} := \frac{\partial \hat{\mathbf{x}}}{\partial \mathbf{x}} = \frac{\partial \hat{\mathbf{x}}}{\partial \mathbf{s}}\frac{\partial \mathbf{s}}{\partial \mathbf{x}} = \frac{\partial \hat{\mathbf{x}}}{\partial \mathbf{s}}(\frac{\partial \mathbf{x}}{\partial \mathbf{s}})^{-1} = \hat{\mathbf{J}}(\mathbf{J})^{-1} \tag{1}$$

of a deformed hex element with vertices $\mathbf{v}^i$ can be measured at the center point $\mathbf{o} := \sum_i \mathbf{v}^i/2^3$ and called $\mathbf{F_o}$ – or at one of the vertices i and named $\mathbf{F}_i$. [2] proposes to increase accuracy by blending for hex k and vertex i the gradients as $\overline{\mathbf{F}}_i^k := (\mathbf{F}_i^k + \mathbf{F_o}^k)/2$.

Plasticity Decomposition. Recent approaches [18–20] recommend multiplica-tive decomposition of the deformation gradient into elastic and plastic parts: $\mathbf{F} = \mathbf{F}^e\mathbf{F}^p$ both for better numerical stability, to support compressibility and because the classic additive (strain) decomposition $\epsilon = \epsilon^e + \epsilon^p$ is only accu-rate for infinitesimal strains but fails for large deformations (or to easily model incompressibility). Starting with the singular value decomposition $\mathbf{F} = \mathbf{U}\mathbf{D}\mathbf{V}^T$ where $\mathbf{D}$ is the diagonal and $\mathbf{U}, \mathbf{V}$ are orthogonal (rotations or reflections), the total deformation is factored into a (polar, orthogonal) rotation tensor $\mathbf{R}$ and a symmetric positive-definite tensor called the (right) stretch tensor $\mathbf{S}$:

$$\mathbf{F} = (\mathbf{U}\mathbf{V}^T)(\mathbf{V}\mathbf{D}\mathbf{V}^T) = \mathbf{R}\mathbf{S}. \tag{2}$$

Following [19] we determine the elastic component. Then we extract the plastic components from both the rotation and the stretch tensor to obtain the factoring

$$\mathbf{F} = \mathbf{R}^e\mathbf{R}^p\mathbf{S}^e\mathbf{S}^p \tag{3}$$

into elastic rotation, plastic rotation, elastic stretch and plastic stretch as follows.

Plastic Stretching. Following [20] the linearized strain $\epsilon = \mathbf{V}(\mathbf{D} - \mathbf{I})\mathbf{V}^T$ is derived from $\mathbf{S}$ and converted into the first Piola–Kirchhoff stress σ by the classic stress-strain relation. Then the plastic stretching deformation is

$$\mathbf{S}^p = \mathbf{V}(\frac{\mathbf{D}}{(\det \mathbf{D})^{1/3}})^\gamma \mathbf{V}^T \quad \gamma := \min\{\nu \Delta t \frac{\|\sigma\|_2 - \tau}{\|\sigma\|_2}, 1\} \tag{4}$$

for a plastic flow rate ν, plastic yield threshold τ and time step Δt.

Plastic Rotation. Separating the plastic rotation from the rotation tensor is a challenging problem [21], because the material rotation stems from two sources: shape changing deformation, e.g., shear deformation, and rigid body rotations. When the material changing shape, it is typically not possible to uniquely separate out the rigid body rotation. However, in our surgical simulation context there is no spinning anatomy and we can neglect angular velocity or inertia-related factors. We can therefore assume that plastic rotation depends solely on rotational distortion. We measure rotational distortion as the magnitude of the angle based on geodesics on the unit sphere defined as [22]

$$\Phi(\mathbf{R}_1, \mathbf{R}_2) := \|\log(\mathbf{R}_1 \mathbf{R}_2^T)\| \in [0, \pi]$$

and apply this measure to the rotational component obtained from polar decomposition at the center $\mathbf{F_o} = \mathbf{R_o}\mathbf{U_o}$, respectively blended, $\overline{\mathbf{F}}_i^k = \overline{\mathbf{R}}_i^k \overline{\mathbf{U}}_i^k$. For vertex i and the center $\mathbf{o}$ of hex k this yields the decomposition of the blended vertex rotation $\overline{\mathbf{R}}_i^k$ of (2) so that, dropping the superscript k, and denoting the identity matrix as $\mathbf{I}_3$,

$$\overline{\mathbf{R}}_i^e \, \overline{\mathbf{R}}_i^p := \overline{\mathbf{R}}_i^{(1-\eta)} \, \overline{\mathbf{R}}_i^\eta = \overline{\mathbf{R}}_i, \tag{5}$$

$$\eta := \nu^R \Delta t(\phi_i^k - \tau^R)/(\phi_i^k + \Phi(\mathbf{I}_3, \mathbf{R_o})), \quad \phi_i^k := \Phi(\overline{\mathbf{R}}_i^k, \mathbf{R_o}^k).$$

for a rotation yield-threshold $\tau^R \in [0, \pi]$, rotational plastic flow rate $\nu^R \in [0, 1]$.

The above vertex plastic rotation is based on each vertex's local rotation. This yields flexibility and accuracy but potentially allows strong distortion. Alternatively we propose cell-centered decomposition of rotation to determine a centered rigid rotation $\mathbf{R_o}^p$ that largely preserves the hexahadral element's shape.

$$\mathbf{R_o}^p \mathbf{R_o}^e = \mathbf{R_o}^{\tilde{\eta}} \mathbf{R_o}^{1-\tilde{\eta}}, \tag{6}$$

$$\tilde{\eta} := \nu^R \Delta t(\tilde{\phi}^k - \tau^R)/(\tilde{\phi}^k + \Phi(\mathbf{I}_3, \mathbf{R_o})), \quad \tilde{\phi}^k = \max_{i \in \text{hex}^k} \phi_i^k.$$

Material Hardening. To implement the material hardening, we update the plastic yield threshold by $\tau \leftarrow \tau + \kappa \gamma \|\sigma\|$ and the rotation yield threshold by $\tau^R \leftarrow \tau^R + \kappa \eta \phi_i^k$. The parameter κ controls the amount of work hardening (or softening) per time step.

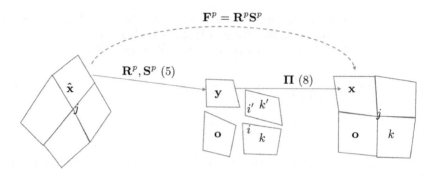

Fig. 4. Representation of a quad-mesh (analogous to a hex-mesh) in (*left*) elastically deformed world space (where the surgery takes place), (*middle*) rest space (isolated elements with strongly deformed domains) and (*right*) material space (relaxation of the deformations to a consistent mesh). The dashed path is not used for iteration.

Plastic Update. At each time step, the plastic stretch $\mathbf{S}_i^p$ first updates the rest position of each vertex i of an isolated rest space element k (see Fig. 4,*middle*):

$$\mathbf{y}_i^k \leftarrow \mathbf{y}_i^k + \mathbf{S}_i^p \mathbf{u}_i, \quad \mathbf{u}_i := \mathbf{R}_i \hat{\mathbf{x}}_i - \mathbf{x}_i, \tag{7}$$

where $\mathbf{u}_\alpha$ is the co-rotational vertex displacement [17]. For vertex plastic rotation, we update the rotation map (see Fig. 4)

$$\boldsymbol{\Pi}_i^k \leftarrow \boldsymbol{\Pi}_i^k \overline{\mathbf{R}}_i^p, \tag{8}$$

and for the alternative cell-center plastic rotation $\boldsymbol{\Pi}^k \leftarrow \boldsymbol{\Pi}^k \mathbf{R}_o^p$.

Even though constructions (4) and (5) imply $\det(\mathbf{S}_i^p) = \det(\mathbf{R}_i^p) = 1$, the volume of each element is not locally preserved, because the plastic offset is applied separately per vertex. Computing the exact volume from the Jacobian $\mathbf{J}$, the ratio β of the deformed volume divided by the original volume can be accurately computed. For element k we update $\mathbf{y}_i^k$ by scaling back to the initial volume:

$$\mathbf{y}_i^k \leftarrow \mathbf{y}_i^k + (\mathbf{y}_i^k - \mathbf{o}^k)(1 - \beta)\Delta t \tag{9}$$

where Δt distributes the adjustment over the iterations so that short time steps animate to slower volume restitution. To combine the vertices $\mathbf{y}_i^k$ of the isolated elements and form a consistently joined 'material', we update $\mathbf{x}_j \leftarrow \sum_{k \in N_j} \boldsymbol{\Pi}_i^k \mathbf{y}_i^k / |N_j|$, where $\mathbf{y}_i^k$ are the $|N_j|$ vertices corresponding to the material space vertex $\mathbf{x}_j$ with global index j in the surrounding cells N_j, see Fig. 4.

4   Results and Discussion

We have incorporated our plasticity decomposition approach by extending linear hexahedral FEM in SOFA 19.12. The examples illustrate some parameter choices. All tests, see also the video, were conducted on a PC with Intel Core.

i7-9700K CPU and 8G RAM running Windows 10. Our code executes at 25–33 hex/ms, For the most expensive computational time steps, immediately after release, our code is ca 20% slower than SOFA's elastic-only corotational code.

Figure 5 compares the stretching of a 9×10-hex bar under different plasticity flow rates ν, plastic yield thresholds τ. The bar is clamped at the middle (shown as black dots) and the stretching force is applied at the central face on the end of the bar. Results show the distribution of the plastic deformations where cross section symmetry is preserved. (The views vary due to perspective projection).

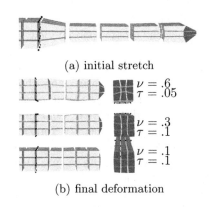

(a) initial stretch

$\nu = .6$
$\tau = .05$

$\nu = .3$
$\tau = .1$

$\nu = .1$
$\tau = .1$

(b) final deformation

Fig. 5. Stretching bars with *top* to *bottom*: high, medium, low stretch plasticity material. Faces between slice 5 and 6 are clamped. (a) initial stretch, (b) side and cross-section views of final deformed rest pose

Figure 6 juxtaposes the blended vertex plastic rotation with the cell-center plastic rotation by twisting the bar with different rotational plasticity flow rates ν^R, and rotational plastic yield thresholds τ^R. The twisting force is applied tangentially at all the four edge centers of the central face of the bar cross-section. Tests show that both methods can capture plastic rotation related to the flow rate ν^R. However, the deformation differs: the vertex approach is better at preserving the cross-sectional shape but distorts each local element more than

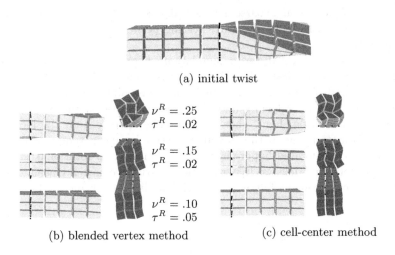

(a) initial twist

$\nu^R = .25$
$\tau^R = .02$

$\nu^R = .15$
$\tau^R = .02$

$\nu^R = .10$
$\tau^R = .05$

(b) blended vertex method   (c) cell-center method

Fig. 6. Twist and plasticity. *top* to *bottom*: high, medium, low rotational plasticity material. (a) initial twist, (b,c) final plastic deformation (side view, *left*; front cross-sectional view, *right*).

the cell-centered approach. The cell-centered approach preserves the local box shape but the cross-sectional view is jagged.

For all configurations of Figs. 5 and 6, the volume of the final deformed bars agrees with the input volume within $< 1\%$. The material hardening parameter is $\kappa = 0.2$ in all test cases. This choice proved effective in preventing ill-shaped material elements due to large plastic deformations: before an element becomes highly distorted its plastic yield threshold is reached and causes fracture.

Twist torques due to rotation of lap surgery instrument heads are extremely low [23]. As a practical solution in the surgical simulation setting, our implementation switches off plasticity when an element is about to become inverted – the remaining elastic FEM solver handles inverted elements robustly [24]. Switching to a purely elastic simulation when hexahedral elements are about to invert favors robustness over physics during flawed, unrealistic high-torque interactions.

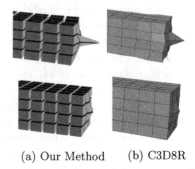

(a) Our Method   (b) C3D8R

Fig. 7. FEM beam simulations using (a) our 8-node hex corotational elasto-plasticity (Young's 5000, Poisson .45), (b) Abaqus 8-node hex element (C3D8R) with hyperelasto-plasticity (Moon-Rivlin, C10 = 1765, C01 = 43, D1 = 1E-05). Initial Stretch, *top*; final plastic deformation, *bottom*, agree for (a) and (b).

We use plasticity for surgical training simulation for a range of anatomical features and laparoscopic surgical procedures. Figure 3a illustrates the plastic deformation of a free-form deformation cage, a piecewise trilinear function on a coarse hexahedral mesh that transform an embedded much finer mesh. Figure 3b illustrates tearing peritoneum. See also the accompanying video.

A comparison of FEM beam simulations between our method and the Abaqus C3D8R hyperelasto-plasticity model is shown in Fig. 7. At ca 7ms per hex Abaqus analysis does not meet real-time constraints. While linear Abaqus C3D8R model stretches locally and visibly less realistic in the 'eyeball norm' of real-time simulation, Fig. 7 shows remarkable agreement on the location of highest deformation and within 5% in the stretch during temporal evolution between the sophisticated hyper-elastic model and our co-rotational extension. As tissue is better modeled as hyper-elastic material, we have no formal analysis to explain the agreement but note that the applications typically have a high Poisson ratio.

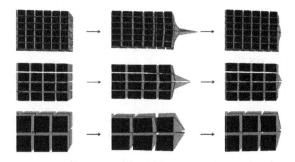

Fig. 8. Refining the stretch test. *left*: initial mesh; *middle*: maximal stretch; *right*: final plastic deformation.

Figure 8 illustrates self-refinement. The three beam models reproduce the same deformation, localized commensurate with the higher resolution. Execution time for refinement scales linearly with the number of hex-elements since no explicit matrix is built or inverted.

Limitations. While generating hex meshes by offsetting natural quad-meshes is highly efficient, the process does not easily lend itself to re-meshing when hex elements become ill-shaped. Our implementation relies on the material hardening to prevent distortion and switches to pure elasticity when elements invert.

To avoid polar rotation ambiguity, we restrict the maximal element rotation to less than π. To increase the total object's rotational angle, the initial mesh has to be designed with sufficiently many pieces (what is usually not a problem in the context of a given surgical procedure).

References

1. Faure, F., et al.: SOFA: a multi-model framework for interactive physical simulation. In: Payan Y. (eds.) Soft Tissue Biomechanical Modeling for Computer Assisted Surgery. Studies in Mechanobiology, Tissue Engineering and Biomaterials, vol. 11, pp. 283–321. Springer, Heidelberg (2012). https://doi.org/10.1007/8415_2012_125
2. James, D.L.: Phong deformation: a better C^0 interpolant for embedded deformation. ACM Trans. Graph. (TOG) **39**(4), 56 (2020)
3. Tadepalli, S.C., Erdemir, A., Cavanagh, P.R.: Comparison of hexahedral and tetrahedral elements in finite element analysis of the foot and footwear. J. Biomech. **44**(12), 2337–2343 (2011)
4. Benzley, S.E., Perry, E., Merkley, K., Clark, B., Sjaardama, G.: A comparison of all hexagonal and all tetrahedral finite element meshes for elastic and elastoplastic analysis. In: Proceedings, 4th International Meshing Roundtable, vol. 17, pp. 179–191. Citeseer (1995)
5. Schneider, T., Hu, Y., Dumas, J., Gao, X., Panozzo, D., Zorin, D.: Decoupling simulation accuracy from mesh quality. ACM Trans. Graph **37**(6), 280:1–280:14 (2018)

6. Schneider, T., Dumas, J., Gao, X., Botsch, M., Panozzo, D., Zorin, D.: Poly-spline finite-element method. ACM Trans. Graph **38**(3), 19:1–19:16 (2019)
7. Shepherd, J.F., Johnson, C.R.: Hexahedral mesh generation constraints. Eng. Comput. **24**(3), 195–213 (2008)
8. Sarrate Ramos, J., Ruiz-Gironés, E., Roca Navarro, F.J.: Unstructured and semi-structured hexahedral mesh generation methods. Comput. Technol. Rev. **10**, 35–64 (2014)
9. Hu, Y., Schneider, T., Wang, B., Zorin, D., Panozzo, D.: Fast tetrahedral meshing in the wild. ACM Trans. Graph **39**(4), 117 (2020)
10. Blacker, T.D.: Automated conformal hexahedral meshing constraints, challenges and opportunities. Eng. Comput. **17**(3), 201–210 (2001)
11. Liu, H., Zhang, P., Chien, E., Solomon, J., Bommes, D.: Singularity-constrained octahedral fields for hexahedral meshing. ACM Trans. Graph **37**(4), 93:1–93:17 (2018)
12. Cherchi, G., Alliez, P., Scateni, R., Lyon, M., Bommes, D.: Selective padding for polycube-based hexahedral meshing. Comput. Graph. Forum **38**(1), 580–591 (2019)
13. Gao, X., Shen, H., Panozzo, D.: Feature preserving octree-based hexahedral meshing. Comput. Graph. Forum **38**(5), 135–149 (2019)
14. Schreck, C., Wojtan, C.: A practical method for animating anisotropic elastoplastic materials. Comput. Graph. Forum **39**, 89–99 (2020). Wiley Online Library
15. Stomakhin, A., Schroeder, C., Chai, L., Teran, J., Selle, A.: A material point method for snow simulation. ACM Trans. Graph. (TOG) **32**(4), 1–10 (2013)
16. Wang, S., et al.: Simulation and visualization of ductile fracture with the material point method. Proc. ACM Comput. Graph. Interact. Tech. **2**(2), 1–20 (2019)
17. Hauth, M., Strasser, W.: Corotational simulation of deformable solids (2004)
18. Irving, G., Teran, J., Fedkiw, R.: Invertible finite elements for robust simulation of large deformation. In: Proceedings of the 2004 ACM SIGGRAPH/Eurographics Symposium on Computer Animation, pp. 131–140 (2004)
19. Bargteil, A.W., Wojtan, C., Hodgins, J.K., Turk, G.: A finite element method for animating large viscoplastic flow. ACM Trans. Graph. (TOG) **26**(3), 16-es (2007)
20. Wicke, M., Ritchie, D., Klingner, B.M., Burke, S., Shewchuk, J.R., O'Brien, J.F.: Dynamic local remeshing for elastoplastic simulation. ACM Trans. Graph. (TOG) **29**(4), 1–11 (2010)
21. Holmedal, B.: Spin and vorticity with vanishing rigid-body rotation during shear in continuum mechanics. J. Mech. Phys. Solids **137**, 103835 (2020)
22. Huynh, D.Q.: Metrics for 3D rotations: comparison and analysis. J. Math. Imaging Vis. **35**(2), 155–164 (2009)
23. Richards, C., Rosen, J., Hannaford, B., Pellegrini, C., Sinanan, M.: Skills evaluation in minimally invasive surgery using force/torque signatures. Surg. Endoscopy **14**(9), 791–798 (2000). https://doi.org/10.1007/s004640000230
24. Irving, G., Teran, J., Fedkiw, R.: Tetrahedral and hexahedral invertible finite elements. Graph. Models **68**(2), 66–89 (2006)

Rapid Treatment Planning
for Low-dose-rate Prostate Brachytherapy
with TP-GAN

Tajwar Abrar Aleef[1]([✉]) [iD], Ingrid T. Spadinger[2], Michael D. Peacock[2],
Septimiu E. Salcudean[3] [iD], and S. Sara Mahdavi[2]

[1] School of Biomedical Engineering, University of British Columbia,
Vancouver, Canada
tajwaraleef@ece.ubc.ca
[2] BC Cancer - Vancouver Centre, Vancouver, Canada
[3] Department of Electrical and Computer Engineering,
University of British Columbia, Vancouver, Canada

Abstract. Treatment planning in low-dose-rate prostate brachytherapy
(LDR-PB) aims to produce arrangement of implantable radioactive seeds
that deliver a minimum prescribed dose to the prostate whilst minimizing
toxicity to healthy tissues. There can be multiple seed arrangements that
satisfy this dosimetric criterion, not all deemed 'acceptable' for implant
from a physician's perspective. This leads to plans that are subjective
where quality of treatment depends on the expertise of the planner. We
propose a method that learns to generate consistent treatment plans
from a large pool of successful clinical data (961 patients). Our model
is based on conditional generative adversarial networks that use a novel
loss function for penalizing the model on spatial constraints of the seeds.
An optional optimizer based on a simulated annealing (SA) algorithm
can be used to further fine-tune the plans if necessary (determined by
the treating physician). Performance analysis was conducted on 150 test
cases demonstrating comparable results to that of the manual plans. On
average, the clinical target volume covered by 100% of the prescribed
dose was 98.9% for our method compared to 99.4% for manual plans.
Moreover, using our model, the planning time was significantly reduced
to an average of 3 s/plan (2.5 min/plan with the optional SA). Compared
to this, manual planning at our centre takes around 20 min/plan.

Keywords: Low-dose-rate brachytherapy · Treatment planning ·
Prostate cancer · Generative adversarial network

1 Introduction

Low-dose-rate prostate brachytherapy (LDR-PB) is considered an effective cura-
tive treatment for men with localized prostate cancer (PCa) [13]. In LDR-PB, a
standard needle template is used to guide and place permanent radioactive seeds

This work was supported by the Canadian Institutes of Health Research (CIHR).

M. de Bruijne et al. (Eds.): MICCAI 2021, LNCS 12904, pp. 581–590, 2021.
https://doi.org/10.1007/978-3-030-87202-1_56

transperineally into the prostate through needles [13,15]. Before the implant, expert planners manually determine the optimal distribution of seeds that deliver a prescribed dose to the target anatomy while minimizing toxicity to other surrounding tissues. Several seed arrangements can fulfil dosimetric constraints and clinical guidelines, although not all are deemed acceptable for implant by the physician due to factors such as their preference on what might make a plan easier to deliver. The selection of optimal plans can hence be subjective where the quality of the outcome depends on the experience of the planner which comes from years of training and practice. This, along with the long planning duration for each new patient, makes manual planning a resource-intensive and time-consuming task. For automating this planning procedure, which can also benefit real-time intra-operative LDR-PB planning, many approaches have been proposed. Common methods focus on meeting the dose-volume criteria with limited constraints on the needle or seed locations (which can determine if a plan is implantable or not). These include various optimization techniques such as: mixed-integer linear programming [2], inverse treatment planning with compressed sensing [5], genetic algorithms [3], and fast simulated annealing [12]. Most of these techniques are highly sensitive to their initialization with their associated optimization costs having multiple minima and a huge search space. The use of machine learning approaches to replicate high-quality plans using respective centre's data is limited in the literature [10,11]. Both these methods initialize a prior plan based on the training data followed by the main optimization step. To initiate the priors from their respective training database, [11] uses a joint Sparse Dictionary Learning approach to learn the relationship between target volumes and seed plans while [10] uses a feature extraction & matching technique to look for similar plans. Both methods use a limited set of features that are selected manually and they rely heavily on the optimization step. Hence, plan characteristics will mostly be those defined within the objective function with less chance of learning plan features that cannot be mathematically expressed. Indeed, [11] reports a significant drop in performance when its optimizer is removed.

In this work, we propose a model based on conditional generative adversarial network (cGAN) where a large set of learnable parameters is trained using retrospective clinical data– capturing implicit clinical features from the data automatically. The proposed model predicts seed plans directly using volumetric data from the anatomy (easily available) and constraints from the needle space (as per standard guidelines). To achieve this, we use a novel loss function that incorporates additional spatial constraints for seed placements. As supervised learning technique is limited to what it has observed in the training dataset, and the problem can have multiple solutions, we also provide an option to fine-tune the results further when required using a simulated annealing (SA) based optimizer. We compare our method with existing and common approaches of automatic plan generation using several plan quality metrics. A further ablation study is conducted to validate the need for the different components in our model. To our best knowledge, this is the first approach that uses deep learning techniques to learn this multi-objective mapping for LDR-PB treatment planning in an end-to-end approach.

2   Methods

2.1   Dataset

With institutional ethics approval, high-quality successful retrospective treatment plan data of 961 patients treated at BC Cancer-Vancouver Centre (Vancouver, BC, Canada) was available for this study. This data includes annotations of the clinical target volumes (CTV, i.e. the prostate) and the planning target volumes (PTV, i.e. the CTV plus a predefined margin). For every patient, 2–4 plan variations were created by expert medical physicists from which one was selected for implantation by the treating physician. These plans consist of seeds located on a 5 mm spaced grid of size $10 \times 13 \times \#axial\ image\ planes$, where 10×13 is the needle template size (see Fig. 2). The data includes prostate volumes ranging from 20–70 cc, all receiving standard I-125 monotherapy with a prescribed dose of 144 Gy. We used 711 cases for training, 100 cases for validation, and the remaining 150 cases for testing.

2.2   Seed Planning with TP-GAN

Multiple seed arrangements can deliver the prescribed dose to the volume of the gland. However, plans are decided not only based on dose coverage but also considering a pool of guidelines that are considered best practices for LDR-PB. Moreover, the expertise of the planners and treating physicians also determines which plan will eventually be selected for the implant. With such a degree of variation, there can be multiple solutions satisfying the planning guidelines, and hence our clinical dataset does not embody a straightforward solution path that can be depicted only from the target volume data. To alleviate this, we narrow down the search space by putting a constrain on the needle locations. Our group previously showed that centre specific needle plans can be automatically & reliably generated given the target volume data (PTV, CTV) using a similar cGAN based approach [1]. The problem then can be considered as an image-to-image mapping task where the inputs are the target volume data with the automatically generated needle plan and the output is the corresponding seed plan. We refer to this proposed model as "TP-GAN" (treatment planning with generative adversarial network).

All the target volume data and plans are first aligned to a reference needle template. The number of axial image planes is set to 14 to cover the maximum prostate length of 14×5 mm, with axial planes zero-padded for shorter prostates. The target volume data is represented as binary masks where all pixels within the margins are set to '1' and the rest of the background are set to '0'. Likewise, the needle and seed plans are also converted to binary matrices following the needle template grid where a pixel with '1' means a presence of a seed/needle in that grid coordinate. Next, a weighted distance transform is performed on the binary volumetric and plan data to provide the network with further distance-based information [8]. For each patient, the corresponding PTV, CTV, and needle plan are axially resized and then stacked in channels to form the input with

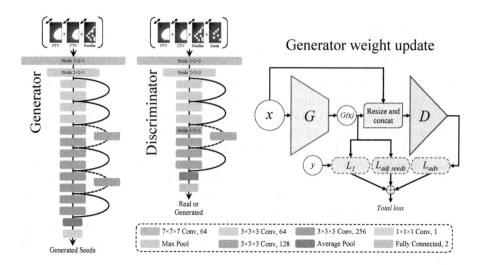

Fig. 1. Architectures of the generator (G) and discriminator (D) of TP-GAN. Figure legend provides kernel and filter sizes of the different layers. The diagram on the right shows how the different losses are calculated for updating weight of G. The input of the model (PTV, CTV, and needle plan) is given by x and the output of the model (seed plan) is given by $G(x)$. y is the corresponding manual plan. The three lightly shaded grey blocks shows the three losses used to optimize G.

dimension $64 \times 64 \times 14 \times 3$. Data augmentation increases the robustness of deep networks and reduces the chance of over-fitting. However, typical geometric or image-based augmentation technique can't be used for our problem as any change in the input space doesn't imply the same change to the output space. Since the plans in our dataset are symmetric (as per our centre's guideline), we utilize the following augmentation technique to double the data for training: the data is vertically split around the center of the template and the left and right sides are treated as two independent samples. After augmentation, the dimension of the input and output data becomes $64 \times 32 \times 14 \times 3$ and $10 \times 6 \times 14$.

We design the TP-GAN model based on [6] - which showed results with great generalizability in terms of learning paired image-to-image transformations for a plethora of applications. TP-GAN consists of a generator (G) that learns to encode the input space (x) to a seed plan ($G(x)$) and a discriminator (D) that learns to recognize between real (y) and generated plans ($G(x)$). The weights of the two networks are updated alternatively [4]. The architecture of G is designed out of a ResNet architecture [14] with 14 layers depth that encodes 4D inputs to 3D seed plans. D is also based on a ResNet architecture with a depth of 10 layers which takes the pair of inputs & outputs and predicts if it is generated or real. A resize and concatenation block is used to resize the real/generated plan and stack it in channels with corresponding x to match the dimension of D's input. Details of the architectures and the loss calculation for updating weights of G are given in Fig. 1.

To minimize hot spots (i.e. high dose regions) in the prostate, a general preference is to avoid implanting adjacent seeds. Although introducing needle plans to the input constrains the model to produce seeds within those needles, without further regularization, the model tends to produce plans with neighbouring adjacent seeds. To mitigate this problem, in addition to the *Adversarial* and L_1 loss from [6], we introduce a new loss function, *Adjacent seed* loss, that imposes a penalty for predicting adjacent seeds during training.

The *Adversarial*, L_1, and *Adjacent seed* losses are given by:

$$\mathcal{L}_{adv}(G, D) = \mathbb{E}_{x,y}[\log D(x, y)] + \mathbb{E}_x[\log(1 - D(x, G(x)))] \tag{1}$$

$$\mathcal{L}_{L_1}(G) = \mathbb{E}_{x,y}[\||y - G(x)\||_1] \tag{2}$$

$$\mathcal{L}_{adj\ seeds}(G) = \mathbb{E}_x[\sum(\max(0, (G(x) * k) - 5)] \tag{3}$$

$$where,\ k = \left[\begin{bmatrix} 0\ 0\ 0 \\ 0\ 1\ 0 \\ 0\ 0\ 0 \end{bmatrix} \begin{bmatrix} 0\ 1\ 0 \\ 1\ 7\ 1 \\ 0\ 1\ 0 \end{bmatrix} \begin{bmatrix} 0\ 0\ 0 \\ 0\ 1\ 0 \\ 0\ 0\ 0 \end{bmatrix} \right]$$

In the *Adjacent seed* loss (Eq. 3), k is a $3 \times 3 \times 3$ kernel used to find the presence of adjacent seeds in the prediction. The adjacent seeds can be localized by convolving k with the output. If the prediction of the model is binary, this convolution will set any pixel with adjacent coordinates to greater than 7. However, since the output is probability values in the range between $[0, 1]$, we lower this threshold to 5. This threshold is a hyperparameter that was selected during the tuning phase. Next, by subtracting 5 from the result of the convolution, we set all pixels with no adjacent seeds to less than or equal to zero. Applying a *ReLu* function $(max(0, X))$ on top of this sets any negative pixels to zero. Now all remaining positive pixels correspond to the adjacent seeds which are then summed to form this loss.

The full objective function of TP-GAN then becomes:

$$\mathcal{L}(G, D) = \alpha(\arg \min_G \max_D \mathcal{L}_{adv}(G, D)) + \beta \mathcal{L}_{L1}(G) + \alpha \mathcal{L}_{adj\ seeds}(G) \tag{4}$$

where, the weights of the different losses were tuned to $\alpha = \frac{1}{3}$ and $\beta = \frac{2}{3}$.

Keras and Tensorflow were used to train, validate, and test our model. For both the G and D, Adam optimizer with a learning rate of 10^{-5} and momentum parameters of $\beta_1 = 0.5$ and $\beta_2 = 0.99$ were used. The model was trained for 1000 epochs with a batch size of 16 samples on a single NVIDIA Tesla V100 GPU (16GB) which took about 25 hr to train. Weights of the model with the lowest validation loss were used for the evaluation of the results. Code of our implementation is available at: https://github.com/tajwarabraraleef/TP-GAN.

2.3   Post Processing Stage and Fine-Tuning with Simulated Annealing

The generated seed plans are then passed to a post-processing stage that initially checks for any remaining adjacent seeds. If found, it first attempts to relocate such seeds by shifting them ±5 mm within their needles. If that doesn't solve the adjacency problem, any remaining adjacent seeds are discarded from the plans. Next, if seeds are not uniformly distributed among neighbouring planes, a uniformization stage from [1] is used to shift seeds among such planes to improve uniformity and reduce potential hot spots. These two stages are computationally inexpensive and don't increase the planning duration. An optional, but beneficial step is the use of an SA optimizer based on [9,12] initialized using results of the post-processing step. SA fine-tunes the results by further enforcing some dosimetric constraints. Hence planning benefits from a combination of learned factors through the training set and explicit constraints determined by clinical guidelines. As we show, this step particularly reduces the unnecessary urethral dose. Furthermore, initializing the SA with our method provides it with a solution that is already close to the global minimum.

3   Results and Discussions

3.1   Performance Analysis

The quality of the plans was assessed using standard dosimetric parameters including: the target volume dose coverage (PTV $V100\%$, $V150\%$ & CTV $V100\%$, $V150\%$), the urethra & rectum (OAR: organs at risk) exposure (URE $V150\%$ & REC $V50\%$), and source usages indicated by the number of seeds & needles used in the plan. Here, $Vx\%$ indicates the percentage of the volume (PTV, CTV, URE, or REC) receiving $x\%$ of the minimum prescribed dose. Table 1 lists the mean and standard deviations of the key quality metrics on the 150 test cases for a number of plan generation techniques which includes: 1) SA initialized using Seattle based planning method ($SA_{seattle}$) [7], 2) joint Sparse Dictionary Learning approach from [11] ($jSDL$), 3) SA initialized with generated needle plans from [1] (SA_{genN}), 4) proposed method (TP-GAN), 5) proposed method with SA tuning (TP-GAN+SA), and 6) Manual plans (*Actual*). To evaluate statistical significance among different plan generation techniques, we ran ANOVA test which resulted in $p < 0.05$ for all key plan quality metrics– indicating the presence of significant difference. To identify these differences, a post hoc analysis was conducted by running paired t-test with p-value adjusted following Bonferroni Correction. The result from this test is included in Table 1, where for each key plan quality metric, significant difference ($p < 0.008$) of TP-GAN and TP-$GAN + SA$ with other techniques are indicated by underlined and bold values, respectively.

Table 1. Comparison of the mean and standard deviation of key plan quality metrics for different techniques on the test set. #N and #S indicates the count of needles and seeds used. Statistical significant difference of $TP\text{-}GAN$ and $TP\text{-}GAN\text{+}SA$ with other techniques are indicated by underlined and bold values, respectively.

	PTV V100%	PTV V150%	CTV V100%	CTV V150%	URE V150%	REC V50%	#N	#S
$SA_{seattle}$	**92.9 ± 2.2**	**47.6 ± 5.1**	**98.2 ± 1.3**	**55.2 ± 7.2**	9.6 ± 7.9	**14.8 ± 4.6**	30 ± 4	100 ± 13
$jSDL$	94.2 ± 3.4	53.6 ± 5.6	98.5 ± 2.0	**61.9 ± 8.3**	22.7 ± 19.0	16.8 ± 7.7	24 ± 3	112 ± 16
SA_{genN}	**94.8 ± 1.7**	**51.0 ± 4.2**	98.9 ± 1.1	58.3 ± 5.4	6.2 ± 5.6	17.3 ± 5.9	28 ± 3	106 ± 16
$TP\text{-}GAN$	94.6 ± 3.9	55.0 ± 11.9	97.8 ± 2.5	60.8 ± 13.7	7.8 ± 11.9	17.6 ± 5.4	24 ± 3	109 ± 17
$TP\text{-}GAN + SA$	95.9 ± 1.6	53.0 ± 3.5	98.8 ± 0.9	59.1 ± 5.0	4.5 ± 3.08	16.7 ± 4.4	27 ± 3	107 ± 15
$Actual$	**96.9 ± 1.2**	**55.9 ± 4.0**	**99.4 ± 0.7**	62.1 ± 4.9	**3.3 ± 5.2**	17.1 ± 4.4	24 ± 3	110 ± 17

Table 2. Mean and standard deviation of the metrics on different ablation settings.

Ablation settings			AUC	Dice coeff	Adj Seeds	Seed Diff
Needle plan	Augmentation	$L_{adj\ seeds}$				
✗	✗	✗	89.1% ± 3.2%	34.6% ± 9.4%	18 ± 7	25 ± 9
✓	✗	✗	97.2% ± 2.2%	71.1% ± 9.1%	50 ± 18	23 ± 9
✓	✓	✗	97.4% ± 2.1%	73.4% ± 11.1%	30 ± 15	13 ± 6
✓	✓	✓	97.8% ± 2.0%	72.3% ± 14.1%	4 ± 5	6 ± 4

From Table 1, $SA_{seattle}$ has inferior performance among all other techniques. We suspect that this is because its initialization is based on a template that may not always converge to the optimal solution. Compared to all other techniques, $jSDL$ has the highest OAR toxicity as its objective function does not consider OAR dosage. The SA we use explicitly includes OAR dosimetry in its cost function while $TP\text{-}GAN$ implicitly learns to avoid OAR from training data. Although, results from SA_{genN} are similar to those of $TP\text{-}GAN$, the latter can produce results instantly in 3 s and it also learns to generate seed plans directly from the training data compared to just generating needle plans. Even though $TP\text{-}GAN$ can produce favourable plans on its own, its standard deviation across the metrics are high in general. Due to the limited dataset and the possibility of multiple optimal solutions, this technique works best on similar prostate shapes it has observed in the training dataset. A general trend observed for $TP\text{-}GAN$ is that the number of seeds is overestimated for smaller prostate volumes and underestimated for larger prostate volumes; resulting in over/under dose coverage on smaller/larger prostates. With SA tuning, $TP\text{-}GAN + SA$ produces the best results among the automated techniques and is comparable to manually created plans ($p < 0.008$ for most metrics when compared with the other techniques). However, this increases the planning duration from 3 s to an average of 2.5 min/plan which is still significantly lower than the average manual planning duration of 20 min. Figure 2 shows results of plans generated using $TP\text{-}GAN$ and $TP\text{-}GAN + SA$ vs actual historical plan for two test patients (P1 and P2). P1 represents a good case where $TP\text{-}GAN$ produces comparable dose coverage to the manual plan and $TP\text{-}GAN + SA$ produces even better coverage than the

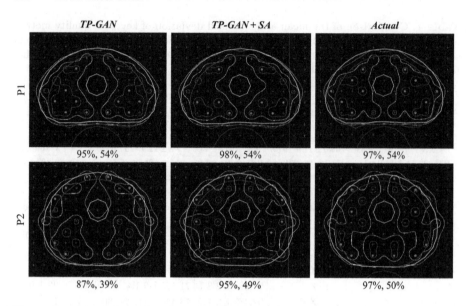

Fig. 2. Generated plans of two patients (P1 and P2) using *TP-GAN*, *TP-GAN + SA*, and *Actual* manual plan. The numbers below each plan indicate the respective PTV $V100\%$ and $V150\%$ of their full plan. Here, the template, seeds, urethra, rectum, CTV, PTV, $V100\%$, $V150\%$, and $V200\%$ is represented by orange (+), pink, yellow, blue, purple, cyan, green, white, and red, respectively. (Color figure online)

manual plan. P2 represents a case where *TP-GAN* fails to produce sufficient dose coverage. This is because P2 is an out of distribution sample with a rather large prostate volume (70 cc; largest in our dataset). With SA tuning, such cases can be dealt with as seen by the significant improvement in plan generated by *TP-GAN + SA* for P2.

3.2 Ablation Study

We analyzed the importance of the different components of TP-GAN which includes the use of: needle plans in the input, augmentation, and *Adjacent seed* loss. These components were added sequentially, meaning improvement of a latter component is based on having the preceding components in the model. These different configurations, as indicated in Table 2, were trained using the same hyper-parameters and evaluated on the validation set. As we are interested in replicating the manual plans, we evaluated plan similarity by calculating the area under the receiver operating characteristic curve (AUC) and Dice coefficient between the predicted and manual seed locations. We also compared spatial characteristics by measuring the number of adjacent seeds in predicted plans and the difference between the number of seeds used in predicted and real plans. From Table 2, a clear improvement is seen with each additional component. The adjacent seeds are low for no needle plan because the model is not predicting enough seeds as can be indicated by the very low Dice coefficient and high seed difference. Adding needle plan improves the Dice coefficient significantly, but now with no constraint on the output, the model predicts a large number of adjacent seeds. Adding augmentation improves the results as the model over-fits less and sees more variation in the input– indicated by the increase in Dice coefficient and decrease in adjacent seeds & seed difference. Finally, adding the proposed loss keeps the Dice coefficient in the same range while significantly lowering adjacent seeds and seed difference in the prediction.

4 Conclusion

We proposed a novel end-to-end method called TP-GAN for the automatic generation of LDR-PB treatment plans. To our best knowledge, this is the first method using fully automatic feature extraction to learn implicit clinical factors which are not possible to manually determine. A novel loss function is proposed to penalize unacceptable seed placements. Comprehensive evaluation was made between our proposed model with other automatic approaches for seed plan generation using pertinent clinical measures. From the results, we suggest the use of $TP\text{-}GAN$ for rapid plan generation and the use of $TP\text{-}GAN + SA$ when further fine-tuning is required. Both of these methods can significantly save crucial time and resources for brachytherapy clinicians. Centres with less experience can utilize model trained using data from experienced centres to improve their quality of treatment. Furthermore, such a rapid planning method can be used for real-time plan generation for centres providing intra-operative LDR-PB treatment.

References

1. Aleef, T.A., Spadinger, I.T., Peacock, M.D., Salcudean, S.E., Mahdavi, S.S.: Centre-specific autonomous treatment plans for prostate brachytherapy using CGANs. Int. J. Comput. Assist. Radiol. Surg., 1–10 (2021)
2. D'Souza, W.D., Meyer, R., Thomadsen, B.R., Ferris, M.: An iterative sequential mixed-integer approach to automated prostate brachytherapy treatment plan optimization. Phys. Med. Biol. **46**(2), 297 (2001)
3. Ferrari, G., Kazareski, Y., Laca, F., Testuri, C.E.: A model for prostate brachytherapy planning with sources and needles position optimization. Oper. Res. Health Care **3**(1), 31–39 (2014)
4. Goodfellow, I., et al.: Generative adversarial nets. In: Advances in Neural Information Processing Systems, pp. 2672–2680 (2014)
5. Guthier, C., Aschenbrenner, K., Buergy, D., Ehmann, M., Wenz, F., Hesser, J.: A new optimization method using a compressed sensing inspired solver for real-time LDR-brachytherapy treatment planning. Phys. Med. Biol. **60**(6), 2179 (2015)
6. Isola, P., Zhu, J.Y., Zhou, T., Efros, A.A.: Image-to-image translation with conditional adversarial networks. In: Proceedings of the IEEE Conference on Computer Vision and Pattern Recognition, pp. 1125–1134 (2017)
7. John, S.: The seattle prostate institute approach to treatment planning for permanent implants. In: Dicker, A.P., Merrick, G., Gomella, L., Valicenti, R.K., Waterman, F. (eds.) Basic and Advanced Techniques in Prostate Brachytherapy, chap. 15, pp. 178–201. CRC Press, London (2005)

8. Karimi, D., Salcudean, S.E.: Reducing the hausdorff distance in medical image segmentation with convolutional neural networks. IEEE Trans. Med. Imaging **39**(2), 499–513 (2019)

9. Mahdavi, S.S., Peacock, M.D., Morris, W.J., Spadinger, I.T.: Automatic dual air kerma strength treatment planning for focal low-dose-rate prostate brachytherapy boost using dosimetric and geometric constraints. arXiv preprint arXiv:2010.12617 (2020)

10. Nicolae, A., et al.: Evaluation of a machine-learning algorithm for treatment planning in prostate low-dose-rate brachytherapy. Int. J. Radiat. Oncol. Biol. Phys. **97**(4), 822–829 (2017)

11. Nouranian, S., Ramezani, M., Spadinger, I., Morris, W.J., Salcudean, S.E., Abolmaesumi, P.: Automatic prostate brachytherapy preplanning using joint sparse analysis. In: Navab, N., Hornegger, J., Wells, W.M., Frangi, A.F. (eds.) MICCAI 2015. LNCS, vol. 9350, pp. 415–423. Springer, Cham (2015). https://doi.org/10.1007/978-3-319-24571-3_50

12. Pouliot, J., Tremblay, D., Roy, J., Filice, S.: Optimization of permanent 125I prostate implants using fast simulated annealing. Int. J. Radiat. Oncol. Biol. Phys. **36**(3), 711–720 (1996)

13. Stish, B.J., Davis, B.J., Mynderse, L.A., McLaren, R.H., Deufel, C.L., Choo, R.: Low dose rate prostate brachytherapy. Transl. Androl. Urol. **7**(3), 341 (2018)

14. Szegedy, C., Ioffe, S., Vanhoucke, V., Alemi, A.: Inception-v4, inception-ResNet and the impact of residual connections on learning. In: Proceedings of the AAAI Conference on Artificial Intelligence, vol. 31 (2017)

15. Yu, Y., et al.: Permanent prostate seed implant brachytherapy: report of the American association of physicists in medicine task group no. 64. Med. Phys. **26**(10), 2054–2076 (1999)

Surgical Skill and Work Flow Analysis

Trans-SVNet: Accurate Phase Recognition from Surgical Videos via Hybrid Embedding Aggregation Transformer

Xiaojie Gao[1], Yueming Jin[1], Yonghao Long[1], Qi Dou[1,2(✉)],
and Pheng-Ann Heng[1,2]

[1] Department of Computer Science and Engineering,
The Chinese University of Hong Kong, Hong Kong, China
[2] T Stone Robotics Institute, CUHK, Hong Kong, China
qdou@cse.cuhk.edu.hk

Abstract. Real-time surgical phase recognition is a fundamental task in modern operating rooms. Previous works tackle this task relying on architectures arranged in spatio-temporal order, however, the supportive benefits of intermediate spatial features are not considered. In this paper, we introduce, for the first time in surgical workflow analysis, Transformer to reconsider the ignored complementary effects of spatial and temporal features for accurate surgical phase recognition. Our hybrid embedding aggregation Transformer fuses cleverly designed spatial and temporal embeddings by allowing for active queries based on spatial information from temporal embedding sequences. More importantly, our framework processes the hybrid embeddings in parallel to achieve a high inference speed. Our method is thoroughly validated on two large surgical video datasets, i.e., Cholec80 and M2CAI16 Challenge datasets, and outperforms the state-of-the-art approaches at a processing speed of 91 fps.

Keywords: Surgical phase recognition · Transformer · Hybrid embedding aggregation · Endoscopic videos

1 Introduction

With the developments of intelligent context-aware systems (CAS), the safety and quality of modern operating rooms have significantly been improved [19]. One underlying task of CAS is surgical phase recognition, which facilitates surgery monitoring [2], surgical protocol extraction [33], and decision support [21]. However, purely vision-based recognition is quite tricky due to similar inter-class appearance and scene blur of recorded videos [13,21]. Essentially, online recognition is even more challenging because future information

Electronic supplementary material The online version of this chapter (https:// doi.org/10.1007/978-3-030-87202-1_57) contains supplementary material, which is available to authorized users.

© Springer Nature Switzerland AG 2021
M. de Bruijne et al. (Eds.): MICCAI 2021, LNCS 12904, pp. 593–603, 2021.
https://doi.org/10.1007/978-3-030-87202-1_57

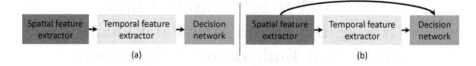

Fig. 1. (a) Previous methods extract spatio-temporal features successively for surgical phase recognition; (b) We propose to reuse extracted spatial features together with temporal features to achieve more accurate recognition.

is not allowed to assist current decision-making [30]. Moreover, processing high-dimensional video data is still time-consuming, given the real-time application requirement.

Temporal information has been verified as a vital clue for various surgical video analysis tasks, such as robotic gesture recognition [7,8], surgical instrument segmentation [12,32]. Initial methods for surgical workflow recognition, utilized statistical models, such as conditional random field [3,23] and hidden Markov models (HMMs) [5,22,26]. Nevertheless, temporal relations among surgical frames are highly complicated, and these methods show limited representation capacities with pre-defined dependencies [13]. Therefore, long short-term memory (LSTM) [11] network was combined with ResNet [10] in SV-RCNet [13] to model spatio-temporal dependences of video frames in an end-to-end fashion. Yi et al. [30] suggested an Online Hard Frame Mapper (OHFM) based on ResNet and LSTM to focus on the pre-detected rigid frames. Gao et al. [8] devised a tree search algorithm to consider future information from LSTM for surgical gesture recognition. With additional tool presence labels, multi-task learning methods are proposed to boost phase recognition performance. Twinanda [27] replaced the HMM of EndoNet [26] with LSTM to enhance its power of modeling temporal relations. MTRCNet-CL [14], the best multi-task framework, employed a correlation loss to strengthen the synergy of tool and phase predictions. To overcome limited temporal memories of LSTMs, Convolutional Neural Networks (CNN) are leveraged to extract temporal features. Funke et al. [7] used 3D CNN to learn spatial and temporal features jointly for surgical gesture recognition. Zhang et al. [31] devised a Temporal Convolutional Networks (TCN) [17,18] bridged with a self-attention module for offline surgical video analysis. Czempiel et al. [4] designed an online multi-stage TCN [6] called TeCNO to explore long-term temporal relations in pre-computed spatial features. TMRNet [15], a concurrent work, integrated multi-scale LSTM outputs via non-local operations. However, these methods process spatial and temporal features successively, as shown in Fig. 1 (a), which leads to losses of critical visual attributes.

Transformer [28] allows concurrently relating entries inside a sequence at different positions rather than in recurrent computing styles, which facilitates the preservation of essential features in overlong sequences. Therefore, it can enable the discovery of long-term clues for accurate phase recognition in surgical videos whose average duration spans minutes or hours. Moreover, thanks to its parallel computing fashion, high speed in both training and inference stages is realized.

Besides strong capacity in sequence learning, Transformer also demonstrates outstanding ability in visual feature representation [9,16]. Recently, Transformer was employed to fuse multi-view elements in point clouds and illustrated excellent outcomes [29], which implies its potential to promote the synergy of spatial and temporal features in surgical videos.

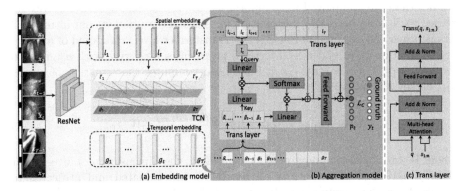

Fig. 2. Overview of our proposed Trans-SVNet for surgical phase recognition. (a) Extracted spatial embeddings enable the generation of temporal embeddings, (b) and are fused with the temporal information for refined phase predictions in our aggregation model, with the architecture of (c) Transformer layer presented in detail. (Color figure online)

In this paper, we propose a novel method, named Trans-SVNet, for accurate phase recognition from surgical videos via Hybrid Embedding Aggregation **Trans**former. As shown in Fig. 1 (b), we reconsider the spatial features as one of our hybrid embeddings to supply missing appearance details during temporal feature extracting. Specifically, we employ ResNet and TCN to generate spatial and temporal embeddings, respectively, where representations with the same semantic labels cluster in the embedding space. Then, we introduce Transformer, for the first time, to aggregate the hybrid embeddings for accurate surgical phase recognition by using spatial embeddings to attend supporting information from temporal embedding sequences. More importantly, our framework is parameter-efficient and shows extraordinary potential for real-time applications. We extensively evaluate our Trans-SVNet on two large public[1] surgical video datasets. Our approach outperforms all the compared methods and achieves a real-time processing speed of 91 fps.

2   Method

Figure 2 presents an overview of our proposed Trans-SVNet, composed of embedding and aggregation models. Our embedding model first represents surgical

---

[1] http://camma.u-strasbg.fr/datasets.

video frames with spatial embeddings l and temporal embeddings g. The aggregation model fuses the hybrid embeddings by querying l from g to explore their synergy for accurate phase recognition.

2.1   Transformer Layer

Rather than only employed for temporal feature extraction, spatial features are reused to discover necessary information for phase recognition via our introduced Transformer. As depicted in Fig. 2 (c), a Transformer layer, composed of a multi-head attention layer and a feed-forward layer, fuses a query q with a temporal sequence $s_{1:n} = [s_1, \ldots, s_{n-1}, s_n]$. Each head computes the attention of q with $s_{1:n}$ as key and value:

$$\text{Attn}(q, s_{1:n}) = \text{softmax}(\frac{W_q q (W_k s_{1:n})^{\text{T}}}{\sqrt{d_k}}) W_v s_{1:n}, \tag{1}$$

where W are linear mapping matrices and d_k is the dimension of q after linear transformation. The outputs of all heads are concatenated and projected to enable the residual connection [10] with q followed by a layer normalization [1]. Since each attention head owns different learnable parameters, they concentrate on respective features of interest and jointly represent crucial features. We find it necessary to utilize multiple heads rather than a single head to produce a much faster convergence speed. The feed-forward layer is made up of two fully connected layers connected with a ReLU activation. The residual connection and layer normalization are applied in a similar way as the multi-head attention layer. Finally, the output of the Transformer layer is denoted as $\text{Trans}(q, s_{1:n})$, which contains synthesized information of q and $s_{1:n}$.

2.2   Video Embedding Extraction

Given the discrete and sequential nature of video frames, we suggest two kinds of embeddings to represent their spatial and temporal information, which extends the spirit of word embeddings [20] to surgical video analysis. Let $x_t \in \mathbb{R}^{H \times W \times C}$ and $y_t \in \mathbb{R}^N$ denote the t-th frame of a surgical video with T frames in total and the corresponding one-hot phase label, respectively. We first employ a very deep ResNet50 [10] to extract discriminative spatial embeddings, which is realized by training a frame-wise classifier using the cross-entropy loss. Note that we only utilize phase labels because additional annotations like tool presence labels are not widely available, and single-task methods are more practical in real-world applications. Then, outputs of the average pooling layer of ResNet50 are made as our spatial embeddings, i.e., $l_t \in \mathbb{R}^{2048}$, and high-dimensional video data are converted into low-dimensional embeddings.

To save memory and time, temporal embeddings are directly extracted from the spatial embeddings generated by the trained and fixed ResNet50. We first adjust the dimension of l_t with a 1×1 convolutional layer and generate $l'_t \in \mathbb{R}^{32}$. Then, we exploit TCN to process the embedding sequence of a whole video without touching future information as illustrated in Fig. 2 (a). For easy comparison, we employ TeCNO [4], a two-stage TCN model, to generate temporal embeddings using $l'_{1:T}$. Owing to multi-layer convolutions and dilated kernels, its temporal receptive field is increased to several minutes. Since l_t is not updated, spatial embeddings of a whole video could be processed in a single forward computation, and the network converges quickly. Moreover, the outputs of the last stage of the TeCNO are used as our temporal embedding $g_t \in \mathbb{R}^N$.

2.3 Hybrid Embedding Aggregation

Our aggregation model, consisting of two Transformer layers, aims to output the refined prediction p_t of frame x_t by fusing the pre-computed hybrid video embeddings only available at time step t. The intuition is that a fixed-size representation encoded with spatio-temporal details is insufficient to express all critical features in both spatial and temporal dimensions, thus information loss is inevitably caused. Hence, we propose to look for supportive information based on a spatial embedding l_t from an n-length temporal embedding sequence $g_{t-n+1:t}$ (see Sect. 3 for ablation study), which allows for the rediscovery of missing yet crucial details during temporal feature extraction. In other words, our aggregation model learns a function $\mathbb{R}^{2048} \times \mathbb{R}^{n \times N} \to \mathbb{R}^N$.

Before synthesizing the two kinds of embeddings, they first conduct internal aggregation, respectively. On the one hand, dimension reduction is executed for the temporal embedding l_t to generate $\tilde{l}_t \in \mathbb{R}^N$ by

$$\tilde{l}_t = \tanh(W_l l_t), \tag{2}$$

where $W_l \in \mathbb{R}^{N \times 2048}$ is a parameter matrix. On the other hand, the temporal embedding sequence $g_{t-n+1:t}$ is processed by one of our Transformer layer to capture self-attention and an intermediate sequence $\tilde{g}_{t-n+1:t} \in \mathbb{R}^{n \times N}$ is produced.[2] Specifically, each entry in $[g_{t-n+1}, \ldots, g_{t-1}, g_t]$ attends all entries of the sequence, which is denoted as

$$\tilde{g}_i = \mathrm{Trans}(g_i, g_{t-n+1:t}), \quad i = t-n+1, \ldots, t. \tag{3}$$

Given self-aggregated embeddings $\tilde{l}$ and $\tilde{g}$, we employ the other Transformer layer to enable $\tilde{l}_t$ to query pivotal information from $\tilde{g}_{t-n+1:t}$ as key and value while fuse with the purified temporal features through residual additions (red arrow in Fig. 2 (b)). Next, the output of the second Transformer layer is activated with the Softmax function to predict phase probability:

$$p_t = \mathrm{Softmax}(\mathrm{Trans}(\tilde{l}_t, \tilde{g}_{t-n+1:t})). \tag{4}$$

Although the fused embeddings have a dimension of N, they still contain rich information for further processing. Lastly, our aggregation model is trained using the cross-entropy loss:

$$\mathcal{L}_C = -\sum_{t=1}^{T} y_t \log(p_t). \tag{5}$$

---

[2] Zero padding is applied if necessary.

3   Experiments

Datasets. We extensively evaluate our Trans-SVNet on two challenging surgical video datasets of cholecystectomy procedures recorded at 25 fps, i.e., Cholec80 [26] and M2CAI16 Challenge dataset [25]. Cholec80 includes 80 laparoscopic videos with 7 defined phases annotated by experienced surgeons. Its frame resolution is either 1920×1080 or 854×480. This dataset also provides tool presence labels to allow for multi-task learning. We follow the same evaluation procedure of previous works [13,26,30] by separating the dataset into the first 40 videos for training and the rest for testing. The M2CAI16 dataset consists of 41 videos that are segmented into 8 phases by expert physicians. Each frame has a resolution of 1920×1080. It is divided into 27 videos for training and 14 videos for testing, following the split of [13,24,30]. All videos are subsampled to 1 fps following previous works [13,26], and frames are resized into 250×250.

Table 1. Phase recognition results (%) of different methods on the Cholec80 and M2CAI16 datasets. The best results are marked in bold. Note that the * denotes methods based on multi-task learning that requires extra tool labels.

Method	Cholec80				M2CAI16				#param
	Accuracy	Precision	Recall	Jaccard	Accuracy	Precision	Recall	Jaccard	
EndoNet* [26]	81.7 ± 4.2	73.7 ± 16.1	79.6 ± 7.9	—	—	—	—	—	58.3M
EndoNet+LSTM* [27]	88.6 ± 9.6	84.4 ± 7.9	84.7 ± 7.9	—	—	—	—	—	68.8M
MTRCNet-CL* [14]	89.2 ± 7.6	86.9 ± 4.3	88.0 ± 6.9	—	—	—	—	—	29.0M
PhaseNet [24,26]	78.8 ± 4.7	71.3 ± 15.6	76.6 ± 16.6	—	79.5 ± 12.1	—	—	64.1 ± 10.3	58.3M
SV-RCNet [13]	85.3 ± 7.3	80.7 ± 7.0	83.5 ± 7.5	—	81.7 ± 8.1	81.0 ± 8.3	81.6 ± 7.2	65.4 ± 8.9	28.8M
OHFM [30]	87.3 ± 5.7	—	—	67.0 ± 13.3	85.2 ± 7.5	—	—	68.8 ± 10.5	47.1M
TeCNO [4]	88.6 ± 7.8	86.5 ± 7.0	87.6 ± 6.7	75.1 ± 6.9	86.1 ± 10.0	85.7 ± 7.7	$\mathbf{88.9 \pm 4.5}$	74.4 ± 7.2	24.7M
Trans-SVNet (ours)	$\mathbf{90.3 \pm 7.1}$	$\mathbf{90.7 \pm 5.0}$	$\mathbf{88.8 \pm 7.4}$	$\mathbf{79.3 \pm 6.6}$	$\mathbf{87.2 \pm 9.3}$	$\mathbf{88.0 \pm 6.7}$	87.5 ± 5.5	$\mathbf{74.7 \pm 7.7}$	24.7M

Evaluation Metrics. We employ four frequently-used metrics in surgical phase recognition for comprehensive comparisons. These measurements are accuracy (AC), precision (PR), recall (RE), and Jaccard index (JA), which are also utilized in [13,30]. The AC is calculated at the video level, defined as the percentage of frames correctly recognized in the entire video. Since the video classes are imbalanced, the PR, RE, and JA are first computed towards each phase and then averaged over all the phases. We also count the number of parameters to indicate the training and inference speed to a certain degree.

Implementation Details. Our embedding and aggregation models are trained one after the other on PyTorch using an NVIDIA GeForce RTX 2080 Ti GPU. We initialize the parameters of the ResNet from a pre-trained model on the ImageNet [10]. It employs an SGD optimizer with a momentum of 0.9 and a learning rate of 5e-4 except for its fully connected layers with 5e-5. Its batch size is set to 100, and data augmentation is applied, including 224×224 cropping, random mirroring, and color jittering. We re-implement TeCNO [4] based on their released code with only phase labels and directly make outputs of its second stage as our temporal embeddings. We report the re-implemented results of

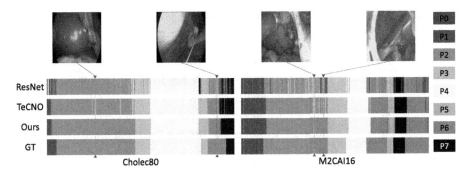

Fig. 3. Color-coded ribbon illustration for two complete surgical videos. The time axes are scaled for better visualization.

TeCNO, and this well-trained model directly generates our temporal embeddings without further tuning. Our aggregation model is trained by Adam optimizer with a learning rate of 1e-3 and utilizes a batch size identical to the length of each video. The number of attention heads is empirically set to 8, and the temporal sequence length n is 30. N is set to the dimension of the one-hot phase label. Our code is released at: https://github.com/xjgaocs/Trans-SVNet.

Comparison with State-of-the-Arts. Table 1 presents comparisons of our Trans-SVNet with seven existing methods without a post-processing strategy. Using extra tool presence annotations of the Cholec80 dataset, multi-task learning methods [14, 26, 27] generally achieve high performances, and MTRCNet-CL beats all single-task models except ours. As for methods using only phase labels, PhaseNet is far behind all other models due to its shallower network. Thus the much deeper ResNet50 becomes a standard visual feature extractor since SV-RCNet [13]. As a multi-step learning framework like OHFM, our approach gains a significant improvement by 6%-12% in JA with a much simpler training procedure. Compared to the state-of-the-art TeCNO with the same backbones, our Trans-SVNet gains a boost by 4% in PR and JA on the larger Cholec80 dataset with a negligible increase in parameters (~30k). In a word, our Trans-SVNet outperforms all the seven compared methods, especially on the enormous Cholec80 dataset. Our method is observed to achieve a more remarkable improvement on the Cholec80 dataset than the M2CAI16 dataset. The underlying reason is that the M2CAI16 dataset is smaller and contains less challenging videos. The robustness of our method yields a better advantage on the more complicated Cholec80 dataset. Thanks to the designed low-dimensional video embeddings, our model generates predictions at 91 fps with one GPU, which vastly exceeds the video recording speed.

Qualitative Comparison. In Fig. 3, we show the color-coded ribbon of two complete laparoscopic videos from the two datasets. Due to the lack of temporal relations, ResNet suffers from noisy patterns and generates frequently jumped

Table 2. Ablative testing results (%) for increasing length of our temporal embedding sequence on the Cholec80 dataset.

Length (n)	Accuracy	Precision	Recall	Jaccard
0	82.1 ± 7.8	78.0 ± 6.4	78.5 ± 10.8	61.7 ± 11.3
10	89.9 ± 7.2	89.6 ± 5.2	88.4 ± 7.9	78.4 ± 6.6
20	90.2 ± 7.1	90.2 ± 5.1	$\mathbf{88.8 \pm 7.7}$	79.1 ± 6.6
30	$\mathbf{90.3 \pm 7.1}$	90.7 ± 5.0	$\mathbf{88.8 \pm 7.4}$	$\mathbf{79.3 \pm 6.6}$
40	$\mathbf{90.3 \pm 7.0}$	$\mathbf{90.8 \pm 4.9}$	88.5 ± 7.2	79.0 ± 6.8

Table 3. Phase recognition results (%) of different architectures and their P-values in JA towards our proposed method on the Cholec80 dataset.

Architecture			Accuracy	Precision	Recall	Jaccard	P-values
PureNet	ResNet		82.1 ± 7.8	78.0 ± 6.4	78.5 ± 10.8	61.7 ± 11.3	2e-8
	TeCNO		88.6 ± 7.8	86.5 ± 7.0	87.6 ± 6.7	75.1 ± 6.9	2e-7
	ResNet *cat* TeCNO		87.9 ± 7.5	86.6 ± 5.9	85.3 ± 8.2	73.0 ± 7.8	2e-8
Transformer	Query	Key	Accuracy	Precision	Recall	Jaccard	P-values
	l_t	$l_{t-n+1:t}$	81.9 ± 9.2	78.0 ± 12.5	78.3 ± 12.8	60.8 ± 12.4	2e-8
	g_t	$g_{t-n+1:t}$	89.1 ± 7.8	87.6 ± 6.3	87.7 ± 6.9	76.2 ± 6.6	4e-7
	g_t	$l_{t-n+1:t}$	89.2 ± 7.5	87.7 ± 6.7	87.7 ± 7.0	76.1 ± 7.0	3e-7
	l_t	$g_{t-n+1:t}$	$\mathbf{90.3 \pm 7.1}$	$\mathbf{90.7 \pm 5.0}$	$\mathbf{88.8 \pm 7.4}$	$\mathbf{79.3 \pm 6.6}$	—

predictions. TeCNO achieves smoother results by relating long-term temporal information in spatial embeddings generated by ResNet. However, its predictions for P2 in both videos still need to be improved. We also visualize some of the misclassified frames of TeCNO and find they are negatively influenced by excessive reflection, where bright but trivial parts might dominate the extracted spatial features, making it easy to miss pivotal information. Aggregating embeddings from ResNet and TeCNO elegantly, our Trans-SVNet contributes to more consistent and robust predictions of surgical phases, which highlights its promotion towards the synergy between the hybrid embeddings.

Ablation Study. We first analyze the effect of different length n of our temporal embedding sequence on the Cholec80 dataset, and the results are reported in Table 2. It is observed that our design of temporal sequence is undoubtedly necessary to gain a notable boost relative to not using temporal embeddings, i.e., $n = 0$. We also notice that gradually increasing the temporal sequence length produces improvements towards all metrics, and our approach behaves almost equally well with the length $n \in [20, 40]$. The boost tends to be slower because adding n by one increases the temporal sequence span by one second (only for $n > 0$) and over-long sequences bring too much noise. Therefore, we choose $n = 30$ as the length of our temporal embedding sequence.

Table 3 lists the results of different network structures, i.e., letting embeddings from ResNet and TeCNO be query or key in every possible combination, to identify which one makes the best use of information. We first show baseline methods without Transformer denoted as PureNet. ResNet *cat* TeCNO employs a superficial linear layer to process concatenated l_t and g_t, whose performance unsurprisingly falls between ResNet and TeCNO. As for Transformer-based networks, there are no advancements to use spatial embedding l_t to query $l_{t-n+1:t}$. The reason is that spatial embeddings cannot indicate their orders in videos and bring ambiguity in the aggregation stage. Better performances are achieved than PureNet by letting TeCNO embeddings g with sequential information be either query or key, which justifies Transformer rediscovers necessary details neglected by temporal extractors. Our Trans-SVNet uses l_t to query $g_{t-n+1:t}$ and generates the best outcomes with a clear margin, which confirms the effectiveness of our proposed architecture. We also calculate P-values in JA using Wilcoxon signed-rank test for compared settings towards our Trans-SVNet. It is found that P-values are substantially less than 0.05 in all cases, which indicates that our model learns a formerly non-existent but effective policy.

4   Conclusion

We propose a novel framework to fuse different embeddings based on Transformer for accurate real-time surgical phase recognition. Our novel aggregation style allows the retrieval of missing but critical information with rarely additional cost. Extensive experimental results demonstrate that our method consistently outperforms the state-of-the-art models while maintains a breakneck processing speed. The excellent performance and parameter efficiency of our method justify its promising applications in real operating rooms.

Acknowledgements. This work was supported by Hong Kong RGC TRS Project T42-409/18-R, National Natural Science Foundation of China with Project No. U1813204, and Shenzhen-HK Collaborative Development Zone.

References

1. Ba, J.L., Kiros, J.R., Hinton, G.E.: Layer normalization. arXiv preprint arXiv:1607.06450 (2016)
2. Bricon-Souf, N., Newman, C.R.: Context awareness in health care: a review. Int. J. Med. Informatics **76**(1), 2–12 (2007)
3. Charrière, K., et al.: Real-time analysis of cataract surgery videos using statistical models. Multimedia Tools Appl., 1–19 (2017). https://doi.org/10.1007/s11042-017-4793-8
4. Czempiel, T., et al.: TeCNO: surgical phase recognition with multi-stage temporal convolutional networks. In: Martel, A.L., et al. (eds.) MICCAI 2020. LNCS, vol. 12263, pp. 343–352. Springer, Cham (2020). https://doi.org/10.1007/978-3-030-59716-0_33

5. Dergachyova, O., Bouget, D., Huaulmé, A., Morandi, X., Jannin, P.: Automatic data-driven real-time segmentation and recognition of surgical workflow. Int. J. Comput. Assist. Radiol. Surg. **11**(6), 1081–1089 (2016). https://doi.org/10.1007/s11548-016-1371-x

6. Farha, Y.A., Gall, J.: MS-TCN: multi-stage temporal convolutional network for action segmentation. In: Proceedings of the IEEE/CVF Conference on Computer Vision and Pattern Recognition, pp. 3575–3584 (2019)

7. Funke, I., et al.: Using 3D convolutional neural networks to learn spatiotemporal features for automatic surgical gesture recognition in video. In: Shen, D., et al. (eds.) MICCAI 2019. LNCS, vol. 11768, pp. 467–475. Springer, Cham (2019). https://doi.org/10.1007/978-3-030-32254-0_52

8. Gao, X., Jin, Y., Dou, Q., Heng, P.A.: Automatic gesture recognition in robot-assisted surgery with reinforcement learning and tree search. In: IEEE International Conference on Robotics and Automation, pp. 8440–8446. IEEE (2020)

9. Han, K., et al.: A survey on visual transformer. arXiv preprint arXiv:2012.12556 (2020)

10. He, K., Zhang, X., Ren, S., Sun, J.: Deep residual learning for image recognition. In: Proceedings of the IEEE Conference on Computer Vision and Pattern Recognition, pp. 770–778 (2016)

11. Hochreiter, S., Schmidhuber, J.: Long short-term memory. Neural Comput. **9**(8), 1735–1780 (1997)

12. Jin, Y., Cheng, K., Dou, Q., Heng, P.-A.: Incorporating temporal prior from motion flow for instrument segmentation in minimally invasive surgery video. In: Shen, D., et al. (eds.) MICCAI 2019. LNCS, vol. 11768, pp. 440–448. Springer, Cham (2019). https://doi.org/10.1007/978-3-030-32254-0_49

13. Jin, Y., et al.: SV-RCNet: workflow recognition from surgical videos using recurrent convolutional network. IEEE Trans. Med. Imaging **37**(5), 1114–1126 (2018)

14. Jin, Y., et al.: Multi-task recurrent convolutional network with correlation loss for surgical video analysis. Med. Image Anal. **59**, 101572 (2020)

15. Jin, Y., Long, Y., Chen, C., Zhao, Z., Dou, Q., Heng, P.A.: Temporal memory relation network for workflow recognition from surgical video. IEEE Trans. Med. Imaging (2021)

16. Khan, S., Naseer, M., Hayat, M., Zamir, S.W., Khan, F.S., Shah, M.: Transformers in vision: a survey. arXiv preprint arXiv:2101.01169 (2021)

17. Lea, C., Flynn, M.D., Vidal, R., Reiter, A., Hager, G.D.: Temporal convolutional networks for action segmentation and detection. In: Proceedings of the IEEE Conference on Computer Vision and Pattern Recognition, pp. 156–165 (2017)

18. Lea, C., Vidal, R., Reiter, A., Hager, G.D.: Temporal convolutional networks: a unified approach to action segmentation. In: Hua, G., Jégou, H. (eds.) ECCV 2016. LNCS, vol. 9915, pp. 47–54. Springer, Cham (2016). https://doi.org/10.1007/978-3-319-49409-8_7

19. Maier-Hein, L., et al.: Surgical data science for next-generation interventions. Nat. Biomed. Eng. (2017)

20. Mikolov, T., Sutskever, I., Chen, K., Corrado, G., Dean, J.: Distributed representations of words and phrases and their compositionality. In: Advances in Neural Information Processing Systems (2013)

21. Padoy, N.: Machine and deep learning for workflow recognition during surgery. Minimally Invasive Therapy Allied Technol. **28**(2), 82–90 (2019)

22. Padoy, N., Blum, T., Feussner, H., Berger, M.O., Navab, N.: On-line recognition of surgical activity for monitoring in the operating room. In: Proceedings of the AAAI Conference on Artificial Intelligence, pp. 1718–1724 (2008)

23. Quellec, G., Lamard, M., Cochener, B., Cazuguel, G.: Real-time segmentation and recognition of surgical tasks in cataract surgery videos. IEEE Trans. Med. Imaging **33**(12), 2352–2360 (2014)
24. Twinanda, A.P., Mutter, D., Marescaux, J., de Mathelin, M., Padoy, N.: Single- and multi-task architectures for surgical workflow challenge at M2CAI 2016. arXiv preprint arXiv:1610.08844 (2016)
25. Twinanda, A.P., Shehata, S., Mutter, D., Marescaux, J., De Mathelin, M., Padoy, N.: MICCAI modeling and monitoring of computer assisted interventions challenge. http://camma.u-strasbg.fr/m2cai2016/
26. Twinanda, A.P., Shehata, S., Mutter, D., Marescaux, J., De Mathelin, M., Padoy, N.: EndoNet: a deep architecture for recognition tasks on laparoscopic videos. IEEE Trans. Med. Imaging **36**(1), 86–97 (2017)
27. Twinanda, A.P.: Vision-based approaches for surgical activity recognition using laparoscopic and RBGD videos. Ph.D. thesis, Strasbourg (2017)
28. Vaswani, A., et al.: Attention is all you need. In: Advances in Neural Information Processing Systems, pp. 5998–6008 (2017)
29. Wang, Y., Solomon, J.M.: Deep closest point: learning representations for point cloud registration. In: Proceedings of the IEEE/CVF International Conference on Computer Vision, pp. 3523–3532 (2019)
30. Yi, F., Jiang, T.: Hard frame detection and online mapping for surgical phase recognition. In: Shen, D., et al. (eds.) MICCAI 2019. LNCS, vol. 11768, pp. 449–457. Springer, Cham (2019). https://doi.org/10.1007/978-3-030-32254-0_50
31. Zhang, J., et al.: Symmetric dilated convolution for surgical gesture recognition. In: Martel, A.L., et al. (eds.) MICCAI 2020. LNCS, vol. 12263, pp. 409–418. Springer, Cham (2020). https://doi.org/10.1007/978-3-030-59716-0_39
32. Zhao, Z., Jin, Y., Gao, X., Dou, Q., Heng, P.-A.: Learning motion flows for semi-supervised instrument segmentation from robotic surgical video. In: Martel, A.L., et al. (eds.) MICCAI 2020. LNCS, vol. 12263, pp. 679–689. Springer, Cham (2020). https://doi.org/10.1007/978-3-030-59716-0_65
33. Zisimopoulos, O., et al.: DeepPhase: surgical phase recognition in CATARACTS videos. In: Frangi, A.F., Schnabel, J.A., Davatzikos, C., Alberola-López, C., Fichtinger, G. (eds.) MICCAI 2018. LNCS, vol. 11073, pp. 265–272. Springer, Cham (2018). https://doi.org/10.1007/978-3-030-00937-3_31

OperA: Attention-Regularized Transformers for Surgical Phase Recognition

Tobias Czempiel[1(✉)], Magdalini Paschali[1], Daniel Ostler[2], Seong Tae Kim[3], Benjamin Busam[1], and Nassir Navab[1,4]

[1] Computer Aided Medical Procedures, Technische Universität München, Munich, Germany
Tobias.czempiel@tum.de
[2] MITI, Klinikum Rechts der Isar, Technische Universität München, Munich, Germany
[3] Department of Computer Science and Engineering, Kyung Hee University, Yongin-si, South Korea
[4] Computer Aided Medical Procedures, Johns Hopkins University, Baltimore, USA

Abstract. In this paper we introduce OperA, a transformer-based model that accurately predicts surgical phases from long video sequences. A novel attention regularization loss encourages the model to focus on high-quality frames during training. Moreover, the attention weights are utilized to identify characteristic high attention frames for each surgical phase, which could further be used for surgery summarization. OperA is thoroughly evaluated on two datasets of laparoscopic cholecystectomy videos, outperforming various state-of-the-art temporal refinement approaches.

Keywords: Surgical workflow analysis · Surgical phase recognition · Transformers · Self-attention · Cholecystectomy

1 Introduction

Surgical workflow analysis is a crucial task for the operating room (OR) of the future [1]. Specifically, automatic detection of surgical phases is one of its most essential components. An efficient surgical phase recognition system will build the foundation for automated surgical assistance and cognitive guidance [2,3]. Online analysis during an ongoing intervention can provide feedback to surgeons and alarm the staff in case of erroneous or adverse events [4]. Additionally, extracting surgical phases during an operation and group different procedures based on their unique characteristics plays an important role for modern surgical training. Since automatic extraction of surgical phases is particularly challenging, advanced Machine Learning (ML) methodologies [5] have been employed towards solving it.

Electronic supplementary material The online version of this chapter (https://doi.org/10.1007/978-3-030-87202-1_58) contains supplementary material, which is available to authorized users.

© Springer Nature Switzerland AG 2021
M. de Bruijne et al. (Eds.): MICCAI 2021, LNCS 12904, pp. 604–614, 2021.
https://doi.org/10.1007/978-3-030-87202-1_58

However, factors such as variability of patient anatomy, surgeon style [6] as well as limited availability and quality of training data present problems for modern ML algorithms.

A recent development in ML that could help overcome these challenges in surgical workflow analysis is transformer networks [7]. Transformers have shown their vast potential for sequential modeling in Natural Language Processing (NLP) [8] and have quickly become the gold standard in this area. Transformer networks have the capability to create temporal relationships between current and previous frames using self-attention, much like frequently used LSTM methods [9]. However, self-attention enables learning in long sequences without forgetting of previous information which often hampers LSTM-based methods.

An additional advantage of transformer networks and self-attention over other approaches used for surgical phase recognition is their ability to visualize the attention weights for a sequence, which could yield further insights into the decision-making process of a model.

1.1 Related Work

Automatic extraction of surgical phases was initially performed using binary surgical signals [10], where a comparison with an average surgery determined the surgical phase. Hidden Markov Models (HMM), provided an extension of this idea capable of online predictions [11]. In EndoNet, Twinanda et al. [5] utilized image features extracted with a Convolutional Neural Network (CNN) to predict the surgical phase and surgical tool presence directly from surgical images. EndoLSTM [12] additionally performed temporal refinement with LSTMs [9], which improved the results substantially. A variety of works combined pre-trained CNNs as feature extractors, followed by temporal refinement with LSTMs [13,14]. In MTRCNet-CL [15], Jin et al. proposed a CNN/LSTM model to refine the prediction over short sequences in an end-to-end fashion including a correlation loss to identify phase and tool correlations in an explicit manner. Czempiel et al. [16] proposed TeCNO, which combined Temporal Convolutional Networks (TCN) with a ResNet-50 [17] feature extractor. Transformer models were first introduced for NLP [7] where they quickly became the state-of-the-art in a plethora of downstream tasks [8,18]. Furthermore, the versatility of transformers has been showcased not only for vision tasks such as image classification [19] and text-to-image generation [20] but also in biology for the challenging protein folding problem with *AlphaFold* [21]. In surgical data sciences, transformers have been explored only for surgical tool [22] classification.

An additional aspect of transformers and self-attention is the fact that their attention weights could be used for model insights and explanation. Some works have claimed that attention has limited explanation capabilities [23]. However, this assumption has been challenged [24] suggesting that each work should define their notion of explanation since it could be dependent on the task at hand.

In this paper, we introduce for the first time, OperA, a transformer-based method for online surgical phase prediction for laparoscopic operations. Our contributions are:

- We successfully leverage a transformer-based model for surgical phase recognition that outperforms other temporal refinement methods.
- We propose a novel attention regularizer, that improves the automatic extraction of the most relevant frames with high feature-quality.
- We utilize the attention weights to extract and visualize characteristic frames.
- We carefully evaluate OperA on two challenging surgical video datasets.

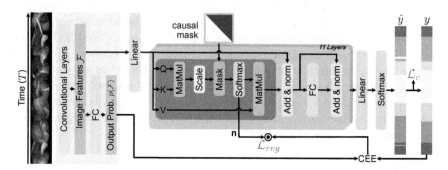

Fig. 1. Overview of the proposed OperA model. Image features $\mathcal{F}$ are used as input for the transformer. The output logits $p(\mathcal{F})$ of the feature extraction backbone are used in combination with the normalized frame-wise attention weights **n** to regularize the attention.

2 Methodology

Our proposed model, OperA, consists of a CNN for visual feature extraction followed by multiple self-attention layers. The attention map is regularized during training to focus on reliable CNN image features. The full network architecture is visualized in Fig. 1.

For our feature extraction backbone we trained a ResNet-50 [17] frame-wise CNN without sequential modeling. We trained this model on phase recognition and additionally on surgical tool detection, if tool information was available in the dataset. The result of the feature extraction backbone are per frame image features $\mathcal{F} \in \mathbb{R}^{2048}$ and their corresponding class probabilities $p(\mathcal{F}) \in [0,1]^c$ with c the number of classes.

2.1 Sequential Transformer Network

Our model expands on the well-known Transformer architecture [7] with the addition of our attention regularization that will be discussed below. Transformers have the capabilities to model long sequences in a parallel manner using self-attention by relating every input feature with other input features regardless of their distance in the sequence [25]. Visualized in Fig. 1, we first calculate the query Q, key K and value V, the inputs for the scaled dot product attention using a linear layer such that $(Q, K, V) = \text{Linear}(\mathcal{F}) \in \mathbb{R}^{3d}$ with $d = 64$.

$$\text{AttentionWeights}(Q, K) = \text{softmax}\left(\text{mask}\left(\frac{QK^T}{\sqrt{d}}\right)\right) \qquad (1)$$

$$\text{Attention}(Q, K, V) = \text{AttentionWeights}(Q, K)V \tag{2}$$

Our architecture uses 11 consecutive layers each consisting of a linear layer, a scaled dot-product attention layer, a layer normalization [26] and residual connections [17]. Similar to the architecture of the Vision Tranformer [19] after the last encoding layer, a linear layer followed by a softmax is used to estimate the class-wise output probabilities y for each frame of the sequence. For the training of the model we use a median frequency balanced cross-entropy loss [27] $\mathcal{L}_c$. Causal masking [28] with the binary mask $M \ni \{0, 1\}$ is performed on the attention map of the model, to prevent information leakage from future frames to the current frame prediction. This allows us to use OperA for real-time surgical phase prediction.

2.2   Normalized Frame-Wise Attention

For each frame, each layer of our transformer generates one column in $A = \text{AttentionWeights}(Q, K)$. Due to the softmax activation function (Eq. 1) each row of the attention map A sums up to 1. Column-wise summation of the attention weights in A results in the total attention value for each frame at time t. Due to the causal mask visualized in Fig. 1, the first frame has the opportunity to contribute T times, where T is the number of frames in a video, while the last frame is considered only once. We therefore need to normalize the frame-wise attention by dividing the total attention of each frame with the number of times this frame is considered, thus equally weighting the attention of all frames regardless of their position in the video. The normalized frame-wise attention is calculated by: $\mathbf{n} = (n_1, \ldots, n_T)$ with $n_j = \dfrac{\sum_i A_{ij}}{\sum_i M_{ij}}$, $M_{ij} = 0$ if $j > i$ else 1.

A_{ij} quantifies how much attention is being paid by frame i (query) to frame j (key). M_{ij} will be zero if the key index j is larger than the query index i as we want to restrict our model to only respect previous events in the video.

2.3   Attention Regularization

Different from NLP and Visual Transformers, the input of OperA is generated by a CNN backbone network. The quality of each frame embedding of the CNN can vary drastically, especially for frames where the CNN predictions are incorrect. To this end, OperA should focus on higher quality CNN features, that were correctly classified by the backbone CNN. Such features have higher softmax probabilities, or confidence, and lower cross-entropy values.

We learn this relationship by comparing the normalized frame-wise attention weights with the prediction error of our CNN. The regularization then reads:

$$\mathcal{L}_{reg} = \langle \mathbf{n}, \text{CEE}\left(p(\mathcal{F}), y\right)\rangle \tag{3}$$

The Cross Entropy Evaluation value (CEE) describes the residual error of $p(\mathcal{F})$ compared to the ground truth label y. It should be noted that the weights of the backbone CNN remain frozen and CEE is only used for the optimization of the attention weights. Multiplying CEE with the normalized frame-wise attention $\mathbf{n}$ explicitly penalizes the model if a high attention value was generated for a feature with low CNN confidence. We apply the proposed regularization to the first attention layer as it has a direct relationship with the input visual features. The final loss function used for model training is denoted as: $\mathcal{L} = \mathcal{L}_c + \lambda \cdot \mathcal{L}_{reg}$

Summing up all the normalized attention weights for each layer we generate a final attention value for each frame. In order to interpret whether OperA focuses on highly-informative frames that correctly represent each phase, we extract the frames with Highest Attention (HA) and the ones with Lowest Attention (LA).

3 Experimental Setup

Datasets. For the evaluation of OperA we use two challenging surgical workflow intra-operative video datasets of laparoscopic cholecystectomy procedures. The publicly available Cholec80 [5] includes 80 videos with a resolutions of 1920 × 1080 or 854 × 480 pixels recorded at 25 frames-per-second (fps). For this work, the dataset was sub-sampled to 1fps. Every frame in the surgical video has been manually assigned to one out of seven surgical phases. Additionally, seven different tool annotation labels sampled at 1 fps are provided.

Contrary to previous Cholec80 splits that used 32 videos for training, 8 for validation and 40 for testing, we increased the number of videos in the validation set from 8 to 12, as we believed that 8 videos is not a big enough sample size to find the models that generalize best to the test set. For the same reason we also increased the number of training videos from 32 to 48. Among those 48 training and 12 validation videos we performed 5-fold cross validation. Following all previous works [12, 14–16] we kept an unseen test-set of 20 videos that was not part of the cross-validation. In the test phase we used the best model of each cross-validation split and averaged the results of those 5 models on the unseen test set. All the results reported, including the baselines, were created using the same split and cross-validation approach.

CSW (Cholecystectomy Surgical Workflow) is an in-house dataset of 85 laparoscopic cholecystectomy videos with resolution 1920 × 1080 pixels and sampling rate of 1 fps. CSW includes the 7 surgical phases of Cholec80, shown in Fig. 3, along with one additional phase Pre-preparation, used to describe frames before the Preparation phase. The phases have been annotated by expert physicians with no additional tool-presence information. 20 videos are utilized for testing and the remaining 65 videos for training (52) and validation (13). For all the experiments 5-fold cross validation is performed. To balance our combined loss function we set λ to 1.

Model Training. OperA was trained for the task of surgical phase recognition using the Adam optimizer with an initial learning rate of 1e−5 for 30 epochs. We report the test results extracted by the model that performed best on the validation set for each fold. The batch size is identical to the length of each video. Our method was implemented in PyTorch and our models were trained on an

Table 1. Ablative testing results for 6 and 11 transformer layers and with the addition of Attention Regularization (Reg). Average metrics over 5 folds are reported (%) with the corresponding standard deviation ($\pm$).

Layers	Reg	Accuracy	Precision	Recall
		Cholec80		
6	–	90.35 ± 0.71	80.64 ± 1.41	86.48 ± 0.61
6	✓	90.49 ± 0.70	81.38 ± 0.29	**86.98 ± 0.61**
11	-	90.37 ± 0.86	81.60 ± 0.40	86.23 ± 0.34
11	✓	**91.26 ± 0.64**	**82.19 ± 0.70**	86.92 ± 0.86
		CSW		
6	-	84.88 ± 1.43	82.76 ± 1.43	87.20 ± 1.02
6	✓	85.41 ± 0.95	83.00 ± 1.34	87.41 ± 1.66
11	-	85.02 ± 1.01	82.89 ± 1.20	**87.82 ± 0.75**
11	✓	**85.77 ± 0.95**	**83.32 ± 1.52**	87.68 ± 1.08

NVIDIA Titan V 12 GB GPU using Polyaxon[1]. The source code for OperA along with the Evaluation scripts is publicly available[2].

Evaluation Metrics and Baselines. To comprehensively measure the results we report the video-level Accuracy (Acc) together with the Precision and Recall [11] and average the results over the 5 splits. We perform ablative testing to identify the most suitable number of attention layers and to test the effect of the regularization term (Sect. 2.3). Finally, we compare OperA with a variety of surgical phase recognition baselines.

4   Results and Discussion

Effect of Layers and Regularization. In Table 1 we compare models trained with 6 and 11 attention layers and evaluate the impact of the attention regularization. 11 attention layers is the highest amount of layers that we could fit in the VRAM. For both datasets, the results slightly increase for 11 attention layers by ~1% both in terms of Accuracy and Precision with similar Recall values. Regarding the attention regularization, a ~1% improvement is reported for both datasets and number of layers. As we will discuss later, the attention regularization does not only marginally increases the model performance but also the quality of the highest attention video frames.

Baseline Comparison. We compare OperA with various methods for surgical phase recognition. ResNet-50 is the feature extraction backbone, ResLSTM [14] and MTRCNet-CL [15] utilize LSTMs for the temporal refinement. Different to

[1] https://polyaxon.com/
[2] https://github.com/tobiascz/OperA/

Table 2. Baseline comparisons for Cholec80 and CSW. MTRCNet-CL requires tool information, thus cannot be used for CSW. We report the average metrics over 5-fold cross validation along with their respective standard deviation ($\pm$).

	Accuracy	Precision	Recall
	Cholec80		
ResNet-50	81.21 ± 1.16	68.35 ± 1.61	78.31 ± 1.14
ResLSTM	87.94 ± 0.80	80.26 ± 1.12	84.43 ± 0.85
MTRCNET-CL	85.64 ± 0.21	79.31 ± 0.97	82.67 ± 0.114
TeCNO	89.05 ± 0.79	80.90 ± 0.75	**87.44 ± 0.64**
OperA + PE	90.20 ± 1.45	80.78 ± 1.42	86.08 ± 0.89
OperA	**91.26 ± 0.64**	**82.19 ± 0.70**	86.92 ± 0.86
	CSW		
ResNet-50	73.90 ± 1.89	69.06 ± 1.38	74.20 ± 1.63
ResLSTM	82.97 ± 1.18	82.08 ± 1.57	86.15 ± 0.94
MTRCNET-CL	–	–	–
TeCNO	85.09 ± 1.67	82.03 ± 0.20	86.50 ± 0.43
OperA + PE	83.67 ± 1.54	81.34 ± 1.60	86.94 ± 0.98
OperA	**85.77 ± 0.95**	**83.32 ± 1.10**	**87.68 ± 0.71**

the other models MTRCNet-CL is trained in an end-to-end fashion combining a CNN feature extraction with LSTM training. One of the downsides to this approach is that due to memory constraint only a limited sequence of the video can be used per-batch. ResLSTM, TeCNO and OperA use pre-trained image features and can therefore analyze a full video sequence at once.

First, we see that temporal refinement achieves a substantial improvement over ResNet-50 ranging from 4–10% for Cholec80 and 9–12% for CSW, showcasing its advantage. MTRCNet-CL is outperformed by the other temporal models by 2–6% potentially due to the limited sequence length that can be processed in every batch. OperA with or without positional encoding (PE) [7] outperforms the other temporal models by 2–6% in terms of accuracy for Cholec80 showcasing the abilities of transformers to model long temporal dependencies. Regarding CSW, OperA without PE has increased accuracy by 0.6–3% and an increase for Precision and Recall over all baselines. For both datasets it can be seen that PE marginally decreases the performance, potentially due to the increased sequence length in surgical videos in comparison to NLP tasks (Table 2).

Predictions and Attention Values. In Fig. 2 we visualize the ground truth, predictions and attention values for video 66 of Cholec80. First, we see that the predictions of OperA are smoother and more consistent than the ones of the CNN. Moreover, the HA frames (denoted with $\triangle$) in most cases correspond with frames, where the CNN and OperA predictions were correct, while LA frames (denoted with $\triangledown$) are at positions where the CNN predictions were wrong, confirming that attention regularization works as intended.

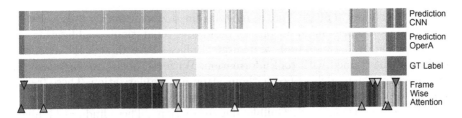

Fig. 2. Qualitative results of the predictions per phase for video 66 from Cholec80 using the feature extraction CNN and OperA compared to the ground truth labels. In the frame-wise attention, brighter (yellow) color corresponds to higher attention, darker (blue) color to lower attention. The position of the LA frames for each phase is denoted with ▽ and the HA frames with △. (Color figure online)

Highest and Lowest Attention Frames. In Fig. 3 we visualize the HA and LA frames per phase for the models trained with and without attention regularization. Visual inspection revealed that LA frames are generally less descriptive for the respective surgical phase. However, as we can see highlighted by the blue boxes, the model trained without attention regularization has minimum attention for frames containing surgical tools that are quite characteristic of their respective phase. Regarding the HA frames, they are more diverse and represen-

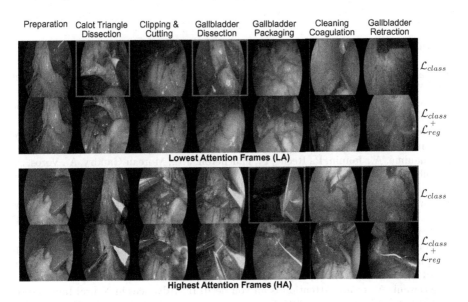

Fig. 3. Visualization of frames of video 66 of Cholec80 with highest (HA) and lowest (LA) attention per phase for the models with and without attention regularization. Blue and red boxes denote frames of the model without regularization that have low attention, while they are descriptive of their phase and high attention, while they are not. (Color figure online)

tative of their surgical phase. For the model trained with attention regularization surgical tools are present in all phases besides "Preparation", highlighting the strong correlation between tools and surgical phases even though the attention model was not trained with tool information. With red boxes, we showcase the HA frames for the model trained without attention regularization. These frames were not descriptive of their phase and very similar to each other in the case of "Cleaning Coagulation" and "Gallbladder Retraction". These findings highlight the benefits of the proposed attention regularization and its potential for surgery video summarization.

5   Conclusion

In this paper we introduced OperA, a transformer-based model that accurately predicted surgical phases of cholecystectomy procedures, outperforming a variety of baselines on two challenging laparoscopic video datasets. Additionally, our novel attention regularizer enabled OperA to extract characteristic high attention frames. Future work includes applying our method to different laparoscopic procedures and explore its potential for surgical video summarization.

Acknowledgements. Our research is partly funded by the DFG research unit PLAFOKON (FKZ 620/33-2) and BMBF research project ARTEKMED (FKZ 16SV8088) in collaboration with the Minimal-invasive Interdisciplinary Intervention Group.

References

1. Maier-Hein, L., et al.: Surgical data science: a consensus perspective. arXiv preprint arXiv:1806.03184 (2018)
2. Garrow, C.R., et al.: Machine learning for surgical phase recognition: a systematic review. Ann. Surg. **273**, 684–693 (2020)
3. Padoy, N.: Machine and deep learning for workflow recognition during surgery. Minim. Invasive Ther. Allied Technol. **28**, 82–90 (2019)
4. Huaulmé, A., Jannin, P., Reche, F., Faucheron, J.L., Moreau-Gaudry, A., Voros, S.: Offline identification of surgical deviations in laparoscopic rectopexy. Artif. Intell. Med. **104**(May), 2020 (2019)
5. Twinanda, A.P., Shehata, S., Mutter, D., Marescaux, J., De Mathelin, M., Padoy, N.: EndoNet: a deep architecture for recognition tasks on laparoscopic videos. IEEE Trans. Med. Imaging **36**(1), 86–97 (2017)
6. Funke, I., Mees, S.T., Weitz, J., Speidel, S.: Video-based surgical skill assessment using 3D convolutional neural networks. Int. J. Comput. Assist. Radiol. Surg. **14**(7), 1217–1225 (2019)
7. Vaswani, A., et al.: Attention is all you need. In: Advances in Neural Information Processing Systems, vol. 2017-Decem, no. Nips, pp. 5999–6009 (2017)
8. Devlin, J., Chang, M.W., Lee, K., Toutanova, K.: BERT: pre-training of deep bidirectional transformers for language understanding. In: NAACL HLT 2019–2019 Conference of the North American Chapter of the Association for Computational Linguistics: Human Language Technologies - Proceedings of the Conference, vol. 1, no. Mlm, pp. 4171–4186 (2019)

9. Hochreiter, S., Schmidhuber, J.: Long short-term memory. Neural Comput. **9**(8), 1735–1780 (1997)

10. Ahmadi, S.-A., Sielhorst, T., Stauder, R., Horn, M., Feussner, H., Navab, N.: Recovery of surgical workflow without explicit models. In: Larsen, R., Nielsen, M., Sporring, J. (eds.) MICCAI 2006. LNCS, vol. 4190, pp. 420–428. Springer, Heidelberg (2006). https://doi.org/10.1007/11866565_52

11. Padoy, N., Blum, T., Ahmadi, S.A., Feussner, H., Berger, M.O., Navab, N.: Statistical modeling and recognition of surgical workflow. Med. Image Anal. **16**(3), 632–641 (2012)

12. Twinanda, A.P., Padoy, N., Troccaz, M.J., Hager, G.: Vision-based approaches for surgical activity recognition using laparoscopic and RBGD videos, Thesis, no. Umr 7357 (2017)

13. Yengera, G., Mutter, D., Marescaux, J., Padoy, N.: Less is more: surgical phase recognition with less annotations through self-supervised pre-training of CNN-LSTM networks. arXiv preprint arXiv:1805.08569 (2018)

14. Jin, Y., et al.: SV-RCNet: workflow recognition from surgical videos using recurrent convolutional network. IEEE Trans. Med. Imaging **37**(5), 1114–1126 (2018)

15. Jin, Y., et al.: Multi-task recurrent convolutional network with correlation loss for surgical video analysis. Med. Image Anal. **59**, 101572 (2020)

16. Czempiel, T., et al.: TeCNO: surgical phase recognition with multi-stage temporal convolutional networks. In: Martel, A.L., et al. (eds.) MICCAI 2020. LNCS, vol. 12263, pp. 343–352. Springer, Cham (2020). https://doi.org/10.1007/978-3-030-59716-0_33

17. He, K., Zhang, X., Ren, S., Sun,J.: Deep residual learning for image recognition. In: 2016 IEEE Conference on Computer Vision and Pattern Recognition (CVPR). IEEE (2016)

18. Brown, T.B., et al.: Language models are few-shot learners. arXiv (2020)

19. Dosovitskiy, A., et al.: An image is worth 16 × 16 words: transformers for image recognition at scale. arXiv preprint arXiv:2010.11929 (2020)

20. Ramesh, A., et al.: Zero-shot text-to-image generation. arXiv preprint arXiv:2102.12092 (2021)

21. Heo, L., Feig, M.: High-accuracy protein structures by combining machine-learning with physics-based refinement. Proteins **88**, 637–642 (2020)

22. Kondo, S.: Lapformer: surgical tool detection in laparoscopic surgical video using transformer architecture. In: Computer Methods in Biomechanics and Biomedical Engineering: Imaging & Visualization, pp. 1–6 (2020)

23. Jain, S., Wallace, B.C.: Attention is not explanation. In: Proceedings of the 2019 Conference of the North American Chapter of the Association for Computational Linguistics: Human Language Technologies, NAACL-HLT 2019, Minneapolis, MN, USA, 2–7 June 2019, vol. 1, pp. 3543–3556. Association for Computational Linguistics (2019)

24. Wiegreffe, S., Pinter, Y.: Attention is not not explanation. In: Proceedings of the 2019 Conference on Empirical Methods in Natural Language Processing and the 9th International Joint Conference on Natural Language Processing (EMNLP-IJCNLP), Hong Kong, China, pp. 11–20. Association for Computational Linguistics (2019)

25. Kim, Y., Denton, C., Hoang, L., Rush, A.M.: Structured attention networks. In: International Conference on Learning Representations, pp. 1–21 (2017)

26. Ba, J.L., Kiros, J.R., Hinton, G.E.: Layer normalization. arXiv preprint arXiv:1607.06450 (2016)

27. Eigen, D., Fergus, R.: Predicting depth, surface normals and semantic labels with a common multi-scale convolutional architecture. In: 2015 IEEE International Conference on Computer Vision (ICCV). IEEE (2015)
28. Al-Rfou, R., Choe, D., Constant, N., Guo, M., Jones, L.: Character-level language modeling with deeper self-attention. In: Proceedings of the AAAI Conference on Artificial Intelligence, vol. 33, July 2019

Surgical Workflow Anticipation Using Instrument Interaction

Kun Yuan[1(✉)], Matthew Holden[2], Shijian Gao[3], and Won-Sook Lee[1]

[1] Faculty of Engineering, University of Ottawa, Ottawa, Canada
{kyuan033,wslee}@uottawa.ca
[2] School of Computer Science, Carleton University, Ottawa, Canada
matthew.holden@carleton.ca
[3] Department of Electrical and Computer Engineering, University of Minnesota, Minneapolis, USA
gao00379@umn.edu

Abstract. Surgical workflow anticipation, including surgical instrument and phase anticipation, is essential for an intra-operative decision-support system. It deciphers the surgeon's behaviors and the patient's status to forecast surgical instrument and phase occurrence before they appear, providing support for instrument preparation and computer-assisted intervention (CAI) systems. We investigate an unexplored surgical workflow anticipation problem by proposing an Instrument Interaction Aware Anticipation Network (IIA-Net). Spatially, it utilizes rich visual features about the context information around the instrument, i.e., instrument interaction with their surroundings. Temporally, it allows for a large receptive field to capture the long-term dependency in the long and untrimmed surgical videos through a causal dilated multi-stage temporal convolutional network. Our model enforces an online inference with reliable predictions even with severe noise and artifacts in the recorded videos. Extensive experiments on Cholec80 dataset demonstrate the performance of our proposed method exceeds the state-of-the-art method by a large margin (1.40 v.s. 1.75 for inMAE and 2.14 v.s. 2.68 for eMAE). The code is published on https://github.com/Flaick/Surgical-Workflow-Anticipation.

Keywords: Surgical workflow analysis · Anticipation · Temporal convolutional networks · Endoscopic videos · Instrument detection

1 Introduction

Context-aware assistance is integral for CAI systems, of which the most crucial task is surgical workflow anticipation. It anticipates the occurrence of surgical instruments and phases before they appear, enabling the efficient instrument preparation and intelligent robot assistance system design [7,23]. The benefit

Electronic supplementary material The online version of this chapter (https://doi.org/10.1007/978-3-030-87202-1_59) contains supplementary material, which is available to authorized users.

© Springer Nature Switzerland AG 2021
M. de Bruijne et al. (Eds.): MICCAI 2021, LNCS 12904, pp. 615–625, 2021.
https://doi.org/10.1007/978-3-030-87202-1_59

of anticipation is three-fold [20]. Firstly, instrument anticipation offers a useful reference for decision making in a robotic assistance system. It helps to identify instrument usage triggering so that a robotic system can decide when to intervene. Also, for context-aware assistance, anticipating instruments such as irrigator can help early detection and prevention of potential complications, e.g., massive haemorrhage. Thirdly, it allows real-time instruction for automated surgical coaching therefore increasing patient safety and reducing surgical errors. The anticipation of surgical phases can also provide vital input for optimizing communication in the operating room (OR).

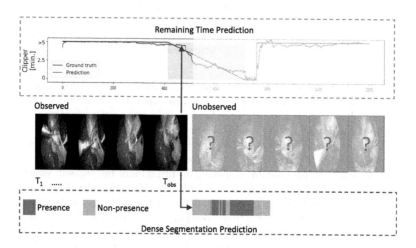

Fig. 1. Anticipation frameworks. Given an observed sequence and current time instant, T_{obs}, bottom part shows the conventional anticipation works that predicts dense segmentations. The upper part shows our strategy handling anticipation task as a real-time remaining time prediction task.

Recent models [23,28] for surgical workflow anticipation possess spatial-temporal limitations. Spatially, they use AlexNet[17], VGG[25] and similar architectures to extract a feature vector, representing instrument/phase presence for each frame. However, they ignore the task-specific combinations present in surgical anticipation applications, i.e., instrument-instrument and instrument-surrounding interactions. This information precisely reflects the surgeon's intention and patient's anatomy status, helping models generalize to the low-quality input materials [16] and variability of patient's anatomy and surgeon style [8]. Novelly, our IIA-Net addresses instrument-instrument interaction in the form of a correlation matrix and designed geometric relations among instruments. Also, the instrument-surrounding interaction is included via the semantic segmentation map. This makes our extracted feature to be representative enough to identify the trigger event for the next instrument and phase occurrence.

Temporally, existing works have difficulty handling non-stationary time series. Especially for surgical workflow whose laparoscopic surgery transitions among instruments and phases are ambiguous and various. This requires the temporal modeling method to integrate recent observations with the long-range context in a computationally efficient way. However, widely used RNNs [11] learn a pattern from shorts segments of time series and apply it to other parts to get predictions, losing the distant observation information. Therefore, we opt for dilated temporal convolutions to handle the full resolution of time series. This aids temporal pattern modeling and does not require complex computational resources.

Initial works [1,4,7,9,15,19] handle anticipation as a dense segmentation prediction task, shown in the bottom of Fig. 1. They require a pre-loading process before performing anticipation, limiting their usage in online surgical applications. Specifically, [1] needs to observe at least 10%/20% of the video before it starts the prediction. Also, Fig. 1 shows an example where the predicted dense segmentation usually contains the short segments, which are ambiguous to determine the trending of the instrument's presence.

Our contribution is four-fold: (1) Spatially, we propose a novel instrument interaction module (IIM) for the feature extraction process. (2) Temporally, we apply, for the first time, the causal dilated multi-stage temporal convolutional network (MSTCN) structure to surgical workflow anticipation, with an accurate and fast online inference. (3) We combine spatial and temporal information to form a two-step IIA-Net for surgical workflow anticipation. (4) We propose a multi-task learning schema to jointly anticipate instrument and phase occurrence, which are important challenges in surgical workflow anticipation.

2   Methodology

Our IIA-Net composes of two parts, a feature extractor with an Instrument Interaction Module (IIM) and a temporal model using MSTCN. Spatially, our IIA-Net models the surgeon's intention through extracting rich geometric features of the instrument-instrument interactions and semantic features of instrument-surrounding interactions. Motivated by the recognition methods [13,14,21,27], we introduce tool and phase signal to boost the feature extraction process. Temporally, we utilize causal dilated MSTCN [5] to capture long-term patterns with a large receptive field. Unlike the dense segmentation prediction, shown in Fig. 1, our IIA-Net follows [23] to handle anticipation as a real-time remaining time regression problem without any latency or pre-loading process.

2.1   Task Formulation

We process the anticipation task as a regression problem both for instrument and phase anticipation. Given a frame i from video x, we firstly extract semantic map s_i and instrument bounding boxes b_i. At the same time, we obtain the instrument presence signal t_i and phase signal p_i from the manual annotations. Given the observed sequence $\{(x_1, s_1, b_1, t_1, p_1), ... (x_{T_{obs}}, s_{T_{obs}}, b_{T_{obs}}, t_{T_{obs}}, p_{T_{obs}})\}$ from time 1 to T_{obs}, our model predicts the remaining time until the occurrence of τ/α for instrument/phase. The ground truth $r(x_{T_{obs}}, \tau/\alpha)$ for current time instant T_{obs} ranges $[0, h]$, where 0 denotes that the τ/α is currently happening and h denotes that τ/α will not happen within next h minutes.

2.2   Network Architecture

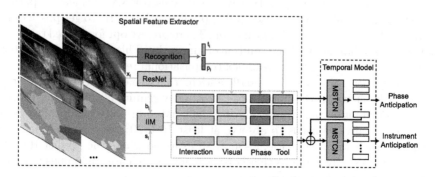

Fig. 2. Overview of the proposed model. For each frame observed, its estimated seman-tic map and tool detection are forwarded to instrument interaction module (IIM) to extract interaction feature. The manual annotations for phase and tool signal are fed into temporal model jointly with interaction and visual features.

Figure 2 shows the overall network architecture of our IIA-Net. It is a two-step model with a feature extractor and a temporal model. The feature extractor takes five inputs x_i, s_i, b_i, t_i, p_i mentioned in Sect. 2.1, of which the s_i and b_i are used for IIM to model instrument-instrument interactions and the instrument-surrounding interactions. The frame x_i is encoded by ResNet50 [10] into visual features, and the tool signal t_i, phase signal p_i are provided by the manual annotations from Cholec80 dataset. They are embedded into the feature space and concatenated with interaction feature and visual feature jointly for the input of the next temporal model.

For the temporal pattern modeling, we apply a multi-stage temporal convo-lutional network firstly for phase anticipation. Then we concatenate the above five feature vectors and the prediction of phase anticipation together as the input for the instrument anticipation. In the rest of this section, we will introduce the above modules in details.

2.3   Instrument Interaction Module

In this module, we model surgeons' intention by analyzing the instrument-instrument interaction and instrument-surrounding interaction, shown in Fig. 3. We assume each frame is processed to obtain the spatial coordinates and bound-ing boxes of all instruments. Also, we extract the categorical prior for each frame, which characterizes the semantic class of a region in an image. (e.g., liver, gall-bladder).

Instrument-Instrument Encoder. This encoder explicitly models the geomet-ric relation among instruments. Here we only consider the interaction between grasper and other instruments because the grasper is the most frequently used

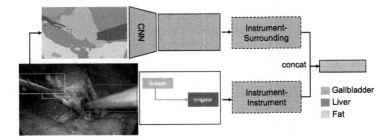

Fig. 3. Instrument interaction module. Upper: instrument-surrounding modeling uses pooled scene semantic features to encode features; Bottom: instrument-instrument modeling extracts the spatial relations between the grasper and the other instruments.

instrument. It provides the primary support for the other instruments during the surgery.

We encode the geometric relation $G \in \mathbb{R}^{M \times 4}$ using Eq. 1 that is proven effective in object detection [18]. Specifically, at any time instant, given the bounding box of grasper (x_g, y_g, w_g, h_g) and M other instruments in the scene $\{(x_m, y_m, w_m, h_m)|m \in [1, M]\}$, we encode the geometric relation into $G \in \mathbb{R}^{M \times 4}$, the m-th row of which equals to:

$$G_m = [log(\frac{|x_g - x_m|}{w_g}), log(\frac{|y_g - y_m|}{h_g}), log(\frac{w_m}{w_g}), log(\frac{h_m}{h_g})] \qquad (1)$$

This encoding computes the geometric relation in terms of the geometric distance and the fraction box size. We then embed this geometric feature at each time instant into $\mathbb{R}^{T_{obs} \times C_1}$ where C_1 is the embedding size.

Instrument-Surrounding Encoder. To encode an instrument's nearby anatomical surroundings, we first extract pixel-level scene semantic classes for each frame. Here, we use totally $N_s = 7$ scene classes (i.e., background, liver, fat, abdominal wall, tool Shaft, tool tip, gallbladder). Then we transform the integer semantic map into N_s binary masks of the size $T_{obs} \times h \times w$, where h, w are spatial resolution. We apply two convolutional layers on the binary masks with a stride of 2 to get the scene CNN features. We then average the scene feature along the spatial dimensions and generate a feature vector as the encoder's output.

The generated feature vector is in $\mathbb{R}^{T_{obs} \times C_2}$, where C_2 is the number of channels in the convolution layers. After combining the feature vectors from instrument-instrument encoder and instrument-surrounding encoder, the final feature vector outputed from IIM is in $\mathbb{R}^{T_{obs} \times (C_1 + C_2)}$.

2.4 Multi-stage Temporal Convolutional Network

To model temporal patterns in the anticipation task, we modify the MSTCN [5] to build a lightweight temporal network. The network is constructed fully with dilated temporal convolutions without neither pooling layers nor fully connected layers. This design keeps the model processing the full resolution temporal sequence and reduces the number of parameters.

To apply our model in an online mode, we use causal convolutions in the network. Instead of the acausal convolutions in [5] with predictions depend on both n past and n future frames, the causal convolutions ensure the prediction of current instant not relying on any n future frames but only depends on the current and previous frames.

3   Experiment Setup

3.1   Datasets and Preprocessing

Anticipation Dataset. We evaluate our method on publicly available surgical workflow intraoperative video dataset, Cholec80 [26], which contains laparoscopic cholecystectomy procedures for the resection of the gallbladder. Cholec80 dataset consists of 80 videos ranging from 15 min to 90 min. We follow the same split as [23], separating the dataset to 60 videos for training and 20 for testing. We resize the videos spatial resolution to 224×224 to dramatically reduce the computational cost. Also, we resample the video from 25 fps to 1 fps.

Detection and Segmentation Dataset. As mentioned above, we need to extract instrument bounding boxes and semantic maps for the Cholec80 dataset. However, annotating such dataset is manually unfeasible. Therefore, we opt for training the segmentation model [24] on a synthesized dataset [22], which utilizes conditional GAN [12] to generate Cholec80 style laparoscopic images from simulation images. Then, we apply the trained model to infer on the Cholec80 dataset. The segmentation result can be found in the supplementary materials.

To detect the surgical instrument bounding boxes on Cholec80, we leverage the dataset from [13] to train a YOLO [2,6] detector. The trained model is proven to detect surgical instruments on Cholec80 dataset effectively [13].

3.2   Evaluation Metrics

Automatic instrument preparation is one of the primary tasks that benefits from surgical workflow anticipation. It does not require tools or phases to be anticipated too far in advance. Also, the preparation system should only react to the signals that indicate tool/phase is anticipating. Therefore, we measure the performance of 'anticipating' frames ($0 < r(x_{T_{obs}}, \tau/\alpha) < h$, using inMAE. Also, we propose to use eMAE to evaluate intervals ($0 < r(x_{T_{obs}}, \tau/\alpha) < 0.1h$) that provides the most effective support to the computer-assistance system. Also, we utilize the pMAE in [23] to measure the precision performance of our model.

Table 1. Effect of IIM and MSTCN on different feature extraction models for instrument anticipation. We report the inMAE/eMAE averaging over instrument types in minutes per metric when $h = 5$ min. T: tool signal feature; P: phase signal feature; IIM: interaction feature from instrument interaction module.

	ResNet50	ResNet50+T+P	ResNet50+T+P+IIM	Baseline
No MSTCN	1.99/4.06	1.79/3.58	1.57/2.51	1.75/2.68
1 Stage	1.62/3.74	1.57/3.29	1.42/2.22	
2 Stages	1.59/3.67	1.45/3.23	**1.40/2.14**	
3 Stages	1.53/3.64	1.60/3.31	1.48/2.15	

4   Results and Discussions

4.1   Effect of IIM and Stages in MSTCN

We conduct ablative testing to compare different feature extraction models, ResNet50 [10], ResNet50 with instrument and phase features, ResNet50 with all added features, to identify a suitable feature extractor for our model. Additionally, we conduct experiments with different numbers of MSTCN stages to determine which architecture is best able to capture temporal patterns.

As shown in Table 1, the ResNet50 with all features outperforms ResNet50 across the board with improvements ranging from 1.99 to 1.57 in inMAE and 4.06 to 2.51 in pMAE. This increase can be attributed to the improved representation by our designed features. Among the features that we added, the IIM makes more contribution than the instrument and phase signals. This suggests that modeling the interactions signifies the surgeon's intention and the occurrence of the next situation. Interestingly, ResNet50 with all added feature achieves a comparable result with the baseline model [23] even without any temporal modeling.

Table 1 also highlights the substantial performance improvement achieved by the MSTCN refinement stages. Those results demonstrate the ability of MSTCN to improve the performance of any feature extractor. All feature extractors achieve higher performance when only 1 stage is used. However, 2 stages model outperforms 3 stages model. This could indicate that 3 stages of refinement lead to overfitting on the training set for the limited amount of data.

4.2   Anticipation Results

We evaluate the model for instrument and phase anticipation on horizons of 2, 3, and 5 min. We remove the horizon setting of 7 min since anticipating the surgical workflow too early is unnecessary for instrument preparation and robot assistance. We re-implement methods from [23] and retrain them as the baseline methods for phase anticipation.

Table 2 shows that IIA-Net achieves lower inMAE and pMAE error compared to the previous methods. Regarding the pMAE error, the margin increases even

Table 2. inMAE/pMAE comparison. We report the mean over instrument types in minutes per metric. Ours 2D: our feature extractor without temporal training.

	Instrument			Phase		
	$h = 2$ min	$h = 3$ min	$h = 5$ min	$h = 2$ min	$h = 3$ min	$h = 5$ min
MeanHist	1.09/0.93	1.62/1.34	2.64/2.14	–	–	–
OracleHist (offline)	(0.92/0.83)	(1.31/1.18)	(2.01/1.73)	–	–	–
Baseline [23]	0.77/0.64	1.13/0.92	1.80/1.49	0.63/0.62	0.86/0.85	1.17/1.37
Ours 2D	0.70/0.52	1.07/0.77	1.65/**1.16**	0.70/0.53	1.04/0.76	1.40/**1.12**
Ours	**0.66/0.42**	**0.97/0.69**	**1.48**/1.28	**0.62/0.49**	**0.81/0.73**	**1.08**/1.22

Table 3. eMAE comparison. We report the mean over instrument types in minutes.

	Instrument			Phase		
	$h = 2$ min	$h = 3$ min	$h = 5$ min	$h = 2$ min	$h = 3$ min	$h = 5$ min
MeanHist	1.85	2.72	4.35	–	–	–
OracleHist (offline)	(1.36)	(1.93)	(2.96)	–	–	–
Baseline [23]	1.12	1.65	2.68	**1.02**	1.47	1.54
Ours 2D	1.07	1.65	2.51	1.38	1.85	2.42
Ours	**1.01**	**1.46**	**2.14**	1.18	**1.42**	**1.09**

further. Even though [23] is trained in an end-to-end fashion, it is also outperformed by our IIA-Net, which is trained in a two-step process. Interestingly, our model trained without temporal context (Ours 2D) achieved a lower pMAE when $h = 5$. This is because the 2D model has difficulty foreseeing long-horizon occurrence and easily predicts the value that is close to $0/h$, making its prediction unsmooth. Also, our model achieves the lowest eMAE error, seen from Table 3. This suggests that our model can effectively identify instrument or phase occurrence a few seconds ahead. In real-world scenarios, this is typically the most critical time for accurate anticipation.

To further verify the feasibility of our model in the real-world surgical scenario, we test our model's running time performance given the online video stream. Based on the Pytorch 1.4 framework and single RTX 2080 ti, our model is able to process each frame within 0.0293 s when deploying the spatial feature extractor and temporal model parallely. The running time is 10% faster than [23] taking 0.0328 s, indicating our model is applicable in a real-world setup.

4.3   Limitations

The primary limitation of our current experimental setup is the incorporation of tool and phase signals. Specifically, we train the network using the signals from human annotation instead of recognition models. In real-world scenarios where this ground-truth is not available, the model's performance will likely be reduced. We conjecture, however, the degradation will be minimal using recent models which show superior performance for tool and phase recognition (95% and 88% of accuracy for tool and phase recognition). Also, [3] shows that the predicted phase signal is consistent and smooth not only within one phase, but also for the often ambiguous phase transitions. This means our IIA-Net will likely make a reliable prediction even with predicted tool and phase signals.

5 Conclusion

In this paper, we propose the IIA-Net, which incorporates existing surgical workflow analysis methods, i.e., tool detection, phase recognition, laparoscopic image segmentation, and outperforms previous works. It shows that the interaction relationship during spatial feature extraction is effective to resolve surgical workflow anticipation. Without temporal training, our model is a strong baseline for the following 2D works. Furthermore, temporal modelling using a MSTCN with causal and dilated convolution handles full temporal resolution of time series, fitting extreme long laparoscopic workflow well. Its large receptive field captures distant as well recent observations. Our multi-task learning schema provides a potential direction to jointly perform instrument and phase anticipation. Future work includes evaluation of our method on real-world phase and tool signals.

References

1. Abu Farha, Y., Richard, A., Gall, J.: When will you do what?-anticipating temporal occurrences of activities. In: Proceedings of the IEEE Conference on Computer Vision and Pattern Recognition, pp. 5343–5352 (2018)
2. Bochkovskiy, A., Wang, C.Y., Liao, H.Y.M.: Yolov4: optimal speed and accuracy of object detection. arXiv preprint arXiv:2004.10934 (2020)
3. Czempiel, T., et al.: TeCNO: surgical phase recognition with multi-stage temporal convolutional networks. In: Martel, A.L., et al. (eds.) MICCAI 2020. LNCS, vol. 12263, pp. 343–352. Springer, Cham (2020). https://doi.org/10.1007/978-3-030-59716-0_33
4. Du, N., Dai, H., Trivedi, R., Upadhyay, U., Gomez-Rodriguez, M., Song, L.: Recurrent marked temporal point processes: embedding event history to vector. In: Proceedings of the 22nd ACM SIGKDD International Conference on Knowledge Discovery and Data Mining, pp. 1555–1564 (2016)
5. Farha, Y.A., Gall, J.: MS-TCN: multi-stage temporal convolutional network for action segmentation. In: Proceedings of the IEEE/CVF Conference on Computer Vision and Pattern Recognition, pp. 3575–3584 (2019)
6. Farhadi, A., Redmon, J.: Yolov3: an incremental improvement. Computer Vision and Pattern Recognition, cite as (2018)
7. Forestier, G., Petitjean, F., Riffaud, L., Jannin, P.: Automatic matching of surgeries to predict surgeons' next actions. Artif. Intell. Med. **81**, 3–11 (2017)
8. Funke, I., Mees, S.T., Weitz, J., Speidel, S.: Video-based surgical skill assessment using 3D convolutional neural networks. Int. J. Comput. Assist. Radiol. Surg. **14**(7), 1217–1225 (2019)
9. Gao, J., Yang, Z., Nevatia, R.: Red: reinforced encoder-decoder networks for action anticipation. arXiv preprint arXiv:1707.04818 (2017)

10. He, K., Zhang, X., Ren, S., Sun, J.: Deep residual learning for image recognition. In: Proceedings of the IEEE Conference on Computer Vision and Pattern Recognition, pp. 770–778 (2016)
11. Hochreiter, S., Schmidhuber, J.: Long short-term memory. Neural Comput. **9**(8), 1735–1780 (1997)
12. Isola, P., Zhu, J.Y., Zhou, T., Efros, A.A.: Image-to-image translation with conditional adversarial networks. In: Proceedings of the IEEE Conference on Computer Vision and Pattern Recognition, pp. 1125–1134 (2017)
13. Jin, A., et al.: Tool detection and operative skill assessment in surgical videos using region-based convolutional neural networks. In: 2018 IEEE Winter Conference on Applications of Computer Vision (WACV), pp. 691–699. IEEE (2018)
14. Jin, Y., et al.: SV-RCNet: workflow recognition from surgical videos using recurrent convolutional network. IEEE Trans. Med. Imaging **37**(5), 1114–1126 (2017)
15. Ke, Q., Fritz, M., Schiele, B.: Time-conditioned action anticipation in one shot. In: Proceedings of the IEEE/CVF Conference on Computer Vision and Pattern Recognition, pp. 9925–9934 (2019)
16. Klank, U., Padoy, N., Feussner, H., Navab, N.: Automatic feature generation in endoscopic images. Int. J. Comput. Assist. Radiol. Surg. **3**(3), 331–339 (2008)
17. Krizhevsky, A., Sutskever, I., Hinton, G.E.: ImageNet classification with deep convolutional neural networks. In: Advances in Neural Information Processing Systems, vol. 25, pp. 1097–1105 (2012)
18. Liang, J., Jiang, L., Niebles, J.C., Hauptmann, A.G., Fei-Fei, L.: Peeking into the future: Predicting future person activities and locations in videos. In: Proceedings of the IEEE/CVF Conference on Computer Vision and Pattern Recognition, pp. 5725–5734 (2019)
19. Mahmud, T., Hasan, M., Roy-Chowdhury, A.K.: Joint prediction of activity labels and starting times in untrimmed videos. In: Proceedings of the IEEE International Conference on Computer Vision, pp. 5773–5782 (2017)
20. Maier-Hein, L., et al.: Surgical data science for next-generation interventions. Nature Biomed. Eng. **1**(9), 691–696 (2017)
21. Padoy, N.: Machine and deep learning for workflow recognition during surgery. Minimally Invasive Ther. Allied Technol. **28**(2), 82–90 (2019)
22. Pfeiffer, M., et al.: Generating large labeled data sets for laparoscopic image processing tasks using unpaired image-to-image translation. In: Shen, D., et al. (eds.) MICCAI 2019. LNCS, vol. 11768, pp. 119–127. Springer, Cham (2019). https://doi.org/10.1007/978-3-030-32254-0_14
23. Rivoir, D., et al.: Rethinking anticipation tasks: uncertainty-aware anticipation of sparse surgical instrument usage for context-aware assistance. In: Martel, A.L., et al. (eds.) MICCAI 2020. LNCS, vol. 12263, pp. 752–762. Springer, Cham (2020). https://doi.org/10.1007/978-3-030-59716-0_72
24. Ronneberger, O., Fischer, P., Brox, T.: U-Net: convolutional networks for biomedical image segmentation. In: Navab, N., Hornegger, J., Wells, W.M., Frangi, A.F. (eds.) MICCAI 2015. LNCS, vol. 9351, pp. 234–241. Springer, Cham (2015). https://doi.org/10.1007/978-3-319-24574-4_28
25. Simonyan, K., Zisserman, A.: Very deep convolutional networks for large-scale image recognition. arXiv preprint arXiv:1409.1556 (2014)
26. Twinanda, A.P., Mutter, D., Marescaux, J., de Mathelin, M., Padoy, N.: Single- and multi-task architectures for surgical workflow challenge at m2cai 2016. arXiv preprint arXiv:1610.08844 (2016)

27. Twinanda, A.P., Shehata, S., Mutter, D., Marescaux, J., De Mathelin, M., Padoy, N.: EndoNet: a deep architecture for recognition tasks on laparoscopic videos. IEEE Trans. Med. Imaging **36**(1), 86–97 (2016)
28. Twinanda, A.P., Yengera, G., Mutter, D., Marescaux, J., Padoy, N.: RSDNet: learning to predict remaining surgery duration from laparoscopic videos without manual annotations. IEEE Trans. Med. Imaging **38**(4), 1069–1078 (2018)

Multi-view Surgical Video Action Detection via Mixed Global View Attention

Adam Schmidt[1]([✉]), Aidean Sharghi[2], Helene Haugerud[2], Daniel Oh[2],
and Omid Mohareri[2]

[1] University of British Columbia, Vancouver, Canada
adamschmidt@ece.ubc.ca
[2] Intuitive Surgical Inc., Sunnyvale, USA

Abstract. Automatic surgical activity detection in the operating room can enable intelligent systems that potentially lead to more efficient surgical workflow. While real-world implementations of video activity detection in the OR most likely rely on multiple video feeds observing the environment from different view points to handle occlusion and clutter, the research on the matter has been left under-explored. This is perhaps due to the lack of a suitable dataset, thus, as our first contribution, we introduce the first large-scale multi-view surgical action detection dataset that includes over 120 temporally annotated robotic surgery operations, each recorded from 4 different viewpoints, resulting in 480 full-length surgical videos. As our second contribution, we design a novel model architecture that can detect surgical actions by utilizing multiple time-synchronized videos with shared field of view to better detect the activity that is taking place at any time. We explore early, hybrid, and late fusion methods for combining data from different views. We settle on a late fusion model that remains insensitive to sensor locations and feeding order, improving over single-view performance by using a mixing in the style of attention. Our model learns how to dynamically weight and fuse information across all views. We demonstrate improvements in mean Average Precision across the board using our new model.

1   Introduction

Automatic action detection is the task of recognizing both the type of each action instance and its corresponding start and end times in a given video. Such algorithms enable a variety of intelligent applications in different environments. In the case of operating rooms, they highly facilitate workflow analysis and annotation, surgical team skill assessment, smart scheduling systems, and potentially improve surgical outcomes by detecting human errors.

Video activity recognition approaches can be divided into three categories depending on their input. Most existing models fall under the category of single-view unimodal approaches. In multi-modal activity recognition models, two or more videos from different modalities are used to classify actions. Finally, multi-view activity recognition

Electronic supplementary material The online version of this chapter (https://doi.org/10.1007/978-3-030-87202-1_60) contains supplementary material, which is available to authorized users.

© Springer Nature Switzerland AG 2021
M. de Bruijne et al. (Eds.): MICCAI 2021, LNCS 12904, pp. 626–635, 2021.
https://doi.org/10.1007/978-3-030-87202-1_60

models operate on multiple videos observing the scene from different viewpoints. Existing datasets for multi-view action classification on short clips mainly include a single instance of an action [11,20], but the more complex task of multi-view action detection in long and complex videos captured in the wild has no suitable dataset, leaving the task fairly unexplored.

Since operating rooms are often cluttered, relying on a single video feed to recognize activities is unrealistic. Therefore, in this paper, we delve into multi-view action detection. As our first contribution, we prepare the first-of-its-kind *"Multi-view Robot-Assisted Operation Dataset"* for action detection. This dataset includes over 120 full-length surgery cases, each recorded via four time-synchronized time-of-flight sensors installed in strategic locations within different operating rooms of a medical institution. These videos average 2 h long, and are individually annotated with 10 clinically significant activity classes. We look to improve activity detection results in surgical environments as it can increase efficiency, reduce cost, and improve care delivery.

For our second contribution, we design a new model architecture that utilizes multiple views to improve activity detection. Our model takes in long videos captured from different viewpoints and recognizes the activity taking place at any time. We introduce a flexible late fusion model that is capable of taking in multiple views in any order as input and outputting a classification decision that is better than that of any individual view. Our model fuses decisions in a latent space using a method similar to Attention is All You Need introduced by Vaswani *et al.* [23], except instead of operating on a per-element basis, we define a global key in order to be able to regress a global estimate. This type of model can prove useful in any application where there are multiple data-sources with similar modality and different views. We also explore early and hybrid fusion methods, but notice no improvement. Thus we opt for our late fusion method.

2 Related Work

Action detection has been studied in medical computer assisted interventions, although the task is typically known as *surgical phase segmentation* [27,30]. Given an endoscopic video of a surgical case, the objective is to find the start and end times of each surgical sub-task. These videos are often hours long but the actions of interest are fairly coarse and limited. Patient mobility detection in ICU [16,28] is another relevant area where similar works could be found. Perhaps the most relevant work to our efforts is that of Sharghi *et al.* [21] who study the same problem as ours, but use single views.

Most of the work in the computer vision community is focused on classifying trimmed videos. The conventional action classification models are based on hand-crafted visual features [10,24] while newer models are designed to learn discriminative spatio-temporal features to better represent videos [4,6]. Most of these methods assume trimmed videos, i.e., the action of interest takes nearly the entire video duration. Thus, for action detection—detecting instances of actions in long videos—components are added to the base action classification models.

Automatic action detection has applications in surveillance and sport, and workflow and interaction analysis. Recent works [12,29], use an action proposal module that is responsible for selecting video segments that are likely to have an action of interest occurring in them. The features of such segments are then fed to a classifier to determine the action class.

One way to improve practicality and performance of action detection models is through the use of complementary input signals. Multimodal detection models can exploit signals including visual, pose, WiFi footprint, etc. simultaneously [3,9]. Another way to make robust action detection models is to deploy multiple cameras. This is perhaps the most effective approach to reliably account for clutter, occlusion, and camera malfunction.

We define multi-view as a dataset where the data modality is the same, but it is collected from different locations e.g. audio from multiple microphones or video from multiple cameras. While there exist datasets for multi-view action classification [2,11, 18,20,25], i.e., classifying a trimmed clip including a single instance of one action, only a few were collected for multi-view action detection as well (please refer to the supplementary material for a table summarizing multi-view datasets). In the medical space, the CATARACTS 2020 dataset [2] looks to classify steps in cataract surgery using two non-overlapping views could benefit from our same approach, although we deal with Time-of-Flight data. The small amount of data available for action detection highlights the importance of our efforts in this work to collect such a dataset and develop a model that can take advantage of multiple video feeds that share a common field of view.

3   Dataset

The dataset presented in this work has been collected from real clinical cases as part of an Institutional Review Board (IRB) approved study. Our data collection system includes four time-synchronized Time-of-Flight (ToF) sensors and is deployed in two robotic operating rooms. We acquired data from 120 full-length surgical cases from four viewpoints (hence 480 videos in total) and annotated every video with a predefined set of clinical activities. See Fig. 1a and 1b for example activities and layout, respectively.

All surgical cases in our dataset were captured in a single medical institution with two robotic operating rooms. We cover 28 different types of procedure across multiple specialties performed by 16 surgeons all using the daVinci Xi surgical system. Such

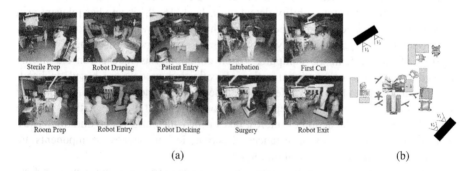

Sterile Prep   Robot Draping   Patient Entry   Intubation   First Cut

Room Prep   Robot Entry   Robot Docking   Surgery   Robot Exit

(a)   (b)

Fig. 1. Dataset: (a) Example surgical activities and annotations (b) Operating room layout and sensor placement

dataset characteristics enable studying a model's generalization capabilities from different aspects. By dividing the videos into train and test data according to which OR they were captured in, we can determine the sensitivity of the trained model to the OR layout changes. Testing the model on unseen operations, we assess if the model can generalize to novel procedures. We can also train on operations performed by a group of surgeons and evaluate on the remaining surgeons that the model has not seen during training. Finally, to obtain an upper-bound on model's performance, we can split the data into train and test randomly.

4 Model Architecture

This section details our approach to action detection in long videos. We first describe our single-view model in Sect. 4.1. Then, using the single-view model as a building block, we design a smart fusion model that fuses data from multiple views to improve performance in Sect. 4.2.

Before delving in, we briefly cover our notation for the dataset. Videos for a surgery recorded with N cameras are denoted as: $V^i, i \in \{1, ..., N\}$. Each video is defined as a set of frames of grayscale ToF images, $V^i := \{I_1^i, ..., I_{l_i}^i\}$. l_i is the number of frames that video V^i has, and I_n^i is a single ToF frame.

Each video has a set of timestamps $T^i := \left(t_n^i\right)_{n \in \{1, 2, ..., l_i\}}$ as well. Each timestamp $t_n^i \in \mathbb{R}$ is a real number, synced up to a base-time with other sensors. In order to learn a model that can run online and classify an action at time t_n^i, we want to use the current view in addition to prior frames of all other views. We align frames temporally to their nearest prior frames in order to implement a neural network on top of multiple views.

For an annotated set of videos of the same surgery the ground truth may vary between views because videos are annotated individually. To deal with this, we treat the current view (the view with the most recent timestamp) as the output label to regress against. The current view, as expected, varies over time. This formulation of using the current view allows us to have the most recent data to estimate the most recent ground truth annotation.

We first focus on improving the single-view model presented in [21] by increasing the modeling capacity, and then adapt it to work on any number of video feeds. We propose a few changes to their model. Firstly we add a skip layer around the RNN to allow the loss to propagate more easily to earlier pieces of the model. We then add layers to the model and increase channel sizes: allowing the model to compute in a space larger than the class size. Finally, we develop a model to fuse detections over multiple views. Our final method is both flexible and extensible: working with any number of views in any order; and any other clip-level backbones such as SlowFast [6], or X3D [5] could be inserted to achieve similar results.

4.1 Single-View RNN Model

Our single-view recurrent neural network (RNN) is a model that takes in a set of clips for a view and returns an activity class for said view. Since the videos are hours long, we use a clip-level model I3D [4] to obtain compact, yet representative latent features for each input clip. These features are passed through an RNN to obtain the class probabilities. We train I3D first on clips, and then train an action detection model on top of the clip features. This allows us to avoid excess costs that result from having to back-propagate all the way through a CNN model.

We denote C_t^{ij} as the synchronized clip of size l_{clip} ending at time t. i denotes the camera that we are using as the current view (the one we regress against), while j represents the view we are aligning temporally. In our single-view formulation $i = j$ as there is only one view available. We leave the notation in this more general manner for our multi-view model that we introduce later.

First, we train an I3D model f on all clips C^{ii}. The model f takes in a clip and outputs an action label. After training this model we fix the weights in f and use all the layers but the last to obtain a representative latent vector from f. To estimate the action detection label we take prior seen frames, and transform them with the I3D model, f. This results in a set of latent vectors: $z_s^{ij} = \left(f(C_{16}^{ij}), \ldots, f(C_{s*16}^{ij}) \right)$. The single-view model only uses the z^{ii} in our notation, meaning clips from video i aligned to video i.

We use these latent vectors as inputs for an RNN model g_{single} which uses a few fully connected layers and an RNN to estimate an output classification: $\hat{y}_s^i = fc\left(g_{single}\left(z_s^{ii} \right) \right)$. $\hat{y}_s^i$ is the estimated logit probability for clip s from view i. We omit frame numbers through the rest of the paper; frames being lined up using the frame number s is implied.

The d_{latent}-dimensional output of this model is denoted as $v_{single}^i = g_{single}(z^{ii})$. This is used as an input for our multi-view model. This single-view model takes in all prior frames as inputs for a single view, i, and outputs a feature $v_{single}^i \in \mathbb{R}^{d_{latent}}$ that is turned into a logit probability with a fully connected layer $\hat{y}_{single}^i = fc(v_{single}^i)$. We use binary cross entropy as our loss, with input as logit values ($LogitBCE$). Loss for the single-view model is defined as $\mathcal{L}_{singleview} = LogitBCE(y, \hat{y}_{single})$, Where $\hat{y}_{single}$ is estimated using a single RNN model following the structure shown in a single sensor branch of Fig. 2. We add a residual layer around the RNN to allow the network to propagate gradients and still perform well on single-clip inputs.

4.2   Smart Fusion Model

We use our single-view model as a building block for the multi-view case. All the weights in g_{single} are shared through all views. The multi-view model is designed to run on top of this as a form of late fusion, taking in the outputs of g_{single}. We do not freeze the weights of the model g_{single}, the only weights frozen throughout this paper are those of the backbone clip-level model f (I3D in our case) after training on clips.

The smart fusion model mixes the latent output vectors from single-view RNNs:

$$g_{multi} = mix\left(g_{single}\left(z^{i0} \right), \ldots, g_{single}\left(z^{iN} \right) \right) \tag{1}$$

The mix function takes in a set of d_{latent} sized vectors and fuses them. We use a fully connected layer to output the final prediction result: $\hat{y} = fc(g_{multi})$ We define mix as a function that sums over a set of latent vectors:

$$\sum_j \omega_j g_{single}\left(z^{ij} \right) \tag{2}$$

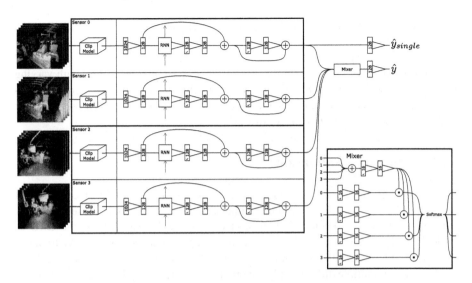

Fig. 2. Our multi-view action detection architecture takes in an arbitrary number of views, mixes them, and outputs a decision. Both the all-view output and the single-view output are used for the loss. Here we show the model with four inputs. The mixer module outputs scalar weight values used for mixing each view in latent space.

We could also define the mix operation more simply as a concatenation operation or an averaging operation, and we run this as an ablation experiment. The mixing weights ω can be determined as a function of inputs, or fixed beforehand. If we define $\omega_j = \frac{1}{4}$ then we have an average pooling model. We can define functions for similar procedures such as argmax (picking most confident), or fixed weightings based on validation performance as in Machado *et al.* [17].

In a similar essence to Attention is All You Need [23], we use scaled dot product attention. This gives us a data-dependent way to decide how much each view should contribute, enabling the model to adapt to the situation. Our model differs in that the query vector q is a globally estimated value, rather than having a query vector for each input value. We do this as we want the model to make a global decision about which action is happening.

We opt to learn the weight vector that is used according to an attention model:

$$\omega^\top = softmax(\frac{q^\top K}{\sqrt{d_k}}), K \in \mathbb{R}^{d_k \times N} \tag{3}$$

with K being the latent view feature vectors formed into a matrix, q is a global vector estimated using average pooling of the latent vectors, and d_k is the mixer dimension. The final model, shown in Fig. 2, is then defined as:

$$\hat{y} = fc(g_{multi}). \tag{4}$$

We define the multiview loss as a binary cross entropy $\mathcal{L}_{multiview} = LogitBCE(y, \hat{y})$ This is combined with the single view loss to ease training and enable the network to maintain good performance on single-view inputs in case of loss of a view: $\mathcal{L} = \mathcal{L}_{multiview} + \mathcal{L}_{singleview}$.

In our case, we apply this model to $(1, 2, 4)$ views, but there is no limit to the input size, and our model remains adaptive. For instance, we could even train with varying numbers of views in the training set, and we enable this in our implementation for cases where we do not have data for all cameras throughout. When given just one view our model decays to a single-view model by design.

5   Experimental Validation

For our experiments, we set our model's latent dimension d_{latent} to 128, the I3D output dimension d_{clip} to 1024, and the mixer dimension d_k to 16. We test on sets of views from $N \in (1, 2, 4)$. After testing both LSTMs and GRUs, we settle on the GRU as our RNN of choice. Results of this test are in the supplementary material. We train our model using PyTorch with an Adam optimizer and learning rate of $lr = 0.0015$.

We train on multiple different splits in order to comprehensively evaluate and understand generalization to different environments. For all splits, we group videos of the same individual surgery into the same split in order to prevent overtraining. A brief description of the splits follows.

Random. A random 80% of the videos is selected for training.

Operating Room. This split divides test and train between the two ORs we have in our dataset, with OR1 (57.5% of cases) for training, and OR2 for testing.

Procedure Type. Data is split based on procedure types. We select 12 of 28 surgery types for testing, and the remainder for training, resulting in a 80%–20% split.

Surgeon. Data is split based on surgeons, selected to maintain a 80%–20% split.

Results on each of these splits helps inform us about generalization ability of the models. Operating room changes, for example, result in a drastically different layout of the scene. Likewise, surgeon and procedure splits have different workflows based on surgeon tendencies and procedure requirements.

We evaluate our pipeline using mean Average Precision (mAP). To showcase our model leveraging additional views, we re-generate splits, and re-train I3D and our model 5 times, reporting standard deviation and mean in Table 1. We additionally test the model using a bidirectional model in order to evaluate how well the model would work in an offline environment.

Using our fusion model we show that increasing views results in a large improvement in mAP as shown in Table 1. In short, going from one view to four views with our model increases the mAP by 6.33% for OR, 3.19% for surgeons, 3.74% for procedure, and 2.53% for random. Incorporating additional views results in the largest relative improvement for the OR split, likely because weighting confident views is most useful in cases where we are moving to novel view directions.

We also train our model with a bidirectional RNN and compare it to prior work. Our model has a clear improvement over prior work on all splits. This is shown in Table 2. Even in the single-view case of (N = 1) our model still outperforms the prior work noticeably. The reasons for this lie in the 20 surgical case videos we add to the dataset

Table 1. Our multi-view smart fusion model's results in mean Average Precision over 5 runs for (1, 2, 4) views.

Split	$1 \pm \sigma_1$	$2 \pm \sigma_2$	$4 \pm \sigma_4$
OR	69.2 ± 1.40	72.2 ± 1.65	75.5 ± 1.35
Surgeons	78.1 ± 0.08	79.6 ± 1.23	81.3 ± 1.75
Procedure	80.6 ± 0.56	82.8 ± 0.73	84.3 ± 0.61
Random	87.9 ± 2.00	89.3 ± 1.82	90.5 ± 1.86

Table 2. mAP for bidirectional action detection on different splits. In 'n-view', n represents the number of videos used for training/testing our smart fusion model.

	1-view	2-views	4-views	[21]
OR	77.10	79.23	**81.84**	58.4
Surgeons	85.08	86.70	**87.72**	78.77
Procedure	87.07	89.27	**90.36**	78.04
Random	91.50	92.56	**93.09**	88.81

in addition to the changes we make to their single-view model. We increase the amount of layers, channel depth, and add a skip layer around the RNN.

To show the importance of the mixer module, we ran an ablation experiment that demonstrates its improvement compared to averaging or concatenating as shown in Table 3. We also experimented with fusion in the clip model by integrating our mixer module into I3D (diagram in supplemental material). Integrating a learned global descriptor using this method did not perform better than just using our model in Fig. 2.

Table 3. Ablation comparison using mAP for fusion given four views. Our smart fusion method outperforms the rest.

Fusion type	OR	Surgeons	Procedure	Random
Smart fusion	**76.91**	**83.50**	**84.39**	**90.26**
Average	76.25	81.01	83.75	89.13
Concatenate	76.49	81.88	83.34	90.19

6 Conclusion

In this paper, we investigate the benefits of using multiple views for activity detection in the operating room. We introduce the first-of-its-kind *"Multi-view Robot-Assisted Operation Dataset"* that includes over 120 surgical cases captured in operating rooms, all with multiple views.

Then we implement an attention based model to mix multiple views into a global representation. We demonstrate that our model outperforms the prior state-of-the-art. We design our model to work with varying view location and quality, dynamically focusing the attention towards informative views. Our model is flexible and portable for performing in other multi-view environments where similar datasets can be created, such as endoscopic surgery.

Acknowledgements. This work was completed while Adam Schmidt was an intern at Intuitive Surgical.

References

1. University of Central Florida-aerial camera, rooftop camera and ground camera dataset. https://www.crcv.ucf.edu/data/UCF-ARG.php
2. Al Hajj, H., et al.: CATARACTS: challenge on automatic tool annotation for cataract surgery. Med. Image Anal. **52**, 24–41 (2019)
3. Baltrušaitis, T., Ahuja, C., Morency, L.P.: Multimodal machine learning: a survey and taxonomy. IEEE Trans. Pattern Anal. Mach. Intell. **41**(2), 423–443 (2018)
4. Carreira, J., Zisserman, A.: Quo Vadis, action recognition? A new model and the kinetics dataset. In: Proceedings of the IEEE Conference on Computer Vision and Pattern Recognition, pp. 6299–6308 (2017)
5. Feichtenhofer, C.: X3D: Expanding architectures for efficient video recognition. arXiv:2004.04730 [cs], April 2020
6. Feichtenhofer, C., Fan, H., Malik, J., He, K.: SlowFast networks for video recognition. In: Proceedings of the IEEE International Conference on Computer Vision, pp. 6202–6211 (2019)
7. Gkalelis, N., Kim, H., Hilton, A., Nikolaidis, N., Pitas, I.: The i3DPost multi-view and 3D human action/interaction database. In: 2009 Conference for Visual Media Production, London, United Kingdom, pp. 159–168. IEEE, November 2009. https://doi.org/10.1109/CVMP.2009.19
8. Home Office Scientific Development Branch: Imagery library for intelligent detection systems (i-LIDS). In: 2006 IET Conference on Crime and Security, pp. 445–448, June 2006
9. Joze, H.R.V., Shaban, A., Iuzzolino, M.L., Koishida, K.: MMTM: multimodal transfer module for CNN fusion. In: Proceedings of the IEEE/CVF Conference on Computer Vision and Pattern Recognition, pp. 13289–13299 (2020)
10. Laptev, I.: On space-time interest points. Int. J. Comput. Vis. **64**(2–3), 107–123 (2005)
11. Li, W., Wong, Y., Liu, A.A., Li, Y., Su, Y.T., Kankanhalli, M.: Multi-camera action dataset for cross-camera action recognition benchmarking. In: 2017 IEEE Winter Conference on Applications of Computer Vision (WACV), pp. 187–196, March 2017. https://doi.org/10.1109/WACV.2017.28
12. Lin, T., Liu, X., Li, X., Ding, E., Wen, S.: BMN: boundary-matching network for temporal action proposal generation. In: Proceedings of the IEEE International Conference on Computer Vision, pp. 3889–3898 (2019)
13. Liu, A., Su, Y., Jia, P., Gao, Z., Hao, T., Yang, Z.: Multiple/single-view human action recognition via part-induced multitask structural learning. IEEE Trans. Cybern. **45**(6), 1194–1208 (2015). https://doi.org/10.1109/TCYB.2014.2347057
14. Liu, A., Xu, N., Nie, W., Su, Y., Wong, Y., Kankanhalli, M.: Benchmarking a multimodal and multiview and interactive dataset for human action recognition. IEEE Trans. Cybern. **47**(7), 1781–1794 (2017). https://doi.org/10.1109/TCYB.2016.2582918

15. Liu, J., Shahroudy, A., Perez, M., Wang, G., Duan, L.Y., Kot, A.C.: NTU RGB+D 120: a large-scale benchmark for 3D human activity understanding. IEEE Trans. Pattern Anal. Mach. Intell. **42**(10), 2684–2701 (2020). https://doi.org/10.1109/TPAMI.2019.2916873

16. Ma, A.J., et al.: Measuring patient mobility in the ICU using a novel noninvasive sensor. Crit. Care Med. **45**(4), 630 (2017)

17. Machado, G., Ferreira, E., Nogueira, K., Oliveira, H., Gama, P., dos Santos, J.A.: AiRound and CV-BrCT: novel multi-view datasets for scene classification. arXiv:2008.01133 [cs], August 2020

18. Murtaza, F., Yousaf, M.H., Velastin, S.A.: Multi-view human action recognition using 2D motion templates based on MHIs and their HOG description. IET Comput. Vis. **10**(7), 758–767 (2016). https://doi.org/10.1049/iet-cvi.2015.0416

19. Rybok, L., Friedberger, S., Hanebeck, U.D., Stiefelhagen, R.: The KIT Robo-kitchen data set for the evaluation of view-based activity recognition systems. In: 2011 11th IEEE-RAS International Conference on Humanoid Robots, Bled, Slovenia, pp. 128–133. IEEE, October 2011. https://doi.org/10.1109/Humanoids.2011.6100854

20. Shahroudy, A., Liu, J., Ng, T.T., Wang, G.: NTU RGB+ D: a large scale dataset for 3D human activity analysis. In: Proceedings of the IEEE Conference on Computer Vision and Pattern Recognition, pp. 1010–1019 (2016)

21. Sharghi, A., Haugerud, H., Oh, D., Mohareri, O.: Automatic operating room surgical activity recognition for robot-assisted surgery. In: Martel, A.L., et al. (eds.) MICCAI 2020. LNCS, vol. 12263, pp. 385–395. Springer, Cham (2020). https://doi.org/10.1007/978-3-030-59716-0_37

22. Sigurdsson, G.A., Gupta, A., Schmid, C., Farhadi, A., Alahari, K.: Actor and observer: joint modeling of first and third-person videos. arXiv:1804.09627 [cs], April 2018

23. Vaswani, A., et al.: Attention is all you need. arXiv:1706.03762 [cs], December 2017

24. Wang, H., Schmid, C.: Action recognition with improved trajectories. In: Proceedings of the IEEE International Conference on Computer Vision, pp. 3551–3558 (2013)

25. Wang, J., Nie, X., Xia, Y., Wu, Y., Zhu, S.C.: Cross-view action modeling, learning, and recognition. In: 2014 IEEE Conference on Computer Vision and Pattern Recognition, Columbus, OH, USA, pp. 2649–2656. IEEE, June 2014. https://doi.org/10.1109/CVPR.2014.339

26. Weinland, D., Ronfard, R., Boyer, E.: Free viewpoint action recognition using motion history volumes. Comput. Vis. Image Underst. **104**(2–3), 249–257 (2006). https://doi.org/10.1016/j.cviu.2006.07.013

27. Yengera, G., Mutter, D., Marescaux, J., Padoy, N.: Less is more: surgical phase recognition with less annotations through self-supervised pre-training of CNN-LSTM networks. arXiv preprint arXiv:1805.08569 (2018)

28. Yeung, S., et al.: A computer vision system for deep learning-based detection of patient mobilization activities in the ICU. NPJ Digit. Med. **2**(1), 1–5 (2019)

29. Zhao, Y., Xiong, Y., Wang, L., Wu, Z., Tang, X., Lin, D.: Temporal action detection with structured segment networks. In: Proceedings of the IEEE International Conference on Computer Vision, pp. 2914–2923 (2017)

30. Zia, A., Hung, A., Essa, I., Jarc, A.: Surgical activity recognition in robot-assisted radical prostatectomy using deep learning. In: Frangi, A.F., Schnabel, J.A., Davatzikos, C., Alberola-López, C., Fichtinger, G. (eds.) MICCAI 2018. LNCS, vol. 11073, pp. 273–280. Springer, Cham (2018). https://doi.org/10.1007/978-3-030-00937-3_32

Interhemispheric Functional Connectivity in the Primary Motor Cortex Distinguishes Between Training on a Physical and a Virtual Surgical Simulator

Anirban Dutta[3]([⊠]), Anil Kamat[1], Basiel Makled[4], Jack Norfleet[4], Xavier Intes[1,2], and Suvranu De[1,2]

[1] Center for Modeling, Simulation and Imaging in Medicine, Rensselaer Polytechnic Institute, Troy, NY, USA
[2] Department of Biomedical Engineering, Rensselaer Polytechnic Institute, Troy, NY, USA
[3] Department of Biomedical Engineering, University at Buffalo, Buffalo, NY, USA
anirband@buffalo.edu
[4] U.S. Army Futures Command, Combat Capabilities Development Command Soldier Center STTC, Orlando, FL, USA

Abstract. Functional brain connectivity using functional near-infrared spectroscopy (fNIRS) during a pattern cutting (PC) task was investigated in physical and virtual simulators. 14 right-handed novice medical students were recruited and divided into separate cohorts for physical (N = 8) and virtual (N = 6) PC training. Functional brain connectivity measured were based on wavelet coherence (WCOH) from task-related oxygenated hemoglobin (HBO2) changes from baseline at left and right prefrontal cortex (LPFC, RPFC), left and right primary motor cortex (LPMC, RPMC), and supplementary motor area (SMA). HBO2 changes within the neurovascular frequency band (0.01–0.07 Hz) from long-separation channels were used to compute average inter-regional WCOH metrics during the PC task. The coefficient of variation (CoV) of WCOH metrics and PC performance metrics were compared. WCOH metrics from short-separation fNIRS time-series were separately compared. Partial eta squared effect size (Bonferroni correction) between the physical versus virtual simulator cohorts was found to be highest for LPMC-RPMC connectivity. Also, the percent change in magnitude-squared WCOH metric was statistically ($p < 0.05$) different for LPMC-RPMC connectivity between the physical and the virtual simulator cohorts. Percent change in WCOH metrics from extracerebral sources was not different at the 5% significance level. Also, higher CoV for both LPMC-RPMC magnitude-squared WCOH metric and PC performance metrics were found in physical than a virtual simulator. We conclude that interhemispheric connectivity of the primary motor cortex is the distinguishing functional brain connectivity feature between the physical versus the virtual simulator cohorts. Brain-behavior relationship based on CoV between

---

Electronic supplementary material The online version of this chapter (https://doi.org/10.1007/978-3-030-87202-1_61) contains supplementary material, which is available to authorized users.

M. de Bruijne et al. (Eds.): MICCAI 2021, LNCS 12904, pp. 636–644, 2021.
https://doi.org/10.1007/978-3-030-87202-1_61

the LPMC-RPMC magnitude-squared WCOH metric and the FLS PC performance metric provided novel insights into the neuroergonomics of the physical and virtual simulators that are crucial for validating Virtual Reality technology.

Keywords: fNIRS · Fundamental Laparoscopic Surgery · Physical surgical simulator · Virtual surgical simulator

1   Introduction

Fundamentals of Laparoscopic Surgery (FLS) is a pre-requisite for board certification in general surgery in the USA. As a part of the FLS program, five psychomotor tasks with increasing task complexity are used for training and evaluation: (i) pegboard transfers, (ii) pattern cutting, (iii) placement of a ligating loop, (iv) suturing with extracorporeal knot tying, and (v) suturing with intracorporeal knot tying. An important aspect of learning laparoscopic surgery compared to open surgery entails the sensorimotor and visuomotor adaptation of the 3D surgical field to 2D visualization, reduced tactile perception, and bimanual hand-eye coordination using ergonomically altered surgical instruments [1]. Prior works have shown the regional extent of brain activation in novices during laparoscopic surgery training in the functional magnetic resonance imaging (MRI) environment that depended on the task complexity [1] where tasks ii–v have shown activation of the Brodmann area (BA) 5 and BA 7 of the parietal cortex related to visual-motor coordination besides expected activation of BA 4 for bimanual coordination that includes the primary motor cortex (M1). While mainly the lateral portion of BA 6 of the frontal cortex showed activation for tasks i-iii, both the lateral and the medial portions of BA 6, including the premotor cortex and the supplementary motor area (SMA), showed activation in the most complex bimanually coordinated task v. In order to investigate these neural correlates (brain aspect) of laparoscopic surgery training, functional near infrared spectroscopy (fNIRS) can provide a better temporal resolution and ease of administration when compared to fMRI; however, at a lower spatial resolution and signal-to-noise ratio (SNR) [2]. Many prior works [3–7] have assessed surgery training using fNIRS mainly to compare skill levels; however, we could not find prior works on portable neuroimaging to compare surgery training in physical versus virtual simulators. This is crucial since surgical training field has seen a rapid emergence of new virtual training technologies which are gradually replacing physical training simulators with virtual training simulators. Here, virtual simulators been shown to improve acquisition of skills [8]; although, the perception-action neural correlates of skill acquisition in physical versus virtual simulators are unknown.

In this study, we investigated fNIRS to capture brain activation during the pattern cutting (PC) task which may be considered a transitional task from the purely unimanual task i to the completely bimanual task v in terms of task complexity. In the PC task, the trainees cut a marked gauze and are given a score based on the accuracy of cutting and the time spent completing the activity. During the PC activity, the trainee provides traction to the gauze with one hand to place it in the best possible orientation to the cutting hand, and uses endoscopic scissors in the other hand to cut into the gauze along the pre-marked circle until it is completely removed. Therefore, modulation of interhemispheric inhibition of the M1 for bimanual coordination is expected during the PC task. Our prior work [9] investigated the brain functional connectivity based on wavelet coherence metrics for the following brain regions in right-handed subjects: left lateral prefrontal cortex, medial prefrontal cortex, right lateral prefrontal cortex, left medial primary motor cortex, and SMA, and found only in the case of FLS box trainer (not for virtual basic laparoscopic skills trainer: VBLaST) a correspondence to the surgical motor skills. Specifically, the partial eta squared effect size for the inter-regional magnitude-squared wavelet coherence metric between the medial prefrontal cortex and the SMA was found to be highest for expert versus novice comparison in the FLS box trainer cohort. However, a lack of investigation of the correspondence of brain functional connectivity to the surgical motor skills in VBLaST cohort needed further investigation since VBLaST has demonstrated face and construct validity [10, 11] including skill transfer [12]. Learning bimanual coordination of finger movements by novices for PC task performance (speed and accuracy) is postulated to involve learning to modulate interhemispheric inhibition of the primary motor cortices so we included fNIRS of bilateral primary motor cortices in the current study. Here, wavelet coherence [9] and cross-spectrum can provide a measure of the time-varying inter-regional association that may be sensitive to PC bimanual task-coupling to the interhemispheric coordination in physical versus virtual simulators. Therefore, we investigated fNIRS oxy-hemoglobin (HbO2)-based wavelet coherence metric within the neurovascular frequency band (0.01–0.07 Hz) related to neuronal activation, which was used to compare changes in the PC task-related brain functional connectivity from pre-task resting-state baseline in novices training in physical versus virtual simulators.

2   Materials and Methods

2.1   Subjects and Experimental Design

The study was approved by the Institutional Review Board of the Massachusetts General Hospital and the University at Buffalo. A priori power analysis, based on two-sample t-tests, determined the minimum number of samples required for this study. With a 95% confidence interval and a minimum power of 0.80, a minimum of eight subjects each for the expert and novice surgeon cohort group, four subjects for the FLS training group, three subjects for the VBLaST training group, and four subjects for the control group was estimated in our prior work [9]. In this study, we analyzed the data from 14 healthy novice medical students from prior work [9] where they performed performing the PC task in the physical and virtual simulators (demographics in the Table 1 of the Supplementary Materials). The novice medical students did not perform any laparoscopic procedures earlier. All right-handed subjects were divided into two cohorts; the physical simulator group consisted of 8 subjects whereas the virtual simulator group had 6 subjects. All the subjects were instructed verbally with a standard set of instructions on how to perform the PC task, goal, and the rule of the task completion. The optical probes or optodes held by a standard electroencephalography (EEG) cap (www.easycap.de) were mounted on the scalp of each participant by avoiding any hair in between the source/detector and the scalp. During the trial, the subjects were asked to perform the PC task, where the goal was to cut along the circular mark on a piece of gauze as accurately and as quickly as possible (details in our prior work [9]). After a baseline rest period of 1 min, the trial was started and the maximum time provided for doing the PC task was 5 min. The scoring of the gauze cut on the physical simulator were subjectively assessed by the proctor, whereas for the virtual simulator the scores were automatically generated (details in prior works [3, 9]).

2.2   Equipment

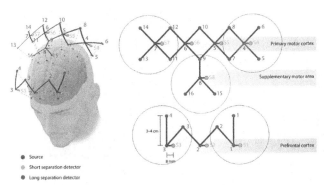

Fig. 1. Selected optode positions over the PFC, M1, and SMA (from our prior work [3]) for non-overlapping cortical regions, Left PFC (LPFC), Right PFC (RPFC), Left M1 (LPMC), Right M1 (RPMC), and SMA, based on anatomical guidance in AtlasViewer. Red dots indicate infrared sources, blue dots indicate long separation detectors, and light blue dots indicate short separation detectors. In the current study, PFC has two sources (1 and 3), two short separation detectors (S1 and S3), and four long separation detectors (1 to 4). The M1 has 2 sources (4 and 7), 2 short separation detectors (S4 and S7), and 8 long detectors (5 to 8 and 11 to 14). The SMA has one source (8), one short separation detector (S8), and three long separation detectors (9, 15, and 16).

A 32-channel continuous-wave near-infrared spectrometer (CW6 system, TechEn Inc.) was used for this study that delivered infrared light at 690 nm and 830 nm. Optode montage consisted of eight long-distance and eight-short distance sources coupled to 16 detectors. 25 long-distance (30–40 mm) channels and 8 short-distance (~8 mm) channels measured brain activation at the bilateral prefrontal cortex (PFC), primary motor cortex (M1), and SMA that were assessed using Monte Carlo simulations in AtlasViewer

(https://github.com/BUNPC/AtlasViewer). Based on the EEG cap locations and the optode sensitivity profile (Supplementary Materials – Fig. S1 we selected fNIRS channels that measured from the non-overlapping cortical regions (see Fig. 1), Left PFC (LPFC), Right PFC (RPFC), Left M1 (LPMC), Right M1 (RPMC), and SMA, based on anatomical guidance in AtlasViewer – an open-source software in Matlab (Mathworks Inc., USA) [9].

2.3   Data Processing for Oxy-Hemoglobin Time-Series

Motion artifact detection and correction were performed using combined spline interpolation and Savitzky-Golay filtering [10] in HOMER3 (https://github.com/BUNPC/Homer3), which is an open-source software in Matlab (Mathworks Inc., USA). Then, modified Beer-Lambert law was used to convert the detectors' raw optical data into optical density. Then, the conversion of optical density to changes in HbO2 concentrations with differential path-length factors of 6.4 (690 nm) and 5.8 (830 nm) based on our prior work [3]. The mean HbO2 changes for each brain region [9], LPFC, RPFC, LPMC, RPMC, SMA, from long separation channels (inter-optode distance of 30–40 mm) specific to the cortical activity in those regions was analyzed within neurovascular frequency band (0.01–0.07 Hz) based on our prior work [9]. Short separation channels (inter-optode distance of 8 mm) captured the systemic physiology originating from non-cortical superficial regions which was separately analyzed to investigate the effect of noise using the wavelet transform [13].

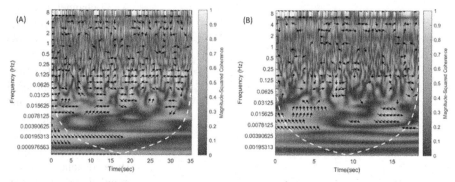

Fig. 2. Illustrative examples of the magnitude-squared wavelet coherence between signals from the supplementary motor area and right primary cortex in the physical simulator (A) and virtual simulator (B). The color bar represents the squared magnitude.

2.4 Mean and the Coefficient of Variation of Functional Connectivity Metrics from fNIRS HbO2 Time Series and FLS Performance Scores

Pair-wise inter-regional functional connectivity between LPFC, RPFC, LPMC, RPMC, and SMA were estimated using wavelet coherence from the fNIRS HbO2 time series. In this study, wavelet coherence based pair-wise inter-regional functional connectivity was computed using the analytic Morlet wavelets that has shown good performance in time-frequency analysis by many prior works [9, 14]. Analytic Morlet wavelets were centered within the neurovascular frequency band (0.01–0.07 Hz) and applied using Wavelet Toolbox in Matlab (Mathworks Inc., USA). Due to the complex nature of the Morlet wavelet, the wavelet transform for each time and scale is a complex value. We performed the time-frequency analysis of the magnitude-squared wavelet coherence (WCOH), as shown in Fig. 2, and found the mean and the variance of WCOH along the time domain within the neurovascular frequency band (0.01–0.07 Hz). Also, for a normalized measure of WCOH, the corresponding sine and cosine of the phase difference was averaged in time yielding wavelet phase coherence (WPCO) between the inter-regional HbO2 time series.

The mean and the variance (also, coefficient of variation) of WCOH and WPCO were computed for the baseline resting state (1 min) and the PC task state separately from the long-separation and the short separation fNIRS channels. Here, brain-behavior relationship can be elucidated based on their coefficient of variation (CoV) [15], i.e., a higher CoV in the brain metrics (WCOH and WPCO) is postulated to be related to a higher CoV in the behavior metrics (FLS performance scores). Here, mean WCOH and WPCO values from the long-separation channels at the baseline and during the PC task estimated the functional connectivity for the following inter-regions: LPFC-RPFC, LPFC-LPMC, LPFC-RPMC, LPFC-SMA, RPFC-LPMC, RPFC-RPMC, RPFC-SMA, LPMC-RPMC, LPMC-SMA, RPMC-SMA.

The pre-task pair-wise mean WCOH and WPCO values were considered as the baseline connectivity and the percentage change from baseline was computed for the PC task state separately for the long-separation and the short-separation fNIRS channels. As a control measure, the percentage change from baseline during the PC task in the mean WCOH and WPCO values from the short separation channels were computed to determine systemic noise effects. After checks for normality (Shapiro–Wilk test, quantile-quantile plot) and homogeneity of variance (Levene test), we performed one-way multivariate analysis of variance (one-way MANOVA) in SPSS version 27 (IBM) to determine whether there is any significant difference in the inter-regional (i.e., LPFC-RPFC, LPFC-LPMC, LPFC-RPMC, LPFC-SMA, RPFC-LPMC, RPFC-RPMC, RPFC-SMA, LPMC-RPMC, LPMC-SMA, RPMC-SMA) functional connectivity metrics (average WCOH and WPCO values) using Wilks' Lambda between the physical and the virtual simulator cohorts. Then, to determine how the dependent variables (i.e., inter-regional functional connectivity) differ for the independent variable (physical versus virtual simulator), partial eta squared effect size was used with alpha correction with Bonferroni correction. Then, one-way analysis of variance (ANOVA) was performed to determine whether the independent variable, the physical (FLSbox) and the virtual (VBLaST) simulators have different effects on individual inter-regional functional connectivity (percent change in magnitude-squared WCOH metric) separately for the long-separation and the short-separation fNIRS channels.

3 Results

3.1 Physical Versus Virtual Simulator Effect on the Functional Connectivity

An illustrative example of the magnitude-squared wavelet coherence plot for RPMC-SMA in the physical and virtual simulators is shown in the Fig. 2. Grouped plot of the matrix of percentage change from baseline in average WCOH and WPCO during PC task in physical (FLSbox) and virtual (VBLaST) simulators across ten inter-regional LPFC-RPFC, LPFC-LPMC, LPFC-RPMC, LPFC-SMA, RPFC-LPMC, RPFC-RPMC, RPFC-SMA, LPMC-RPMC, LPMC-SMA, RPMC-SMA are shown in the Supplementary materials (Fig. S2). One-way multivariate ANOVA did not find significant differences at the 1% level in the averages of ten (inter-regional) WCOH and ten (inter-regional) WPCO measures between the physical (FLSbox) and the virtual (VBLaST) simulators. Partial eta squared effect size (Bonferroni correction) was found to be highest for LPMC-RPMC connectivity.

3.2 Statistical Analysis on Physical Versus Virtual Simulator Effect on the Individual Inter-regional Functional Connectivity

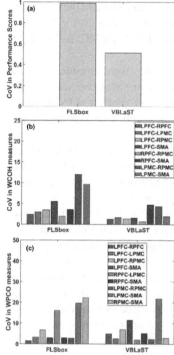

Shapiro-Wilk test of normality conducted on the functional brain connectivity metrics (WCOH, WPCO) found that while all the metrics in the virtual simulator cohort passed the test at the 5% level, few metrics in the physical simulator cohort digressed from the normal distribution. Therefore, we performed both the one-way ANOVA and the Wilcoxon Rank Sum Test to determine whether the independent variable, the physical (FLSbox) and the virtual (VBLaST) simulators, have different effects on the individual response variables, WCOH and WPCO measures. Results are summarized in the Fig. S3 (Supplementary Materials), where the percent change from baseline in the WCOH metric for LPMC – RPMC functional connectivity was found to be statistically ($p < 0.05$) different between the physical and the virtual simulators for the long separation channels. However, no significant differences at the 5% level in the percent change from baseline in the percent change from baseline in the WCOH metric were found for the short separation channels. No significant differences at the 5% level in the percent change from baseline in the WPCO metrics between the physical and the virtual simulator cohorts were found for the long and the short separation channels.

Fig. 3. Coefficient of variation (CoV) in brain (WCOH, WPCO) and behavior (FLS performance) measures. [WCOH, WPCO outliers removed based on 'isoutlier' in Matlab] (A) CoV for the FLS performance scores in the physical and the virtual simulators. (B) CoV for the percent change from baseline in inter-regional magnitude-squared wavelet coherence (WCOH) measures in simulators. (C) CoV for percent change from baseline in inter-regional wavelet phase coherence (WPCO) measures in simulators.

3.3 Brain – Behavior Correspondence

Figure 3 shows that the physical simulator resulted in higher CoV than the virtual simulator in both the brain (WCOH, WPCO) and the behavior metrics (FLS performance scores). Specifically, CoV for the WCOH measure (see Fig. 3B) for the LPMC-RPMC and the CoV for the WPCO measure for the RPMC-SMA (see Fig. 3C) were highest in the physical simulator.

4 Discussion

WCOH metric for the LPMC-RPMC functional connectivity was the best marker between FLS PC training in the physical and the virtual simulators in terms of both partial eta squared effect size and the CoV-based brain-behavior analysis (see Fig. 3). Importantly, higher CoV for the LPMC-RPMC functional connectivity in the physical simulator corresponded with higher CoV in FLS performance score in the physical simulator, as shown in Fig. 3. Even with our limitation due to a small sample size, our partial eta squared effect size and the CoV-based brain-behavior analysis highlighted the relevance of LPMC-RPMC functional connectivity during FLS PC training. Here, nomological validity can be derived from prior work [15] that showed that cortical functional MRI variability explained the movement extent variability. Also, the relevance of inter-hemispheric LPMC-RPMC functional connectivity is supported by the relevance of corpus callosum in bimanual coordination [16]. Our brain – behavior approach based on the correspondence of CoV for brain and behavior metrics can be used to find neural correlates (brain aspect) to assess and improve laparoscopic surgical skills. Therefore, fNIRS-based investigation of the variability in behavior is crucial in neuroergonomics for studying the brain and behavior at work [17] and its facilitation with neuroimaging guided transcranial electrical stimulation [18–22].

5 Conclusion

We showed that inter-hemispheric primary motor cortex functional connectivity based on wavelet coherence was most dissimilar in novices between the PC task performed in the physical versus virtual simulators.

Acknowledgment. Medical Technology Enterprise Consortium (MTEC) award #W81XWH 2090019 (2020-628), U.S. Army Futures Command, Combat Capabilities Development Command Soldier Center STTC cooperative research agreement #W912CG-21-2-0001.

Disclosures: No authors have neither relevant financial or competing interests nor other potential conflicts of interests.

References

1. Bahrami, P., et al.: Neuroanatomical correlates of laparoscopic surgery training. Surg. Endosc. **28**(7), 2189–2198 (2014). https://doi.org/10.1007/s00464-014-3452-7
2. Cui, X., Bray, S., Bryant, D.M., Glover, G.H., Reiss, A.L.: A quantitative comparison of NIRS and fMRI across multiple cognitive tasks. Neuroimage **54**, 2808–2821 (2011). https://doi.org/10.1016/j.neuroimage.2010.10.069
3. Nemani, A., et al.: Assessing bimanual motor skills with optical neuroimaging. Sci. Adv. **4**, eaat3807 (2018). https://doi.org/10.1126/sciadv.aat3807
4. Nemani, A., Kruger, U., Cooper, C.A., Schwaitzberg, S.D., Intes, X., De, S.: Objective assessment of surgical skill transfer using non-invasive brain imaging. Surg. Endosc. **33**(8), 2485–2494 (2018). https://doi.org/10.1007/s00464-018-6535-z
5. Khoe, H.C.H., et al.: Use of prefrontal cortex activity as a measure of learning curve in surgical novices: results of a single blind randomised controlled trial. Surg. Endosc. **34**(12), 5604–5615 (2020). https://doi.org/10.1007/s00464-019-07331-7

6. Leff, D.R., et al.: Functional prefrontal reorganization accompanies learning-associated refinements in surgery: a manifold embedding approach. Comput. Aided Surg. **13**, 325–339 (2008). https://doi.org/10.3109/10929080802531482

7. Keles, H.O., Cengiz, C., Demiral, I., Ozmen, M.M., Omurtag, A.: High density optical neuroimaging predicts surgeons's subjective experience and skill levels. PLoS ONE **16**, e0247117 (2021). https://doi.org/10.1371/journal.pone.0247117

8. Jordan, J.A., Gallagher, A.G., McGuigan, J., McClure, N.: Virtual reality training leads to faster adaptation to the novel psychomotor restrictions encountered by laparoscopic surgeons. Surg. Endosc. **15**(10), 1080–1084 (2001). https://doi.org/10.1007/s004640000374

9. Nemani, A., et al.: Functional brain connectivity related to surgical skill dexterity in physical and virtual simulation environments. NPh. **8**, 015008 (2021). https://doi.org/10.1117/1.NPh.8.1.015008

10. Sankaranarayanan, G., et al.: Preliminary face and construct validation study of a virtual basic laparoscopic skill trainer. J. Laparoendosc. Adv. Surg. Tech. A. **20**, 153–157 (2010). https://doi.org/10.1089/lap.2009.0030

11. Linsk, A.M., et al.: Validation of the VBLaST pattern cutting task: a learning curve study. Surg. Endosc. **32**(4), 1990–2002 (2017). https://doi.org/10.1007/s00464-017-5895-0

12. Nemani, A., Ahn, W., Cooper, C., Schwaitzberg, S., De, S.: Convergent validation and transfer of learning studies of a virtual reality-based pattern cutting simulator. Surg. Endosc. **32**(3), 1265–1272 (2017). https://doi.org/10.1007/s00464-017-5802-8

13. Duan, L., Zhao, Z., Lin, Y., Wu, X., Luo, Y., Xu, P.: Wavelet-based method for removing global physiological noise in functional near-infrared spectroscopy. Biomed. Opt. Exp. **9**, 3805–3820 (2018). https://doi.org/10.1364/BOE.9.003805

14. Zhang, X., et al.: Activation detection in functional near-infrared spectroscopy by wavelet coherence. J Biomed Opt. **20**, 016004 (2015). https://doi.org/10.1117/1.JBO.20.1.016004

15. Haar, S., Donchin, O., Dinstein, I.: Individual movement variability magnitudes are explained by cortical neural variability. J. Neurosci. **37**, 9076–9085 (2017). https://doi.org/10.1523/JNEUROSCI.1650-17.2017

16. Gooijers, J., Swinnen, S.P.: Interactions between brain structure and behavior: the corpus callosum and bimanual coordination. Neurosci. Biobehav. Rev. **43**, 1–19 (2014). https://doi.org/10.1016/j.neubiorev.2014.03.008

17. Dehais, F., Lafont, A., Roy, R., Fairclough, S.: A neuroergonomics approach to mental workload, engagement and human performance. Front Neurosci. **14** (2020). https://doi.org/10.3389/fnins.2020.00268

18. Guhathakurta, D., Dutta, A.: Computational pipeline for NIRS-EEG joint imaging of tDCS-evoked cerebral responses—an application in ischemic stroke. Front. Neurosci. **10** (2016). https://doi.org/10.3389/fnins.2016.00261

19. Ashcroft, J., Patel, R., Woods, A.J., Darzi, A., Singh, H., Leff, D.R.: Prefrontal transcranial direct-current stimulation improves early technical skills in surgery. Brain Stimul. **13**, 1834–1841 (2020). https://doi.org/10.1016/j.brs.2020.10.013

20. Gao, Y., Cavuoto, L., Schwaitzberg, S., Norfleet, J.E., Intes, X., De, S.: The effects of transcranial electrical stimulation on human motor functions: a comprehensive review of functional neuroimaging studies. Front Neurosci. **14** (2020). https://doi.org/10.3389/fnins.2020.00744

21. Otal, B., et al.: Opportunities for guided multichannel non-invasive transcranial current stimulation in poststroke rehabilitation. Front Neurol. **7** (2016). https://doi.org/10.3389/fneur.2016.00021

22. Rezaee, Z., et al.: Feasibility of combining functional near-infrared spectroscopy with electroencephalography to identify chronic stroke responders to cerebellar transcranial direct current stimulation—a computational modeling and portable neuroimaging methodological study. Cerebellum (2021). https://doi.org/10.1007/s12311-021-01249-4

Surgical Visualization and Mixed, Augmented and Virtual Reality

Image-Based Incision Detection for Topological Intraoperative 3D Model Update in Augmented Reality Assisted Laparoscopic Surgery

Tom François[1,2(✉)], Lilian Calvet[1,3], Callyane Sève-d'Erceville[1],
Nicolas Bourdel[1], and Adrien Bartoli[1]

[1] CHU Clermont-Ferrand, CNRS, SIGMA Clermont, Institut Pascal.,
Université Clermont Auvergne, Clermont-Ferrand, France
[2] Be-Ys Research, 123 route de Meyrin, 1219 Châtelaine, Suisse
`tom.francois@etu.uca.fr`
[3] Université de Toulouse, Toulouse, France

Abstract. Augmented Reality (AR) is a promising way to precisely locate the internal structures of an organ in laparoscopy. Several methods have been proposed to register a preoperative 3D model reconstructed from MRI or CT to the intraoperative laparoscopy 2D images. These methods assume a fixed topology of the 3D model. They thus quickly fail once the organ is cut to remove pathological internal structures. We propose to add image-based incision detection in the registration pipeline, in order to update the topology of the organ model. Whenever an incision is detected, it is transferred to the 3D model, whose topology is then updated accordingly, and registration started. We trained a UNet as incision detector from 181 labelled incision images, collected from 10 myomectomy procedures. It obtains a mean precision, recall and f1 score of 0.05, 0.36, and 0.08 from 10-fold cross-validation. Overall, topology updating improves 3D registration accuracy by 5% on average.

Keywords: Incision detection · Laparoscopic surgery · Registration

1 Introduction

A large number of procedures are nowadays performed by laparoscopy, during which localising the internal anatomical structures such as tumours is a major challenge. An active research subject in Computer Assisted Intervention (CAI) is hence to develop laparoscopic surgery guidance systems using Augmented Reality (AR). AR is achieved by overlaying the laparoscopic video stream in realtime with the internal anatomical structures extracted as a preoperative 3D model from MRI or CT. AR thus relies on the ability to compute the geometric transformation between the preoperative 3D model and the laparoscopic video stream. This is the *registration* problem, which represents a strong technical challenge because of the organ deformation. The current realtime registration methods

© Springer Nature Switzerland AG 2021
M. de Bruijne et al. (Eds.): MICCAI 2021, LNCS 12904, pp. 647–656, 2021.
https://doi.org/10.1007/978-3-030-87202-1_62

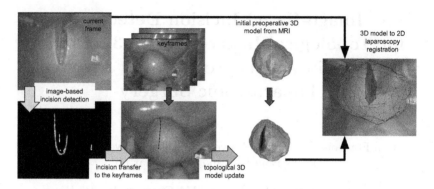

Fig. 1. Description of the complete pipeline. (column 1) Image-based incision detection. (column 2) Incision transfer to the keyframes. (column 3) Topological 3D model update. (column 4) Registered 3D model projected onto the current frame.

work in two steps [7, 12, 14]. First, they deform the preoperative 3D model to fit the intraoperative deformation state of the organ. This step uses an intraoperative 3D model reconstructed by means of Structure-from-Motion (SfM), SLAM or stereovision. Second, they track the organ using the intraoperative 3D model. This process works well as long as the organ is mobilised but does not change its shape too much. However, when the surgeon starts incising the organ, it obviously affects its shape in much more dramatic ways than mobilisation, causing existing registration methods to break down. Consequently, AR is only available at the very early steps of surgery and stops as soon as organ incision starts. Our main objective is to develop a solution to this major limitation.

We propose a framework which both tracks the organ and adapts the intraoperative 3D organ model to the transformations undergone by the organ during surgery. Concretely, this means that we update the intraoperative 3D organ model. The adaptation handles changes in the 3D model shape due to deformation of the organ and changes in the 3D model topology due to incision of the organ. A closely related work to ours is [16], which adapts the 3D model topology using a *geometric criterion*. Their key idea is to compute the registration and then detect the incisions. Concretely, an excess of extension of a mesh's edge triggers the deletion of the edge. This is an interesting idea but requires the registration to be highly accurate: noise in the registration, even temporary and mild, will cause spurious edge deletion, with no possibility of later recovery. Because deformable registration is a difficult problem, and because the model topology is wrong before registration, we cannot expect registration to be always highly accurate.

Our framework introduces a novel key idea in intraoperative 3D model update: *image-based incision detection*. Concretely, we detect the presence of an incision and its visible boundaries in the laparoscopy image. Importantly, this detection is independent of the registration. As opposed to the geometric criterion of [16], which depends on the registration, our image-based incision detection can be exploited to strengthen registration computation. Concretely, we use an incision detection DNN to obtain the image boundaries of the incision, which we transfer to the intraoperative 3D model, and update the model topology prior to solving for the non-rigid registration. Transferring the incision boundaries to the intraoperative 3D model represents a key step, which we achieve by means of special deformable image warps exploiting the incision boundaries explicitly.

To summarise, our contributions are three-fold. Our first contribution is a novel framework for topological intraoperative 3D model update, in order to facilitate AR in the presence of organ incisions by exploiting the geometric and image-based incision criteria. This framework has two new core components. Our second contribution is a labelled database and an incision detection DNN. Our third contribution is an incision transfer method, from the laparoscopy image to the intraoperative 3D model, prior to registration. We evaluate our framework in three parts. First, we evaluate our incision detector using edge detection metrics. Second, we evaluate the benefit of updating the 3D model topology on an ex-vivo experiment using 4 ex-vivo pig kidneys. We perform an ablation study for our framework and compare it with previous work. Third, we evaluate on postoperative surgical patient images.

2 Related Work

Registration of Deformable Organs. The registration of deformable organs uses priors including deformation smoothness [4,7] and bio-mechanics [3,13]. Recent works based on deep learning [19] are yet inapplicable to laparoscopy. These methods all assume that the model topology is given and fixed.

Image-Based Detection. Image-based incision detection has not been specifically addressed in the literature, but many other detection and segmentation tasks have been explored [15]. In laparoscopy, tool detection and segmentation [9,11] and organ-specific contour detection [8,14] are active subjects. Existing works train an encoder-decoder derived from the U-Net [18] with dedicated datasets.

Incision Simulation. Cutting a virtual 3D model can be achieved in several ways [20]. We have chosen *element deleting*, which is simple and runs fast.

Geometric Incision Detection. [16] uses a Finite Element Method to simulate the strain energy of the 3D model and visually controls it using SURF point correspondences. It uses a metric to measure how much has the distance changed between the deformed points. If some point pairs move apart significantly, a cut point is detected. The ex-vivo experiments described in [16] present large cuts with limited deformation but large piecewise rigid motion of the incised organ. The method requires that the registration is solved perfectly from the SURF points, which is very unlikely in practice.

3 Methodology

We base our pipeline on the uterus registration method [7]. We consider that the intraoperative 3D model has been reconstructed successfully using SfM [1,2]. We denote the input data as $\{\mathcal{T}_i, \mathbf{p}_i, \mathbf{S}_0, \mathcal{I}, \mathbf{q}\}$, with $\mathcal{T}_i$ one keyframe, $\mathbf{p}_i$ the feature points in $\mathcal{T}_i$ corresponding to the feature points $\mathbf{q}$ in the current frame $\mathcal{I}$, and $\mathbf{S}_0$ the 3D model reconstructed from the N keyframes. In particular, $\mathbf{S}_0$ represents the reconstructed *intact model*, as a closed surface. The output is $\{\mathbf{S}_t, \mathbf{Q}\}$, with $\mathbf{S}_t$ the incised 3D model and $\mathbf{Q}$ the registered model in the target frame.

3.1   Overview

Updating the 3D model according to the laparoscopic video stream involves to detect the incision both spatially and temporally. To simplify the problem, we make the following hypotheses. *(i)* A single incision is observed during the video sequence. *(ii)* The length of the incision only grows in time, which is always true. *(iii)* There is at least one frame where the incision is visible entirely, which is realistic as the surgeon checks the incision once done. These assumptions allow us to only detect the incision in the current frame and update the 3D model when the length of the detected incision has grown. Our pipeline can be summarized in 4 steps. (1) We detect the location of the incision in the current frame. (2) The detected incision location is transferred independently to multiple keyframes $\mathcal{T}_i$, where the organ is still intact. These keyframes have been used in the SfM reconstruction beforehand. This provides the transformation to both merge all transferred incisions in one reference keyframe, and to backproject the incision onto the reconstructed 3D model. (3) The 3D model is virtually cut according to the detected incision. (4) The registration is estimated between the updated 3D model and the current frame. In this pipeline, incision under-detection is very well-handled and hence preferred to over-detection.

3.2   Image-Based Incision Detection

We collected and annotated a total of 181 images from 10 myomectomy procedure videos. All enrolled participants gave their written informed consent following the IRB approval 2018-A03130-55. We trained a UNet [18] architecture to predict the visible boundaries of the incision. Our network produces an output $P(x,y)$ representing the probability of having an incision boundary at (x,y). The actual detected incision is obtained by selecting the main connected component, which is thinned and converted to a polyline. This post-processing removes many false positive pixels, contributing to the method's accuracy. The dataset contains various incision techniques that used different types of tools. The first type is the regular scissors. The boundaries of the incision made by these are difficult to perceive if the incision is not bleeding. The second type of tools are the heat cutting tools. The boundaries of the incision are burned while cutting which makes them easier to perceive but the cutting may also produce smoke.

3.3   Image to Model Incision Transfer

Image Deformation Model. The image deformation model we use is a parametric warp. There is a large variety of such functions in the literature. We use

Fig. 2. Image-based incision detection on test images. The groundtruth is in blue, the detection in red and the overlap in green. (Color figure online)

the generic model [5]. The warp $\mathcal{W} : \mathbb{R}^2 \times \mathbb{R}^{l \times 2} \mapsto \mathbb{R}^2$ maps 2D points from the current image to the reference image and depends on a set of l 2D control points $c_1, \ldots, c_l$ stacked in the parameter matrix $L \in \mathbb{R}^{l \times 2}$. The parameters are chosen by minimizing a cost function ε, composed of a data term ε_d, based on the average distance between the warped points in $\mathbf{q}$ and the matched points in $\mathbf{p}_i$, a smoothing term ε_s that controls the smoothness of the motion field, and a shrinking term ε_f that forces the warp to close the incision area. The first two terms are described in [17]:

$$\varepsilon_d(L) = \frac{1}{n_c} \sum_{j=1}^{n_c} \|\mathcal{W}(\mathbf{q}_j, L) - \mathbf{p}_{i,j}\|_2^2 \qquad \varepsilon_s(L) = \frac{1}{m_2}\|ZL\|_{\mathcal{F}}^2, \qquad (1)$$

with n_c the number of correspondences, Z the second derivatives of $\mathcal{W}$ stacked, evaluated at m_2 points. Minimizing the data and smoothing terms has a closed-form solution [5], which we use to initialise our solution.

Closing the Incision. In the keyframe $\mathcal{T}_i$, the boundaries of the incision should be superimposed to each other, as the organ is still intact. Thus, the estimated warp should close the detected incision area, denoted $\mathcal{H}$. Several works on warp estimation [5, 10, 17] have tackled this issue for self-occlusions, which occur when some parts of an observed surface are occluded by itself. In a self-occluded area, the data term gives absolutely no constraint. The best method from [10] uses a shrinking term:

$$\varepsilon_f(L) = \frac{1}{m_1} \sum_{\mathbf{a} \in \mathcal{H}} \min_{d \in \mathbb{S}^1} \left\| \frac{\partial_d \mathcal{W}(\mathbf{a}, L)}{\partial \mathbf{a}} \right\|_{\mathcal{F}}^2, \qquad (2)$$

with $\mathbb{S}^1$ the unit circle, and $\frac{\partial_d \cdot}{\partial \mathbf{a}}$ the directional derivatives along direction d, evaluated at m_1 points. The incision area is geometrically similar to a self-occlusion of the target image $\mathcal{I}$. We thus use Levenberg-Marquardt to minimise:

$$\varepsilon(L) = \varepsilon_d(L) + \lambda_s \varepsilon_s(L) + \lambda_f \varepsilon_f(L). \qquad (3)$$

Keyframes Mutual Agreement. The image to keyframe transfer is computed for every keyframe $\mathcal{T}_i$ that has been used in SfM to reconstruct $\mathbf{S}_0$. SfM provides the relative pose of every keyframe, which allows us to accurately transfer any point of the organ across the keyframes. For every estimated warp $\mathcal{W}_i$ from the set of correspondences $(\mathcal{I} \leftrightarrow \mathcal{T}_i)$, we can thus transfer the warped incision boundaries in a reference keyframe $\mathcal{T}_r$. As all the resulting warped incisions should align in the reference frame $\mathcal{T}_r$, we form a robust average using the median of the transferred incisions, see Fig. 3. The final transferred incision is obtained by thinning.

Fig. 3. Incision transfer to a reference keyframe. Left: current frame with the detected incision boundaries. Middle: incision boundaries transferred to the reference keyframe. Right: median warp (yellow) and final transferred incision (green). (Color figure online)

3.4   Topological Model Update

Whenever the detected incision length increases, we create a 3D model that represents the incision. Once the incision is backprojected on the intact model surface, we create the outline of the incision. The bottom of the incision is defined by moving at a fixed depth along the surface normal. The depth and width of the incision are manually set for each procedure.

3.5   3D-2D Registration

We formulate the problem as a non-linear energy-minimisation problem, with energies coming from prior and data terms. The prior term encodes the model's internal energy, which is used to regularise the problem. The data term enforces the projected registered 3D model to match the 2D correspondences $\mathbf{q}$ in the target frame. The energy function $E \in \mathbb{R}^+$ is given by:

$$E(x) = E_c(x, \mathbf{p}, \mathbf{q}) + \lambda_i E_i(x), \tag{4}$$

where λ_i is a scalar weight to control the non-rigid deformation. We optimize E by iterative non-linear optimization using a *stiff-to-flexible* strategy [6].

4   Results

Acquiring valid and precise groundtruth is a common challenge in surgical AR. The image-based incision detection can be evaluated very efficiently on a test set of labelled clinical images. The rest of the pipeline is evaluated on ex-vivo organs, for which the image-based detection is controlled. Qualitative results of the complete pipeline on patient data are shown in Fig. 4.

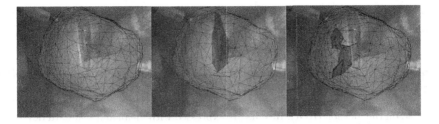

Fig. 4. Visual comparison on patient data: from left to right, Intact (baseline), ImageDet (ours), and GeoDet [16]. The detected incision is colored in black on the mesh.

4.1   Image-Based Incision Detection

For evaluation, we applied 10-fold cross-validation using the 10 procedures forming 10 image sets. For every permutation, the trained model is applied on the left-out images, considered as test set. The mean results for the concatenated test sets are 0.049, 0.356 and 0.084 for precision, recall and f1 score respectively. We notice that the quality of the predictions differs greatly between procedures (see Fig. 2). We obtained the best results with the regular Weighted Cross Entropy (WCE), compared to training strategies with the combination of WCE with Tversky Loss. We initialize the UNet with weights pretrained for occluding contours detection of the uterus [8] and train using a step learning rate strategy for 80 epochs, scaling down every 10 epochs by 0.1. The initial learning rate is set to 10^{-4} and the model stops improving after 50 epochs. The class weights for WCE are set to 10 and 0.1 for incision and non-incision respectively.

4.2   Ex-Vivo Registration

Description. We conducted an ablation study over 4 ex-vivo pig kidneys. We reconstruct the 3D scene before and after the incision using SfM, obtaining respectively the intact and groundtruth incised models. We recorded the video while the kidney was incised using a scalpel. Kidneys 1–3 are deeply incised, namely throughout the organ, while kidney 4 undergoes a shallow incision (6–10 mm). The average kidney dimensions are $140 \times 75 \times 25$ mm. Our experiments are similar to [16] but with less displacement, which is more realistic. For warp estimation, we used $\lambda_s = 0.5, \lambda_f = 5$ and a 5×5 grid of control points. For the registration, λ_i is set to 10^{-10}.

Image to Model Incision Transfer. The transferred incision has been evaluated on its own with a manually annotated ground-truth. We obtain an Hausdorff distance and symmetric distance of 15.8px and 6.1px respectively. The average RMSR over the different kidneys is 8.5px.

Geometric Incision Detection. We have reimplemented [16], which we call GeoDet. The main difference is that we use a different registration method. For our two last experiments, we have not been able to make GeoDet work.

The method either detects many false cut points or detects no incision. We use GeoDet with $r_{P_F^0} = 30$ mm, $\tau = 4.5$, $n = 7$.

Table 1. Evaluation of ex-vivo kidney data for the tested approaches. '–' in GeoDet means that no incision was detected. '2D' is the reprojection distance error (in px). 3D distances are presented for the external registered surface (A) and a cropped region centered on the incision (ROI). Results are expressed with mean and standard deviation (the lower, the better; best in bold, second best underlined).

		2D (px)	A (i → GT)	A (GT → i)	ROI (i → GT)	ROI (GT → i)
K1	Intact	8.92 ± 7,05	5 ± 2.11	5.14 ± 2.7	4.95 ± 2.06	5.42 ± 2.99
	ImageDet	**8.85 ± 7.23**	4.73 ± 2.89	4.11 ± 2.14	4.95 ± 3.56	**3.57 ± 1.92**
	ImageDet-50%	9.02 ± 7.45	**4.15 ± 1.85**	4.11 ± 2.11	**4.02 ± 1.95**	3.8 ± 2.05
	GeoDet	8.88 ± 7.03	5.12 ± 2.88	4.56 ± 2.48	5.12 ± 3.34	3.88 ± 2.32
K2	Intact	6.09 ± 4.03	10.82 ± 5.6	9.23 ± 5.48	14.47 ± 5.02	12.1 ± 5.66
	ImageDet	6.07 ± 3.97	**10.7 ± 5.34**	**8.6 ± 4.84**	13.62 ± 4.65	**10.41 ± 4.59**
	ImageDet-50%	**6.07 ± 3.98**	10.8 ± 5.43	9.03 ± 5.15	14.08 ± 4.73	11.55 ± 5.06
	GeoDet	6.1 ± 4.04	10.81 ± 5.5	8.89 ± 5.18	**13.44 ± 5.44**	11.02 ± 5.51
K3	Intact	8.84 ± 7.96	13.45 ± 6.51	11.56 ± 6.21	15.26 ± 6.52	14.13 ± 6.32
	ImageDet	8.5 ± 8.42	**11.44 ± 5.77**	**9.35 ± 5.33**	**11.46 ± 5.95**	**10.07 ± 5.62**
	ImageDet-50%	**8.21 ± 7.6**	12.85 ± 6.36	10.79 ± 5.94	13.04 ± 6.36	11.97±6.27
	GeoDet	–	–	–	–	–
K4	Intact	5.69 ± 5.68	2.4 ± 1.9	2.05 ± 1.27	1.65 ± 1.33	1.63 ± 1.2
	ImageDet	5.68 ± 5.62	2.4 ± 1.91	1.99 ± 1.25	1.81 ± 1.48	1.56 ± 1.17
	ImageDet-50%	**5.68 ± 5.61**	**2.32 ± 1.88**	**1.97 ± 1.23**	**1.61 ± 1.28**	1.54 ± 1.14
	GeoDet	–	–	–	–	–

Evaluation. In the ablation study, we compare the methods Intact [7], ImageDet (ours), ImageDet-50% (ours), and GeoDet [16]. ImageDet-50% simply uses half of the incision to simulate misdetection. In Table 1, we compare the mean reprojection error between the registered model and the manual correspondences. In order to compare the different methods, we measure the sampled distance between the external surface of the registered model with the groundtruth model. We also provide the same distances on a Region of Interest centered on the incision location. the registration examples from Figs. 4 and 5 show how ImageDet manages to consistently update and register the 3D model, and outperforms GeoDet.

5   Discussion and Future Work

We have given the first pipeline which detects incisions from the images in endoscopic surgery and uses the information to perform topological preoperative 3D model registration. Our dataset for incision detection is still small. We expect a significant performance boost upon expanding it. We expect a combination of

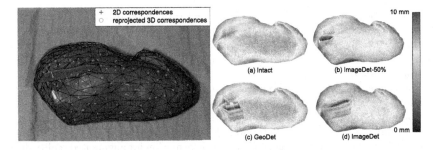

Fig. 5. Visual results for registration on kidney 1. Left: Reprojection error between correspondences (white line) for ImageDet. Right: Distance from registered model to groundtruth with methods Intact (a), ImageDet-50% (b), GeoDet (c), ImageDet (d).

the ImageDet (ours) and GeoDet [16] methods to improve precision and robustness. Finally, we plan to include new terms in the registration loss, to associate the image detected incision boundaries to the 3D model.

References

1. Griwodz, C., et al.: Alicevision meshroom: an open-source 3D reconstruction pipeline. In: Proceedings of the 12th ACM Multimedia Systems Conference - MMSys 2021. ACM Press (2021). https://doi.org/10.1145/3458305.3478443
2. Photoscan (2015). https://www.agisoft.com
3. Adagolodjo, Y., Trivisonne, R., Haouchine, N., Cotin, S., Courtecuisse, H.: Silhouette-based pose estimation for deformable organs application to surgical augmented reality. In: IROS, pp. 539–544. IEEE (2017)
4. Amir-Khalili, A., Nosrati, M.S., Peyrat, J.-M., Hamarneh, G., Abugharbieh, R.: Uncertainty-encoded augmented reality for robot-assisted partial nephrectomy: a phantom study. In: Liao, H., Linte, C.A., Masamune, K., Peters, T.M., Zheng, G. (eds.) AE-CAI/MIAR -2013. LNCS, vol. 8090, pp. 182–191. Springer, Heidelberg (2013). https://doi.org/10.1007/978-3-642-40843-4_20
5. Bartoli, A.: Maximizing the predictivity of smooth deformable image warps through cross-validation. JMIV **31**(2–3), 133–145 (2008). https://doi.org/10.1007/s10851-007-0062-1
6. Collins, T., et al.: A system for augmented reality guided laparoscopic tumour resection with quantitative ex-vivo user evaluation. In: Peters, T., et al. (eds.) CARE 2016. LNCS, vol. 10170, pp. 114–126. Springer, Cham (2017). https://doi.org/10.1007/978-3-319-54057-3_11
7. Collins, T., et al.: Augmented reality guided laparoscopic surgery of the uterus. IEEE Trans. Med. Imaging **40**(1), 371–380 (2020)
8. François, T., et al.: Detecting the occluding contours of the uterus to automatise augmented laparoscopy: score, loss, dataset, evaluation and user study. Int. J. Comput. Assist. Radiol. Surg. **15**(7), 1177–1186 (2020). https://doi.org/10.1007/s11548-020-02151-w
9. Garcia-Peraza-Herrera, L.C., et al.: ToolNet: holistically-nested real-time segmentation of robotic surgical tools. In: IROS, pp. 5717–5722. IEEE (2017)

10. Gay-Bellile, V., Bartoli, A., Sayd, P.: Direct estimation of nonrigid registrations with image-based self-occlusion reasoning. TPAMI **32**(1), 87–104 (2008)
11. Han, L., Wang, H., Liu, Z., Chen, W., Zhang, X.: Vision-based cutting control of deformable objects with surface tracking. IEEE/ASME Transactions on Mechatronics (2020)
12. Haouchine, N., Dequidt, J., Berger, M.O., Cotin, S.: Monocular 3D reconstruction and augmentation of elastic surfaces with self-occlusion handling. IEEE Trans. Vis. Comput. Graph. **21**(12), 1363–1376 (2015)
13. Haouchine, N., Dequidt, J., Peterlik, I., Kerrien, E., Berger, M.O., Cotin, S.: Image-guided simulation of heterogeneous tissue deformation for augmented reality during hepatic surgery. In: ISMAR, pp. 199–208. IEEE (2013)
14. Hattab, G., et al.: Kidney edge detection in laparoscopic image data for computer-assisted surgery. Int. J. Comput. Assist. Radiol. Surg. **15**(3), 379–387 (2019). https://doi.org/10.1007/s11548-019-02102-0
15. Litjens, G.: A survey on deep learning in medical image analysis. Med. Image Anal. **42**, 60–88 (2017)
16. Paulus, C.J., Haouchine, N., Kong, S.-H., Soares, R.V., Cazier, D., Cotin, S.: Handling topological changes during elastic registration. Int. J. Comput. Assist. Radiol. Surg. **12**(3), 461–470 (2016). https://doi.org/10.1007/s11548-016-1502-4
17. Pizarro, D., Bartoli, A.: Feature-based deformable surface detection with self-occlusion reasoning. IJCV **97**(1), 54–70 (2012). https://doi.org/10.1007/s11263-011-0452-0
18. Ronneberger, O., Fischer, P., Brox, T.: U-Net: convolutional networks for biomedical image segmentation. In: Navab, N., Hornegger, J., Wells, W.M., Frangi, A.F. (eds.) MICCAI 2015. LNCS, vol. 9351, pp. 234–241. Springer, Cham (2015). https://doi.org/10.1007/978-3-319-24574-4_28
19. de Vos, B.D., Berendsen, F.F., Viergever, M.A., Sokooti, H., Staring, M., Išgum, I.: A deep learning framework for unsupervised affine and deformable image registration. Med. Image Anal. **52**, 128–143 (2019)
20. Wu, J., Westermann, R., Dick, C.: A survey of physically based simulation of cuts in deformable bodies. Comput. Graph. Forum **34**(6), 161–187 (2015)

Using Multiple Images and Contours for Deformable 3D-2D Registration of a Preoperative CT in Laparoscopic Liver Surgery

Yamid Espinel[1]([✉]), Lilian Calvet[1,2], Karim Botros[1], Emmanuel Buc[2], Christophe Tilmant[1], and Adrien Bartoli[1]

[1] EnCoV, Institut Pascal, UMR 6602 CNRS/Université Clermont-Auvergne, Clermont-Ferrand, France
{yamid.espinel,karim.botros,christophe.tilmant,adrien.bartoli}@uca.fr
[2] University Hospital of Clermont-Ferrand, Clermont-Ferrand, France
ebuc@chu-clermontferrand.fr

Abstract. Deformable registration is required to achieve laparoscopic augmented reality but still is an open problem. Some of the existing methods reconstruct a preoperative model and register it using anatomical landmarks from a single image. This is not accurate due to depth ambiguities. Other methods require of non-standard devices unadapted to the clinical practice. A reasonable way to improve accuracy is to combine multiple images from a monocular laparoscope. We propose three novel registration methods exploiting information from multiple images. The first two are based on rigidly-related images (MV-B and MV-C) and the third one on non-rigidly-related images (MV-D). We evaluated registration accuracy quantitatively on synthetic and phantom data, and qualitatively on patient data, comparing our results with state of the art methods. Our methods outperforms, reducing the partial visibility and depth ambiguity issues of single-view approaches. We characterise the improvement margin, which may be slight or significant, depending on the scenario.

Keywords: Laparoscopy · Liver · Registration · Augmented reality

1 Introduction

Laparoscopic liver surgery is less invasive than open surgery, reducing patient trauma and recovery time. However, it is difficult for the surgeon to localise the inner structures such as vessels and tumours. This is mainly caused by the large size of the liver and its proximity to the laparoscope. Augmented Reality (AR) can provide aid to surgery by showing these internal structures, hence improving resection planning. Concretely, AR overlays a preoperative 3D model reconstructed from CT data onto the laparoscopic images. AR requires one to register the preoperative 3D model to the 2D laparoscopic images. This is done using anatomical landmarks present on the liver's surface. Because the liver deforms

M. de Bruijne et al. (Eds.): MICCAI 2021, LNCS 12904, pp. 657–666, 2021.
https://doi.org/10.1007/978-3-030-87202-1_63

significantly between the preoperative and intraoperative states, the registration must be a 3D-2D deformable one. This is a difficult and currently open problem. Existing works have addressed the cases of monocular laparoscopes [1–3] and of stereoscopic laparoscopes, possibly with external tracking [4–10]. Methods [1,2] are single-view. They use sparse landmarks, leaving ambiguities on the registration. Method [3] is multi-view and reconstructs an intraoperative shape by Structure-from-Motion (SfM), solving registration by ICP and 3D landmarks. This method was however only tested with animal data as the required correspondences are not available in real surgery cases. Finally, methods [4–10] perform rigid registration only. The monocular case is important because it forms the standard in many operating rooms. Therefore, there is a need for an unambiguous monocular registration method compatible with the clinical constraints and the desired clinical outcomes, studied in [13].

We propose to use multiple monocular laparoscopic images to solve deformable registration without unrealistic requirements. Our methods solve registration for cases where images are rigidly-related and non-rigidly-related. We assume that the images show the liver from multiple viewpoints. Our methods are inspired by [1], a method named SV (for Single-View) hereinafter. Concretely, we deform the preoperative 3D model using a particle system with biomechanical properties and position-based dynamics [15]. For the rigidly-related case, we propose two methods. The first one, named MV-B (for Multi-View Base), guides deformation using the landmarks from all the images and the inter-image rigidity as constraints. The second one, named MV-C (for Multi-View Correspondences), extends MV-B by exploiting the liver's texture information via inter-image keypoint correspondences. For the non-rigidly-related case, we propose the MV-D method (for Multi-View Deformable Correspondences) which, compared to MV-C, does not use the inter-image rigidity and produces several registered shapes. These methods are illustrated in Fig. 1. By exploiting multiple images, our methods solve the partial visibility and depth ambiguity issues inherent to SV. We evaluated our methods on synthetic, phantom and patient data. For the synthetic and phantom case, target registration errors (TRE) were measured on uniformly-distributed control points inside the liver models. For the patient case, we measured the liver's 2D reprojection errors on control views. We aim to have a TRE lower than 1 cm, as it is the resection margin advised for Hepato-Cellular Carcinoma (HCC) interventions [17]. We do not measure TRE on patient data as having a reliable ground-truth is not possible with the available devices in the surgery room.

2   Background

SV uses a volumetric deformable model M to predict image contours C. M is modeled using the isotropic Neo-Hookean elastic model [15] with generic values for the human liver's mechanical parameters [16]. We use this model as it can simulate large deformations in a computationally inexpensive way and is validated for the hepatic tissue [11,12]. The constraints given by the elastic model

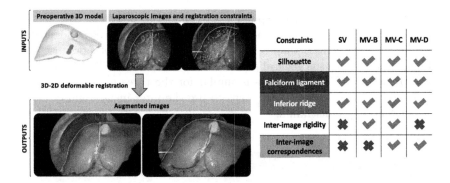

Fig. 1. Characteristics of the state-of-the-art and proposed registration methods.

are denoted as Ω_{def}. The predicted contours C should match the observed image contours C^*. As shown in Fig. 1, SV uses three types of contours: the ridge contour, which has a very distinctive profile; the falciform ligament contour; and the silhouette contour which corresponds to the occluding boundaries of the liver. The constraints given by this set of contours are denoted as Ω_{ctr}. SV uses a convergent algorithm of alternating projections to solve the registration problem. As such, it finds the closest model M to the constraints Ω_{def} and Ω_{ctr}:

$$\min_{M} dist(M, \Omega_{def}) + dist(M, \Omega_{ctr}) \tag{1}$$

where $dist$ is the distance between the model and the constraints.

3   Methodology

The common principle of our methods is to use several laparoscopic images to solve registration. These images should present a noticeable change in camera pose, while keeping some overlap. This can be done by tilting and panning the laparoscope. Views from different trocars can also be used, should they sufficiently overlap. Using multiple images is especially useful for the liver case as it helps to overcome the partial visibility and, thus, the low precision issues inherent to single-view methods.

These methods try to solve registration in two situations: when the liver does not deform across images, e.g. during the exploratory phase (rigidly-related case) and when it does, e.g. during surgical manipulation (non-rigidly-related case). For the rigidly-related case, MV-B and MV-C take advantage of the liver's rigidity to constrain registration. Thus, they produce a single deformed shape, along with the corresponding camera poses for each of the views. In order to have suitable images for this case, the surgeon can pause the artificial ventilation system for a very short period of time (about 10 s), while the scene is filmed. Compared to MV-B, MV-D makes use of inter-image correspondences which helps to improve the registration accuracy. However, MV-B is still useful in cases where enough valid correspondences cannot be obtained. For the non-rigidly-related case, MV-D also makes use of inter-image correspondences and produces several deformed shapes, according to the number of views used.

Figure 2 illustrates the general pipeline followed by MV-B, MV-C and MV-D to solve registration. The principle is similar for the three methods, taking SV as basis and adding the necessary steps to use the inter-image rigidity and the correspondences as constraints.

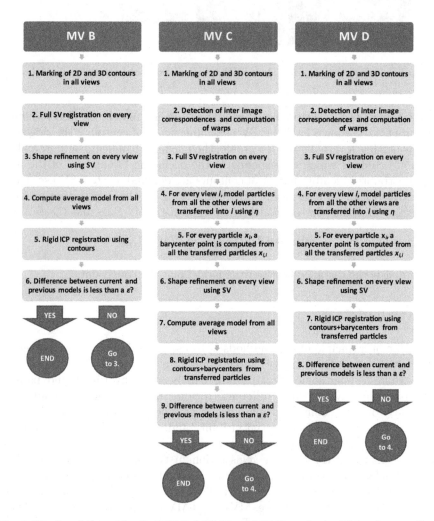

Fig. 2. Pipeline followed by the MV-B, MV-C and MV-D methods. (Color figure online)

3.1   Multi-View Rigid Base (MV-B)

MV-B extends SV to solve registration on N rigidly-related laparoscopic images simultaneously. At every iteration, it computes an average model M from all the views after an SV refinement has been done on every view. A single general model is thus produced from each view's individual contribution. Equation (1) becomes:

$$\min_M \sum_{i=1}^{N} dist(M, \Omega_{def_i}) + \sum_{i=1}^{N} dist(M, \Omega_{ctr_i}) \tag{2}$$

The algorithm that solves this problem is composed by the blue and green stages from Fig. 2. The computation of the average model is the key to impose the inter-image rigidity. It converges when the difference between models from previous and current iterations $\Delta M < 10^{-3}$. This criterion was found after running our methods for a large number of iterations on a variety of synthetic, phantom and patient data.

3.2   Multi-View Rigid with Inter-image Correspondences (MV-C)

MV-C extends MV-B to exploit inter-image correspondences. These correspondences are obtained using SIFT and mismatches rejected using FBDSD [18]. Our key idea is to measure the consistency between the registration and the correspondences using the P particles $x_p, p \in [1, P]$, representing the deformable model. These correspondences are related to each other by the warp transformation function η_{ji}, which is based on the Rigid-Perspective Thin-Plate Spline [14]. Such function η_{ji} lets us transfer any point from image j to image i, as shown in Fig. 3. Like MV-B, MV-C generates a single general model from each view's individual contribution by using the inter-image rigidity and the available inter-image correspondences Ω_{corr} as constraints:

$$\min_M \sum_{i=1}^{N} dist(M, \Omega_{def_i}) + \sum_{i=1}^{N} dist(M, \Omega_{ctr_i}) + \sum_{i=1}^{N} dist(M, \Omega_{corr_i}) \tag{3}$$

The algorithm that solves this problem is composed by the blue, yellow and green stages from Fig. 2. It adds the necessary steps to use the keypoint correspondences as constraints, and it also converges when $\Delta M < 10^{-3}$.

3.3   Multi-View Deformable Correspondences (MV-D)

MV-D modifies MV-C to avoid using the inter-image rigidity to solve registration, as well as to use the inter-image correspondences in the SV-refinement stage. In this way, MV-D produces multiple shapes M_N according to the number of images used and the deformations exerted by the liver on each of them. This is especially useful in cases where the liver is being manipulated by tools or other external forces. Consequently, MV-D can be seen as:

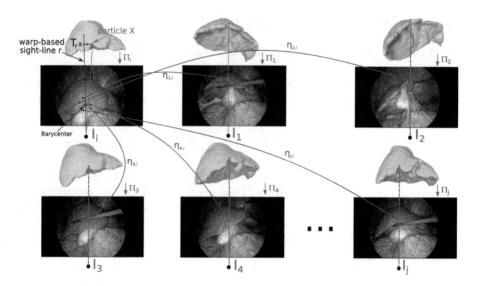

Fig. 3. Warp-based sight-line for particle location update. The particle locations are updated as follows: pixel motion is preliminary computed for all image-pairs i, j and the warps robustly estimated; during an iteration of the optimization algorithm, for every image i, a particle x_p is updated such that its new location corresponds to its orthogonal projection onto its warp-based sight line, namely the backprojections of the barycenter of the imaged particles $\eta_{ji}(\Pi_j(x_p))$ through the warps.

$$\min_{M_1,\ldots,M_N} \sum_{i=1}^{N} dist(M_i, \Omega_{def_i}) + \sum_{i=1}^{N} dist(M_i, \Omega_{ctr_i}) + \sum_{i=1}^{N} dist(M_i, \Omega_{corr_i}) \quad (4)$$

The algorithm that solves this problem is composed by the blue and yellow stages from Fig. 2, avoiding the usage of the inter-image rigidity. It also converges when $\Delta M < 10^{-3}$.

4   Experimental Results

4.1   Rigidly-Related Views

Synthetic Data: We reconstructed a virtual 3D liver model from a patient's CT and synthesized 10 virtual deformations using Abaqus [19], by simulating gravity, the pneumoperitoneum and the action of surgical instruments. We generated 10 images of each deformation by simulating a virtual moving laparoscope. We estimated registration for a varying number of images, going from 1 to 8, and measured TRE as the average prediction error for uniformly sampled points within the virtual liver. It should be noted that, while the contour marking is done manually, it does not take more than 5 min to mark all the 8 views used as the maximum case for our experiments. Thus, the impact on the surgical workflow is minimal. We use 8 views as a maximum to keep the computation time reasonable. We repeated the estimation 10 times for each number of images, randomly selecting the images being used. The results are shown in Fig. 4(a). We observe that the TRE for SV is steady around (9.57 ± 7.99) mm, while for MV-B and MV-C it consistently decreases as the number of views increases. The decrease is notable in both average and standard deviation, reaching (7.99 ± 7.34) mm and (5.95 ± 4.46) mm respectively for MV-B and MV-C.

Phantom Data: We 3D printed the synthetic deformations generated in Abaqus using PLA (Polylactic Acid). We then used a surgical laparoscope and a pelvitrainer box to take 10 pictures of each printed model. Similarly to the synthetic case, 10 combinations of 8 views were generated per deformation and experiments were run using 1 to 8 views for every combination. Distances were also measured between the registered and ground truth control points, for which the average and standard deviation are shown in Fig. 4(b). As for the synthetic case, SV remains steady around (11.96 ± 7.72) mm, while for MV-B and MV-C it consistently decreases as the number of views increases. We can see a decrease in both average and standard deviation, especially for MV-C. MV-B remains close to SV, reaching (11.59 ± 10.95) mm, while MV-C decreases to and (8.30 ± 4.59) mm.

Patient Data: We collected data for 5 patients, for which we had IRB approval (IRB8526-2019-CE58, CPP Sud-Est VI). We kept 9 images per patient. Out of these images, we singled one out to serve as a control view. The control view is not used to compute registration but as a means to verify registration, using the landmark prediction error expressed in px (number of pixels). The control view error is a weak measure of TRE because it only concerns the visible liver surface. We measure such reprojection errors due to the difficulty of having a reliable groundtruth to evaluate the registration accuracy in 3D. It is shown in Fig. 5(a). We run both MV-B and MV-C with 8 images. We observe a clear benefit of using multiple images and of using the inter-image correspondences.

4.2 Non-rigidly-related Views

Synthetic Data: From the previously generated synthetic data, 10 combinations of 8 views were taken to assess MV-D, with every view corresponding to a different deformation. As for the rigidly-related case, we estimated registration for a varying number of images and measured TRE as the average prediction error for uniformly sampled points within the virtual liver. The results are shown in Fig. 4(c). For the experiments using 8 views, SV has a TRE of (21.83 ± 32.61) mm on the control points, while that for MV-D is (20.87 ± 32.96) mm.

Phantom Data: From the generated phantom data, and similarly to the non-rigidly-related synthetic case, experiments were done on 10 combinations of 8 views, with every view corresponding to a different deformation. TRE results on a varying number of images are shown in Fig. 4(d). For the experiments using 8 views, SV has a TRE of (11.59 ± 6.95) mm on the control points, while MV-D's one is (10.71 ± 6.35) mm.

Patient Data: From the previously acquired patient data we have selected 9 images per patient. Here, the liver exerts significant deformation across the images. As for the rigidly-related case, we singled one image out to serve as a control view. Landmark prediction error is computed on the control views and expressed in px (number of pixels). It is shown in Fig. 5(b). We run both MV-B and MV-C with 8 images. We observe a slight benefit of using multiple images to solve registration.

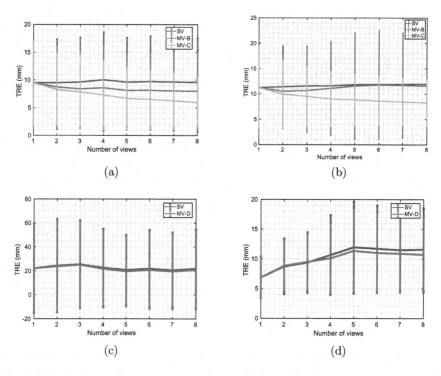

Fig. 4. Mean TRE and standard deviations on (a) synthetic data usisng the rigidly-related methods, (b) phantom data using the rigidly-related methods, (c) synthetic data using the non-rigidly-related methods, (d) phantom data using the non-rigidly-related methods.

Patient	SV	MV-B	MV-C
1	26.07	_21.98_	**19.49**
2	58.40	_10.28_	**09.49**
3	42.78	_32.24_	**29.75**
4	**17.13**	21.72	_21.13_
5	26.71	**15.46**	_16.95_
average	34.28	_20.33_	**19.36**

(a)

Patient	SV	MV-D
1	32.47	**30.14**
2	63.08	**57.62**
3	**46.24**	49.79
4	25.07	**21.31**
5	31.85	**29.75**
average	39.74	**37.73**

(b)

Fig. 5. Control view errors (px) on patient data for (a) rigidly-related methods with best results in bold and second best underlined and (b) non-rigidly-related methods with best results in bold.

5 Conclusions

We proposed 3 multi-view methods for 3D-2D deformable registration of preoperative CT data into intraoperative images for laparoscopy of liver. They aim to solve registration on rigidly- and non-rigidly-related views. Results on synthetic data show that MV-B improves the mean registration accuracy by 1.58 mm, while MV-C improves it by 3.62 mm compared to SV. MV-D shows a slight improvement of 1 mm compared to SV. On phantom data, MV-B has a similar performance to SV with a difference of 0.37 mm, while MV-C improves registration by 3.66 mm. MV-D also shows a similar performance to SV with a difference of 0.88 mm. On patient data, the reprojection error measured on control views is improved by 13.95 px, 14.92 px and 2.01 px for MV-B, MV-C and MV-D respectively. It means that, for the rigidly-related case, we can see an improvement with respect to SV as we increment the number of views, with registration errors below 1 cm for MV-C. For the non-rigidly-related case, MV-D behaves similarly to SV, with a slight improvement of 1 mm. Visual inspection on the error distribution shows that TRE follows an unimodal distribution with positive skewness. It is worth noting that TRE is measured on the whole 3D model, including the visible and hidden regions. Hidden regions can increase the global TRE as they are less accurately registered than the visible ones. As future work, we will focus on (i) performing extended validation on clinical patient data, (ii) developing a real-time non-rigidly tracking strategy taking the multi-view initial registration as basis, and (iii) using other visual information with improved registration approaches to increase the registration accuracy, such as surgical tools pose and Structure-from-Motion.

References

1. Koo, B., Özgür, E., Le Roy, B., Buc, E., Bartoli, A.: Deformable registration of a preoperative 3D liver volume to a laparoscopy image using contour and shading cues. In: Descoteaux, M., Maier-Hein, L., Franz, A., Jannin, P., Collins, D.L., Duchesne, S. (eds.) MICCAI 2017. LNCS, vol. 10433, pp. 326–334. Springer, Cham (2017). https://doi.org/10.1007/978-3-319-66182-7_38

2. Adagolodjo, Y., Trivisonne, R., Haouchine, N., Cotin, S., Courtecuisse, H.: Silhouette-based pose estimation for deformable organs application to surgical augmented reality. In: IROS (2017)
3. Modrzejewski, R., Collins, T., Seeliger, B., Bartoli, A., Hostettler, A., Marescaux, J.: An in vivo porcine dataset and evaluation methodology to measure soft-body laparoscopic liver registration accuracy with an extended algorithm that handles collisions. Int. J. Comput. Assist. Radiol. Surg. 14(7), 1237–1245 (2019). https://doi.org/10.1007/s11548-019-02001-4
4. Chen, L., Tang, W., John, N.W., Wuan, T.R., Zhang, J.J.: SLAM-based dense surface reconstruction in monocular minimally invasive surgery and its application to augmented reality. Comput. Meth. Prog. Biomed. 158, 135–146 (2018)
5. Haouchine, N., Roy, F., Untereiner, L., Cotin, S.: Using contours as boundary conditions for elastic registration during minimally invasive hepatic surgery. In: IROS (2016)
6. Robu, M.R., et al.: Global rigid registration of CT to video in laparoscopic liver surgery. Int. J. Comput. Assist. Radiol. Surg. 13(6), 947–956 (2018). https://doi.org/10.1007/s11548-018-1781-z
7. Thompson, S., et al.: Accuracy validation of an image guided laparoscopy system for liver resection. Proc. SPIE - Int. Soc. Opt. Eng. 9415(09), 1–12 (2015)
8. Plantefeve, R., Peterlik, I., Haouchine, N., Cotin, S.: Patient-specific biomechanical modeling for guidance during minimally-invasive hepatic surgery. Ann. Biomed. Eng. 44, 139–153 (2016)
9. Clements, L., Collins, J., Weis, J., Simpson, A., Kingham, T., Jarnagin, W., Miga, M.: Deformation correction for image guided liver surgery: an intraoperative fidelity assessment. Surgery 162(3), 537–547 (2017)
10. Bernhardt, S., Nicolau, S., Bartoli, A., Agnus, V., Soler, L., Doignon, C.: Using shading to register an intraoperative CT scan to a laparoscopic image. In: Computer-Assisted and Robotic Endoscopy, CARE (2015)
11. Chui, C., Kobayashi, E., Chen, X., Hisada, T., Sakuma, I.: Combined compression and elongation experiments and non-linear modelling of liver tissue for surgical simulation. Med. Biol. Eng. Comput. 44, 787–798 (2004)
12. Shi, H., Farag, A., Fahmi, R., Chen, D.: Validation of finite element models of liver tissue using Micro-CT. IEEE Trans. Biomed. Eng. 55, 978–984 (2008)
13. Thompson, S., Hu, M., Johnsen, S., Gurusamy, K., Davidson, B., Hawkes, D.: Towards Image Guided Laparoscopic Liver Surgery, Defining the System Requirement. LIVIM (2011)
14. Bartoli, A., Perriollat, M., Chambon, S.: Generalized thin-plate spline warps. Int. J. Comput. Vis. 88, 85–110 (2010)
15. Bender, J., Koschier, D., Charrier, P., Weber, D.: Position-based simulation of continuous materials. Comput. Graph. 44, 1–10 (2014)
16. Nava, A., Mazza, E., Furrer, M., Villiger, P., Reinhart, W.H.: In vivo mechanical characterization of human liver. Med. Image Anal. 12(2), 203–216 (2008)
17. Zhong, F.P., Zhang, Y.J., Liu, Y., Zou, S.B.: Prognostic impact of surgical margin in patients with hepatocellular carcinoma: a meta-analysis. Medicine 96(37), e8043 (2017)
18. Pizarro, D., Bartoli, A.: Feature-based deformable surface detection with self-occlusion reasoning. Int. J. Comput. Vis. 97, 54–70 (2010)
19. 3DS Abaqus. http://edu.3ds.com/en/software/abaqus-student-edition. Accessed 2 Mar 2021

SurgeonAssist-Net: Towards Context-Aware Head-Mounted Display-Based Augmented Reality for Surgical Guidance

Mitchell Doughty[1,2(✉)], Karan Singh[3], and Nilesh R. Ghugre[1,2,4]

[1] Department of Medical Biophysics, University of Toronto, Toronto, Canada
mitchell.doughty@mail.utoronto.ca
[2] Schulich Heart Program, Sunnybrook Health Sciences Centre, Toronto, Canada
[3] Department of Computer Science, University of Toronto, Toronto, Canada
[4] Physical Sciences Platform, Sunnybrook Research Institute, Toronto, Canada

Abstract. We present SurgeonAssist-Net: a lightweight framework making action-and-workflow-driven virtual assistance, for a set of predefined surgical tasks, accessible to commercially available optical see-through head-mounted displays (OST-HMDs). On a widely used benchmark dataset for laparoscopic surgical workflow, our implementation competes with state-of-the-art approaches in prediction accuracy for automated task recognition, and yet requires 7.4× fewer parameters, 10.2× fewer floating point operations per second (FLOPS), is 7.0× faster for inference on a CPU, and is capable of near real-time performance on the Microsoft HoloLens 2 OST-HMD. To achieve this, we make use of an efficient convolutional neural network (CNN) backbone to extract discriminative features from image data, and a low-parameter recurrent neural network (RNN) architecture to learn long-term temporal dependencies. To demonstrate the feasibility of our approach for inference on the HoloLens 2 we created a sample dataset that included video of several surgical tasks recorded from a user-centric point-of-view. After training, we deployed our model and cataloged its performance in an online simulated surgical scenario for the prediction of the current surgical task. The utility of our approach is explored in the discussion of several relevant clinical use-cases. Our code is publicly available at https://github.com/doughtmw/surgeon-assist-net.

Keywords: Augmented reality · Machine learning · Surgical task prediction · Head-mounted display · Microsoft HoloLens · Neural networks

1 Introduction

There has been significant interest in adoption of augmented reality (AR) for surgical guidance in the medical field, due to its ability to enhance task performance when effectively implemented [1]. The use of a see-through head-mounted

© Springer Nature Switzerland AG 2021
M. de Bruijne et al. (Eds.): MICCAI 2021, LNCS 12904, pp. 667–677, 2021.
https://doi.org/10.1007/978-3-030-87202-1_64

display (HMD) as the visualization medium, as opposed to a monitor, has been demonstrated to provide a further benefit to efficiency by eliminating the visual disconnect between the monitor and the surgical scene [2].

Though recent work has indicated the applicability of AR in laparoscopic and endoscopic procedures [3,4], neurosurgery [5], orthopedic surgery [6], and general surgery [7], there remains the concern of overloading the user with too much additional information, distracting them from the task at hand and resulting in reduced performance over routine standardized techniques [8].

Motivation. The bulk of research work into AR systems for medical image-guidance has centered around technical developments, including calibration [9], alignment [1], and visualization [10,11] and focused on achieving quantitative metrics like speed and accuracy [12]. Due to a lack of optimized virtual workflow and information representation, these advances do not guarantee the effective translation of guidance systems to clinical practice; this is evidenced by the absence of widely used commercial see-through HMD navigation systems [13].

To bridge the gap towards an effective OST-HMD based AR guidance system, our aim was to address these pitfalls by creating a framework capable of understanding the current action of a user and supporting them with only the critical information that is relevant to the task at hand, thus reducing the information overload and the need for manual control of displayed virtual content.

Related Work. Context-aware surgery involves the interpretation of the large amount of information created during a surgical procedure, with focus on recognizing/predicting key tasks [14], monitoring incidents [15], and highlighting adverse events. Surgical task prediction in an off-line context has been recently investigated in neurosurgery [16], laparoscopy [17], and cataract surgery [18] and has proposed the use of a secondary monitor to display virtual assistance. These applications have relied on various types of input features to predict the current surgical phase, such as radio-frequency identification chips attached to surgical instruments [17], instrument signals [19], robot kinematic data [20], external infrared measurement systems [21], or laparoscopic video [14].

Recent applications to context-aware surgery have remained limited to procedures where high-performance computing resources and video data are readily accessible [14,22,23]. If these systems are to be generalized to other surgical procedures, the display of context-aware information on a secondary monitor could introduce a potentially disorienting visual disconnect for the user.

To address these challenges, Katić et al. propose a context-aware system, based on an OST-HMD, for intraoperative AR in dental implant surgery. In a porcine study, the authors demonstrated an improved task completion time and acceptance of their system by dental surgeons [21]. However, the reliance on additional sensors, computing power and specialized markers for task prediction makes their method challenging to incorporate into a typical operating room workflow and incapable of generalization to different surgical scenarios.

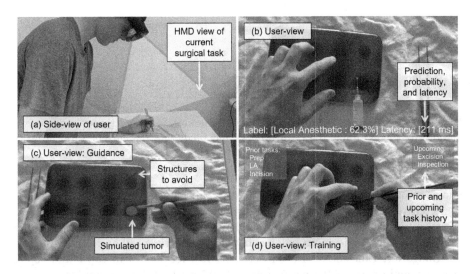

Fig. 1. Example images from online inference with the SurgeonAssist-Net app on a HoloLens 2. (a) Side-view of the user and phantom. User-view with (b) information on the current surgical phase prediction, (c) minimal virtual models for task-specific guidance, and (d) information on prior and upcoming surgical tasks for user-training.

In contrast to these systems, we propose SurgeonAssist-Net, a novel framework to predict the current surgical task that a user is performing and, using the context-aware predictions, ensure that the virtual augmentation meets the current information needs of the user. Our approach does not rely on the use of external sensors, custom/specialized hardware, or additional computing power. We are the first to demonstrate the implementation of a context-aware platform on a commercially available OST-HMD with near real-time performance, eliminating the visual disconnect between a monitor and the patient (Fig. 1).

2   Methods

2.1   SurgeonAssist-Net: Surgical Task Prediction

To provide context-awareness to the wearer of an OST-HMD from user-centric input video, we have created the SurgeonAssist-Net framework, composed of an EfficientNet-Lite-B0 [24,25] backbone for feature extraction and a gated recurrent unit (GRU) RNN framework [26] for storing long term dependency information (Fig. 2). These networks are jointly trained in an end-to-end manner, generating features that encode both spatial and temporal information.

Spatial Feature Extraction. Deep learning and the introduction of deep CNNs has led to vast improvements in interpreting high-dimensional data over traditional approaches, finding successful applications in object detection and

Fig. 2. Overview of the deep learning-based framework for extracting relevant information from video frame data (10 frames) and predicting the current surgical task.

image recognition [27]. With EfficientNet, Tan et al. overcame scaling issues common to increasingly deep CNNs through a compound scaling method that optimally adjusted the width, depth, and resolution of the network by using fixed coefficients, thus achieving a balance between network speed and accuracy [24]. On ImageNet, EfficientNet-B0 outperforms ResNet-50 [28] in top-1 and top-5 accuracy and offers a 4.9× parameter and 10.5× FLOPS reduction [24].

With EfficientNet-Lite-B0 [25], modifications to the EfficientNet-B0 model were implemented to optimize performance for mobile CPU applications; we use pre-trained weights from ImageNet to serve as an initial starting point for training. The final fully connected layer at the end of the EfficientNet-Lite-B0 network was removed and replaced with a global average-pooling layer to output a $7 \times 7 \times 320$ tensor of high-level discriminative features that was reshaped to a vector of length 1280 to serve as an input to the GRU framework (Fig. 2).

Temporal Information Modeling. Recurrent neural networks can handle variable-length sequence inputs by using a hidden state augmented with non-linear mechanisms whose activation at each time step is reliant on that of the prior frame [29]. Both GRU and long short-term memory (LSTM) components have been used for time-series forecasting tasks like analysis of video data for activity recognition, image captioning, and surgical task prediction [14]. It has been demonstrated that GRUs perform similarly to LSTM units [30], but have fewer total parameters and are more well-suited to real-time inference applications.

For our RNN architecture, we found optimal results using a single GRU cell with 128 hidden units, followed by a decision network with a ReLU activation, dropout layer with probability of 0.2, and a fully connected output layer with C output nodes—corresponding to the C potential surgical tasks. The parameters of the GRU cell and dense layer were initialized using Xavier normal initialization [31]. During inference, we used an online recognition mode accessing only current and prior frames. After performing initial hyperparameter evaluation experiments, we found that a sequence length of 10 video frames provided an optimal trade-off for system performance and speed (Fig. 2).

2.2   Integrating SurgeonAssist-Net for Online Inference

We used the Microsoft HoloLens 2 OST-HMD for the recording of user-centric video and visualization of context-aware surgical task predictions. The HoloLens 2 is capable of visualization of three-dimensional (3D) virtual models through stereoscopic vision via two-dimensional (2D) laser beam scanning displays, offering a field of view of 43×29 degrees (horizontal $\times$ vertical) to the wearer.

Input frames of size 896×504 px were requested from the HoloLens 2 photo-video camera at a rate of 30 frames per second (FPS), resized to 256×256 px using nearest-neighbor interpolation [32], and center cropped to a final resolution of 224×224 px; these served as input to the prediction framework. We leveraged the Windows Machine Learning and Open Neural Net Exchange (ONNX) [33] libraries within a C# Universal Windows Platform (UWP) app to perform inference using the SurgeonAssist-Net model. The OpenCV library [32] was included within a C++ UWP runtime component to prepare input frame data for prediction. The network output, the predicted task, was then used to optimize the virtual content shown to the user based on their current information needs.

2.3   Cholec80 Dataset

Cholec80 contains 80 videos (1920×1080 px or 854×480 px at 25 FPS) of cholecystectomy surgeries performed by 13 surgeons, complete with phase annotations of the 7 surgical phases for a procedure (25 FPS) defined by a senior surgeon [14]. The original videos were down-sampled from 25 FPS to 1 FPS to match the temporal resolution used by other groups [23]. We use nearest-neighbor interpolation to scale the input video frames from the original resolution to 256×256 px to improve computational efficiency [32]. For all tests using the Cholec80 dataset, 32 videos were used for a train split, 40 videos for a test split, and the remaining 8 videos for a validation split, as in prior work [14].

Our framework was implemented using the PyTorch [34] deep learning library. We trained our network for 25 epochs with a batch size of 32 on 2× NVIDIA V100 GPUs, each with 32 GB HBM2 memory, and reported the average network performance across three training runs. For optimization, we used stochastic gradient descent (SGD), an initial learning rate of $5e^{-4}$, and a momentum of 0.9. Sequence-wise data augmentation was performed on each training batch of image data, including random cropping of input images to 224×224 px, horizontal and vertical flipping, and random color augmentation.

To evaluate the performance of the SurgeonAssist-Net for task recognition, we employed the widely used metrics of accuracy (AC), precision (PR), and recall (RE) [14] and compared the results to other recent approaches. Furthermore, we included an estimate of the total number of parameters in each model, the model size, the inference time (latency), and the FLOPS for an input image sequence of size $(t \times 224 \times 224 \times 3)$, where t is the input sequence length of that specific approach. A single core of an AMD Ryzen 3600 CPU was used to measure the average latency for each network over 10 runs on a subset of the testing data.

2.4   User-Centric Surgical Tasks Dataset

Due to the lack of available video of surgical tasks from a user-centric point of view, we created our own dataset for training the SurgeonAssist-Net framework and evaluating its online performance on the HoloLens 2 device. The dataset included a total of five surgical tasks performed by three novice users on a gelatin phantom as they worked to remove a simulated subsurface tumor. During the task, the typical suite of surgical tools: scalpel, forceps, scissors, clamp, and syringe, was made available to the users. We recorded the dataset using the photo-video camera on the HoloLens 2 OST-HMD (1280 × 720 px at 30 FPS). A total of 52,500 frames were extracted from the videos at a rate of 30 FPS. Details of the dataset including tasks and duration are included in Table 1.

Table 1. Details of the user-centric surgical tasks dataset.

Phase	Preparation	Local anes.	Incision	Excision	Inspection
Duration (sec)	27.2 ± 11.5	39.2 ± 10.0	15.1 ± 9.9	32.3 ± 23.5	12.3 ± 8.9
Annotations	4,110	5,880	13,800	21,360	7,350

As with the Cholec80 benchmark dataset, we resized input video frames of the user-centric surgical tasks dataset to a resolution of 256 × 256 px [32] and used an input sequence length of 10 frames. We segmented the dataset such that 3 videos were used for a train split, 1 video for a test split, and the remaining 1 video for a validation split.

3   Results and Discussion

3.1   Cholec80: Surgical Task Prediction on a Benchmark Dataset

Table 2 compares the performance and latency of each approach on the Cholec80 dataset. Twinanda et al. have presented surgical task prediction results using learned visual features and temporal dependencies [14] based on (1) the single-task PhaseNet with features extracted from a modified AlexNet backbone [35]; and (2) the multi-task EndoNet framework which makes use of a modified AlexNet backbone for feature extraction and tool classification; both approaches use a single image frame as the input sequence. The single-task SV-RCNet [22] and multi-task MTRCNet-CL [23] networks share a similar ResNet-50 architecture for feature extraction and an LSTM cell with 512 hidden nodes for phase prediction. Additionally, both the SV-RCNet and MTRCNet-CL approaches use an input sequence length of 10 frames for prediction. As we were only interested in single-task performance, we did not report the results of multi-task approaches like EndoNet and MTRCNet-CL in our evaluation.

SurgeonAssist-Net outperformed the PhaseNet [14] framework across AC, PR, and RE metrics, required 46× fewer parameters for operation, and used

45× less memory for model deployment. When compared to SV-RCNet [22], SurgeonAssist-Net scored better in AC and PR metrics and required 7.4× fewer model parameters, achieved 10.2× faster FLOPS, and used 3× less time for inference. Due to its performance efficiency, low parameter count, and compact model size, SurgeonAssist-Net can be effectively operated in computationally restricted environments for real-time inference. Figure 3 provides a qualitative representation of the performance of SurgeonAssist-Net, without any form of post-processing, across a subset of the Cholec80 dataset (Video41).

Table 2. Results versus state-of-the-art using the Cholec80 surgical tasks dataset.

Method	Frames (t)	AC (%)	PR (%)	RE (%)	Size (MB)	Params (M)	FLOPS (B)	Latency (ms)
Ours	10	**85.8**	**81.5**	81.4	**15.9**	**3.91**	4.04	532.4
SV-RCNet [22]	10	85.3	80.7	**83.5**	115.3	28.76	41.25	1593.8
PhaseNet [14]	1	73.0	67.0	63.4	718.8	179.71	**0.83**	**99.3**

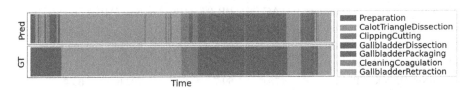

Fig. 3. SurgeonAssist-Net phase prediction (Pred) performance visualized relative to the ground truth (GT) phase labels on Video41 of the Cholec80 dataset (51 m 43 s in duration). The legend indicates the color coding of each individual phase.

3.2 User-Centric Surgical Tasks Dataset: Task Prediction and Online Performance on the HoloLens 2

To make the SurgeonAssist-Net model available to the HoloLens 2 through the UWP interface, we converted our trained model from its PyTorch implementation to ONNX format [33]. Table 3 includes the relative performance of the SurgeonAssist-Net framework in PyTorch and ONNX formats compared with ONNX converted implementations of PhaseNet [14] and SV-RCNet [22] when evaluated on the user-centric surgical tasks dataset.

A small decrease in AC, PR, RE and model size was recorded following conversion of the SurgeonAssist-Net model to ONNX format. However, we also measured a 5.2× decrease in CPU inference time by the ONNX model when compared to its PyTorch implementation; this speedup was due to the compilation of an efficient graph model during ONNX model export. Similar relative performance across AC, PR, and RE by the SurgeonAssist-Net model as compared to the ONNX converted PhaseNet [14] and SV-RCNet [22] was observed.

HoloLens 2 Performance and Feasibility. To evaluate the real-world performance of the SurgeonAssist-Net model on the HoloLens 2 headset, we created a sample application that displayed the prediction, with its associated probability and latency, for the current surgical task being performed. In Fig. 1 we include a sample image from an experiment where a user was presented with the same gelatin phantom and surgical tools as in the user-centric surgical tasks dataset and tasked with removing a subsurface tumor. Across the test, there was good agreement between the predicted and user-performed tasks.

The latency of the SurgeonAssist-Net ONNX model on the HoloLens 2 CPU, averaged across 30 s of online predictions, was measured to be 219.2 ms, or roughly 5 FPS, with a single image input sequence. To measure the feasibility of online inference with other networks on the HoloLens 2, we repeated the above experiment with ONNX converted PhaseNet [14] and SV-RCNet [22] models; however, we were unable to successfully load or operate either model on the HoloLens 2 CPU due to the large model size and/or high FLOPS requirements.

Table 3. Results versus state-of-the-art using the user-centric surgical tasks dataset.

Method	Frames (t)	AC (%)	PR (%)	RE (%)	Size (MB)	Latency (ms)
Ours PyTorch	10	**85.5**	**88.4**	76.5	15.9	538.6
Ours ONNX	10	85.1	87.5	75.3	**15.2**	103.1
SV-RCNet ONNX [22]	10	84.9	87.3	**77.5**	112.2	721.2
PhaseNet ONNX [14]	1	70.6	69.7	60.1	702.0	**64.7**

3.3   Clinical Significance

Aside from accuracy, a primary limitation of AR-guided approaches is the reliance upon a user to manually control the appearance and presentation of virtually augmented entities, thereby adapting the visualization to their current surgical context. This is tedious and detracts from their focus on the task. Our work on surgical task prediction is thus critical and foundational in ensuring that the automated augmentation of virtual models meets the current information needs. In this work, the predicted surgical task serves as a prerequisite to displaying the optimal virtual information to the user. We will now briefly discuss three clinical scenarios involving surgical guidance, user-training, and performance evaluation, where the SurgeonAssist-Net could be readily incorporated.

Guidance. The predicted task context can control the choice and presentation of the augmented virtual models. For example, in general surgery, as a surgeon picks up a scalpel and the incision phase of a procedure is detected, the HMD would display a relevant virtual model indicating the target site for surgical entry (Fig. 1). Throughout the procedure, our surgical phase detection would enable different virtual models and information relevant to the surgical phase to be optimally selected and presented, without user intervention.

Training. Task prediction can be used to guide a student, wearing the HMD, in practicing a general surgery task on a cadaver. Their active learning can be reinforced by presenting them with a task history of phases performed, or of upcoming surgical actions, given their present surgical step (Fig. 1). Additional relevant information, in the form of visual cues, text, or audio, could be presented in tandem with the detected task to enhance the training experience.

Evaluation. Task analytics can provide surgeons with quantitative data on a surgical procedure. For example, a surgeon performing a less frequent procedure could wear the HMD while re-training and be provided with chronology and analytics of the time spent in each surgical phase (Fig. 3). This information, when compared to peers, could serve to suggest focus areas for improvement.

4   Conclusions and Future Work

The focus of this work was to create a lightweight framework capable of understanding user-centric activities and providing virtual real-time workflow assistance for a pre-defined set of surgical tasks. By training the SurgeonAssist-Net framework on the user-centric surgical tasks dataset, we were able to use it effectively as an event-detection heuristic to associate known events in an online scenario by using the on-board computing resources of an OST-HMD (in this case, the HoloLens 2). For future investigation, we envision that an in-depth user-study to evaluate a manually-controlled AR experience versus the context-aware approach could further demonstrate the benefits of action-driven virtual assistance. We also expect that a larger user-centric training dataset would result in better consistency in predictions from the SurgeonAssist-Net framework. Nonetheless, we have demonstrated the potential capabilities of an online context-aware surgical guidance platform and brought attention to its capacity to overcome issues which had previously plagued AR-based image-guidance systems.

Acknowledgements. This work was supported by the Natural Sciences and Engineering Research Council of Canada (NSERC) Discovery program (RGPIN-2019-06367).

References

1. Peters, T.M.: Image-guidance for surgical procedures. Phys. Med. Biol. **51**(14), R505 (2006)
2. Liu, D., Jenkins, S.A., Sanderson, P.M., Fabian, P., Russell, W.J.: Monitoring with head-mounted displays in general anesthesia: a clinical evaluation in the operating room. Anesth. Analg. **110**(4), 1032–1038 (2010)
3. Bernhardt, S., Nicolau, S.A., Soler, L., Doignon, C.: The status of augmented reality in laparoscopic surgery as of 2016. Med. Image Anal. **37**, 66–90 (2017)
4. Zorzal, E.R., et al.: Laparoscopy with augmented reality adaptations. J. Biomed. Inform. **107**, 103463 (2020)

5. Meola, A., Cutolo, F., Carbone, M., Cagnazzo, F., Ferrari, M., Ferrari, V.: Augmented reality in neurosurgery: a systematic review. Neurosurg. Rev. **40**(4), 537–548 (2016). https://doi.org/10.1007/s10143-016-0732-9
6. Jud, L., et al.: Applicability of augmented reality in orthopedic surgery-a systematic review. BMC Musculoskelet. Disord. **21**(1), 1–13 (2020)
7. Rahman, R., Wood, M.E., Qian, L., Price, C.L., Johnson, A.A., Osgood, G.M.: Head-mounted display use in surgery: a systematic review. Surg. Innov. **27**(1), 88–100 (2020)
8. Dixon, B.J., Daly, M.J., Chan, H., Vescan, A.D., Witterick, I.J., Irish, J.C.: Surgeons blinded by enhanced navigation: the effect of augmented reality on attention. Surg. Endosc. **27**(2), 454–461 (2013)
9. Grubert, J., Itoh, Y., Moser, K., Swan, J.E.: A survey of calibration methods for optical see-through head-mounted displays. IEEE Trans. Visual Comput. Graphics **24**(9), 2649–2662 (2017)
10. Kersten-Oertel, M., Jannin, P., Collins, D.L.: The state of the art of visualization in mixed reality image guided surgery. Comput. Med. Imaging Graph. **37**(2), 98–112 (2013)
11. Hong, J., et al.: Three-dimensional display technologies of recent interest: principles, status, and issues [invited]. Appl. Opt. **50**(34), H87–H115 (2011)
12. Cleary, K., Peters, T.M.: Image-guided interventions: technology review and clinical applications. Annu. Rev. Biomed. Eng. **12**, 119–142 (2010)
13. Eckert, M., Volmerg, J.S., Friedrich, C.M.: Augmented reality in medicine: systematic and bibliographic review. JMIR mHealth uHealth **7**(4), e10967 (2019)
14. Twinanda, A.P., Shehata, S., Mutter, D., Marescaux, J., De Mathelin, M., Padoy, N.: Endonet: a deep architecture for recognition tasks on laparoscopic videos. IEEE Trans. Med. Imaging **36**(1), 86–97 (2016)
15. Suzuki, T., Sakurai, Y., Yoshimitsu, K., Nambu, K., Muragaki, Y., Iseki, H.: Intraoperative multichannel audio-visual information recording and automatic surgical phase and incident detection. In: 2010 Annual International Conference of the IEEE Engineering in Medicine and Biology, pp. 1190–1193. IEEE (2010)
16. Forestier, G., et al.: Multi-site study of surgical practice in neurosurgery based on surgical process models. J. Biomed. Inform. **46**(5), 822–829 (2013)
17. Navab, N., Traub, J., Sielhorst, T., Feuerstein, M., Bichlmeier, C.: Action-and workflow-driven augmented reality for computer-aided medical procedures. IEEE Comput. Graphics Appl. **27**(5), 10–14 (2007)
18. Quellec, G., Lamard, M., Cochener, B., Cazuguel, G.: Real-time task recognition in cataract surgery videos using adaptive spatiotemporal polynomials. IEEE Trans. Med. Imaging **34**(4), 877–887 (2014)
19. Padoy, N., Blum, T., Ahmadi, S.A., Feussner, H., Berger, M.O., Navab, N.: Statistical modeling and recognition of surgical workflow. Med. Image Anal. **16**(3), 632–641 (2012)
20. Lea, C., Vidal, R., Hager, G.D.: Learning convolutional action primitives for fine-grained action recognition. In: 2016 IEEE International Conference on Robotics and Automation (ICRA), pp. 1642–1649. IEEE (2016)
21. Katić, D., et al.: A system for context-aware intraoperative augmented reality in dental implant surgery. Int. J. Comput. Assist. Radiol. Surg. **10**(1), 101–108 (2014). https://doi.org/10.1007/s11548-014-1005-0
22. Jin, Y., et al.: SV-RCNet: workflow recognition from surgical videos using recurrent convolutional network. IEEE Trans. Med. Imaging **37**(5), 1114–1126 (2017)
23. Jin, Y., et al.: Multi-task recurrent convolutional network with correlation loss for surgical video analysis. Med. Image Anal. **59**, 101572 (2020)

24. Tan, M., Le, Q.: Efficientnet: rethinking model scaling for convolutional neural networks. In: International Conference on Machine Learning, pp. 6105–6114. PMLR (2019)
25. Liu, R.: Higher accuracy on vision models with efficientnet-lite. TensorFlow Blog (2020). https://blog.tensorflow.org/2020/03/higher-accuracy-on-visionmodels-with-efficientnet-lite.html. Accessed 30 Apr 2020
26. Cho, K., Van Merriënboer, B., Bahdanau, D., Bengio, Y.: On the properties of neural machine translation: encoder-decoder approaches. arXiv preprint arXiv:1409.1259 (2014)
27. LeCun, Y., Bengio, Y., Hinton, G.: Deep learning. Nature **521**(7553), 436–444 (2015)
28. He, K., Zhang, X., Ren, S., Sun, J.: Deep residual learning for image recognition. In: Proceedings of the IEEE Conference on Computer Vision and Pattern Recognition, pp. 770–778 (2016)
29. Hochreiter, S., Schmidhuber, J.: Long short-term memory. Neural Comput. **9**(8), 1735–1780 (1997)
30. Chung, J., Gulcehre, C., Cho, K., Bengio, Y.: Empirical evaluation of gated recurrent neural networks on sequence modeling. arXiv preprint arXiv:1412.3555 (2014)
31. Glorot, X., Bengio, Y.: Understanding the difficulty of training deep feedforward neural networks. In: Proceedings of the Thirteenth International Conference on Artificial Intelligence and Statistics, pp. 249–256. JMLR Workshop and Conference Proceedings (2010)
32. Bradski, G.: The opencv library. Dr. Dobb's J. Softw. Tools **25**, 120–125 (2000)
33. Bai, J., Lu, F., Zhang, K., et al.: ONNX: open neural network exchange (2019). https://github.com/onnx/onnx
34. Paszke, A., et al.: Pytorch: an imperative style, high-performance deep learning library. arXiv preprint arXiv:1912.01703 (2019)
35. Krizhevsky, A., Sutskever, I., Hinton, G.E.: Imagenet classification with deep convolutional neural networks. Adv. Neural. Inf. Process. Syst. **25**, 1097–1105 (2012)

Author Index

Printed in the United States
by Baker & Taylor Publisher Services

Printed in the United States
by Baker & Taylor Publisher Services